Benchmark Papers in Acoustics

Series Editor: R. Bruce Lindsay
Brown University

PUBLISHED VOLUMES

UNDERWATER SOUND / *Vernon M. Albers*
ACOUSTICS: Historical and Philosophical Development / *R. Bruce Lindsay*
SPEECH SYNTHESIS / *James L. Flanagan and L. R. Rabiner*
PHYSICAL ACOUSTICS / *R. Bruce Lindsay*
MUSICAL ACOUSTICS, PART I: Violin Family Components / *Carleen M. Hutchins*
MUSICAL ACOUSTICS, PART II: Violin Family Functions / *Carleen M. Hutchins*
ULTRASONIC BIOPHYSICS / *Floyd Dunn and William D. O'Brien*
VIBRATION: Beams, Plates, and Shells / *Arturs Kalnins and Clive L. Dym*
MUSICAL ACOUSTICS: Piano and Wind Instruments / *Earle L. Kent*

Additional volumes in preparation

Benchmark Papers
in Acoustics / 8

A BENCHMARK® Books Series

VIBRATION
Beams, Plates, and Shells

Edited by

ARTURS KALNINS
Lehigh University

CLIVE L. DYM
Bolt Beranek and Newman Inc.

Dowden, Hutchinson & Ross, Inc.
STROUDSBURG, PENNSYLVANIA

Distributed by
HALSTED PRESS
A Division of
John Wiley & Sons, Inc.

Benchmark Papers in Acoustics, Volume 8
Library of Congress Catalog Card Number: 76-18764
ISBN: 0-87933-243-3

78 77 76 1 2 3 4 5
Manufactured in the United States of America.

LIBRARY OF CONGRESS CATALOGING IN PUBLICATION DATA
Main entry under title:
Vibration: beams, plates, and shells
(Benchmark papers in acoustics ; 8)
Includes indexes.
1. Vibration—Addresses, essays, lectures.
I. Kalnins, Arturs. II. Dym, Clive L.
TA355.V52 620.3'08 76-18764
ISBN 0-87933-243-3

Exclusive Distributor: **Halsted Press**
A Division of John Wiley & Sons, Inc.
ISBN: 0-470-98943-2

SERIES EDITOR'S FOREWORD

The "Benchmark Papers in Acoustics" constitute a series of volumes that make available to the reader in carefully organized form important papers in all branches of acoustics. The literature of acoustics is vast in extent and much of it, particularly the earlier part, is inaccessible to the average acoustical scientist and engineer. These volumes aim to provide a practical introduction to this literature, since each volume offers an expert's selection of the seminal papers in a given branch of the subject, that is, those papers which have significantly influenced the development of that branch in a certain direction and introduced concepts and methods that possess basic utility in modern acoustics as a whole. Each volume provides a convenient and economical summary of results as well as a foundation for further study for both the person familiar with the field and the person who wishes to become acquainted with it.

Each volume has been organized and edited by an authority in the area to which it pertains. In each volume there is provided an editorial introduction summarizing the technical significance of the field being covered. Each article is accompanied by editorial commentary, with necessary explanatory notes, and an adequate index is provided for ready reference. Articles in languages other than English are either translated or abstracted in English. It is the hope of the publisher and editor that these volumes will constitute a working library of the most important technical literature in acoustics of value to students and research workers.

The present volume *Vibration: Beams, Plates, and Shells* has been edited by Arturs Kalnins of Lehigh University and Clive L. Dym of Bolt Beranek and Newman. In its 41 seminal articles it covers progress in the theory of vibrations of the principal varieties of solid configurations as well as wave motion through such media. The principal emphasis is on developments during the twentieth century, but there is adequate reference to the historical background in the associated editorial commentaries. The book contains an important section on nonlinear vibrations, which have been assuming a greater role in current applications. The material in this volume is of great significance for the application of acoustical theory to a wide variety of engineering problems.

R. BRUCE LINDSAY

PREFACE

This volume is intended to serve as an introduction to the fundamental aspects of the dynamic behavior of beams, plates, and shells, as reflected in the research literature. The selection of papers has been guided by the notion that we wanted to put together a *tutorial* volume, or a research textbook, that would give the reader the necessary starting information to carry out further work on the subject. Thus we have chosen papers not necessarily for their originality, but for their clarity. Happily, in many instances, we were able to realize both objectives within a single paper. Occasionally, however, we have chosen not to reprint a paper that originated an idea or a technique, but rather a later work that seemed to use to represent a clearer exposition.

We would like to thank Professor R. Bruce Lindsay for his interest, his encouragement, and his patience with our effort.

ARTURS KALNINS
CLIVE L. DYM

Contents

PART III: FREE VIBRATION OF SHELLS

PART VI. DAMPING

CONTENTS BY AUTHOR

VIBRATION

INTRODUCTION

The second half of the twentieth century has been, in many ways, a fruitful period in the development of applied mechanics. Spurred on by many external events, including both the vast support of scientific research by the various agencies of the U.S. Department of Defense and the increased awareness of the role of engineering science in engineering education, the discipline of engineering mechanics has emerged as an important and interesting field in its own right. Among the many problems that would come into the purview of mechanics would be that of the dynamic response of elastic structural systems, and elements of those systems, to dynamic loading. And, as we shall see quite often in this Volume, there is a prescient quality to much of the work published in the early 1950s, for much of it was concerned with choosing appropriate models to represent the dynamic response of certain structural elements that would assume proportions of considerable significance in the era of Sputnik, ballistic missiles, and nuclear power plants.

An important element in a structural system is an element which has one dimension, called thickness, much smaller than other dimensions. Beams, plates, and shells form a class of such elements. The property of thickness being considerably smaller than any other length justifies the construction of two-dimensional mathematical models which are simpler but more limited than the corresponding three-dimensional models. Consequently, the mathematical models of beams, plates, and shells can recognize only those features in the input that are averaged over distances comparable to thickness. Similarly, the output represents quantities that are averaged over the same distances. Thickness in beam, plate, and shell models can be looked upon as a length-scale that sets the limit of resolution of the model.

In vibration problems, the length-scale determines also a time-scale, which equals the propagation time of a disturbance across one thickness. Changes in input that occur within the time-scale cannot be recognized by the models; a meaningful prediction of a transient response can be made only after a time interval that is comparable to the time-scale has elapsed.

The study of vibration of beam, plate, and shell elements is important in acoustics because they are capable of vibrating in modes that resemble flexure. The general trend of the resonant frequencies of such flexural modes is that they decrease with decreasing thickness, and, therefore, for sufficiently small thicknesses, the infinite set of flexural modes begins at the lower end of the frequency spectrum. The major portion of strain energy of a flexural mode is in flexural strain energy, which depends only on the curvature changes and is produced largely by normal deflections. It follows that the flexural modes with the largest amplitudes are produced by loads that are normal to the bounding surfaces of beams, plates, and shells.

The flexural modes are important in applications for which (1) the forcing frequency is in the lower frequency spectrum; (2) the response of the normal deflection is desired; and (3) the loads are normal to a bounding surface. The last two cases are particularly important in acoustics.

In addition to the flexural modes, there are separate (infinite) sets of modes of other types that begin in higher frequency ranges. Many mathematical models predict also the extensional modes; and more refined models can predict the thickness shear and stretch modes, which comprise the higher frequency mode sets. Practically all types of vibration of beams, plates, and shells have found applications in acoustics.

Part I

VIBRATION AND WAVES IN BEAMS

Editors' Comments on Papers 1 Through 5

1 **TIMOSHENKO**
On the Correction for Shear of the Differential Equation for Transverse Vibrations of Prismatic Bars

2 **MINDLIN and DERESIEWICZ**
Timoshenko's Shear Coefficient for Flexural Vibrations of Beams

3 **LEONARD and BUDIANSKY**
On Traveling Waves in Beams

4 **MIKLOWITZ**
Flexural Wave Solutions of Coupled Equations Representing the More Exact Theory of Bending

5 **BOLEY**
An Approximate Theory of Lateral Impact on Beams

The set of papers that comprises this section is an interesting reflection of the generalizations discussed in the Introduction. With the exception of the very first paper, which has been included largely for historical interest, all of the remaining papers are concerned with the analysis of the response of simple beams; all of them contain attempts to decide how complex a beam model is required to accurately portray the response of a beam to a sharp (spatially) or sudden (temporally) load; and, in fact, all of these papers were published in the early 1950s.

In the first paper, one of the classics in the field, the great mechanician S. P. Timoshenko outlines how the effects of transverse shear deformation may be incorporated into a model of a bent beam that already includes the effect of rotatory inertia. Although not explicitly stated, two important points stand out in this paper. The first is that the two higher-order effects, rotatory inertia and transverse shear deformation, are of the same order of magnitude, viz., equation (13). In that equation it can be seen that both terms are of the order of the square of the ratio of the radius of gyration to the beam length. This can be brought out by a more detailed study of the roots of the Timoshenko equations (see, for example, C. L. Dym and I. H. Shames, *Solid Mechanics: A Variational Approach,* New York: McGraw-Hill, pp. 370–377, 1973).

The second point of interest is the introduction of the *shape factor*, λ, in equation (4). Again, with almost no discussion of the origin of or the calculation of this factor, Timoshenko simply introduces it as if it was obvious that it belonged. The purpose of this factor is, in fact, to account for the Timoshenko beam theory assumption of a shear stress that is uniform through the beam thickness, rather than having the proper parabolic distribution. A number of approaches to properly defining this shape factor have appeared [e.g., G. R. Cowper, *J. Appl. Mech.*, **33,** 335–340 (1966), and some of the references listed therein] and no doubt the subject is not entirely closed even now.

The second paper in this section is devoted to discussion and calculation of this shape factor. The paper develops the technique made famous by the first author (see Mindlin's paper on higher-order plate theory, No. 9 in the next section) wherein an approximate frequency from the Timoshenko theory is matched to a corresponding limiting value obtained from a three-dimensional elastodynamic solution. In this case the constant is obtained by comparing simple thickness-shear mode frequencies, for a variety of cross-sections.

The paper by Mindlin and Deresiewicz is also of interest because of a historical remark that is inserted in the first reference. That is, they point out that although Rayleigh is credited with developing the rotatory inertia correction, and Timoshenko is credited with the shear deformation correction, both of these corrections may be found in a French treatise, due to M. Bresse, that was published in 1859!

The next three papers are intended to illustrate the reasons for the development and use of a higher-order beam theory. That is, there are circumstances in which the elementary beam theory is inadequate. The inadequacies arise, of course, from the assumptions that are incorporated into the elementary beam theory. Leonard and Budiansky credit W. Flügge (see Ref. 6 of their paper) with being the first to recognize that a traveling wave solution could not be constructed for the elementary theory, as that theory predicts that a disturbance will propagate at an infinite velocity. Only if the shear and rotatory inertia correction factors are included can discontinuities in shear and moment propagate at finite (and generally different) speeds. The approaches taken in the three papers (Leonard and Budiansky, Miklowitz, and Boley) are rather different, and they deserve further comment.

Leonard and Budiansky make use of the method of characteristics to demonstrate quite clearly that, if the Timoshenko equations are used, shear discontinuities will propagate at a certain characteristic shear wave speed, while moment discontinuities will propagate with a bending wave speed. This demonstrated, the authors then go on to develop some numerical results by integrating modal solutions and some appropriately applied exact solutions are obtained for infinitely long beams.

The comparisons between solutions indicate that the modal analyses cannot replicate the discontinuities introduced in these particular problems. Further, even the numerical integration scheme loses accuracy as multiple reflections build up behind the discontinuities. This latter problem, however, may not be difficult to resolve in the current era of high-speed computers with large memory banks.

Miklowitz's paper presents a methodology, based on the Laplace transform, for treating both infinite and finite beams via a traveling wave solution. This paper also makes explicit the propagation, in the Timoshenko model, of disturbances at two different speeds, one reflecting a shear wave speed, the other a bending-related wave speed. The solutions are developed in closed form; and, although no numerical results are presented, a very interesting observation is made regarding beams of finite length. That is, Miklowitz follows Uflyand (see Ref. 2 of Miklowitz's paper) and notes that a fairly ordinary solution for the bending deflection of a beam of finite length, equation (22), can be interpreted as a superposition of wave solutions, as in equation (23). Then one can see the development of standing waves as a process of repeated reflection of traveling waves. A very similar analysis is given in great detail by Meirovitch for the propagation of longitudinal waves in rods (see L. Meirovitch, *Analytical Methods in Vibrations,* New York: Macmillan, pp. 343–352, 1967).

In the final paper of this section, Boley uses a traveling wave methodology from the outset. However, the paper he presents has the more practical goal of determining how important are the shear and rotatory inertia correction factors, and when they need to be applied. The answer that arises is that, in the immediate neighborhood of a propagating shear discontinuity, there is a small boundary layer wherein these correction factors are very important. The highlight of this paper is an order-of-magnitude analysis [equations (24) through (27)] that demonstrates that the additional terms introduced in the Timoshenko model are required only if information is required about the *very short-time* response of a beam, and in the *immediate* neighborhood of the propagating discontinuities. An extension of this work, which is able to predict the beam response outside of the boundary layer, was subsequently published by Boley and C.-C. Chao [*J. Appl. Mech.,* **25,** 31–36 (1958)].

1

Reprinted from *Phil. Mag.*, **41**, 744–746 (1921)

LXVI. *On the Correction for Shear of the Differential Equation for Transverse Vibrations of Prismatic Bars.* By Prof. S. P. Timoshenko *.

IN studying the transverse vibrations of prismatic bars, we usually start from the differential equation

$$\mathrm{EI}\frac{\partial^4 y}{\partial x^4}+\frac{\rho\Omega}{g}\frac{\partial^2 y}{\partial t^2}=0, \quad . \quad . \quad . \quad . \quad . \quad (1)$$

in which EI denotes the flexural rigidity of the bar,
Ω the area of the cross-section,
and $\frac{\rho}{g}$ the density of the material.

When the "rotatory inertia" is taken into consideration, the equation takes the form

$$\mathrm{EI}\frac{\partial^4 y}{\partial x^4}-\frac{\mathrm{I}\rho}{g}\frac{\partial^4 y}{\partial x^2\partial t^2}+\frac{\rho\Omega}{g}\frac{\partial^2 y}{\partial t^2}=0. \quad . \quad . \quad . \quad (2)$$

I now propose to show how the effect of the shear may be taken into account in investigating transverse vibrations, and I shall deduce the general equation of vibration, from which equations (1) and (2) may be obtained as special cases.

Fig. 1.

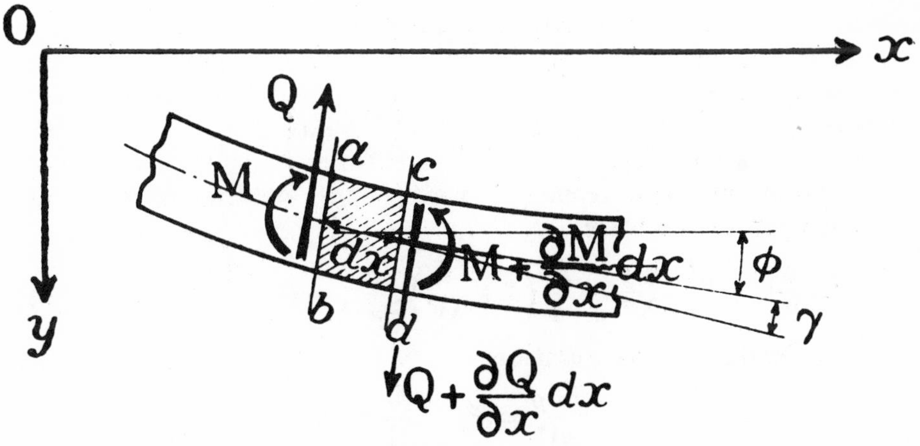

Let *a b c d* (fig. 1) be an element bounded by two adjacent cross-sections of a prismatic bar. M and Q denote respectively the bending moment and the shearing force.

* Communicated by Mr. R. V. Southwell, M.A. Translated from the Russian by Prof. M. G. Yatsevitch.

The position of the element during vibration will be determined by the displacement of its centre of gravity and by the angular rotation ϕ in the (x, y) plane: the axis Ox may be taken as coinciding with the initial position of the axis of the bar.

The angle at which the tangent to the curve into which the axis of the bar is bent (the curve of deflexion) is inclined to the axis Ox will differ from the angle ϕ by the angle of shear γ. Hence, for very small deflexions, we may write

$$\frac{\partial y}{\partial x}=\phi+\gamma. \quad . \quad . \quad . \quad . \quad . \quad . \quad (3)$$

For determining M and Q we have the familiar expressions

$$M=-EI\frac{\partial\phi}{\partial x}, \quad Q=\lambda C\Omega\gamma=\lambda C\Omega\left(\frac{\partial y}{\partial x}-\phi\right), \quad . \quad (4)$$

where C denotes the modulus of rigidity, for the material of the bar, and λ is a constant which depends upon the shape of the cross-section.

The equations of motion will now be :—

for the rotation—

$$-\frac{\partial M}{\partial x}dx+Qdx=\frac{\rho I}{g}\frac{\partial^2\phi}{\partial t^2}dx,$$

or
$$EI\frac{\partial^2\phi}{\partial x^2}+\lambda C\Omega\left(\frac{\partial y}{\partial x}-\phi\right)-\frac{\rho I}{g}\frac{\partial^2\phi}{\partial t^2}=0, \quad . \quad . \quad . \quad (5)$$

if we substitute from equations (4);

for translation in the direction of Oy—

$$\frac{\partial Q}{\partial x}dx=\frac{\rho\Omega}{g}\frac{\partial^2 y}{\partial t^2}dx,$$

or
$$\frac{\rho\Omega}{g}\frac{\partial^2 y}{\partial t^2}-\lambda C\Omega\left(\frac{\partial^2 y}{\partial x^2}-\frac{\partial\phi}{\partial x}\right)=0. \quad . \quad . \quad . \quad . \quad (6)$$

Eliminating ϕ from (5) and (6), we obtain the required equation in the form

$$EI\frac{\partial^4 y}{\partial x^4}+\frac{\rho\Omega}{g}\frac{\partial^2 y}{\partial t^2}-\frac{\rho I}{g}\left(1+\frac{E}{\lambda C}\right)\frac{\partial^4 y}{\partial x^2\partial t^2}+\frac{\rho^2 I}{g^2\lambda C}\frac{\partial^4 y}{\partial t^4}=0. \quad (7)$$

Introducing the notation

$$\frac{EIg}{\rho\Omega}=a^2, \quad \frac{I}{\Omega}=k^2,$$

we may write equation (7) in the form

$$a^2\frac{\partial^4 y}{\partial x^4}+\frac{\partial^2 y}{\partial t^2}-k^2\left(1+\frac{E}{\lambda C}\right)\frac{\partial^4 y}{\partial x^2\partial t^2}+\frac{k^2\rho}{g\lambda C}\frac{\partial^4 y}{\partial t^4}=0. \quad (8)$$

In order to estimate the influence of the shear upon the frequency of the vibrations, let us consider the case of a prismatic bar with supported ends. The type of the vibrations may be assumed to be given by

$$y = \mathrm{Y} \sin \frac{m\pi x}{l} \cos p_m t, \quad . \quad . \quad . \quad . \quad . \quad (9)$$

where l represents the length of the bar, and p_m is the required frequency. By substitution from (9) in equation (8), we obtain the following equation for the frequency:

$$\alpha^2 \frac{m^4\pi^4}{l^4} - p_m{}^2 - \frac{m^2\pi^2k^2}{l^2}\left(1 + \frac{\mathrm{E}}{\lambda \mathrm{C}}\right)p_m{}^2 + \frac{k^2\rho}{g\lambda\mathrm{C}}p_m{}^4 = 0. \quad (10)$$

If only the first two terms on the left side of this equation are retained (this will correspond to the equation (1)), we have

$$p_m = \alpha \frac{m^2\pi^2}{l^2} = \frac{\alpha\pi^2}{\mathrm{L}^2}, \quad . \quad . \quad . \quad . \quad . \quad . \quad (11)$$

where $\mathrm{L} = \frac{l}{m}$ represents the length of a wave.

By retaining the first three terms of equation (10) (*i. e.* by neglecting the terms which involve λ), we find

$$p_m = \frac{\alpha\pi^2}{\mathrm{L}^2}\left(1 - \frac{1}{2}\frac{\pi^2k^2}{\mathrm{L}^2}\right) \quad . \quad . \quad . \quad . \quad . \quad (12)$$

approximately: this result corresponds to equation (2), where the rotatory inertia is taken into consideration.

By using the complete equation (10), and neglecting small quantities of the second order, we find

$$p_m = \frac{\alpha\pi^2}{\mathrm{L}^2}\left[1 - \frac{1}{2}\frac{\pi^2k^2}{\mathrm{L}^2}\left(1 + \frac{\mathrm{E}}{\lambda\mathrm{C}}\right)\right] \quad . \quad . \quad . \quad (13)$$

approximately.

Assuming the values

$$\lambda = \tfrac{2}{3}, \qquad \mathrm{E} = \tfrac{8}{3}\mathrm{C},$$

we have

$$\frac{\mathrm{E}}{\lambda\mathrm{C}} = 4,$$

and hence we see that the correction for shear is four times greater than the correction for rotatory inertia. The value of the correction of course increases with a decrease in the wave-length L, *i. e.*, with an increase in m.

Yougoslavia, Videm.
Summer 1920.

2

Reprinted from *Proc. 2nd U.S. Natl. Cong. Appl. Mech.*, 175–178 (1955)

TIMOSHENKO'S SHEAR COEFFICIENT FOR FLEXURAL VIBRATIONS OF BEAMS

R. D. Mindlin
H. Deresiewicz

Professor of Civil Engineering,
Columbia University, New York. N. Y.
Assistant Professor of Civil Engineering,
Columbia University, New York, N. Y.

It is pointed out that Timoshenko's shear coefficient, in his equations of flexural vibrations of beams, depends on both the shape of the cross-section and the mode of motion. It is shown how the latter may be taken into account by matching solutions, of Timoshenko's equations and the three-dimensional equations, for simple thickness-shear motions of infinite beams of various cross-sectional shapes.

Introduction

The range of applicability of the one-dimensional theory of flexural vibrations of beams was extended to higher frequencies by Timoshenko when he took into account the effect of transverse shear deformation. He arrived at the free-vibration equations [1]*

$$EI\,\frac{\partial^2\psi}{\partial x^2} + k\left(\frac{\partial y}{\partial x} - \psi\right) AG - \frac{I\gamma}{g}\,\frac{\partial^2\psi}{\partial t^2} = 0$$

$$\frac{\gamma A}{g}\,\frac{\partial^2 y}{\partial t^2} - k\left(\frac{\partial^2 y}{\partial x^2} - \frac{\partial\psi}{\partial x}\right) AG = 0 \qquad (1)$$

governing the transverse deflection, y, and the slope, ψ, of the deflection curve when the shear is neglected. The coefficient k is defined as the ratio of the average shear stress on a section to the product of the shear modulus and the angle of shear at the neutral axis. This ratio depends upon the distribution of shear stress on the section and, hence, k depends upon the shape of the section, as Timoshenko observed. However, the distribution of shear stress on a section depends also on the mode of motion of the beam. For example, for low modes of motion of a slender beam, the shear stress has a maximum at the neutral axis, while, for very high modes, the shear stress has a minimum at the same place. Thus, k depends both on the shape of the section and the frequency of vibration.

*Numbers in brackets refer to Bibliography at the end of the paper.

In the solution of Equations (1), the simplest interpretation of k is that it is a constant. For a beam of given cross-sectional shape, then, k can have the correct value for only one frequency. In the past, several calculations of k have been made, for various cross-sectional shapes, on the basis of statical considerations, that is, at zero frequency. These values are satisfactory for the low modes of motion of slender beams, where, in fact, the influence of transverse shear deformation is small in comparison with its influence at high frequencies.

As the frequency is increased, a point is reached at which a drastic change occurs in the spectrum. This is the frequency, ω', of the first thickness-shear mode of a beam of infinite length, i.e., the lowest frequency at which the infinite beam can vibrate with no transverse deflection, the displacement being entirely parallel to the axis of the beam.

The reason for the change in character of the frequency spectrum is that, at the frequencies of the first thickness-shear mode and its overtones, there is strong coupling between the flexural and thickness-shear modes of motion. Examples of this phenomenon have been described in detail in previous papers [2, 3] for special cases of crystal plates for which the equations of motion are the same as those for a Timoshenko beam.

It is desirable, then, that the thickness-shear frequency calculated from Equations (1) should be the same as that calculated from the equations to which (1) are an approximation, namely, the three-dimensional equations of small vibrations of an elastic body. In this connection it must be recognized that, whereas Timoshenko interpreted ψ as the slope of the deflection curve when the shear deformation is neglected, the product of ψ and the depth of the beam may be interpreted as the maximum axial displacement of a transverse section. Thus, ω' is calculated from (1) by setting $y = 0$ and ψ proportional to $\exp(i\omega' t)$, following which ω' is set equal to the corresponding frequency calculated from the three-dimensional equations. Since ω' depends upon k, the result is a formula for k.

It has been shown [4] that, for a rectangular section, this procedure leads to $k = \pi^2/12 = 0.822$. Comparison with experiments [2,3] shows that, when k is calculated in this manner, Timoshenko's equations give good results for both low and high frequencies of beams with free ends. This good agreement may be expected to hold, for slender beams, regardless of the end conditions, even for high modes, since the frequencies of high modes are

This investigation was supported by the Office of Naval Research under Contract Nonr-266(09) with Columbia University.

insensitive to the boundary conditions.

In the present paper, k is calculated in the same manner for a variety of cross-sectional shapes. To do this, it is necessary to solve the three-dimensional equations of elasticity for the appropriate frequency and equate it with the value of ω' obtained from Equations (1). In the case of motion parallel to the axis of the bar, the three-dimensional equations and boundary conditions reduce to equations governing a familiar hydrodynamical problem for which many solutions are known. The determination of k is thus reduced to an interpretation of these solutions and some additional computations. Results are given for the following sections: circle, ellipse, orthogonal parabolas and a variety of ovaloids. The values of k computed for these sections all lie within about 10 per cent of that for the rectangular section.

Thickness-Shear Motion: Timoshenko Theory

If y is set equal to zero in Equations (1), the second of the equations gives $\partial\psi/\partial x = 0$, so that the first of (1) reduces to

$$\frac{d^2\psi}{dt^2} + k\left(\frac{c}{r}\right)^2 \psi = 0 \tag{2}$$

where $c = (Gg/\gamma)^{1/2}$, i.e., the velocity of shear waves in an infinite, isotropic, elastic medium, and $r = (I/A)^{1/2}$, i.e., the radius of gyration of the cross section. Hence, the frequency of pure thickness-shear vibration of a Timoshenko beam is

$$\omega' = ck^{1/2}/r \tag{3}$$

This is the frequency which is to be equated to the one calculated from the solution of the three-dimensional equations for each section.

Thickness-Shear Motion: Three-Dimensional Theory

Let the neutral axis of the beam be the z-axis of a rectangular coordinate system and let u, v and w be components of displacement in the x, y and z directions, respectively. For pure thickness-shear motion,

$$u = v = 0$$
$$w = \zeta(x,y)\, e^{i\omega t} \tag{4}$$

Then the three-dimensional equations of motion [5] reduce to

$$\nabla^2\zeta + \delta^2\zeta = 0 \tag{5}$$

where $\delta = \omega/c$ and $\nabla^2 = \partial^2/\partial x^2 + \partial^2/\partial y^2$. The condition that the traction vanish on the cylindrical surface of the beam reduces to

$$\partial\zeta/\partial n = 0 \tag{6}$$

on the boundary, where n is the normal to the boundary.

The differential equation (5) and the boundary condition (6) are the same as those governing the small oscillations of a fluid in a basin of uniform depth [6] and the small vibrations of a gas in a rigid cylindrical container [7]. Solutions are available for a variety of boundaries.

Rectangle

In the case of a rectangular section, the frequency is independent of the width. If the depth is $2a$, the frequency is [6]

$$\omega = \pi c/2a \tag{7}$$

Equating (7) and (3), and noting that $r^2 = a^2/3$, the shear constant, k, is found to be $\pi^2/12$, or, approximately, 0.822.

Circle

For a circular section of radius a, the lowest antisymmetric mode has a frequency [6]

$$\omega = 1.841\, c/a \tag{8}$$

The radius of gyration of a circle about a diameter is $a/2$. Hence, equating (8) and (3), $k = 0.847$.

Ellipse

The elliptic section was treated by Jeffreys [8], who computed frequencies for two values of the eccentricity

$$e = (1 - b^2/a^2)^{1/2} \tag{9}$$

where a and b are the semi-major and semi-minor axes, respectively. Additional values were computed and are listed in the columns headed $\omega a/c$ and $\omega b/c$ in Table I. The corresponding values of ω were then equated to that in Equation (3) to obtain the values of k listed in Table I.

It is interesting to notice that the frequency (1.886 c/a) of the first antisymmetric mode of a very narrow ellipse, about its minor axis, is greater than that (1.571 c/a) of a

TABLE I

	Motion antisymmetric about minor axis		Motion antisymmetric about major axis	
	$\omega a/c$	k	$\omega b/c$	k
Rectangle	1.571	.822	1.571	.822
Ellipse				
e = 1.0	1.886*	.889	**	
e = 0.9	1.878	.882	**	
e = 0.8	1.87 *	.87	1.78 *	.79
e = 0.7	1.858	.863	1.814	.823
e = 0.6	1.856	.861	1.823	.831
e = 0.3	1.845	.851	1.837	.844
e = 0 (circle)	1.841	.847	1.841	.847
Orthogonal Parabolas	2.117	.896		

* Jeffreys (8)
** Convergence slow

rectangle of depth equal to the major diameter of the ellipse. Jeffreys observed that this is due to the concentration of motion near the center of the ellipse. On the other hand, the corresponding frequency for a very wide ellipse appears to approach that of the rectangle. However, the corresponding values of k are not the same because of the difference in radii of gyration.

Parabolas

The hydrodynamic problem for a symmetric section bounded by a pair of orthogonal parabolas was studied by Hidaka [9]. Such a section has the property $a/b = 2$, where a and b are the semi-major and semi-minor axes, respectively. For motion antisymmetric with respect to the minor axis, Hidaka found a secular equation which can be reduced to

$$J_{-\frac{1}{4}}(\delta a/2) = 0 \qquad (10)$$

where $J(x)$ is the Bessel function of the first kind. The lowest root of Equation (10) is [10]

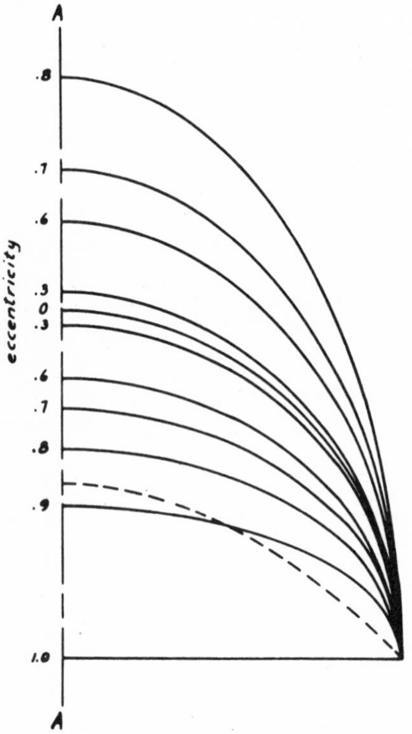

FIGURE 1. ELLIPTIC AND PARABOLIC SECTIONS FOR WHICH THICKNESS-SHEAR FREQUENCY (ω) AND TIMOSHENKO'S CONSTANT (k) ARE GIVEN IN TABLE I. DASHED LINE IS PARABOLIC SECTION. THICKNESS-SHEAR MOTION IS ANTISYMMETRIC ABOUT AXIS AA.

$$\delta a/2 = 1.0585 \qquad (11)$$

so that

$$\omega = 2.117\, c/a \qquad (12)$$

Equating (12) and (3) and noting that $r^2 = a^2/5$, the result is $k = 0.896$.

This completes the calculations for sections for which exact solutions are available. The results are assembled in Table I and the sections are illustrated in Fig. 1.

Ovaloids

Approximate frequencies, good for narrow sections, may be obtained by neglecting the variation of displacement across the width. Hidaka [9] did this for sections bounded by the curves

$$\frac{x^2}{a^2} + \left(\frac{y^2}{b^2}\right)^{\frac{1}{2}m} = 1$$
$$m = 2^{-\mu}, \quad \mu = 0, 1, 2, \ldots \qquad (13)$$

A set of these curves, for the case $a = 2b$, is shown in Fig. 2.

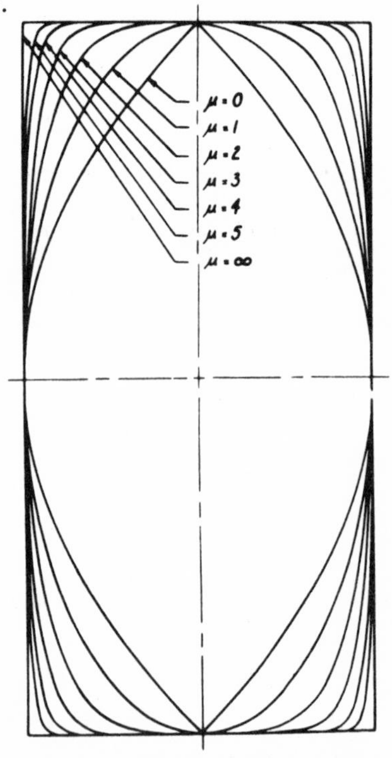

FIGURE 2. OVALOID SECTIONS FOR WHICH APPROXIMATE VALUES OF ω and k ARE GIVEN IN TABLE II.

TABLE II

μ	m	$\omega a/c$	r^2/a^2	k
0	1 *	2.150	1/5	.924
1	1/2 **	1.886	1/6	.889
2	1/4	1.717	2/7	.862
3	1/8	1.655	4/13	.843
4	1/16	1.616	8/25	.836
5	1/32	1.591	16/49	.826
∞	0 ***	1.571	1/3	.822

*parabolas; **ellipse; ***rectangle.

Hidaka's results for the lowest mode, antisymmetric about the horizontal axis of Fig. 2, are given in Table II, along with the corresponding values of k. In finding k, the radius of gyration (r), defined by

$$r^2 = \frac{\int_0^a y x^2 \, dx}{\int_0^a y \, dx} \tag{14}$$

must be computed. Inserting y from Equation (13) into Equation (14), and making the substitution $x = a \sin\theta$ in the integrals, we find

$$\frac{r^2}{a^2} = 1 - \frac{\int_0^{\pi/2} \cos^{(3+2m)}\theta \, d\theta}{\int_0^{\pi/2} \cos^{(1+2m)}\theta \, d\theta} \tag{15}$$

By aid of the formula [11]

$$\int_0^{\pi/2} \cos^n x \, dx = \frac{\sqrt{\pi}}{2} \frac{\Gamma\left[\frac{n+1}{2}\right]}{\Gamma\left[\frac{n}{2}+1\right]}, \quad n > -1$$

and the recursion formula for the Gamma function, Equation (15) reduces to

$$r^2 = a^2/(3 + 2m) \tag{16}$$

The results in Table II are the same as in Table I for the rectangle ($\mu = \infty$) for any b/a since the assumption of uniform displacement across the width is exact in pure thickness-shear motion. For the ellipse ($\mu = 1$), Table II gives the same value as Table I only for $b/a = 0$ (i.e., $e = 1$), since here, too, there is no variation of displacement across the width. As the section becomes wider, there is some discrepancy, but it is not great. Even for $a = b$ (the circle), Table I gives $k = 0.847$, while the approximation yields 0.889. The approximation is not intended, of course, to apply to $b > a$. Another comparison may be made for the parabola with $a = 2b$. The value of k in Table I is 0.896, while the approximation gives 0.924. Thus it may be expected that the approximate solution of Equation (5) will give results good to within a few per cent for sections at least twice as deep as they are wide.

Bibliography

1. S. Timoshenko, "Vibration Problems in Engineering," D. Van Nostrand Company, New York, 2nd Edition, 1937, p. 338.

It is of some historical interest that both the rotatory inertia correction, usually attributed to Rayleigh ("Theory of Sound," Cambridge, England, first edition, 1877; current edition, reference [7], art. 162), and the transverse shear correction, usually attributed to Timoshenko (Philosophical Magazine, Ser. 6, Vol. 41, 1921, pp. 744–746, and Vol. 43, 1922, pp. 125–131), are given by M. Bresse in his "Cours de Mécanique Appliquée," Mallet-Bachelier, Paris, 1859, p. 126.

2. R. D. Mindlin, "Thickness-Shear and Flexural Vibrations of Crystal Plates," J. Appl. Phys., Vol. 22, 1951, pp. 316–323.

3. R. D. Mindlin, "Forced Thickness-Shear and Flexural Vibrations of Piezoelectric Crystal Plates," J. Appl. Phys., Vol. 23, 1952, pp. 83–88.

4. R. D. Mindlin, "Influence of Rotatory Inertia and Shear on Flexural Motions of Isotropic, Elastic Plates," J. Appl. Mech., Vol. 18, 1951, pp. 31–38.

5. S. Timoshenko and J. N. Goodier, "Theory of Elasticity," McGraw-Hill Book Co., New York, 2nd Edition, 1951, p. 452.

6. H. Lamb, "Hydrodynamics," Dover Publications, New York, 6th Edition, 1945, pp. 282–290.

7. Lord Rayleigh, "Theory of Sound," Dover Publications, New York, 2nd Edition, 1945, art. 339.

8. H. Jeffreys, "The Free Oscillations of Water in an Elliptic Lake," Proc. London Math. Soc., Ser. 2, Vol. 23, 1924, pp. 455–476.

9. K. Hidaka, "The Oscillations of Water in Spindle-Shaped and Elliptic Basins as well as the Associated Problems," Mem. Imp. Marine Observ., Kobe, Japan, Vol. 4, 1931, pp. 99–219.

10. "Tables of Bessel Functions of Fractional Order," Columbia University Press, New York, 1948, Vol. 1, p. 384.

11. B. O. Peirce, "A Short Table of Integrals," Ginn and Company, Boston, 3rd revised edition, 1929, No. 483.

3

Reprinted from *Natl. Adv. Comm. Aeron. Report 1173,* 1–27 (1954)

ON TRAVELING WAVES IN BEAMS [1]

By ROBERT W. LEONARD and BERNARD BUDIANSKY

SUMMARY

The basic equations of Timoshenko for the motion of vibrating nonuniform beams, which allow for effects of transverse shear deformation and rotary inertia, are presented in several forms, including one in which the equations are written in the directions of the characteristics. The propagation of discontinuities in moment and shear, as governed by these equations, is discussed.

Numerical traveling-wave solutions are obtained for some elementary problems of finite uniform beams for which the propagation velocities of bending and shear discontinuities are taken to be equal. These solutions are compared with modal solutions of Timoshenko's equations and, in some cases, with exact closed solutions.

INTRODUCTION

The theoretical analysis of transient stresses in aircraft wings and fuselages subjected to impact loadings has generally been performed by means of a mode-superposition method that uses the natural modes of vibration predicted by the elementary engineering theory of beam bending. (See, for example, refs. 1 to 3.) For very sharp impact loadings, however, this approach is known to have certain shortcomings: For sharp impacts of short duration, many modes are often required for a satisfactory degree of convergence (see, for example, ref. 4); in addition, the use of elementary beam theory in the calculation of the higher modes of vibration is inaccurate because of the neglect of, among other factors, the effects of transverse shear deformation and rotary inertia which become increasingly important for higher and higher modes (ref. 5).

A classically recognized alternative to the modal method of calculating transient stresses in elastic bodies is the traveling-wave approach, which seeks to trace directly the propagation of stresses through the body (ref. 3). Although the traveling-wave concept has been successfully used to treat such simple problems as longitudinal and torsional impact of rods, only recently have serious attempts been made to study the transient bending response of beams by this approach. Flügge (ref. 6) was apparently the first to point out that elementary beam theory could not serve as an adequate basis for the traveling-wave approach since the elementary theory predicts that disturbances propagate with infinite velocity; he showed, however, that a traveling-wave theory could be constructed by modifying the elementary theory, as Timoshenko (ref. 7) did, to include first-order effects of transverse shear deformation and rotary inertia. On the basis of this more accurate theory, Flügge found that discontinuities in moment and shear travel along the beam with finite, and generally distinct, velocities. A similar analysis was carried out independently by Robinson (ref. 8) who, exploiting the hyperbolic nature of Timoshenko's equations, proposed the use of approximate methods of solution and gave some numerical results for a particular example. Pfeiffer (ref. 9) also suggested the possibility of step-by-step solutions by the method of characteristics. In reference 10, Uflyand attempted an analytical solution of Timoshenko's equations for the case of a simply supported beam subjected to a sudden application of load; however, as was shown by Dengler and Goland (ref. 11), Uflyand's work is marred by the fact that he applied boundary conditions that are incorrect for Timoshenko's theory. The only known example of an exact traveling-wave solution based on Timoshenko's theory was presented by Dengler and Goland for the case of an infinitely long beam subjected to a concentrated impulse.

Thus, although the use of Timoshenko's theory as a basis for the transient-stress analysis of beams has been seriously considered, few problems have actually been solved. Additional basic studies of Timoshenko's equations and their solution, particularly for finite-length beams, constitute necessary prerequisites to the development of practical methods of dynamic-stress analysis based on the traveling-wave approach. In the present report, several specific problems of transient loading of uniform beams of finite length are considered and their solutions by various procedures, all based on the Timoshenko theory, are presented. These procedures are (a) numerical step-by-step integration—the "method of characteristics," (b) mode superposition, and (c) exact closed-form solution. The examples are all for the special case of equal propagation velocities of shear and bending disturbances; only for this case have exact solutions been found in closed form. For the sake of completeness, the presentation of these solutions is preceded by an exposition of the basic equations of Timoshenko's theory, a derivation of the characteristic lines and characteristic forms of these equations, and a discussion of their implications concerning propagation of disturbances.

[1] Supersedes NACA TN 2874, "On Traveling Waves in Beams" by Robert W. Leonard and Bernard Budiansky, 1953.

SYMBOLS

A	cross-sectional area
E	Young's modulus of elasticity
G	shear modulus of elasticity
I	cross-sectional moment of inertia
L	length of finite beam (arbitrary length for infinite beam)
M	internal bending moment (see fig. 1)
$\overline{M}$	dimensionless internal bending moment, ML/EI_B
$R=\left(\frac{c_2}{c_1}\right)^2\left(\frac{L}{r_i}\right)^2$	
V	vertical shear force on a cross section (see fig. 1)
$\overline{V}$	dimensionless vertical shear force, VL^2/EI_B
c_1	propagation velocity of bending discontinuities, $\sqrt{\frac{EI_B}{\rho I_i}}$
c_2	propagation velocity of shear discontinuities, $\sqrt{\frac{A_SG}{\rho A_i}}$
p	operator used in the Laplace transformation
q	intensity of distributed external loading
$\overline{q}$	dimensionless intensity of external loading, qL^3/EI_B
r	cross-sectional radius of gyration
t	time
v	velocity of deflection, y_t
$\overline{v}$	dimensionless velocity of deflection, v/c_2
x	coordinate along beam
y	deflection (see fig. 1)
$\overline{y}$	dimensionless deflection, y/L
δ	denotes a discontinuity in quantity immediately following
$\lambda=\frac{1}{2}\frac{L}{r_i}$	
ξ	dimensionless coordinate along beam, x/L
ρ	density of beam material
τ	dimensionless time, c_1t/L
ψ	rotation of cross section about neutral axis (see fig. 1)
ω	velocity of rotation of cross section, ψ_t
$\overline{\omega}$	dimensionless velocity of rotation, $\omega L/c_1$

Subscripts:

B	contributing to resistance of beam to bending
i	contributing to inertia
S	contributing to resistance of beam to shearing
x,t,ξ,τ	indicate partial derivatives with respect to those quantities

BASIC EQUATIONS

NONUNIFORM BEAMS

Timoshenko's equations.—The Timoshenko theory of beam bending (ref. 7) constitutes a modification of elementary beam theory that attempts to account for the effects of transverse shear deformation and rotary inertia; the basic assumption of elementary theory—that plane sections remain plane—is retained. The moment M, shear V, deflection y, and cross-sectional rotation ψ (see fig. 1) of a nonuniform beam subject to a dynamic lateral loading q are governed, according to this theory, by the following four simultaneous partial-differential equations:

$$M+EI_B\psi_x=0 \tag{1a}$$

$$V-A_SG(y_x-\psi)=0 \tag{1b}$$

$$M_x-V+\rho I_i\psi_{tt}=0 \tag{1c}$$

$$V_x-\rho A_iy_{tt}+q=0 \tag{1d}$$

The first two equations constitute elastic laws relating the deformations to the internal loading. Equation (1a) expresses the same relationship between moment and cross-sectional rotation as that given by elementary beam theory Equation (1b) stipulates a linear relationship between the shear V and the shear angle $y_x-\psi$ at the neutral axis; A_S is the so-called "effective" shear-carrying area, different from the total area A_i since the true shear angle actually varies over the cross section. Equations (1c) and (1d) prescribe rotational and translational equilibrium, respectively, with the term $\rho I_i\psi_{tt}$ representing the contribution of rotary inertia.

The moment and shear may be eliminated from equations (1) to yield two simultaneous partial-differential equations in y and ψ:

$$\left.\begin{array}{l}(EI_B\psi_x)_x+A_SG(y_x-\psi)-\rho I_i\psi_{tt}=0\\ [A_SG(y_x-\psi)]_x-\rho A_iy_{tt}+q=0\end{array}\right\} \tag{2}$$

This form is convenient for finding the normal modes and frequencies of free vibration ($q=0$) predicted by Timoshenko's theory and for carrying out modal analyses that make use of these modes. In this theory, a natural mode is described by a pair of functions $[y(x), \psi(x)]$ rather than a single function $y(x)$, as is the case in elementary beam theory.

For a traveling-wave analysis, however, it is advantageous to return to the original system of equations (eqs. (1)) but to put them in a more convenient form by differentiating

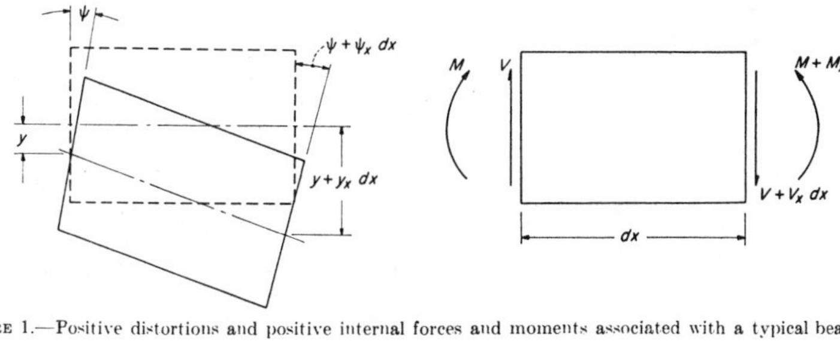

FIGURE 1.—Positive distortions and positive internal forces and moments associated with a typical beam element.

equations (1a) and (1b) with respect to time. The equations become:

$$\omega_x+\frac{1}{EI_B}M_t=0 \tag{3a}$$

$$v_x-\frac{1}{A_SG}V_t-\omega=0 \tag{3b}$$

$$M_x+\rho I_t\omega_t-V=0 \tag{3c}$$

$$V_x-\rho A_t v_t+q=0 \tag{3d}$$

where the new variables, linear velocity v and angular velocity ω, have been introduced for y_t and ψ_t, respectively.

Equations (3) comprise four first-order linear partial-differential equations in the four variables v, ω, M, and V. Furthermore, equations (3a) and (3c) contain derivatives of only M and ω, whereas equations (3b) and (3d) contain derivatives of only v and V. These facts are exploited in the next section in seeking characteristic lines and characteristic forms of these equations.

Characteristics and the characteristic form of the equations.—In equations (3a) and (3c), the variables M and ω are differentiated with respect to both space and time; it would be advantageous to replace them by equivalent equations each involving only total derivatives (or differentials) in a particular direction in the space-time plane. The lines in the space-time plane having these particular directions—the characteristic lines or so-called "characteristics"—and the equivalent equations written in these directions are found as follows (ref. 12).

A linear combination of equations (3a) and (3c),

$$\mu(M_t+EI_B\omega_x)+M_x+\rho I_t\omega_t-V=0 \tag{4}$$

is formed, where the function μ is to be determined in such a way that the partial derivatives in equation (4) combine to give total derivatives $\frac{dM}{d\sigma}$ and $\frac{d\omega}{d\sigma}$ in the direction of an as yet unknown characteristic line $[x(\sigma),t(\sigma)]$. In order that the terms involving derivatives of M combine in the form

$$M_x\frac{dx}{d\sigma}+M_t\frac{dt}{d\sigma}=\frac{dM}{d\sigma}$$

the function μ must satisfy the following equation:

$$\mu=\frac{\frac{dt}{d\sigma}}{\frac{dx}{d\sigma}}=\frac{dt}{dx}$$

where $\frac{dt}{dx}$ is the required slope of the characteristic line. Similarly, in order that the terms involving derivatives of ω combine in the form

$$\omega_x\frac{dx}{d\sigma}+\omega_t\frac{dt}{d\sigma}=\frac{d\omega}{d\sigma}$$

the function μ must also satisfy the equation

$$\frac{\rho I_t}{\mu EI_B}=\frac{\frac{dt}{d\sigma}}{\frac{dx}{d\sigma}}=\frac{dt}{dx}$$

Since the characteristic slope must be the same in both cases, μ is defined by

$$\mu^2=\frac{\rho I_t}{EI_B}$$

Thus, the two values of μ and the corresponding characteristic slopes $\frac{dt}{dx}$ are:

$$\left.\begin{aligned}\mu&=\frac{1}{c_1}\\ \frac{dt}{dx}&=\frac{1}{c_1}\end{aligned}\right\} \tag{5}$$

$$\left.\begin{aligned}\mu&=-\frac{1}{c_1}\\ \frac{dt}{dx}&=-\frac{1}{c_1}\end{aligned}\right\} \tag{6}$$

where $c_1=\sqrt{\frac{EI_B}{\rho I_t}}$. Then, multiplying equation (4) by dt and using equations (5) yields

$$\frac{1}{c_1}\,dM+\rho I_t\,d\omega-V\,dt=0$$

when

$$dt=\frac{1}{c_1}\,dx$$

Similarly, using equation (6) gives

$$-\frac{1}{c_1}\,dM+\rho I_t\,d\omega-V\,dt=0$$

when

$$dt=-\frac{1}{c_1}\,dx$$

In an analogous fashion, it can be shown, from equations (3b) and (3d), that

$$\frac{1}{c_2}\,dV-\rho A_t\,dv+\left(q+\frac{A_SG}{c_2}\,\omega\right)dt=0$$

when

$$dt=\frac{1}{c_2}\,dx$$

and

$$-\frac{1}{c_2}\,dV-\rho A_t\,dv+\left(q-\frac{A_SG}{c_2}\,\omega\right)dt=0$$

when

$$dt=-\frac{1}{c_2}\,dx$$

where $c_2=\sqrt{\frac{A_SG}{\rho A_t}}$.

Then, the system of equations (3) has associated with it four real characteristic directions and is thus "totally hyperbolic" (ref. 12). A network of characteristics can readily be constructed without prior knowledge of the unknowns M, V, ω, and v since their slopes depend only on the material and geometrical properties of the beam. For uniform beams, as well as for tapered beams having uniform material properties and geometrically similar cross sections, c_1 and c_2 are constant, and the characteristics will therefore be straight; in general, however, the characteristics may be

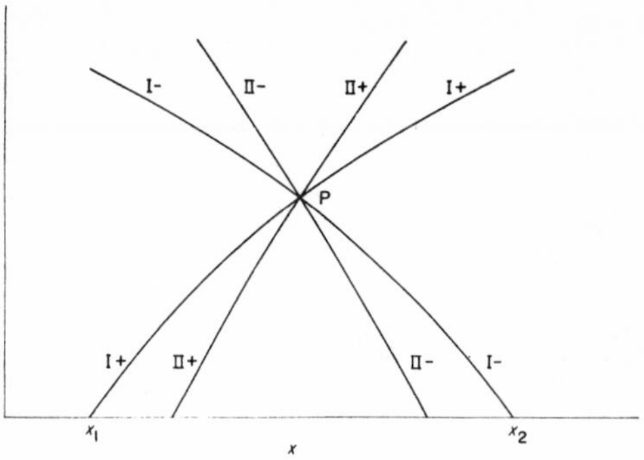

FIGURE 2.—The characteristics of Timoshenko's equations for a point in the x, t plane.

curved. Figure 2 illustrates the four characteristics passing through a point P in the space-time plane with the characteristics designated as

I+: $$\frac{dt}{dx}=\frac{1}{c_1}$$

I−: $$\frac{dt}{dx}=-\frac{1}{c_1}$$

II+: $$\frac{dt}{dx}=\frac{1}{c_2}$$

II−: $$\frac{dt}{dx}=-\frac{1}{c_2}$$

It is known that, by virtue of the totally hyperbolic character of the basic equations, the values of the unknowns M, V, ω, and v at the point P depend only on their initial values at $t=0$ between the points x_1 and x_2 on the beam (ref. 12). Furthermore, these values at P can, in turn, have influence only on points lying in the region above P enveloped by the I+ and I− characteristics through P. Thus no signal can proceed along the beam with a velocity greater than c_1 (which is generally larger than c_2, as illustrated in fig. 2).

For the sake of easy reference, the four characteristic differential forms of the basic equations are grouped below.

Along I+: $$\frac{1}{c_1}\,dM+\rho I_t\,d\omega-V\,dt=0 \tag{7a}$$

Along I−: $$\frac{1}{c_1}\,dM-\rho I_t\,d\omega+V\,dt=0 \tag{7b}$$

Along II+: $$\frac{1}{c_2}\,dV-\rho A_t\,dv+(\rho A_t c_2\omega+q)dt=0 \tag{7c}$$

Along II−: $$\frac{1}{c_2}\,dV+\rho A_t\,dv+(\rho A_t c_2\omega-q)dt=0 \tag{7d}$$

Related forms of these characteristic equations have been written by Robinson (ref. 8) and Pfeiffer (ref. 9).

Propagation of discontinuities.—Characteristics are lines across which discontinuities may exist (ref. 12); indeed, this property is often taken as the basic definition of a characteristic. In the present problem, discontinuities (or jumps) in M and ω can therefore exist across the I+ and I− characteristics, and discontinuities in V and v can exist across the II+ and II− characteristics. Hence, a jump in M or ω will propagate with the velocity c_1, whereas a jump in V or v must proceed with the velocity c_2. It should be remarked that such discontinuities would appear only in the limiting case of a beam subjected to an instantaneous loading. The solution of such idealized problems, which are often instructive, requires a knowledge of the laws governing the variations in strength of these discontinuities as they propagate through the beam. These laws are determined below for nonuniform beams for which it is assumed that the condition $c_1=c_2$ does not hold over any portion of the beam; in other words, the I and II characteristics are distinct. The special case where $c_1=c_2$ is considered subsequently when uniform beams are discussed.

Let a and b be two points on a I− characteristic on either side of a particular I+ characteristic

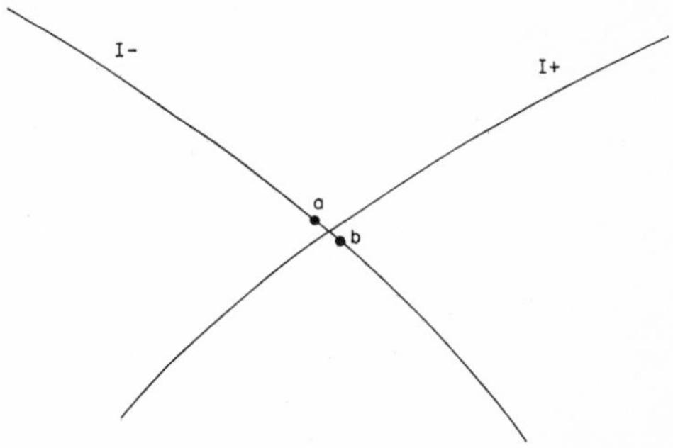

If M is discontinuous across the I+ characteristic, then M_a-M_b retains a finite value δM as a and b are allowed to approach the I+ characteristic from either side. Consequently, from equation (7b) written along the I− characteristic,

$$\frac{1}{c_1}\,\delta M-\rho I_t\,\delta\omega=0$$

since dt approaches zero as a and b approach each other. Thus, everywhere *along* a I+ characteristic, jumps δM and $\delta\omega$ *across* this characteristic are related by

$$\delta M=c_1\rho I_t\delta\omega \tag{8}$$

Similarly, the jumps across the other characteristics can be readily shown to satisfy

Along I−: $$\delta M=-c_1\rho I_t\delta\omega \tag{9}$$

Along II+: $$\delta V=-c_2\rho A_t\delta v \tag{10}$$

Along II−: $$\delta V=c_2\rho A_t\delta v \tag{11}$$

A jump in M is thus always accompanied by a definite jump in ω; similarly, jumps in V and v are always coupled together.

The variations in the magnitude of a discontinuity as one proceeds along a characteristic may be determined in the following manner. Equation (7a) is written for the upper side and the lower side of the I+ characteristic; then, since

V is continuous across a I+ characteristic, the difference of the two equations yields

$$d(\delta M)+c_1\rho I_t d(\delta\omega)=0$$

along I+. Eliminating $\delta\omega$ by using equation (8) gives

$$d(\delta M)+c_1\rho I_t d\left(\frac{\delta M}{c_1\rho I_t}\right)=0$$

By carrying out the indicated differentiation in the second term and dividing by $2\delta M$, the following result is obtained:

$$\frac{d(\delta M)}{\delta M}=\frac{1}{2}\frac{d(c_1\rho I_t)}{c_1\rho I_t}$$

Solution of this equation gives

$$(\delta M)_2=(\delta M)_1\sqrt{\frac{(c_1\rho I_t)_2}{(c_1\rho I_t)_1}} \tag{12}$$

as the relationship between the magnitude of the jumps in M at two points 1 and 2 on the I+ characteristic.

It can be shown that the identical relationship holds between jumps in M at two points on a I− characteristic. Similarly, it can be found that, on II+ and II− characteristics,

$$(\delta V)_2=(\delta V)_1\sqrt{\frac{(c_2\rho A_t)_2}{(c_2\rho A_t)_1}} \tag{13}$$

for any two points 1 and 2 on a given characteristic. The corresponding variations of the jumps $\delta\omega$ and δv are, of course, readily determined from equations (8) to (11).

UNIFORM BEAMS

Nondimensional form of the equations.—The examples to be presented in this report are all concerned with beams having uniform cross-sectional size and shape and uniform material properties throughout their length. For such beams, it is convenient to express Timoshenko's equations in nondimensional form. Thus, equations (3) may be written in terms of $\overline{M}=\frac{ML}{EI_B}$, $\overline{V}=\frac{VL^2}{EI_B}$, $\overline{\omega}=\frac{\omega L}{c_1}$, and $\overline{v}=\frac{v}{c_2}$ as

$$\overline{\omega}_\xi+\overline{M}_\tau=0 \tag{14a}$$

$$\frac{c_2}{c_1}\overline{v}_\xi-\frac{1}{R}\overline{V}_\tau-\overline{\omega}=0 \tag{14b}$$

$$\overline{M}_\xi+\overline{\omega}_\tau-\overline{V}=0 \tag{14c}$$

$$\overline{V}_\xi-R\frac{c_1}{c_2}\overline{v}_\tau+\overline{q}=0 \tag{14d}$$

where $\xi=\frac{x}{L}$, $\tau=\frac{c_1 t}{L}$, $\overline{q}=\frac{qL^3}{EI_B}$, and $R=\left(\frac{c_2}{c_1}\right)^2\left(\frac{L}{r_t}\right)^2$. The quantity L refers to the beam length for all beams except those of infinite length, in which case any convenient arbitrary length may be chosen for L.

The characteristics of Timoshenko's equations for a uniform beam are defined in the ξ,τ plane by the families of straight lines

I+: $$\frac{d\tau}{d\xi}=1 \tag{15a}$$

I−: $$\frac{d\tau}{d\xi}=-1 \tag{15b}$$

II+: $$\frac{d\tau}{d\xi}=\frac{c_1}{c_2} \tag{15c}$$

II−: $$\frac{d\tau}{d\xi}=-\frac{c_1}{c_2} \tag{15d}$$

The nondimensionalized characteristic forms of the basic equations become

Along I+: $$d\overline{M}+d\overline{\omega}-\overline{V}\,d\tau=0 \tag{16a}$$

Along I−: $$d\overline{M}-d\overline{\omega}+\overline{V}\,d\tau=0 \tag{16b}$$

Along II+: $$d\overline{V}-R\,d\overline{v}+\left(R\overline{\omega}+\frac{c_2}{c_1}\overline{q}\right)d\tau=0 \tag{16c}$$

Along II−: $$d\overline{V}+R\,d\overline{v}+\left(R\overline{\omega}-\frac{c_2}{c_1}\overline{q}\right)d\tau=0 \tag{16d}$$

In addition to the restriction to uniform beams, for which c_1 and c_2 are constants, the examples presented herein are further limited to those beams for which the propagation velocities c_1 and c_2 are equal. This assumption has been made because the simplifications that result not only permit the ready attainment of numerical solutions but also, in particular cases, permit the attainment of exact closed solutions for comparison. Since, for this very special case, the characteristics II coincide with the characteristics I, equations (16) may now be written, for $\overline{q}=0$, as

Along I+: $$\left\{\begin{matrix} d\overline{M}+d\overline{\omega}-\overline{V}\,d\tau=0 \\ d\overline{V}-4\lambda^2\,d\overline{v}+4\lambda^2\overline{\omega}\,d\tau=0\end{matrix}\right\} \tag{17a}$$

Along I−: $$\left\{\begin{matrix} d\overline{M}-d\overline{\omega}+\overline{V}\,d\tau=0 \\ d\overline{V}+4\lambda^2\,d\overline{v}+4\lambda^2\overline{\omega}\,d\tau=0\end{matrix}\right\} \tag{17b}$$

where $\lambda=\frac{L}{2r_t}$.

Propagation of discontinuities when $c_1=c_2$.—Equations (8) to (13), which describe the behavior of discontinuities in a nonuniform beam, also describe, as a special case, the behavior of discontinuities in a uniform beam for which $c_1\neq c_2$. They show that such discontinuities propagate with constant magnitude.

However, when the beam has properties such that $c_1=c_2$, these equations are no longer valid. Equations which are valid in this case may be derived in precisely the same way by using equations (17) instead of equations (7). Discontinuities in such beams can be shown to be related by

Along I+: $$\left\{\begin{matrix} \delta\overline{M}=\delta\overline{\omega} \\ \delta\overline{V}=-4\lambda^2\delta\overline{v}\end{matrix}\right\} \tag{18a}$$

Along I−: $$\left\{\begin{matrix} \delta\overline{M}=-\delta\overline{\omega} \\ \delta\overline{V}=4\lambda^2\delta\overline{v}\end{matrix}\right\} \tag{18b}$$

as they propagate, and they can be shown to vary in magnitude according to the equations

$$\text{Along I+ and I}-:\begin{Bmatrix}\frac{d}{d\tau}(\delta\bar{\omega})-\frac{1}{2}\,\delta\bar{V}=0\\ \frac{d}{d\tau}(\delta\bar{V})+2\lambda^2\delta\bar{\omega}=0\end{Bmatrix} \tag{19}$$

Equations (19) may be solved simultaneously to obtain

$$\text{Along I+ and I}-:\begin{Bmatrix}\delta\bar{\omega}=A\cos(\lambda\tau-B)\\ \delta\bar{V}=-2\lambda A\sin(\lambda\tau-B)\end{Bmatrix} \tag{20}$$

where A and B are arbitrary constants which must be evaluated by using known values of $\delta\bar{\omega}$ and $\delta\bar{V}$ at some point on the characteristic. The variations in $\delta\bar{M}$ and $\delta\bar{v}$ can then be readily found by using equations (18).

Thus, for the case $c_1=c_2$, discontinuities in a uniform beam do not propagate unchanged but vary in magnitude sinusoidally as they progress through the beam.

LIMITATIONS OF THE THEORY

It may be well to insert a word of caution about the applicability of Timoshenko's theory. The investigations of Prescott (ref. 13) and Cooper (ref. 14) have shown that, when the response of a beam includes components whose wave length is small compared to the depth of the beam, the assumption of Timoshenko's theory that plane sections remain plane after bending is, as might be expected, too restrictive. Since applied disturbances which could give rise to discontinuities would obviously excite even the shortest wave length in the spectrum of the response, the results obtained by application of Timoshenko's theory to such hypothetical problems cannot, in themselves, have practical significance. However, carrying out solutions involving discontinuities is a useful means of testing methods of solution of the Timoshenko equations with a view to applying these methods to problems in which discontinuities do not exist. Furthermore, the admittedly inaccurate response to an infinitely abrupt disturbance may be used to obtain the correct response to disturbances of a more practical nature through the application of Duhamel's superposition integral.

SPECIFIC EXAMPLES—FINITE UNIFORM BEAMS WITH $c_1=c_2$

Three specific examples are considered; they are: a cantilever beam given a step velocity at the root, a simply supported beam subjected to a step moment at one end, and a simply supported beam subjected to a ramp-platform moment at one end.

METHODS OF SOLUTION

In the examples, the results of calculation by the following three methods are compared: (a) numerical step-by-step integration along the characteristics, (b) normal-mode superposition, and (c) exact closed-form solution. The first two procedures are approximate in character, but they could conceivably be generalized sufficiently to be applied to practical structures; the last procedure, although exact, would rarely be useful in practice and is introduced herein chiefly as a check on the accuracy of the first two.

The detailed descriptions and applications of the three methods are contained in appendixes A, B, and C. In brief, the numerical procedure exploits a grid of characteristic lines as shown in figure 3 (a). For the case $c_1=c_2$ that is under consideration, this grid consists of two families of lines in the ξ,τ plane with slopes 1 and -1. A recurrence formula is developed in appendix A that gives the values of $\bar{\omega}$ and $\bar{V}$ at station 1 of a typical interior mesh (see fig. 3 (b)) in terms of the values of $\bar{\omega}$ and $\bar{V}$ at stations 2, 3, and 4. Repeated application of this formula, together with the use of special formulas for the half-meshes at either end of the beam and the knowledge of the magnitudes of jumps in $\bar{\omega}$ and $\bar{V}$ across characteristics where they occur, permits $\bar{\omega}$ and $\bar{V}$ to be calculated throughout the ξ,τ plane. Subsequent determination of $\bar{M}$ and $\bar{v}$ is achieved by means of simple addition formulas utilizing these calculated values of $\bar{\omega}$ and $\bar{V}$.

Although the solutions derived in appendix B have actually been obtained by Laplace transform techniques, they have

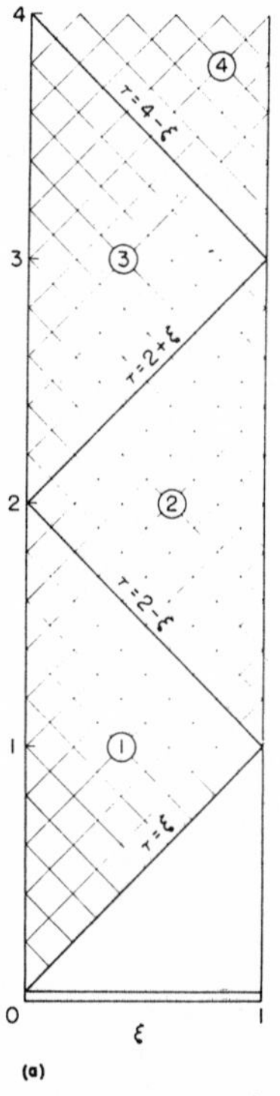

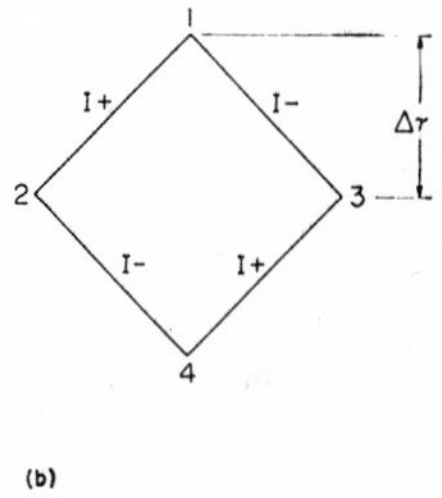

(a) Space-time plane. (b) Typical interior mesh.

FIGURE 3.—Characteristic grid for application of numerical procedure to uniform beams with disturbance applied at $\xi=0$. $c_1=c_2$.

been termed "modal solutions" because they are exactly those which would result from an application of the usual mode-superposition process. The exact closed solutions in appendix C have also been obtained through the use of Laplace transforms.

CANTILEVER BEAM GIVEN A STEP VELOCITY AT THE ROOT

The first example to be considered, the response of a uniform cantilever beam given a step velocity $\overline{v}=1$ at the root, is the equivalent of the most severe idealized landing problem, the instantaneous arrest of the root of a moving cantilever beam. Computed results for the shear and moment at the root of such a beam having a ratio of length to radius of gyration of 10 ($\lambda=5$) and properties such that $c_1=c_2$ are presented in figures 4 (a) and 4 (b), respectively. Two separate curves obtained by the numerical procedure are shown—one from a grid that divides the beam into 10 segments and the other from a 20-segment solution. The modal solution includes the contributions of the first eight modes. Results given by an exact closed solution are shown for both the shear and moment at the root up to the time $\tau=2$. These exact results are actually those for an infinitely long beam, since the influence of the free end is not felt at the root until $\tau=2$. After $\tau=2$, the influence of the free end is felt and, in this case, an exact solution is not feasible. To illustrate the time range covered in the plots, the point corresponding to half the period of the first mode of vibration of the beam is indicated on the time scale of each plot.

In figure 4 (a), the shear discontinuities evident in the numerical solutions occur each time the discontinuous wave front returns to the root after being reflected at the free end. The beam is seen to react violently to each of these boosts by the wave front, with more and more oscillations occurring after each succeeding jump. The frequency of these oscillations tends to increase with each succeeding boost until limited by the finite time interval. In these regions of violent oscillation the accuracy of the numerical solutions is obviously questionable; indeed the question arises as to whether these oscillations are really predicted by the theory or are merely the result of some instability in the numerical process. This question is resolved in the next section in which the simply supported beam is considered. At any rate, away from the regions of violent oscillations, the comparisons with the modal solution are favorable, and, for $\tau<2$, the fine-grid numerical solution almost coincides with the exact closed solution valid in this region.

The comparisons between the numerical and modal solutions are very good in figure 4 (b) where the time history of the moment at the root is plotted. Again, the fine-grid numerical solution nearly coincides with the exact closed solution for $\tau<2$.

SIMPLY SUPPORTED BEAM WITH AN APPLIED END MOMENT

Step moment.—A simply supported uniform beam with a ratio of length to radius of gyration of 10 ($\lambda=5$) and properties such that $c_1=c_2$ is subjected to a step moment $\overline{M}=1$ at the end $\xi=0$. Computed results for the shear at $\xi=0$ are presented in figure 5 (a) and the time history of moment at the center of the beam is presented in figure 5 (b). The point corresponding to the full period of the first mode is marked on the time scale of each plot. The numerical curves for both shear and moment were obtained by a 20-segment solution. The modal curves were obtained by adding dynamic corrections to the static solutions, the dynamic corrections being calculated with six modes for both the shear and the moment.

This example affords an answer to the question raised in the preceding section with regard to the stability of the numerical procedure following the passage of a discontinuous wave front. The fact is that the violent oscillations that occur after the discontinuity actually appear in the exact solution (fig. 5 (a)) and are hence inherently present in the theory.

In figure 5 (a) the scale of dimensionless vertical shear happens to be precisely the dynamic overshoot factor; it is of interest to note that values at least 15 times the static shear are predicted when the moment is applied suddenly.

For shear (fig. 5 (a)), the inaccuracies in the numerical solution just after the discontinuities are evident; however, the numerical results approximate the exact solution very well elsewhere. A similar observation may be made for the modal solution, which, as would be expected, ignores the discontinuities and violent oscillations caused by the wave front. The numerical and modal curves for the moment in figure 5 (b) follow the same pattern, agreeing well everywhere except in the regions immediately following the discontinuities.

Ramp-platform moment.—Computed results for the shear at the end $\xi=0$ and the moment at the center of the same simply supported uniform beam subjected to the applied ramp-platform moment

$$\overline{M}(0,\tau)=\tau \qquad (0\leqq\tau\leqq1)$$

$$\overline{M}(0,\tau)=1 \qquad (\tau>1)$$

are presented in figures 6 (a) and 6 (b). Again, the period of the first mode of vibration is marked on the time scales.

The numerical curves were recomputed with a 20-segment grid for the new forcing function; however, the results for the modal and exact solutions were obtained by means of a superposition of the preceding results. This superposition was carried out analytically for the complete modal solution and for the exact solution in the range $\tau<2$. In the range $\tau>2$, it was necessary to carry out the superposition for the exact solution numerically.

In figure 6 the time to peak value of the applied moment is seen to be approximately one-seventh the period of the first mode; predicted shear values approximately three and one-half times the static response occur.

With the removal of the discontinuity, there remain no high-frequency oscillations which the numerical solution might be unable to represent. In fact, all three solutions for the shear and both moment solutions are seen to be in excellent agreement.

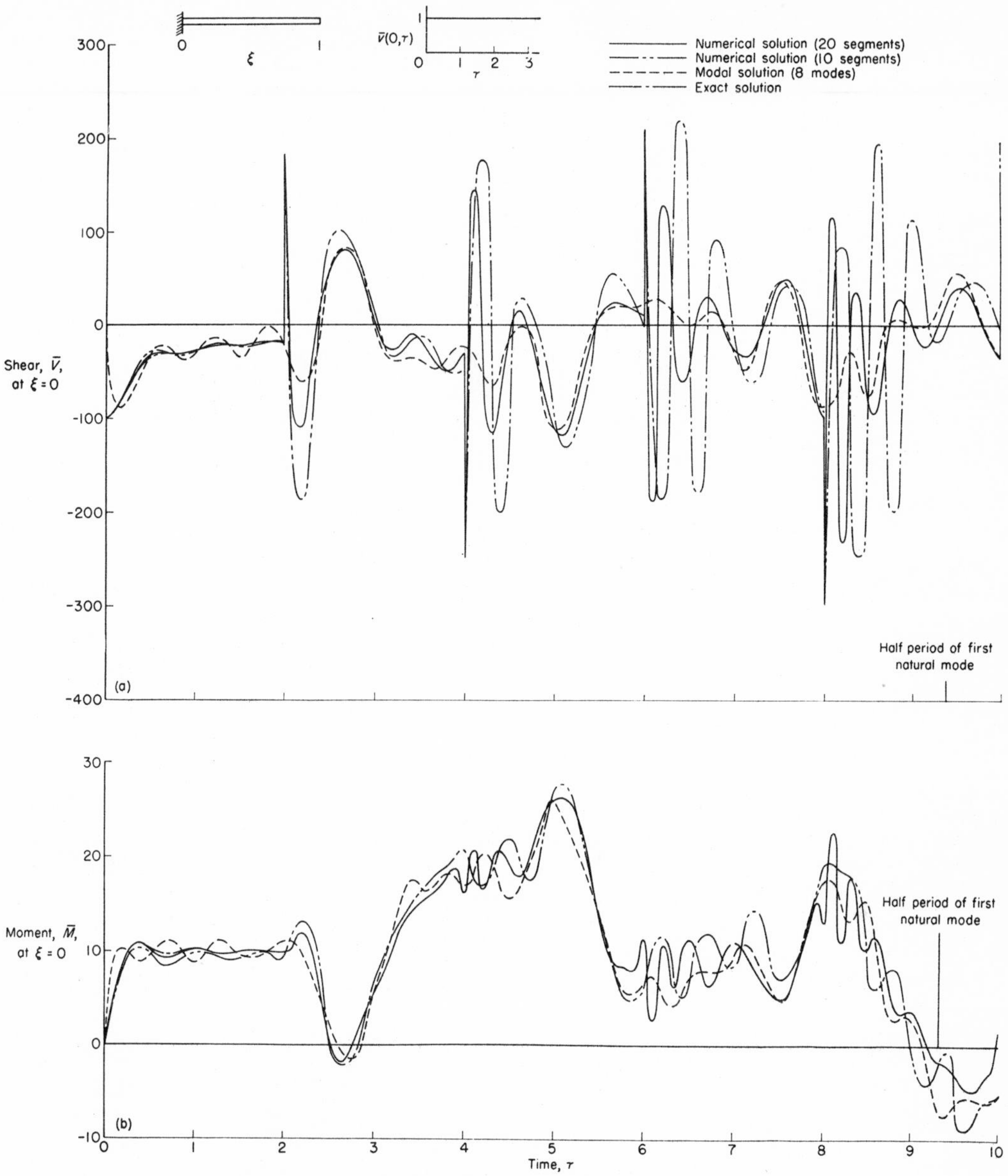

(a) Time history of the nondimensional vertical shear at the root.
(b) Time history of the nondimensional moment at the root.

FIGURE 4.—Response of a uniform cantilever beam subjected to a step velocity at the root.

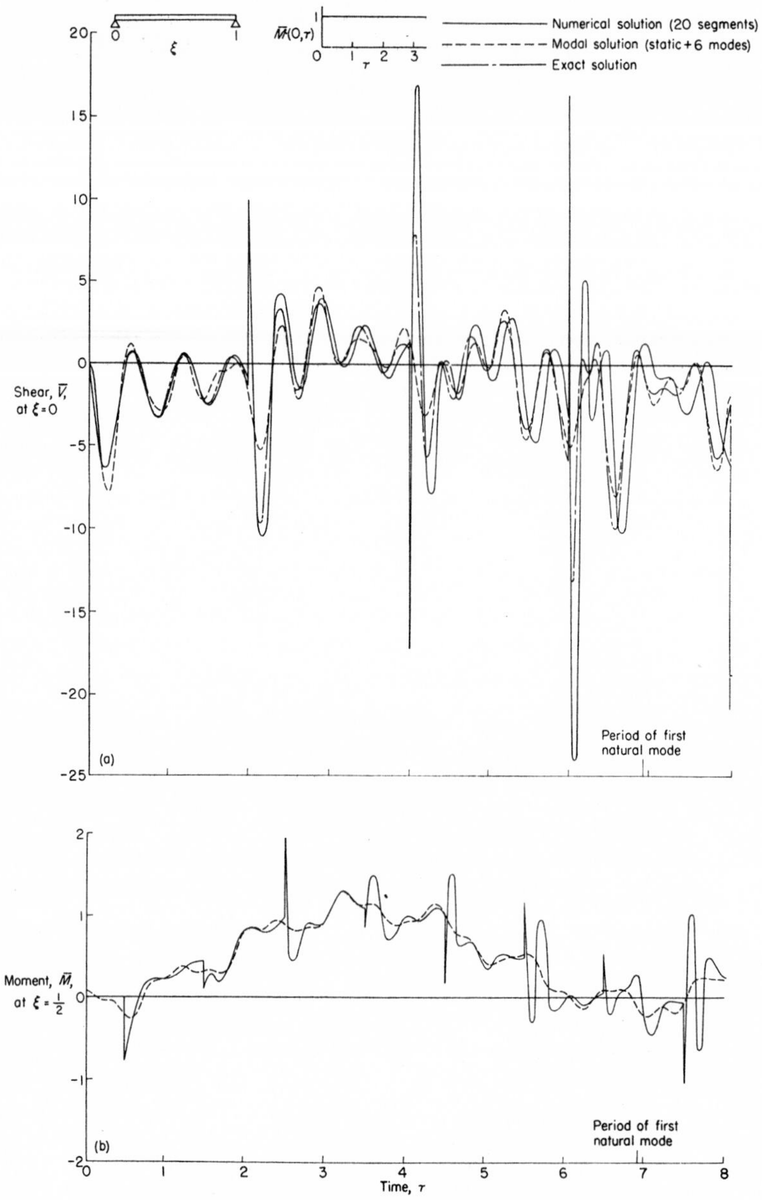

(a) Time history of the nondimensional vertical shear at the end $\xi=0$.
(b) Time history of the nondimensional moment at the center.

FIGURE 5.—Response of a uniform simply supported beam subjected to a step moment at $\xi=0$.

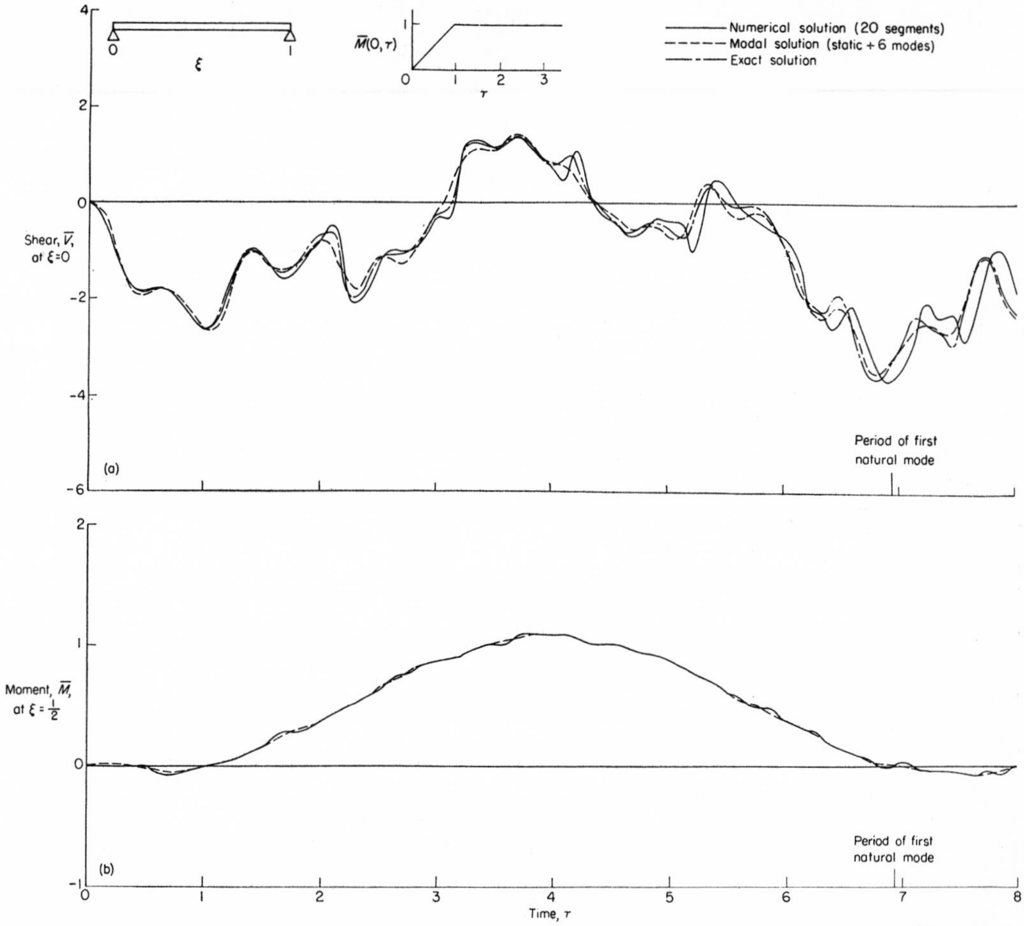

(a) Time history of the nondimensional vertical shear at the end $\xi=0$. (b) Time history of the nondimensional moment at the center.

FIGURE 6.—Response of a uniform simply supported beam subjected to a ramp-platform moment at $\xi=0$.

CONCLUDING REMARKS

Timoshenko's equations for the motion of vibrating nonuniform beams may be written in a characteristic form which appears to be well-suited to solution by numerical methods. In the examples presented in this report, all of which are for uniform beams with equal propagation velocities of bending and shear discontinuities ($c_1=c_2$), the solutions by the numerical and modal methods generally agree well with each other as well as with exact closed solutions where these have been obtained. However, the modal method, of course, fails entirely to reproduce discontinuities in shear or moment, and the numerical procedure, although it yields these discontinuities, represents the initial oscillations which follow with decreasing accuracy as more and more reflections of the wave front occur. In the more practical situations where discontinuities do not exist, these difficulties will, of course, not arise.

The results of the examples carried out by the numerical traveling-wave procedure encourage the viewpoint that such traveling-wave analyses may eventually be of practical usefulness. This kind of procedure is inherently simple and straightforward and has the advantage that the bulk of the labor involved is routine computation rather than mathematical analysis. It should be emphasized, however, that numerical solutions of Timoshenko's equations have been demonstrated only for uniform beams in which the propagation velocities c_1 and c_2 are equal; numerical procedures for the general case where they are unequal remain to be developed and tested.

LANGLEY AERONAUTICAL LABORATORY,
NATIONAL ADVISORY COMMITTEE FOR AERONAUTICS,
LANGLEY FIELD, VA., *October 28, 1952.*

APPENDIX A

NUMERICAL SOLUTIONS FOR UNIFORM BEAMS WITH $c_1=c_2$

MATRIX FORMULATION

Let the differential equations (17) be replaced by the finite-difference equations

$$\text{Along I+:}\quad \begin{cases} \Delta\overline{M}+\Delta\overline{\omega}-\overline{V}\Delta\tau=0 & \text{(A1a)} \\ \Delta\overline{V}-4\lambda^2\Delta\overline{v}+4\lambda^2\overline{\omega}\Delta\tau=0 & \text{(A1b)} \end{cases}$$

$$\text{Along I}-:\quad \begin{cases} \Delta\overline{M}-\Delta\overline{\omega}+\overline{V}\Delta\tau=0 & \text{(A1c)} \\ \Delta\overline{V}+4\lambda^2\Delta\overline{v}+4\lambda^2\overline{\omega}\Delta\tau=0 & \text{(A1d)} \end{cases}$$

and consider a closely spaced network of I+ and I− characteristics in the space-time plane as shown in figure 3 (a). Let the corners of a typical interior mesh of this grid be designated as shown in figure 3 (b). A step-by-step integration formula for $\overline{\omega}$ and $\overline{V}$ may now be derived in the following manner.

Equations (A1) may be written along the upper sides of the typical mesh to obtain

$$\overline{M}_1-\overline{M}_2+\overline{\omega}_1-\overline{\omega}_2-\frac{\Delta\tau}{2}(\overline{V}_1+\overline{V}_2)=0 \qquad \text{(A2)}$$

$$\overline{M}_1-\overline{M}_3-\overline{\omega}_1+\overline{\omega}_3+\frac{\Delta\tau}{2}(\overline{V}_1+\overline{V}_3)=0 \qquad \text{(A3)}$$

$$\overline{V}_1-\overline{V}_2-4\lambda^2(\overline{v}_1-\overline{v}_2)+2\lambda^2\Delta\tau(\overline{\omega}_1+\overline{\omega}_2)=0 \qquad \text{(A4)}$$

$$\overline{V}_1-\overline{V}_3+4\lambda^2(\overline{v}_1-\overline{v}_3)+2\lambda^2\Delta\tau(\overline{\omega}_1+\overline{\omega}_3)=0 \qquad \text{(A5)}$$

where $\overline{\omega}$ and $\overline{V}$ have been assumed to vary linearly in the small intervals between the corners. Obviously equations (A2) to (A5) may be solved simultaneously to obtain the four quantities $\overline{V}$, $\overline{\omega}$, $\overline{v}$, and $\overline{M}$ at point 1 in terms of their values at points 2 and 3. However, it is noted that, if $\overline{V}$, $\overline{\omega}$, $\overline{v}$, and $\overline{M}$ at points 2 and 3 were determined by a similar process for the preceding meshes, they already satisfy the equations

$$\overline{M}_2-\overline{M}_4-\overline{\omega}_2+\overline{\omega}_4+\frac{\Delta\tau}{2}(\overline{V}_2+\overline{V}_4)=0 \qquad \text{(A6)}$$

$$\overline{M}_3-\overline{M}_4+\overline{\omega}_3-\overline{\omega}_4-\frac{\Delta\tau}{2}(\overline{V}_3+\overline{V}_4)=0 \qquad \text{(A7)}$$

$$\overline{V}_2-\overline{V}_4+4\lambda^2(\overline{v}_2-\overline{v}_4)+2\lambda^2\Delta\tau(\overline{\omega}_2+\overline{\omega}_4)=0 \qquad \text{(A8)}$$

$$\overline{V}_3-\overline{V}_4-4\lambda^2(\overline{v}_3-\overline{v}_4)+2\lambda^2\Delta\tau(\overline{\omega}_3+\overline{\omega}_4)=0 \qquad \text{(A9)}$$

Equations (A6) and (A7) may be added to equations (A2) and (A3), respectively, to obtain

$$\overline{M}_1-\overline{M}_4+\overline{\omega}_1-2\overline{\omega}_2+\overline{\omega}_4-\frac{\Delta\tau}{2}(\overline{V}_1-\overline{V}_4)=0$$

and

$$\overline{M}_1-\overline{M}_4-\overline{\omega}_1+2\overline{\omega}_3-\overline{\omega}_4+\frac{\Delta\tau}{2}(\overline{V}_1-\overline{V}_4)=0$$

which may, in turn, be subtracted to obtain the single equation

$$\overline{\omega}_1-\overline{\omega}_2-\overline{\omega}_3+\overline{\omega}_4-\frac{\Delta\tau}{2}(\overline{V}_1-\overline{V}_4)=0 \qquad \text{(A10)}$$

Similarly, equations (A8) and (A9) may be subtracted from equations (A4) and (A5), respectively, to obtain

$$\overline{V}_1-2\overline{V}_2+\overline{V}_4-4\lambda^2(\overline{v}_1-\overline{v}_4)+2\lambda^2\Delta\tau(\overline{\omega}_1-\overline{\omega}_4)=0$$

and

$$\overline{V}_1-2\overline{V}_3+\overline{V}_4+4\lambda^2(\overline{v}_1-\overline{v}_4)+2\lambda^2\Delta\tau(\overline{\omega}_1-\overline{\omega}_4)=0$$

and these may be added to derive

$$\overline{V}_1-\overline{V}_2-\overline{V}_3+\overline{V}_4+2\lambda^2\Delta\tau(\overline{\omega}_1-\overline{\omega}_4)=0 \qquad \text{(A11)}$$

Equations (A10) and (A11) may now be solved simultaneously to obtain the step-by-step integration formula

$$\begin{vmatrix}\overline{\omega}_1\\ \overline{V}_1\end{vmatrix}=\frac{1}{(\lambda\,\Delta\tau)^2+1}\left\{\begin{bmatrix}A_{23}\end{bmatrix}\begin{vmatrix}\overline{\omega}_2+\overline{\omega}_3\\ \overline{V}_2+\overline{V}_3\end{vmatrix}+\begin{bmatrix}A_4\end{bmatrix}\begin{vmatrix}\overline{\omega}_4\\ \overline{V}_4\end{vmatrix}\right\} \qquad \text{(A12)}$$

where

$$[A_{23}]=\begin{bmatrix}1 & \frac{\Delta\tau}{2}\\ -2\lambda^2\Delta\tau & 1\end{bmatrix}$$

and

$$[A_4]=\begin{bmatrix}(\lambda\,\Delta\tau)^2-1 & -\Delta\tau\\ 4\lambda^2\Delta\tau & (\lambda\,\Delta\tau)^2-1\end{bmatrix}$$

Thus $\overline{\omega}$ and $\overline{V}$ may be determined at every interior mesh point in the space-time plane by the repeated application of formula (A12) to each mesh as it is encountered. The half-meshes which occur at the vertical left and right boundaries of the plane (fig. 3 (a)) require special formulas incorporating the known boundary data. These formulas may be derived from equations (A1) by a procedure similar to that used in obtaining equation (A12). Boundary formulas for some specific examples are presented in subsequent sections of this appendix.

Besides boundary data, initial data must be provided in order that a step-by-step solution may be begun. In all the examples considered, the beam is initially at rest and has a disturbance applied at the point $\xi=0$ beginning at time $\tau=0$. The region $\tau<\xi$ (without grid lines in fig. 3 (a)) therefore is one of zero stress and motion and $\overline{\omega}$ and $\overline{V}$ are given along the line $\tau=\xi$ by the known conditions at the wave front. It is with these values that each numerical solution is begun.

Discontinuities in $\overline{\omega}$ and $\overline{V}$ offer no special difficulties since they propagate so as to be always located on characteristics which, of course, can be made part of the basic grid. Thus they are simply added as they are encountered.

Once $\bar{\omega}$ and $\bar{V}$ have been determined at the grid intersections by the step-by-step process, $\bar{M}$ and $\bar{v}$ may be found by repeated application of equation (A2) or (A3) and equation (A4) or (A5) along the proper grid lines from boundaries where $\bar{M}$ and $\bar{v}$ are known. For example, if $\bar{M}$ is known along the left boundary, it may be found at a point g by

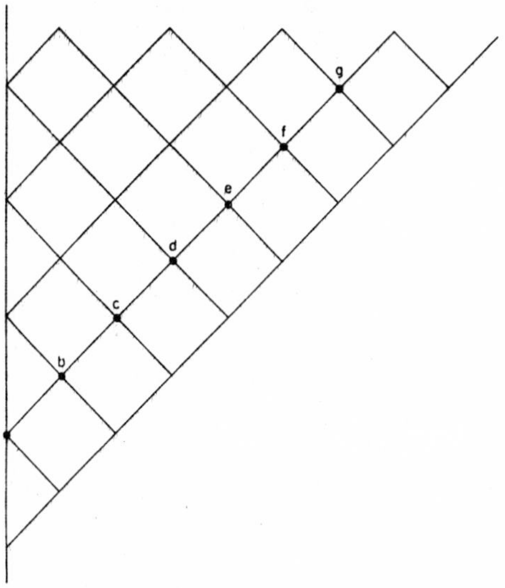

applying equation (A2) successively to the intervals ab, bc, cd, . . . fg. The resulting expression for $\bar{M}_g$ becomes

$$\bar{M}_g=\bar{M}_a+\bar{\omega}_a-\bar{\omega}_g+\Delta\tau\left(\frac{1}{2}\bar{V}_a+\bar{V}_b+\ldots+\bar{V}_f+\frac{1}{2}\bar{V}_g\right) \qquad \text{(A13)}$$

This procedure is seen to correspond to integration of the first of differential equations (17a) by means of the trapezoidal rule.

SPECIFIC PROBLEMS

Cantilever beam given a step velocity at the root.—If the root $\xi=0$ of a uniform cantilever beam is given a step velocity $\bar{v}=1$ at time $\tau=0$, the boundary conditions may be written

$$\left.\begin{aligned}\bar{v}(0,\tau)&=1\\ \bar{\omega}(0,\tau)&=0\\ \bar{M}(1,\tau)&=0\\ \bar{V}(1,\tau)&=0\end{aligned}\right\} \qquad \text{(A14)}$$

Typical left-boundary and right-boundary half-meshes are

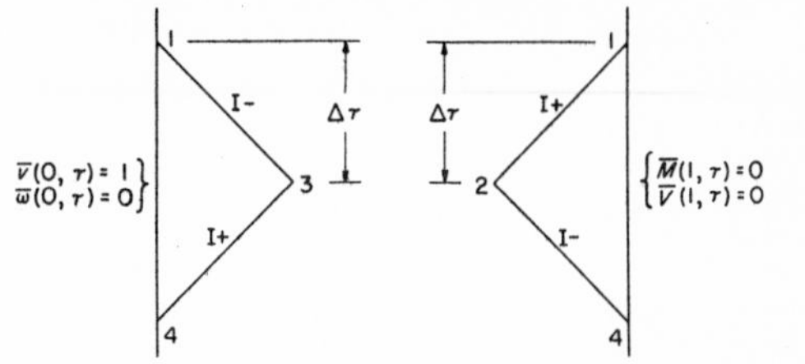

Application of equations (A1b) and (A1d) to the sides of the left-boundary half-mesh and proper combination of the resulting equations to eliminate $\bar{v}$ produces

$$\bar{V}_1=2\bar{V}_3-\bar{V}_4 \qquad \text{(A15)}$$

Similarly, application of equations (A1a) and (A1c) to a right-boundary half-mesh produces

$$\bar{\omega}_1=2\bar{\omega}_2-\bar{\omega}_4 \qquad \text{(A16)}$$

so that a complete set of integration formulas for determining $\bar{\omega}$ and $\bar{V}$ for this problem is now available. In addition, if the unused equations (eqs. (A1a) and (A1c) for the left-boundary half-mesh and eqs. (A1b) and (A1d) for the right) are applied and combined to eliminate $\bar{M}$ and $\bar{v}$ at points 3 and 2, the recurrence relations

$$\bar{M}_1\doteq\bar{M}_4-2\bar{\omega}_3-\frac{\Delta\tau}{2}(\bar{V}_1-\bar{V}_4) \qquad \text{(A17)}$$

for left-boundary half-meshes and

$$\bar{v}_1=\bar{v}_4-\frac{1}{2\lambda^2}\bar{V}_2+\frac{\Delta\tau}{2}(\bar{\omega}_1-\bar{\omega}_4) \qquad \text{(A18)}$$

for right-boundary half-meshes are obtained. These equations may be used to compute $\bar{M}$ at the root and $\bar{v}$ at the tip after $\bar{\omega}$ and $\bar{V}$ have been determined everywhere.

Since the applied disturbance is initially discontinuous, $\bar{\omega}$ and $\bar{V}$ will be discontinuous along the line $\tau=\xi$ (see fig. 3 (a)). In order to begin the step-by-step solution for $\bar{\omega}$ and $\bar{V}$, these discontinuities must be determined in advance. Furthermore, since the wave front is reflected back into the beam at either end, $\bar{\omega}$ and $\bar{V}$ will also be discontinuous along the lines $\tau=2-\xi$, $\tau=2+\xi$, . . ., and $\delta\bar{\omega}$ and $\delta\bar{V}$ must be determined along each of these lines before the step-by-step solution can be extended beyond it. The discontinuities at the wave front are determined as follows.

From the boundary conditions (A14), it is seen that $\delta\bar{\omega}(0,0)=0$ and $\delta\bar{v}(0,0)=1$. Thus, from equation (18a),

$\delta\overline{V}(0,0)=-4\lambda^2$, and equations (20) become, for the line $\tau=\xi$,

$$\left.\begin{aligned}\delta\overline{\omega}(\xi,\xi)&=-2\lambda\sin\lambda\xi\\ \delta\overline{V}(\xi,\xi)&=-4\lambda^2\cos\lambda\xi\end{aligned}\right\}\qquad\text{(A19)}$$

At $\xi=1$, the discontinuities across $\tau=\xi$ are $\delta\overline{\omega}(1,1)=\delta\overline{M}(1,1)=-2\lambda\sin\lambda$ and $\delta\overline{V}(1,1)=-4\lambda^2\cos\lambda$, so that, if the boundary conditions $\overline{M}(1,\tau)=\overline{V}(1,\tau)=0$ are to be satisfied, jumps must occur across $\tau=2-\xi$ at the point (1,1) with the magnitudes $\delta\overline{\omega}(1,1)=-\delta\overline{M}(1,1)=-2\lambda\sin\lambda$ and $\delta\overline{V}(1,1)=4\lambda^2\cos\lambda$. In view of these initial conditions at $\tau=1$, equations (20) become, for the line $\tau=2-\xi$,

$$\left.\begin{aligned}\delta\overline{\omega}(\xi,2-\xi)&=-2\lambda\sin\lambda\xi\\ \delta\overline{V}(\xi,2-\xi)&=4\lambda^2\cos\lambda\xi\end{aligned}\right\}\qquad\text{(A20)}$$

Initial jump values $\delta\overline{\omega}(0,2)=0$ and $\delta\overline{V}(0,2)=4\lambda^2$ for the line $\tau=2+\xi$ may be found from equation (A20) by satisfying the boundary conditions $\overline{\omega}(0,\tau)=0$ and $\overline{v}(0,\tau)=1$. From these initial values, the discontinuities across $\tau=2+\xi$ are found to be the negative of those across $\tau=\xi$, or

$$\left.\begin{aligned}\delta\overline{\omega}(\xi,2+\xi)&=2\lambda\sin\lambda\xi\\ \delta\overline{V}(\xi,2+\xi)&=4\lambda^2\cos\lambda\xi\end{aligned}\right\}\qquad\text{(A21)}$$

Then, it must be true that

$$\left.\begin{aligned}\delta\overline{\omega}(\xi,4-\xi)&=2\lambda\sin\lambda\xi\\ \delta\overline{V}(\xi,4-\xi)&=-4\lambda^2\cos\lambda\xi\end{aligned}\right\}\qquad\text{(A22)}$$

and so forth, with the values on each succeeding line $\tau=n\pm\xi$ repeating the values on the line $\tau=(n-4)\pm\xi$. The variations in the magnitudes of $\delta\overline{\omega}$ and $\delta\overline{V}$ as the wave front propagates back and forth through the beam are thus clearly defined, and, since $\tau<\xi$ is a region of zero stress and motion, equations (A19) define the values of $\overline{\omega}$ and $\overline{V}$ along the line $\tau=\xi$.

With $\overline{\omega}$ and $\overline{V}$ known along $\tau=\xi$, the solution for $\overline{\omega}$ and $\overline{V}$ may be begun by applying formula (A15) to the half-mesh in the lower corner of triangular region ① (fig. 3 (a)). It is continued step by step throughout the triangle, with formula (A12) being used for all interior meshes. When the first triangle is complete, the known jumps along the line $\tau=2-\xi$ may be added to the values computed for the under side of this line, and the solution may then be carried out in triangular region ② beginning again at the lower corner this time with formula (A16). In this way the solution may be carried through as many triangular regions as desired.

Two sets of computations, with time intervals $\Delta\tau=0.1$ and $\Delta\tau=0.05$, have been made for a cantilever beam for which $\lambda=5$. These time intervals correspond to grids dividing the beam into 10 and 20 segments, respectively. (The computations for the 20-segment solution were made on the Bell Telephone Laboratories X–66744 relay computer at the Langley Laboratory.) The computed time histories of shear and moment at the root have been plotted in figures 4 (a) and 4 (b), respectively, up to $\tau=10$. In order to obtain these time histories, the computations for $\overline{\omega}$ and $\overline{V}$ had to be carried through the first nine triangular regions.

Simply supported beam with an applied end moment.—If the end $\xi=0$ of a uniform simply supported beam for which $c_1=c_2$ is subjected to an applied bending moment $\overline{M}_0(\tau)$ beginning at time $\tau=0$, the boundary conditions are

$$\left.\begin{aligned}\overline{M}(0,\tau)&=\overline{M}_0(\tau)\\ \overline{v}(0,\tau)&=0\\ \overline{M}(1,\tau)&=0\\ \overline{v}(1,\tau)&=0\end{aligned}\right\}\qquad\text{(A23)}$$

Typical left-boundary and right-boundary half-meshes for this beam are

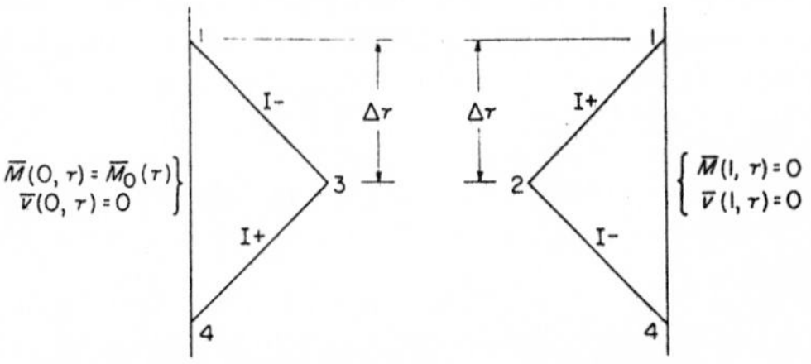

Equations (A1c), (A1a), (A1d), and (A1b) may be written along the sides of the left-boundary half-mesh to give, respectively,

$$(\overline{M}_0)_1-\overline{M}_3-\overline{\omega}_1+\overline{\omega}_3+\frac{\Delta\tau}{2}(\overline{V}_1+\overline{V}_3)=0$$

$$\overline{M}_3-(\overline{M}_0)_4+\overline{\omega}_3-\overline{\omega}_4-\frac{\Delta\tau}{2}(\overline{V}_3+\overline{V}_4)=0$$

$$\overline{V}_1-\overline{V}_3-4\lambda^2\overline{v}_3+2\lambda^2\Delta\tau(\overline{\omega}_1+\overline{\omega}_3)=0$$

$$\overline{V}_3-\overline{V}_4-4\lambda^2\overline{v}_3+2\lambda^2\Delta\tau(\overline{\omega}_3+\overline{\omega}_4)=0$$

Adding the first two equations and subtracting the last two equations produces

$$(\overline{M}_0)_1-(\overline{M}_0)_4-\overline{\omega}_1+2\overline{\omega}_3-\overline{\omega}_4+\frac{\Delta\tau}{2}(\overline{V}_1-\overline{V}_4)=0$$

and

$$\overline{V}_1-2\overline{V}_3+\overline{V}_4+2\lambda^2\Delta\tau(\overline{\omega}_1-\overline{\omega}_4)=0$$

which may be solved simultaneously for $\overline{\omega}_1$ and $\overline{V}_1$ to obtain

$$\begin{vmatrix}\overline{\omega}_1\\ \overline{V}_1\end{vmatrix}=\frac{1}{(\lambda\,\Delta\tau)^2+1}\left\{\begin{bmatrix}\Lambda_{23}\end{bmatrix}\begin{vmatrix}2\overline{\omega}_3\\ 2\overline{V}_3\end{vmatrix}+\begin{bmatrix}\Lambda_4\end{bmatrix}\begin{vmatrix}\overline{\omega}_4\\ \overline{V}_4\end{vmatrix}+((\overline{M}_0)_1-(\overline{M}_0)_4)\begin{vmatrix}1\\ -2\lambda^2\Delta\tau\end{vmatrix}\right\} \tag{A24}$$

where $[\Lambda_{23}]$ and $[\Lambda_4]$ have been defined in conjunction with equation (A12).

A similar process produces, for the right-boundary half-mesh, the formula

$$\begin{vmatrix}\overline{\omega}_1\\ \overline{V}_1\end{vmatrix}=\frac{1}{(\lambda\,\Delta\tau)^2+1}\left\{\begin{bmatrix}\Lambda_{23}\end{bmatrix}\begin{vmatrix}2\overline{\omega}_2\\ 2\overline{V}_2\end{vmatrix}+\begin{bmatrix}\Lambda_4\end{bmatrix}\begin{vmatrix}\overline{\omega}_4\\ \overline{V}_4\end{vmatrix}\right\} \tag{A25}$$

Now consider the case in which the applied moment is a step moment; that is, $\overline{M}_0(\tau)=1$. Equation (A24) reduces to

$$\begin{vmatrix}\overline{\omega}_1\\ \overline{V}_1\end{vmatrix}=\frac{1}{(\lambda\,\Delta\tau)^2+1}\left\{\begin{bmatrix}\Lambda_{23}\end{bmatrix}\begin{vmatrix}2\overline{\omega}_3\\ 2\overline{V}_3\end{vmatrix}+\begin{bmatrix}\Lambda_4\end{bmatrix}\begin{vmatrix}\overline{\omega}_4\\ \overline{V}_4\end{vmatrix}\right\} \tag{A26}$$

for left-boundary half-meshes. A similarity between the boundary formulas (A25) and (A26) and the interior formula (A12) is apparent; indeed, formula (A12) may be used everywhere if the sums of the values at points 2 and 3 are simply replaced by twice the values at the inboard point of each boundary mesh.

For this problem, as in the preceding one, the magnitudes of the discontinuities $\delta\overline{\omega}$ and $\delta\overline{V}$ at the wave front must be determined in advance for every point on the grid lines defining its position. From the boundary conditions and equations (18a) it is apparent that $\delta\overline{\omega}(0,0)=\delta\overline{M}(0,0)=1$ and $\delta\overline{V}(0,0)=0$. Thus, for the line $\tau=\xi$, equations (20) become

$$\left.\begin{aligned}\delta\overline{\omega}&=\cos\lambda\tau\\ \delta\overline{V}&=-2\lambda\sin\lambda\tau\end{aligned}\right\} \tag{A27}$$

Furthermore, it is found that, when the wave front is reflected at either of the simply supported ends, the signs as well as the magnitudes of the discontinuities remain unchanged, in contrast to the behavior of a discontinuous wave front reflected at a free or fixed end as in the preceding case. Thus equation (A27) must be valid for the entire zigzag path defining the position of the wave front.

With the discontinuities known, the step-by-step solution can be begun as before. Again, it is convenient to complete each traingular region before proceeding to the next.

Computations for this problem were made on the Bell computer, with a 20-segment grid, for a beam with $\lambda=5$. The quantities $\overline{\omega}$ and $\overline{V}$ were found in the first eight triangular regions; in addition, the moment at the center was computed. The time histories of shear at the end $\xi=0$ and of moment at the center $\xi=\frac{1}{2}$ have been plotted in figures 5(a) and 5(b), respectively.

Now consider the case in which a ramp-platform moment is applied to the beam; that is, $\overline{M}_0(\tau)$ is defined by

$$\left.\begin{aligned}\overline{M}_0(\tau)&=\tau &&(0\leqq\tau\leqq 1)\\ \overline{M}_0(\tau)&=1 &&(\tau>1)\end{aligned}\right\} \tag{A28}$$

Equation (A24) reduces to

$$\begin{vmatrix}\overline{\omega}_1\\ \overline{V}_1\end{vmatrix}=\frac{1}{(\lambda\,\Delta\tau)^2+1}\left\{\begin{bmatrix}\Lambda_{23}\end{bmatrix}\begin{vmatrix}2\overline{\omega}_3\\ 2\overline{V}_3\end{vmatrix}+\begin{bmatrix}\Lambda_4\end{bmatrix}\begin{vmatrix}\overline{\omega}_4\\ \overline{V}_4\end{vmatrix}+2\Delta\tau\begin{vmatrix}1\\ -2\lambda^2\Delta\tau\end{vmatrix}\right\} \tag{A29}$$

for the left-boundary half-meshes for which $\tau_1\leqq 1$ and reduces to equation (A26) for all the rest. Equations (A12) and (A25) are, of course, still valid for interior meshes and right-boundary half-meshes, respectively.

Since $\delta\overline{\omega}(0,0)=\delta\overline{V}(0,0)=0$ for this case, the solution begins with initial values $\overline{\omega}=\overline{V}=0$ along the line $\tau=\xi$ and there are no discontinuities to be added. Thus the solution may progress in any convenient manner without regard for the position of the wave front.

The same quantities were computed for this problem that were obtained in the preceding case. As before, it was assumed that $\lambda=5$, and the calculations were performed on the Bell computer with a 20-segment grid. The resulting time histories of shear at $\xi=0$ and moment at $\xi=\frac{1}{2}$ are presented in figures 6(a) and 6(b), respectively.

APPENDIX B

MODAL SOLUTIONS FOR UNIFORM BEAMS

The solutions carried out in this and the following appendix make considerable use of Laplace transform techniques; all the Laplace transforms used, most of which were taken directly from references 15 and 16, are given in table I for the sake of easy reference.

TABLE I.—LAPLACE TRANSFORMS

Number	$F(\tau)$ (a)		$f(p)=\int_0^\infty e^{-p\tau}F(\tau)d\tau$	Reference
1	$\int_0^\tau F_1(\tau-\theta)F_2(\theta)d\theta$	$(\tau>0)$	$f_1(p)f_2(p)$	15
2	$F(\tau-\xi)$	$(\tau>\xi)$	$e^{-\xi p}f(p)$	15
3	$e^{i\lambda\tau}F(\tau)$	$(\tau>0)$	$f(p-i\lambda)$	15
4	$\int_0^\tau F(\theta)d\theta$	$(\tau>0)$	$\frac{1}{p}f(p)$	15
5	$\tau F(\tau)$	$(\tau>0)$	$-f'(p)$	15
6	$F^n(\tau)=\frac{d^n}{d\tau^n}[F(\tau)]$	$(\tau>0)$	$p^nf(p)-p^{n-1}F(+0)-p^{n-2}F'(+0)-\ldots-F^{n-1}(+0)$	15
7	1	$(\tau>0)$	$\frac{1}{p}$	15
8	τ	$(\tau>0)$	$\frac{1}{p^2}$	15
9	$J_0(\lambda\tau)$	$(\tau>0)$	$\frac{1}{\sqrt{p^2+\lambda^2}}$	15
10	$J_1(\lambda\tau)$	$(\tau>0)$	$\frac{1}{\lambda}\left(1-\frac{p}{\sqrt{p^2+\lambda^2}}\right)$	15
11	$e^{i\lambda\tau}[J_0(\lambda\tau)-iJ_1(\lambda\tau)]$	$(\tau>0)$	$\frac{i}{\lambda}\left(\sqrt{\frac{p-i2\lambda}{p}}-1\right)$	--
12	$e^{-i\lambda\tau}[J_0(\lambda\tau)+iJ_1(\lambda\tau)]$	$(\tau>0)$	$\frac{i}{\lambda}\left(1-\sqrt{\frac{p+i2\lambda}{p}}\right)$	--
13	$\cos\lambda\tau\, J_0(\lambda\tau)+\sin\lambda\tau\, J_1(\lambda\tau)$	$(\tau>0)$	$\frac{i}{2\lambda}\left(\sqrt{\frac{p-i2\lambda}{p}}-\sqrt{\frac{p+i2\lambda}{p}}\right)$	--
14	$\frac{2\lambda}{\tau}J_1(2\lambda\tau)$	$(\tau>0)$	$\sqrt{p^2+4\lambda^2}-p$	15
15	$J_0(\lambda\sqrt{\tau^2-\xi^2})$	$(\tau>\xi)$	$\frac{e^{-\xi\sqrt{p^2+\lambda^2}}}{\sqrt{p^2+\lambda^2}}$	15
16	$\frac{\lambda\xi J_1(\lambda\sqrt{\tau^2-\xi^2})}{\sqrt{\tau^2-\xi^2}}$	$(\tau>\xi)$	$e^{-\xi p}-e^{-\xi\sqrt{p^2+\lambda^2}}$	15
17	$\frac{\tau J_1(\lambda\sqrt{\tau^2-\xi^2})}{\sqrt{\tau^2-\xi^2}}+iJ_0(\lambda\sqrt{\tau^2-\xi^2})$	$(\tau>\xi)$	$\frac{1}{\lambda}\left(e^{-\xi p}-\sqrt{\frac{p-i\lambda}{p+i\lambda}}e^{-\xi\sqrt{p^2+\lambda^2}}\right)$	16

[a] $F(\tau)=0$ for all values of τ not specified.

EQUATIONS

The equations of motion (2) may be written for a uniform beam vibrating in the absence of external distributed loading in nondimensional form as

$$\left.\begin{aligned}(\bar{y}_\xi-\psi)_\xi-\left(\frac{c_1}{c_2}\right)^2\bar{y}_{\tau\tau}=0\\ \psi_{\xi\xi}+R(\bar{y}_\xi-\psi)-\psi_{\tau\tau}=0\end{aligned}\right\} \quad \text{(B1)}$$

where $\bar{y}=\frac{y}{L}$. The dimensionless bending moment $\bar{M}$ and vertical shear $\bar{V}$ are given by (see eqs. (1))

$$\bar{M}=-\psi_\xi \quad \text{(B2)}$$

$$\bar{V}=R(\bar{y}_\xi-\psi) \quad \text{(B3)}$$

If the beam is initially at rest ($\bar{y}(\xi,0)=\bar{y}_\tau(\xi,0)=\psi(\xi,0)=\psi_\tau(\xi,0)=0$), equations (B1) may be transformed to (where transform 6 in table I has been used)

$$\left.\begin{aligned}(Y_\xi-\Psi)_\xi-\left(\frac{c_1}{c_2}\right)^2p^2Y=0\\ \Psi_{\xi\xi}+R(Y_\xi-\Psi)-p^2\Psi=0\end{aligned}\right\} \quad \text{(B4)}$$

in terms of the Laplace transforms $Y(\xi,p)=\int_0^\infty e^{-p\tau}\bar{y}(\xi,\tau)d\tau$ and $\Psi(\xi,p)=\int_0^\infty e^{-p\tau}\psi(\xi,\tau)d\tau$. The solutions of equations (B4) have the form

$$\left.\begin{aligned}Y(\xi,p)=A(p)e^{m\xi}\\ \Psi(\xi,p)=B(p)e^{m\xi}\end{aligned}\right\} \quad \text{(B5)}$$

Substituting these relations into equations (B4) leads to a biquadratic equation in m

$$m^4-p^2\left[1+\left(\frac{c_1}{c_2}\right)^2\right]m^2+\left(\frac{c_1}{c_2}\right)^2p^2(p^2+R)=0 \quad \text{(B6)}$$

and gives the following relationship between B and A:

$$B=\frac{m^2-\left(\frac{c_1}{c_2}\right)^2p^2}{m}A \quad \text{(B7)}$$

Let the solutions of equation (B6) be written $m=\pm im_1$ and $\pm m_2$. Then

$$m_1=p\sqrt{\frac{-\left[1+\left(\frac{c_1}{c_2}\right)^2\right]+\sqrt{\left[1-\left(\frac{c_1}{c_2}\right)^2\right]^2-\frac{4R}{p^2}\left(\frac{c_1}{c_2}\right)^2}}{2}}$$

$$m_2=p\sqrt{\frac{1+\left(\frac{c_1}{c_2}\right)^2+\sqrt{\left[1-\left(\frac{c_1}{c_2}\right)^2\right]^2-\frac{4R}{p^2}\left(\frac{c_1}{c_2}\right)^2}}{2}}$$

and the general solution of equations (B4) is

$$Y(\xi,p)=C_1\cos m_1\xi+C_2\sin m_1\xi+C_3\cosh m_2\xi+C_4\sinh m_2\xi \quad \text{(B8)}$$

$$\begin{aligned}\Psi(\xi,p)=&\frac{1}{m_1}\left[m_1^2+\left(\frac{c_1}{c_2}\right)^2p^2\right](C_2\cos m_1\xi-C_1\sin m_1\xi)+\\ &\frac{1}{m_2}\left[m_2^2-\left(\frac{c_1}{c_2}\right)^2p^2\right](C_4\cosh m_2\xi+C_3\sinh m_2\xi)\end{aligned} \quad \text{(B9)}$$

where C_1, C_2, C_3, and C_4 are functions of p which must be chosen so that the boundary conditions are satisfied.

SPECIFIC PROBLEMS

Cantilever beam given a step velocity at the root.—If the root $\xi=0$ of a cantilever beam is given a step velocity $\bar{v}=\frac{c_1}{c_2}\bar{y}_\tau=1$ at time $\tau=0$, the boundary conditions may be written

$$\left.\begin{aligned}\bar{y}(0,\tau)&=\frac{c_2}{c_1}\tau\\ \psi(0,\tau)&=0\\ \psi_\xi(1,\tau)&=0\\ \psi(1,\tau)&=\bar{y}_\xi(1,\tau)\end{aligned}\right\} \quad \text{(B10)}$$

These boundary conditions transform to (transform 8, table I)

$$\left.\begin{aligned}Y(0,p)&=\frac{c_2}{c_1}\frac{1}{p^2}\\ \Psi(0,p)&=0\\ \Psi_\xi(1,p)&=0\\ \Psi(1,p)&=Y_\xi(1,p)\end{aligned}\right\} \quad \text{(B11)}$$

The constants C_1, C_2, C_3, and C_4 are determined by substituting equations (B8) and (B9) into equations (B11); the resulting expressions for $Y(\xi,p)$ and $\Psi(\xi,p)$ may be written in the forms

$$Y(\xi,p)=\frac{c_2}{c_1}\frac{U_Y(\xi,p)}{p^2D(p)} \quad \text{(B12)}$$

and

$$\Psi(\xi,p)=\frac{c_2}{c_1}\frac{U_\Psi(\xi,p)}{p^2D(p)} \quad \text{(B13)}$$

where

$$\begin{aligned}U_Y(\xi,p)=&\left(1-\frac{m_1}{m_2}\sin m_1\sinh m_2+\frac{1}{Q}\cos m_1\cosh m_2\right)\cos m_1\xi+\\ &\left(1+\frac{m_2}{m_1}\sin m_1\sinh m_2+Q\cos m_1\cosh m_2\right)\cosh m_2\xi+\\ &\left(\frac{1}{Q}\sin m_1\cosh m_2+\frac{m_1}{m_2}\cos m_1\sinh m_2\right)\sin m_1\xi-\\ &\left(\frac{m_2}{m_1}\sin m_1\cosh m_2+Q\cos m_1\sinh m_2\right)\sinh m_2\xi\end{aligned}$$

$$U_{\Psi}(\xi,p)=\frac{m_2^2-\left(\frac{c_1}{c_2}\right)^2p^2}{m_2}\left[\left(\frac{m_2}{m_1}\sin m_1\cosh m_2+\right.\right.$$

$$\left.Q\cos m_1\sinh m_2\right)(\cos m_1\xi-\cosh m_2\xi)-$$

$$Q\frac{m_2}{m_1}\left(1-\frac{m_1}{m_2}\sin m_1\sinh m_2+\frac{1}{Q}\cos m_1\cosh m_2\right)\sin m_1\xi+$$

$$\left.\left(1+\frac{m_2}{m_1}\sin m_1\sinh m_2+Q\cos m_1\cosh m_2\right)\sinh m_2\xi\right]$$

$$D(p)=2+\left(Q+\frac{1}{Q}\right)\cos m_1\cosh m_2+$$

$$\left(\frac{m_2}{m_1}-\frac{m_1}{m_2}\right)\sin m_1\sinh m_2$$

$$Q=\frac{m_1^2+\left(\frac{c_1}{c_2}\right)^2p^2}{m_2^2-\left(\frac{c_1}{c_2}\right)^2p^2}$$

The inverse transforms $\bar{y}(\xi,\tau)$ and $\psi(\xi,\tau)$ are determined by substitution of $Y(\xi,p)$ and $\Psi(\xi,p)$ into the complex inversion integral (ref. 15). The singularities of the resulting integrands $e^{p\tau}Y(\xi,p)$ and $e^{p\tau}\Psi(\xi,p)$ must therefore be examined. Although the functions m_1 and m_2 are, in themselves, multiple-valued functions of p with branch points at $p=0$, $\pm i\sqrt{R}$, and $\pm\frac{2\sqrt{R}\frac{c_1}{c_2}}{1-\left(\frac{c_1}{c_2}\right)^2}$, it follows from a consideration of the fundamental theorem of the uniqueness of solutions of ordinary linear differential equations that $Y(\xi,p)$ and $\Psi(\xi,p)$ must nevertheless be single valued. The integrands $e^{p\tau}Y(\xi,p)$ and $e^{p\tau}\Psi(\xi,p)$ thus have no singularities other than poles. The inverse transforms $\bar{y}(\xi,\tau)$ and $\psi(\xi,\tau)$ will therefore be taken as the sum of the residues at the poles of $e^{p\tau}Y(\xi,p)$ and $e^{p\tau}\Psi(\xi,p)$, respectively (see ref. 15).

It can be shown that no singularities occur in the numerator functions $U_Y(\xi,p)$ and $U_{\Psi}(\xi,p)$; therefore, all the poles of $Y(\xi,p)$ and $\Psi(\xi,p)$ must be introduced by the zeros of the denominator $p^2D(p)$. Consider first the equation

$$D(p)=2+\left(Q+\frac{1}{Q}\right)\cos m_1\cosh m_2+\left(\frac{m_2}{m_1}-\frac{m_1}{m_2}\right)\sin m_1\sinh m_2=0 \tag{B14}$$

This equation has an infinite number of roots $p=\pm p_n$, $n=1, 2, \ldots \infty$, all of which are, in general, simple poles of both $Y(\xi,p)$ and $\Psi(\xi,p)$. In addition, $p=0$ can be shown to be a double pole of $Y(\xi,p)$ but not a pole of $\Psi(\xi,p)$.

The sum of the residues of $e^{p\tau}Y(\xi,p)$ at the poles $p=0$ and $p=\pm p_n$, $n=1, 2, 3, \ldots \infty$, is

$$\bar{y}(\xi,\tau)=\frac{c_2}{c_1}\left\{\tau+\sum_{n=1}^{\infty}\frac{1}{p_n^2}\left[\frac{U_Y(\xi,p_n)}{D'(p_n)}e^{p_n\tau}+\frac{U_Y(\xi,-p_n)}{D'(-p_n)}e^{-p_n\tau}\right]\right\}$$

and, similarly, the sum of the residues of $e^{p\tau}\Psi(\xi,p)$ at $p=\pm p_n$, $n=1, 2, 3, \ldots \infty$, may be written

$$\psi(\xi,\tau)=\frac{c_2}{c_1}\sum_{n=1}^{\infty}\frac{1}{p_n^2}\left[\frac{U_{\Psi}(\xi,p_n)}{D'(p_n)}e^{p_n\tau}+\frac{U_{\Psi}(\xi,-p_n)}{D'(-p_n)}e^{-p_n\tau}\right]$$

where

$$D'(p)=\frac{d}{dp}[D(p)]$$

$$=\frac{Rp\left(\frac{c_1}{c_2}\right)^2}{m_1m_2}\left(\frac{1}{m_2^2}-\frac{1}{m_1^2}\right)\sin m_1\sinh m_2-\frac{2}{p}\left[\frac{(m_1^2+m_2^2)^2}{Rp^2\left(\frac{c_1}{c_2}\right)^2}+\right.$$

$$\left.4\right]\cos m_1\cosh m_2+\frac{m_2}{p}\left\{\left(Q+\frac{1}{Q}\right)\left[1+\frac{Rp^2\left(\frac{c_1}{c_2}\right)^2}{m_2^2(m_1^2+m_2^2)}\right]+\right.$$

$$\left.\frac{m_2^2-m_1^2}{m_2^2}\left[1+\frac{Rp^2\left(\frac{c_1}{c_2}\right)^2}{m_1^2(m_1^2+m_2^2)}\right]\right\}\cos m_1\sinh m_2-$$

$$\frac{m_1}{p}\left\{\left(Q+\frac{1}{Q}\right)\left[1+\frac{Rp^2\left(\frac{c_1}{c_2}\right)^2}{m_1^2(m_1^2+m_2^2)}\right]-\frac{m_2^2-m_1^2}{m_1^2}\left[1+\right.\right.$$

$$\left.\left.\frac{Rp^2\left(\frac{c_1}{c_2}\right)^2}{m_2^2(m_1^2+m_2^2)}\right]\right\}\sin m_1\cosh m_2$$

However, it is seen that $U_Y(\xi,p)=U_Y(\xi,-p)$, $U_{\Psi}(\xi,p)=U_{\Psi}(\xi,-p)$, and $D'(p)=-D'(-p)$ so that these equations reduce to

$$\bar{y}(\xi,\tau)=\frac{c_2}{c_1}\left[\tau+2\sum_{n=1}^{\infty}\frac{U_Y(\xi,p_n)}{p_n^2D'(p_n)}\sinh p_n\tau\right] \tag{B15}$$

and

$$\psi(\xi,\tau)=2\frac{c_2}{c_1}\sum_{n=1}^{\infty}\frac{U_{\Psi}(\xi,p_n)}{p_n^2D'(p_n)}\sinh p_n\tau \tag{B16}$$

If, in equation (B14), p is replaced by ik, the result may be written

$$2+\left[\frac{(\alpha^2+\beta^2)^2}{Rk^2\left(\frac{c_1}{c_2}\right)^2}-2\right]\cosh\alpha\cos\beta-\left(\frac{\beta}{\alpha}-\frac{\alpha}{\beta}\right)\sinh\alpha\sin\beta=0 \tag{B17}$$

where

$$\alpha=k\sqrt{\frac{-\left[1+\left(\frac{c_1}{c_2}\right)^2\right]+\sqrt{\left[1-\left(\frac{c_1}{c_2}\right)^2\right]^2+\frac{4R}{k^2}\left(\frac{c_1}{c_2}\right)^2}}{2}}$$

$$\beta=k\sqrt{\frac{1+\left(\frac{c_1}{c_2}\right)^2+\sqrt{\left[1-\left(\frac{c_1}{c_2}\right)^2\right]^2+\frac{4R}{k^2}\left(\frac{c_1}{c_2}\right)^2}}{2}}$$

and $k=\frac{\Omega L}{c_1}$, Ω being the circular frequency of vibration. Equation (B17) is the frequency equation for a uniform cantilever beam (see ref. 5) and its solutions $k=k_n$, $n=1, 2, 3, \ldots \infty$, are the nondimensional natural vibration frequencies of such a beam. Equations (B15) and (B16) may be written in terms of these natural frequencies simply by replacing p_n by ik_n. The results are

$$\bar{y}(\xi,\tau)=\frac{c_2}{c_1}\left[\tau+2\sum_{n=1}^{\infty}X_{\bar{y}_n}(\xi)\sin k_n\tau\right] \tag{B18}$$

and

$$\psi(\xi,\tau)=2\frac{c_2}{c_1}\sum_{n=1}^{\infty}X_{\psi_n}(\xi)\sin k_n\tau \tag{B19}$$

where

$$\begin{aligned}X_{\bar{y}_n}(\xi)=\frac{1}{k_n\Phi_n}\Bigg[&\Big(1+\frac{\alpha_n}{\beta_n}\sin\beta_n\sinh\alpha_n+\frac{1}{\phi_n}\cos\beta_n\cosh\alpha_n\Big)\cosh\alpha_n\xi+\\&\Big(1-\frac{\beta_n}{\alpha_n}\sin\beta_n\sinh\alpha_n+\phi_n\cos\beta_n\cosh\alpha_n\Big)\cos\beta_n\xi-\\&\Big(\frac{\alpha_n}{\beta_n}\cosh\alpha_n\sin\beta_n+\frac{1}{\phi_n}\sinh\alpha_n\cos\beta_n\Big)\sinh\alpha_n\xi+\\&\Big(\frac{\beta_n}{\alpha_n}\sinh\alpha_n\cos\beta_n+\phi_n\cosh\alpha_n\sin\beta_n\Big)\sin\beta_n\xi\Bigg]\end{aligned}$$

$$\begin{aligned}X_{\psi_n}(\xi)=\frac{\beta_n^2-\left(\frac{c_1}{c_2}\right)^2k_n^2}{k_n\beta_n\Phi_n}\Bigg[&\Big(\frac{\beta_n}{\alpha_n}\sinh\alpha_n\cos\beta_n+\phi_n\cosh\alpha_n\sin\beta_n\Big)(\cos\beta_n\xi-\\&\cosh\alpha_n\xi)+\phi_n\frac{\beta_n}{\alpha_n}\Big(1+\frac{\alpha_n}{\beta_n}\sin\beta_n\sinh\alpha_n+\\&\frac{1}{\phi_n}\cos\beta_n\cosh\alpha_n\Big)\sinh\alpha_n\xi-\Big(1-\frac{\beta_n}{\alpha_n}\sin\beta_n\sinh\alpha_n+\\&\phi_n\cos\beta_n\cosh\alpha_n\Big)\sin\beta_n\xi\Bigg]\end{aligned}$$

$$\begin{aligned}\Phi_n=&\frac{Rk_n^2\left(\frac{c_1}{c_2}\right)^2}{\alpha_n\beta_n}\left(\frac{1}{\beta_n^2}-\frac{1}{\alpha_n^2}\right)\sin\beta_n\sinh\alpha_n+2\Bigg[\frac{(\alpha_n^2+\beta_n^2)^2}{Rk_n^2\left(\frac{c_1}{c_2}\right)^2}-\\&4\Bigg]\cos\beta_n\cosh\alpha_n-\beta_n\Bigg\{\Bigg[\frac{(\alpha_n^2+\beta_n^2)^2}{Rk_n^2\left(\frac{c_1}{c_2}\right)^2}-2\Bigg]\Bigg[1-\frac{Rk_n^2\left(\frac{c_1}{c_2}\right)^2}{\beta_n^2(\alpha_n^2+\beta_n^2)}\Bigg]+\\&\frac{\beta_n^2-\alpha_n^2}{\beta_n^2}\Bigg[1-\frac{Rk_n^2\left(\frac{c_1}{c_2}\right)^2}{\alpha_n^2(\alpha_n^2+\beta_n^2)}\Bigg]\Bigg\}\cosh\alpha_n\sin\beta_n+\\&\alpha_n\Bigg\{\Bigg[\frac{(\alpha_n^2+\beta_n^2)^2}{Rk_n^2\left(\frac{c_1}{c_2}\right)^2}-2\Bigg]\Bigg[1-\frac{Rk_n^2\left(\frac{c_1}{c_2}\right)^2}{\alpha_n^2(\alpha_n^2+\beta_n^2)}\Bigg]-\\&\frac{\beta_n^2-\alpha_n^2}{\alpha_n^2}\Bigg[1-\frac{Rk_n^2\left(\frac{c_1}{c_2}\right)^2}{\beta_n^2(\alpha_n^2+\beta_n^2)}\Bigg]\Bigg\}\sinh\alpha_n\cos\beta_n\end{aligned}$$

$$\phi_n=\frac{\alpha_n^2+\left(\frac{c_1}{c_2}\right)^2k_n^2}{\beta_n^2-\left(\frac{c_1}{c_2}\right)^2k_n^2}$$

The coefficients $X_{\bar{y}_n}(\xi)$ and $X_{\psi_n}(\xi)$ are seen to be the natural mode shapes associated with the frequency k_n. As a check on the validity of the Laplace transform procedure these quantities may be shown to satisfy the differential equations (B1) and the boundary conditions (B10).

Substituting $\bar{y}(\xi,\tau)$ and $\psi(\xi,\tau)$ from equations (B18) and (B19) into equations (B2) and (B3) gives, for the moment and shear,

$$\overline{M}(\xi,\tau)=-2\frac{c_2}{c_1}\sum_{n=1}^{\infty}X_{\overline{M}_n}(\xi)\sin k_n\tau \tag{B20}$$

where

$$\begin{aligned}X_{\overline{M}_n}(\xi)=\frac{\beta_n^2-\left(\frac{c_1}{c_2}\right)^2k_n^2}{k_n\Phi_n}\Bigg[&\phi_n\Big(1+\frac{\alpha_n}{\beta_n}\sin\beta_n\sinh\alpha_n+\\&\frac{1}{\phi_n}\cosh\alpha_n\cos\beta_n\Big)\cosh\alpha_n\xi-\Big(1-\frac{\beta_n}{\alpha_n}\sin\beta_n\sinh\alpha_n+\\&\phi_n\cosh\alpha_n\cos\beta_n\Big)\cos\beta_n\xi-\Big(\frac{\beta_n}{\alpha_n}\sinh\alpha_n\cos\beta_n+\\&\phi_n\cosh\alpha_n\sin\beta_n\Big)\Big(\sin\beta_n\xi+\frac{\alpha_n}{\beta_n}\sinh\alpha_n\xi\Big)\Bigg]\end{aligned}$$

and

$$\overline{V}(\xi,\tau)=-2R\frac{c_1}{c_2}\sum_{n=1}^{\infty}X_{\overline{V}_n}(\xi)\sin k_n\tau \tag{B21}$$

where

$$\begin{aligned}X_{\overline{V}_n}(\xi)=\frac{k_n}{\beta_n\Phi_n}\Bigg[&\frac{\beta_n}{\alpha_n}\Big(1+\frac{\alpha_n}{\beta_n}\sin\beta_n\sinh\alpha_n+\\&\frac{1}{\phi_n}\cos\beta_n\cosh\alpha_n\Big)\sinh\alpha_n\xi+\Big(1-\frac{\beta_n}{\alpha_n}\sin\beta_n\sinh\alpha_n+\\&\phi_n\cos\beta_n\cosh\alpha_n\Big)\sin\beta_n\xi-\Big(\frac{\beta_n}{\alpha_n}\sinh\alpha_n\cos\beta_n+\\&\phi_n\cosh\alpha_n\sin\beta_n\Big)\Big(\cos\beta_n\xi+\frac{1}{\phi_n}\cosh\alpha_n\xi\Big)\Bigg]\end{aligned}$$

The first eight terms in equations (B20) and (B21) have been used to compute the quantities $\overline{M}(0,\tau)$ and $\overline{V}(0,\tau)$. These results have been plotted in the range $0\leqq\tau\leqq 10$ in figures 4(a) and 4(b).

Simply supported beam with an applied step end moment.—If a step moment $\overline{M}=1$ is applied to the end $\xi=0$ of a simply supported beam at time $\tau=0$, the boundary conditions are

$$\left.\begin{aligned}\bar{y}(0,\tau)&=0\\\bar{y}(1,\tau)&=0\\\psi_\xi(0,\tau)&=-1\\\psi_\xi(1,\tau)&=0\end{aligned}\right\} \tag{B22}$$

These equations transform to (transform 7, table I)

$$\left.\begin{aligned}Y(0,p)&=0\\Y(1,p)&=0\\\Psi_\xi(0,p)&=-\frac{1}{p}\\\Psi_\xi(1,p)&=0\end{aligned}\right\} \tag{B23}$$

If conditions (B23) are used to evaluate the functions C_1,

C_2, C_3, and C_4 in equations (B8) and (B9), the quantities $Y(\xi,p)$ and $\Psi(\xi,p)$ become

$$Y(\xi,p)=\frac{1}{p(m_1^2+m_2^2)}(\cos m_1\xi-\cot m_1\sin m_1\xi-\cosh m_2\xi+\coth m_2\sinh m_2\xi) \tag{B24}$$

$$\Psi(\xi,p)=-\frac{1}{p(m_1^2+m_2^2)}\left[\frac{m_1^2+\left(\frac{c_1}{c_2}\right)^2p^2}{m_1}(\sin m_1\xi+\cot m_1\cos m_1\xi)+\frac{m_2^2-\left(\frac{c_1}{c_2}\right)^2p^2}{m_2}(\sinh m_2\xi-\coth m_2\cosh m_2\xi)\right] \tag{B25}$$

or they may be written in the forms

$$Y(\xi,p)=\frac{N_Y(\xi,p)}{D_Y(p)} \tag{B26}$$

where

$$N_Y(\xi,p)=\sin m_1\sinh m_2(\cos m_1\xi-\cosh m_2\xi)-\cos m_1\sinh m_2\sin m_1\xi+\sin m_1\cosh m_2\sinh m_2\xi$$

$$D_Y(p)=p(m_1^2+m_2^2)\sin m_1\sinh m_2$$

and

$$\Psi(\xi,p)=\frac{N_\Psi(\xi,p)}{D_\Psi(p)} \tag{B27}$$

where

$$N_\Psi(\xi,p)=m_1\left[m_2^2-\left(\frac{c_1}{c_2}\right)^2p^2\right]\sin m_1(\cosh m_2\cosh m_2\xi-\sinh m_2\sinh m_2\xi)-m_2\left[m_1^2+\left(\frac{c_1}{c_2}\right)^2p^2\right]\sinh m_2(\cos m_1\cos m_1\xi+\sin m_1\sin m_1\xi)$$

$$D_\Psi(p)=m_1m_2D_Y(p)$$

Here again, the functions $Y(\xi,p)$ and $\Psi(\xi,p)$ are single valued, and the inverse transforms $\bar{y}(\xi,\tau)$ and $\psi(\xi,\tau)$ of the quantities $Y(\xi,p)$ and $\Psi(\xi,p)$ are taken as the sums of the residues at the poles of $e^{p\tau}Y(\xi,p)$ and $e^{p\tau}\Psi(\xi,p)$, respectively. Consider all the roots of the denominators $D_Y(p)$ and $D_\Psi(p)$. For $m_1=\pm n\pi$ or $m_2=\pm in\pi$, $n=1, 2, 3, \ldots \infty$ (any of which are roots of both denominators), equation (B6) becomes

$$p^4+\left\{R+n^2\pi^2\left[1+\left(\frac{c_2}{c_1}\right)^2\right]\right\}p^2+\left(\frac{c_2}{c_1}\right)^2n^4\pi^4=0 \qquad (n=1, 2, 3, \ldots \infty) \tag{B28}$$

The solutions of equation (B28) are $p=\pm ia_n$ and $\pm ib_n$, $n=1, 2, 3, \ldots \infty$, where

$$a_n=\sqrt{\frac{R+n^2\pi^2\left[1+\left(\frac{c_2}{c_1}\right)^2\right]-\sqrt{\left\{R+n^2\pi^2\left[1+\left(\frac{c_2}{c_1}\right)^2\right]\right\}^2-4\left(\frac{c_2}{c_1}\right)^2n^4\pi^4}}{2}}$$

$$b_n=\sqrt{\frac{R+n^2\pi^2\left[1+\left(\frac{c_2}{c_1}\right)^2\right]+\sqrt{\left\{R+n^2\pi^2\left[1+\left(\frac{c_2}{c_1}\right)^2\right]\right\}^2-4\left(\frac{c_2}{c_1}\right)^2n^4\pi^4}}{2}}$$

and these points are seen to be the locations of simple poles of both $Y(\xi,p)$ and $\Psi(\xi,p)$. The roots $p=\pm\frac{2\sqrt{R}\frac{c_2}{c_1}}{1-\left(\frac{c_2}{c_1}\right)^2}$ of the equation $m_1^2+m_2^2=0$ are zeros of the denominator, but these are also roots of the numerators $N_Y(\xi,p)$ and $N_\Psi(\xi,p)$ and are not poles of either $Y(\xi,p)$ or $\Psi(\xi,p)$. The root $p=0$, on the other hand, is a pole of both $Y(\xi,p)$ and $\Psi(\xi,p)$ for, although both numerators also go to zero at this point, the denominators vanish more rapidly. The equation $m_1=0$ has three roots: $p=0$, which has already been discussed, and $p=\pm i\sqrt{R}$. The latter points are not poles of $Y(\xi,p)$ since they are roots of $N_Y(\xi,p)=0$, but they are simple poles of $\Psi(\xi,p)$.

First consider the residues of $e^{p\tau}Y(\xi,p)$ and $e^{p\tau}\Psi(\xi,p)$ at the poles $\pm ia_n$ and $\pm ib_n$. The sums of the residues at these poles provide the results

$$\bar{y}_d(\xi,\tau)=\sum_{n=1}^{\infty}\left[\frac{N_Y(\xi,ia_n)}{D_Y'(ia_n)}e^{ia_n\tau}+\frac{N_Y(\xi,-ia_n)}{D_Y'(-ia_n)}e^{-ia_n\tau}+\frac{N_Y(\xi,ib_n)}{D_Y'(ib_n)}e^{ib_n\tau}+\frac{N_Y(\xi,-ib_n)}{D_Y'(-ib_n)}e^{-ib_n\tau}\right] \tag{B29}$$

and

$$\psi_{d_1}(\xi,\tau)=\sum_{n=1}^{\infty}\left[\frac{N_\Psi(\xi,ia_n)}{D_\Psi'(ia_n)}e^{ia_n\tau}+\frac{N_\Psi(\xi,-ia_n)}{D_\Psi'(-ia_n)}e^{-ia_n\tau}+\frac{N_\Psi(\xi,ib_n)}{D_\Psi'(ib_n)}e^{ib_n\tau}+\frac{N_\Psi(\xi,-ib_n)}{D_\Psi'(-ib_n)}e^{-ib_n\tau}\right] \tag{B30}$$

where the subscripts on $\bar{y}$ and ψ indicate that they constitute only parts—so-called "dynamic" parts—of the complete solutions for $\bar{y}(\xi,\tau)$ and $\psi(\xi,\tau)$. In order to obtain the values of $\frac{N_Y}{D_Y'}$ and $\frac{N_\Psi}{D_\Psi'}$ at $p=\pm ia_n$ and $\pm ib_n$, it is necessary to use the fact that, when p equals one of the roots $\pm ia_n$ and $\pm ib_n$, one of four equations $m_1=n\pi$, $m_1=-n\pi$, $m_2=in\pi$, or $m_2=-in\pi$

is satisfied. The question of which value of p corresponds to each equation need not be answered since the following relationships can be derived:

$$\left(\frac{N_Y}{D_Y'}\right)_{m_1=n\pi}=\left(\frac{N_Y}{D_Y'}\right)_{m_1=-n\pi}=\left(\frac{N_Y}{D_Y'}\right)_{m_2=in\pi}=\left(\frac{N_Y}{D_Y'}\right)_{m_2=-in\pi}=-\frac{\left(\frac{c_2}{c_1}\right)^2 n\pi \sin n\pi\xi}{2a_n^2b_n^2+p^2(a_n^2+b_n^2)} \tag{B31}$$

$$\left(\frac{N_\Psi}{D_\Psi'}\right)_{m_1=n\pi}=\left(\frac{N_\Psi}{D_\Psi'}\right)_{m_1=-n\pi}=\left(\frac{N_\Psi}{D_\Psi'}\right)_{m_2=in\pi}=\left(\frac{N_\Psi}{D_\Psi'}\right)_{m_2=-in\pi}=-\frac{\left(\frac{c_2}{c_1}\right)^2\left[n^2\pi^2+\left(\frac{c_1}{c_2}\right)^2p^2\right]\cos n\pi\xi}{2a_n^2b_n^2+p^2(a_n^2+b_n^2)} \tag{B32}$$

As an example, let $p=ia_n$. Then one of the four possibilities $m_1=n\pi$, $m_1=-n\pi$, $m_2=in\pi$, and $m_2=-in\pi$ must occur. But for any of these possibilities, the ratios $\frac{N_Y}{D_Y'}$ and $\frac{N_\Psi}{D_\Psi'}$ have the forms given by equations (B31) and (B32). Thus,

$$\left.\begin{aligned}\frac{N_Y(\xi,ia_n)}{D_Y'(ia_n)}&=\frac{\left(\frac{c_2}{c_1}\right)^2 n\pi\sin n\pi\xi}{a_n^2(a_n^2-b_n^2)}\\ \frac{N_\Psi(\xi,ia_n)}{D_\Psi'(ia_n)}&=\frac{\left[\left(\frac{c_2}{c_1}\right)^2\frac{n^2\pi^2}{a_n^2}-1\right]\cos n\pi\xi}{a_n^2-b_n^2}\end{aligned}\right\} \tag{B33}$$

The values of the other necessary ratios can be found in a similar manner and equations (B29) and (B30) reduce to

$$\bar{y}_d(\xi,\tau)=2\left(\frac{c_2}{c_1}\right)^2\sum_{n=1}^{\infty}\frac{n\pi\sin n\pi\xi}{a_n^2-b_n^2}\left(\frac{\cos a_n\tau}{a_n^2}-\frac{\cos b_n\tau}{b_n^2}\right) \tag{B34}$$

and

$$\psi_{d_1}(\xi,\tau)=2\sum_{n=1}^{\infty}\frac{\cos n\pi\xi}{a_n^2-b_n^2}\left[\left(\frac{b_n^2}{n^2\pi^2}-1\right)\cos a_n\tau-\left(\frac{a_n^2}{n^2\pi^2}-1\right)\cos b_n\tau\right] \tag{B35}$$

Next, consider the contributions of the pole at $p=0$. Equations (B24) and (B25) may be expanded in a Laurent series about $p=0$. The resulting equations for $Y(\xi,p)$ and $\Psi(\xi,p)$ may be written

$$Y(\xi,p)=\frac{1}{p}\left[\frac{\xi^3}{6}-\frac{\xi^2}{2}+\frac{\xi}{3}+0(p)\right] \tag{B36}$$

$$\Psi(\xi,p)=\frac{1}{p}\left[\frac{\xi^2}{2}-\xi+\frac{1}{3}+\frac{1}{R}+0(p)\right] \tag{B37}$$

where $0(p)$ signifies terms of order p or higher. Thus the quantities $e^{p\tau}Y(\xi,p)$ and $e^{p\tau}\Psi(\xi,p)$ have simple poles at $p=0$ with the residues

$$\bar{y}_{st}(\xi,\tau)=\frac{\xi^3}{6}-\frac{\xi^2}{2}+\frac{\xi}{3} \tag{B38}$$

and

$$\psi_{st}(\xi,\tau)=\frac{\xi^2}{2}-\xi+\frac{1}{3}+\frac{1}{R} \tag{B39}$$

respectively. It can be readily verified that equations (B38) and (B39) actually constitute the solution of the problem when the beam is considered to be loaded statically.

Lastly, consider the points $p=\pm i\sqrt{R}$, which are simple poles of $\Psi(\xi,p)$. It can be shown that

$$\lim_{p\to\pm i\sqrt{R}}\frac{N_\Psi}{D_\Psi'}=-\frac{1}{2R} \tag{B40}$$

so that the sum of the residues of $e^{p\tau}\Psi(\xi,p)$ at these poles provides the additional dynamic contribution

$$\psi_{d_2}(\xi,\tau)=-\frac{1}{R}\cos\sqrt{R}\,\tau \tag{B41}$$

Summing the contributions of all the poles gives

$$\bar{y}(\xi,\tau)=\frac{\xi^3}{6}-\frac{\xi^2}{2}+\frac{\xi}{3}+2\left(\frac{c_2}{c_1}\right)^2\sum_{n=1}^{\infty}\frac{n\pi\sin n\pi\xi}{a_n^2-b_n^2}\left(\frac{\cos a_n\tau}{a_n^2}-\frac{\cos b_n\tau}{b_n^2}\right) \tag{B42}$$

$$\psi(\xi,\tau)=\frac{\xi^2}{2}-\xi+\frac{1}{3}+\frac{1}{R}-\frac{1}{R}\cos\sqrt{R}\,\tau+2\sum_{n=1}^{\infty}\frac{\cos n\pi\xi}{a_n^2-b_n^2}\left[\left(\frac{b_n^2}{n^2\pi^2}-1\right)\cos a_n\tau-\left(\frac{a_n^2}{n^2\pi^2}-1\right)\cos b_n\tau\right] \tag{B43}$$

The responses $\bar{y}(\xi,\tau)$ and $\psi(\xi,\tau)$ are seen to have the same form as solutions obtained by Williams' method (ref. 2); that is, they are the sum of a static part, $\bar{y}_{st}(\xi,\tau)$ and $\psi_{st}(\xi,\tau)$, and a dynamic correction, $\bar{y}_d(\xi,\tau)$ and $\psi_{d_1}(\xi,\tau)+\psi_{d_2}(\xi,\tau)$, which is expanded in a series of the natural mode shapes. The $\bar{y}$- and ψ-components of the natural modes have the shapes $\sin n\pi\xi$ and $\cos n\pi\xi$, respectively, and there are seen to be two natural frequencies, a_n and b_n, associated with each integer n so that each integer corresponds to two modes. From equations (B2) and (B3),

$$\overline{M}(\xi,\tau)=1-\xi+2\sum_{n=1}^{\infty}\frac{n\pi\sin n\pi\xi}{a_n^2-b_n^2}\left[\left(\frac{b_n^2}{n^2\pi^2}-1\right)\cos a_n\tau-\left(\frac{a_n^2}{n^2\pi^2}-1\right)\cos b_n\tau\right] \tag{B44}$$

and

$$\overline{V}(\xi,\tau)=\cos\sqrt{R}\,\tau-1+2R\sum_{n=1}^{\infty}\frac{\cos n\pi\xi}{a_n^2-b_n^2}(\cos a_n\tau-\cos b_n\tau) \tag{B45}$$

Six terms ($n=3$) have been used in the summations of equations (B44) and (B45) in computing the quantities $\overline{M}(1/2,\tau)$ and $\overline{V}(0,\tau)$. These results have been plotted in figures 5 (a) and 5 (b) up to $\tau=8$.

Simply supported beam with an applied ramp-platform end moment.—Let the response to unit step moment at $\xi=0$, obtained above, be designated $\overline{M}_1(\xi,\tau)$ and $\overline{V}_1(\xi,\tau)$. Then the response to an arbitrary applied moment $\overline{M}(0,\tau)$ may be obtained from Duhamel's superposition integral as

$$\left.\begin{aligned}\overline{M}(\xi,\tau)&=\overline{M}(0,0)\overline{M}_1(\xi,\tau)+\int_0^\tau \overline{M}_1(\xi,\tau-\theta)\overline{M}_\theta(0,\theta)d\theta\\ \overline{V}(\xi,\tau)&=\overline{M}(0,0)\overline{V}_1(\xi,\tau)+\int_0^\tau \overline{V}_1(\xi,\tau-\theta)\overline{M}_\theta(0,\theta)d\theta\end{aligned}\right\}\qquad\text{(B46)}$$

where the subscript θ indicates differentiation with respect to that variable. Let the applied moment have the ramp-platform time history

$$\left.\begin{aligned}\overline{M}(0,\tau)&=\tau &&(0\leqq\tau\leqq1)\\ \overline{M}(0,\tau)&=1 &&(\tau>1)\end{aligned}\right\}\qquad\text{(B47)}$$

Then equations (B46) become

$$\left.\begin{aligned}\overline{M}(\xi,\tau)&=\int_0^\tau \overline{M}_1(\xi,\tau-\theta)d\theta &&(0\leqq\tau\leqq1)\\ \overline{M}(\xi,\tau)&=\int_0^1 \overline{M}_1(\xi,\tau-\theta)d\theta &&(\tau>1)\end{aligned}\right\}\qquad\text{(B48)}$$

and

$$\left.\begin{aligned}\overline{V}(\xi,\tau)&=\int_0^\tau \overline{V}_1(\xi,\tau-\theta)d\theta &&(0\leqq\tau\leqq1)\\ \overline{V}(\xi,\tau)&=\int_0^1 \overline{V}_1(\xi,\tau-\theta)d\theta &&(\tau>1)\end{aligned}\right\}\qquad\text{(B49)}$$

Substituting $\overline{M}_1$ and $\overline{V}_1$ from equations (B44) and (B45) gives for $\overline{M}$ and $\overline{V}$

$$\left.\begin{aligned}\overline{M}(\xi,\tau)&=(1-\xi)\tau+2\sum_{n=1}^{\infty}\frac{n\pi\sin n\pi\xi}{a_n^2-b_n^2}\left[\left(\frac{b_n^2}{n^2\pi^2}-1\right)\frac{\sin a_n\tau}{a_n}-\left(\frac{a_n^2}{n^2\pi^2}-1\right)\frac{\sin b_n\tau}{b_n}\right] &&(0\leqq\tau\leqq1)\\ \overline{M}(\xi,\tau)&=1-\xi+2\sum_{n=1}^{\infty}\frac{n\pi\sin n\pi\xi}{a_n^2-b_n^2}\left[\left(\frac{b_n^2}{n^2\pi^2}-1\right)\frac{\sin a_n\tau-\sin a_n(\tau-1)}{a_n}-\left(\frac{a_n^2}{n^2\pi^2}-1\right)\frac{\sin b_n\tau-\sin b_n(\tau-1)}{b_n}\right] &&(\tau>1)\end{aligned}\right\}\qquad\text{(B50)}$$

and

$$\left.\begin{aligned}\overline{V}(\xi,\tau)&=\frac{1}{\sqrt{R}}\sin\sqrt{R}\,\tau-\tau+2R\sum_{n=1}^{\infty}\frac{\cos n\pi\xi}{a_n^2-b_n^2}\left(\frac{\sin a_n\tau}{a_n}-\frac{\sin b_n\tau}{b_n}\right) &&(0\leqq\tau\leqq1)\\ \overline{V}(\xi,\tau)&=\frac{1}{\sqrt{R}}\left[\sin\sqrt{R}\,\tau-\sin\sqrt{R}\,(\tau-1)\right]-1+2R\sum_{n=1}^{\infty}\frac{\cos n\pi\xi}{a_n^2-b_n^2}\left[\frac{\sin a_n\tau-\sin a_n(\tau-1)}{a_n}-\frac{\sin b_n\tau-\sin b_n(\tau-1)}{b_n}\right] &&(\tau>1)\end{aligned}\right\}\qquad\text{(B51)}$$

As in the preceding case, computations using terms up to $n=3$ in equations (B50) and (B51) have been made for $\overline{M}(1/2,\tau)$ and $\overline{V}(0,\tau)$ in the range $0\leqq\tau\leqq8$. The results are plotted in figures 6(a) and 6(b).

APPENDIX C

EXACT CLOSED SOLUTIONS FOR UNIFORM BEAMS WITH $c_1=c_2$

INTRODUCTION

In this appendix some exact closed solutions of Timoshenko's equations are derived for infinitely long uniform beams for which $c_1=c_2$. The results for beams of infinite length are then utilized to obtain results for beams of finite length. This is done either by simply restricting the infinite-beam solution to a time interval in which it coincides with the corresponding finite-beam solution or by superposing infinite-beam solutions in such a way that the result satisfies the boundary conditions for a finite beam.

EQUATIONS

If the propagation velocities c_1 and c_2 are taken to be equal and if $\overline{q}=0$, equations (14) reduce to

$$\overline{\omega}_\xi+\overline{M}_\tau=0 \tag{C1a}$$

$$\overline{V}_\xi-4\lambda^2\overline{v}_\tau=0 \tag{C1b}$$

$$\overline{M}_\xi+\overline{\omega}_\tau-\overline{V}=0 \tag{C1c}$$

$$\overline{v}_\xi-\frac{1}{4\lambda^2}\overline{V}_\tau-\overline{\omega}=0 \tag{C1d}$$

where $\lambda=\frac{L}{2r_i}$. This system may be further reduced by writing a linear combination of equations (C1a) and (C1b) and a linear combination of (C1c) and (C1d) to obtain

$$Z_\xi+2\lambda iK_\tau=0 \tag{C2a}$$

$$Z_\tau+2\lambda iK_\xi-2\lambda iZ=0 \tag{C2b}$$

in terms of the complex variables $Z=\overline{V}+2\lambda i\overline{\omega}$ and $K=\overline{M}+2\lambda i\overline{v}$.

For a beam initially at rest, equations (C2) may be transformed to (transform 6, table I)

$$\zeta_\xi+2\lambda ip\kappa=0 \tag{C3a}$$

$$2\lambda i\kappa_\xi+(p-2\lambda i)\zeta=0 \tag{C3b}$$

in terms of the Laplace transforms $\zeta(\xi,p)=\int_0^\infty e^{-p\tau}Z(\xi,\tau)d\tau$ and $\kappa(\xi,p)=\int_0^\infty e^{-p\tau}K(\xi,\tau)d\tau$. Eliminating κ produces the equation

$$\zeta_{\xi\xi}-p(p-2\lambda i)\zeta=0$$

which has the solution

$$\zeta(\xi,p)=A(p)e^{-\xi\sqrt{p(p-2\lambda i)}}+B(p)e^{\xi\sqrt{p(p-2\lambda i)}} \tag{C4}$$

where A and B are governed by the boundary conditions. If the beam extends to infinity in the positive ξ-direction, boundary conditions stipulating that the beam remain undisturbed at $\xi=\infty$ must be satisfied. Thus, equation (C4) reduces to

$$\zeta(\xi,p)=A(p)e^{-\xi\sqrt{p(p-2\lambda i)}} \tag{C5}$$

Substituting this expression into equation (C3a) produces

$$\kappa(\xi,p)=\frac{A(p)}{2\lambda ip}\sqrt{p(p-2\lambda i)}\,e^{-\xi\sqrt{p(p-2\lambda i)}} \tag{C6}$$

The quantity $A(p)$ may now be determined from the remaining boundary conditions in conjunction with equation (C5) or (C6), or both.

SPECIFIC PROBLEMS

Infinite cantilever given a step velocity at the root.—If the end $\xi=0$ of an infinitely long beam is restrained from rotating and is given a unit step velocity at time $\tau=0$, the boundary conditions are $\overline{v}(0,\tau)=1$ and $\overline{\omega}(0,\tau)=0$. In terms of Z and K these conditions may be written

$$\text{I.P.}[Z(0,\tau)]=0$$

and

$$\text{I.P.}[K(0,\tau)]=2\lambda$$

where I.P. designates the imaginary part of the quantity in brackets. The boundary conditions may be transformed to (transform 7, table I)

$$\text{I.P.}[\zeta(0,p)]=0 \tag{C7a}$$

and

$$\text{I.P.}[\kappa(0,p)]=\frac{2\lambda}{p} \tag{C7b}$$

respectively.

Because of the nature of these conditions, it will be convenient to proceed in the following manner to determine the function $A(p)$. From equation (C5), it may be seen that $A(p)=\zeta(0,p)$; thus condition (C7a) establishes at once that $A(p)$ for this problem is a real quantity. At $\xi=0$ equation (C6) becomes

$$\kappa(0,p)=\frac{A(p)}{2\lambda ip}\sqrt{p(p-2\lambda i)}$$

or, in view of condition (C7b),

$$\text{R.P.}[\kappa(0,p)]+\frac{2\lambda i}{p}=\frac{A(p)}{2\lambda ip}\sqrt{p(p-2\lambda i)} \tag{C8}$$

where R.P.$[\kappa(0,p)]$ is the real part of $\kappa(0,p)$. Since $A(p)$ and p are real, the conjugate equation, which must also hold, is

$$\text{R.P.}[\kappa(0,p)]-\frac{2\lambda i}{p}=-\frac{A(p)}{2\lambda ip}\sqrt{p(p+2\lambda i)} \tag{C9}$$

Eliminating R.P.$[\kappa(0,p)]$ between equations (C8) and (C9) results in

$$A(p)=-2\lambda i\left(\sqrt{\frac{p-2\lambda i}{p}}-\sqrt{\frac{p+2\lambda i}{p}}\right) \tag{C10}$$

If $A(p)$ is substituted into equation (C5), the result may be written in the form

$$\zeta(\xi,p)=-[4\lambda^2+2\lambda i(p-\sqrt{p^2+4\lambda^2})]\frac{e^{-\xi\sqrt{p(p-2\lambda i)}}}{\sqrt{p(p-2\lambda i)}}$$

which has the inverse transform (transforms 1, 3, 14, and 15, table I)

$$Z(\xi,\tau)=-4\lambda^2\left\{e^{i\lambda\tau}J_0(\lambda\sqrt{\tau^2-\xi^2})-i\int_\xi^\tau e^{i\lambda\theta}J_0(\lambda\sqrt{\theta^2-\xi^2})\frac{J_1[2\lambda(\tau-\theta)]}{\tau-\theta}d\theta\right\}$$

$$(\tau>\xi)\quad\text{(C11)}$$

where J_0 and J_1 are Bessel functions of the first kind. From equation (C11) the shear and angular velocity are obtained as

$$\overline{V}(\xi,\tau)=-4\lambda^2\left\{\cos\lambda\tau J_0(\lambda\sqrt{\tau^2-\xi^2})+\int_\xi^\tau \sin\lambda\theta J_0(\lambda\sqrt{\theta^2-\xi^2})\frac{J_1[2\lambda(\tau-\theta)]}{\tau-\theta}d\theta\right\}$$

$$(\tau>\xi)\quad\text{(C12)}$$

and

$$\overline{\omega}(\xi,\tau)=-2\lambda\left\{\sin\lambda\tau J_0(\lambda\sqrt{\tau^2-\xi^2})-\int_\xi^\tau \cos\lambda\theta J_0(\lambda\sqrt{\theta^2-\xi^2})\frac{J_1[2\lambda(\tau-\theta)]}{\tau-\theta}d\theta\right\}$$

$$(\tau>\xi)\quad\text{(C13)}$$

in terms of integrals that apparently defy evaluation by analytical methods.

The time history of the shear $\overline{V}(0,\tau)$ at the root of the beam may be obtained as a special case of equation (C12); however, it will be noted that $L\{\overline{V}(0,\tau)\}=\zeta(0,p)=A(p)$ (where $L\{\overline{V}(0,\tau)\}$ denotes the Laplace transform of $\overline{V}(0,\tau)$) so that the inverse of equation (C10) gives $\overline{V}(0,\tau)$ directly. Thus the shear at the root is found explicitly in terms of tabulated functions as (transform 13, table I)

$$\overline{V}(0,\tau)=-4\lambda^2[\cos\lambda\tau\; J_0(\lambda\tau)+\sin\lambda\tau\; J_1(\lambda\tau)]\quad(\tau>0)\quad\text{(C14)}$$

Since $A(p)$ may be replaced by $L\{\overline{V}(0,\tau)\}$, the inverse transform of equation (C6) may be written for this problem in the form (transforms 1, 2, 3, and 17, table I)

$$K(\xi,\tau)=-\frac{i}{2}\left\{\frac{1}{\lambda}e^{i\lambda\xi}\overline{V}(0,\tau-\xi)-\int_\xi^\tau \overline{V}(0,\tau-\theta)e^{i\lambda\theta}\left[\frac{\theta J_1(\lambda\sqrt{\theta^2-\xi^2})}{\sqrt{\theta^2-\xi^2}}+iJ_0(\lambda\sqrt{\theta^2-\xi^2})\right]d\theta\right\}$$

$$(\tau>\xi)$$

or, after integration by parts,

$$K(\xi,\tau)=2\lambda ie^{i\lambda\tau}J_0(\lambda\sqrt{\tau^2-\xi^2})-\int_\xi^\tau\left\{\overline{V}(0,\tau-\theta)+\frac{2\lambda i}{\tau-\theta}\sin\lambda(\tau-\theta)J_1[\lambda(\tau-\theta)]\right\}e^{i\lambda\theta}J_0(\lambda\sqrt{\theta^2-\xi^2})\,d\theta$$

$$(\tau>\xi)\quad\text{(C15)}$$

where $\overline{V}(0,\tau)$ is defined by equation (C14). From equation (C15) the moment and linear velocity may be written as

$$\overline{M}(\xi,\tau)=-2\lambda\sin\lambda\tau J_0(\lambda\sqrt{\tau^2-\xi^2})-2\lambda\int_\xi^\tau\left\{\frac{1}{2\lambda}\overline{V}(0,\tau-\theta)\cos\lambda\theta-\frac{\sin\lambda\theta}{\tau-\theta}\sin\lambda(\tau-\theta)J_1[\lambda(\tau-\theta)]\right\}J_0(\lambda\sqrt{\theta^2-\xi^2})d\theta$$

$$(\tau>\xi)\quad\text{(C16)}$$

and

$$\overline{v}(\xi,\tau)=\cos\lambda\tau J_0(\lambda\sqrt{\tau^2-\xi^2})-\int_\xi^\tau\left\{\frac{1}{2\lambda}\overline{V}(0,\tau-\theta)\sin\lambda\theta+\frac{\cos\lambda\theta}{\tau-\theta}\sin\lambda(\tau-\theta)J_1[\lambda(\tau-\theta)]\right\}J_0(\lambda\sqrt{\theta^2-\xi^2})d\theta$$

$$(\tau>\xi)\quad\text{(C17)}$$

again in terms of integrals requiring evaluation by approximate methods.

As in the case of the shear, it will be more convenient to determine the moment $\overline{M}(0,\tau)$ at the root by means other than a reduction of the general relation, equation (C16). Since, for this problem, R.P.$[\kappa(0,p)]=L\{\overline{M}(0,\tau)\}$, equation (C8) may be written, after substitution for $A(p)$, in the form

$$L\{\overline{M}(0,\tau)\}=\frac{1}{p}(\sqrt{p^2+4\lambda^2}-p)$$

The inverse transform of this equation is (transforms 4 and 14, table I)

$$\overline{M}(0,\tau)=2\lambda\int_0^\tau\frac{J_1(2\lambda\theta)}{\theta}d\theta\qquad(\tau>0)\quad\text{(C18)}$$

But, from reference 17,

$$\frac{J_1(2\lambda\tau)}{\tau}=2\lambda J_0(2\lambda\tau)-\frac{d}{d\tau}[J_1(2\lambda\tau)]$$

and

$$\int_0^\tau J_0(2\lambda\theta)d\theta=\tau\left\{J_0(2\lambda\tau)+\frac{\pi}{2}[J_1(2\lambda\tau)H_0(2\lambda\tau)-J_0(2\lambda\tau)H_1(2\lambda\tau)]\right\}$$

where H_0 and H_1 are Struve functions. Thus, equation (C18) becomes

$$\overline{M}(0,\tau)=2\lambda\{\lambda\tau[2-\pi H_1(2\lambda\tau)]J_0(2\lambda\tau)-[1-\pi\lambda\tau H_0(2\lambda\tau)]J_1(2\lambda\tau)\}$$

$$(\tau>0)\quad\text{(C19)}$$

The Struve functions H_0 and H_1 are tabulated in reference 17 in the range $0\leqq 2\lambda\tau\leqq 15.9$. For larger values of the argument it is convenient to use the approximations (ref. 17)

$$H_0(2\lambda\tau)\approx Y_0(2\lambda\tau)+\frac{1}{\pi\lambda\tau}\qquad(2\lambda\tau>15.9)$$

$$H_1(2\lambda\tau)\approx Y_1(2\lambda\tau)+\frac{2}{\pi}\left(1+\frac{1}{4\lambda^2\tau^2}\right)\qquad(2\lambda\tau>15.9)$$

where Y_0 and Y_1 are Bessel functions of the second kind. From reference 18,

$$Y_0(2\lambda\tau)J_1(2\lambda\tau)-Y_1(2\lambda\tau)J_0(2\lambda\tau)=\frac{1}{\pi\lambda\tau}$$

so that

$$\overline{M}(0,\tau)\approx 2\lambda\left[1-\frac{J_0(2\lambda\tau)}{2\lambda\tau}\right]\quad(2\lambda\tau>15.9)\qquad\text{(C20)}$$

Computed values of $\overline{V}(0,\tau)$ and $\overline{M}(0,\tau)$, obtained from equations (C14), (C19), and (C20) for a beam with $\lambda=5$, have been plotted in figures 4 (a) and 4 (b), respectively, in the range $0\leqq\tau\leqq 2$. In this range, the root of a finite cantilever beam behaves as if the beam were infinite, since the effect of the free end is not felt until the return of the wave front at $\tau=2$.

Infinite simply supported beam with an applied step end moment.—If the end $\xi=0$ of an infinitely long beam is simply supported and is subjected to a unit step bending moment at time $\tau=0$, the boundary conditions are $\overline{M}(0,\tau)=1$ and $\overline{v}(0,\tau)=0$, or $K(0,\tau)=1$. This condition transforms to (transform 7, table I)

$$\kappa(0,p)=\frac{1}{p}\qquad\text{(C21)}$$

Substituting equation (C6) into condition (C21) reveals that

$$A(p)=\frac{2\lambda i}{\sqrt{p(p-2\lambda i)}}\qquad\text{(C22)}$$

so that equations (C5) and (C6) become

$$\zeta(\xi,p)=2\lambda i\frac{e^{-\xi\sqrt{p(p-2\lambda i)}}}{\sqrt{p(p-2\lambda i)}}\qquad\text{(C23)}$$

and

$$\kappa(\xi,p)=\frac{1}{p}e^{-\xi\sqrt{p(p-2\lambda i)}}\qquad\text{(C24)}$$

respectively. The inverse transform of equation (C23) is (transforms 3 and 15, table I)

$$Z(\xi,\tau)=2\lambda i e^{i\lambda\tau}J_0(\lambda\sqrt{\tau^2-\xi^2})\quad(\tau>\xi)\qquad\text{(C25)}$$

and the inverse of equation (C24) may be written (transforms 2, 3, 4, and 16, table I) as

$$K(\xi,\tau)=e^{i\lambda\xi}-\lambda\xi\int_{\xi}^{\tau}e^{i\lambda\theta}\frac{J_1(\lambda\sqrt{\theta^2-\xi^2})}{\sqrt{\theta^2-\xi^2}}d\theta\quad(\tau>\xi)\qquad\text{(C26)}$$

From these equations, the shear, angular velocity, moment, and linear velocity are obtained as

$$\overline{V}(\xi,\tau)=-2\lambda\sin\lambda\tau J_0(\lambda\sqrt{\tau^2-\xi^2})\quad(\tau>\xi)\qquad\text{(C27)}$$

$$\overline{\omega}(\xi,\tau)=\cos\lambda\tau J_0(\lambda\sqrt{\tau^2-\xi^2})\quad(\tau>\xi)\qquad\text{(C28)}$$

$$\overline{M}(\xi,\tau)=\cos\lambda\xi-\lambda\xi\int_{\xi}^{\tau}\cos\lambda\theta\frac{J_1(\lambda\sqrt{\theta^2-\xi^2})}{\sqrt{\theta^2-\xi^2}}d\theta\quad(\tau>\xi)\qquad\text{(C29)}$$

$$\overline{v}(\xi,\tau)=\sin\lambda\xi-\lambda\xi\int_{\xi}^{\tau}\sin\lambda\theta\frac{J_1(\lambda\sqrt{\theta^2-\xi^2})}{\sqrt{\theta^2-\xi^2}}d\theta\quad(\tau>\xi)\qquad\text{(C30)}$$

Thus, $\overline{V}$ and $\overline{\omega}$ are determined everywhere in closed form in terms of tabulated functions, but the relations for $\overline{M}$ and $\overline{v}$ contain integrals which must be evaluated by approximate methods.

The results of this section are utilized in the next section where a finite simply supported beam is considered.

Finite simply supported beam with an applied step end moment.—If the end $\xi=0$ of a finite simply supported beam is subjected to a unit step bending moment at time $\tau=0$, the boundary conditions may be written

$$\left.\begin{aligned}&\overline{M}(0,\tau)=1\\&\overline{v}(0,\tau)=\overline{M}(1,\tau)=\overline{v}(1,\tau)=0\end{aligned}\right\}\qquad\text{(C31)}$$

or, in terms of K,

$$\left.\begin{aligned}K(0,\tau)=1\\K(1,\tau)=0\end{aligned}\right\}\qquad\text{(C32)}$$

Although a direct solution of equations (C2) is possible in this case, the response of the finite beam may be obtained by a somewhat simpler procedure in which the responses of infinite beams to the same disturbance are superposed. This procedure is described below.

A series of semi-infinite simply supported uniform beams extending in opposite directions are shown in figure 7. The beams have been positioned in the figure so that the origins of the space coordinate ξ lie on the same vertical line, and the segments of greatest interest, $0\leqq\xi\leqq 1$, have been outlined with solid lines while the rest of each beam is defined by dashed lines. For each beam the origin of the coordinate ξ is seen to lie at a different position relative to the end where a unit positive moment is suddenly applied at time $\tau=0$. An infinite number of these beams is assumed to exist.

The response of the top beam has already been determined as (eqs. (C25) and (C26))

$$Z(\xi,\tau)=2\lambda i e^{i\lambda\tau}J_0(\lambda\sqrt{\tau^2-\xi^2})\quad(\tau>\xi)$$

and

$$K(\xi,\tau)=G(\xi,\tau)\quad(\tau>\xi)$$

where

$$G(\xi,\tau)=e^{i\lambda\xi}-\lambda\xi\int_{\xi}^{\tau}e^{i\lambda\theta}\frac{J_1(\lambda\sqrt{\theta^2-\xi^2})}{\sqrt{\theta^2-\xi^2}}d\theta$$

The response of the second beam may be determined from these equations simply by accounting for the changed

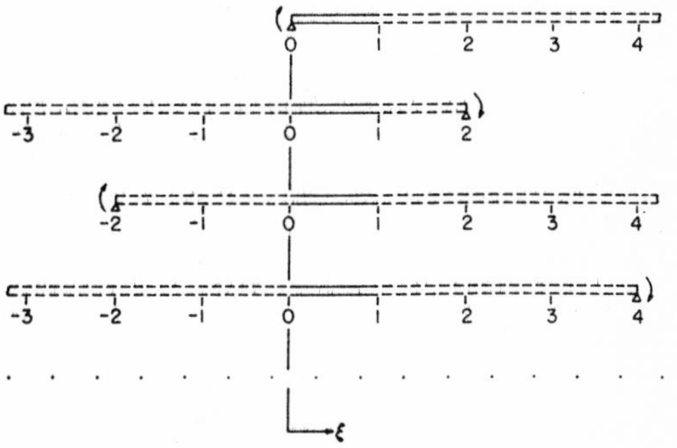

FIGURE 7.—Semi-infinite uniform beams superposed to obtain the response of a finite simply supported beam.

direction and shifted origin of the coordinate ξ. This process gives

$$Z(\xi,\tau)=2\lambda ie^{i\lambda\tau}J_0[\lambda\sqrt{\tau^2-(2-\xi)^2}] \qquad (\tau>2-\xi)$$

and

$$K(\xi,\tau)=-G(2-\xi,\tau) \qquad (\tau>2-\xi)$$

In order to obtain the response of the third beam from the response of the first, only the shifted origin need be accounted for. Thus, for this beam,

$$Z(\xi,\tau)=2\lambda ie^{i\lambda\tau}J_0[\lambda\sqrt{\tau^2-(2+\xi)^2}] \qquad (\tau>2+\xi)$$

$$K(\xi,\tau)=G(2+\xi,\tau) \qquad (\tau>2+\xi)$$

Similarly, for the fourth beam,

$$Z(\xi,\tau)=2\lambda ie^{i\lambda\tau}J_0[\lambda\sqrt{\tau^2-(4-\xi)^2}] \qquad (\tau>4-\xi)$$

$$K(\xi,\tau)=-G(4-\xi,\tau) \qquad (\tau>4-\xi)$$

and so forth, for all the rest. In each case, the region where the response is not specified is a region of zero response.

Let the responses of all these beams to their respective disturbances be superposed and consider only the response of the segment $0\leqq\xi\leqq1$ of the resulting composite beam. Since the wave fronts of all the disturbances travel at the same velocity $\frac{d\xi}{d\tau}=1$, the response of this segment may be written

$$\left.\begin{aligned}
&Z(\xi,\tau)=0 && (0\leqq\tau\leqq\xi)\\
&Z(\xi,\tau)=2\lambda ie^{i\lambda\tau}J_0(\lambda\sqrt{\tau^2-\xi^2}) && (\xi\leqq\tau\leqq2-\xi)\\
&Z(\xi,\tau)=2\lambda ie^{i\lambda\tau}\{J_0(\lambda\sqrt{\tau^2-\xi^2})+J_0[\lambda\sqrt{\tau^2-(2-\xi)^2}]\} && (2-\xi\leqq\tau\leqq2+\xi)\\
&Z(\xi,\tau)=2\lambda ie^{i\lambda\tau}\{J_0(\lambda\sqrt{\tau^2-\xi^2})+J_0[\lambda\sqrt{\tau^2-(2-\xi)^2}]+J_0[\lambda\sqrt{\tau^2-(2+\xi)^2}]\} && (2+\xi\leqq\tau\leqq4-\xi)\\
&\qquad\cdot\quad\cdot\quad\cdot\quad\cdot\quad\cdot
\end{aligned}\right\}\quad\text{(C33)}$$

and

$$\left.\begin{aligned}
&K(\xi,\tau)=0 && (0\leqq\tau\leqq\xi)\\
&K(\xi,\tau)=G(\xi,\tau) && (\xi\leqq\tau\leqq2-\xi)\\
&K(\xi,\tau)=G(\xi,\tau)-G(2-\xi,\tau) && (2-\xi\leqq\tau\leqq2+\xi)\\
&K(\xi,\tau)=G(\xi,\tau)-G(2-\xi,\tau)+G(2+\xi,\tau) && (2+\xi\leqq\tau\leqq4-\xi)\\
&\qquad\cdot\quad\cdot\quad\cdot\quad\cdot\quad\cdot
\end{aligned}\right\}\quad\text{(C34)}$$

Since $G(0,\tau)=1$, equations (C34) are seen to satisfy the boundary conditions (C32) and the response obtained above must be that of a simply supported uniform beam to a unit dimensionless step moment applied at the support $\xi=0$.

The vertical shear $\overline{V}$ and bending moment $\overline{M}$ may be obtained from equations (C33) and (C34) as

$$\left.\begin{aligned}
&\overline{V}(\xi,\tau)=0 && (0\leqq\tau\leqq\xi)\\
&\overline{V}(\xi,\tau)=-2\lambda\sin\lambda\tau\ J_0(\lambda\sqrt{\tau^2-\xi^2}) && (\xi\leqq\tau\leqq2-\xi)\\
&\overline{V}(\xi,\tau)=-2\lambda\sin\lambda\tau\{J_0(\lambda\sqrt{\tau^2-\xi^2})+J_0[\lambda\sqrt{\tau^2-(2-\xi)^2}]\} && (2-\xi\leqq\tau\leqq2+\xi)\\
&\overline{V}(\xi,\tau)=-2\lambda\sin\lambda\tau\{J_0(\lambda\sqrt{\tau^2-\xi^2})+J_0[\lambda\sqrt{\tau^2-(2-\xi)^2}]+J_0[\lambda\sqrt{\tau^2-(2+\xi)^2}]\} && (2+\xi\leqq\tau\leqq4-\xi)\\
&\qquad\cdot\quad\cdot\quad\cdot\quad\cdot\quad\cdot
\end{aligned}\right\}\quad\text{(C35)}$$

and

$$\left.\begin{aligned}
&\overline{M}(\xi,\tau)=0 && (0\leqq\tau\leqq\xi)\\
&\overline{M}(\xi,\tau)=H(\xi,\tau) && (\xi\leqq\tau\leqq2-\xi)\\
&\overline{M}(\xi,\tau)=H(\xi,\tau)-H(2-\xi,\tau) && (2-\xi\leqq\tau\leqq2+\xi)\\
&\overline{M}(\xi,\tau)=H(\xi,\tau)-H(2-\xi,\tau)+H(2+\xi,\tau) && (2+\xi\leqq\tau\leqq4-\xi)\\
&\qquad\cdot\quad\cdot\quad\cdot\quad\cdot\quad\cdot
\end{aligned}\right\}\quad\text{(C36)}$$

where

$$H(\xi,\tau)=\cos\lambda\xi-\lambda\xi\int_{\xi}^{\tau}\cos\lambda\theta\frac{J_1(\lambda\sqrt{\theta^2-\xi^2})}{\sqrt{\theta^2-\xi^2}}d\theta$$

The angular and linear velocities $\bar{\omega}$ and $\bar{v}$ may, of course, also be obtained from equations (C33) and (C34).

Computations have been made for the quantity $\bar{V}(0,\tau)$ from equation (C35) with $\lambda=5$. The results have been plotted in figure 5(a) in the range $0\leqq\tau\leqq 8$.

Finite simply supported beam with an applied ramp-platform end moment.—Let the applied moment at the end $\xi=0$ of a simply supported uniform beam have the time history

$$\left.\begin{aligned}\bar{M}(0,\tau)&=\tau\qquad(0\leqq\tau\leqq 1)\\ \bar{M}(0,\tau)&=1\qquad(\tau>1)\end{aligned}\right\}\qquad\text{(C37)}$$

The response of the beam to this disturbance may be obtained from the response to a unit step by using Duhamel's superposition integral. Thus, if the response to a unit step, as given by equation (C33), is designated $Z_1(\xi,\tau)$, the response $Z(\xi,\tau)$ to the applied moment (C37) may be written

$$\left.\begin{aligned}Z(\xi,\tau)&=\int_0^{\tau}Z_1(\xi,\tau-\theta)d\theta\qquad(0\leqq\tau\leqq 1)\\ Z(\xi,\tau)&=\int_0^{1}Z_1(\xi,\tau-\theta)d\theta\qquad(\tau>1)\end{aligned}\right\}\qquad\text{(C38)}$$

Substituting from equations (C33) and letting $\xi=0$ results in the following expressions for the end response $Z(0,\tau)$:

$$\left.\begin{aligned}Z(0,\tau)&=2\lambda i\int_0^{\tau}e^{i\lambda(\tau-\theta)}J_0[\lambda(\tau-\theta)]d\theta\qquad(0\leqq\tau\leqq 1)\\ Z(0,\tau)&=2\lambda i\int_0^{1}e^{i\lambda(\tau-\theta)}J_0[\lambda(\tau-\theta)]d\theta\qquad(1\leqq\tau\leqq 2)\\ Z(0,\tau)&=2\lambda i\int_0^{1}e^{i\lambda(\tau-\theta)}J_0[\lambda(\tau-\theta)]d\theta+\\ &\quad 4\lambda i\int_0^{\tau-2}e^{i\lambda(\tau-\theta)}J_0\left[\lambda\sqrt{(\tau-\theta)^2-4}\right]d\theta\qquad(2\leqq\tau\leqq 3)\\ Z(0,\tau)&=2\lambda i\int_0^{1}e^{i\lambda(\tau-\theta)}J_0[\lambda(\tau-\theta)]d\theta+\\ &\quad 4\lambda i\int_0^{1}e^{i\lambda(\tau-\theta)}J_0\left[\lambda\sqrt{(\tau-\theta)^2-4}\right]d\theta\qquad(3\leqq\tau\leqq 4)\\ Z(0,\tau)&=2\lambda i\int_0^{1}e^{i\lambda(\tau-\theta)}J_0[\lambda(\tau-\theta)]d\theta+\\ &\quad 4\lambda i\int_0^{1}e^{i\lambda(\tau-\theta)}J_0\left[\lambda\sqrt{(\tau-\theta)^2-4}\right]d\theta+\\ &\quad 4\lambda i\int_0^{\tau-4}e^{i\lambda(\tau-\theta)}J_0\left[\lambda\sqrt{(\tau-\theta)^2-16}\right]d\theta,\qquad(4\leqq\tau\leqq 5)\\ &\quad\cdot\qquad\cdot\qquad\cdot\qquad\cdot\qquad\cdot\end{aligned}\right\}\qquad\text{(C39)}$$

It can be seen that (transforms 3, 4, and 9, table I)

$$L\left\{\int_0^{\tau}e^{i\lambda\theta}J_0(\lambda\theta)d\theta\right\}=\frac{1}{p\sqrt{p(p-2\lambda i)}}=-\frac{i}{\lambda}\frac{d}{dp}\left(\sqrt{\frac{p-2\lambda i}{p}}-1\right)$$

or (transforms 5 and 11, table I)

$$\int_0^{\tau}e^{i\lambda\theta}J_0(\lambda\theta)d\theta=\tau e^{i\lambda\tau}[J_0(\lambda\tau)-iJ_1(\lambda\tau)]$$

Then, $Z(0,\tau)$ may be reduced in the region $0\leqq\tau\leqq 2$ to

$$Z(0,\tau)=2\lambda i\int_0^{\tau}e^{i\lambda\theta}J_0(\lambda\theta)d\theta$$

$$Z(0,\tau)=2\lambda i\tau e^{i\lambda\tau}[J_0(\lambda\tau)-iJ_1(\lambda\tau)]\qquad(0\leqq\tau\leqq 1)\qquad\text{(C40)}$$

and

$$\begin{aligned}Z(0,\tau)&=2\lambda i\int_{\tau-1}^{\tau}e^{i\lambda\theta}J_0(\lambda\theta)d\theta\\ &=2\lambda i\left[\int_0^{\tau}e^{i\lambda\theta}J_0(\lambda\theta)d\theta-\int_0^{\tau-1}e^{i\lambda\theta}J_0(\lambda\theta)d\theta\right]\\ Z(0,\tau)&=2\lambda i\tau e^{i\lambda\tau}[J_0(\lambda\tau)-iJ_1(\lambda\tau)]-\\ &\quad 2\lambda i(\tau-1)e^{i\lambda(\tau-1)}\{J_0[\lambda(\tau-1)]-iJ_1[\lambda(\tau-1)]\}\\ &\qquad(1\leqq\tau\leqq 2)\qquad\text{(C41)}\end{aligned}$$

From these equations, the shear at $\xi=0$ may be written in this range as

$$\left.\begin{aligned}\bar{V}(0,\tau)&=2\lambda\tau[\cos\lambda\tau\, J_1(\lambda\tau)-\sin\lambda\tau\, J_0(\lambda\tau)]\qquad(0\leqq\tau\leqq 1)\\ \bar{V}(0,\tau)&=2\lambda\tau[\cos\lambda\tau\, J_1(\lambda\tau)-\sin\lambda\tau\, J_0(\lambda\tau)]-\\ &\quad 2\lambda(\tau-1)\{\cos\lambda(\tau-1)J_1[\lambda(\tau-1)]-\\ &\quad\sin\lambda(\tau-1)J_0[\lambda(\tau-1)]\}\qquad(1\leqq\tau\leqq 2)\end{aligned}\right\}\qquad\text{(C42)}$$

In the range $\tau>2$, however, the integrals in equations (C39) apparently cannot be evaluated analytically. If the shear $\bar{V}_1(0,\tau)$ due to a unit step moment has been computed, it may be convenient to write, for the shear at the point $\xi=0$ resulting from the disturbance defined by equation (C37),

$$\left.\begin{aligned}\bar{V}(0,\tau)&=\int_0^{\tau}\bar{V}_1(0,\tau-\theta)d\theta\qquad(0\leqq\tau\leqq 1)\\ \bar{V}(0,\tau)&=\int_0^{1}\bar{V}_1(0,\tau-\theta)d\theta\qquad(\tau>1)\end{aligned}\right\}\qquad\text{(C43)}$$

and then evaluate the integral $\int_0^1\bar{V}_1(0,\tau-\theta)d\theta$ numerically to obtain $\bar{V}(0,\tau)$ in the range $\tau>2$.

The quantity $\bar{V}(0,\tau)$, for a beam with $\lambda=5$, has been computed in the range $0\leqq\tau\leqq 2$ from equations (C42) and has been obtained in the range $2\leqq\tau\leqq 8$ by a numerical integration of the exact curve of figure 5 (a) in accordance with equation (C43). These results are presented in figure 6 (a).

REFERENCES

1. Biot, M. A., and Bisplinghoff, R. L.: Dynamic Loads on Airplane Structures During Landing. NACA WR W-92, 1944. (Formerly NACA ARR 4H10.)
2. Williams, D.: Displacements of a Linear Elastic System Under a Given Transient Load. British S.M.E. C/7219/DW/19, Aug. 1946.
3. Isakson, G.: A Survey of Analytical Methods for Determining Transient Stresses in Elastic Structures. Contract No. N5 ori-07833, Office Naval Res. (Project NR-035-259), M.I.T., Mar. 3, 1950.
4. Ramberg, Walter: Transient Vibration in an Airplane Wing Obtained by Several Methods. Res. Paper RP1984, Nat. Bur. of Standards Jour. Res., vol. 42, no. 5, May 1949, pp. 437–447.
5. Kruszewski, Edwin T.: Effect of Transverse Shear and Rotary Inertia on the Natural Frequency of a Uniform Beam. NACA TN 1909, 1949.
6. Flügge, W.: Die Ausbreitung von Biegungswellen in Stäben. Z.a.M.M., Bd. 22, Nr. 6, Dec. 1942, pp. 312–318.
7. Timoshenko, S.: Vibration Problems in Engineering. Second ed., D. Van Nostrand Co., Inc., 1937, pp. 337–338.
8. Robinson, A.: Shock Transmission in Beams. R. & M. No. 2265, British A.R.C., 1945.
9. Pfeiffer, F.: Über die Differentialgleichung der transversalen Stabschwingungen. Z.a.M.M., Bd. 25/27, Nr. 3, June 1947, pp. 83–91.
10. Uflyand, Y. S.: Rasprostranenie voln pri poperechnykh kolebaniyakh sterzhney i plastin. (Propagation of Waves in Transverse Vibrations of Beams and Plates.) Prikladnaya Matematika i Mekhanika (Moscow, Leningrad), vol. XII, no. 3, May–June 1948, pp. 287–300.
11. Dengler, M. A., and Goland, M.: Transverse Impact of Long Beams, Including Rotatory Inertia and Shear Effects. Proc. First U. S. Nat. Cong. Appl. Mech. (Chicago, Ill., 1951), A.S.M.E., 1952, pp. 179–186.
12. Courant, R., and Friedrichs, K. O.: Supersonic Flow and Shock Waves. Pure & Appl. Math., vol. I, Interscience Publishers, Inc. (New York), 1948, pp. 38–59.
13. Prescott, John: Elastic Waves and Vibrations of Thin Rods. Phil. Mag., ser. 7, vol. 33, no. 225, Oct. 1942, pp. 703–754.
14. Cooper, J. L. B.: The Propagation of Elastic Waves in a Rod. Phil. Mag., ser. 7, vol. 38, no. 276, Jan. 1947, pp. 1–22.
15. Churchill, Ruel V.: Modern Operational Mathematics in Engineering. McGraw-Hill Book Co., Inc., 1944.
16. Campbell, George A., and Foster, Ronald M.: Fourier Integrals for Practical Applications. Monograph B-584, Bell Telephone System, 1942.
17. McLachlan, N. W.: Bessel Functions for Engineers. Clarendon Press (Oxford), 1934.
18. Jahnke, Eugene, and Emde, Fritz: Tables of Functions. Fourth Ed., Dover Publications, 1945, p. 144.

Reprinted from *J. Appl. Mech.*, **20,** 511–514 (1953)

Flexural Wave Solutions of Coupled Equations Representing the More Exact Theory of Bending

By JULIUS MIKLOWITZ,[2] PASADENA, CALIF.

Presented here is a new method for deriving flexural wave solutions for the Timoshenko bending theory. The method is based on a breakdown of the total deflection into its bending and shear components. Instead of treating the full Timoshenko equation (1)[3] an equivalent set of coupled equations, representing the rotational and translatory motions of the beam element, is solved. The advantages of this method stem from (*a*) the simplicity of the associated expressions for the moment and shear force, which are the elementary bending theory relations, and (*b*) the well-defined nature of the related boundary conditions. The latter is particularly important since it is difficult to define the proper boundary conditions associated with the full Timoshenko equation. This is evidenced in the works of Uflyand (2) and Dengler and Goland (3), both of which are concerned with wave solutions for the infinite beam under the action of a concentrated transverse load. The quoted work (3) points out the erroneous boundary conditions used in the Uflyand work (2). The present method is applied to the same case treated in the works (2, 3). Agreement is shown with the Dengler and Goland solution. The Uflyand solution is shown to have meaning when interpreted properly. The derivation of transforms for other beam cases, both finite and infinite, by the present method has also been included in this work.

Nomenclature

The following nomenclature is used in the paper:

A, A_j = arbitrary functions of p

A_0 = cross-sectional area of beam, a constant

B, B_j = arbitrary functions of p

$C = \dfrac{A_0\gamma}{EIg}$, where E = modulus of elasticity; I = moment of inertia of beam cross section with respect to neutral axis of bending; g = acceleration due to gravity; γ = specific weight of beam material

C_j = arbitrary functions of p

D_j = arbitrary functions of p

$M(x, t)$ = moment

$S(x, t)$ = shear force

c_1 = a limiting wave velocity, defined by $\sqrt{\dfrac{Eg}{\gamma}}$

c_2 = a limiting wave velocity, defined by $\sqrt{\dfrac{k'Gg}{\gamma}}$, where

$k'G$ = effective shear modulus, relating average shear stress acting over a beam cross section to shear deformation at neutral plane in accord with elementary bending

l = length

$m(x, p)$ = Laplace transform of $M(x, t)$

n, n_j = roots of characteristic equation

p = transformation parameter

$s(x, p)$ = Laplace transform of $S(x, t)$

t = time

$v(x, p)$, $v_b(x, p)$, $v_s(x, p)$ = Laplace transforms of $y(x, t)$, $y_b(x, t)$, and $y_s(x, t)$

x = co-ordinate along beam

$y(x, t)$ = total lateral deflection of beam element

$y_b(x, t)$ = lateral deflection of beam element due to bending

$y_s(x, t)$ = lateral deflection of beam element due to shear

Introduction

As a means of studying the propagation of flexural waves in beams, recent work in the literature has been focused on the derivation of wave solutions of the more exact equation of bending due to Timoshenko (1). Uflyand (2) showed that this differential equation, the derivation of which includes considerations of the shear force and rotatory inertia effects, had wave character as against the nonwave character of the elementary bending equation. He was the first to derive wave solutions, treating both the infinite and finite beams subjected to a transverse impact. Later Dengler and Goland (3), focusing on the infinite beam, derived wave solutions of the Timoshenko equation which are related to Uflyand's solution. In this work (3) it is pointed out that the boundary conditions associated with the Uflyand solution (2) are incorrect. With this as an example, Dengler and Goland comment on the difficulty of defining the proper boundary conditions (at a cut) associated with the Timoshenko equation. In their solutions a definite loading is applied at $x = 0$ and no cut is required.

Instead of working with the total deflection equation as was done in the quoted works (2, 3), Anderson (4), concerned with classical solutions (nonwave), used the equivalent set of coupled equations

$$EI\frac{\partial^3 y_b}{\partial x^3} + k'A_0G\frac{\partial y_s}{\partial x} - \frac{I\gamma}{g}\frac{\partial^3 y_b}{\partial x\partial t^2} = 0 \quad\ldots\ldots[1a]$$

$$\frac{\gamma A_0}{g}\frac{\partial^2 y_b}{\partial t^2} + \frac{\gamma A_0}{g}\frac{\partial^2 y_s}{\partial t^2} - k'A_0G\frac{\partial^2 y_s}{\partial x^2} = 0\ldots\ldots[1b]$$

[1] The work reported here was supported by the U. S. Naval Ordnance Test Station through the Exploratory and Foundational Research program.

[2] Consultant, Naval Ordnance Test Station.

[3] Numbers in parentheses refer to the Bibliography at the end of the paper.

Contributed by the Applied Mechanics Division and presented at the Semi-Annual Meeting, Los Angeles, Calif., June 28–July 2, 1953, of The American Society of Mechanical Engineers.

Discussion of this paper should be addressed to the Secretary, ASME, 29 West 39th Street, New York, N. Y., and will be accepted until January 11, 1954, for publication at a later date. Discussion received after the closing date will be returned.

Note: Statements and opinions advanced in papers are to be understood as individual expressions of their authors and not those of the Society. Manuscript received by ASME Applied Mechanics Division, February 12, 1953. Paper No. 53—SA-6.

These equations[4] result by employing the transformation

$$\left.\begin{aligned} y &= y_b + y_s \\ \psi &= \frac{\partial y_b}{\partial x} \end{aligned}\right\} \qquad [2]$$

at the appropriate point in the Timoshenko derivation. Equation [1a] is the differential equation for the rotation, and Equation [1b] for the translatory motion of the beam element.[5] The advantages in using the coupled equations in deriving solutions for the Timoshenko bending mechanism stem from the simplicity of associated expressions. The moment and shear force are those of the elementary bending theory

$$M(x, t) = EI \frac{\partial^2 y_b}{\partial x^2} \qquad [3a]$$

$$S(x, t) = -k' A_0 G \frac{\partial y_s}{\partial x} \qquad [3b]$$

In addition the quoted work (4) points out the well-defined nature of the associated boundary conditions.[6]

This paper presents a method of deriving wave solutions for the Timoshenko theory based on the coupled-equation approach. As might be expected, the simplicity of Equations [3], and particularly the well-defined nature of the boundary conditions associated with Equations [1], in view of the difficulty discussed in the quoted works (2, 3), are definite aids in the wave treatment. The present method is applied to the infinite beam case treated in the literature (2, 3), and agreement is shown with the Dengler and Goland solution (3). The Uflyand solution (2) is shown to have meaning when interpreted properly.

A section of the paper is devoted to a discussion of wave solutions for finite beams.

The Method

The method presented offers a convenient means of obtaining separate wave solutions for any or all of the basic items of interest in beam action; the deflections $y_b(x, t)$ and $y_s(x, t)$, the moment $M(x, t)$ and the shear force $S(x, t)$. In view of the nature of Equations [1] interest initially focuses on the deflections. With the initial conditions

$$y_b(x, 0) = y_s(x, 0) = \frac{\partial y_b(x, 0)}{\partial t} = \frac{\partial y_s(x, 0)}{\partial t} = 0$$

the Equations [1] transform to

$$\left.\begin{aligned} EI \frac{d^3 v_b}{dx^3} + k' A_0 G \frac{dv_s}{dx} - \frac{I\gamma}{g} p^2 \frac{dv_b}{dx} &= 0 \\ \frac{\gamma}{g} p^2 v_b + \frac{\gamma}{g} p^2 v_s - k' G \frac{d^2 v_s}{dx^2} &= 0 \end{aligned}\right\} \qquad [4]$$

according to the Laplace transform theory (9).

Equations [4] constitute a system of two homogeneous linear ordinary differential equations. A possible solution of these equations is of the type

$$\left.\begin{aligned} v_b(x, p) &= Ae^{-nx} \\ v_s(x, p) &= Be^{-nx} \end{aligned}\right\} \qquad [5]$$

The system of Equations [4] is satisfied therefore if

$$\left.\begin{aligned} \left(\frac{I\gamma}{g} p^2 n - EIn^3\right) A - k' A_0 G n B &= 0 \\ \frac{\gamma}{g} p^2 A + \left(\frac{\gamma}{g} p^2 - k' G n^2\right) B &= 0 \end{aligned}\right\} \qquad [6]$$

and these equations are compatible if the determinant

$$\begin{vmatrix} \left(\frac{I\gamma}{g} p^2 n - EIn^3\right) & -k' G A_0 n \\ \frac{\gamma}{g} p^2 & \left(\frac{\gamma p^2}{g} - k' G n^2\right) \end{vmatrix} = 0 \qquad [7]$$

Equation [7], of fourth degree in n, is the characteristic equation of the system of Equations [4]. Introducing the quantities c_1, c_2, and C, Equation [7] may be written as

$$n^4 - p^2 \left(\frac{1}{c_2^2} + \frac{1}{c_1^2}\right) n^2 + p^2 \left(\frac{p^2}{c_1^2 c_2^2} + C\right) = 0 \qquad [7a]$$

Assuming that Equation [7a] has four distinct roots n_1, n_2, n_3, and n_4, then the general solution of the system of Equations [4] is

$$v_b(x, p) = \sum_{j=1}^{4} A_j e^{-n_j x} \qquad [8a]$$

$$v_s(x, p) = \sum_{j=1}^{4} B_j e^{-n_j x} \qquad [8b]$$

where A_j, B_j, and n_j, all functions of p, satisfy Equations [6].

Equations [6] must hold for all values of n_j, therefore

$$\frac{B_j}{A_j} = \frac{\frac{p^2}{c_1^2} - n_j^2}{c_2^2 C} = -\frac{p^2}{p^2 - c_2^2 n_j^2} \qquad [9]$$

and eliminating the B_j-values we have for Equations [8]

$$v_b(x, p) = \sum_{j=1}^{4} A_j e^{-n_j x} \qquad [10a]$$

$$v_s(x, p) = \sum_{j=1}^{4} \phi_j A_j e^{-n_j x} \qquad [10b]$$

where
$$\phi_j = \frac{\frac{p^2}{c_1^2} - n_j^2}{c_2^2 C}$$

Equations [10] contain the four required arbitrary functions $A_j(p)$ which are determined by the boundary conditions of a particular problem.[7] It may be noted that Equation [7a] produces the same biquadratic expression for n_j as in the Uflyand work (2).[8] In terms of his nomenclature then

$$n_j = \pm M \sqrt{p} \sqrt{p \mp N \sqrt{p^2 - a^2}} \qquad [11]$$

where

$$M = \frac{1}{\sqrt{2}} \sqrt{\frac{1}{c_2^2} + \frac{1}{c_1^2}}, \quad N = \frac{\frac{1}{c_2^2} - \frac{1}{c_1^2}}{\frac{1}{c_2^2} + \frac{1}{c_1^2}}, \quad a = \frac{2\sqrt{C}}{\frac{1}{c_2^2} - \frac{1}{c_1^2}}$$

[4] These equations are also stated in the works of Davidson and Meier (5), and Robinson (6), and suggested in the work of Mindlin (7).

[5] Equations [c], [d] reference (1), p. 338.

[6] The author learned later of the work of Boley, Heninger, and Zimnoch (8), which contains a variational derivation of the coupled Equations [1] and related boundary conditions similar to that associated with the work (4).

[7] This method of producing Equations [10] is discussed in von Kármán and Biot (10).

[8] See equations [1.6] and [1.7] in reference (2).

Of primary interest in engineering problems are the stresses, hence the importance of the transforms for the moment and shear force. They may be obtained from Equations [10] by making use of Equations [3]

$$m(x, p) = EI \sum_{j=1}^{4} n_j^2 A_j e^{-n_j x} \qquad [12a]$$

$$s(x, p) = k' A_0 G \sum_{j=1}^{4} n_j \phi_j A_j e^{-n_j x} \qquad [12b]$$

Note that each of the Equations [10*a*], [10*b*], [12*a*], and [12*b*] bears marked resemblance to Uflyand's transform for the total deflection in a beam (2).[9] Depending on the information desired, the general procedure Uflyand used in getting his inverse transform could be followed for any or all of the inverse transforms associated with Equations [10*a*], [10*b*], [12*a*], and [12*b*]. The method employs the complex inversion integral and contour integration after the functions $A_j(p)$ are properly defined for the particular problem being solved. The details are given in the quoted work (2).

Infinite-Beam Solutions

The present method will now be applied to the problem

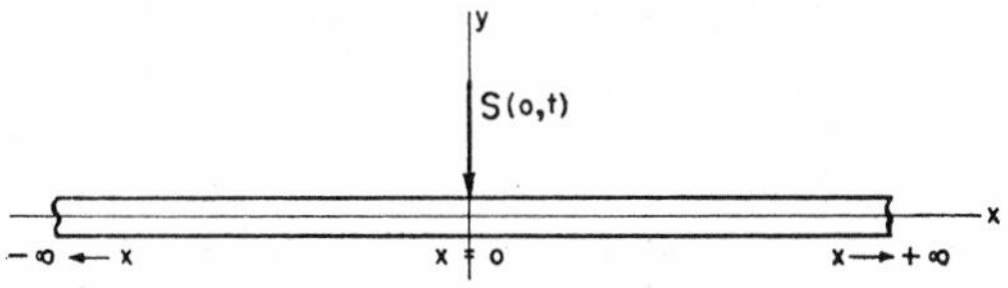

Fig. 1 Infinite-Beam Problem
(Concentrated transverse load at $x = 0$.)

treated in the quoted works (2, 3), that of an infinite beam with a concentrated transverse load applied at the station $x = 0$, Fig. 1. The transform derived agrees with that Dengler and Goland derived (3). Further, the Uflyand solution (2), previously shown to be in error because of incorrect boundary conditions (3), is given meaning.

The boundary conditions[10] at $x = 0$ are taken as

$$\left.\begin{aligned} \frac{\partial y_s(0, t)}{\partial x} &= -\frac{S(0, t)}{k' A_0 G} \\ \frac{\partial y_b(0, t)}{\partial x} &= 0 \end{aligned}\right\} \qquad [13]$$

and the $x > 0$ portion of the beam is considered. The transformation of these yields

$$\left.\begin{aligned} \frac{dv_s(0, p)}{dx} &= -\frac{s(0,p)}{k' A_0 G} \\ \frac{dv_b(0, p)}{dx} &= 0 \end{aligned}\right\} \qquad [14]$$

Applying these to the Equations [10*a*] and [10*b*] the latter become

$$v_b(x, p) = A_1 e^{-n_1 x} + A_2 e^{-n_2 x} \qquad [15a]$$

$$v_s(x, p) = \phi_1 A_1 e^{-n_1 x} + \phi_2 A_2 e^{-n_2 x} \qquad [15b]$$

where

$$A_1(p) = \frac{s(0, p)}{EI n_1 (n_2^2 - n_1^2)}$$

$$A_2(p) = -\frac{s(0, p)}{EI n_2 (n_2^2 - n_1^2)}$$

and

$$n_{1,2} = M \sqrt{p} \sqrt{p \mp N \sqrt{p^2 - a^2}}$$

these two values of n necessarily being chosen to satisfy the Bromwich contour condition of Re $n_1 > 0$, Re $n_2 > 0$. The expression for the moment, and hence bending stress, can be obtained directly from Equation [12*a*]

$$m(x, p) = EI [n_1^2 A_1 e^{-n_1 x} + n_2^2 A_2 e^{-n_2 x}] \qquad [16]$$

Equation [16] agrees with the transform Dengler and Goland derived for the moment (3).[11] Interesting is the fact that Equation [15*a*] and the transform Uflyand derived for the total deflection (2)[12] are the same. Hence it is clear that his solution may be used if it is interpreted as just that of the bending deflection.

In concluding this section of the paper the author would like to comment briefly on another basic case he has treated—that

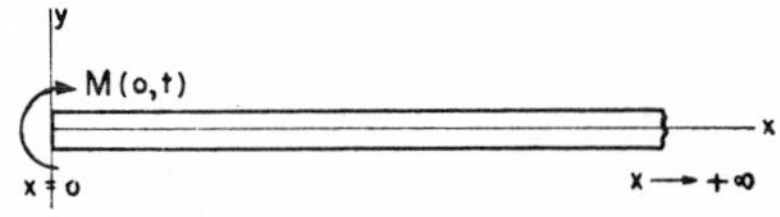

Fig. 2 Semi-Infinite-Beam Problem
(Pure moment at $x = 0$.)

of a pure moment applied to the free end of a semi-infinite beam, Fig. 2. The $x = 0$ boundary conditions for this case are

$$\left.\begin{aligned} \frac{\partial^2 y_b(0, t)}{\partial x^2} &= \frac{M(0, t)}{EI} \\ \frac{\partial y_s(0, t)}{\partial x} &= 0 \end{aligned}\right\} \qquad [17]$$

Interest in this work was in the bending stresses; hence attention was focused on the moment transform. Equations [17] yield for this case the Equation [16]

where

$$\left.\begin{aligned} A_1 &= \frac{c_2^2 C m(0, p)\, \phi_2}{EI\,(n_1 - n_2)\left(\dfrac{p^2}{c_1^2} + n_1 n_2\right) n_1} \\ A_2 &= -\frac{c_2^2 C m(0, p)\, \phi_1}{EI\,(n_1 - n_2)\left(\dfrac{p^2}{c_1^2} + n_1 n_2\right) n_2} \end{aligned}\right\} \qquad [18]$$

and where the same comments on n apply, as in the previous case. The complete solution for this case, with $M(0, t)$ taken as a sharp half-sine input, will be presented in the near future. Curves representing the wave shapes will be included.

Finite-Beam Solutions

It is of interest engineering-wise to consider wave solutions for the finite-beam cases. The general transformed solutions for the finite beam are contained in Equations [10*a*], [10*b*], [12*a*], and [12*b*]. They can be written in terms of n_1 and n_2 since $n_3 = -n_1$, and $n_4 = -n_2$ follow from Equation [11]; hence the general transformed solutions for the deflections may be written as

[9] See equation [1.8] in reference (2).

[10] These were considered by the present author to be the proper conditions for this case. They are actually one of four possible sets of associated end conditions generated in the variational derivation mentioned in footnote 6.

[11] See equation [39] of reference (3).

[12] See equations [1.8] and [4.3] of reference (2).

$$v_b(x, p) = C_1 \cosh n_1x + D_1 \sinh n_1x + C_2 \cosh n_2x + D_2 \sinh n_2x \quad [19a]$$

$$v_s(x, p) = \phi_1[C_1 \cosh n_1x + D_1 \sinh n_1x] + \phi_2[C_2 \cosh n_2x + D_2 \sinh n_2x] \quad [19b]$$

and for the moment and shear force as

$$m(x, p) = EI[n_1^2(C_1 \cosh n_1x + D_1 \sinh n_1x) + n_2^2(C_2 \cosh n_2x + D_2 \sinh n_2x)] \quad [20a]$$

$$s(x, p) = -k'A_0G[n_1\phi_1(C_1 \sinh n_1x + D_1 \cosh n_1x) + n_2\phi_2(C_2 \sinh n_2x + D_2 \cosh n_2x)] \quad [20b]$$

where C and D are functions of p, and are determined by the boundary conditions of the problem. The case treated by Uflyand, the simply supported beam with a transverse force at $x = 0$, will now be treated by the present method as an example,

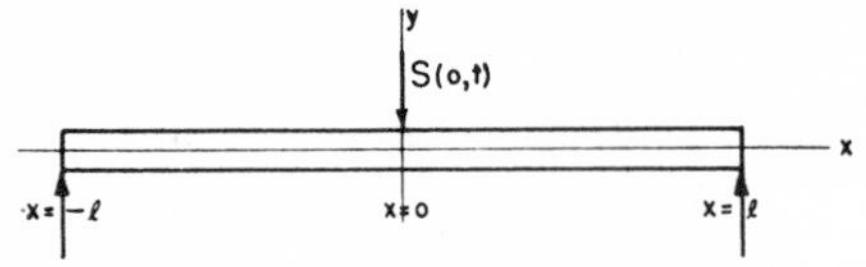

Fig. 3 Finite-Beam Problem
(Concentrated transverse load at $x = 0$.)

Fig. 3. For this case, the following conditions at $x = l$ (4) are needed in addition to the conditions at $x = 0$.[13]

$$\left.\begin{aligned} y_b(l, t) + y_s(l, t) &= 0 \\ \frac{\partial^2 y_b(l, t)}{\partial x^2} &= 0 \end{aligned}\right\} \quad [21]$$

These conditions applied to Equations [19a] and [19b] produce the following transform for the bending deflection

$$v_b(x, p) = \frac{s(0, p)}{EI(n_2^2 - n_1^2)}\left[\frac{\sinh n_2(x - l)}{n_2 \cosh n_2l} - \frac{\sinh n_1(x - l)}{n_1 \cosh n_1l}\right] \quad [22]$$

and similar expressions for the shear deflection, moment, and shear force. Note here again, as might be expected, that Equation [22] is the same as the transform for the total deflection in the Uflyand work (2).[14] Hence the remainder of his finite-beam solution may be used if the results are interpreted as just the bending deflection. As Uflyand points out, the form of Equation [22] may be interpreted in its wave sense, i.e.

$$\frac{\sinh n(x - l)}{\cosh nl} = -e^{-nx} + e^{-n(2l-x)} + e^{-n(2l+x)} - e^{-n(4l-x)} - e^{-n(4l+x)} + \ldots \quad [23]$$

the terms after the first being related to wave reflections from the beam ends. A transform in this form simplifies the finite-beam case since the complete solution may be written from the infinite-beam solution. The latter is based on the first term on the right-hand side of Equation [23]. Therefore the reflection contributions can be obtained from the infinite-beam solution by substituting $2l - x$, $2l + x$, $4l - x$, and so on, for x, according to Equation [23].[15] It should be pointed out that wave solutions of finite-beam problems with boundary conditions involving an odd and an even derivative are much more difficult since the transform cannot be written in the form of Equation [22]. A fixed or free end gives this difficulty, the conditions on the former being

$$\left.\begin{aligned} y_b + y_s &= 0 \\ \frac{\partial y_b}{\partial x} &= 0 \end{aligned}\right\} \quad [24]$$

and on the latter

$$\left.\begin{aligned} \frac{\partial y_s}{\partial x} &= 0 \\ \frac{\partial^2 y_b}{\partial x^2} &= 0 \end{aligned}\right\} \quad [25]$$

The case shown in Fig. 4 would give a convenient solution. The solution for this case would be of interest since it would

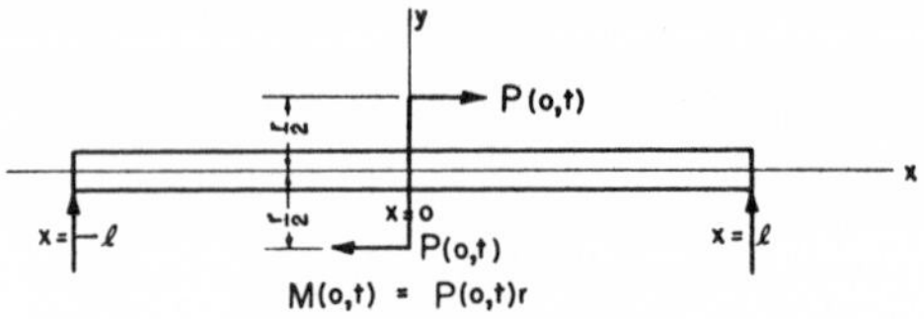

Fig. 4 Finite-Beam Problem
(Moment at $x = 0$.)

yield information on the propagation of waves in a finite beam due to a moment source. The boundary conditions at $x = 0$ for this case[10] would be

$$\left.\begin{aligned} \frac{\partial^2 y_b(0, t)}{\partial x^2} &= \frac{M(0, t)}{EI} \\ y_b(0, t) + y_s(0, t) &= 0 \end{aligned}\right\} \quad [26]$$

and the conditions at $x = l$ are again Equations [21].

Acknowledgment

Acknowledgment is made to the U. S. Naval Ordnance Test Station for permission to publish this work.

BIBLIOGRAPHY

1 "Vibration Problems in Engineering," by S. Timoshenko, D. Van Nostrand Company, Inc., New York, N. Y., 1937, p. 338.

2 "The Propagation of Waves in the Transverse Vibration of Bars and Plates," by Ya. S. Uflyand, *Prikladnaia Matematika i Mekhanika*, vol. 12, May–June, 1948, pp. 287–300 (in Russian). See Rev. 33, AMR 3.

3 "Transverse Impact of Long Beams, Including Rotatory Inertia and Shear Effects," by M. A. Dengler and M. Goland, Proceedings of First U. S. National Congress of Applied Mechanics, 1951.

4 "Flexural Vibrations of Uniform Beams According to the Timoshenko Theory," by R. A. Anderson, published in this issue, pp. 504–510.

5 "Impact on Prismatical Bars," by T. Davidson and J. H. Meier, Proceedings of the SESA, vol. 4, 1946, p. 106.

6 "Shock Transmission in Beams," by A. Robinson, Aeronautical Research Council Technical Report No. 2265, Ministry of Supply, London, England, 1950, p. 16.

7 "Influence of Rotatory Inertia and Shear on Flexural Motions of Isotopic, Elastic Plates," by R. D. Mindlin, Journal of Applied Mechanics, Trans. ASME, vol. 73, 1951, pp. 31–38. Reference is to p. 33.

8 "An Energy Theory of Transverse Impact on Beams," by B. A. Boley, R. E. Heninger, and V. P. Zimnoch, ONR Report, Project NR-064-355, January, 1952.

9 "Modern Operational Mathematics in Engineering," by R. V. Churchill, McGraw-Hill Book Co., Inc., New York, N. Y., 1944.

10 "Mathematical Methods in Engineering," by Th. von Kármán and M. A. Biot, McGraw-Hill Book Co., Inc., New York, N. Y., 1940, p. 41.

[13] See Equations [13].

[14] See equation [7.6] in reference (2). There is a misprint in this equation: $u(x, t)$ should be replaced by $v(x, p)$.

[15] Equations [7.6] to [7.13] in reference (2).

5

Reprinted from *J. Appl. Mech.*, **22**, 69–76 (1955)

An Approximate Theory of Lateral Impact on Beams[1]

By B. A. BOLEY,[2] NEW YORK, N. Y.

The approximate theory derived in this paper describes, by means of a "traveling-wave" approach, the behavior of beams under transverse impact. Lateral impact is considered in detail, namely, one in which a section of the beam undergoes a sudden change in velocity or shear force. The theory considers the effects of shear deformations and of rotatory inertia according to Timoshenko's model, and that of lateral contraction as suggested by Love. The governing equations and the boundary conditions are developed with the aid of an energy-variation technique. Numerical examples are given in which the behavior of the boundary layer near the point of impact is examined. For one of these the exact solution is available and is in agreement with the present approximate results. Some general considerations concerning the velocity of propagation also are discussed.

Introduction

THE behavior of beams under various types of dynamic lateral loads has been found, in some experimental investigations, to be well approximated by the Bernoulli-Euler theory of bending (2, 3, 4, 5).[3] This is not true, however, of the transmission of pulses with very short wave lengths; in such cases it becomes necessary to consider the shearing deformations and the rotatory inertia of the beam. Timoshenko's beam model (6) takes these quantities into account and has been shown by Mindlin (7, 8) to give results which are in good agreement with those obtained either experimentally or from the three-dimensional theory of elasticity (9, 10, 11, 12). A similarly favorable comparison is given by Kolsky (13).

Some exact solutions of the Timoshenko beam equations (which result from the ones used in this paper when lateral contraction terms are neglected) have been obtained for certain special cases (14, 15, 17, 28) but, in general, are quite cumbersome. An attempt is presented here to develop simpler, if approximate, solutions; it employs a traveling-wave approach and is based on Timoshenko's model with an additional correction for lateral contraction analogous to that introduced by Pochhammer (16) in the case of longitudinal impact. The present paper represents a first effort in this direction and consequently is limited in many respects. For example, attention has been concentrated on the behavior of a short portion of the beam close to the point of impact, with generalizations indicated briefly.

Consider a bar, to one end of which is suddenly imparted, for example, a lateral velocity; the ensuing motion of the beam will then certainly contain high-frequency components and therefore will require consideration of the shear and rotatory inertia effects. These effects will be of great importance in the neighborhood of the sudden jump in velocity, but the experimental evidence previously cited [and in particular the work of Vigness (reference 5) who considers the present loading condition] indicates that at some distance from the disturbance front they are not essential for a prediction of the bending stresses.

A consideration of the orders of magnitude of the various terms in the basic equation is carried out in this paper and shows that near the point of impact the shearing deformations and rotatory inertia effect are important, principally in a short "boundary layer" immediately following the disturbance front. The length of this layer is of the order of the cross-sectional dimensions of the bar. Therefore the solution can be set up in two parts; i.e., within the boundary layer the Timoshenko model must be used, but for the remainder of the beam the Bernoulli-Euler theory is adequate.

The behavior of the boundary layer, arising in beams under lateral impact, is described in this paper with an approximate energy theory. Expressions are assumed, in terms of some arbitrary parameters, for the deflection of the beam and for the distance traveled by the disturbance. The parameters are then determined with the aid of the principle of virtual displacements. Four illustrative examples are presented. The exact solution for one of these is known (28) and is in good agreement with the present results.

Basic Assumptions and Relations

Consider a bar, not necessarily uniform in cross section, of length L. Let x measure the distance along the bar, and let y and z be the principal axes of the cross section. Assume that the centroids of all cross sections lie on the x-axis.

The bar is struck at the end where $x = 0$ with a force $P = P(t)$ in the z-direction, the time $t = 0$ being chosen as the beginning of impact. Let the displacements of any point (x, y, z) be u_1, v_1, and w_1, in the x, y, and z-directions, respectively, and denote by $w = w(x, t)$ the displacement, in the z-direction, of points on the x-axis. Assume that displacement w consists of two parts; one, denoted by w_b, taking place with bending, but no shearing, deformations; the other, denoted by w_s, taking place with shearing, but no bending, deformations. Assume further that u_1 and v_1 vanish on the x-axis.

The displacements of points off the x-axis may be related to those of points on this axis, as a first approximation, by means of the formulas for the case of pure bending (18, 19); then

$$\left.\begin{aligned} u_1 &= -w_b'z \\ v_1 &= \nu w_b''zy \\ w_1 &= w_s + w_b + (\nu/2)w_b''(z^2 - y^2) \end{aligned}\right\} \quad \ldots\ldots\ldots [1]$$

where ν is Poisson's ratio and primes indicate differentiation with respect to x. The strains and stresses at any point in the

[1] The work reported here was sponsored by the Office of Naval Research, U. S. Navy, under contract N6-onr-225-29, project NR-064-355 (1).[3]

[2] Associate Professor of Civil Engineering, Columbia University.

[3] Numbers in parentheses refer to the Bibliography at the end of the paper.

Contributed by the Applied Mechanics Division and presented at the Annual Meeting, New York, N. Y., November 28–December 3, 1954, of The American Society of Mechanical Engineers.

Discussion of this paper should be addressed to the Secretary, ASME, 29 West 39th Street, New York, N. Y., and will be accepted until April 11, 1955, for publication at a later date. Discussion received after the closing date will be returned.

Note: Statements and opinions advanced in papers are to be understood as individual expressions of their authors and not those of the Society. Manuscript received by ASME Applied Mechanics Division January 12, 1953. Paper No. 54—A-24.

bar are then, in the usual notation

$$\left.\begin{aligned} \epsilon_x &= -(\epsilon_y/\nu) = -(\epsilon_z/\nu) = (\sigma_x/E) = -w_b''z \\ \gamma_{yz} &= \sigma_y = \sigma_z = \tau_{yz} = 0 \\ \gamma_{xy} &= (\tau_{xy}/G) = w_{bxy}''' \\ \gamma_{xz} &= (\tau_{xz}/G) = w_s' + (\nu/2)\, w_b'''\,(z^2 - y^2) \end{aligned}\right\} \quad \ldots [2]$$

where E is Young's modulus and $G = E/(2 + 2\nu)$ the shear modulus.

The striking of the bar gives rise to a disturbance which travels along the bar; let the distance traveled in a time t be $L_1 = L_1(t)$; then $L_1(0) = 0$. It follows that

$$w = w_b = w_s = w_b' = 0, \text{ if } x \geq L_1 \ldots\ldots\ldots\ldots [3]$$

is necessary for continuity of the bar, and because sudden changes in slope can only occur with the action of shear. From these equations the kinematical conditions to be satisfied at the head of the disturbance wave (18, art. 205) are easily derived to be

$$w'\dot{L}_1 + \dot{w} = w_b'\dot{L}_1 + \dot{w}_b = w_s'\dot{L}_1 + \dot{w}_s$$
$$= w_b''\dot{L}_1 + \dot{w}_b' = 0 \ldots\ldots [4]$$

where dots indicate differentiation with respect to time and $\dot{L}_1$ represents the velocity of advance of the wave front.

Derivation of Basic Equations

The basic equations of the problem of transverse impact on a bar will now be developed from the principle of virtual displacements, with the aid of D'Alembert's principle. In the present problem

$$\delta(U + V + K_1) = 0 \ldots\ldots\ldots\ldots\ldots\ldots [5]$$

where the variation of the strain energy is

$$\delta U = \delta(1/2) \int_0^{L_1} \int \int (\sigma_x\epsilon_x + \sigma_y\epsilon_y + \sigma_z\epsilon_z + \tau_{xy}\gamma_{xy} + \tau_{yz}\gamma_{yz} + \tau_{xz}\gamma_{xz})\, dx\, dy\, dz \ldots\ldots [6]$$

the variation of the potential of the external forces is

$$\delta V = -P(t)\delta w(0, t) + \int_0^{L_1} \int \int \rho(\ddot{u}_1\, \delta u_1 + \ddot{v}_1\, \delta v_1 + \ddot{w}_1\, \delta w_1)\, dx\, dy\, dz \ldots\ldots [7]$$

the increment in kinetic energy due to the variation of the length L_1 is

$$\delta K_1 = [K_{x = L_1}]\delta L_1 \ldots\ldots\ldots\ldots\ldots\ldots [8]$$

and where the kinetic energy K per unit of length is

$$K = (\rho/2) \int \int [(\dot{u}_1)^2 + (\dot{v}_1)^2 + (\dot{w}_1)^2]\, dy\, dz \ldots\ldots [9]$$

In the foregoing equations $\rho = \rho(x)$ is the mass density of the bar, and the limits of integrations in the y and z-directions must be chosen so as to cover the entire cross section.

The last three terms in Equation [6], which represent the strain energy of shear, cannot be calculated directly from Equations [2] because the shear stress due to w_s is incorrectly given there as constant in any one cross section. A numerical coefficient N is therefore introduced, which differs from the usual one (20, 21) because of the inclusion, in its calculation, of the effect of lateral contraction. The determination of N is carried out in the Appendix for a rectangular section and the results are plotted in Fig. 1. The final form of the strain-energy variation is then

$$\delta U = \delta \int_0^{L_1} (EI_y/2)(w_b'')^2\, dx + \delta \int_0^{L_1} (G/N)(A/2)(w' - w_b')^2\, dx \ldots\ldots [10]$$

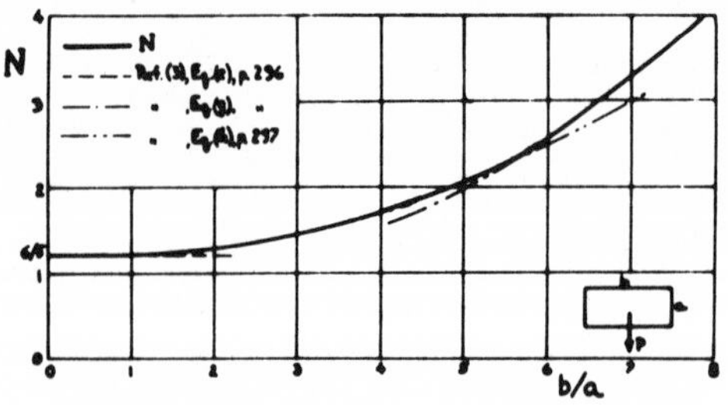

Fig. 1 Values of N for Rectangular Sections

where I_y is the moment of inertia of the cross section about the y-axis.

The various energy quantities can now be evaluated with the aid of Equations [1] and the variations indicated in Equation [5] carried out according to the rules of the calculus of variations. The variation in the length L_1, appearing in the upper limits of the integrals of Equation [10], must be taken into account with the aid of Leibnitz's Formula (22)

$$\delta \int_{f_1}^{f_2} F(x)dx = \int_{f_1}^{f_2} \delta F(x)dx + F(f_2)\,\delta f_2 - F(f_1)\delta f_1 \ldots [11]$$

The final form of Equation [5], correct to terms of the fourth order in the cross-sectional dimensions, is then

$$\left.\begin{aligned} &\int_0^{L_1} \{-(EI_y w_b'')' - (GA/N)(w' - w_b') + \rho I_y \ddot{w}_b' - (\nu\rho I_y k\ddot{w}/2)'\}\, \delta w_b' dx \\ &+ \int_0^{L_1} \{-[(GA/N)(w' - w_b')]' + \rho A\ddot{w} + (\nu\rho I_y k\ddot{w}_b''/2)\}\, \delta w\, dx \\ &+ \{[(EI_y w_b'' + (\nu\rho I_y k\ddot{w}/2)]\, \delta w_b' + (GA/N)(w' - w_b')\,\delta w + [(EI_y/2)(w_b'')^2 \\ &+ (G/N)(A/2)(w' - w_b')^2 + (\rho A/2)(\dot{w})^2 + (\rho I_y/2)(\dot{w}_b')^2 \\ &+ (\nu\rho I_y k\dot{w}\,\dot{w}_b''/2)]\, \delta L_1\}_{\text{at } x = L_1} \\ &- \{[EI_y w_b'' + (\nu\rho I_y k\ddot{w}/2)]\, \delta w_b' + [(GA/N)(w' - w_b') + P]\, \delta w\}_{\text{at } x = 0} = 0 \end{aligned}\right\} \ldots [12]$$

where

$$k = 1 - (I_z/I_y) \ldots\ldots\ldots\ldots\ldots\ldots [13]$$

and where I_z is the moment of inertia of the cross section about the z-axis. Equation [12] must be satisfied for any arbitrary variation of the displacements; hence the governing equations for the transverse motion of a beam with variable cross section are

$$-(EI_y w_b'')' - (GA/N)(w' - w_b') + \rho I_y \ddot{w}_b' - (\nu\rho I_y k\ddot{w}/2)' = 0$$
$$-[(GA/N)(w' - w_b')]' + \rho A\ddot{w} + (\nu\rho I_y k\ddot{w}_b''/2) = 0 \ldots\ldots [14]$$

In the special case of a uniform bar, w_b and its derivatives may be eliminated; the result is (D indicating a differential operator)

$$Dw = EI_y w^{iv} + \rho A\ddot{w} - \rho I_y[1 + (EN/G) - \nu k]\ddot{w}'' + (\rho^2 I_y N/G)\ddddot{w} + [(\nu\rho I_y k)^2 N/(4GA)]\ddddot{w}'' = 0 \ldots [15]$$

It may similarly be shown that this equation takes the form

$$Dw_b' = Dw_s' = 0 \qquad [16]$$

when written in terms of the bending and shear displacements, respectively. If the terms containing Poisson's ratio are neglected, Equation [15] becomes identical with Timoshenko's equation (6) and very similar to Prescott's (23). If shear deformations are neglected Love's equation (18) results; if both shear and Poisson's ratio terms are neglected the well-known equation for the transverse motion of beams including the effect of rotatory inertia (24) is obtained. The relative importance of the various terms of Equation [15] will be discussed later.

Boundary Conditions

Some of the boundary conditions to be satisfied at the wave front ($x = L_1$) are given by Equations [3]. The remaining conditions there and at $x = 0$ can be obtained from Equation [12]. At $x = 0$ they are either

$$EI_y w_b'' + (\nu\rho I_y k \ddot{w}/2) = 0 \quad \text{or} \quad \delta w_b' = 0 \qquad [17a]$$

and either

$$(GA/N)(w - w_b') + P = 0 \quad \text{or} \quad \delta w = 0 \qquad [17b]$$

Thus at the struck end of the bar either the bending moment (modified by a term which vanishes if $I_y = I_z$) or the slope due to bending may be prescribed, and either the shear force or the total displacement.

Consider now the quantities δL_1 and $\delta w_b'$, contained in the boundary conditions at the head of the wave. Since w and $(w + \delta w)$ must vanish at $x = L_1$ and $x = L_1 + \delta L_1$, respectively, it follows that

$$\delta L_1 = -(1/w')\delta w \quad \text{at } x = L_1 \qquad [18]$$

and similarly in terms of the bending and shear displacements.

The quantity $\delta w_b'$ stands for the variation of the slope (due to bending) at $x = L_1$, the variation being taken without changing x. It differs from the variation $\delta(w_b')_{\text{total}}$ of the slope at the head of the wave, inasmuch as the latter must be calculated taking into account a change in x equal to δL_1. The relation between the two slope changes is then, with the aid of Equation [18]

$$\delta w_b' = \delta(w_b')_{\text{total}} + (w_b''/w')\delta w \quad \text{at } x = L_1 \qquad [19]$$

The bracket of Equation [12] labeled $x = L_1$ indicates that the work done by the shear force and by the moment at $x = L_1$ during the virtual variation of the displacements must balance the total strain plus kinetic energies stored in the portion of the beam defined by $L_1 \leq x \leq (L_1 + \delta L_1)$. It corresponds to the dynamical conditions given by Love (18, art. 206). Use of Equations [18] and [19] gives the boundary conditions as either

$$EI_y w_b'' + (\nu\rho I_y k \ddot{w}/2) = 0 \quad \text{or} \quad \delta(w_b')_{\text{total}} = 0 \text{ at } x = L_1 \qquad [20]$$

where, because of Equation [3], the latter condition must always be satisfied, and

$$(1 - \alpha^2)(w_b'')^2 - [\alpha^2 - (1/\gamma)](1/r_y)^2(w')^2 + (\nu k/a^2)(\ddot{w} w_b'' - \dot{w}\dot{w}_b'') = 0 \qquad [21]$$

at $x = L_1$ with the following notation

$$\alpha = \dot{L}_1/a; \; a = \sqrt{E/\rho}; \; \gamma = EN/G; \; r_y^2 = I_y/A \qquad [22]$$

Equations [3], [17a], [17b], and [21] are the boundary conditions of the problem.

Remarks on Role of Shear, Rotatory Inertia, and Lateral Contraction

It is well known (7, 23) that if the effects of shear and rotatory inertia are not considered in the analysis of a beam under lateral impact, the physically impossible result of an infinite velocity of propagation is obtained for very short waves. However, a theory developed by Timoshenko (25) on the basis of the Bernoulli-Euler equation

$$EI_y w^{iv} + \rho A \ddot{w} = 0 \qquad [23]$$

which results from Equation [15] when all shear, rotatory inertia, and Poisson's ratio effects are neglected, gives stresses and displacements in close agreement with some experiments (2, 3, 4, 5). It therefore appears that shear and rotatory inertia, though essential to a study of the velocity of propagation, play a somewhat limited role. This may be clarified by the following study of the orders of magnitude of the various terms of Equation [15] pertaining to positions of the wave front close to the struck end of the beam.

Consider a portion of the beam, originally at rest, near to and including the end struck laterally. Let order of magnitude of the displacement in this portion of the bar at a time t be denoted by w; in the case of lateral impact w is a finite and nonvanishing quantity. The order of magnitude (denoted by 0) of its time derivative can then be approximated by

$$0(\partial w/\partial t) = w/t \qquad [24a]$$

The order of magnitude of its x-derivative may be taken as

$$0(\partial w/\partial x) = w/(\lambda L) \qquad [24b]$$

where the dimensionless coefficient λ appearing in the characteristic length λL will be determined later. The magnitudes of higher derivatives may then be estimated by repeated application of Equations [24]. The orders of magnitude of the various terms in Equation [15] are then (if a common multiplier $[EI_y w/(\lambda L)^4]$ is omitted), respectively

$$\left.\begin{aligned} &0(w^{iv} - \text{term}) = 1 \\ &0(\ddot{w} - \text{term}) = \lambda^4 S_y^2 (L/at)^2 \\ &0(\ddot{w}'' - \text{term}) = (1 + \gamma - \nu k)\lambda^2(L/at)^2 \\ &0(\ddddot{w} - \text{term}) = \lambda^4\gamma(L/at)^4 \\ &0(\ddddot{w}'' - \text{term}) = (\lambda\nu k/S_y)^2(\gamma/4)(L/at)^4 \end{aligned}\right\} \qquad [25]$$

In view of the restriction to cases of lateral impact, it is reasonable to suppose that the second of these, which represents the translatory inertia, is important at all times. It is now necessary to determine which of the other terms of Equation [15] are of equal importance.

Assume first, that the w^{iv} — term and the $\ddot{w}$ — term are of the same order of magnitude (as they would have to be if Equation [23] is to hold in good approximation); then

$$0(\lambda^2) = (at/L)(1/S_y) \qquad [26a]$$

The orders of magnitude of the various terms of Equation [15] are then

$$\left.\begin{aligned} &0(w^{iv} - \text{term}) = 0(\ddot{w} - \text{term}) = 1 \\ &0(\ddot{w}'' - \text{term}) = (1 + \gamma - \nu k)(1/S_y)(L/at) \\ &0(\ddddot{w} - \text{term}) = (\gamma/S_y^2)(L/at)^2 \\ &0(\ddddot{w}'' - \text{term}) = (\gamma/4)(\nu^2k^2/S_y^3)(L/at)^3 \end{aligned}\right\} \qquad [26b]$$

The third and fourth terms represent mainly the shear and rotatory inertia effects, and the last term is due principally to lateral contraction. Equations [26b] show that the latter effect is small unless νk is large. The orders of magnitude of these three terms is small compared to 1 unless the time t is very small: specifically, it is more than 1 if, respectively

$$\left.\begin{aligned} &t < [(1 + \gamma - \nu k)/S_y](L/a) \\ &t < [(\gamma)^{1/2}/S_y](L/a) \\ &t < [\gamma\nu^2k^2/(4S_y^4)]^{1/3}(L/a) \end{aligned}\right\} \qquad [26c]$$

The largest of the foregoing limiting values is (unless νk is very large) the first; it follows that if t is large compared with the quantity $[(1 + \gamma - \nu k)/S_y](L/a)$, the first two terms of Equation [15] predominate and the others are negligible. If on the other hand, t is small compared with that quantity, it is found from Equations [26b] that the $w''^{\cdot\cdot}$ — term (for instance) is large compared to unity. This is contrary to the hypothesis that the $w^{\cdot\cdot}$ term always predominates; hence it is necessary to repeat the whole argument on the basis of

$$O(\lambda^2) = (1 + \gamma - \nu k)/S_y^2 \qquad [27a]$$

namely, the relation necessary to make the second and third terms of Equations [25] of the same order of magnitude. Equations [26b] are then replaced by

$$\left.\begin{aligned} &O(w^{\text{iv}} - \text{term}) = 1 \\ &O(w^{\cdot\cdot} - \text{term}) = O(w''^{\cdot\cdot} - \text{term}) \\ &\qquad = (1 + \gamma - \nu k)^2(1/S_y)^2(L/at)^2 \\ &O(w^{\cdot\cdot\cdot\cdot} - \text{term}) = (1 + \gamma - \nu k)^2(1/S_y)^4(L/at)^4 \\ &O(w''^{\cdot\cdot\cdot\cdot} - \text{term}) = (1 + \gamma - \nu k)(\nu k/S_y^4)(\gamma/4)(L/at)^4 \end{aligned}\right\} \qquad [27b]$$

Here the $w''^{\cdot\cdot}$ term predominates if t is small; similar results may be obtained by equating the order of magnitude of the $w^{\cdot\cdot\cdot\cdot}$ term with that of the w'' term. Considering only the largest of the limiting values of Equations [26c], one may then conclude that if t is very small compared with the quantity $[(1 + \gamma - \nu k)/S_y](L/a)$ the shear and rotatory inertia effects are of paramount importance, and the bending terms secondary. On the other hand, if t is large compared with that quantity, only bending terms need be considered; thus, except for a short interval, Equation [23] will describe with good accuracy the behavior of the bar. If t is of the same order of magnitude as that quantity, all terms should be considered; but because of the large value of a this time is very small. For example, for a 1.7-in-square aluminum or steel bar with $\nu = 0.3$, shear and rotatory inertia may be neglected if $t \gg 10^{-5}$ sec. In the numerical examples which follow, the time t_0 (at which the shear deformations at $x = 0$ become negligible) is 2.1×10^{-5} sec, 6.2×10^{-5} sec, 3.9×10^{-5} sec, in examples I, II(a), and II(b), respectively, for this same bar, and hence in good agreement with that predicted here.

To determine how far the disturbance has traveled in 10^{-5} sec it is necessary to find first the speed of propagation. Immediately after impact the shear (and rotatory inertia) terms predominate, and it is, therefore, reasonable to expect that for a short time the bulk of the disturbance will propagate with the speed

$$L_1^{\cdot} = \sqrt{G/(N\rho)} \qquad [28]$$

With this velocity the disturbance will have traveled in the bar approximately 1.1 in. in 10^{-5} sec, that is, less than the cross-sectional dimensions of the bar.

The velocity of propagation given by Equation [28] is identical with one given by the characteristic solution of Equation [15], lateral contraction terms being omitted (15). Such a solution shows that there are two distinct velocities with which a sudden disturbance may be propagated; the slower of these is that given by Equation [28] and corresponds to a sudden change in velocity and in shear force; the other is

$$L_1^{\cdot} = a = \sqrt{E/\rho} \qquad [28a]$$

and corresponds to a sudden change in bending moment and in the angular velocity $w'^{\cdot}_b$.

The initial shear force at the struck end will now be calculated. The shear force Q is, in general

$$Q = (GA/N)(w' - w_b') \qquad [29]$$

At $x = L_1$ the quantity w_b' is zero; at the beginning of impact $L_1 = 0$ and $w^{\cdot} = v_0$, where v_0 is the velocity with which the beam is struck. The initial force is then (see Equation [4])

$$Q_0 = -GAv_0/(L_1^{\cdot}N) = -Av_0\sqrt{G\rho/N} \qquad [30]$$

and the average shear stress

$$\tau_0 = -Gv_0/L_1^{\cdot} \qquad [30a]$$

This equation has the same form as that which gives (19) the initial stress in a bar struck axially with a velocity v_0. Equations [30] hold whenever the one point of the beam undergoes a sudden change in velocity of magnitude v_0. The foregoing value for the initial shear force is identical to that given by an exact solution of Equation [15], lateral contraction terms being neglected (28).

The argument has been developed for a region close to the struck end of the bar but could be extended to any position of the wave front at which a sudden change in velocity occurred. Such a wave front travels with a velocity given by Equation [28]; for a wave front traveling with the velocity of Equation [28a] a similar argument could be developed by replacing the displacement w by the rotation w_b', and the shear-force discontinuity by one in the bending moment.

Remarks Concerning the Speed of Propagation

Consider first a bar with $I_y = I_z$, for which all terms containing Poisson's ratio vanish. In particular, Condition [21] becomes

$$(1 - \alpha^2)r_y^2(w_b'')^2 = [\alpha^2 - (1/\gamma)](w')^2 \text{ at } x = L_1 \qquad [31]$$

Inspection shows that the terms $(1 - \alpha^2)$ and $[\alpha^2 - (1/\gamma)]$ must be of the same sign. Since $\gamma > 1$ they must be positive and must lie in the following region

$$(1/\gamma)^{1/2} \leq \alpha \leq 1 \qquad [32]$$

It may be verified that the displacements given by

$$w = c \sin \omega[t - (x/L_1^{\cdot})] \qquad [33]$$

satisfy Equation [15] with $k = 0$ if

$$(\omega r_y/a)^2 = \alpha^4/[(1 - \alpha^2)(1 - \alpha^2\gamma)] \qquad [33a]$$

Comparison of Equations [32] and [33a] shows that ω must be imaginary. A real value of ω would imply that the condition of continuity of the slope due to bending at $x = L_1$ is violated. Let, in fact, the condition $w_b'' = 0$ be chosen in Relations [20], rather than $\delta(w_b')_{\text{total}} = 0$; then Equation [31] would be replaced by

$$(w_b')^2 + (r_y w_b''/a)^2 = (1 - \gamma\alpha^2)(w')^2 \qquad [34]$$

This requires

$$(1/\gamma)^{1/2} \geqslant \alpha \qquad [34a]$$

which, when substituted into Equation [33a] gives $\omega^2 > 0$.

The limiting values of α given by Equation [32] are identical with those given by the characteristic solution of the governing equations (see Equations [28]). They correspond to disturbances with very short wave length, i.e. $(1/\omega) = 0$, or, more specifically, to the propagation of discontinuities in bending moment (with $w_s' = w' = 0$) and in shear force (with $w_b'' = 0$), respectively, for the upper and lower limit.

To investigate the general case ($k \neq 0$), let the displacement of Equation [33] be again substituted into Equation [15]; then

$$(\omega r_y k\alpha/a)^2(\gamma/2) = b \pm \sqrt{b^2 - \nu^2k^2\gamma\alpha^4} \qquad [35]$$

where

$$b = (1 - \alpha^2)(1 - \alpha^2\gamma) + \alpha^2\nu k \ldots\ldots\ldots [35a]$$

The quantity $\nu^2k^2\gamma\alpha^4$ is always positive; therefore, ω^2 can be negative (as shall be assumed) only if $b < 0$. However, ω^2 will be real only if $|b| > |\nu k \sqrt{\gamma}\alpha^2|$. These two conditions on b will be satisfied if, and only if

$$(1 - \alpha^2)(1 - \alpha^2\gamma) + \alpha^2\nu k + |\nu k \sqrt{\gamma}\alpha^2| < 0 \ldots [35b]$$

The limiting values of α (namely, those which cause the left-hand side of Relation [35b] to vanish) are plotted in Fig. 2 for rectangular sections of height a and width b, and with $\nu = 0.3$.

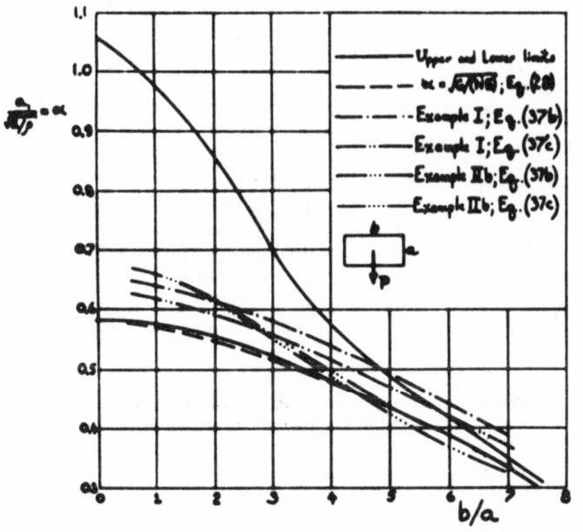

FIG. 2 VARIATION OF VELOCITY OF PROPAGATION WITH SIDE RATIO OF A RECTANGULAR SECTION
(Values for examples I and IIb were calculated for $t = t_0$.)

Both the upper and lower limits of α decrease as b/a increases and for large values of b/a they lie so close that they practically establish immediately the value of α. In the numerical examples discussed later the calculated velocity of propagation (whose meaning is discussed later) also decreases with increasing (b/a) and always lies close to the lower limit.

APPROXIMATE METHOD OF SOLUTION

The foregoing theoretical development suggests an approximate method of solution of problems of lateral impact of bars. In this method, expressions are assumed, in terms of some arbitrary parameters, for the deflections of the beam. The parameters are then adjusted so as to approximate a solution of Equation [12]. In general, two wave fronts will arise, each followed by a short region in which shear deformations and rotatory inertia are important. It will then be convenient to assume the displacements in two parts, one vanishing when $x > at$ (see Equation [28a]), the other when $x > at\sqrt{1/\gamma}$ (see Equation [28]). As mentioned in the introduction, however, it is desired to obtain solutions valid for short times only, that is, when the disturbance has affected only a short portion of the beam. During such a period the bulk of the disturbance travels with velocity close to that given by Equation [28]. For such a limited solution it is therefore possible to simplify the problem by assuming that only one wave front exists, traveling with a "mean" velocity L_1' to be calculated from Equation [12]. The details of this method of solution are then as follows:

1 Assume a function $L_1 = L_1(t)$ for the distance traveled by the disturbance at any time t and functions $w = w(x, t)$ and $w_b = w_b(x, t)$ for the displacements of the bar, in terms of n arbitrary parameters. These functions must satisfy the boundary conditions given by Equations [3], [17a], [17b], and [21], and the condition $L_1(0) = 0$.

2 Substitute the assumed functions in Equation [12]. The brackets labeled $x = 0$ and $x = L_1$ will vanish because the boundary conditions already have been satisfied. Note that

$$\delta w = \sum_{i=1}^{n} (\partial w/\partial p_i)\delta p_i \ldots\ldots\ldots\ldots\ldots [36]$$

and similarly for w_b', where p_i denotes one of the arbitrary parameters.

3 The result of step 2 is an expression of the form

$$\sum_{i=1}^{n} F_i(t, p_1, p_2, \ldots\ldots, p_n)\,\delta p_i = 0 \ldots\ldots\ldots [37]$$

Since the δp_i quantities are arbitrary, the exact solution requires that F_i vanish for any i and all t; but in an approximate solution one must be satisfied with adjusting the available parameters to obtain the best possible approximation. The following methods are suggested to obtain an approximate solution in the range $0 < t < T$:

(a) By the method of least squares, that is, by minimizing the sum of the squares of the total errors in each function F_i; then

$$(\partial/\partial p_j) \int_0^T \sum_{i=1}^{n} F_i^2 dt = 0 \quad j = 1, 2, \ldots\ldots, n \ldots\ldots [37a]$$

(b) By setting equal to zero (or minimizing) the average total error in each function F_i; then

$$\int_0^T F_i dt = 0 \quad i = 1, 2, \ldots\ldots, n \ldots\ldots\ldots\ldots [37b]$$

(c) By substituting $t = T$ in each function F_i; then

$$F_i(T, p_1, p_2, \ldots\ldots, p_n) = 0 \quad i = 1, 2, 3, \ldots\ldots, n \ldots\ldots [37c]$$

In the numerical examples these three methods were found to yield very similar results. In example I the results of methods (a) and (b) are, for all practical purposes, the same. The results of methods (b) and (c) for the propagation velocity in examples I, II(a), and II(b), respectively, differ by 3.5, 6.8, and 0.0 per cent for a square bar, and by 4.5, 12.0, and 5.3 per cent for a rectangular bar with a width-to-height ratio of 20 to 3. The values given are percentages of the smaller value in each case.

It is probably possible to assume fairly accurate functions for the quantities L_1, w, and w_b, without an excessive number of parameters, if the solution is restricted to the "boundary layer" previously discussed. If, on the other hand, a description of a longer disturbed portion is sought, it may be found that the number of parameters required becomes excessive. In such cases it is suggested that different displacement functions be assumed near the disturbance front and at some distance from it so as to conform with the previous discussion concerning the importance of the shearing deformations and the rotatory inertia effect. This method is illustrated in two of the numerical examples but it should be emphasized that the small number of parameters used still makes the results unreliable outside the boundary layer. The discussion of Fig. 4 indicates, however, that here the Bernoulli-Euler theory provides an adequate approximation.

Expressed in general terms, the choice for the displacement functions is as follows:

Region 1: $(L_1 - L_0) \leq x \leq L_1$

$$\left.\begin{aligned} w &= w_1(x, t) \\ w_b &= w_{b_1}(x, t) \end{aligned}\right\} \quad [38a]$$

Region 2; $0 \leq x \leq (L_1 - L_0)$

$$\left.\begin{aligned} w &= w_2(x, t) \\ w_b &= w_2(x, t) - w_1(L_1 - L_0, t) + w_{b_1}(L_1 - L_0, t) \end{aligned}\right\} \quad [38b]$$

These functions must satisfy the conditions of continuity at $x = L_1 - L_0$, namely

$$\left.\begin{aligned} w_1(L_1 - L_0, t) &= w_2(L_1 - L_0, t) \\ w_{b_1}'(L_1 - L_0, t) &= w_2'(L_1 - L_0, t) \end{aligned}\right\} \quad [38c]$$

as well as the following requirements on the shear force and bending moment

$$\left.\begin{aligned} (GA/N)[w_1'(L_1 - L_0, t) - w_{b_1}'(L_1 - L_0, t)] &= 0 \\ EI_y w_{b_1}''(L_1 - L_0, t) &= EI_y w_2''(L_1 - L_0, t) \end{aligned}\right\} \quad [38d]$$

Within the distance $(L_1 - L_0)$, which varies as the disturbance progresses along the bar, the effects of shear, rotatory inertia, and lateral contraction are important. The shear-strain energy vanishes in Region 2. The distance L_0 may be taken as a parameter or may be expressed in terms of other parameters. If the time elapsed after the start of impact is very small ($t \leq t_0$), only Region 1 will be present. The numerical examples show that for large values of time $(L_1 - L_0)$ is small compared with L_1, and that in Region 2 the (constant) shearing displacements constitute only a small fraction of the total displacements. In general, the order of magnitude of $(L_1 - L_0)$ is $(1 + \gamma - \nu k)(L_1^{\cdot}/a)(L/S_y)$, where the value for the propagation velocity $L_1^{\cdot}$ may be taken in good approximation from Equation [28].

Numerical Examples

Four numerical examples, described in Table 1, were solved by the foregoing procedure. The details of the solutions, omitted here, have been presented more fully in reference (1). The assumed displacements are given in Table 2, together with the expression for the time $t = t_0$ at which the disturbance has traveled a distance equal to the length of Region 1. Typical results of the calculations are plotted in Figs. 2 to 5.

In each example the velocity of propagation $L_1^{\cdot} = a_1$ was taken as one of the parameters, and therefore, the values obtained represent mean velocities within the range of validity of the solution. In other words, a_1 is a weighted average of the velocities of propagation of all the component wave lengths contained in the disturbance and, as expected, was close to the velocity, given by Equation [28], for the propagation of sudden jumps in velocity and shear force.

TABLE 1 DESCRIPTION OF NUMERICAL EXAMPLES

	Example I	Example II(a)	Example II(b)	Example III
Boundary conditions at $x = 0$	$w = v_0t$ $w_b'' = 0$	$w = v_0t$ $w_b' = 0$		$w = v_0t + v_1t^2$ $w_b'' = 0$
Arbitrary parameters	α	α, c		α
Equations used in solution	[37a], [37b], [37c]	[37b], [37c]		[37b]
Restrictions on calculations	Lateral contraction neglected	Lateral contraction neglected	Lateral contraction considered, $t \leq t_0$	$k = 0$ $t \leq t_0$

Notes: (a) $\alpha = a_1/\sqrt{E/\rho}$, where distance traveled by disturbance is $L_1 = a_1t$. (b) Quantities v_0 and v_1 are constants: v_1 is given in Equation [40a]. (c) At $t = t_0$ disturbance has traveled a distance equal to length of Region 1 of Equations [38a]

The variation of the shear force at the struck end of the bar of example I is plotted in Fig. 4 together with that obtained from the exact solution, Equation [28]. It may be seen that the two curves are in good agreement within the boundary layer previously discussed. It should be noted that (taking as an example a rectangular section), a wave traveling with the velocity of Equation [28] will have traveled a distance equal to the height of the bar when the abscissa of Fig. 4 equals approximately 6.2.

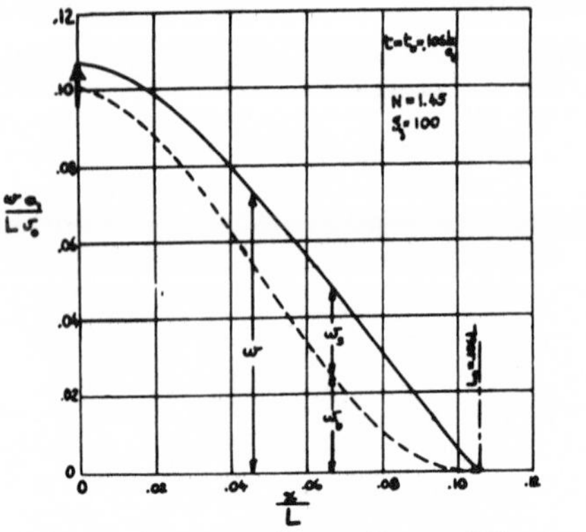

Fig. 3 Deflected Shape of the Beam of Example II(b) for $t = t_0$, Calculated From Equation [37c]

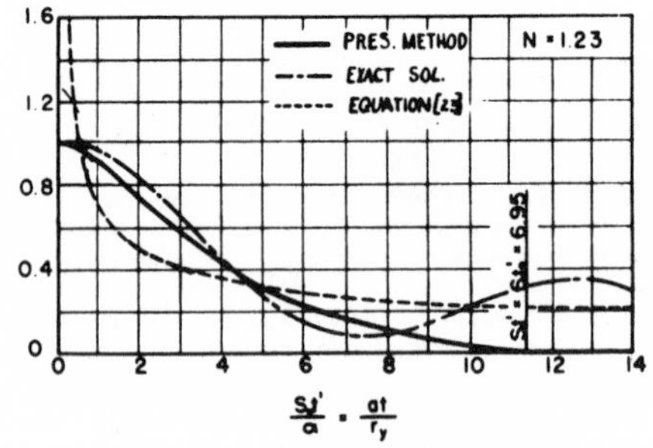

Fig. 4 Variation of Shear Force at Point of Impact for Example I

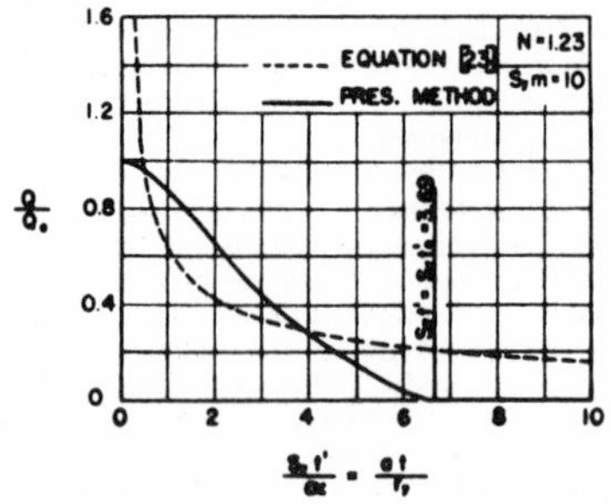

Fig. 5 Variation of Shear Force at Point of Impact for Example III

TABLE 2 EXPRESSIONS FOR ASSUMED DEFLECTIONS AND FOR t_0

Example No.	Displacements of Region 1 $(a_2t-t_0 \le x \le a_2t)$	Displacements of Region 2 $(0 \le x \le a_2t-t_0)$	Expressions for $t_0=L_0/a_2$ when $t \le t_0$	Expressions for $t_0=L_0/a_2$ when $t \ge t_0$
I	$w=v_0\left(t-\frac{x}{a_2}\right)$ $w_b=\frac{v_0K}{6r_y}\left(2a_2t-3xt+\frac{x^3}{a_2^2t}\right)$	$w=w_s=v_0\left(t-\frac{x}{a_2}\right)-\frac{v_0K}{r_y}\left[\frac{x^4-2a_2(t-t_0)x^3+a_2^2(t-t_0)^2x^2}{2a_2^4t(t-t_0)^2}\right]$ $w_b=w_s-v_0t_0+\frac{v_0a_2t_0^2K(3t-t_0)}{6r_yt}$	$t_0=\frac{2r_y}{Ka_2}$	$t_0=t-\sqrt{t^2-\frac{2tr_y}{Ka_2}}$
IIa	$w=v_0\left(t-\frac{x}{a_2}\right)+\frac{v_0c}{aL}\left(a_2tx-2x^2+\frac{x^3}{a_2t}\right)$ $w_b=\frac{v_0K}{6r_y}\left(a_2t-\frac{3x^2}{a_2}+\frac{2x^3}{a_2^2t}\right)$	$w=w_s=v_0\left[t+Px^2+\frac{Nx^3}{a_2(t-t_0)}+\frac{Mx^4}{a_2^2(t-t_0)^2}\right]$ $w_b=w_s-v_0\left[t_0+\frac{ca_2^2t_0^2(t-t_0)}{aLt}-\frac{Ka_2t_0^2(3t-2t_0)}{6r_yt}\right]$	$t_0=\frac{aL}{ca_2^2}$	$t_0=\frac{1}{2}\left[t-\sqrt{t^2-\frac{4tr_y}{Ka_2}}\right]$
IIb	$w=v_0\left(t-\frac{x}{a_2}\right)-\frac{v_0Rc}{aL\sqrt{k}}\left(a_2tx-2x^2+\frac{x^3}{a_2t}\right)$ $w_b=\frac{v_0c}{aL}\left(a_2^2t^2-4x^2+\frac{4x^3}{a_2t}-\frac{x^4}{a_2^2t^2}\right)$	—	$t_0=-\frac{\sqrt{k}aL}{Rca_2^2}$	—
III	$w=v_0\left(t-\frac{x}{a_2}\right)+v_1\left(t-\frac{x}{a_2}\right)^2$ $w_b=\frac{v_0K}{6r_y}\left(2a_2t^2-3xt+\frac{x^3}{a_2^2t}\right)$	—	$t_0=\frac{2v_0r_y}{v_0a_2K-4r_yv_1}$	—
Notation for this Table	$K=\frac{\sqrt{2\gamma-1}}{\sqrt{\gamma(1-\alpha^2)}}$; $M=\frac{KS_y}{a_2Lt}\left(\frac{t}{2}-t_0\right)-\frac{1}{a_2^2(t-t_0)}+\frac{at_0(4t-3t_0)}{aLt(t-t_0)}$; $N=-\frac{2KS_y}{a_2Lt}\left(\frac{t}{2}-t_0\right)+\frac{3}{a_2^2(t-t_0)}+\frac{ct_0(-10t+7t_0)}{aLt(t-t_0)}$; $P=\frac{ct_0^2\alpha a_2-L}{a_2^2L(t-t_0)}-M-N$; $R=\frac{2(1-\alpha^2)}{\alpha^2}+\frac{S_y^2(1-\alpha^2)}{8c\alpha^2\gamma}$			

Beyond that point the Bernoulli-Euler theory is a sufficiently good approximation since it represents the mean value about which the shear force oscillates. The amplitude of these oscillations is small compared with the maximum shear force and it was therefore thought permissible to neglect it. It should be mentioned, however, that these oscillations arise because of the presence of shear deformations, which therefore in reality affect the entire beam. Only if they are neglected is it possible to carry out the order-of-magnitude considerations as in Equation [24]; consequently, the use of the Bernoulli-Euler theory in this region is consistent with the accuracy employed throughout this development.

Examples II(a) and II(b) represent solutions of the same problem using two different deflected shapes, respectively, neglecting and considering the effect of lateral contraction. For k = 0, however, all lateral contraction terms vanish; hence, in this particular case, the discrepancy between the results of those examples is due to the variation in deflected shape alone. The difference between the two velocities of propagation is only 4.7 per cent of the smaller value, indicating that, in this case at least, a change in assumed shape does not influence greatly the results.

For other values of k examples II(a) and II(b) give widely varying results, indicating that the importance of lateral contraction increases with k. Thus, for a rectangular bar with a width-to-height ratio of 20 to 3, the foregoing error increases to 46.4 per cent of the smaller velocity.

In example III the quantity v_0 was determined by assuming that the beam is struck at $x = 0$ by a rigid mass M moving with an initial velocity v_0. The assumption was made that the beam and the mass remain in contact; therefore this solution holds only within the duration of the subimpacts described in references (2, 3, 4). However, the time t_0 is considerably smaller than the measured duration of the subimpacts. It may be remarked, incidentally, that it is difficult to find direct experimental proof for these solutions within the boundary layer because of the extremely short distance covered by it. The only verification possible appears to be that concerning the accuracy of the basic equations at high frequencies (8, 13).

Returning to example III, let the displacements of the struck end of the beam be expressed as

$$w(0, t) = \sum_{i=0}^{\infty} v_i t^{(i+1)} \quad \text{[39]}$$

where v_0 is the initial velocity, and $2v_1$ the initial acceleration. The equation of motion of the striking mass is

$$(GA/N)(w' - w_b') = M\ddot{w} \text{ at } x = 0 \quad \text{[40]}$$

When the various displacement functions are substituted in this equation, the quantities v_i can be determined in terms of v_0 by equating coefficients of like powers of t. If only the two lowest powers are kept, the result is

$$v_1/v_0 = -GA/(2NMa_1) \quad \text{[40a]}$$

The results of example III depend on the parameter $(S_y m)$, where S_y is the slenderness ratio of the bar, and m is the ratio of the mass of the striking body to the mass of the bar. If m is very large, or the bar very slender, the end of the bar moves with constant velocity, thus approaching the conditions of example I. The results of the two examples are, in fact, identical if $(1/S_y m)$ vanishes. The variation of the shear force of the struck end of the beam of example III is shown in Fig. 5 for a value $S_y m$ = 10. The variation is similar to that of Fig. 4 but the shear force of course decreases more rapidly than in example I. Again, the Bernoulli-Euler theory may be used for regions beyond the boundary layer.

In conclusion, it is felt that the present approach is in certain respects different from the methods usually employed in this type of problem and therefore may be of some interest. Of course the applications presented are limited in scope; it is, nevertheless, hoped that they may prove of value as a basis for further work.

ACKNOWLEDGMENT

The author gratefully acknowledges the contributions to the development described of Robert E. Heninger, Vincent P.

Zimnoch, and Chi-Chang Chao and wishes to thank the Office of Naval Research, U. S. Navy, for its support of the investigation.

BIBLIOGRAPHY

1 "An Energy Theory of Transverse Impact on Beams, Including the Effects of Shear, Rotatory Inertia and Lateral Contraction," by B. A. Boley, R. E. Heninger, and V. P. Zimnoch, Technical Report submitted to the Office of Naval Research, U. S. Navy, under contract N6-onr-225-29, project NR-064-355, January, 1952.

2 "Photoelastic Study of Stresses Due to Impact," by F. Tusi and M. Nisida, *Philosophical Magazine*, vol. 21, 1936, p. 448.

3 "Impact Stresses in a Freely Supported Beam," by R. N. Arnold, Proceedings of The Institution of Mechanical Engineers, London, England, vol. 137, 1937, pp. 217–281.

4 "Impact on Beams," by H. L. Mason, Journal of Applied Mechanics, Trans. ASME, vol. 58, 1936, p. A-55.

5 "Transverse Waves in Beams," by I. Vigness, Proceedings of the Society for Experimental Stress Analysis, vol. 8, 1951, pp. 69–82.

6 "On the Correction for Shear of the Differential Equation for Transverse Vibrations of Prismatic Bars," by S. Timoshenko, *Philosophical Magazine*, vol. 41, 1921, pp. 744–746.

7 "Influence of Rotatory Inertia and Shear on Flexural Motions of Isotropic Elastic Plates," by R. D. Mindlin, Journal of Applied Mechanics, Trans. ASME, vol. 73, 1951, pp. 31–38.

8 "Timoshenko's Shear Coefficient for Flexural Vibrations of Beams," by R. D. Mindlin and H. Deresiewicz, Technical Report No. 10, Office of Naval Research Contract Nonr-266(09), June, 1953.

9 "The Dispersion of Supersonic Waves in Cylindrical Rods of Polycrystalline Silver, Nickel and Magnesium," by S. Shear and A. B. Focke, *Physical Review*, vol. 57, 1940, p. 532.

10 "The Velocity of Longitudinal Waves in Cylindrical Bars," by D. Bancroft, *Physical Review*, vol. 59, 1941, p. 588.

11 "Dispersion of Elastic Waves in Solid Circular Cylinders," by G. E. Hudson, *Physical Review*, vol. 63, 1943, p. 46.

12 "A Critical Study of the Hopkinson Pressure Bar," by R. M. Davies, Philosophical Transactions of the Royal Society of London, England, vol. 240, 1948, p. 375.

13 "Stress Waves in Solids," by H. Kolsky, Clarendon Press, Oxford, England, 1953.

14 "Transverse Impact of Long Beams, Including Rotatory Inertia and Shear Effects," by M. A. Dengler and M. Goland, Proceedings of the First U. S. National Congress of Applied Mechanics, 1952, p. 179.

15 "On Traveling Waves in Beams," by R. W. Leonard and B. Budiansky, NACA TN 2874, January, 1953.

16 "Über Fortpflanzungsgeschwindigkeiten Kleiner Schwingungen in Einem Unbegrensten Isotropen Kreiscylinder," by L. Pochhammer, *Crelle Journal für Mathematik*, vol. 81, 1876, pp. 324–336.

17 "Flexural Waves in Beams According to More Exact Theory of Bending," by J. Miklowitz, NAYORD Report 2049, 1953.

18 "The Mathematical Theory of Elasticity," by A. E. H. Love, fourth edition, Dover Publications, New York, N. Y., 1944.

19 "Theory of Elasticity," by S. Timoshenko and J. N. Goodier, second edition, McGraw-Hill Book Company, Inc., New York, N. Y., 1951.

20 "Theory of Elastic Stability," by S. Timoshenko, McGraw-Hill Book Company, Inc., New York, N. Y., 1936.

21 "Numerical Procedures for the Calculation of the Stresses in Monocoques," by N. J. Hoff, B. Klein, and P. A. Libby, NACA TN 999.

22 "Higher Mathematics for Engineers and Physicists," by I. S. Sokolnikoff and E. S. Sokolnikoff, McGraw-Hill Book Company, Inc., New York, N. Y., 1941.

23 "Elastic Waves and Vibrations of Thin Rods," by J. Prescott, *Philosophical Magazine*, vol. 33, seventh series, 1942, pp. 703–754.

24 "Theory of Sound," by J. W. S. Rayleigh, Dover Publications, New York, N. Y., 1945.

25 "Vibration Problems in Engineering," by S. Timoshenko, second edition, D. Van Nostrand Company, Inc., New York, N. Y., 1937.

26 "The Application of Saint-Venant's Principle in Some Problems of Impact," by B. A. Boley, Technical Report No. 2, U. S. Navy Contract Nonr-266(20), Columbia University, 1953. Also, paper to be published in the Journal of Applied Mechanics.

27 "The Propagation of Elastic Waves in a Rod," by J. L. B. Cooper, *Philosophical Magazine*, vol. 38, seventh series, 1953, p. 147.

28 "Some Solutions of the Timoshenko-Beam Equations," by B. A. Boley and Chi-Chang Chao, Technical Report No. 3, U. S. Navy Contract Nonr-266(20), Columbia University, 1954.

Appendix

Determination of N

The quantity N is defined by the equation

$$2GU_s = \int\int (\tau_{xz}^2 + \tau_{yz}^2)\,dy\,dz = (N/A)\left(\int\int \tau_{xz}\,dy\,dz\right)^2$$

for a beam bent by a force in the z-direction. In this equation U_s is the shear-strain energy per unit of length, and the limits of integration must be chosen so as to cover the entire cross section. For a rectangular section, sufficiently accurate values of the shear stresses may be obtained from the approximate formulas given by Timoshenko.[4] The results are plotted in Fig. 1.

These values of N correspond to the stress distribution at some distance from the point of load application, and there is some question as to whether they are applicable in the portion of the beam immediately behind the disturbance front (Region 1 of Equation [38a]). For very small values of time, Region 1 is close to the point of impact, and therefore (according to Saint Venant's principle) strongly affected by the details of load application. Whether this is true of Region 1 for other values of time is the subject of an investigation in progress at present (26).

The coefficient N cannot be used to account for normal stress distributions radically different from that of Equations [3], such as arise when skin waves become prominent (19). In such case a more accurate theory, similar to Prescott's (23, 27), would be required, or the present approach might be modified by the introduction of some arbitrary parameters in the stress variation within the cross section of the bar. The coefficient N was calculated in the foregoing from statical considerations; the correction suggested by Mindlin (8, 14) was not included in the calculations.

[4] Reference (19), p. 327 ff.

Part II

VIBRATION AND WAVES IN PLATES

Editor's Comments on Papers 6 Through 15

Although it is a commonplace that beams are one-dimensional plates, and that plates are two-dimensional generalizations of beams, it is also true that much of what we know about elastodynamics derives from investigations of plate theories, with concommitant attempts to match plate results to those obtained from three-dimensional elasticity theory. We have collected in this section a set of papers that will, hopefully, trace some early applications of conventional plate theory, trace the development of improved higher-order plate theories and place them in the context of elasticity theory, and then go on to some modern applications.

The first two papers, those by Airey and Young, might seem elementary to us now, for after all, they are "only" the eigenvalue problems of the free vibrations of plates, using the classical theory. And that theory, incidentally, while variously attributed to Lagrange and to Sophie Germain, had been around since just after the beginning of the nineteenth century. However, it is important to remember the times in which these investigations were carried out. Airey's paper predates the First World War. And yet, by careful use of classical mathematical analysis, he calculates frequencies and nodal diameters to four-decimal-place accuracy. That is to say, rather than working with existing tables of Bessel functions, as did the previous investigators of these types of problems, Airey applies some results of "old-fashioned" analysis [viz., the steps leading from equation (4a) to equation (7)] to get series expansions for the roots of the transcendental frequency equations.

Young applied the Ritz approach to an investigation of the free vibration frequencies of some rectangular plates. And although it is now a garden-variety sort of tool, ideas stemming from the calculus of variations were then only rarely applied to mechanics problems, and then usually in more theoretical investigations. In this paper, Young successfully applied what is now termed the Rayleigh-Ritz procedure together with some judicious choices of beam functions (see Ref. 10 of the paper) to get some very accurate and useful frequencies and mode shapes.

The third paper in this group is also historical in nature. Lamb's analysis, together with previously obtained results of Pochhammer and Rayleigh, provides one of the touchstones (or benchmarks) for much of the later development of the theory of waves and vibrations in plates. This is in part because of the lucidity of the analysis, and in part because of the array of results that are uncovered. These include the symmetrical extensional modes in plane stress [equation (15)], symmetrical shear modes [equation (45)], and the "ordinary" antisymmetrical bending modes [equation (50)]. Of course, not all of these results are immediately recognizable to us today, because of the changes in notation and style. In addition to a discussion of the effects of compressibility, Lamb also gives a nice discussion of Rayleigh surface waves. And, as befits a man who wrote a classic treatise on hydrodynamics, Lamb also compares the penetration of the Rayleigh waves in an elastic solid with the much smaller penetration of hydrodynamic surface waves (viz., pp. 79–80).

Mindlin's paper may be reasonably characterized as the centerpiece of this section. While due credit must be (and is) assigned to the work of Rayleigh, Timoshenko, and Reissner* [E. Reissner, *J. Appl. Mech.,* **12**, 69–77 (1945); and *Quar. Appl. Math.,* **5,** 55–68 (1947)], it is true that this paper of Mindlin's had perhaps more impact on the way people thought about the vibrations of elastic structures than almost any other paper of our time. Contained in the paper is a complete study of the foundations of higher-order plate theory, including the derivation of the fundamental equations from the three-dimensional theory of elasticity; the verification of these equations and the development of the proper boundary conditions via an energy principle; considerations of compatibility; and a discussion of the propagation of bending and shear waves that both justifies the theory developed and puts it into its proper context.

Mindlin demonstrates (Figure 1) that in order to predict correctly the flexural wave speed, and thus the flexural frequency, for a high-frequency (short wave length) wave, both the transverse shear and rotatory inertia corrections to classical plate theory are required. He states (p. 95) that most of the correction comes from the shear correction, and plots the results for Poisson's ratio as one-half. However, it is not difficult to combine equations (45) and (48) to show that the high-frequency limit is dependent on the value of Poisson's ratio, and that for smaller values of Poisson's ratio the rotatory inertia must contribute as well.

Also included in this paper is the notion of picking the shear correction factor to suit the problem at hand. Thus that factor can be chosen to match the speed of Rayleigh surface waves [equation (47)], or to match the thickness-shear frequencies [equation (51)]. And, in a small way, the discussion of these possibilities leads us to the next three papers (Mindlin, Deresiewicz and Mindlin, and Mindlin and Medick).

That is because all of these papers, and many other papers by Mindlin and his colleagues, represent attempts to obtain from plate theories replications of the results of three-dimensional elastodynamics,

*In his classic monograph (R. D. Mindlin, *An Introduction to the Mathematical Theory of Vibrations of Elastic Plates,* U.S. Army Signal Corps Engineering Laboratories, 1955), Mindlin notes that "The important conclusion that three edge conditions are required, rather than the two of the classical theory of plates, was reached by E. Reissner." In this context it is also worth suggesting a second look at Miklowitz's paper on beams, reprinted in the first section, and his comments on Uflyand's work and boundary conditions.

and so they also represent attempts to validate those plate theories. And validation is one of the principal points of the next paper, Mindlin on crystal plates. This paper is of interest because it presents for the first time the *mathematical* structure of the terrace-like structure of the high-frequency spectrum, and it also presents a very detailed comparison of the theoretical spectra with experimental data obtained for the vibration of AT quartz plates. Further, some simple formulas are presented for various branches of the frequency spectrum.

The next paper is an axisymmetric version of a problem hinted at previously, that is, the coupling of flexural waves with thickness-shear waves. That this does occur is depicted quite clearly in this paper (Figure 1). This effect is a high-frequency effect, however: the ordinates of the curves in that figure indicate that the lowest thickness-shear frequency has been exceeded when coupling does occur, and it is limited to thick plates. Deresiewicz and Mindlin have also added some interesting remarks on modal density and on plate sizes as they are affected by using (or not using) a higher-order theory.

In a companion paper [R. D. Mindlin and H. Deresiewicz, *J. Appl. Phys.*, **25,** 1329–1332 (1954)] they remove the restriction of axisymmetric behavior, and find that not only are the flexural and thickness-shear modes coupled; there is also a third set, the thickness-twist modes, that are also highly coupled into the system. The coupling here is degenerate, though, for while the flexural modes are coupled with each of the thickness modes, the thickness modes are not coupled to each other. Also, it must be borne in mind that these coupling phenomena occur only at high frequencies, in very thick plates.

The paper by Mindlin and Medick has the same general focus as the previous two; it is, however, concerned with motion that is symmetric about the middle plane. In this class of motion, termed extensional, there appear thickness-shear modes and thickness-stretch modes. Here the thickness modes are coupled to each other, in contradistinction to the motion (discussed earlier) that is antisymmetric with respect to the thickness. A complete accounting of all the modes and their coupling is given in Mindlin's monograph [see preceeding footnote and also R. D. Mindlin, "Waves and vibrations in isotropic elastic plates," in J. N. Goodier and N. J. Hoff (eds.), *Structural Mechanics,* Elmsford, N.Y.: Pergamon, pp. 199–232, 1960]. It might be noted in passing that the methodology applied in this paper to obtain the approximate solution—i.e., the series expansion in powers of the thickness coordinate—dates back to the early part of the nineteenth century; it will appear once more when we turn to the analysis of composite plates.

Miklowitz's paper is an application of the higher-order plate theory in the solution of a forced motion problem. As the plate is infinite in extent, traveling waves that die out at infinity are generated. Thus, while the reflections that produce standing waves in a finite system do not appear here, we could use these results for the very early response of a finite plate (see, for example, the paper of Leonard and Budiansky in the previous section). Of particular interest in the results are the arrival times (or travel speeds) of the various waves and discontinuities. It can be seen (Figures 3 through 5) that the moment waves propagate at a speed consistent with the "plate speed" [equations following equations (9)], and that there are no discontinuities in the moments. The shear waves travel at a speed proportional to the shear-wave speed, modified by a Mindlin-type constant, and they also propagate the discontinuity due to the concentrated load applied at the center of the plate. Note, however, that the pulse input leads to the propagation of a second discontinuity, while the step input sends out only a single discontinuity. Other interesting features of this paper include comparisons of the long-time solutions with corresponding static solutions [equations (53)], and a nice physical explanation of Kelvin's stationary-phase method (page 123).

All of the plate analyses we have discussed so far pertain only to homogeneous, isotropic elastic plates. The paper by Sun, Achenbach, and Herrmann represents a serious attempt to deal with the very complicated problem of wave propagation in a stack of plates that are laminated to each other. It is vastly different from the elementary theory of composites, wherein the overall thickness of the stack is assumed to be sufficiently small so that one does not have to deal with displacements on a layer-by-layer basis (e.g., J. E. Ashton, J. C. Halpin, and P. H. Petit, *Primer on Composite Materials: Analysis,* Westport, Conn.: Technomic Publishing, pp. 30–34, 1969). Here, the individual plate displacements must be treated [viz., equations (3)–(5)]; and even though an averaging procedure is introduced [equations (23) and (24)], and the displacement series are truncated, the resulting equations of motion are quite complex. Some results are given, and some reasonable agreement with available exact results are demonstrated, particularly for flexural modes (Figure 3). Again, many types of propagation are possible, and the approximate theory based on averaging the stack thickness does not reproduce all the results found from the exact (not averaged) solution.

In the final paper, that by Feit, a different sort of coupling is considered. For waves propagating through a dense fluid and exciting an elastic plate, or for a mechanically excited plate vibrating in a dense medium, there is coupling and interaction between the fluid pressure loading and the plate response. In the present analysis, the coupling term is the last term in equation (3). One of the interesting findings here is a shift in the coincidence frequency (that frequency for which the speed of sound in the surrounding medium is equal to the bending-wave speed in the plate) due to the inclusion of shear and rotatory inertia correction factors. It is worth observing that this shift must be evaluated for a plate in water because in a dense medium the sound speed is high, so that the coincidence frequency would be high. For a less dense medium, e.g., air, the coincidence frequency would be lower because the speed of sound is lower; and for a plate vibrating in air one can use the classical theory of plates. In fact, in such a case, the coupling term may also be disregarded. These matters, particularly for dense media, have been treated extensively by Junger and Feit (M. C. Junger and D. Feit, *Sound, Structures, and Their Interaction,* Cambridge, Mass.: M.I.T. Press, 1972).

6

Reproduced by permission of the Institute of Physics, London, from *Proc. Phys. Soc., London.*, **23**, 225–232 (1911)

XXI. *The Vibrations of Circular Plates and their Relation to Bessel Functions.* By John R. Airey, *M.A., B.Sc., late Scholar of St. John's College, Cambridge.*

First Received February 15, 1911. Received in Final Form March 7, 1911.

The vibrations of circular plates were first investigated by Poisson * in a celebrated memoir read before the French Academy of Sciences in 1829. Three cases were considered: (*a*) when the circumference was fixed; (*b*) when the plate was "supported"; (*c*) when the plate was free. The ratios of the radii of the nodal circles to the radius of the plate were calculated when the vibrating plate had no nodal diameter and one or two nodal circles. Kirchhoff † extended Poisson's results for the free plate by calculating six ratios of the radii when the plate vibrated with one, two or three nodal diameters, whilst Schulze ‡ found eight more values of the ratios for a plate with fixed circumference.

The calculation of these ratios required the determination of the roots of equations involving Bessel functions with real and imaginary arguments. These appear to have been found by a "trial and error" method or by interpolation from tables of these functions.

The object of the present Paper is to give a general method of solving these equations—viz., equation (4A) for a circular plate with fixed circumference (Table I.), and equation (9A) for a free circular plate (Tables II. and III.). From the roots so calculated, the radii of the nodal circles and the times of vibration in any given mode are readily found.

(A) *Vibrations of a Circular Plate with Fixed Circumference.*

The displacement of a point on the plate from its position of equilibrium is given by

$$w = A\cos(p\theta)\{J_p(\kappa r)+\lambda J_p(i\kappa r)\}\cos(qt-\epsilon). \quad . \quad . \quad (1)$$

The boundary conditions in this case are $w=0$ and $\frac{dw}{dr}=0$ when $r=a$, or, if the radius of the plate is equal to unity, when $r=1$.

* "Mémoires de l'Académie royale des Sciences de l'Institut de France," tome VIII., 1829.

† Kirchhoff, "Pogg. Annalen," 1850. Strehlke, "Pogg. Annalen," 1855.

‡ Schulze, "Ann. der Physik," XXIV., 1907.

Hence (1) gives

$$J_p(\kappa a)+\lambda J_p(i\kappa a)=0, \quad (2)$$

$$J_p'(\kappa a)+\lambda J_p'(i\kappa a)=0. \quad (3)$$

Eliminating λ and writing x for κa, we get

$$\frac{J_p'(x)}{J_p(x)}=\frac{J_p'(ix)}{J_p(ix)} \quad (4)$$

or

$$\frac{J_{p+1}(x)}{J_p(x)}+\frac{I_{p+1}(x)}{I_p(x)}=0. \quad (4A)$$

This becomes, after the substitution of the semi-convergent series for $J_p(x)$, &c.,

$$\tan\left(x-\frac{2p+1}{4}\pi\right)=\frac{I_{p+1}\,.\,P_p+I_p\,.\,Q_{p+1}}{I_{p+1}\,.\,Q_p-I_p\,.\,P_{p+1}}. \quad (5)$$

or $$\tan\left(x-\frac{p\pi}{2}-n\pi\right)=\frac{I_p(P_{p+1}-Q_{p+1})-I_{p+1}(P_p+Q_p)}{I_p(P_{p+1}+Q_{p+1})+I_{p+1}(P_p-Q_p)} \quad (5A)$$

$$=-ay-8ay^2-\frac{a}{3}(m^2+2m+93)y^3-8a^3y^4$$

$$-\frac{2a}{15}(m^4+6m^3+744m^2-10726m+56055)y^5\ldots,$$

where $a=4p^2-1$, $m=4p^2$ and $y=\frac{1}{8x}$.

Then, by Gregory's series,

$$x=n\pi+\frac{p\pi}{2}-a\{y+8y^2+\tfrac{4}{3}(m+23)y^3$$

$$+\tfrac{1}{15}(96m^2-18624m+110688)y^5\ldots\}$$

$$=\beta-\frac{a}{8x}-\frac{8a}{(8x)^2}-\frac{4a(m+23)}{3(8x)^3}$$

$$-\frac{a}{15}\cdot\frac{(96m^2-18624m+110688)}{(8x)^5}\ldots. \quad (6)$$

By Lagrange's theorem, if

$$x=\beta+\frac{p}{x}+\frac{q}{x^2}+\frac{r}{x^3}+\frac{s}{x^4}+\frac{t}{x^5}\ldots,$$

$$x=\beta+\frac{p}{\beta}+\frac{q}{\beta^2}+\frac{r-p^2}{\beta^3}+\frac{s-3pq}{\beta^4}+\frac{t-2q^2-4pr+2p^3}{\beta^5}\ldots.$$

Hence

$$x=\beta-(m-1)\left[\frac{1}{8\beta}+\frac{8}{(8\beta)^2}+\frac{4(7m+17)}{3(8\beta)^3}+\frac{192(m-1)}{(8\beta)^4}+\frac{32(83m^2+218m+2579)}{15(8\beta)^5}\cdots\right], \quad . (7)$$

where

$$\beta=\left(\frac{2n+p}{2}\right)\pi.$$

Two roots of equation (4A) were given by Poisson when $p=0$, viz., 3·196 and 6·292, whilst Lord Rayleigh gave the values 3·20 and 6·3. For other values of p, Schulze found the following roots: When $p=1$, 4·612, 7·80, 10·95; and when $p=2$, 5·904 and 9·40.

The roots of equation (4) have been calculated from the expression given in (7), when $p=0$, 1, 2 and 3, some of the earlier roots by interpolation.

TABLE I.

Roots of $\frac{J_{p+1}(x)}{J_p(x)}+\frac{I_{p+1}(x)}{I_p(x)}=0.$

No. of root.	$p=0.$	$p=1.$	$p=2.$	$p=3.$
1	3·1955	4·611	5·906	7·144
2	6·3064	7·799	9·197	10·536
3	9·4395	10·958	12·402	13·795
4	12·5771	14·109	15·579	17·005
5	15·7164	17·256	18·745	20·192
6	18·8565	20·401	21·901	23·366
7	21·9971	23·545	25·055	26·532
8	25·1379	26·689	28·205	29·693
9	28·2790	29·832	31·354	32·849
10	31·4200	32·975	34·502	36·003

By substituting one of these values in (2), the value of λ corresponding to this root can be found—*e.g.*, when the plate is vibrating with one nodal diameter and three nodal circles, $\kappa a=x=10{\cdot}958$, and

$$\lambda=-\frac{J_1(10{\cdot}958)}{J_1(10{\cdot}958i)}$$

or

$$i\lambda=0{\cdot}0000255\ldots.$$

Equation (2) becomes

$$J_1(\kappa r)+0{\cdot}0000255 I_1(\kappa r)=0.$$

The roots of this equation are 3·8312, 7·0024 and 10·958.

Therefore, the ratios of the radii of the nodal circles to that of the plate are

$$\frac{3{\cdot}8312}{10{\cdot}958}=0{\cdot}3496 \text{ and } \frac{7{\cdot}0024}{10{\cdot}958}=0{\cdot}6390,$$

Schulze ("Ann. der Physik.," 1907) gives the values
0·350 and 0·640.

The expression for the frequency of vibration of the plate is

$$\mathrm{N}=\frac{ax^2}{\sqrt{1-\mu^2}},$$

where a is constant for the same plate (Lord Rayleigh, "Theory of Sound," Vol. I., § 217), μ is Poisson's ratio, and x is a root of equation (4). Since x is independent of μ, N is only affected by a change in the assumed value of μ through the factor $\frac{1}{\sqrt{1-\mu^2}}$.

(B) *Vibrations of a Free Circular Plate.*

The boundary conditions require

$$-\lambda=\frac{p^2(\mu-1)\{x\mathrm{J}_p'(x)-\mathrm{J}_p(x)\}-x^3\mathrm{J}_p'(x)}{p^2(\mu-1)\{ix\mathrm{J}_p'(ix)-\mathrm{J}_p(ix)\}+ix^3\mathrm{J}_p'(ix)} \quad . \quad . \quad (8)$$

and

$$-\lambda=\frac{(\mu-1)\{x\mathrm{J}_p'(x)-p^2\mathrm{J}_p(x)\}-x^2\mathrm{J}\ (x)}{(\mu-1)\{ix\mathrm{J}_p'(x)-p^2\mathrm{J}_p(ix)\}+x^2\mathrm{J}_p(ix)}, \quad . \quad . \quad (9)$$

where x is written for κa.

Eliminating λ, the expressions on the right of (8) and (9) are equal. . . . . (9A)

(i.) When $p=0$, *i.e.*, when there is no nodal diameter, equation (9A) becomes

$$2(1-\mu)+ix\frac{\mathrm{J}_0(ix)}{\mathrm{J}_0'(ix)}+x\frac{\mathrm{J}_0(x)}{\mathrm{J}_0'(x)}=0. \quad . \quad . \quad . \quad (10)$$

or, using Poisson's value for $\mu(\mu=\frac{1}{4})$

$$\frac{\mathrm{J}_0(x)}{\mathrm{J}_1(x)}=\frac{3}{2x}-\frac{\mathrm{I}_0(x)}{\mathrm{I}_1(x)}. \quad . \quad . \quad . \quad . \quad . \quad (10\mathrm{A})$$

Substituting the semi-convergent series for $\mathrm{J}_0(x)$, $\mathrm{J}_1(x)$, &c., we get

$$\frac{\mathrm{P}_0 \cdot \cos\left(x-\frac{\pi}{4}\right)-\mathrm{Q}_0 \sin\left(x-\frac{\pi}{4}\right)}{\mathrm{P}_1 \cdot \sin\left(x-\frac{\pi}{4}\right)+\mathrm{Q}_1 \cos\left(x-\frac{\pi}{4}\right)}=\frac{3}{2x}-\frac{\mathrm{I}_0(x)}{\mathrm{I}_1(x)}=a \quad . \quad (11)$$

or
$$\tan\left(x-\frac{\pi}{4}\right)=\frac{P_0-aQ_1}{Q_0+aP_1}. \quad . \quad . \quad . \quad . \quad . \quad (12)$$

Putting $y=\frac{1}{8x}$, $a=12y-\frac{I_0(x)}{I_1(x)}$
$$=-1+8y-24y-192y^3-2016y^4-27648y\ldots$$

This value substituted in (12) gives, after simplification,

$$\tan\left(x-\frac{\pi}{4}\right)=-1-10y-10y^2+320y^3+3650y^4+59840y^5\ldots$$

$$\therefore\quad \tan(x-n\pi)=-5y+20y^2+85y^3+500y^4+21070y^5\ldots.$$

Expressing the angle in terms of the tangent, we find

$$x-n\pi=-5y+20y^2+\frac{380}{3}y^3+20320y^5\ldots$$

or
$$x=n\pi-\frac{5}{8x}+\frac{20}{(8x)^2}+\frac{380}{3(8x)^3}+\frac{20320}{(8x)^5}\ldots.$$

Then, as before, by Lagrange's theorem,

$$x=\beta-\frac{5}{8\beta}+\frac{20}{(8\beta)^2}-\frac{220}{3(8\beta)^3}+\frac{2400}{(8\beta)^4}+\frac{54560}{3(8\beta)^5}\cdots$$

(ii.) When $p=1$—*i.e.*, when the vibrating plate has one nodal diameter, equation (9A) gives

$$\frac{(6+4x^2)J_1-(3x+4x^3)J_0}{(6-4x^2)J_1-3xJ_0}=\frac{(6-4x^2)I_1-(3x-4x^3)I_0}{(6+4x^2)I_1-3xI_0}. \quad (13)$$

This reduces to

$$\tan\left(\frac{\pi}{2}-x+n\pi\right)=9y-84y^2+639y^3-6804y^4+168714y^5\ldots.$$

Hence
$$x=(2n+1)\frac{\pi}{2}-\frac{9}{8x}+\frac{84}{(8x)^2}-\frac{396}{(8x)^3}-\frac{65261}{(8x)^5}\cdots$$

Writing β for $(2n+1)\frac{\pi}{2}$, Lagrange's theorem then gives

$$x=\beta-\frac{9}{8\beta}+\frac{84}{(8\beta)^2}-\frac{1044}{(8\beta)^3}+\frac{18144}{(8\beta)^4}-\frac{385877}{(8\beta)^5}\cdots.$$

(iii.) The general expression for the roots of equation (9A), as far as the term containing $1/(8\beta)^4$ can be obtained from a result of Kirchhoff's, viz.,

$$\tan\left(x-n\pi-\frac{p\pi}{2}\right)=\frac{B/(8x)+C/(8x)^2-D/(8x)^3\ldots}{A+B/(8x)+D/(8x)^3\ldots}\ldots,$$

where $A=\gamma=\frac{4}{3}$, using Poisson's value of $\mu=\frac{1}{4}$,

$$B=\gamma\,(1-4p^2)-8,$$

$$C=\gamma\,(1-4p^2)(9-4p^2)+48(1+4p^2),$$

$$D=-\frac{\gamma}{3}\{(1-4p^2)(9-4p^2)(13-4p^2)\}+8(9+136p^2+80p^4)$$

$$\tan\left(x-n\pi-\frac{p\pi}{2}\right)=-(4p^2+5)\,y+4(16p^2+5)y^2$$

$$-\tfrac{1}{3}(64p^6+304p^4+1804p^2-255)y^3\ldots.$$

Writing β for $(2n+p)\frac{\pi}{2}$ and m for $4p^2$, it is easily shown that

$$x=\beta-\frac{m+5}{8x}+\frac{4(4m+5)}{(8x)^2}-\frac{4(m-1)(m+95)}{(8x)^3}+\frac{0}{(8x)^4}\ldots$$

$$\therefore\; x=\beta-\frac{m+5}{8\beta}+\frac{4(4m+5)}{(8\beta)^2}-\frac{4(7m^2+154m+55)}{3(8\beta)^3}$$

$$+\frac{96(m+5)(4m+5)}{(8\beta)^4}\ldots. \qquad (14)$$

This series is not convergent enough to give the earlier roots of equation (9A). These can be obtained without difficulty from tables of Bessel functions.

Poisson (Mém. Acad., 1829) found the first two roots of equation (9A), when $p=0$, viz., 2·9815 and 6·1936.

Kirchhoff calculated some of the roots of the general equation (9A) by expressing it in the form

$$0=1-\frac{x^4}{A_1}+\frac{x^8}{A_2}-\frac{x^{12}}{A_3}+\ldots$$

and finding the roots by "trial." Only the first two roots, $(\lambda l)^4$, in each case ($p=0$, 1, 2, 3) could be calculated from the table of values of A_1, A_2, &c., given in the Paper. Kirchhoff's roots are readily expressed in the same form as those calculated from (14). Several of these values have been verified and are included in the following table :—

TABLE II.

Roots of Equation (9A).

When $p=0, 1, 2, 3$. $\mu=\frac{1}{4}$ (Poisson's value).

n.	$p=0$.	$p=1$.	$p=2$.	$p=3$.
0	...	...	2·348	3·571
1	2·982	4·518	5·940	7·291
2	6·192	7·729	9·186	10·600
3	9·362	10·903	12·381	13·821
4	12·519	14·024	15·556	17·015
5	15·669	17·218	18·721	20·203
6	18·817	20·368	21·880	23·363
7	21·963	23·516	25·035	26·526
8	25·108	26·663	28·187	29·685
9	28·253	29·809	31·337	32·841

The values of $\lambda(p=0, \mu=\frac{1}{4})$ can be found from (8)

$$\lambda=\frac{J_0'(x)}{iJ_0'(ix)}=-\frac{J_1(x)}{I_1(x)},$$

and x has one of the values in column 2 of Table II. When $x=\kappa a=9{\cdot}362$, for example, $\lambda=-0{\cdot}0001299\ldots$. The radii of the nodal circles are readily found from the roots of the equation

$$J_0(\kappa r)-0{\cdot}0001299-J_0(i\kappa r)=0. \quad . \quad . \quad . \quad (15)$$

The four roots of (15) are 2·40406, 5·5369, 8·3662 and 9·3620. Hence the radius of the plate being taken as unity,

$r_1=0{\cdot}25679$ 0·25679,
$r_2=0{\cdot}59143$ 0·59147,
$r_3=0{\cdot}89365$ 0·89381.

For comparison, Kirchhoff's calculated results* are given in the second column.

The roots of equation (9A) vary with the assumed value of μ, but, with the exception of some of the earlier roots, the variations are comparatively small. The value of the roots for any given value of μ can be obtained without difficulty. For example, if $\mu=\frac{1}{3}$ (Wertheim's value),

$$x=\beta-\frac{3m+13}{3(8\beta)}+\frac{40(3m+5)}{9(8\beta)^2}\ldots$$

* Strehlke, "Pogg. Ann.," 1855. Lord Rayleigh, "Theory of Sound," Vol. I.

If this expression be divided by the value of x in (14), the new series of roots can be found with considerable accuracy by multiplying the roots in Table II. by this quotient, viz. :—

$$1+\frac{2}{24\beta^2}-\frac{4\beta(6m-5)-6(m+5)}{(24\beta^2)^2}\ldots.$$

TABLE III.
Roots of Equation (9A).
$\mu=\frac{1}{3}$ (Wertheim's value).

n.	$p=0$.	$p=1$.	$p=2$.	$p=3$.
0	...	...	2·292	3·496
1	3·013	4·530	5·936	7·274
2	6·206	7·737	9·188	10·595
3	9·371	10·910	12·386	13·820
4	12·526	14·029	15·559	17·015

The first two roots in each column are those given by Kirchhoff (Crelle, 1850, § 85).

The calculated values of the radii of the nodal circles vary very little for different values of μ. Taking the case where the variation is greatest, viz., when $p=3$ and $n=0$, the change in the value of the radius when μ is changed from $\frac{1}{4}$ to $\frac{1}{3}$ is less than 1 in 500. (Lord Rayleigh, "Theory of Sound," Vol. I., p. 363.)

The change in the calculated value of the frequency of vibration of a "free" plate for a given change in μ is easily found from a consideration of the expression for the frequency, viz. :—

$$N=\frac{ax^2}{\sqrt{1-\mu^2}},$$

a being constant for the same plate and x one of the roots of equation (9A). When the value of μ is changed x also changes and both contribute to the variation in the value of N. The factor $1/\sqrt{1-\mu^2}$ introduces a change of about 2·7 per cent. in the frequency when the value of μ is changed from $\frac{1}{4}$ to $\frac{1}{3}$. Since N varies as the square of the root (x), the variation due to this is easily found from Tables II. and III. For example, when $p=3$, a change in μ from $\frac{1}{4}$ to $\frac{1}{3}$ diminishes the first root by about 2 per cent., and, therefore, the calculated value of the frequency is about 4 per cent. less. In the second mode of vibration, the decrease in the frequency is less than 1 in 200, in the third mode less than 1 in 1,000, and so on.

7

Reprinted from *J. Appl. Mech.*, **17,** 448–453 (1950)

Vibration of Rectangular Plates by the Ritz Method

By DANA YOUNG,[1] AUSTIN, TEXAS

Ritz's method is one of several possible procedures for obtaining approximate solutions for the frequencies and modes of vibration of thin elastic plates. The accuracy of the results and the practicability of the computations depend to a great extent upon the set of functions that is chosen to represent the plate deflection. In this investigation, use is made of the functions which define the normal modes of vibration of a uniform beam. Tables of values of these functions have been computed as well as values of different integrals of the functions and their derivatives. With the aid of these data, the necessary equations can be set up and solved with reasonable effort. Solutions are obtained for three specific plate problems, namely, (*a*) square plate clamped at all four edges, (*b*) square plate clamped along two adjacent edges and free along the other two edges, and (*c*) square plate clamped along one edge and free along the other three edges.

Nomenclature

THE following nomenclature is used in the paper. Any consistent set of units can be used for the physical quantities. For reference purposes the units are given in terms of the engineering inch-pound-second system:

V = elastic strain energy of bending of a plate, in-lb
E = modulus of elasticity, psi
μ = Poisson's ratio
h = thickness of plate, in.
D = $Eh^3/12(1 - \mu^2)$ = bending stiffness of a plate, lb-in.
w = lateral deflection of plate, in.
x, y = rectangular co-ordinates
ρ = mass density of plate material (lb sec²/in.⁴)
f = frequency, cycles per sec (cps)
ω = $2\pi f$ = angular frequency, radians/sec
λ = $\omega^2 \rho h a^3 b/D$ = characteristic value
a, b = lateral dimensions of plate, in.; see Fig. 1
l = length of beam, in.
m, n; i, k; p, q; r, s = positive integers
A_{mn} = coefficient used in series representation of deflection
X_m = a function of x alone
Y_n = a function of y alone
φ_r = characteristic function of a vibrating beam, as defined by Equations [5], [6], [7]
α_r = parameter in expressions for φ_r; values given in Table 1
ϵ_r = parameter in expressions for φ_r; values given in Table 1
$E_{mi}, F_{kn}, H_{im}, K_{kn}$ = definite integrals defined by Equations [10], [11], [12]
$C_{mn}^{(ik)}$ = coefficients defined by Equations [15] and [16]
δ_{mn} = Kronecker delta, defined by Equation [14]

Introduction

An exact solution of the differential equation of a vibrating plate is known for the case of a rectangular plate which is simply supported at all four edges (1, 2),[2] and also for a rectangular plate which is simply supported along one pair of opposite edges with any conditions at the other two edges (3). For other combinations of edge conditions the solutions are more complicated, and it has been necessary to resort to various approximate methods. The procedure developed by Ritz (4) has been found to be useful in such problems and, in fact, was used by Ritz himself (5) to calculate in great detail the frequencies and nodal patterns of a square plate with all four edges free.

The convergence and accuracy of Ritz's method have been discussed by various authors including E. Trefftz (6), R. Courant (7), and L. Collatz (8). It is known that this method gives upper bounds for the frequencies, that is, the frequencies calculated by Ritz's procedure are always higher than the exact values. Also, the accuracy of the results cannot be estimated with certainty in most cases. In spite of these limitations, the method has yielded satisfactory solutions for numerous problems in equilibrium, buckling, and vibration.

While Ritz's method is well known, it has not been used as much as might be expected for plate-vibration problems. There appears to be little published data for the vibration of rectangular plates by Ritz's method except for a square plate with free edges (5). This is probably due, at least in part, to the great amount of computational labor which is required both to set up and to solve the necessary equations. The amount of computation involved depends to a large extent upon the set of functions that is used to represent the plate deflection. For these functions some investigators have taken a series of polynomials while others have used combinations of the characteristic functions which define the normal modes of a vibration of a uniform beam. For example, an application of the polynomial functions to a plate equilibrium problem has been given by G. Pickett (9), while the use of the beam-vibration functions is illustrated by Ritz's solution (5). It is these latter types of functions that have been selected for use in this study.

In order to simplify the computations, values of different integrals of the functions and their derivatives have been calculated, and the results are given herein. In addition, tables of

---

[1] Professor of Applied Mechanics, University of Texas. Mem. ASME.

For presentation at the Annual Conference of the Applied Mechanics Division, Purdue University, Lafayette, Ind., June 22–24, 1950, of THE AMERICAN SOCIETY OF MECHANICAL ENGINEERS.

Discussion of this paper should be addressed to the Secretary, ASME, 29 West 39th Street, New York, N. Y., and will be accepted until one month after final publication of the paper itself in the JOURNAL OF APPLIED MECHANICS.

NOTE: Statements and opinions advanced in papers are to be understood as individual expressions of their authors and not those of the Society. Manuscript received at ASME Headquarters, March 27, 1950. Paper No. 50—APM-18.

[2] Numbers in parentheses refer to the Bibliography at the end of the paper.

values for the functions are now available (10). With the aid of these data the work required to set up the necessary equations is considerably reduced. A simple iteration procedure can be used to solve the equations.

It is to be observed that the analysis herein is for a homogeneous plate of uniform thickness and is based upon the ordinary theory of thin plates.

Ritz's Method

For a uniform plate which is vibrating harmonically with amplitude $w(x, y)$ and angular frequency ω, the maximum potential energy is given by

$$V = \frac{D}{2}\int\int\left[\left(\frac{\partial^2 w}{dx^2}\right)^2 + \left(\frac{\partial^2 w}{\partial y^2}\right)^2 + 2\mu\frac{\partial^2 w}{\partial x^2}\frac{\partial^2 w}{\partial y^2} + 2(1-\mu)\left(\frac{\partial^2 w}{\partial x \partial y}\right)^2\right]dx\,dy \quad [1]$$

and the maximum kinetic energy is

$$\frac{1}{2}\rho h\omega^2\int\int w^2\,dx\,dy$$

where the integrations are to be taken over the domain of the plate surface. Equating these two expressions, we have

$$\omega^2 = \frac{2}{\rho h}\frac{V}{\int\int w^2 dxdy} \quad [2]$$

It is known, see (2), (5), or (11), that the natural frequencies are determined by finding expressions for w that satisfy the boundary conditions and minimize the expression, Equation [2]. The direct application of the calculus of variations to minimize Equation [2] leads to the partial differential equation for a vibrating plate. Instead of following such a procedure, Ritz's method consists of assuming the deflection $w(x, y)$ as a linear series of "admissible" functions and adjusting the coefficients in the series so as to minimize Equation [2].

For rectangular plates, with the edges parallel to the x- and y-axes, it is expedient to take the series approximation for w in the form

$$w(x, y) = \sum_{m=1}^{p}\sum_{n=1}^{q} A_{mn}X_m(x)\cdot Y_n(y) \quad [3]$$

Each function X_mY_n must be "admissible," that is, it must satisfy the so-called "artificial boundary conditions," but need not satisfy any "natural boundary conditions;" see (8) or (11). In the case of plates, prescribed values for the deflection and also for the slope constitute artificial boundary conditions, while the requirement that second or third derivatives or combinations thereof vanish at the boundary is a natural condition. From a practical consideration of the rate of convergence, it is desirable to satisfy natural boundary conditions if possible.

When $w(x, y)$ as given by Equation [3] is substituted in Equation [2], the right-hand side becomes a function of the coefficients A_{mn}. This is minimized by taking the partial derivative with respect to each coefficient and equating to zero. Thus we arrive at a set of equations each of which has the form

$$\frac{\partial V}{\partial A_{ik}} - \frac{\omega^2\rho h}{2}\frac{\partial}{\partial A_{ik}}\int\int w^2 dxdy = 0 \quad [4]$$

where A_{ik} is any one of the coefficients A_{mn}. Equation [4] represents a system of linear homogeneous equations in the unknowns A_{mn}. The natural frequencies $\omega_1, \omega_2, \ldots$ are determined from the condition that the determinant of the system must vanish.

As discussed in the introduction, the appropriate characteristic functions for vibrating beams will be used for X_m and Y_n. A summary of the properties of these functions is given in the next section.

Characteristic Functions for Vibrating Beam

The different types of beams will be identified by a compound adjective which describes the end conditions. Thus a "clamped-clamped" beam is one which is rigidly clamped at both ends; a "clamped-free" beam is clamped at the end $x = 0$ and free at the end $x = l$; a "free-free" beam is free at both ends.

For each type of beam there is an infinite number of normal modes in which the beam can vibrate laterally. The method of determining the set of characteristic functions which define the normal modes for any type of beam is given in standard references such as (1) and (2). The characteristic functions for the three types of beams used in this paper are as follows:

Clamped-Clamped Beam

$$\varphi_r = \cosh\frac{\epsilon_r x}{l} - \cos\frac{\epsilon_r x}{l} - \alpha_r\left(\sinh\frac{\epsilon_r x}{l} - \sin\frac{\epsilon_r x}{l}\right) \quad [5]$$

Clamped-Free Beam

$$\varphi_r = \cosh\frac{\epsilon_r x}{l} - \cos\frac{\epsilon_r x}{l} - \alpha_r\left(\sinh\frac{\epsilon_r x}{l} - \sin\frac{\epsilon_r x}{l}\right) \quad [6]$$

Free-Free Beam

$$\varphi_1 = 1 \quad [7a]$$

$$\varphi_2 = \sqrt{3}\,(1 - 2x/l) \quad [7b]$$

$$\varphi_r = \cosh\frac{\epsilon_r x}{l} + \cos\frac{\epsilon_r x}{l} - \alpha_r\left(\sinh\frac{\epsilon_r x}{l} + \sin\frac{\epsilon_r x}{l}\right), \quad (r = 3, 4, 5, \ldots) \quad [7c]$$

In each of the foregoing expressions $r = 1, 2, 3, \ldots$; hence each expression defines an infinite set of functions. The numerical values of α_r and ϵ_r for each set of functions are given in Table 1. It should be noted that, while Equations [5] and [6] have the same general form, the values of α_r and ϵ_r are different. The functions have significance only in the interval $0 \le x \le l$.

Tables of values of these functions are given in (10) to five decimal places and at intervals of the argument $x/l = 0.02$.

Equation [7c] is the usual expression for the characteristic functions of a free-free beam; when $r = 3$ we have the first mode of free vibration. The functions φ_1 and φ_2 represent a rigid-body translation and rotation and are included in order to obtain a complete orthogonal set.

The boundary conditions satisfied by the functions in each set are the same as the end conditions of the corresponding beam. That is, for the clamped-clamped functions $\varphi_r = d\varphi_r/dx = 0$ at $x = 0$ and $x = l$; for the clamped-free functions $\varphi_r = d\varphi_r/dx = 0$ at $x = 0$ and $d^2\varphi_r/dx^2 = d^3\varphi_r/dx^3 = 0$ at $x = l$; for the free-free functions

$$d^2\varphi_r/dx^2 = d^3\varphi_r/dx^3 = 0 \text{ at } x = 0 \text{ and } x = l$$

Each of the characteristic functions except Equations [7a] and [7b] satisfies the differential equation $d^4\varphi_r/dx^4 = \epsilon_r^4\varphi_r/l^4$. Each set of the functions is orthogonal in the interval 0 to l, that is, for any two functions φ_r and φ_s in the same set, the following relations hold

TABLE 2 INTEGRALS OF CHARACTERISTIC FUNCTIONS OF CLAMPED-CLAMPED BEAM

Values of $\ell \int_0^{\ell} \frac{d\phi_r}{dx}\frac{d\phi_s}{dx}\,dx$

r \ s	1	2	3	4	5	6
1	12.30262	0	- 9.73079	0	- 7.61544	0
2	0	46.05012	0	- 17.12892	0	- 15.19457
3	- 9.73079	0	98.90480	0	- 24.34987	0
4	0	- 17.12892	0	171.58566	0	- 31.27645
5	- 7.61544	0	- 24.34987	0	263.99798	0
6	0	- 15.19457	0	- 31.27645	0	376.15008

Note: $\int_0^{\ell} \phi_r \frac{d^2\phi_s}{dx^2}\,dx = -\int_0^{\ell} \frac{d\phi_r}{dx}\frac{d\phi_s}{dx}\,dx$

TABLE 1 VALUES OF α_r AND ϵ_r

Type of Beam	r	α_r	ϵ_r	ϵ_r^4
Clamped-Clamped	1	0.9825 0222	4.7300 408	500.564
	2	1.0007 7731	7.8532 046	3 803.537
	3	0.9999 6645	10.9956 078	14 617.630
	4	1.0000 0145	14.1371 655	39 943.799
	5	0.9999 9994	17.2787 596	89 135.407
	6	1.0000 0000	20.4203 522	173 881.316
	r > 6	1.0	(2r + 1)π/2	
Clamped-Free	1	0.7340 955	1.8751 041	12.362
	2	1.0184 6644	4.6940 911	485.519
	3	0.9992 2450	7.8547 574	3 806.546
	4	1.0000 3355	10.9955 407	14 617.273
	5	0.9999 9855	14.1371 684	39 943.832
	r > 5	1.0	(2r - 1)π/2	
Free-Free	1		0	0
	2		0	0
	3	0.9825 0222	4.7300 408	500.564
	4	1.0007 7731	7.8532 046	3 803.537
	5	0.9999 6645	10.9956 078	14 617.630
	6	1.0000 0145	14.1371 655	39 943.799
	7	0.9999 9994	17.2787 596	89 135.407
	r > 7	1.0	(2r - 3)π/2	

$$\int_0^l \varphi_r \varphi_s \, dx \begin{array}{l} = l \quad (\text{for } r = s) \\ = 0 \quad (\text{for } r \neq s) \end{array} \Bigg\} \ldots\ldots\ldots\ldots [8]$$

The second derivatives of the function in each set are also orthogonal and satisfy the relations

$$\int_0^l \frac{d^2\varphi_r}{dx^2}\frac{d^2\varphi_s}{dx^2}\,dx \begin{array}{l} = \dfrac{\epsilon_r^4}{l^3} \quad (\text{for } r = s) \\ = 0 \quad (\text{for } r \neq s) \end{array} \Bigg\} \ldots\ldots\ldots [9]$$

with the exception of φ_1 and φ_2 for the free-free functions, Equations [7(*a*), (*b*)], for which

$$\int_0^l \left(\frac{d^2\varphi_1}{dx^2}\right)^2 dx = \int_0^l \left(\frac{d^2\varphi_2}{dx^2}\right)^2 dx = 0$$

For convenience in handling these exceptions, it is expedient to define $\epsilon_1 = \epsilon_2 = 0$ arbitrarily for the free-free functions; with this notation the relations, Equations [9], are valid for all functions in each set. Numerical values of ϵ_r^4 are given in Table 1.

In addition to the integrals defined by Equations [8] and [9], it is necessary to evaluate

$$\int_0^l \varphi_r \frac{d^2\varphi_s}{dx^2}\,dx, \text{ and } \int_0^l \frac{d\varphi_r}{dx}\frac{d\varphi_s}{dx}\,dx$$

TABLE 3 INTEGRALS OF CHARACTERISTIC FUNCTIONS OF CLAMPED-FREE BEAM

Values of $\ell \int_0^{\ell} \frac{d\phi_r}{dx}\frac{d\phi_s}{dx}\,dx$

r \ s	1	2	3	4	5
1	4.64778	- 7.37987	3.94151	- 6.59339	4.59198
2	- 7.37987	32.41735	- 22.35243	13.58245	- 22.83952
3	3.94151	- 22.35243	77.29889	- 35.64827	20.16203
4	- 6.59339	13.58245	- 35.64827	1[illegible].90185	- 48.71964
5	4.59198	- 22.83952	20.16203	- 48.71964	228.13325

Values of $\ell \int_0^{\ell} \phi_r \frac{d^2\phi_s}{dx^2}\,dx$

r \ s	1	2	3	4	5
1	0.85824	- 11.74322	27.45315	- 37.39025	51.95662
2	1.87385	- 13.29425	- 9.04222	30.40119	- 33.70907
3	1.56451	3.22933	- 45.90423	- 8.33537	36.38656
4	1.08737	5.54065	4.25360	- 98.91821	- 7.82895
5	0.91404	3.71642	11.23264	4.73605	-171.58466

TABLE 4 INTEGRALS OF CHARACTERISTIC FUNCTIONS OF FREE-FREE BEAM

Values of $\ell \int_0^{\ell} \frac{d\phi_r}{dx}\frac{d\phi_s}{dx}\,dx$

r \ s	1	2	3	4	5	6	7
1	0	0	0	0	0	0	0
2	0	12.00000	0	13.85641	0	13.85641	0
3	0	0	49.48082	0	35.37751	0	36.60752
4	0	13.85641	0	108.92459	0	57.58881	0
5	0	0	35.37751	0	186.86671	0	78.10116
6	0	13.85641	0	57.58881	0	284.68314	0
7	0	0	36.60752	0	78.10116	0	402.22805

Values of $\ell \int_0^{\ell} \phi_r \frac{d^2\phi_s}{dx^2}\,dx$

r \ s	1	2	3	4	5	6	7
1	0	0	18.58910	0	43.98096	0	69.11504
2	0	0	0	40.59448	0	84.08889	0
3	0	0	- 12.30262	0	52.58440	0	101.62255
4	0	0	0	- 46.05012	0	55.50868	0
5	0	0	1.80069	0	- 98.90480	0	60.12891
6	0	0	0	5.28566	0	-171.58566	0
7	0	0	0.57069	0	9.86075	0	-263.99798

when using these functions in Ritz's method. Values of these integrals have been computed and are given in Tables 2, 3, and 4, for the clamped-clamped, clamped-free, and free-free functions, respectively.

Application of Ritz's Method

The characteristic functions discussed previously are the functions that will be used for X_m and Y_n in Equation [3]. The particular sets to be used in any problem will depend upon the boundary conditions of the plate. Consider a rectangular plate, Fig. 1, bounded by the lines $x = 0, x = a, y = 0, y = b$. Assume, for example, that the plate is clamped along the edge $x = 0$ and free along the other three edges. In this case the clamped-free functions, Equation [6], should be used for X_m, and the free-free functions, Equation [7], should be used for Y_n. We observe that when any one of the sets of functions, Equation [5], [6], or [7], is used for X_m, we take $l = a$; if used for Y_n, we take $l = b$ and replace x by y. Appropriate changes of the subscripts r and s to either m and i or to n and k are to be made in each case. With the three sets of functions given herein, solutions can be obtained for rectangular plates having any combination of free and clamped edges.

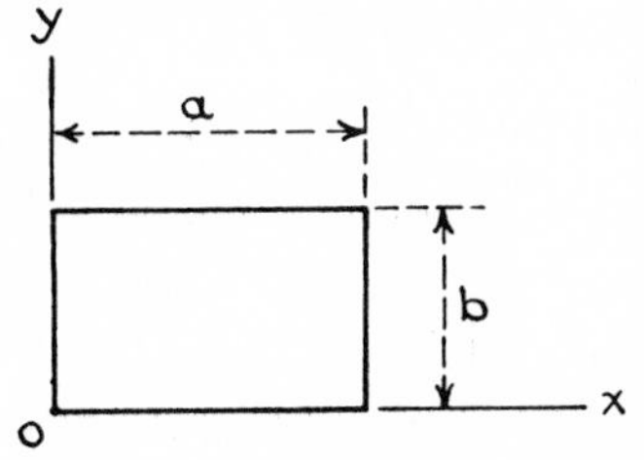

Fig. 1 Co-Ordinate System for Plate

It is convenient to introduce the following notation

$$E_{im} = a\int_0^a X_i \frac{d^2X_m}{dx^2}dx \qquad E_{mi} = a\int_0^a X_m \frac{d^2X_i}{dx^2}dx \ldots [10]$$

$$F_{kn} = b\int_0^b Y_k \frac{d^2Y_n}{dy^2}dy \qquad F_{nk} = b\int_0^b Y_n \frac{d^2Y_k}{dy^2}dy \ldots [11]$$

$$H_{im} = a\int_0^b \frac{dX_i}{dx}\frac{dX_m}{dx}dx \qquad K_{kn} = b\int_0^b \frac{dY_k}{dy}\frac{dY_n}{dy}dy \ldots [12]$$

Since the appropriate φ-functions are to be used for X_m and Y_n the numerical values of the foregoing integrals can be taken directly from the data given in Tables 2, 3, and 4.

Using Equations [3] and [1], and taking into account the orthogonality relations, Equations [8] and [9], the set of Equations [4] can be reduced to the form

$$\sum_{m=1}^{p}\sum_{n=1}^{q} [C_{mn}^{(ik)} - \lambda\delta_{mn}] A_{mn} = 0 \ldots [13]$$

where

$$\lambda = \omega^2 pha^3b/D \ldots [14]$$

$$\delta_{mn} = 1 \text{ for } mn = ik$$
$$= 0 \text{ for } mn \neq ik$$

and

$$C_{mn}^{(ik)} = \mu\frac{a}{b}[E_{mi}F_{kn} + E_{im}F_{nk}] + 2(1-\mu)\frac{a}{b}H_{im}K_{kn} \ldots [15]$$

which is valid for $mn \neq ik$. For $mn = ik$, the coefficient is

$$C^{(ik)}_{ik} = \frac{b}{a}\epsilon_i^4 + \frac{a^3}{b^3}\epsilon_k^4 + 2\mu\frac{a}{b}E_{ii}F_{kk} + 2(1-\mu)\frac{a}{b}H_{ii}K_{kk} \ldots [16]$$

In Equation [16], ϵ_i is to be taken from the data in Table 1 corresponding to the φ-function that represents X_m, while ϵ_k is to be taken from data for the φ function that represents Y_n.

There will be one equation of the type [13] for each of the $p \cdot q$ combinations of ik. The characteristic values λ are found from the condition that determinant of this system of equations must vanish. If there are more than three or four equations in the system, the mathematical labor of expanding the determinant and solving for roots of the polynomial in λ is prohibitive. In such cases it is expedient to solve for λ by one of the known iterative procedures. One of the advantages of using the φ-functions for X_m and Y_n is that the diagonal terms in the determinant are large compared to the others, and as a result the characteristic values and modes can be found by the simple iteration procedure used by Ritz (5).

Square Cantilever Plate

Consider the case of a square plate which is clamped along one edge and free along the other three edges. Using the co-ordinate system in Fig. 1, let us take $x = 0$ as the clamped edge. For X_m we use the clamped-free functions, Equation [6], and for Y_n we use the free-free functions, Equation [7]. Taking $a = b$ and assuming Poisson's ratio $\mu = 0.3$, the coefficients $C_{mn}^{(ik)}$ as defined by Equations [15] and [16] have been computed for an 18-term series based on taking $m = 1, 2, 3$ and $n = 1, 2, 3, 4, 5, 6$. It is found that the corresponding system of 18 Equations [13] divides into two independent groups of 9 equations each. One of these groups includes only $n = 1, 3, 5$ and represents deflections which are symmetrical about the line $y = b/2$. The other group includes only $n = 2, 4, 6$, and represents deflections which are antisymmetrical with respect to the line $y = b/2$.

The coefficients for the symmetrical group are given in Table 5. The coefficients along the principal diagonal are large compared to the others, and, consequently, the system can be solved quite simply by the iteration procedure which was used by Ritz (5). A brief explanation of this procedure will now be given. The equations corresponding to the coefficients in Table 5 are

$$(12.36 - \lambda) A_{11} + 4.79 A_{13} + 11.33 A_{15} + \ldots = 0$$
$$4.79 A_{11} + (828.51 - \lambda) A_{13} + 244.21 A_{15} + \ldots = 0$$
$$11.33 A_{11} + 244.21 A_{13} + (15794.99 - \lambda) A_{15} + \ldots = 0$$

Only the first three equations have been indicated here in order to save space; the six other equations are to be understood as following these.

Assume that we wish to find the lowest frequency. It may be inferred from examination of the equations that A_{11} is the predominant amplitude coefficient. For convenience, take $A_{11} = 1.0$. The first equation can be written as

$$\lambda = 12.36 + 4.79 A_{13} + 11.33 A_{15} + \ldots [a]$$

and the others can be put in the form

$$A_{13} = -(4.79 + 244.21 A_{15} + \ldots)/(828.51 - \lambda) \ldots [b]$$

$$A_{15} = -(11.33 + 244.21 A_{13} + \ldots)/(15794.99 - \lambda) \ldots [c]$$

and similarly for the other six equations. For the first trial, assume values for $A_{13}, A_{15}, \ldots$; these may be taken as zero if no better guess is apparent. Using these first trial values, calculate λ from Equation [a] and then calculate A_{13} from Equation [b]. Using this improved value for A_{13} and the first trial values for the other A_{mn}, calculate A_{15} from Equation [c]. Continue the procedure for the remaining equations.

After these improved values for A_{mn} are calculated, they are substituted back into Equation [a] and a second trial value for λ is calculated. Following the same steps as before, we now recalculate A_{13}, A_{15}, ... The procedure is repeated until the values of λ and A_{mn} in successive iterations are close enough to give the desired accuracy.

The same procedure can be used for the higher frequencies. For example, examination of Table 5 indicates that A_{21} is the predominant amplitude coefficient for the next higher frequency. Accordingly, we take $A_{21} = 1.0$ and start by rewriting the equation corresponding to the fourth row of coefficients in Table 5 in the form

$$\lambda = 485.52 - 65.49\, A_{13} - 154.94\, A_{15} + \ldots .$$

The other equations are written in a form similar to Equations [b] and [c], and then the equations are solved in the same manner as before.

The first three frequencies of the symmetrical modes and the first two frequencies of the antisymmetrical modes have been calculated and the results are given in Table 6. The relative magnitudes of the amplitude coefficients A_{mn} corresponding to each frequency are also given in the table. In each case the magnitude of the predominant amplitude coefficient is taken equal to unity.

Knowing the relative values of A_{mn} and the values of the φ-functions, the shape of each mode of vibration can be calculated from Equation [3]. In this particular example, the predominant amplitude coefficient is so large compared to the others that the shape of each mode is practically determined by the term in Equation [3] corresponding to that one coefficient. Thus it is seen that the lowest symmetrical mode is primarily a cantilever beam deflection in the x-direction and practically a constant in the y-direction. In the lowest antisymmetrical mode, each section x = const rotates essentially as a straight line about the axis $y = b/2$. The nodal lines for each mode of vibration are shown by the sketches in Table 6.

TABLE 5 COEFFICIENTS FOR VIBRATION OF SQUARE CANTILEVER PLATE SYMMETRIC MODES

i	k	$c^{(ik)}_{11}$	$c^{(ik)}_{13}$	$c^{(ik)}_{15}$	$c^{(ik)}_{21}$	$c^{(ik)}_{23}$	$c^{(ik)}_{25}$	$c^{(ik)}_{31}$	$c^{(ik)}_{33}$	$c^{(ik)}_{35}$
1	1	12.36	4.79	11.33	0	30.95	24.72	0	8.72	20.64
1	3	4.79	828.51	244.21	-65.49	-474.80	-342.30	153.10	165.95	234.73
1	5	11.33	244.21	15794.99	-154.94	-549.75	-1637.83	362.23	629.15	170.16
2	1	0	-65.49	-154.94	485.52	-74.14	-175.41	0	18.01	42.61
2	3	10.45	-474.80	-549.75	-74.14	3329.87	1388.69	-50.43	-1526.98	-1061.02
2	5	24.72	-342.30	-1637.83	-175.41	1388.69	24372.88	-119.31	-1247.98	-5675.23
3	1	0	153.10	362.23	0	-50.43	-119.31	3806.55	-256.00	-605.67
3	3	8.72	165.95	629.15	18.01	-1526.98	-1247.98	-256.00	10000.69	3079.54
3	5	20.64	234.73	170.16	42.61	-1061.02	-5675.23	-605.67	3079.54	41370.70

TABLE 6 FREQUENCIES AND MODES OF VIBRATION OF SQUARE CANTILEVER PLATE

	1st Mode	2nd Mode	3rd Mode	4th Mode	5th Mode
$\dfrac{\omega}{\sqrt{D/\rho h a^4}}$	3.494	8.547	21.44	27.46	31.17
Amplitude coefficients	$A_{11} = 1.0000$	$A_{12} = 1.0000$	$A_{11} = 0.0054$	$A_{11} = 0.0090$	$A_{12} = -0.1201$
	$A_{13} = -0.0087$	$A_{14} = -0.0134$	$A_{13} = 0.2731$	$A_{13} = 1.0000$	$A_{14} = 0.0627$
	$A_{15} = -0.0008$	$A_{16} = -0.0011$	$A_{15} = 0.0092$	$A_{15} = -0.0120$	$A_{16} = 0.0080$
	$A_{21} = -0.0026$	$A_{22} = 0.1212$	$A_{21} = 1.0000$	$A_{21} = -0.2866$	$A_{22} = 1.0000$
	$A_{23} = -0.0050$	$A_{24} = 0.0044$	$A_{23} = 0.0713$	$A_{23} = 0.1786$	$A_{24} = -0.0388$
	$A_{25} = -0.0011$	$A_{26} = 0.0006$	$A_{25} = 0.0079$	$A_{25} = 0.0009$	$A_{26} = -0.0013$
	$A_{31} = 0.0001$	$A_{32} = -0.0020$	$A_{31} = -0.0118$	$A_{31} = -0.0451$	$A_{32} = 0.0776$
	$A_{33} = -0.0014$	$A_{34} = -0.0011$	$A_{33} = 0.0050$	$A_{33} = 0.0125$	$A_{34} = 0.0086$
	$A_{35} = -0.0006$	$A_{36} = -0.0006$	$A_{35} = -0.0003$	$A_{35} = -0.0023$	$A_{36} = 0.0024$
Nodal lines	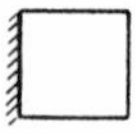	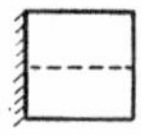	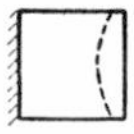	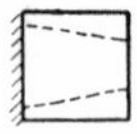	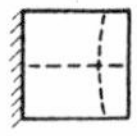

Clamped Square Plate

Consider a square plate which is clamped along all four edges. For this case we use the clamped-clamped functions Equation [5] for both X_m and Y_n. The calculations for this problem have been carried out for a 36-term series based on taking both m and n equal to 1, 2, 3, 4, 5, 6. The 36 equations of the type [13] divide in four independent groups of 9 equations each. If we denote by x' and y' the co-ordinates referred to a set of axes through the middle point of the plate, then one group contains only functions which are even in both x' and y'; the second group contains only functions which are odd in both x' and y'; the third group contains functions which are even in x' and odd in y'; the fourth group contains functions which are odd in x' and even in y'. The last two groups are essentially the same if we interchange x and y, and lead to identical frequencies; hence there are only 3 groups to solve. Further, each of the first two groups can be rearranged to form two independent subgroups, one of which has 6 equations and the other 3 equations; the first subgroup includes solutions for which $A_{mn} = A_{nm}$ while the second subgroup includes solutions for which $A_{mn} = -A_{nm}$.

The frequencies and relative amplitudes have been calculated for the first 6 modes and the results are presented in Table 7. The magnitude of the predominant amplitude coefficient is taken as unity for each mode.

The fundamental frequency of a vibrating clamped square plate under a uniform tension in both the x and y-directions has been computed by A. Weinstein and W. Z. Chien (12). The calculations are based upon a variational method developed by A. Weinstein. This method gives a "lower" bound for the frequency as contrasted to Ritz's method which gives an "upper" bound. As a check, the Ritz method calculations for a clamped square plate were extended to include the effect of a uniform tension T per unit length. In order to obtain a direct comparison with the results in (12), the calculations were carried through for two different values of T, namely, $T_1 = 10\pi^2 D/a^2$ and $T_2 = 100\, \pi^2 D/a^2$. A 9-term series, corresponding to taking both m and n equal to 1, 3, 5 was used; this leads to a system of only 6 equations since $A_{mn} = A_{nm}$ for the fundamental frequency. The fundamental frequency calculated on this basis was found to be higher than the lower bound given in (12) by only 0.1 per cent for the tension T_1 and by 2.5 per cent for the tension T_2. Results are given in (12) for a Ritz method solution based upon the two-term series

$$w = A \cos^2 \frac{\pi x'}{a} \cos^2 \frac{\pi y'}{b} + B \cos^3 \frac{\pi x'}{a} \cos^3 \frac{\pi y'}{b}$$

The upper bound for the frequency in this case is higher than the lower bound by 0.8 per cent for the tension T_1, and by 4.5 per cent for the tension T_2.

TABLE 7 FREQUENCIES AND MODES OF VIBRATION OF CLAMPED SQUARE PLATE

	1st Mode	2nd Mode	3rd Mode	4th Mode	5th Mode	6th Mode
$\frac{\omega}{\sqrt{D/\rho h a^4}}$	35.99	73.41	108.27	131.64	132.25	165.15
Amplitude coefficients	A_{11} = 1.0000	A_{12} = 1.0000	A_{22} = 1.0000	A_{13} = 1.0000	A_{11} = -0.0280	A_{12} = -0.0406
	A_{13} = 0.0142	A_{14} = 0.0101	A_{24} = 0.0326	A_{15} = 0.0085	A_{13} = 1.0000	A_{14} = -0.0105
	A_{15} = 0.0020	A_{16} = 0.0020	A_{26} = 0.0073	A_{31} = -1.0000	A_{15} = 0.0055	A_{16} = -0.0017
	A_{31} = 0.0142	A_{32} = 0.0406	A_{42} = 0.0326	A_{35} = -0.0141	A_{31} = 1.0000	A_{32} = 1.0000
	A_{33} = -0.0031	A_{34} = -0.0022	A_{44} = -0.0019	A_{51} = -0.0085	A_{33} = 0.1267	A_{34} = 0.0560
	A_{35} = -0.0009	A_{36} = -0.0007	A_{46} = -0.0010	A_{53} = 0.0141	A_{35} = 0.0118	A_{36} = 0.0141
	A_{51} = 0.0020	A_{52} = 0.0070	A_{62} = 0.0073		A_{51} = 0.0055	A_{52} = 0.0238
	A_{53} = -0.0009	A_{54} = -0.0011	A_{64} = -0.0010		A_{53} = 0.0118	A_{54} = -0.0011
	A_{55} = -0.0004	A_{56} = -0.0005	A_{66} = -0.0006		A_{55} = -0.0018	A_{56} = -0.0009
Nodal lines	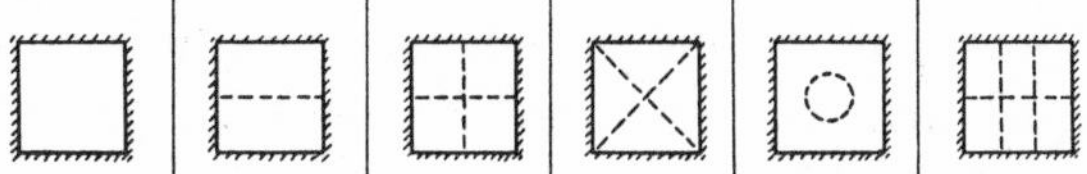					

rechtteckiger Platten," by W. Voight, *Göttingen Nachrichten*, 1893, pp. 225–230.

Square Plate Clamped at Two Adjacent Edges and Free at the Other Edges

For this case assume that the plate is clamped along the two edges $x = 0$ and $y = 0$, and free along the other two edges. The clamped-free functions, Equation [6], are used for both X_m and Y_n. Calculations for this problem have been carried out for a 9-term series based on taking both m and n equal to 1, 2, 3. The set of 9 equations can be divided in two independent groups. One group which contains 6 equations represents modes which are symmetrical about the diagonal line $x = y$, while the other which contains 3 equations represents modes which are antisymmetrical about the line $x = y$.

The frequencies, relative amplitudes, and nodal patterns for the first 5 modes are presented in Table 8. As before, the magnitude of the predominant amplitude coefficient is taken as unity for each mode.

TABLE 8 FREQUENCIES AND MODES OF VIBRATION OF SQUARE PLATE HAVING TWO ADJACENT EDGES CLAMPED AND OTHER TWO EDGES FREE

	1st Mode	2nd Mode	3rd Mode	4th Mode	5th Mode
$\frac{\omega}{\sqrt{D/\rho h a^4}}$	6.958	24.08	26.80	48.05	63.14
Amplitude coefficients	A_{11} = 1.0000	A_{11} = 0	A_{11} = -0.1172	A_{11} = 0.0286	A_{11} = 0
	A_{12} = 0.0604	A_{12} = 1.0000	A_{12} = 1.0000	A_{12} = -0.1566	A_{12} = 0.0030
	A_{13} = -0.0030	A_{13} = 0.00003	A_{13} = 0.0553	A_{13} = -0.0825	A_{13} = 1.0000
	A_{21} = 0.0604	A_{21} = -1.0000	A_{21} = 1.0000	A_{21} = -0.1566	A_{21} = -0.0030
	A_{22} = -0.0101	A_{22} = 0	A_{22} = 0.3223	A_{22} = 1.0000	A_{22} = 0
	A_{23} = -0.0003	A_{23} = -0.0221	A_{23} = 0.0111	A_{23} = 0.1458	A_{23} = 0.1330
	A_{31} = -0.0030	A_{31} = -0.00003	A_{31} = 0.0553	A_{31} = -0.0825	A_{31} = -1.0000
	A_{32} = -0.0003	A_{32} = 0.0221	A_{32} = 0.0111	A_{32} = 0.1458	A_{32} = -0.1330
	A_{33} = -0.0017	A_{33} = 0	A_{33} = 0.0022	A_{33} = -0.0019	A_{33} = 0
Nodal lines	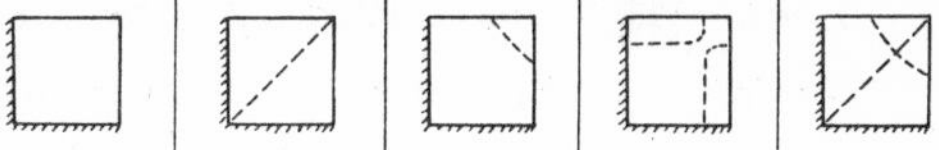				

ACKNOWLEDGMENT

A portion of this work was carried out as a part of the program of the Defense Research Laboratory of the University of Texas and was sponsored by the Navy Department, Bureau of Ordnance.

BIBLIOGRAPHY

1 "Theory of Sound," by Lord Rayleigh, second American edition, Dover Publications, New York, N. Y., 1945.

2 "Vibration Problems in Engineering," by S. Timoshenko, second edition, D. Van Nostrand Company, Inc., New York, N. Y., 1937.

3 "Bermerkung zu dem Problem der transversalen Swingungen rechtteckiger Platten," by W. Voight, *Göttingen Nachrichten*, 1893, pp. 225–230.

4 "Über eine neue Method zur Losung gewisser Variationsprobleme der Mathematische Physik," by W. Ritz, *Journal für reine and angewandte Mathematik*, vol. 135, 1909, pp. 1–61.

5 "Theorie der Transversalschwingungen einer quadratischen Platten mit freien Randern," by W. Ritz, *Annalen der Physik*, Vierte Folge, vol. 28, 1909, pp. 737–786.

6 "Konvergenz und Fehlerschätzung beim Ritzchen Verfahren," by E. Trefftz, *Mathematische Annalen*, vol. 100, 1928, p. 503.

7 "Variational Methods for the Solution of Problems of Equilibrium and Vibration," by R. Courant, *Bulletin of the American Mathematical Society*, New York, N. Y., vol. 49, 1939, pp. 1–23.

8 "Eigenwertprobleme," by Lothar Collatz, Chelsea Publishing Company, New York, N. Y., 1948.

9 "Solution of Rectangular Clamped Plate With Lateral Lead by Generalized Energy Method," by Gerald Pickett, Journal of Applied Mechanics, Trans. ASME, vol. 61, 1931, p. A-178.

10 "Table of Characteristic Functions Representing the Normal Modes of Vibration of a Beam," by D. Young and R. P. Felgar, Engineering Research Series, No. 44, University of Texas, Austin, Texas, July 1, 1949.

11 "Advanced Methods in Applied Mathematics," by R. Courant, New York University, New York, N. Y., 1941.

12 "On the Vibrations of a Clamped Plate Under Tension," by A. Weinstein and W. Z. Chien, *Quarterly of Applied Mathematics*, vol. 1, April, 1943, pp. 61–68.

13 "The Solution of Small Displacement Stability or Vibration Problems Concerning a Flat Rectangular Panel When the Edges Are Either Clamped or Simply Supported," by H. G. Hopkins, Reports and Memoranda, Aeronautical Research Council, London, England, no. 2234, June, 1945.

8

Reprinted from *Proc. Roy. Soc. London*, Ser. A, **93**, 114–128 (1917)

On Waves in an Elastic Plate.

By HORACE LAMB, F.R.S.

(Received July 10, 1916.)

The theory of waves in an infinitely long cylindrical rod was discussed by Pochhammer in 1876 in a well-known paper.* The somewhat simpler problem of two-dimensional waves in a solid bounded by parallel planes was considered by Lord Rayleigh† and by the present writer‡ in 1889. The main object in these various investigations was to verify, or to ascertain small corrections to, the ordinary theory of the vibrations of thin rods or plates, and the wave-length was accordingly assumed in the end to be great in comparison with the thickness.

It occurred to me some time ago that a further examination of the two-dimensional problem was desirable for more than one reason. In the first place, the number of cases in which the various types of vibration of a solid, none of whose dimensions is regarded as small, have been studied is so restricted that any addition to it would have some degree of interest, if merely as a contribution to elastic theory. Again, modern seismology has suggested various questions relating to waves and vibrations in an elastic stratum imagined as resting on matter of a different elasticity and density.§ These questions naturally present great mathematical difficulties, and it seemed unpromising to attempt any further discussion of them unless the comparatively simple problem which forms the subject of this paper should be found to admit of a practical solution. In itself it has, of course, no bearing on the questions referred to.

Even in this case, however, the period-equation is at first sight rather intractable, and it is only recently that a method of dealing with it (now pretty obvious) has suggested itself. The result is to give, I think, a fairly complete view of the more important modes of vibration of an infinite plate, together with indications as to the character of the higher modes, which are of less interest. I may add that the numerical work has been greatly simplified by the help of the very full and convenient Tables of hyperbolic and circular functions issued by the Smithsonian Institution.||

* 'Crelle,' vol. 81, p. 324 ; see also Love, 'Elasticity,' 1906, p. 275.

† 'Proc. Lond. Math. Soc.,' vol. 20, p. 225 ; 'Scientific Papers,' vol. 3, p. 249.

‡ See 'Proc. Lond. Math. Soc.,' vol. 21, p. 85.

§ Love, 'Some Problems of Geodynamics,' 1911, Chap. XI.

|| 'Hyperbolic Functions,' Washington, 1909. I have to thank Mr. J. E. Jones for kindly verifying and, where necessary, correcting my calculations.

1. The motion is supposed to take place in two dimensions x, y, the origin being taken in the medial plane, and the axis of y normal to this. The thickness of the plate is denoted by $2f$. The stress-strain equations are, in the usual notation,

$$\left.\begin{aligned} p_{xx} &= \lambda(\partial u/\partial x + \partial v/\partial y) + 2\mu\, \partial u/\partial x, \\ p_{xy} &= \mu(\partial v/\partial x + \partial u/\partial y), \\ p_{yy} &= \lambda(\partial u/\partial x + \partial v/\partial y) + 2\mu\, \partial v/\partial y. \end{aligned}\right\} \quad (1)$$

It is known that the solution of the equations of motion is of the type

$$u = \partial\phi/\partial x + \partial\psi/\partial y, \qquad v = \partial\phi/\partial y - \partial\psi/\partial x, \quad (2)$$

the functions ϕ, ψ being subject to the equations

$$\rho\frac{\partial^2\phi}{\partial t^2} = (\lambda+2\mu)\nabla^2\phi, \qquad \rho\frac{\partial^2\psi}{\partial t^2} = \mu\nabla^2\psi, \quad (3)$$

where $$\nabla^2 = \partial^2/\partial x^2 + \partial^2/\partial y^2.$$

We now assume a time-factor $e^{i\sigma t}$ (omitted in the sequel), and write

$$h^2 = \frac{\rho\sigma^2}{\lambda+2\mu}, \qquad k^2 = \frac{\rho\sigma^2}{\mu}, \quad (4)$$

so that $$(\nabla^2+h^2)\phi = 0, \qquad (\nabla^2+k^2)\psi = 0. \quad (5)$$

We further assume, for the present purpose, periodicity with respect to x. This is most conveniently done by means of a factor $e^{i\xi x}$, the wave-length being accordingly

$$\lambda' = 2\pi/\xi. \quad (6)$$

Writing $$\alpha^2 = \xi^2 - h^2, \qquad \beta^2 = \xi^2 - k^2, \quad (7)$$

we have $$\frac{\partial^2\phi}{\partial y^2} = \alpha^2\phi, \qquad \frac{\partial^2\psi}{\partial y^2} = \beta^2\psi. \quad (8)$$

The equations (1) now give

$$\left.\begin{aligned} \frac{p_{yy}}{\mu} &= (\xi^2+\beta^2)\phi - 2i\xi\frac{\partial\psi}{\partial y}, \\ \frac{p_{xy}}{\mu} &= 2i\xi\frac{\partial\phi}{\partial y} + (\xi^2+\beta^2)\psi. \end{aligned}\right\} \quad (9)$$

Symmetrical Vibrations.

2. When the motion is symmetrical with respect to the plane $y = 0$ we assume, in accordance with (8),

$$\phi = \mathrm{A}\cosh\alpha y\, e^{i\xi x}, \qquad \psi = \mathrm{B}\sinh\beta y\, e^{i\xi x}. \quad (10)$$

This gives, for the stresses on the faces $y = \pm f$,

$$\left.\begin{aligned} p_{yy}/\mu &= \{\mathrm{A}(\xi^2+\beta^2)\cosh\alpha f - \mathrm{B}\, 2i\xi\beta\cosh\beta f\}e^{i\xi x}, \\ p_{xy}/\mu &= \pm\{\mathrm{A}\, 2i\xi\alpha\sinh\alpha f + \mathrm{B}(\xi^2+\beta^2)\sinh\beta f\}e^{i\xi x} \end{aligned}\right\} \quad (11)$$

Equating these to zero, and eliminating the ratio A/B, we have the period-equation*

$$\frac{\tanh \beta f}{\tanh \alpha f} = \frac{4\xi^2 \alpha\beta}{(\xi^2+\beta^2)^2}. \quad (12)$$

In the most important type of *long* waves, ξf, αf, βf, are all small, and the limiting form of the equation is

$$(\xi^2+\beta^2)^2 - 4\xi^2\alpha^2 = 0, \quad (13)$$

whence

$$\frac{k^2}{\xi^2} = 4\left(1-\frac{h^2}{k^2}\right) = \frac{4(\lambda+\mu)}{\lambda+2\mu}, \quad (14)$$

in virtue of (4). Hence if V be the wave-velocity

$$V^2 = \frac{\sigma^2}{\xi^2} = \frac{4(\lambda+\mu)}{\lambda+2\mu}\frac{\mu}{\rho}. \quad (15)$$

This is in agreement with the ordinary theory, where the thickness is treated as infinitely small and the influence of lateral inertia is neglected.

For waves which are very *short*, on the other hand, as compared with the thickness $2f$, the quantities ξf, αf, βf are large, and the equation tends to the form

$$(2\xi^2-k^2)^2 - 4\xi^2\alpha\beta = 0. \quad (16)$$

This is easily recognised as the equation to determine the period of "Rayleigh waves" on the surface of an elastic solid.† It is known that if the substance be incompressible ($h = 0$, $\alpha = \xi$) the wave-velocity is

$$V = 0{\cdot}9554\,(\mu/\rho)^{\frac{1}{2}}, \quad (17)$$

whilst on Poisson's hypothesis of $\lambda = \mu$,

$$V = 0{\cdot}9194\,(\mu/\rho)^{\frac{1}{2}}. \quad (18)$$

These results will, in fact, present themselves later.

3. In virtue of the relations (4), (7), the equation (12) may be regarded as an equation to find σ when ξ is given, *i.e.* to determine the periods of the various modes corresponding to any given wave-length; but from this point of view it is difficult to handle. It is easier to determine the values of ξ corresponding to given values of the ratio β/α. This is equivalent to finding the wave-lengths corresponding to a given wave-velocity. Putting, in fact, $\beta = m\alpha$, we find from (7),

$$\xi^2 = \frac{k^2 - m^2h^2}{1-m^2}, \quad (19)$$

and therefore

$$V^2 = \frac{\sigma^2}{\xi^2} = \frac{k^2}{\xi^2}\frac{\mu}{\rho} = \frac{(\lambda+2\mu)(1-m^2)}{\lambda+2\mu-m^2\mu}\frac{\mu}{\rho}, \quad (20)$$

so that V depends upon m only.

* An equivalent equation is given by Rayleigh.

† Rayleigh, 'Proc. Lond. Math. Soc.,' vol. 17, p. 3 (1887); 'Scientific Papers,' vol. 2, p. 441.

4. For simplicity we will suppose in the first instance that the substance is incompressible, so that $\lambda = \infty$, $h = 0$, $\alpha = \xi$. Putting

$$\beta = m\xi, \qquad \xi f = \omega, \tag{21}$$

the equation (12) becomes

$$\frac{\tanh m\omega}{\tanh \omega} = \frac{4m}{(1+m^2)^2}, \tag{22}$$

whilst
$$V^2 = (1-m^2)\mu/\rho. \tag{23}$$

It is evident that real values of m must be less than unity. Moreover we have

$$\frac{d}{d\omega} \log \frac{\tanh m\omega}{\tanh \omega} = \frac{2}{\omega}\left(\frac{m\omega}{\sinh 2m\omega} - \frac{\omega}{\sinh 2\omega}\right), \tag{24}$$

which latter expression is easily seen to be positive so long as $m < 1$. Hence as ω increases from 0, the first member of (22) increases from m to its asymptotic value unity. There is therefore one and only one value of ω corresponding to any assigned value of m which makes the second member of (22) less than unity. And since

$$(1+m^2)^2 - 4m = (m-1)(m^3+m^2+3m-1), \tag{25}$$

the right-hand member of (22) is less than unity (for $m < 1$) only so long as $m < 0{\cdot}2956$, which is the positive root of the second factor in (25). The admissible real values of m therefore range from 0·2956 to 0. The former of these makes $\omega = \infty$, $\lambda' = 0$, and gives to V the value (17).

The values of ω corresponding to a series of values of m between the above limits, together with the corresponding values ($\lambda'/2f$) of the ratio of wave-length to thickness, and the corresponding wave-velocities, are given in Table I (p. 118).

So far β has been taken to be real, or $\xi < k$. In the opposite case we may write

$$\beta_1^2 = k^2 - \xi^2, \tag{26}$$

and assume
$$\phi = A \cosh \alpha y\, e^{i\xi x}, \qquad \psi = B \sin \beta_1 y\, e^{i\xi x}. \tag{27}$$

The period-equation is then found to be

$$\frac{\tan \beta_1 f}{\tanh \alpha f} = \frac{4\xi^2\alpha\beta_1}{(\xi^2-\beta_1^2)^2}, \tag{28}$$

α being still supposed to be real.

In the case of incompressibility, writing

$$\xi f = \omega, \qquad \beta_1 = n\xi, \tag{29}$$

we have
$$\frac{\tan n\omega}{\tanh \omega} = \frac{4n}{(1-n^2)^2}, \tag{30}$$

whilst
$$V^2 = (1+n^2)\mu/\rho. \tag{31}$$

If as n increases from 0 we take always the lowest root of (30), we obtain a series of values of ω continuous with the roots of (22) and diminishing down to zero, when $n = \sqrt{3}$. This latter value makes $V = 2\sqrt{(\mu/\rho)}$, in agreement with the general formula (15) for long waves. Numerical results for a series of values of n ranging from 0 to $\sqrt{3}$ are included in Table I.

Table I.—Symmetrical Type. $\lambda = \infty$.

[The unit of V is $\sqrt{(\mu/\rho)}$.]

m.	n.	ω.	$\lambda'/2f$.	V.
0·2956	—	∞	0·0	0·9554
0·29	—	8·67	0·362	0·957
0·28	—	7·09	0·442	0·960
0·27	—	6·38	0·492	0·963
0·26	—	5·93	0·530	0·966
0·25	—	5·61	0·560	0·968
0·20	—	4·75	0·662	0·979
0·15	—	4·35	0·722	0·989
0·10	—	4·14	0·759	0·995
0·05	—	4·03	0·779	0·999
0·0	0·0	4·00	0·785	1·0
—	0·1	3·872	0·811	1·005
—	0·2	3·570	0·880	1·020
—	0·3	3·218	0·976	1·044
—	0·4	2·883	1·090	1·077
—	0·5	2·587	1·214	1·118
—	0·6	2·331	1·348	1·166
—	0·7	2·108	1·490	1·221
—	0·8	1·912	1·643	1·281
—	0·9	1·732	1·814	1·345
—	1·0	$\frac{1}{2}\pi$	2·0	$\sqrt{2}$
—	1·1	1·417	2·217	1·487
—	1·2	1·269	2·476	1·562
—	1·3	1·121	2·803	1·640
—	1·4	0·967	3·25	1·721
—	1·5	0·798	3·94	1·808
—	1·6	0·594	5·29	1·887
—	1·7	0·292	10·8	1·972
—	1·71	0·241	13·0	1·981
—	1·72	0·175	18·0	1·989
—	1·73	0·081	38·8	1·998
—	$\sqrt{3}$	0·0	∞	2·0

The relation between wave-length and wave-velocity for the whole series of modes investigated in this and in the preceding section is shown by the curve A on the opposite page. The unit of the horizontal scale is $\lambda'/2f$; that of the vertical scale is $V/\sqrt{(\mu\rho^{-1})}$.

5. On the present hypothesis of incompressibility there is a displacement function Ψ, analogous to the stream-function of hydrodynamics, viz., we have:—

$$u = \partial\Psi/\partial y, \qquad v = -\partial\Psi/\partial x. \tag{32}$$

The lines $\Psi =$ const. give the directions in which the particles oscillate, whilst if they are drawn for small equidistant values of the constant their

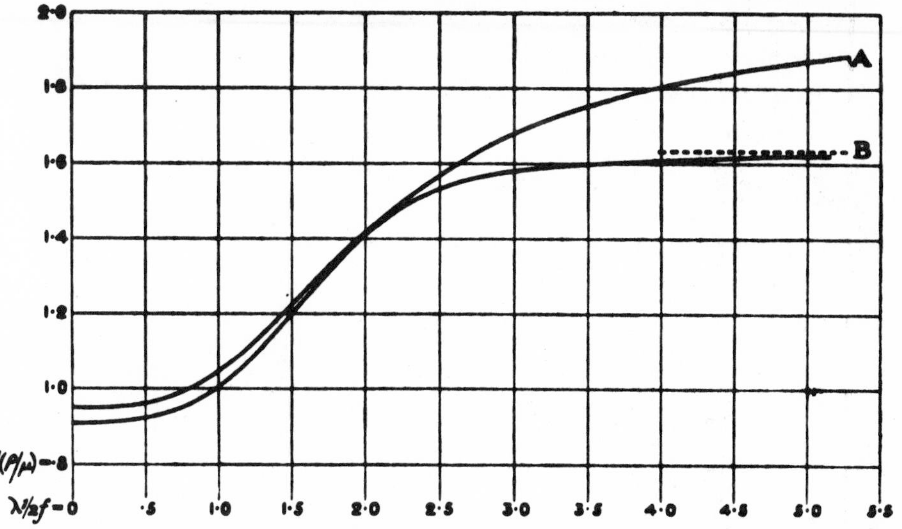

FIG. 1.

greater or less degree of closeness indicates the relative amplitudes. In the case of standing waves the system of lines retains its position in space; in the case of a progressive wave-train it must be imagined to advance with the waves.

If m be real we find, omitting a constant factor,

$$\Psi = \{2m \cosh m\varpi \sinh \xi y - (1+m^2) \cosh \varpi \sinh m\xi y\} e^{i\xi x}. \tag{33}$$

In the opposite case we may write

$$\Psi = \{2n \cos n\varpi \sinh \xi y - (1-n^2) \cosh \varpi \sin n\xi y\} e^{i\xi x}. \tag{34}$$

I have thought it worth while to make diagrams illustrating the configuration of the lines of displacement in the class of modes of vibration which have so far been obtained.

The case of the Rayleigh waves, corresponding to $\varpi = \infty$, is, of course, of independent interest. In the present problem their wave-length is infinitely short; but if we transfer the origin to the surface $y = f$, and then make $f = \infty$, we get, omitting a numerical factor,

$$\Psi = (e^{m\xi y} - 0{\cdot}5437\, e^{\xi y})\, e^{i\xi x}, \tag{35}$$

where $m = 0{\cdot}2956$. On this scale the wave-length may have any value. The diagram (fig. 2) shows the great difference in the character of the motion, and the slow diminution of amplitude with increasing depth, as compared with the "surface waves" of hydrodynamics.* It follows, in fact,

* Lamb, 'Hydrodynamics,' 4th ed., art. 228.

from (35) that the ratio of the vertical amplitude to that at the surface (in the same vertical) at depths of $\frac{1}{4}$, $\frac{1}{2}$, 1 wave-length, is 1·30, 0·814, -0·340, respectively, whereas in the hydrodynamical problem the corresponding numbers are 0·208, 0·043, 0·002.

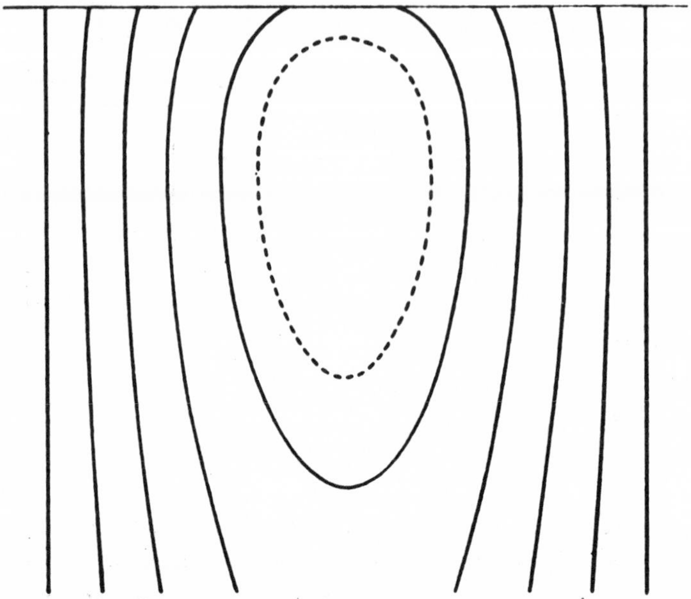

FIG. 2.

In the case of $\xi < k$, the displacement function is given by (34). I have chosen for illustration, as giving a wave-length neither too great nor too small, the case of $n = 1{\cdot}6$, which makes

$$\omega = 0{\cdot}594, \qquad n\omega = 0{\cdot}950, \qquad \lambda'/2f = 5{\cdot}29,$$

and, accordingly,

$$\Psi = (1{\cdot}010 \sinh \xi y + \sin n\xi y)\, e^{i\xi x}. \tag{36}$$

The result is shown in fig. 3, which, like the former diagram, covers half a wave-length. The diagram may be taken to illustrate also, in a general

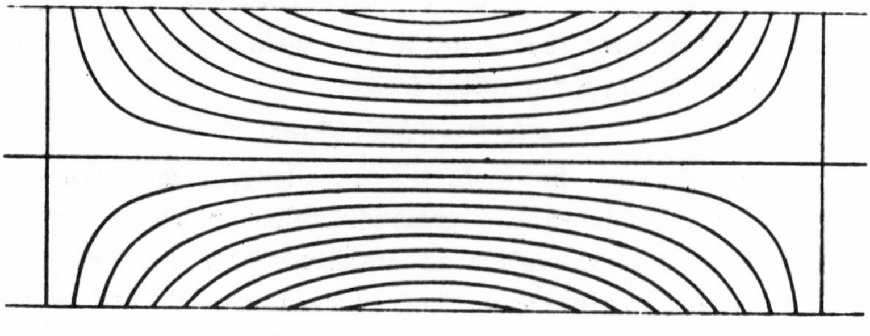

FIG. 3.

way, the most important type of longitudinal vibrations in a cylindrical rod.

6. The modes of vibration so far obtained include all the more interesting types, from the physical point of view, of the symmetrical class; but there are, of course, an infinity of others. These correspond to the higher roots of the equation (30). Thus when $n = 0{\cdot}1$ we easily find the approximate solutions

$$\omega/\tfrac{1}{2}\pi = 22{\cdot}47, \quad 42{\cdot}47, \quad 62{\cdot}47, \ldots. \tag{37}$$

For $n = 0{\cdot}2$

$$\omega/\tfrac{1}{2}\pi = 12{\cdot}28, \quad 22{\cdot}28, \quad 32{\cdot}48, \ldots. \tag{38}$$

For $n = 0{\cdot}3$

$$\omega/\tfrac{1}{2}\pi = 8{\cdot}72, \quad 15{\cdot}39, \quad 22{\cdot}05, \ldots; \tag{39}$$

and so on.

In these modes the plane xy is mapped out into rectangular compartments whose boundaries are lines of displacement. This may be illustrated by the case of $n = 1$, when the internal compartments are squares. This happens to be particularly simple mathematically. The formula (34) for Ψ is now indeterminate, but is easily evaluated. It is found from (30) that for small concomitant variations of ω and n about $n = 1$ we have $\delta(\omega n) = 0$. This leads to

$$\Psi = \sin \xi y \, e^{i\xi x}, \tag{40}$$

with

$$\xi = (2s+1)\pi/2f, \tag{41}$$

where s is an integer. The configuration is shown in fig. 4 for the case

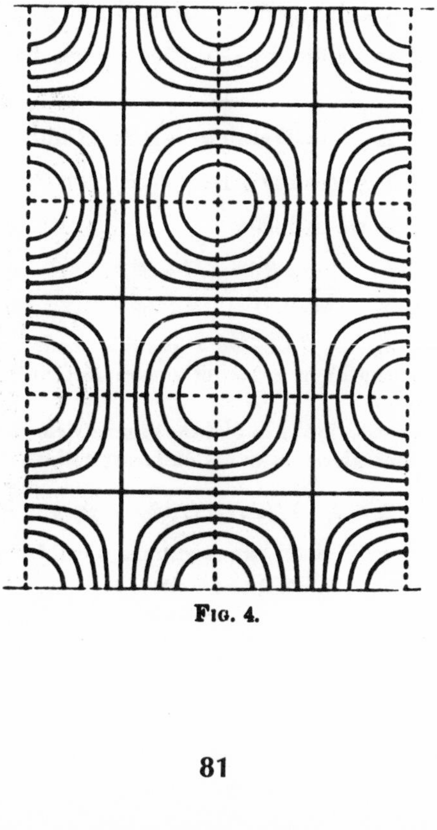

FIG. 4.

$s=1$; but it is to be observed that the dotted lines in the diagram all represent planes which are free from stress, and that consequently any combination of them may be taken to represent free boundaries. This particular solution is, moreover, independent of the hypothesis of incompressibility.* The surface-conditions (11) are, in fact, satisfied by

$$A = 0, \qquad \xi^2 - \beta_1^2 = 0, \qquad \cos \beta_1 f = 0, \tag{42}$$

which lead again to (40) and (41).

As n is increased the compartments referred to become more elongated. For large values of n, and consequently small values of ω or ξf, we have in the limit $n\omega = s\pi$, where s is integral. It is otherwise evident that the surface conditions are satisfied by

$$A = 0, \qquad \xi = 0, \qquad \sin \beta_1 f = 0, \tag{43}$$

whence

$$\phi = 0, \qquad \psi = B \sin (s\pi y/f). \tag{44}$$

The vibration now consists of a shearing motion parallel to x, with $2s$ nodal planes symmetrically situated on opposite sides of $y = 0$. The frequency is given by

$$\sigma^2 = \frac{s^2\pi^2}{f^2}\frac{\mu}{\rho}. \tag{45}$$

Asymmetrical Modes.

7. When the motion is anti-symmetrical with respect to the plane $y=0$ we assume

$$\phi = A \sinh \alpha y\, e^{i\xi x}, \qquad \psi = B \cosh \beta y\, e^{i\xi x}, \tag{46}$$

where α, β are defined as before by (7). This gives for the stresses at the planes $y = \pm f$,

$$\left.\begin{aligned} p_{yy}/\mu &= \pm \{A(\xi^2+\beta^2)\sinh \alpha f - B\, 2i\xi\beta \sinh \beta f\}\, e^{i\xi x}, \\ p_{xy}/\mu &= \{A\, 2i\xi\alpha \cosh \alpha f + B(\xi^2+\beta^2)\cosh \beta f\}\, e^{i\xi x}. \end{aligned}\right\} \tag{47}$$

These surfaces being free, we deduce

$$\frac{\tanh \beta f}{\tanh \alpha f} = \frac{(\xi^2+\beta^2)^2}{4\xi^2\alpha\beta}. \tag{48}$$†

* It was noticed long ago by Lamé as a possible mode of transverse vibration (uniform throughout the length) in a bar of square section, 'Théorie mathématique de l'élasticité,' 2nd ed., p. 170.

There is an analogous solution in the case of the symmetrical vibrations of a cylindrical rod. The surface-conditions given on p. 277 of Love's 'Elasticity' (equation (54)) are satisfied by

$$A = 0, \quad J_1'(\kappa' a) = 0, \quad 2\gamma^2 = p^2\rho/\mu.$$

In the notation of this paper the latter two conditions would be written

$$J_1'(\beta_1 a) = 0, \quad 2\xi^2 = k^2.$$

† Cf. Rayleigh, *loc. cit.*

When the waves are infinitely short this reduces to the form (16) appropriate to Rayleigh waves.

In the case of long flexural waves αf, βf, ξf are small. Writing

$$\tanh \alpha f = \alpha f(1-\tfrac{1}{3}\alpha^2 f^2), \qquad \tanh \beta f = \beta f(1-\tfrac{1}{3}\beta^2 f^2),$$

we find

$$k^2 = \tfrac{4}{3}(1-h^2/k^2)\,\xi^4 f^2, \tag{49}$$

on the supposition that k^2/ξ^2, h^2/ξ^2 are small, which is seen to be verified. This makes

$$V^2 = \tfrac{4}{3}\xi^2 f^2 \frac{\lambda+\mu}{\lambda+2\mu}\frac{\mu}{\rho}, \tag{50}$$

in agreement with the ordinary approximate theory.

It may be pointed out in this connection that Fourier's well-known calculation* of the effect of an arbitrary initial disturbance in an infinitely long bar is physically defective, in that it rests on the assumption that the formula analogous to (50) is valid for all wave-lengths. As a result, it makes the effect of a localised disturbance begin instantaneously at all distances, whereas there is a physical limit, viz. $\sqrt{\{(\lambda+2\mu)/\rho\}}$, to the rate of propagation.

8. For the purpose of a further examination we assume the substance of the plate to be incompressible, so that $\alpha = \xi$, and write $\beta = m\xi$ as before. The equation (48) becomes

$$\frac{\tanh m\omega}{\tanh \omega} = \frac{(1+m^2)^2}{4m}, \tag{51}$$

where $\omega = \xi f$. The wave-velocity is given by (23).

Since m must be less than unity, whilst the second member of (51) exceeds 1 if $m < 0{\cdot}2956$, it appears that for real solutions we are restricted to values of m between 0·2956 and 1. A series of values of ω corresponding to values of m within this range is given on the next page.

The displacement-function is found to be

$$\Psi = \{(1+m^2)\cosh m\omega \cosh \xi y - 2\cosh \omega \cosh m\xi y\}\, e^{i\xi x}. \tag{52}$$

The forms of the lines Ψ = const. for the case of

$$m = 0{\cdot}9, \qquad \omega = 0{\cdot}435, \qquad m\omega = 0{\cdot}392, \qquad \lambda'/2f = 7{\cdot}22,$$

are shown in fig. 5, for a range of half a wave-length. Regarded as belonging to a standing vibration, they indicate a rotation of the matter in the neighbourhood of the nodes, about these points.

* See Todhunter, 'History of the Theory of Elasticity,' vol. 1, p. 112; Rayleigh, Theory of Sound,' vol. 1, art. 192.

Table II.—Asymmetrical Type. $\lambda = \infty$.
[The unit of V is $\sqrt{(\mu/\rho)}$.]

m.	ω.	$\lambda'/2f$.	V.
0·2956	∞	0·0	0·9554
0·30	8·84	0·356	0·954
0·35	4·20	0·748	0·937
0·40	3·028	1·038	0·917
0·45	2·379	1·321	0·893
0·50	1·946	1·614	0·866
0·55	1·627	1·931	0·835
0·60	1·377	2·282	0·800
0·65	1·171	2·683	0·760
0·70	0·995	3·157	0·714
0·75	0·841	3·736	0·661
0·80	0·700	4·45	0·600
0·85	0·563	5·58	0·527
0·90	0·435	7·22	0·436
0·95	0·300	10·5	0·312
1·0	0·0	∞	0·0*

* For *small* values of ξf the value of V is given by equation (50).

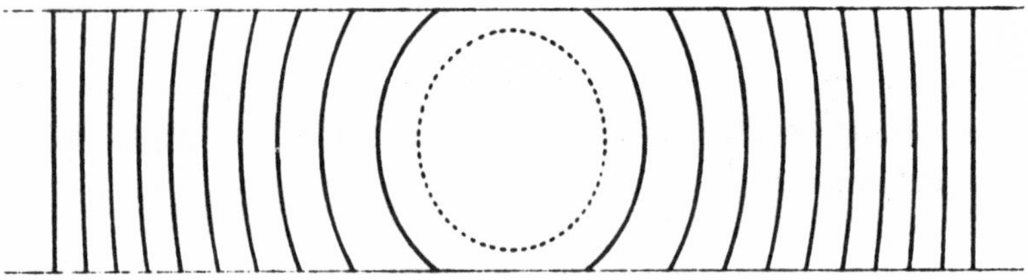

Fig. 5.

So far it has been supposed that $k < \xi$, and consequently that m is real. In the opposite case we assume in place of (46)

$$\phi = \mathrm{A} \sinh \alpha y \, e^{i\xi x}, \qquad \psi = \mathrm{B} \cos \beta_1 y \, e^{i\xi x}, \tag{53}$$

and the period-equation is

$$\frac{\tan \beta_1 f}{\tanh \alpha f} = -\frac{(\xi^2 - \beta_1^2)^2}{4\xi^2 \alpha \beta_1}, \tag{54}$$

where β_1 is defined by (26).

In the case of incompressibility this becomes

$$\frac{\tan n\omega}{\tanh \omega} = -\frac{(1-n^2)^2}{4n}, \tag{55}$$

where $\omega = \xi f$, $n = \beta_1/\xi$. As the preceding investigation (summarised in Table II) evidently covers all the modes in which v has the same sign throughout the thickness, the additional modes to which this equation relates may be dismissed with the remark that they are analogous to those referred to in § 6 above. The particular case $n = 1$ is illustrated by fig. 4 if we imagine the lowest horizontal dotted line to form the lower boundary.

Influence of Compressibility.

9. Although the hypothesis of incompressibility has been adopted for simplicity, the numerical calculations, so far as the more important modes are concerned, are not much more complicated if we abandon this restriction. It will be sufficient to consider the case of the symmetrical types.

We have from (4) and (7)

$$\xi^2 = \frac{k^2\alpha^2 - h^2\beta^2}{k^2 - h^2} = \frac{(\lambda + 2\mu)\,\alpha^2 - \mu\beta^2}{\lambda + \mu}. \tag{56}$$

Hence if we write

$$\alpha f = \omega, \qquad \beta = m\alpha, \tag{57}$$

the equation (12) takes the form

$$\frac{\tanh m\omega}{\tanh \omega} = 4m\,\frac{(\lambda + \mu)(\lambda + 2\mu - m^2\mu)}{(\lambda + 2\mu + m^2\lambda)^2}. \tag{58}$$

The relation of ω to the wave-length is given by

$$\xi^2 f^2 = \frac{(\lambda + 2\mu - m^2\mu)\,\omega^2}{\lambda + \mu}, \tag{59}$$

whilst the wave-velocity is given by (20).

For numerical illustration, we may adopt Poisson's hypothesis as to the relation between the elastic constants. Putting, then, $\lambda = \mu$, we have

$$\frac{\tanh m\omega}{\tanh \omega} = \frac{8m(3 - m^2)}{(3 + m^2)^2}, \tag{60}$$

$$\xi f = \sqrt{(\tfrac{1}{2}(3 - m^2))}\,\omega, \qquad V^2 = \frac{3(1 - m^2)}{3 - m^2}\,\frac{\mu}{\rho}. \tag{61}$$

Real values of m must lie between 0 and 0·4641, this being the positive root of the equation

$$m^3 + 9m^2 + 15m - 9 = 0 \tag{62}$$

obtained by equating the second member of (60) to unity. The wave-velocity corresponding to this latter value of m is

$$V = 0{\cdot}9194\sqrt{(\mu/\rho)}, \tag{63}$$

in accordance with the theory of Rayleigh waves, the wave-length being now infinitely small compared with the thickness.

When m is imaginary ($= in$), we have

$$\frac{\tan n\omega}{\tanh \omega} = \frac{8n(3+n^2)}{(3-n^2)^2}, \qquad (64)$$

$$\xi f = \sqrt{[\tfrac{1}{2}(3+n^2)]}\,\omega, \qquad V^2 = \frac{3(1+n^2)}{3+n^2}\frac{\mu}{\rho}. \qquad (65)$$

The more important modes coming under these formulæ are determined by the lowest root of (64) for values of n ranging from 0 to $\sqrt{15}$, this latter number corresponding to an infinitesimal value of ω, *i.e.* to waves of infinite length.

Numerical results are given in Table III, and the relation between wave-length and wave-velocity is shown by the curve B in fig. 1 (p. 119). The unit of the vertical scale is $V/\sqrt{(\mu\rho^{-1})}$ as before.

Table III.—Symmetrical Type. $\lambda = \mu$.
[The unit of V is $\sqrt{(\mu/\rho)}$.]

m.	n.	ω.	$\lambda'/2f$.	V.
0·4641	—	∞	0·0	0·9194
0·45	—	5·220	0·509	0·925
0·40	—	3·807	0·692	0·942
0·35	—	3·343	0·784	0·956
0·30	—	3·085	0·844	0·969
0·25	—	2·918	0·888	0·978
0·20	—	2·805	0·931	0·986
0·15	—	2·727	0·944	0·992
0·10	—	2·678	0·959	0·997
0·0	0·0	2·640	0·972	1·0
—	0·1	2·604	0·983	1·003
—	0·2	2·506	1·017	1·013
—	0·4	2·214	1·129	1·049
—	0·6	1·911	1·268	1·102
—	0·8	1·649	1·412	1·163
—	1·0	1·432	1·551	1·225
—	1·2	1·253	1·683	1·284
—	1·4	1·105	1·805	1·338
—	1·6	0·979	1·924	1·386
—	$\sqrt{3}$	0·907	2·0	$\sqrt{2}$
—	1·8	0·872	2·04	1·423
—	2·0	0·778	2·16	1·464
—	2·2	0·696	2·28	1·495
—	2·4	0·620	2·42	1·522
—	2·6	0·551	2·58	1·544
—	2·8	0·486	2·78	1·564
—	3·0	0·423	3·04	1·581
—	3·2	0·359	3·40	1·596
—	3·4	0·292	3·99	1·609
—	3·6	0·216	5·15	1·620
—	3·8	0·109	9·8	1·630
—	$\sqrt{15}$	0·0	∞	1·633

As n increases from 0, the modes corresponding to the higher roots of (64) have at first the same general character as in the case of incompressibility

(§ 6). When $n = \sqrt{3}$, we have the division into square compartments, to which fig. 4 refers. When n is infinite, whilst $n\omega$ is finite, we have

$$\alpha = 0, \qquad \beta_1 f = n\omega = s\pi, \qquad \xi f = s\pi/\sqrt{2}. \tag{66}$$

A reference to (11) shows, in fact, that, independently of any special relation between the elastic constants, the boundary conditions are satisfied by $\alpha = 0$, $\sin\beta_1 f = 0$,

and
$$A(\xi^2-\beta_1^2)+2B\xi\beta_1 \cos s\pi = 0. \tag{67}$$

This leads to

$$\left.\begin{aligned} u &= i\left(\cos s\pi + \frac{\lambda}{2\mu}\cos\frac{s\pi y}{f}\right)e^{i\xi x}, \\ v &= \sqrt{\left\{\frac{\lambda^2}{\mu(\lambda+\mu)}\right\}}\sin\frac{s\pi y}{f}e^{i\xi x}. \end{aligned}\right\} \tag{68}$$

There is here a transition to the case where α, as well as β, is imaginary. Writing

$$\alpha_1^2 = h^2-\xi^2, \qquad \beta_1^2 = k^2-\xi^2, \tag{69}$$

and assuming (for the case of symmetry)

$$\phi = A\cos\alpha_1 y\, e^{i\xi x}, \qquad \psi = B\sin\beta_1 y\, e^{i\xi x}, \tag{70}$$

the period-equation is found to be

$$\frac{\tan\beta_1 f}{\tan\alpha_1 f} = -\frac{4\xi^2\alpha_1\beta_1}{(\beta_1^2-\xi^2)^2}. \tag{71}$$

Since

$$\xi^2 = \frac{h^2\beta_1^2-k^2\alpha_1^2}{k^2-h^2} = \frac{\mu\beta_1^2-(\lambda+2\mu)\alpha_1^2}{\lambda+\mu}, \tag{72}$$

we find, writing
$$\alpha_1 f = \omega, \qquad \beta_1 f = q\omega, \tag{73}$$

$$\frac{\tan q\omega}{\tan\omega} = 4(\lambda+\mu)\frac{q(\lambda+2\mu-q^2\mu)}{(q^2\lambda+\lambda+2\mu)^2}. \tag{74}$$

Also

$$\xi^2 f^2 = \frac{q^2\mu-(\lambda+2\mu)}{(\lambda+\mu)}\omega^2,$$

$$V^2 = \frac{(\lambda+2\mu)(q^2-1)}{q^2\mu-(\lambda+2\mu)}\frac{\mu}{\rho}. \tag{75}$$

On Poisson's hypothesis these become

$$\frac{\tan q\omega}{\tan\omega} = \frac{-8q(q^2-3)}{(q^2+3)^2}, \tag{76}$$

$$\xi f = \sqrt{[\tfrac{1}{2}(q^2-3)]}\,\omega, \qquad V^2 = 3\frac{q^2-1}{q^2-3}\frac{\mu}{\rho}. \tag{77}$$

The value of q may range downwards from ∞ to $\sqrt{3}$.

A minute examination of these modes would be laborious, and would hardly repay the trouble. In the extreme case where $q = \sqrt{3}$, the equation (76) is satisfied by either a zero value of tan $q\omega$, or an infinite value of tan ω. The former alternative gives the shearing motions parallel to x, already referred to at the end of § 6 (equations (43) and (44)). The other alternative gives a vibration at right angles to x. It is, in fact, obvious from (11) that the conditions are satisfied by

$$\xi = 0, \qquad B = 0, \qquad \cos \alpha_1 f = 0, \tag{78}$$

whence

$$\phi = \cos (2s+1)\frac{\pi y}{2f}, \qquad \psi = 0, \tag{79}$$

$$\sigma^2 = (s+\tfrac{1}{2})^2 \frac{\pi^2}{f^2} \frac{\lambda+2\mu}{\rho}. \tag{80}$$

When q slightly exceeds $\sqrt{3}$, we have modes of vibration resembling the above types, except for a gradual change of phase in the direction of x. The corresponding values of V, as given by (77), are very great, but it is to be remarked that the notion of "wave-velocity" is in reality hardly applicable (except in a purely geometrical or kinematical sense) to cases of this kind, and that results relating to modes lying outside the limits of the numerical Tables are more appropriately expressed in terms of frequency.* As already remarked, there is a physical limit to the speed of propagation of an initially local disturbance.

* *Cf.* 'Hydrodynamics,' art. 261, where a similar point arises.

9

Reprinted from *J. Appl. Mech.*, **18**, 31–38 (1951)

Influence of Rotatory Inertia and Shear on Flexural Motions of Isotropic, Elastic Plates

By R. D. MINDLIN,[1] NEW YORK, N. Y.

A two-dimensional theory of flexural motions of isotropic, elastic plates is deduced from the three-dimensional equations of elasticity. The theory includes the effects of rotatory inertia and shear in the same manner as Timoshenko's one-dimensional theory of bars. Velocities of straight-crested waves are computed and found to agree with those obtained from the three-dimensional theory. A uniqueness theorem reveals that three edge conditions are required.

Introduction

According to the classical two-dimensional theory of flexural motions of elastic plates, leading to Lagrange's equation

$$D\nabla^4\bar{w} + \rho h \frac{\partial^2 \bar{w}}{\partial t^2} = q(x, y, t) \qquad [1]$$

the wave velocity of straight-crested waves is inversely proportional to the wave length. The exact solution, by Rayleigh (1),[2] Lamb (2), and Timoshenko (3), of the general equations of the linear theory of elasticity, for the case of straight-crested flexural waves, confirms this result only for waves which are long in comparison with the thickness of the plate. As the wave length diminishes, the velocity in the three-dimensional theory has as its upper limit the velocity of Rayleigh surface waves. Hence the classical plate theory cannot be expected to give good results for sharp transients or for the frequencies of modes of vibration of high order. A similar situation exists in regard to the classical one-dimensional theory of flexural motions of elastic bars, leading to the Bernoulli-Euler equation.

In the case of bars, correction terms have been supplied by Rayleigh and Timoshenko. Rayleigh (4) introduced the effect of rotatory inertia but, although the upper limit of the velocity is thereby made finite, its magnitude is too large. It was not until Timoshenko (5, 6) included the effect of transverse shear deformation that a one-dimensional theory of flexural motions of bars was obtained which gives satisfactory results for short waves and high modes of vibration.

In the case of plates, equations of motion analogous to Timoshenko's bar equations have been given by Uflyand (7) and a corresponding theory of plate equilibrium has been given by Hencky (8). One purpose of the present paper is to show how a more comprehensive two-dimensional theory of flexural motions of plates, analogous to Timoshenko's one-dimensional theory for bars, may be deduced directly from the three-dimensional equations of elasticity. Uflyand's equations of motion are reached, but are later modified in the light of a comparison of the velocity of straight-crested waves with the corresponding velocity obtained from the three-dimensional theory. The modification consists of the selection of the value of a constant, appearing in the relation between average transverse shear stress and strain, variously chosen as 2/3 and 8/9 by Timoshenko, and 2/3 by Uflyand. A formula for this constant, involving Poisson's ratio, is suggested (see Equation [47]). By its use, the limiting velocity for very short waves is made identical with the velocity of Rayleigh surface waves, and velocities intermediate between very long and very short waves are brought into close agreement with the three-dimensional theory.

In addition to the derivation of equations of motion, there are derived theorems dealing with kinetic and potential energies, uniqueness, initial and boundary conditions, and single-valuedness of displacements corresponding to similar theorems in the three-dimensional theory.

At various stages of the development, attention is directed to the very close similarities between the theory under discussion and E. Reissner's theory (9, 10) of flexural equilibrium of plates. Of special interest is the conclusion that, as in Reissner's theory, three boundary conditions are to be satisfied rather than the two of classical plate theory.

Plate-Stress Components

The plate is referred to an x, y, z-system of rectangular co-ordinates. The faces of the plate are the planes $z = \pm h/2$ and its cylindrical surfaces are defined by plane curves or polygons C_i parallel to the x, y-plane. The faces of the plate are taken to be free from tangential traction but under normal pressures q_1 and q_2. Thus in the usual notation

$$\tau_{xz}\big]_{z=\pm h/2} = \tau_{yz}\big]_{z=\pm h/2} = 0 \qquad [2]$$

$$\sigma_z\big]_{z=h/2} = -q_1(x, y, t), \quad \sigma_z\big]_{z=-h/2} = -q_2(x, y, t) \qquad [3]$$

For convenience the notation $q = q_2 - q_1$ will be used. Normal pressures are retained on both faces in order that one of them may, if desired, be made proportional to the transverse displacement to simulate the effect of an elastic foundation.

The bending and twisting moments and the transverse shearing forces, all per unit of length, are defined in the customary manner

$$\left.\begin{aligned} (M_x, M_y, M_{yx}) &= \int_{-h/2}^{h/2} (\sigma_x, \sigma_y, \tau_{yx})\, z\,dz \\ (Q_x, Q_y) &= \int_{-h/2}^{h/2} (\tau_{xz}, \tau_{yz})\, dz \end{aligned}\right\} \qquad [4]$$

---

[1] Professor of Civil Engineering, Columbia University; and Consultant, Bell Telephone Laboratories, Murray Hill, N. J. Mem. ASME.

[2] Numbers in parentheses refer to the Bibliography at the end of the paper.

Presented at the Annual Conference of the Applied Mechanics Division, Purdue University, Lafayette, Ind., June 22–24, 1950, of The American Society of Mechanical Engineers.

Discussion of this paper should be addressed to the Secretary, ASME, 29 West 39th Street, New York, N. Y., and will be accepted until April 10, 1951, for publication at a later date. Discussion received after the closing date will be returned.

Note: Statements and opinions advanced in papers are to be understood as individual expressions of their authors and not those of the Society. Manuscript received by the Applied Mechanics Division, January 23, 1950. Paper No. 50—APM-22.

For brevity these will be designated as the "plate-stress components."

Stress-Strain Relations

In three-dimensional elasticity theory there are six components of stress which are expressed in terms of six components of strain through Hooke's law. The stress-strain relations and the expressions for the six components of strain in terms of three components of displacement are then used to reduce to three the number of unknowns in the three equations of motion. In the present theory there are only five plate-stress components and these will be expressed in terms of the same number of "plate-strain components." The latter will then be expressed in terms of three "plate-displacement components."

Consider the general Hooke's law in the form wherein each of the six components of strain is expressed in terms of the six components of stress. Of the six equations, one is ignored, namely, the one containing the unit elongation (ϵ_z) normal to the faces of the plate. The remaining five equations are then solved for σ_x, σ_y, τ_{yz}, τ_{zx}, and τ_{xy} in terms of ϵ_x, ϵ_y, γ_{yz}, γ_{zx}, γ_{xy} and σ_z. Alternatively, one may start with the expressions for the components of stress in terms of the components of strain, solve the one containing σ_z for ϵ_z and use this to eliminate ϵ_z from the remaining five equations. Integrations are now performed on these equations so as to convert the stress components into plate-stress components in accordance with Equations [4]. The results are then altered in two respects: (*a*) The integrals containing σ_z are dropped; (*b*) the coefficients of the integrals containing γ_{zx} and γ_{yz} are replaced by constants whose magnitudes are to be determined later. This procedure is applicable to anisotropic materials. In the case of an isotropic material, we find

$$\left.\begin{aligned} M_x &= D(\Gamma_x + \mu\Gamma_y), \quad M_y = D(\Gamma_y + \mu\Gamma_x), \\ M_{yx} &= (1-\mu)D\Gamma_{yx}/2 \\ Q_x &= G'h\Gamma_{zx}, \quad Q_y = G'h\Gamma_{yz} \end{aligned}\right\}\ldots[5]$$

where the plate modulus D is expressed in terms of Poisson's ratio μ and Young's modulus E, or the modulus of rigidity G, by

$$D = \frac{Eh^3}{12(1-\mu^2)} = \frac{Gh^3}{6(1-\mu)}\ldots\ldots\ldots\ldots[6]$$

and

$$G' = \kappa^2 G\ldots\ldots\ldots\ldots\ldots\ldots\ldots[7]$$

so that G' is the constant which replaces the coefficient, G, of the integrals containing γ_{zx} and γ_{yz} in the integrated stress-strain relations and κ is the constant, referred to in the introduction, for which a formula involving Poisson's ratio will be given presently. The Γ's in Equations [5] are the plate-strain components defined by

$$\left.\begin{aligned} (\Gamma_x, \Gamma_y, \Gamma_{yx}) &= 12h^{-3}\int_{-h/2}^{h/2}(\epsilon_x, \epsilon_y, \gamma_{yx})\,z\,dz \\ (\Gamma_{zx}, \Gamma_{yz}) &= h^{-1}\int_{-h/2}^{h/2}(\gamma_{zx}, \gamma_{yz})\,dz \end{aligned}\right\}\ldots[8]$$

Equations [5] would have been obtained if it had been assumed that $\sigma_z = 0$ at the start. However, the procedure adopted reveals that only a linearly weighted, average effect of σ_z is neglected, rather than σ_z itself.

Plate-Displacement Components

To proceed toward the establishment of plate-strain-displacement relations, the strain-displacement relations of the three-dimensional theory are integrated in accordance with the right-hand sides of Equations [8]. Thus

$$\left.\begin{aligned} \int_{-h/2}^{h/2}(\epsilon_x, \epsilon_y, \gamma_{yx})z\,dz &= \int_{-h/2}^{h/2}\left(\frac{\partial u}{\partial x}, \frac{\partial v}{\partial y}, \frac{\partial v}{\partial x} + \frac{\partial u}{\partial y}\right)z\,dz \\ \int_{-h/2}^{h/2}(\gamma_{zx}, \gamma_{yz})dz &= \int_{-h/2}^{h/2}\left(\frac{\partial u}{\partial z} + \frac{\partial w}{\partial x}, \frac{\partial v}{\partial z} + \frac{\partial w}{\partial y}\right)dz \end{aligned}\right\}\ldots\ldots\ldots[9]$$

It is now assumed that u and v are proportional to z and w is independent of z

$$u = z\psi_x(x, y, t),\ v = z\psi_y(x, y, t),\ w = \bar{w}(x, y, t)\ldots[10]$$

The functions ψ_x and ψ_y correspond to the negative of the function ψ introduced by Timoshenko (6) in his bar theory and to the negatives of the functions ψ/h and χ/h used by Hencky (8).

Questions raised by Goodier (11), in connection with the appearance of Equations [10] in the first version (9) of Reissner's theory of plates, were answered in Reissner's second version (10) by replacing Equations [10] with averages while retaining his assumptions regarding the z-dependence of the components of stress. It appears that the reverse procedure leads to the consideration of shear deformation in the manner adopted by Timoshenko.

From Equations [8], [9], and [10], we have the relations between the plate-strain components and the plate-displacement components ψ_x, ψ_y, $\bar{w}$

$$\left.\begin{aligned} \Gamma_x &= \frac{\partial\psi_x}{\partial x},\ \Gamma_y = \frac{\partial\psi_y}{\partial y},\ \Gamma_{yx} = \frac{\partial\psi_y}{\partial x} + \frac{\partial\psi_x}{\partial y} \\ \Gamma_{zx} &= \psi_x + \frac{\partial\bar{w}}{\partial x},\ \Gamma_{yz} = \psi_y + \frac{\partial\bar{w}}{\partial y} \end{aligned}\right\}\ldots\ldots[11]$$

The analogous relations for classical plate theory are obtained by setting $\Gamma_{zx} = \Gamma_{yz} = 0$.

Substituting Equations [11] into Equations [5], the following relations are obtained between the plate-stress and plate-displacement components

$$\left.\begin{aligned} M_x &= D\left(\frac{\partial\psi_x}{\partial x} + \mu\frac{\partial\psi_y}{\partial y}\right),\ M_y = D\left(\frac{\partial\psi_y}{\partial y} + \mu\frac{\partial\psi_x}{\partial x}\right) \\ M_{yx} &= \frac{1-\mu}{2}D\left(\frac{\partial\psi_y}{\partial x} + \frac{\partial\psi_x}{\partial y}\right) \\ Q_x &= \kappa^2 Gh\left(\frac{\partial\bar{w}}{\partial x} + \psi_x\right),\ Q_y = \kappa^2 Gh\left(\frac{\partial\bar{w}}{\partial y} + \psi_y\right) \end{aligned}\right\}\ldots[12]$$

Equations [12] are similar to the corresponding formulas in E. Reissner's theory.[3] His "equivalent changes of slope," β_x and β_y, are here replaced by ψ_x and ψ_y; his "weighted average displacement" is here replaced by $\bar{w}$. The transverse load q does not appear in the present expressions for M_x and M_y due to the omission of the integrals containing σ_z in the integrated stress-strain relations. The constant $\kappa^2 Gh$ plays a role similar to Reissner's C_s, which has the value $(5/6)Gh$ for isotropic materials and a quadratic variation of transverse shear stress through the thickness of the plate. Hence κ^2 in Reissner's equilibrium theory has the value 5/6. In the present theory the z-dependence of the components of stress is not specified and κ is, as yet, not determined.

Equations of Motion

The stress-equations of motion of three-dimensional elasticity theory

[3] See reference (10), Equations [8], [23], and [28].

$$\left.\begin{aligned}\frac{\partial\sigma_x}{\partial x}+\frac{\partial\tau_{yx}}{\partial y}+\frac{\partial\tau_{zx}}{\partial z}&=\rho\frac{\partial^2u}{\partial t^2}\\ \frac{\partial\tau_{xy}}{\partial x}+\frac{\partial\sigma_y}{\partial y}+\frac{\partial\tau_{zy}}{\partial z}&=\rho\frac{\partial^2v}{\partial t^2}\\ \frac{\partial\tau_{xz}}{\partial x}+\frac{\partial\tau_{yz}}{\partial y}+\frac{\partial\sigma_z}{\partial z}&=\rho\frac{\partial^2w}{\partial t^2}\end{aligned}\right\}\ldots\ldots\ldots\ldots[13]$$

are converted to plate-stress equations of motion in the following manner.[4] The first two of Equations [13] are multiplied by z and integrated over the plate thickness. These become, after making use of Equations [2], [4], and [10]

$$\left.\begin{aligned}\frac{\partial M_x}{\partial x}+\frac{\partial M_{yx}}{\partial y}-Q_x&=\frac{\rho h^3}{12}\frac{\partial^2\psi_x}{\partial t^2}\\ \frac{\partial M_{xy}}{\partial x}+\frac{\partial M_y}{\partial y}-Q_y&=\frac{\rho h^3}{12}\frac{\partial^2\psi_y}{\partial t^2}\end{aligned}\right\}\ldots\ldots\ldots[14]$$

(The right-hand sides of Equations [14] represent the influence of rotatory inertia.) The third of Equations [13] is integrated over the plate thickness and, after using Equations [3], [4], and [10], becomes

$$\frac{\partial Q_x}{\partial x}+\frac{\partial Q_y}{\partial y}+q=\rho h\frac{\partial^2\bar{w}}{\partial t^2}\ldots\ldots\ldots\ldots[15]$$

The three plate-stress equations of motion, Equations [14] and [15], may be expressed in terms of the plate-displacements ψ_x, ψ_y, $\bar{w}$ by using Equations [12], with the result

$$\left.\begin{aligned}&\frac{D}{2}\left[(1-\mu)\nabla^2\psi_x+(1+\mu)\frac{\partial\Phi}{\partial x}\right]-\kappa^2Gh\left(\psi_x+\frac{\partial\bar{w}}{\partial x}\right)=\frac{\rho h^3}{12}\frac{\partial^2\psi_x}{\partial t^2}\\ &\frac{D}{2}\left[(1-\mu)\nabla^2\psi_y+(1+\mu)\frac{\partial\Phi}{\partial y}\right]-\kappa^2Gh\left(\psi_y+\frac{\partial\bar{w}}{\partial y}\right)=\frac{\rho h^3}{12}\frac{\partial^2\psi_y}{\partial t^2}\\ &\kappa^2Gh(\nabla^2\bar{w}+\Phi)+q=\rho h\frac{\partial^2\bar{w}}{\partial t^2}\end{aligned}\right\}\ldots[16]$$

where

$$\Phi=\frac{\partial\psi_x}{\partial x}+\frac{\partial\psi_y}{\partial y}$$

and ∇^2 is Laplace's two-dimensional operator.

Equations [16] are the same as Uflyand's equations of motion (7) except for the value of κ.

Energy Functions

The strain-energy-function W in the three-dimensional theory is given by

$$\left.\begin{aligned}2W&=\sigma_x\epsilon_x+\sigma_y\epsilon_y+\sigma_z\epsilon_z+\tau_{xy}\gamma_{xy}+\tau_{yz}\gamma_{yz}+\tau_{zx}\gamma_{zx}\\ &=\sigma_x\frac{\partial u}{\partial x}+\sigma_y\frac{\partial v}{\partial y}+\sigma_z\frac{\partial w}{\partial z}+\tau_{xy}\left(\frac{\partial v}{\partial x}+\frac{\partial u}{\partial y}\right)\\ &\quad+\tau_{yz}\left(\frac{\partial w}{\partial y}+\frac{\partial v}{\partial z}\right)+\tau_{zx}\left(\frac{\partial u}{\partial z}+\frac{\partial w}{\partial x}\right)\end{aligned}\right\}\ldots[17]$$

[4] This method was employed by Boussinesq (12) to derive plate equations without using the calculus of variations. A. E. Green (13) has shown, recently, how the equations of E. Reissner's plate theory (10) may be obtained in a similar manner.

The analogous plate-strain-energy-function $\bar{W}$ is obtained by substituting Equations [10] into Equations [17] and integrating over the plate thickness, with the result

$$\left.\begin{aligned}2\bar{W}&=M_x\frac{\partial\psi_x}{\partial x}+M_y\frac{\partial\psi_y}{\partial y}+M_{yx}\left(\frac{\partial\psi_y}{\partial x}+\frac{\partial\psi_x}{\partial y}\right)\\ &\quad+Q_x\left(\frac{\partial\bar{w}}{\partial x}+\psi_x\right)+Q_y\left(\frac{\partial\bar{w}}{\partial y}+\psi_y\right)\\ &=M_x\Gamma_x+M_y\Gamma_y+M_{yx}\Gamma_{yx}+Q_x\Gamma_{xz}+Q_y\Gamma_{yz}\end{aligned}\right\}\ldots[18]$$

and this becomes, through Equations [5]

$$4\bar{W}=D(1+\mu)(\Gamma_x+\Gamma_y)^2+2\kappa^2Gh(\Gamma_{yz}^2+\Gamma_{xz}^2)+D(1-\mu)\left[(\Gamma_x-\Gamma_y)^2+\Gamma_{yx}^2\right]\ldots\ldots[19]$$

If $E/(1+\mu)$ and $E/(1-\mu)$ are required to be positive, $\bar{W}$ is positive. Then, if $\bar{W}=0$, all the plate-strain components vanish and, through Equations [5], so do the plate-stress components.

From Equations [19] and [5], we find

$$\left.\begin{aligned}&\frac{\partial\bar{W}}{\partial\Gamma_x}=M_x,\quad\frac{\partial\bar{W}}{\partial\Gamma_y}=M_y,\quad\frac{\partial\bar{W}}{\partial\Gamma_{yx}}=M_{yx}\\ &\frac{\partial\bar{W}}{\partial\Gamma_{xz}}=Q_x,\quad\frac{\partial\bar{W}}{\partial\Gamma_{yz}}=Q_y\end{aligned}\right\}\ldots\ldots[20]$$

The kinetic energy per unit of volume, according to the general linear theory, is

$$\frac{\rho}{2}\left[\left(\frac{\partial u}{\partial t}\right)^2+\left(\frac{\partial v}{\partial t}\right)^2+\left(\frac{\partial w}{\partial t}\right)^2\right]\ldots\ldots\ldots\ldots[21]$$

By using Equations [10] and integrating over the thickness, this becomes, for the plate

$$\frac{\rho h^3}{24}\left[\left(\frac{\partial\psi_x}{\partial t}\right)^2+\left(\frac{\partial\psi_y}{\partial t}\right)^2\right]+\frac{\rho h}{2}\left(\frac{\partial\bar{w}}{\partial t}\right)^2\ldots\ldots\ldots[22]$$

Total Energy and External Work

The kinetic energy in the plate at time t is given by

$$\bar{T}=\iint\left\{\frac{\rho h^3}{24}\left[\left(\frac{\partial\psi_x}{\partial t}\right)^2+\left(\frac{\partial\psi_y}{\partial t}\right)^2\right]+\frac{\rho h}{2}\left(\frac{\partial\bar{w}}{\partial t}\right)^2\right\}dx\,dy\ldots\ldots\ldots\ldots[23]$$

and the potential energy is

$$\bar{V}=\iint\bar{W}\,dx\,dy\ldots\ldots\ldots\ldots\ldots\ldots[24]$$

where the integrations are extended over the surface of the plate.

The total energy at time t is the sum of $\bar{T}$ and $\bar{V}$, which may be written as

$$\begin{aligned}\bar{T}+\bar{V}&=\int_{t_0}^{t}dt\iint\frac{\partial}{\partial t}\left\{\frac{\rho h^3}{24}\left[\left(\frac{\partial\psi_x}{\partial t}\right)^2+\left(\frac{\partial\psi_y}{\partial t}\right)^2\right]+\frac{\rho h}{2}\left(\frac{\partial\bar{w}}{\partial t}\right)^2\right\}dx\,dy\\ &\quad+\int_{t_0}^{t}dt\iint\frac{\partial\bar{W}}{\partial t}dx\,dy+\bar{T}_0+\bar{V}_0\ldots\ldots\ldots\ldots[25]\end{aligned}$$

where $\bar{T}_0$ and $\bar{V}_0$ are the values of $\bar{T}$ and $\bar{V}$ at an initial time t_0.

Performing the operations $\partial/\partial t$ as indicated in Equation [25], the first integrand becomes

$$\frac{\rho h^3}{12}\left[\frac{\partial\psi_x}{\partial t}\frac{\partial^2\psi_x}{\partial t^2}+\frac{\partial\psi_y}{\partial t}\frac{\partial^2\psi_y}{\partial t^2}\right]+\rho h\frac{\partial\bar{w}}{\partial t}\frac{\partial^2\bar{w}}{\partial t^2}\ldots\ldots[26]$$

and the second integrand is

$$\frac{\partial \overline{W}}{\partial t} = \frac{\partial \overline{W}}{\partial \Gamma_x}\frac{\partial \Gamma_x}{\partial t} + \frac{\partial \overline{W}}{\partial \Gamma_y}\frac{\partial \Gamma_y}{\partial t} + \frac{\partial \overline{W}}{\partial \Gamma_{yx}}\frac{\partial \Gamma_{yx}}{\partial t} + \frac{\partial \overline{W}}{\partial \Gamma_{zx}}\frac{\partial \Gamma_{zx}}{\partial t}$$
$$+ \frac{\partial \overline{W}}{\partial \Gamma_{yz}}\frac{\partial \Gamma_{yz}}{\partial t} = \left(M_x \frac{\partial}{\partial x} + M_{yx}\frac{\partial}{\partial y} + Q_x\right)\frac{\partial \psi_x}{\partial t}$$
$$+ \left(M_{yx}\frac{\partial}{\partial x} + M_y \frac{\partial}{\partial y} + Q_y\right)\frac{\partial \psi_y}{\partial t} + \left(Q_x \frac{\partial}{\partial x} + Q_y \frac{\partial}{\partial y}\right)\frac{\ }{\partial t} \qquad \ldots\ldots\ldots [27]$$

by Equations [11] and [20].

In the surface integral of $\partial \overline{W}/\partial t$, Equation [25], the terms containing space derivatives may be integrated by parts to obtain

$$\iint \frac{\partial \overline{W}}{\partial t}\,dx\,dy = \oint \left[\frac{\partial \psi_x}{\partial t}(M_x l + M_{yx} m) + \frac{\partial \psi_y}{\partial t}(M_{yx} l + M_y m) + \frac{\partial \bar{w}}{\partial t}(Q_x l + Q_y m)\right] ds$$
$$- \iint \left[\frac{\partial \psi_x}{\partial t}\left(\frac{\partial M_x}{\partial x} + \frac{\partial M_{yx}}{\partial y} - Q_x\right) + \frac{\partial \psi_y}{\partial t}\left(\frac{\partial M_{yx}}{\partial x} + \frac{\partial M_y}{\partial y} - Q_y\right) + \frac{\partial \bar{w}}{\partial t}\left(\frac{\partial Q_x}{\partial x} + \frac{\partial Q_y}{\partial y}\right)\right] dx\,dy \qquad \ldots\ldots [28]$$

in which l and m are the direction cosines of the normal to the boundary and the line integral is taken around all (external and internal) boundaries. In the integration by parts it is assumed that ψ_x, ψ_y, and $\bar{w}$ are single-valued, otherwise additional line integrals would be required.

In terms of components referred to co-ordinates ν and s measured normal to and along the boundary, respectively, the line integral in Equation [28] may be written as

$$\oint \left(\frac{\partial \psi_\nu}{\partial t} M_\nu + \frac{\partial \psi_s}{\partial t} M_{\nu s} + \frac{\partial \bar{w}}{\partial t} Q_\nu\right) ds$$

Combining these results, we have

$$\overline{T} + \overline{V} = \int_{t_0}^{t} dt \oint \left(\frac{\partial \psi_\nu}{\partial t} M_\nu + \frac{\partial \psi_s}{\partial t} M_{\nu s} + \frac{\partial \bar{w}}{\partial t} Q_\nu\right) ds$$
$$+ \int_{t_0}^{t} dt \iint \left[\frac{\partial \psi_x}{\partial t}\left(\frac{\rho h^3}{12}\frac{\partial^2 \psi_x}{\partial t^2} - \frac{\partial M_x}{\partial x} - \frac{\partial M_{yx}}{\partial y} + Q_x\right)\right.$$
$$+ \frac{\partial \psi_y}{\partial t}\left(\frac{\rho h^3}{12}\frac{\partial^2 \psi_y}{\partial t^2} - \frac{\partial M_{yx}}{\partial x} - \frac{\partial M_y}{\partial y} + Q_y\right)$$
$$\left. + \frac{\partial \bar{w}}{\partial t}\left(\rho h \frac{\partial^2 \bar{w}}{\partial t^2} - \frac{\partial Q_x}{\partial x} - \frac{\partial Q_y}{\partial y}\right)\right] dx\,dy + \overline{T}_0 + \overline{V}_0$$

If Equations [14] and [15] (the plate equations of motion) are satisfied, this reduces to

$$\overline{T} + \overline{V} = \int_{t_0}^{t} dt \oint \left(\frac{\partial \psi_\nu}{\partial t} M_\nu + \frac{\partial \psi_s}{\partial t} M_{\nu s} + \frac{\partial \bar{w}}{\partial t} Q_\nu\right) ds$$
$$+ \int_{t_0}^{t} dt \iint q \frac{\partial \bar{w}}{\partial t}\,dx\,dy + \overline{T}_0 + \overline{V}_0 \qquad \ldots\ldots [29]$$

Equation [29] is simply the form, in the present theory, of the statement that the total energy in the plate at time t is equal to the sum of the energy at time t_0 and the work done by the external forces along the edge and over the surface of the plate during the interval $t - t_0$.

Initial and Boundary Conditions

Appropriate initial and boundary conditions for a system of differential equations are those which are sufficient to assure a unique solution. In the preceding two sections the groundwork has been laid for the application of a plate analog of F. Neumann's uniqueness theorem (14).

Consider two sets of plate displacements and surface loads. If the components of each set satisfy the equations of motion, so will their differences. Then the energies and the plate-stress components calculated from the differences will satisfy an equation of the same form as Equation [29]. If the right-hand side of Equation [29] vanishes, $\overline{T}$ and $\overline{V}$ vanish separately, since they are both positive. If $\overline{T}$ vanishes, the kinetic energy per unit of volume vanishes, since it is positive, and hence the plate velocities vanish. If $\overline{V}$ vanishes, $\overline{W}$ vanishes, since it is positive, and if $\overline{W}$ vanishes, so do the plate-strain and plate-stress components, as long as $E/(1 + \mu)$ and $E/(1 - \mu)$ are required to be positive. Hence, if the right-hand side of Equation [29] (calculated from the differences) vanishes, the two systems must be identical, except possibly for a rigid-body displacement (independent of time, since the velocities vanish) and the latter can be eliminated by requiring the initial displacements to be the same.

Returning, now, to a single set of solutions, appropriate initial and boundary conditions to be specified for the present theory are as follows:

1 On the edges of plate: any combination which contains one member of each of the three pairs of terms in the line integral of Equation [29].

2 Throughout the plate: (a) either q or $\bar{w}$, and (b) the initial values of ψ_x, ψ_y, $\bar{w}$ and their time derivatives.

The relation (see Equations [12])

$$\psi_s = \frac{Q_s}{\kappa^2 G h} - \frac{\partial \bar{w}}{\partial s}$$

is replaced, in the classical theory of plates, by

$$\psi_s = -\frac{\partial \bar{w}}{\partial s}$$

since the transverse shear deformation $Q_s/\kappa^2 G h$ vanishes. The term $M_{\nu s}\,\partial \psi_s/\partial t$ in Equation [29] would then be replaced by $-M_{\nu s}\partial^2 \bar{w}/\partial s\,\partial t$. But

$$-\oint M_{\nu s}\frac{\partial}{\partial s}\left(\frac{\partial \bar{w}}{\partial t}\right) ds = \oint \frac{\partial \bar{w}}{\partial t}\frac{\partial M_{\nu s}}{\partial s}\,ds$$

Hence the second and third terms in the integrand of the line integral of Equation [29] would combine into the single term

$$\frac{\partial \bar{w}}{\partial t}\left(\frac{\partial M_{\nu s}}{\partial s} + Q_\nu\right)$$

leaving only the two edge conditions of Kirchoff. In the present theory, however, three edge conditions are required, as in E. Reissner's theory.

Compatibility Equations

In the integration by parts leading to Equation [29] it was assumed that ψ_x, ψ_y, and $\bar{w}$ are single-valued. If this restriction were relaxed, additional line integrals would appear in Equation [29], and hence there would be additional requirements for uniqueness. If the restriction to single-valued displacements is retained, it is necessary to specify that

$$\oint d\psi_x = 0, \quad \oint d\psi_y = 0, \quad \oint d\bar{w} = 0 \qquad \ldots\ldots [30]$$

where the line integrals are taken around any closed path in the plate.

Usually, single-valuedness may be ascertained by inspection. However, if the differential equations are expressed entirely in terms of plate-stress components, as they may be in equilibrium problems, additional conditions are introduced by Equations [30]. These may be found as follows

$$\oint d\psi_x = \oint \frac{\partial \psi_x}{\partial x} dx + \frac{\partial \psi_x}{\partial y} dy$$

$$= \oint \left(\Gamma_x dx + \frac{1}{2}\Gamma_{yx} dy\right) - \oint \Omega_z dy$$

where

$$\Omega_z = \frac{1}{2}\left(\frac{\partial \psi_y}{\partial x} - \frac{\partial \psi_x}{\partial y}\right)$$

Now

$$-\oint \Omega_z dy = \oint (y - y_0) d\Omega_z$$

where y_0 is the y-co-ordinate of the starting point of integration. Also

$$d\Omega_z = \frac{\partial \Omega_z}{\partial x} dx + \frac{\partial \Omega_z}{\partial y} dy$$

and

$$2\frac{\partial \Omega_z}{\partial x} = \frac{\partial \Gamma_{yx}}{\partial x} - 2\frac{\partial \Gamma_x}{\partial y}, \quad 2\frac{\partial \Omega_z}{\partial y} = 2\frac{\partial \Gamma_y}{\partial x} - \frac{\partial \Gamma_{yx}}{\partial y}$$

Combining these results, we have

$$\oint d\psi_x = \oint (\xi_x dx + \eta_x dy) \ldots\ldots\ldots\ldots [31]$$

where

$$\xi_x = \Gamma_x + (y - y_0)\frac{1}{2}\left(\frac{\partial \Gamma_{yx}}{\partial x} - \frac{\partial \Gamma_x}{\partial y}\right)$$

$$\eta_x = \frac{1}{2}\Gamma_{yx} + (y - y_0)\left(\frac{\partial \Gamma_y}{\partial x} - \frac{1}{2}\frac{\partial \Gamma_{yx}}{\partial y}\right)$$

Similarly

$$\oint d\psi_y = \oint (\xi_y dx + \eta_y dy) \ldots\ldots\ldots\ldots [32]$$

where

$$\xi_y = \frac{1}{2}\Gamma_{yx} - (x - x_0)\left(\frac{1}{2}\frac{\partial \Gamma_{yx}}{\partial x} - \frac{\partial \Gamma_x}{\partial y}\right)$$

$$\eta_y = \Gamma_y - (x - x_0)\left(\frac{\partial \Gamma_y}{\partial x} - \frac{1}{2}\frac{\partial \Gamma_{yx}}{\partial y}\right)$$

For the condition on $\bar{w}$, we have

$$\oint d\bar{w} = \oint\left(\frac{\partial \bar{w}}{\partial x} dx + \frac{\partial \bar{w}}{\partial y} dy\right)$$

$$= \frac{1}{2}\oint (\Gamma_{zx} dx + \Gamma_{yz} dy) - \oint (\Omega_y dx - \Omega_x dy)$$

where

$$\Omega_x = \frac{1}{2}\left(\frac{\partial \bar{w}}{\partial y} - \psi_y\right), \quad \Omega_y = \frac{1}{2}\left(\psi_x - \frac{\partial \bar{w}}{\partial x}\right)$$

(Note that Ω_x, Ω_y, Ω_z are the plate-rotation components.)

Now

$$-\oint (\Omega_y dx - \Omega_x dy) = \oint [(x - x_0) d\Omega_y - (y - y_0) d\Omega_x]$$

where (x_0, y_0) is the starting point of integration. Also

$$d\Omega_x = \frac{\partial \Omega_x}{\partial x} dx + \frac{\partial \Omega_x}{\partial y} dy, \quad d\Omega_y = \frac{\partial \Omega_y}{\partial x} dx + \frac{\partial \Omega_y}{\partial y} dy$$

and

$$2\frac{\partial \Omega_x}{\partial x} = \frac{\partial \Gamma_{zx}}{\partial y} - \Gamma_{yx}, \quad 2\frac{\partial \Omega_x}{\partial y} = \frac{\partial \Gamma_{yz}}{\partial y} - 2\Gamma_y$$

$$2\frac{\partial \Omega_y}{\partial y} = \Gamma_{yx} - \frac{\partial \Gamma_{yz}}{\partial x}, \quad 2\frac{\partial \Omega_y}{\partial x} = 2\Gamma_x - \frac{\partial \Gamma_{zx}}{\partial x}$$

Combining these results, we have

$$\oint d\bar{w} = \oint (\xi_z dx + \eta_z dy) \ldots\ldots\ldots\ldots [33]$$

where

$$\xi_z = \frac{1}{2}\Gamma_{zx} + (x - x_0)\left(\Gamma_x - \frac{1}{2}\frac{\partial \Gamma_{zx}}{\partial x}\right) + \frac{1}{2}(y - y_0)\left(\Gamma_{yx} - \frac{\partial \Gamma_{zx}}{\partial y}\right)$$

$$\eta_z = \frac{1}{2}\Gamma_{yz} + \frac{1}{2}(x - x_0)\left(\Gamma_{yx} - \frac{\partial \Gamma_{yz}}{\partial x}\right) + (y - y_0)\left(\Gamma_y - \frac{1}{2}\frac{\partial \Gamma_{yz}}{\partial y}\right)$$

Equations [31], [32], and [33] constitute the plate analog of Cesàro's theorem[5] for closed paths. Applying Green's theorem to these equations, we find, for reducible paths

$$\left.\begin{aligned}
\oint d\psi_x &= \iint (y - y_0)\left(\frac{\partial^2 \Gamma_y}{\partial x^2} + \frac{\partial^2 \Gamma_x}{\partial y^2} - \frac{\partial^2 \Gamma_{yx}}{\partial x \partial y}\right) dx\, dy \\
\oint d\psi_y &= -\iint (x - x_0)\left(\frac{\partial^2 \Gamma_y}{\partial x^2} + \frac{\partial^2 \Gamma_x}{\partial y^2} - \frac{\partial^2 \Gamma_{yx}}{\partial x \partial y}\right) dx\, dy \\
\oint d\bar{w} &= \frac{1}{2}\iint (y - y_0)\left(2\frac{\partial \Gamma_y}{\partial x} - \frac{\partial \Gamma_{yx}}{\partial y} - \frac{\partial^2 \Gamma_{yz}}{\partial x \partial y} + \frac{\partial^2 \Gamma_{zx}}{\partial y^2}\right) dx\, dy \\
&\quad -\frac{1}{2}\iint (x - x_0)\left(2\frac{\partial \Gamma_x}{\partial y} - \frac{\partial \Gamma_{yx}}{\partial x} - \frac{\partial^2 \Gamma_{zx}}{\partial x \partial y} + \frac{\partial^2 \Gamma_{yz}}{\partial x^2}\right) dx
\end{aligned}\right\} \ldots [34]$$

Hence, three necessary conditions on the plate strains for single-valuedness of ψ_x, ψ_y, $\bar{w}$ are

[5] See reference (14), p. 223.

$$\left.\begin{aligned}&\frac{\partial^2\Gamma_y}{\partial x^2}+\frac{\partial^2\Gamma_x}{\partial y^2}=\frac{\partial^2\Gamma_{xy}}{\partial x\,\partial y}\\&2\frac{\partial\Gamma_x}{\partial y}=\frac{\partial}{\partial x}\left(-\frac{\partial\Gamma_{yz}}{\partial x}+\frac{\partial\Gamma_{zx}}{\partial y}+\Gamma_{xy}\right)\\&2\frac{\partial\Gamma_y}{\partial x}=\frac{\partial}{\partial y}\left(-\frac{\partial\Gamma_{zx}}{\partial y}+\Gamma_{xy}+\frac{\partial\Gamma_{yz}}{\partial x}\right)\end{aligned}\right\}\ldots\ldots[35]$$

and these are also sufficient conditions for a simply connected plate. In an $(n + 1)$ connected plate, each application of Green's theorem introduces n integrals around the n internal boundaries. Hence there are $3n$ additional conditions

$$\oint_{C_i} d\psi_x = 0,\quad \oint_{C_i} d\psi_y = 0,\quad \oint_{C_i} d\bar{w} = 0,\ i = 1, 2 \ldots n$$

where the paths C_i are the internal boundaries.

Equations [35] are the analogs, in the present theory, of the six compatibility equations in 3-dimensional linear elasticity theory. That they are necessary conditions for single-valuedness of ψ_x, ψ_y, and $\bar{w}$ may be determined directly by eliminating these quantities from Equations [11]. The plate-compatibility equations may be expressed in terms of the plate-stress components by using Equations [5].

Equations Governing $\bar{w}$

A single differential equation for $\bar{w}$ may be obtained by eliminating ψ_x and ψ_y from Equations [16] in the following manner: Differentiate the first and second of Equations [16] with respect to x and y, respectively, and add to obtain

$$\left(D\nabla^2 - G'h - \frac{\rho h^3}{12}\frac{\partial^2}{\partial t^2}\right)\Phi = G'h\nabla^2\bar{w}\ldots\ldots.[36]$$

Then eliminate Φ between Equation [36] and the third of Equations [16] with the result

$$\left(\nabla^2 - \frac{\rho}{G'}\frac{\partial^2}{\partial t^2}\right)\left(D\nabla^2 - \frac{\rho h^3}{12}\frac{\partial^2}{\partial t^2}\right)w + \rho h\frac{\partial^2\bar{w}}{\partial t^2} = \left(1 - \frac{D\nabla^2}{G'h} + \frac{\rho h^2}{12G'}\frac{\partial^2}{\partial t^2}\right)q\ldots..[37]$$

Equation [37] is the two-dimensional analog of Timoshenko's beam equation (6).

If the rotatory inertia terms are omitted from Equation [37], it reduces to

$$D\left(\nabla^2 - \frac{\rho}{G'}\frac{\partial^2}{\partial t^2}\right)\nabla^2\bar{w} + \rho h\frac{\partial^2\bar{w}}{\partial t^2} = \left(1 - \frac{D\nabla^2}{G'h}\right)q\ldots[38]$$

If the transverse shear deformation is neglected, but the rotatory inertia terms are retained, Equation [37] reduces to

$$\left(D\nabla^2 - \frac{\rho h^3}{12}\frac{\partial^2}{\partial t^2}\right)\nabla^2\bar{w} + \rho h\frac{\partial^2\bar{w}}{\partial t^2} = q\ldots\ldots..[39]$$

Finally, if both the transverse shear deformation and rotatory inertia terms are omitted, Equation [37] reduces to Equation [1] in which $\nabla^4 = \nabla^2\nabla^2$.

Straight-Crested Waves

The exact solution of the three-dimensional equations of elasticity is known for the case of straight-crested flexural waves in an infinite plate (1, 2, 3). This solution will be used to test the present theory and to complete the equations of motion and the stress-strain-displacement relations by choosing the value of κ.

Three-Dimensional Theory. In the three-dimensional theory, the wave velocity c of wave length λ is given in the form of the transcendental equation (1, 2, 3)

$$\frac{4c_s^2\sqrt{(c_s^2 - \alpha c^2)(c_s^2 - c^2)}}{(2c_s^2 - c^2)^2} = \frac{\tanh\frac{\pi h}{\lambda c_s}\sqrt{c_s^2 - \alpha c^2}}{\tanh\frac{\pi h}{\lambda c_s}\sqrt{c_s^2 - c^2}},\ 0 < \frac{c}{c_s} < 1\ \ldots\ldots\ldots[40]$$

where

$$c_s = \sqrt{G/\rho}$$

i.e., the velocity of shear waves and

$$\alpha = \frac{1 - 2\mu}{2(1 - \mu)}$$

Lamb's values (2) for c/c_s versus h/λ (with $\mu = 1/2$) are plotted

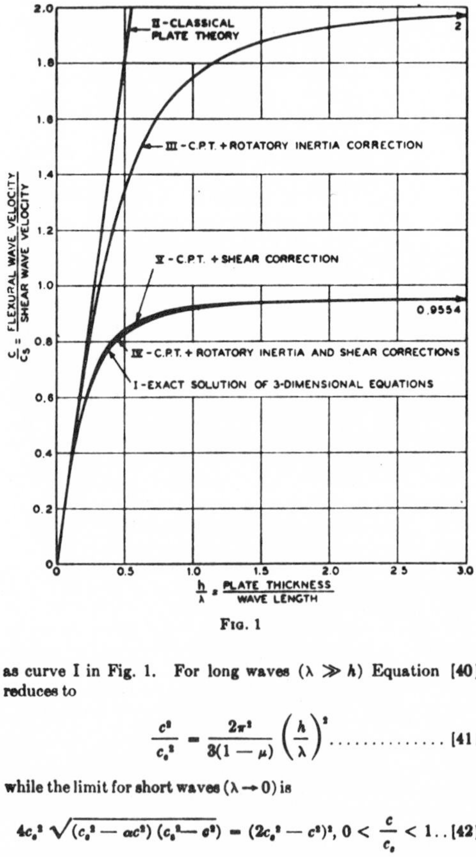

Fig. 1

as curve I in Fig. 1. For long waves ($\lambda \gg h$) Equation [40] reduces to

$$\frac{c^2}{c_s^2} = \frac{2\pi^2}{3(1 - \mu)}\left(\frac{h}{\lambda}\right)^2\ldots\ldots\ldots\ldots..[41]$$

while the limit for short waves ($\lambda \to 0$) is

$$4c_s^2\sqrt{(c_s^2 - \alpha c^2)(c_s^2 - c^2)} = (2c_s^2 - c^2)^2,\ 0 < \frac{c}{c_s} < 1..[42]$$

i.e., the equation for the velocity of Rayleigh surface waves For $\mu = 1/2$, Equation [42] gives $c = 0.9554\,c_s$.

Classical Plate Theory. If the solution for a straight-crested wave

$$\bar{w} = \cos\frac{2\pi}{\lambda}(x - ct) \qquad [43]$$

is substituted into Equation [1], with $q = 0$, we find

$$c^2 = \frac{D}{\rho h}\left(\frac{2\pi}{\lambda}\right)^2 = \frac{2\pi^2 c_s^2}{3(1-\mu)}\left(\frac{h}{\lambda}\right)^2 \qquad [44]$$

which is identical with Equation [41]. Thus, as observed by Rayleigh (1), classical plate theory gives good results for long waves. Equation [44] is plotted (for $\mu = 1/2$) as curve II in Fig. 1. It may be seen that when the wave length becomes less than 5 or 10 times the plate thickness the classical plate theory departs markedly from the three-dimensional theory.

Rotatory Inertia Correction. We consider, next, Equation [39] (with $q = 0$), which contains only the rotatory inertia correction. Insertion of $\bar{w}$ from Equation [43] results in

$$\frac{c^2}{c_s^2} = \frac{2\pi^2}{3(1-\mu)}\left(\frac{h}{\lambda}\right)^2\left[1 + \frac{\pi^2}{3}\left(\frac{h}{\lambda}\right)^2\right]^{-1} \qquad [45]$$

Equation [45], with $\mu = 1/2$, is plotted in Fig. 1 as curve III. For $\lambda \gg h$, Equation [45] reduces to [41] as it should; but the limit of c^2/c_s^2 as $\lambda \rightarrow 0$ is $2/(1-\mu)$, a value that is much too large in comparison with the solution of Equation [42].

Rotatory Inertia and Shear Corrections. In this case the complete left-hand side of Equation [37] is used, with the result

$$\frac{\pi^2}{3}\left(\frac{h}{\lambda}\right)^2\left(1 - \frac{c^2}{\kappa^2 c_s^2}\right)\left(\frac{c_p^2}{c^2} - 1\right) = 1 \qquad [46]$$

where

$$c_p = \sqrt{\frac{E}{\rho(1-\mu^2)}}$$

Accordingly, there are two roots for c^2. Confining attention to the smaller root, Equation [46] reduces to Equation [41] for long waves, as it should. The limit of Equation [46] as $\lambda \rightarrow 0$ is

$$c^2 = \kappa^2 c_s^2$$

According to the three-dimensional theory, this should be the velocity of Rayleigh surface waves. Hence, by Equation [42] the appropriate value of κ^2 is the root of

$$4\sqrt{(1 - \alpha\kappa^2)(1 - \kappa^2)} = (2 - \kappa^2)^2, \quad 0 < \kappa < 1 \qquad [47]$$

For anisotropic plates the analogous constants would be obtained from the algebraic equations whose roots give the velocities of surface waves appropriate to the medium. The dependence of κ^2 on Poisson's ratio is illustrated by noting that it ranges almost linearly from 0.76 for $\mu = 0$ to 0.91 for $\mu = 1/2$.

The smaller root of Equation [46], with $\mu = 1/2$ and $\kappa = 0.9554$, is plotted as curve IV in Fig. 1. The correspondence with the exact solution of the three-dimensional equations is so close that no difference between the two can be detected in computations carried to three figures. Curves IV and I are therefore the same in Fig. 1.

Shear Correction Only. It is interesting to notice that the transverse shear deformation accounts almostly entirely for the discrepancy between classical plate theory and the three-dimensional theory over the whole wave-length spectrum. Equation [38], which contains the shear-correction term only, yields

$$\frac{c^2}{c_s^2} = \frac{2\pi^2}{3(1-\mu)}\left(\frac{h}{\lambda}\right)^2\left[1 + \frac{2\pi^2}{3\kappa^2(1-\mu)}\left(\frac{h}{\lambda}\right)^2\right]^{-1} \qquad [48]$$

This is plotted, for $\mu = 1/2$, as curve V in Fig. 1.

Thickness-Shear Motion

The circular frequency of the first antisymmetric mode of thickness-shear vibration, according to the exact theory, is[6]

$$p = \pi c_s/h \qquad [49]$$

The corresponding solution in the present theory is obtained by setting

$$\psi_y = \bar{w} = 0, \quad \psi_x = e^{ipt}$$

in Equations [16], with the result

$$p = \kappa c_s\sqrt{12}/h \qquad [50]$$

To make Equations [49] and [50] identical requires

$$\kappa^2 = \pi^2/12 \qquad [51]$$

a value very close to Reissner's 5/6.

Substituting $\kappa^2 = \pi^2/12$ into Equation [47] we find

$$\mu = 0.176$$

Thus, in a material with a Poisson's ratio of 0.176, there is no conflict between Equations [51] and [47]. For other values of μ, one must choose or compromise between Equations [51] and [47] in accordance with the relative importance of the two modes of motion.

Reduction to Wave Equations

In the absence of surface loading, the equations of motion may be transformed to an exceptionally simple form with the loss of only trivial solutions.

If two of the plate-displacement components are expressed in terms of the potentials which give rise to the areal dilatation and rotation

$$\psi_x = \partial\varphi/\partial x + \partial H/\partial y, \quad \psi_y = \partial\varphi/\partial y - \partial H/\partial x \qquad [52]$$

and a factor e^{ipt} (hereafter omitted) is assumed, Equations [16] become

$$\left.\begin{aligned} &\frac{\partial}{\partial x}\left[\nabla^2\varphi + (R\delta_0^4 - S^{-1})\varphi - S^{-1}\bar{w}\right] + \frac{1-\mu}{2}\frac{\partial}{\partial y}(\nabla^2 + \omega^2)H = 0 \\ &\frac{\partial}{\partial y}\left[\nabla^2\varphi + (R\delta_0^4 - S^{-1})\varphi - S^{-1}\bar{w}\right] - \frac{1-\mu}{2}\frac{\partial}{\partial x}(\nabla^2 + \omega^2)H = 0 \\ &\nabla^2(\varphi + \bar{w}) + S\delta_0^4\bar{w} = 0 \end{aligned}\right\} \qquad [53]$$

where

$$R = h^2/12, \quad S = D/\kappa^2 Gh, \quad \delta_0^4 = \rho p^2 h/D,$$
$$\omega^2 = 2(R\delta_0^4 - S^{-1})/(1-\mu) \qquad [54]$$

The constants R and S represent the effects of rotary inertia and transverse shear deformation, respectively.

H may be separated from φ and $\bar{w}$ by differentiation, addition, and subtraction of the first two of Equations [53]. These equations become

$$\left.\begin{aligned} &\nabla^2(\nabla^2 + \omega^2)H = 0 \\ &\nabla^2\left[\nabla^2\varphi + (R\delta_0^4 - S^{-1})\varphi - S^{-1}\bar{w}\right] = 0 \end{aligned}\right\} \qquad [55]$$

---

[6] Reference (2), p. 122.

It now remains to separate φ and $\bar{w}$ between the second of Equations [55] and the third of Equations [53]. It will be observed that these equations are satisfied by

$$\varphi = (\sigma - 1)\bar{w} \quad \text{[56]}$$

where σ is a constant, provided

$$\sigma^{-1}S\delta_0^4 = R\delta_0^4 - S^{-1} - [S(\sigma - 1)]^{-1} = \delta^2 \text{ (say)} \quad \text{[57]}$$

and

$$\nabla^2 \bar{w} + \delta^2 \bar{w} = 0 \quad \text{[58]}$$

Solution of Equations [57] yields

$$\delta_1^2, \delta_2^2 = \frac{1}{2}\delta_0^4 \left[R + S \pm \sqrt{(R - S)^2 + 4\delta_0^{-4}}\right] \quad \text{[59]}$$

$$\sigma_1, \sigma_2 = (\delta_2^2, \delta_1^2)(R\delta_0^4 - S^{-1})^{-1} \quad \text{[60]}$$

Hence, two deflection functions ($\bar{w}_1$ and $\bar{w}_2$) are obtained by setting σ, in Equation [56], equal to σ_1 or σ_2 and these functions are governed by separate wave equations obtained by setting δ equal to δ_1 or δ_2 in Equation [58]. Equation [56] is now shown to be nonrestrictive by noting that Equation [37] may be written as

$$(\nabla^2 + \delta_1^2)(\nabla^2 + \delta_2^2)\bar{w} = 0 \quad \text{[61]}$$

when $q = 0$. Finally, in view of Equations [56] and [58], the bracketed terms in Equations [53] vanish, so that the equation governing H reduces to

$$(\nabla^2 + \omega^2) H = 0$$

To sum up, we may write

$$\begin{aligned} \psi_x &= (\sigma_1 - 1)\partial w_1/\partial x + (\sigma_2 - 1)\partial w_2/\partial x + \partial H/\partial y \\ \psi_y &= (\sigma_1 - 1)\partial w_1/\partial y + (\sigma_2 - 1)\partial w_2/\partial y - \partial H/\partial x \quad \text{[62]} \\ w &= w_1 + w_2 \end{aligned}$$

and

$$(\nabla^2 + \omega^2) H = 0 \qquad (\nabla^2 + \delta_1^2)\bar{w}_1 = 0 \qquad (\nabla^2 + \delta_2^2)\bar{w}_2 = 0 \quad \text{[63]}$$

It may be observed that, if $R = S = 0$, H and σ vanish and $\delta^2 = \pm\delta_0^2$, i.e., the present equations all reduce to those of classical plate theory when the effects of rotatory inertia and shear are removed.

BIBLIOGRAPHY

1 "On the Free Vibrations of an Infinite Plate of Homogeneous Isotropic Elastic Matter," by Lord Rayleigh, Proceedings of the London Mathematical Society, London, England, vol. 10, 1889, pp. 225–234.

2 "On Waves in an Elastic Plate," by H. Lamb, Proceedings of the Royal Society of London, England, series A, vol. 93, 1917, pp. 114–128.

3 "On the Transverse Vibrations of Bars of Uniform Cross Section," by S. Timoshenko, *Philosophical Magazine*, series 6, vol. 43, 1922, pp. 125–131.

4 "Theory of Sound," by Lord Rayleigh, second edition, The Macmillan Company, New York, N. Y., vol. 1, p. 258.

5 "On the Correction for Shear of the Differential Equation for Transverse Vibrations of Prismatic Bars," by S. Timoshenko, Philosophical Magazine, series 6, vol. 41, 1921, pp. 744–746.

6 "Vibration Problems in Engineering," by S. Timoshenko, second edition, D. Van Nostrand Company, Inc., New York, N. Y., 1937, p. 337.

7 "The Propagation of Waves in the Transverse Vibrations of Bars and Plates," by Ya. S. Uflyand, Akad. Nauk SSSR. Prikl. Mat. Meh., vol. 12, 1948, pp. 287–300 (Russian).

8 "Über die Berücksichtigung der Schubverzerrung in ebenen Platten," by H. Hencky, Ingenieur Archiv., vol. 16, 1947, pp. 72–76.

9 "The Effect of Transverse Shear Deformation on the Bending of Elastic Plates," by E. Reissner, JOURNAL OF APPLIED MECHANICS, Trans. ASME, vol. 67, 1945, p. A-69.

10 "On Bending of Elastic Plates," by E. Reissner, *Quarterly of Applied Mathematics*, vol. 5, 1947, pp. 55–68.

11 Discussion of Ref. (9), by J. N. Goodier, JOURNAL OF APPLIED MECHANICS, Trans. ASME, vol. 68, 1946, p. A-251.

12 "Étude nouvelle sur l'équilibre et le mouvement des corps solides élastiques dont certaines dimensions sont très-petites par rapport à d'autres," by J. Boussinesq, Journ. de Math., Paris, France, series 2, vol. 16, 1871, pp. 125–274, and series 3, vol. 5, 1879, pp. 329–344

13 "On Reissner's Theory of Bending of Elastic Plates," by A. E. Green, *Quarterly of Applied Mathematics*, vol. 7, 1949, pp. 223–228.

14 "Theory of Elasticity," by A. E. H. Love, fourth edition, The Macmillan Company, New York, N. Y., p. 176.

10

Reprinted from *J. Appl. Phys.*, **22**(3), 316–323 (1951)

Thickness-Shear and Flexural Vibrations of Crystal Plates

RAYMOND D. MINDLIN
Columbia University, New York, and Bell Telephone Laboratories, Murray Hill, New Jersey
(Received September 18, 1950)

The theory of flexural motions of elastic plates, including the effects of rotatory inertia and shear, is extended to crystal plates. The equations are solved approximately for the case of rectangular plates excited by thickness-shear deformation parallel to one edge. Results of computations of resonant frequencies of rectangular, *AT*-cut, quartz plates are shown and compared with experimental data. Simple algebraic formulas are obtained relating frequency, dimensions, and crystal properties for resonances of special interest in design.

INTRODUCTION

EXACT solutions of the three-dimensional equations governing the vibrations of crystals have been obtained for only a few body shapes and modes of motion. Special interest centers around a few shapes and modes which have been found useful as frequency control and filter elements in electric circuits. Among these are the modes excited by forced thickness-shear deformation of plates.

The resonant frequencies of thickness-shear modes in an infinite crystal plate were found by Koga[1] in closed form, but correspondingly simple solutions of the thickness-shear type are not obtainable for a bounded plate. Using Koga's solution as a zero-order approximation, Ekstein[2] applied a perturbation method to obtain approximate frequencies of a plate with a pair of parallel bounding planes normal to the faces of the plate. He arrived at a formula for the resonant frequencies which he showed to have the same form as an empirical formula devised by Sykes.[3] This formula, however, fits Sykes' experimental data only at discrete points whose locations cannot be found from the formula. The curves of frequency *versus* length to thickness ratio obtained from Sykes' and Ekstein's formulas depart markedly from the experimental data in the neighborhood of flexural resonances and become progressively worse both at higher thickness-shear overtones and lower length-to-thickness ratios. Ekstein advanced a qualitative explanation for the discrepancies in the neighborhood of flexural resonances by pointing out that

[1] I. Koga, Physics **3**, 70 (1932).
[2] H. Ekstein, Phys. Rev. **68**, 11 (1945).
[3] R. A. Sykes, Chapter VI, *Quartz Crystals for Electrical Circuits*, edited by R. A. Heising (D. Van Nostrand, New York, 1946), p. 218, Eqs. 6.9.

flexural modes in an infinite plate constitute another zero-order solution, and he had shown previously[4] that "coupling" effects become important when the frequencies of two zero-order modes are nearly equal.

In Ekstein's method of solving plate vibration problems, a start is made with one or more exact solutions (zero-order approximations) of the three-dimensional equations, satisfying some of the boundary conditions (e.g., those on the faces of the plate), and a perturbation theory is applied to satisfy, approximately, the remaining boundary conditions. On the other hand, for bodies having at least one small dimension, the classical procedure in elasticity theory is to reduce the number of variables in the differential equations of the system at the start by omitting certain variables deemed to be unimportant and employing average values or restricted regions of others. In this way an approximate theory is formed (two- or one-dimensional) in which some of the boundary conditions are satisfied identically, and some or all of the remaining ones may be satisfied exactly. Familiar examples are the approximate theories of flexural motions of bars and plates leading to the bernoulli-euler and lagrange equations of motion. In their classical form, these theories are not suitable for the present problem as they contain no shear deformation and, as Rayleigh[5] pointed out, they give good results for flexural wave velocities only when the wavelength is long in comparison with the thickness.

In the case of bars, Rayleigh[6] added a correction term for rotatory inertia to the classical theory with some, but not much, improvement in wave velocities. It was not until Timoshenko[7] introduced a correction for transverse shear deformation that the one-dimensional beam theory was made suitable for short wavelengths and high frequencies.

In the case of isotropic plates it has been shown[8] that the theory corresponding to Timoshenko's beam theory contains the appropriate approximations to the first thickness-shear mode and all the flexure modes. This theory is extended to crystal plates in the present paper. The additional boundaries considered by Ekstein are then introduced and a transcendental frequency equation of simple form is obtained. Notwithstanding its simplicity, the frequency equation contains, in great detail, the elaborate spectrum of frequencies obtained experimentally. The zeros and infinities of the frequency equation correspond to points in the frequency spectrum of special interest in design. For these points explicit solutions are obtained in the form of a pair of algebraic formulas. One of the formulas is similar to that of Sykes and Ekstein but contains an additional term which accommodates the higher thickness-shear overtones and smaller length to thickness ratios. The second formula locates the points at which the curves, computed from the first one, cross the curves computed from the transcendental equation. Thus, both frequency and length to thickness ratio are obtained easily for the points of interest without solving the transcendental equation.

The frequency equation that is obtained is simply the one for flexural resonances according to the improved theory of plates and is quite independent of the notion of thickness shear modes which, of course, do not exist in pure form in a bounded plate. From the point of view of plate theory, shear deformation is always present in flexural motion so that flexural resonance may be excited by forcing shear deformation (piezoelectrically or electrostrictively) at the resonant frequencies of flexure. From this point of view, the resonances in a bounded plate commonly designated as thickness-shear, or thickness-shear overtones, are simply local regions in the spectrum of flexural resonances over which the frequency does not change, as rapidly as elsewhere, with change of plate dimensions.

DERIVATION OF PLATE EQUATIONS

The plate is referred to an x, y, z system of rectangular coordinates. The faces of the plate are the planes $y=\pm h/2$, and its cylindrical surfaces (edges) have their generators normal to the x, z plane. The elastic properties of the material are assumed to have monoclinic symmetry. This includes the crystal classes and cuts commonly employed as thickness-shear resonators. With this symmetry and the x-axis as the principal axis, the stress-displacement relations are (in the I.R.E. Standard notation)

$$\begin{aligned}
T_1&=c_{11}(\partial u_1/\partial x)+c_{12}(\partial u_2/\partial y)+c_{13}(\partial u_3/\partial z)\\
&\qquad+c_{14}[(\partial u_3/\partial y)+(\partial u_2/\partial z)],\\
T_2&=c_{12}(\partial u_1/\partial x)+c_{22}(\partial u_2/\partial y)+c_{23}(\partial u_3/\partial z)\\
&\qquad+c_{24}[(\partial u_3/\partial y)+(\partial u_2/\partial z)],\\
T_3&=c_{13}(\partial u_1/\partial x)+c_{23}(\partial u_2/\partial y)+c_{33}(\partial u_3/\partial z)\\
&\qquad+c_{34}[(\partial u_3/\partial y)+(\partial u_2/\partial z)],\\
T_4&=c_{14}(\partial u_1/\partial x)+c_{24}(\partial u_2/\partial y)+c_{34}(\partial u_3/\partial z)\\
&\qquad+c_{44}[(\partial u_3/\partial y)(\partial u_2/\partial z)],\\
T_5&=c_{55}[(\partial u_1/\partial z)+(\partial u_3/\partial x)]\\
&\qquad+c_{56}[(\partial u_2/\partial x)+(\partial u_1/\partial y)],
\end{aligned} \tag{1}$$

and

$$\begin{aligned}
T_6&=c_{56}[(\partial u_1/\partial z)+(\partial u_3/\partial x)]\\
&\qquad+c_{66}[(\partial u_2/\partial x)+(\partial u_1/\partial y)].
\end{aligned}$$

In the three-dimensional theory, substitution of these expressions for the components of stress into the stress equations of motion

$$\begin{aligned}
&(\partial T_1/\partial x)+(\partial T_6/\partial y)+(\partial T_5/\partial z)=\rho(\partial^2 u_1/\partial t^2),\\
&(\partial T_6/\partial x)+(\partial T_2/\partial y)+(\partial T_4/\partial z)=\rho(\partial^2 u_2/\partial t^2),\\
\text{and}\quad&\\
&(\partial T_5/\partial x)+(\partial T_4/\partial y)+(\partial T_3/\partial z)=\rho(\partial^2 u_3/\partial t^2)
\end{aligned} \tag{2}$$

yields the differential equations governing the displacements.

[4] H. Ekstein, Phys. Rev. **66**, 108 (1944).
[5] Lord Rayleigh, Proc. London Math. Soc. **10**, 225 (1889).
[6] Lord Rayleigh, *Theory of Sound* (Macmillan and Company, London, 1926), second edition, Vol. 1, p. 258.
[7] S. P. Timoshenko, Phil. Mag. **41**, 744 (1921).
[8] R. D. Mindlin, Trans. Am. Soc. Mech. Engr., Paper No. 50-APM-22 (1950).

The corresponding two-dimensional relations of plate theory may be obtained as follows. The normal components of the tractions on planes parallel to the surfaces of the plate are assumed to contribute negligible amounts to the flexural and shear deformations of the plate. This is equivalent to assuming that

$$T_2=0 \tag{3}$$

throughout the plate. The plane faces of the plate are completely traction-free, so that, in addition,

$$T_6=T_4=0 \quad \text{on} \quad y=\pm h/2. \tag{4}$$

The displacements are assumed to have the form

$$u_1=y\psi_x(x,z,t), \quad u_2=\eta(x,z,t), \quad \text{and} \quad u_3=y\psi_z(x,z,t). \tag{5}$$

Equations (3) and (5) serve to eliminate all compressional and face-shear modes and all thickness-shear modes except the first and its overtones. There remain only the possibilities of approximations to plate flexural, torsional, and first thickness-shear modes and their overtones.

Three stress couples and two stress resultants are defined, in the usual manner, by

$$\begin{aligned}(M_x, M_z, M_{xz})&=\int_{-h/2}^{h/2}(T_1, T_3, T_5)y\,dy\\(Q_x, Q_z)&=\int_{-h/2}^{h/2}(T_6, T_4)dy.\end{aligned} \tag{6}$$

By means of (3), the second of Eq. (1) is solved for $\partial u_2/\partial y$, and this expression is used to eliminate $\partial u_2/\partial y$ from the remaining five stress-displacement relations. Equations (5) are now substituted in the remaining five of (1) and integrations are performed so as to convert the left-hand sides of (1) to stress couples and stress resultants in accordance with (6). The formulas for the two stress resultants (Q_x, Q_z) are then altered by introducing a constant multiplying factor whose magnitude is to be determined later by matching an exact solution of the three-dimensional equations. This procedure converts (1) to

$$\begin{aligned}M_x&=D_1(\partial\psi_x/\partial x)+D_2\partial\psi_z/\partial z,\\M_z&=D_2(\partial\psi_x/\partial x)+D_3\partial\psi_z/\partial z,\\M_{xz}&=D_5[(\partial\psi_z/\partial x)+\partial\psi_x/\partial z],\\Q_z=D_4(\psi_z+\partial\eta/\partial z),&\quad\text{and}\quad Q_x=D_6(\psi_x+\partial\eta/\partial x),\end{aligned} \tag{7}$$

where

$$\begin{aligned}&D_1=(c_{11}-c_{12}{}^2/c_{22})h^3/12, \quad D_2=(c_{13}-c_{12}c_{23}/c_{22})h^3/12,\\&D_3=(c_{33}-c_{23}{}^2/c_{22})h^3/12, \quad D_4=\kappa^2(c_{44}-c_{24}{}^2/c_{22})h,\\&D_5=c_{55}h^3/12, \quad \text{and} \quad D_6=\kappa^2c_{66}h,\end{aligned} \tag{8}$$

in which κ^2 is the multiplying factor mentioned above. Of the 13 elastic constants, three (c_{14}, c_{34}, c_{56}) do not contribute, and the remaining ten enter in only six combinations. It may be seen that κ controls the transverse shear correction. For example, if κ is infinite the last two of (7) require

$$\psi_z=-\partial\eta/\partial z, \quad \psi_x=-\partial\eta/\partial x, \tag{9}$$

which are the relations of classical plate theory in which shear deformation is neglected.

The equations of motion (2) are converted, similarly, by integrating over the thickness of the plate after multiplying the first and third by y. If Eqs. (3)–(6) are used, the results are

$$\begin{aligned}&(\partial M_x/\partial x)-Q_x+(\partial M_{xz}/\partial z)=(\rho h^3/12)(\partial^2\psi_x/\partial t^2),\\&(\partial Q_x/\partial x)+(\partial Q_z/\partial z)=\rho h(\partial^2\eta/\partial t^2),\\&\text{and}\\&(\partial M_{xz}/\partial x)-Q_z+(\partial M_z/\partial z)=(\rho h^3/12)(\partial^2\psi_z/\partial t^2).\end{aligned} \tag{10}$$

Finally, inserting (7) into (10), the plate equations of motion are

$$\begin{aligned}&D_1(\partial^2\psi_x/\partial x^2)+D_5(\partial^2\psi_x/\partial z^2)\\&\quad+(D_2+D_5)(\partial^2\psi_z/\partial x\partial z)-D_6(\psi_x+\partial\eta/\partial x)\\&\quad\quad=(\rho h^3/12)(\partial^2\psi_x/\partial t^2),\\&D_6[(\partial\psi_x/\partial x)+(\partial^2\eta/\partial x^2)]\\&\quad+D_4[(\partial\psi_z/\partial z)+(\partial^2\eta/\partial z^2)]=\rho h(\partial^2\eta/\partial t^2),\\&\text{and}\\&D_3(\partial^2\psi_z/\partial z^2)+D_5(\partial^2\psi_z/\partial x^2)\\&\quad+(D_2+D_5)(\partial^2\psi_x/\partial x\partial z)-D_4(\psi_z+\partial\eta/\partial z)\\&\quad\quad=(\rho h^3/12)(\partial^2\psi_z/\partial t^2).\end{aligned} \tag{11}$$

The right-hand sides of the first and third of (11) represent the effects of rotatory inertia. If these are omitted, and if the shear deformation terms (i.e., those containing κ) are eliminated, Eqs. (11), with the help of (9), reduce to the classical theory of orthotropic plates. The retention of the shear and rotatory inertia terms permits the application of the theory of plates to high frequency flexural modes and the accompanying thickness-shear motions.

Equations (7) and (11) replace Hooke's law and the equations of motion of the three-dimensional theory. The associated boundary conditions may be found by converting Neumann's uniqueness theorem[9] to apply to plates. This may be done by using (5) and performing the integrations with respect to y that appear in the volume and surface integrals in Neumann's theorem.[10] In terms of components referred to coordinates ν and s measured, respectively, normal to and along the boundary of the plate, the theorem reveals that, at each point of the edge of the plate, one member of each of the following three products must be specified:

$$M_\nu(\partial\psi_\nu/\partial t), \quad M_{\nu s}(\partial\psi_s/\partial t), \quad Q_\nu(\partial\eta/\partial t). \tag{12}$$

On a free edge, for example,

$$M_\nu=M_{\nu s}=Q_\nu=0. \tag{13}$$

Thus, three boundary conditions are required at each

[9] A. E. H. Love, *Theory of Elasticity* (Cambridge University Press, London, 1927), fourth edition, p. 176.

[10] The development of a plate-analog of Neumann's uniqueness theorem is described in detail in reference 8.

point on the edge, as opposed to two in the classical theory of plates. This is in agreement with E. Reissner's theory of equilibrium of plates[11] in which shear deformation is accounted for in a different manner.

THICKNESS-SHEAR VIBRATION OF AN INFINITE PLATE

The solution of the three-dimensional equations for thickness-shear vibration parallel to the x axis is obtained by setting

$$u_1=u(y)e^{i\omega t}, \quad u_2=u_3=0. \tag{14}$$

The equations of motion (2) then reduce to

$$c_{66}(\partial^2 u/\partial y^2)+\rho\omega^2 u=0.$$

Applying the boundary conditions (4), the usual result[1] is obtained for the circular frequency $\bar{\omega}$ of the first mode:

$$\bar{\omega}^2=\pi^2 c_{66}/\rho h^2. \tag{15}$$

The corresponding solution of the plate, Eqs. (11), is obtained by setting

$$\psi_x=B\exp(i\bar{\omega}t), \quad \eta=\psi_z=0, \tag{16}$$

where B is a constant. The equations of motion (11) are satisfied provided

$$\bar{\omega}^2=12\kappa^2 c_{66}/\rho h^2. \tag{17}$$

The value of the coefficient κ is obtained by equating (15) and (17), with the result

$$\kappa=\pi/(12)^{\frac{1}{2}}. \tag{18}$$

It may be seen that the first antisymmetric mode of thickness-shear motion is contained in plate theory when rotatory inertia and shear terms are included.

THICKNESS-SHEAR EXCITATION OF A RECTANGULAR PLATE

Suppose, now, that the thickness shear vibration

$$\psi_x=Be^{i\omega t}, \quad \eta=\psi_z=0 \tag{19}$$

is forced (piezoelectrically, for example) in a rectangular plate with free edges $x=\pm l/2$, $z=\pm$const. The boundary conditions to be satisfied are, from (13),

$$M_x=M_{xz}=Q_x=0 \quad \text{on} \quad x=\pm l/2 \tag{20}$$

$$M_z=M_{xz}=Q_z=0 \quad \text{on} \quad z=\pm\text{const.} \tag{21}$$

Substitution of (19) into (7) shows that all the boundary conditions are satisfied with the exception that

$$Q_x=D_6Be^{i\omega t} \quad \text{on} \quad x=\pm l/2. \tag{22}$$

Additional modes are required such that, combined with (19), the equations of motion (11) and the boundary conditions (20) and (21) will be satisfied. Since (11) are essentially the equations of flexural motion, it is apparent that the presence of boundaries results, in this case, in the excitation of flexural modes.

[11] E. Reissner, J. Appl. Mech. 12, A-69 (1945); Quart. Appl. Math. 5, 55 (1947).

SOLUTION NEGLECTING WIDTH

If the additional motions introduced to remove (22) are taken as independent of z, the boundary conditions (20) on $x=\pm l/2$ can be satisfied, but small stress couples and resultants will be introduced on $z=\pm$const. Extensive experiments by Sykes on rotated Y-cut quartz plates show that, over a considerable range, the resonant frequencies are practically independent of the width. With the assumption of z-independence, ψ_z disappears from the first two equations of motion (11). These become, after introducing a factor $e^{i\omega t}$, omitted hereafter,

$$\begin{aligned}(\partial^2\psi_x/\partial x^2)+(R\delta_0^4-S^{-1})\psi_x&=S^{-1}\partial\eta/\partial x\\ (\partial\psi_x/\partial x)+[(\partial^2/\partial x^2)+S\delta_0^4]\eta&=0,\end{aligned} \tag{23}$$

where

$$R=h^2/12, \quad S=D_1/D_6,$$

and

$$\delta_0^4=\rho h\omega^2/D_1=\omega^2/(\bar{\omega}^2 RS). \tag{24}$$

The constants R and S represent the effects of rotatory inertia and transverse shear deformation, respectively. The ratio

$$g=R/S=\pi^2 c_{66}/12(c_{11}-c_{12}^2/c_{22}) \tag{25}$$

plays an important role. It is the ratio of the thickness-shear modulus ($\kappa^2 c_{66}$) to the plate flexure modulus. In the case of an isotropic material, c_{66} becomes the shear modulus, and $c_{11}-c_{12}^2/c_{22}$ reduces to $E/(1-\nu^2)$, where E is Young's modulus and ν is the poisson ratio.

Equations (23) are satisfied by

$$\psi_x=(\sigma-1)\partial\eta/\partial x, \tag{26}$$

where σ is a constant, provided

$$\sigma^{-1}S\delta_0^4=R\delta_0^4-S^{-1}-[S(\sigma-1)]^{-1}=\delta^2, \text{ (say)} \tag{27}$$

and

$$(\partial^2\eta/\partial x^2)+\delta^2\eta=0. \tag{28}$$

Solution of (27) yields

$$2(\delta_1^2, \delta_2^2)=\delta_0^4\{R+S\pm[(R-S)^2+4\delta_0^{-4}]^{\frac{1}{2}}\}, \tag{29}$$

and

$$(\sigma_1, \sigma_2)=(\delta_1^{-2}, \delta_2^{-2})S\delta_0^4. \tag{30}$$

Hence, two deflection functions (η_1 and η_2) are obtained by setting δ equal to δ_1 or δ_2 in (28). They are the solutions of the fourth-order equation

$$[(\partial^2/\partial x^2)+\delta_1^2][(\partial^2/\partial x^2)+\delta_2^2]\eta=0, \tag{31}$$

which can be obtained directly by eliminating ψ_x from (23). Equation (31) is, in fact, no more than an anisotropic plate version of Timoshenko's beam equation.[7]

To sum up, we may write

$$\eta=\eta_1+\eta_2, \tag{32}$$

$$\psi_x=(\sigma_1-1)(\partial\eta_1/\partial x)+(\sigma_2-1)\partial\eta_2/\partial x, \tag{33}$$

$$(\partial^2\eta_1/\partial x^2)+\delta_1^2\eta_1=0, \quad (\partial^2\eta_2/\partial x^2)+\delta_2^2\eta_2=0, \tag{34}$$

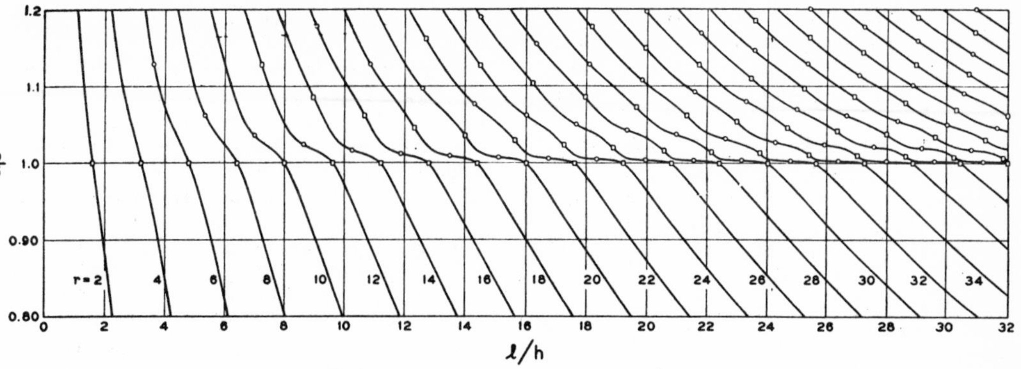

FIG. 1. Resonant frequencies of the even modes of vibration of an *AT*-cut quartz plate as computed from Eqs. (42), (43), and (44). The ordinate is the ratio of the resonant frequency (ω) to the frequency ($\bar{\omega}$) of the first thickness-shear mode of an infinite plate. The abscissa is the ratio of the length (l) to the thickness (h). The circles and squares locate points for which the transcendental frequency equation can be solved explicitly. They are also the points at which the corresponding curves for the odd modes would cross the curves shown.

and, from (7), $M_{xz}=0$,

$$M_x=-D_1[(\sigma_1-1)\delta_1^2\eta_1+(\sigma_2-1)\delta_2^2\eta_2],$$

and (35)

$$Q_x=D_6[(\sigma_1\partial\eta_1/\partial x)+\sigma_2\partial\eta_2/\partial x].$$

The solutions of (34), which are even in x (i.e., the symmetric, or odd order, flexure modes), make Q_x odd in x. Hence, they cannot annul (22), which is even in x, and play no part in satisfying the boundary conditions on $x=\pm l/2$. This is the counterpart of the experimental observation that there is no coupling between the fundamental thickness-shear mode and the odd flexure modes. The remaining solutions of (34) are,

$$\eta_1=A_1\sin\delta_1 x, \quad \text{and} \quad \eta_2=A_2\sin\delta_2 x. \qquad (36)$$

The constants A_1 and A_2 are evaluated from the boundary conditions, i.e., (36) are substituted into (35) and then, at $x=\pm l/2$, $M_x=0$, $Q_x=-D_6B$ to annul (22). This results in the equations

$$A_1(\sigma_1-1)\delta_1^2\sin(\delta_1 l/2)+A_2(\sigma_2-1)\delta_2^2\sin(\delta_2 l/2)=0,$$

and (37)

$$A_1\sigma_1\delta_1\cos(\delta_1 l/2)+A_2\sigma_2\delta_2\cos(\delta_2 l/2)=-B.$$

These equations give A_1 and A_2 in terms of the forcing amplitude B. Resonance occurs when the determinant of (37) vanishes, i.e., when

$$\sigma_2\delta_1(\sigma_1-1)\tan(\delta_1 l/2)=\sigma_1\delta_2(\sigma_2-1)\tan(\delta_2 l/2). \qquad (38)$$

The roots of this equation give the resonant frequencies in terms of the dimensions and elastic constants of the plate.

FREQUENCY EQUATION

To put the frequency Eq. (38) into a form suitable for computing the roots, let

$$\gamma=\delta_1 l/2, \qquad (39)$$

and

$$a=\delta_2/\delta_1=[(1-b)/(1+b)]^{\frac{1}{2}}, \qquad (40)$$

where b is the positive root of

$$b^2=1+4g(\bar{\omega}^2/\omega^2-1)(1+g)^{-2}. \qquad (41)$$

It may be seen, from (29), that δ_1 is always real, while δ_2 is real if $\omega>\bar{\omega}$ and imaginary if $\omega<\bar{\omega}$. From (40), the same applies to a. (The case $\delta_2=0$, $a=0$, and $\omega=\bar{\omega}$ will be considered later along with other special roots.) When a is imaginary, let $a=ia_1$, where a_1 is real. Then (38) may be written as

$$(a^2-g)\tan\gamma=a(1-ga^2)\tan a\gamma, \qquad \omega>\bar{\omega},$$

and (42)

$$(a_1^2+g)\tan\gamma=a_1(1+ga_1^2)\tanh a_1\gamma, \qquad \omega<\bar{\omega}.$$

From (40) and (41),

$$\begin{aligned}\omega/\bar{\omega}&=[1-a^2(1+g)^2/g(1+a^2)^2]^{-\frac{1}{2}}, \quad \omega>\bar{\omega}\\ \omega/\bar{\omega}&=[1+a_1^2(1+g)^2/g(1-a_1^2)^2]^{-\frac{1}{2}}, \quad \omega<\bar{\omega},\end{aligned} \qquad (43)$$

and, from (39),

$$\begin{aligned}l/h&=\gamma(\bar{\omega}/\omega)[(1+a^2)/3(1+g)]^{\frac{1}{2}}, \quad \omega>\bar{\omega}\\ l/h&=\gamma(\bar{\omega}/\omega)[(1-a_1^2)/3(1+g)]^{\frac{1}{2}}, \quad \omega<\bar{\omega}.\end{aligned} \qquad (44)$$

The elastic constants affect the frequencies only through the constants g and c_{66} (the latter is contained in $\bar{\omega}$). Thus, there is a family of roots of (42) for each value of g, i.e., for each crystal class and orientation of cut. Selection of g and a or a_1 fixes $\omega/\bar{\omega}$ according to (43) and defines roots γ of (42). The values of γ are computed and inserted in (44) to give the values of l/h corresponding to the chosen g and a or a_1, i.e., to the chosen $\omega/\bar{\omega}$. For a given g, this procedure is repeated over the desired range of $\omega/\bar{\omega}$ (or a and a_1).

The results of a set of computations (for $g=0.283$) are shown in Fig. 1. The curves will be recognized as being characteristic of the frequency spectrum of resonances excited by thickness-shear vibration of rectangular quartz plates.[12] The circled and squared points in Fig. 1 represent special roots of (42) for which both sides of the equation are infinity and zero, respectively. These roots will be discussed in detail later.

COMPARISON WITH FREQUENCIES OF *AT* QUARTZ PLATES

Mason's values[13] for the elastic constants of quartz, referred to the principal axis of crystal symmetry, are, in units of 10^{10} dynes/cm², $c_{11}^0=86.05$, $c_{12}^0=5.05$, $c_{13}^0=10.45$, $c_{14}^0=18.25$, $c_{33}^0=107.1$, $c_{44}^0=58.65$, and $c_{66}^0=(c_{11}^0-c_{12}^0)/2=40.5$.

The constants which are required, referred to rotated axes, are given by [14]

$$c_{66}=c_{44}^0s^2+(c_{11}^0-c_{12}^0)c^2/2-2c_{14}^0sc,$$
$$c_{22}=c_{11}^0c^4+c_{33}^0s^4+2(2c_{44}^0+c_{13}^0)s^2c^2+4c_{14}^0sc^3,$$
$$c_{12}=c_{12}^0c^2+c_{13}^0s^2-2c_{14}^0sc,\quad c_{11}=c_{11}^0,$$

where $s=\sin\theta$ and $c=\cos\theta$. For the *AT* cut, $\theta=35°15'$. Hence, in units of 10^{10} dynes/cm², $c_{66}=29.34$, $c_{22}=129.9$, $c_{12}=-10.49$, $c_{11}=86.05$, and, from (25) and (18), $g=0.283$. Thus, the curves in Fig. 1 are for the *AT* cut.

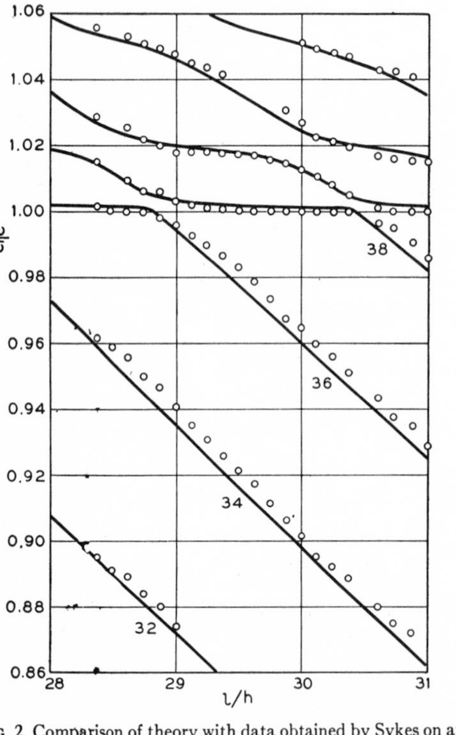

FIG. 2. Comparison of theory with data obtained by Sykes on an *AT* quartz plate initially 25×25×0.807 mm, cut down in decrements of 0.1 mm to 25×21.8×0.807 mm.

Portions of these curves are drawn to a larger scale in Figs. 2 and 3.

Data obtained by Sykes with *AT* plates are plotted as circles in Figs. 2 and 3. The dimensions of the plates were initially 25×25×0.807 mm for Fig. 2 and 22×14 ×1.000 mm for Fig. 3. In successive experiments the x-dimensions (l) were cut down in decrements of 0.1 mm to 21.8 mm and 8 mm, respectively. Hence, l/h varies from 28.38 to 30.98 in Fig. 2 and 8 to 14 in Fig. 3. The recorded resonant frequencies (cps) were converted to $\omega/\bar{\omega}$ by multiplying by $2h/(c_{66}/\rho)^{\frac{1}{2}}=0.485\times10^{-6}$, in which $h=0.807$ and 1.000, $\rho=2.65$, and $c_{66}=29.34\times10^{-10}$. A better fit between experiment and theory would have been obtained in the neighborhood of $\omega/\bar{\omega}=1$ if c_{66} had been calculated from Sykes' data on the *AT* plate rather than from Mason's primary data on the nonrotated constants.

It may be seen that there is a relative displacement, between theory and experiment, of about 0.1 to 0.3 percent along the l/h axis. This could be due to approximations made in the theory or to errors in the elastic constants. In the latter case, for example, a 1 percent error in c_{11} would produce a 0.1 percent displacement of the type observed.

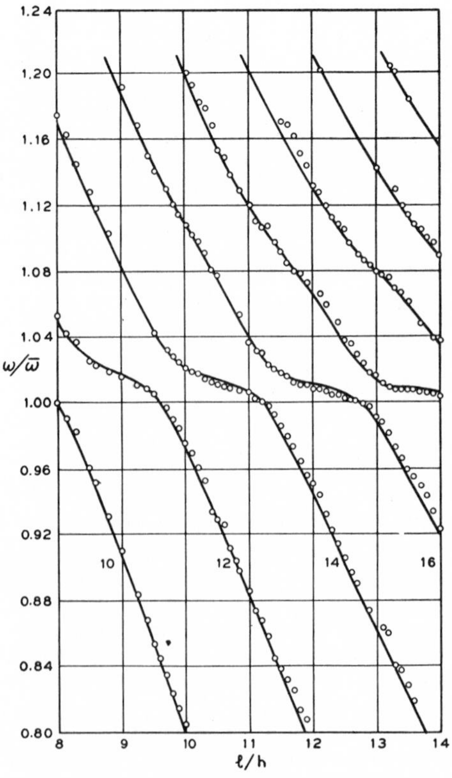

FIG. 3. Comparison of theory with data obtained by Sykes on an *AT* quartz plate, initially 22×14×1.000 mm cut down in decrements of 0.1 mm to 22×8×1.000 mm.

[12] See reference 3, p. 233, Fig. 6.15.

[13] W. P. Mason, *Piezoelectric Crystals and Their Applications to Ultrasonics* (D. Van Nostrand Company, Inc., New York, 1950), p. 84.

[14] See reference 3, p. 247.

SOLUTION OF FREQUENCY EQUATION BY SUCCESSIVE APPROXIMATIONS

Special interest centers around the frequencies of the higher modes in the neighborhood of the thickness-shear frequency $\bar{\omega}$. In this range, the following solution of the frequency equation by successive approximations (suggested by S. A. Schelkunoff) converges very rapidly. Writing the first of (42) in the form

$$\tan\gamma=\varphi(g,a)\tan a\gamma,\quad \omega>\bar{\omega}, \tag{45}$$

let

$$\gamma=r\pi+\beta,\quad -\pi/2<\beta<\pi/2,\quad r=1,2,3\ldots.$$

Since $\tan(r\pi+\beta)=\tan\beta$, (45) may be written

$$\beta=\tan^{-1}[\varphi(g,a)\tan a(r\pi+\beta)],\quad \omega>\bar{\omega}.$$

As a first approximation take

$$\beta_1=\tan^{-1}[\varphi(g,a)\tan ar\pi],\quad \omega>\bar{\omega},$$

and, as successive approximations,

$$\beta_{i+1}=\tan^{-1}[\varphi(g,a)\tan a(r\pi+\beta_i)],\quad \omega>\bar{\omega}.$$

It is usually necessary to take only one or two steps.

Similarly, for $\omega<\bar{\omega}$,

$$\tan\gamma=\varphi_1(g,a_1)\tanh a_1\gamma,\quad \omega<\bar{\omega}, \tag{46}$$

and

$$\beta_{i+1}=\tan^{-1}[\varphi_1(g,a_1)\tanh a_1(r\pi+\beta_i)],\quad \omega<\bar{\omega}.$$

The product $a_1r\pi$ does not have to become very large before this can be written as

$$\beta=\tan^{-1}\varphi_1(g,a_1),$$

which accounts for the monotonic, regularly spaced curves for the low frequency portions of the flexural modes in Fig. 1.

ODD MODES

While only the even flexural modes are excited by the forcing function (19), the odd modes may be excited by forcing a thickness-shear deformation antisymmetrical with respect to x. In this case, the residual stress resultants corresponding to (22) will be odd in x and the even solutions

$$\eta_1=B_1\cos\delta_1x,\quad \eta_2=B_2\cos\delta_2x \tag{47}$$

of (34) will be required to annul them. The only change in the results, therefore, is the interchange of sine and cosine throughout. The transcendental equations are the same as those for the even modes except for the replacement of $\varphi(g,a)$ and $\varphi_1(g,a_1)$ by their reciprocals and a reversal of sign for the case $\omega/\bar{\omega}<1$. The treatment of these equations is then very similar. In particular, the circled and squared points are also roots and hence the odd-mode curves of $\omega/\bar{\omega}$ *vs* l/h, corresponding to Fig. 1, cross the curves of Fig. 1 at these points. Whereas, for the even modes, the circled points are on the flat portions of the curves (slow variation of $\omega/\bar{\omega}$ with l/h) and the squared points are on the steep portions (rapid variation of $\omega/\bar{\omega}$ with l/h), the converse is true for the odd flexural modes.

ALGEBRAIC FORMULAS

The circled points in Fig. 1 are those for which γ and $a\gamma$ in the first of (42) are odd multiples of $\pi/2$. In the case of excitation by thickness-shear, even in x, these points are of special interest, as they are far from the frequencies of neighboring modes and center in regions where the frequency changes slowly with change of dimensions. The squared points, for which γ and $a\gamma$ in the first of (42) are even multiples of $\pi/2$, have the same properties for thickness-shear excitation odd in x. Because of the degeneracy of the transcendental equation for both sets of points, algebraic formulas may be obtained for them.

In the first of (42), the roots under discussion are

$$\begin{aligned}\gamma&=r\pi/2,\quad r=1,2,3\ldots\\ a\gamma&=n\pi/2,\quad n=0,1,2\ldots,\end{aligned} \tag{48}$$

from which

$$a=n/r. \tag{49}$$

Substituting (48) and (49) in the first of (43) and (44), the latter become

$$\omega/\bar{\omega}=g^{\frac{1}{2}}(r^2+n^2)[(gr^2-n^2)(r^2-gn^2)]^{-\frac{1}{2}} \tag{50}$$

$$l/h=(\pi/2)[(gr^2-n^2)(r^2-gn^2)/3g(1+g)(r^2+n^2)]^{\frac{1}{2}}. \tag{51}$$

Equations (50) and (51) give the coordinates ($\omega/\bar{\omega}$, l/h) of the circled points (odd r and n) and squared points (even r and n). These points are plotted in Fig. 4 for the AT plate ($g=0.283$) over the range 0 to 1.2 for $\omega/\bar{\omega}$ and 0 to 32 for l/h.

The equations of the two families of curves passing through the points $n=$const and $r=$const are obtained by eliminating r and n, respectively, between (50) and (51) with the results

$$2(\omega/\bar{\omega})^2=1+(kn^2h^2/l^2)+[(1+kn^2h^2/l^2)^2-4(\kappa nh/l)^4/g]^{\frac{1}{2}}, \tag{52}$$

and

$$2(\omega/\bar{\omega})^2=1+(kr^2h^2/l^2)-[(1+kr^2h^2/l^2)^2-4(\kappa rh/l)^4/g]^{\frac{1}{2}}, \tag{53}$$

where

$$k=\kappa^2(1+g)/g=(\pi^2/12)+(c_{11}c_{22}-c_{12}^2)/c_{22}c_{66}. \tag{54}$$

These curves are illustrated in Fig. 4 for the AT cut.

Equation (52) is to be compared with the empirical equation given by Sykes which, for the present case and in the present notation, is

$$(\omega/\bar{\omega})^2=1+kn^2h^2/l^2, \tag{55}$$

where k is an experimental constant. Sykes' equation was derived by Ekstein, from his theory, with a formula for k more elaborate than (54) but yielding approximately the same numerical values. Equations (55) and (52) differ, in form, only by the term $4(\kappa nh/l)^4/g$ in the

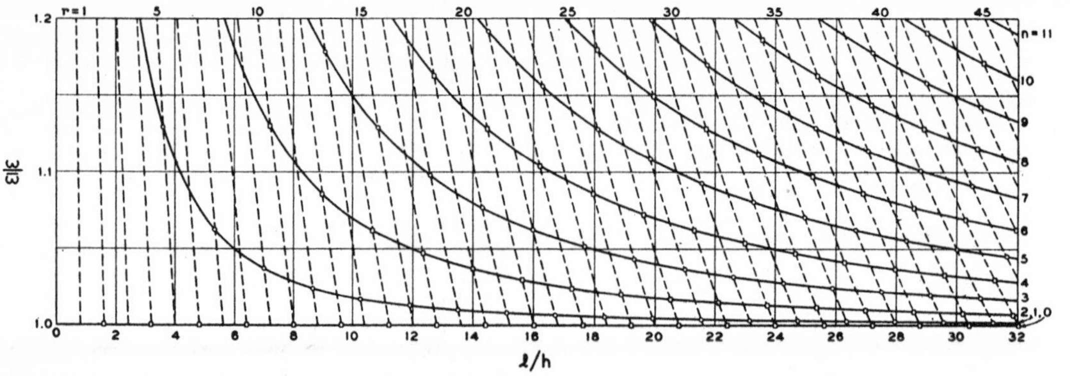

FIG. 4. The circles and squares are those from the upper half of Fig. 1. Their coordinates are given by the algebraic formulas (50) and (51). They are connected by full and dashed curves given by Eqs. (52) and (53), respectively. Points on these curves do not correspond to resonant frequencies except at the intersections marked with a circle or square. The full curves $n=2, 3, 4$ . . . are those commonly designated as the first, second, third . . . overtones of the fundamental thickness-shear mode.

latter. The additional term accommodates the higher overtones of thickness-shear vibration (large n) and small values of l/h.

It should be observed, however, that equations like (52) or (55) have meaning only at discrete points. These may be located by (50) and (51) or by the intersections of the families of curves (52) and (53) as illustrated in Fig. 4.

Regarding Eq. (53), it may be observed that the curves $r=\text{constant}$ are continuations, into the region $\omega/\bar{\omega}>1$, of the flexural mode curves (Fig. 1) from the region $\omega/\bar{\omega}<1$. Alternatively, curves $n+r=\text{constant}$ are close approximations to the curves in the upper region of Fig. 1.

SHAPES OF MODES

The shapes of the modes of vibration may be determined very easily for the frequencies discussed in the preceding section. Noting that $\delta_1=r\pi/l$ and $\delta_2=n\pi/l$, we find, from (32), (33), and (36),

$$\eta=A_1\sin(r\pi x/l)+A_2\sin(n\pi x/l), \tag{56}$$

and

$$h\psi_x=U_1\sin(r\pi x/l)+U_2\sin(n\pi x/l), \tag{57}$$

where

$$U_1=A_1h(\sigma_1-1)\delta_1,\quad U_2=A_2h(\sigma_2-1)\delta_2.$$

These give, respectively, the displacements that would be observed normal and tangential to the surface of the plate. Each contains a short wavelength part, $\sin(r\pi x/l)$, and a long wavelength part, $\sin(n\pi x/l)$. The former is commonly associated with the notion of flexural modes and the latter with thickness-shear modes. Alternatively, each type of mode contains both normal and tangential components of displacement.

The relative amplitudes of the four parts of the surface displacement may be found by using the first of (37), from which

$$\left|\frac{A_2}{A_1}\right|=\left|\frac{(\sigma_1-1)\delta_1^2\sin(\delta_1 l/2)}{(\sigma_2-1)\delta_2^2\sin(\delta_2 l/2)}\right|. \tag{58}$$

Noting that $\sigma_1/\sigma_2=a^2=n^2/r^2$ and $\sigma_1=(1+a^2)/(1+g)$, this becomes, for r and n odd,

$$|A_2/A_1|=(gr^2-n^2)/(r^2-gn^2). \tag{59}$$

For r and n even, the even solutions (47) are used, and the resulting amplitude ratio is the same as (59). Usually r/n is large, and hence $|A_2/A_1|$ is approximately equal to g. Thus, regarding displacements normal to the plate, the part associated with thickness-shear is less than that associated with flexure in the ratio $g:1$ (approximately). The corresponding ratio for the tangential displacements is

$$U_2/U_1=r/n;$$

i.e., the tangential displacement associated with thickness-shear is usually much larger than that associated with the accompanying flexural motion. Similarly, it can be shown that $|U_2/A_2|$ is approximately r/ng times $|U_1/A_1|$; also, $|U_2/A_1|$ is r/n times $|U_1/A_1|$. Thus, in general, for the circled points the predominant motion is tangential to the surface of the plate.

11

Reprinted from *J. Appl. Mech.*, **22**, 86–88 (1955)

Axially Symmetric Flexural Vibrations of a Circular Disk[1]

BY H. DERESIEWICZ[2] AND R. D. MINDLIN,[3] NEW YORK, N. Y.

At high frequencies, the flexural vibrations of a plate are described very poorly by the classical (Lagrange) theory because of neglect of the influence of coupling with thickness-shear vibrations. The latter may be taken into account by inclusion of rotatory inertia and shear-deformation terms in the equations. The resulting frequency spectrum is given, in this paper, for the case of axially symmetric vibrations of a circular disk with free edges and is compared with the spectrum predicted by the classical theory.

INTRODUCTION

THE problem of free vibrations of a thin, isotropic, elastic, circular disk, of constant thickness, was first attacked by Poisson (1)[4] over a century ago. Basing his investigation on the classical (Lagrange) theory of plates, he obtained the lower frequencies of the axially symmetric flexural motions of such a disk with a traction-free boundary.

It is well known that the classical theory satisfactorily predicts actual behavior only for the first few flexural modes of motion of a plate whose thickness is small in comparison with its other dimensions. For the higher flexural modes the influence of coupling with the thickness-shear mode of motion becomes increasingly important. Hence, the classical theory, in which this effect is not taken into account, ceases to yield reliable information (2). In particular, at the frequency of pure thickness-shear vibration of an infinite plate (i.e., displacement constant in direction and parallel to the plane of the plate) a drastic change occurs in the frequency spectrum of a finite plate (3, 4).

The influence of coupling between flexure and shear is taken into account by inclusion of rotatory inertia and shear-deformation terms in the equations. The resulting change in the frequency spectrum has been given previously for the case of a free-free beam (3) and a class of antisymmetric modes of motion of a circular disk (4). The present paper contains a discussion of the axially symmetric, flexural vibrations of a free disk, with emphasis on behavior in the neighborhood of the thickness-shear frequency. The spectrum obtained with consideration of the effects of rotatory inertia and shear deformation is compared with that predicted by the classical theory.

[1] This investigation was supported by the Office of Naval Research under Contract Nonr-266(09).

[2] Assistant Professor of Civil Engineering, Columbia University. Assoc. Mem. ASME.

[3] Professor of Civil Engineering, Columbia University. Mem. ASME.

[4] Numbers in parentheses refer to the Bibliography at the end of the paper.

Contributed by the Applied Mechanics Division for presentation at the Annual Meeting, New York, N. Y., November 28–December 3, 1954, of THE AMERICAN SOCIETY OF MECHANICAL ENGINEERS.

Discussion of this paper should be addressed to the Secretary, ASME, 29 West 39th Street, New York, N. Y., and will be accepted until one month after final publication of the paper itself in the JOURNAL OF APPLIED MECHANICS.

NOTE: Statements and opinions advanced in papers are to be understood as individual expressions of their authors and not those of the Society. Manuscript received by ASME Applied Mechanics Division, February 5, 1954. Paper No. 54—A-15.

PLATE EQUATIONS

If account is taken of rotatory inertia and shear deformation, the plate stress-displacement relations in polar co-ordinates, for the axially symmetric case,[5] become

$$\left.\begin{aligned} M_r &= D\left(\frac{\partial\psi}{\partial r} + \frac{\nu}{r}\psi\right) \\ M_\theta &= D\left(\nu\frac{\partial\psi}{\partial r} + \frac{\psi}{r}\right) \\ Q_r &= \kappa^2\mu h\left(\psi + \frac{\partial w}{\partial r}\right) \\ M_{r\theta} &= Q_\theta = 0 \end{aligned}\right\} \quad [1]$$

where $D = Eh^3/12(1-\nu^2)$, E, μ, ν, and h are Young's modulus, the shear modulus, Poisson's ratio, and the thickness, respectively, and $\kappa^2 = \pi^2/12$. The functions ψ and w are related to the radial and axial components of the displacement according to the approximations

$$\left.\begin{aligned} u_r &\approx z\psi(r,t) \\ u_z &\approx w(r,t) \end{aligned}\right\} \quad [2]$$

u_θ, the circumferential component of the displacement, is zero.

The plate equations of motion,[6] for the present problem, are

$$\left.\begin{aligned} \frac{\partial M_r}{\partial r} + \frac{M_r - M_\theta}{r} - Q_r &= \frac{\rho h^3}{12}\frac{\partial^2\psi}{\partial t^2} \\ \frac{\partial Q_r}{\partial r} + \frac{Q_r}{r} &= \rho h\frac{\partial^2 w}{\partial t^2} \end{aligned}\right\} \quad [3]$$

where ρ is the density of the plate.

If we now insert Equations [1] in [3] we obtain the plate displacement equations of motion. Omitting the time factor e^{ipt}, these become

$$\left.\begin{aligned} &\left(\frac{d^2}{dr^2} + \frac{1}{r}\frac{d}{dr} - \frac{1}{r^2} + \frac{\rho p^2 h^3}{12D}\right)\psi - \frac{k^2\mu h}{D}\left(\psi + \frac{dw}{dr}\right) = 0 \\ &\left(\frac{d}{dr} + \frac{1}{r}\right)\psi + \left(\frac{d^2}{dr^2} + \frac{1}{r}\frac{d}{dr} + \frac{\rho p^2}{\kappa^2\mu}\right)w = 0 \end{aligned}\right\} \quad [4]$$

Equations [4] may be uncoupled by differentiating the second equation once and subtracting the result from the first one. This procedure yields an expression for ψ in terms of w, which, when inserted into the second of Equations [4], gives a single equation on w only

$$\left(\frac{d^2}{dr^2} + \frac{1}{r}\frac{d}{dr} + \delta_1^2\right)\left(\frac{d^2}{dr^2} + \frac{1}{r}\frac{d}{dr} + \delta_2^2\right)w = 0 \quad [5]$$

[5] These are given, in the general case, by equations [3] of reference (4).

[6] For the general case, see equations [2] of reference (4).

where

$$\delta_1^2, \delta_2^2 = \frac{\delta_0^4}{2}\left[R + S \pm \sqrt{(R-S)^2 + 4\delta_0^{-4}}\right]$$

$$S = D/\kappa^2\mu h, \quad \delta_0^4 = \rho p^2 h/D, \quad R = h^2/12$$

The expression for ψ in terms of w is given by

$$\psi = (R\delta_0^4 - S^{-1})^{-1}\frac{d}{dr}\left[\frac{d^2}{dr^2} + \frac{1}{r}\frac{d}{dr} + (S\delta_0^4 + S^{-1})\right]w \quad [6]$$

Solution of Equations of Motion

Equation [5] may be solved for w by noting that

$$w = w_1 + w_2 \quad [7]$$

where w_1 and w_2 satisfy, respectively, the equations

$$\left(\frac{d^2}{dr^2} + \frac{1}{r}\frac{d}{dr} + \delta_i^2\right)w_i = 0, \quad i = 1, 2 \quad [8]$$

Hence, the shear displacement ψ is found to be

$$\psi = (\sigma_1 - 1)\frac{dw_1}{dr} + (\sigma_2 - 1)\frac{dw_2}{dr} \quad [9]$$

where

$$\sigma_1, \sigma_2 = (\delta_2^2, \delta_1^2)(R\delta_0^4 - S^{-1})^{-1}$$

Both of Equations [8] are Bessel equations of zero order. In dealing with a solid disk, their appropriate solutions are

$$\left.\begin{aligned} w_1 &= A_1 J_0(\delta_1 r) \\ w_2 &= A_2 J_0(\delta_2 r) \end{aligned}\right\} \quad [10]$$

where A_1, A_2 are arbitrary constants.

For a plate with a traction-free edge, the boundary conditions are

$$M_r = Q_r = 0 \quad \text{at } r = a \quad [11]$$

where a is the radius of the plate.

Inserting Equations [1], [9], and [10] in Equations [11], the boundary conditions become

$$\left.\begin{aligned} (\sigma_1 - 1)[\delta_1^2 a^2 J_0''(\delta_1 a) + \nu\delta_1 a J_0'(\delta_1 a)]A_1 & \\ + (\sigma_2 - 1)[\delta_2^2 a^2 J_0''(\delta_2 a) + \nu\delta_2 a J_0'(\delta_2 a)]A_2 &= 0 \\ A_1\sigma_1\delta_1 J_0'(\delta_1 a) + A_2\sigma_2\delta_2 J_0'(\delta_2 a) &= 0 \end{aligned}\right\} \quad [12]$$

where primes indicate differentiation with respect to the argument.

Frequency Equation

The secular equation, governing the frequency, is obtained by setting the determinant of Equations [12] equal to zero. The resulting equation may be written in the form

$$\frac{g - \beta^2}{1 + g}\gamma\Gamma_1 + \frac{1 - g\beta^2}{1 + g}\beta\gamma\Gamma_2 - (1 - \nu)(1 - \beta^2) = 0 \quad [13a]$$

where

$$\beta = \delta_2/\delta_1, \quad \gamma = \delta_1 a, \quad g = R/S$$

$$\Gamma_1 = J_0(\gamma)/J_1(\gamma), \quad \Gamma_2 = J_0(\beta\gamma)/J_1(\beta\gamma)$$

It may be observed that g is a material constant, depending only on Poisson's ratio, while the remaining functions in Equation [13a] depend on the frequency.

Since δ_1 is real for all positive values of the frequency p, while δ_2 is real or imaginary, depending on whether p is greater or less than $\bar{p} = \pi(\mu/\rho)^{1/2}/h$, β will be real or imaginary according as $p \gtrless \bar{p}$.[7] For the range $p < \bar{p}$, if we let $\beta = i\beta_1$, the frequency equation may be transformed to the more convenient form

$$\frac{g + \beta_1^2}{1 + g}\gamma\Gamma_1 + \frac{1 + g\beta_1^2}{1 + g}\beta_1\gamma G_2 - (1 - \nu)(1 + \beta_1^2) = 0 \quad [13b]$$

where $G_2 = I_0(\beta_1\gamma)/I_1(\beta_1\gamma)$, and $I_0(x)$, $I_1(x)$ are modified Bessel functions of the first kind.

We may find an explicit formula for the frequency by means of the relation $\beta = \delta_2/\delta_1$ and the expressions for δ_1 and δ_2 which immediately follow Equation [5]. Thus

$$\left.\begin{aligned} p/\bar{p} &= [1 - \beta^2(1 + g)^2/g\,(1 + \beta^2)^2]^{-1/2}, \quad p > \bar{p} \\ p/\bar{p} &= [1 + \beta_1^2(1 + g)^2/g(1 - \beta_1^2)^2]^{-1/2}, \quad p < \bar{p} \end{aligned}\right\} \quad [14]$$

In addition, from the relation $\gamma = \delta_1 a$, we find

$$\left.\begin{aligned} d/h &= \gamma(\bar{p}/p)[(1 + \beta^2)/3(1 + g)]^{1/2}, \quad p > \bar{p} \\ d/h &= \gamma(\bar{p}/p)[(1 - \beta_1^2)/3(1 + g)]^{1/2}, \quad p < \bar{p} \end{aligned}\right\} \quad [15]$$

where d is the diameter of the disk.

The complete solution of the problem is contained in Equations [13], [14], and [15]. With the value of ν specified for the material of the plate, a choice of β or β_1 determines $p/\bar{p}$ by Equations [14] and yields an infinite set of roots γ of Equations [13]. For the chosen β or β_1 each of these roots furnishes a ratio d/h through Equations [15], so that there results an infinite set of values of d/h corresponding to each value of $p/\bar{p}$.

Frequency Spectrum

The resonant frequencies of axially symmetric flexural vibrations of a circular plate with free boundaries, as predicted by the present theory, are shown, for the range $0.80 \leq p/\bar{p} \leq 1.12$, in Fig. 1. The computations were made for $g = 0.283$, corresponding to $\nu = 0.312$. These curves are characteristic of a frequency spectrum arising from the coupling of two infinite systems.[8] The first system consists of the flexural modes, whose frequency spectrum may be visualized as an extension, into the region $p > \bar{p}$, of the curves in the region $p < \bar{p}$. The second system is comprised of the fundamental thickness-shear mode and its overtones. Their spectrum may be discerned as formed by the loci of the flat, nearly horizontal portions of the resonance curves in the region above the thickness-shear frequency. They are, clearly, asymptotic to this frequency.

Comparison With Classical Theory

In order to observe the effect, on the resonant frequencies, of the coupling with thickness-shear motion, it is necessary to make a comparison with the corresponding frequency spectrum obtained from the classical theory, in which this effect is absent. Here, the relation between frequency and diameter to thickness ratio is,[9] in the present notation

$$d/h = \gamma(\bar{p}/p)^{1/2}[8/3\pi^2(1 - \nu)]^{1/4} \quad [16]$$

where the values of γ are given by the roots of the secular equation[10]

$$\gamma(\Gamma_1 + G_1) - 2(1 - \nu) = 0 \quad [17]$$

[7] $\bar{p}$ is the frequency of the first thickness-shear mode of an infinite plate.

[8] See, for example, Fig. 1 of reference (3).

[9] Reference (5), article 217, equations [1] and [4].

[10] Reference (5), article 219, equation [2].

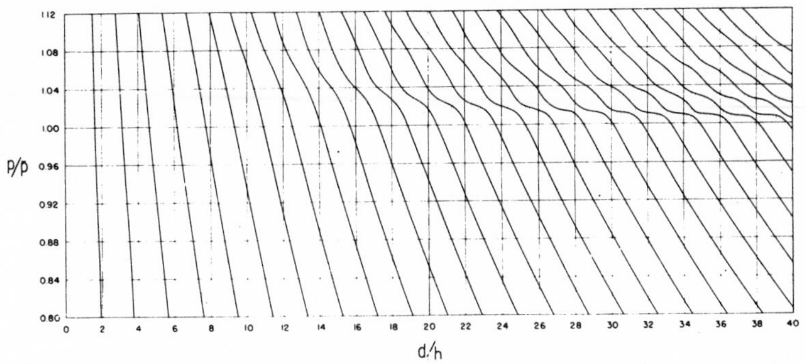

FIG. 1 FREQUENCY SPECTRUM OF AXIALLY SYMMETRIC VIBRATIONS OF A FREE, CIRCULAR DISK, ILLUSTRATING RESULTS OF COUPLING OF FLEXURE AND THICKNESS SHEAR

(d/h = ratio of diameter to thickness; $p/\bar{p}$ = ratio of resonant frequency to frequency of pure thickness-shear vibration of an infinite plate of thickness h; ν = 0.312.)

in which

$$G_1 = I_0(\gamma)/I_1(\gamma)$$

The resulting curves, for $\nu = 0.312$ (corresponding to $g = 0.283$), are shown in Fig. 2.

The presence of the thickness-shear mode is represented, in the present theory, by the rotatory-inertia coefficient R and the shear-deformation coefficient S. (In the sequel, this theory is referred to as RS and the classical theory as C.) It may be verified that the secular equation of RS (Equation [13]) degenerates to that of C (Equation (17]) for the limiting case $R = S = 0$ (so that $\delta_1^2, \delta_2^2 = \pm\delta_0^2$). Hence the curves in Fig. 1, when extended down to $p/\bar{p} = 0$, approach those of Fig. 2 asymptotically. The suppression of shear deformation in C serves as a constraint which raises the frequencies, in that theory, above those of RS. As a result, for a given plate, RS reveals that there are many more resonances, in a given frequency range, than are predicted by C. For example, for $d/h = 40$, there are 25 resonances in the range $0 < p/\bar{p} < 1$, whereas C predicts only 16.

Furthermore, in designing a plate to resonate at a certain frequency, in its fundamental mode, one would be led to choose too large a plate on the basis of C. For example, for $p/\bar{p} = 0.05$, the diameter would be in excess by 2.61 per cent. This discrepancy increases with the ratio of plate thickness to wave length of mode and, hence, with the order of the mode. For example, if the second mode of the plate is to vibrate at $p/\bar{p} = 0.05$, the error in diameter chosen on the basis of C would be 2.86 per cent. The discrepancies are more pronounced at higher frequencies. Thus, the error in diameter of a plate designed, according to C, to resonate at $p/\bar{p} = 0.1$, in the first mode, is 5.33 per cent and, in the second mode, 5.79 per cent.

The most striking difference between the two theories occurs at frequencies above that of the fundamental thickness-shear mode, as may be seen from Figs. 1 and 2.

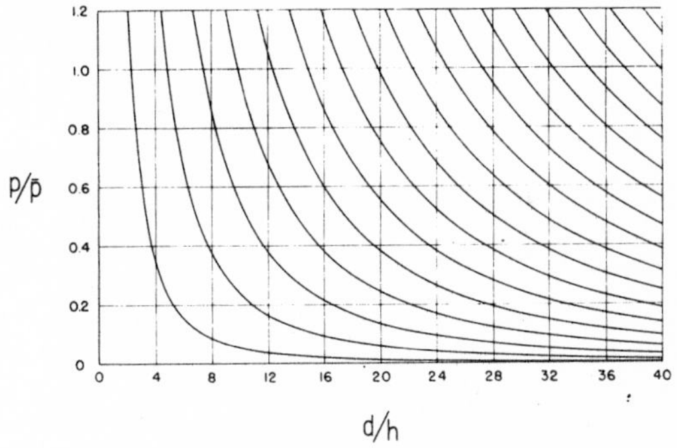

FIG. 2 FREQUENCY SPECTRUM OF AXIALLY SYMMETRIC VIBRATIONS OF A FREE, CIRCULAR DISK ACCORDING TO THE CLASSICAL (LAGRANGE) THEORY OF FLEXURE OF PLATES (ν = 0.312)

BIBLIOGRAPHY

1 "Mémoire sur l'équilibre et le mouvement des corps élastiques," by S. D. Poisson, Paris, Mémoires de l'Académie, vol. 8, 1829; see also "A History of the Theory of Elasticity," by I. Todhunter and K. Pearson, Cambridge University Press, England, 1886, vol. 1, p. 262f.

2 "Influence of Rotatory Inertia and Shear on Flexural Motions of Isotropic, Elastic Plates," by R. D. Mindlin, JOURNAL OF APPLIED MECHANICS, Trans. ASME, vol. 73, 1951, pp. 31–38.

3 "Thickness-Shear and Flexural Vibrations of Crystal Plates," by R. D. Mindlin, *Journal of Applied Physics*, vol. 22, 1951, pp. 316–323.

4 "Thickness-Shear and Flexural Vibrations of an Isotropic, Elastic, Circular Disk," by R. D. Mindlin and H. Deresiewicz, Technical Report No. 11, Office of Naval Research Contract Nonr-266(09), December, 1953.

5 "Theory of Sound," by Lord Rayleigh, Dover Publications, New York, N. Y., 1944.

12

Reprinted from *J. Appl. Mech.*, **26**, 561–569 (1959)

R. D. MINDLIN

Professor of Civil Engineering, Columbia University, New York, N. Y. Mem. ASME

M. A. MEDICK[2]

Senior Staff Scientist, Avco Research and Advanced Development Division, Avco Manufacturing Company, Lawrence, Mass.

Extensional Vibrations of Elastic Plates[1]

A system of approximate, two-dimensional equations of extensional motion of isotropic, elastic plates is derived. The equations take into account the coupling between extensional, symmetric thickness-stretch and symmetric thickness-shear modes and also include two face-shear modes. The spectrum of frequencies for real, imaginary, and complex wave numbers in an infinite plate is explored in detail and compared with the corresponding solution of the three-dimensional equations.

In a previous paper [1][3] a derivation was given of approximate equations of motion that take into account the coupling between face and thickness modes of extensional vibration of elastic plates. The thickness motion considered was that of the lowest thickness-stretch mode, in which the displacement is normal to the middle plane of the plate and the middle plane is the nodal plane. The frequency of this mode depends on Poisson's ratio (ν); the frequency increasing with increasing ν. When $\nu > 1/3$, the frequency of the lowest thickness-stretch mode is higher than the frequency, independent of ν, of the lowest, symmetric thickness-shear mode. In the latter, the displacement is also unidirectional but it is parallel to the middle plane of the plate and there are two nodal planes symmetrically disposed with respect to the middle plane. In the range of Poisson's ratios commonly encountered, both thickness modes can couple with the face modes and with each other. This circumstance has a marked influence on phase and group velocities of waves and on the frequencies and shapes of high-frequency vibrational modes. In the present paper, a derivation is given of approximate equations of motion which include the effects of both thickness modes.

The approximate, two-dimensional equations are deduced from the three-dimensional equations of elasticity by a procedure based on the series-expansion methods of Poisson [2] and Cauchy [3], and the integral method of Kirchhoff [4]. A detailed exposition of the procedure, using a power series, and its application to approximations of orders zero and one are given elsewhere [5]. In approximations of the second and higher orders, awkward mathematical forms are encountered due to the lack of orthogonality of the terms of a power series. At the suggestion of W. Prager, an expansion in a series of Legendre polynomials was studied. In that case, although similar awkward forms appear as a result of the more complicated formula for the derivative, they do not occur, for the most part, until the terms of the third order are reached. Hence the expansion in a series of Legendre polynomials is a convenient one on which to base the present second-order approximation. The method of derivation and the resulting equations (with the inertia terms omitted) are closely related to E. Reissner's theory [6] of three-dimensional corrections for the equations of generalized plane stress.

The expansion in a series of Legendre polynomials of the thickness co-ordinate, followed by an integration across the thickness, converts the three-dimensional equations of elasticity into an infinite series of two-dimensional equations which then are truncated to produce the approximate equations. The full series expressions of displacement, strain, stress, energies, and equations of motion, in conjunction with an understanding of Rayleigh's [7] exact solution of the problem of vibrations of an infinite plate, are of aid in deciding what to include in various orders of approximation and in understanding the implications of what is discarded and what retained. The series expressions and Rayleigh's exact solution also supply both the motivation and the necessary data for making adjustments, of terms that are left after the truncation, in addition to an adjustment of the type made by Poisson in establishing the zero-order equations.

The adjustment analogous to Poisson's serves to uncouple the modes retained from the higher ones without seriously affecting the behavior of the lower ones. The additional adjustments are made to improve the match between the frequency spectra of an infinite plate as obtained from the approximate and exact equations. This is accomplished by the introduction of coefficients analogous to the shear coefficient in the Timoshenko beam equations [8] and the analogous equations for plates [9]. In the latter paper [9] it was shown how the shear coefficient may be chosen to effect a perfect match in one or another part of the spectrum depending upon the frequency range and mode of greatest interest in a particular application of the approximate equations. In the present case, four such coefficients are introduced and, owing to the complexity of the spectrum, several possible combinations of matching points present themselves. The choice is made here to do all the matching at zero wave number. The range of wave numbers and frequencies over which the match remains good is a measure of both the usefulness of the approximate equations and their range of applicability. In this range, solutions of the approximate equations, in the case of finite plates, may be expected to be reliable inasmuch as these solutions are composed essentially of combinations of the modes and overtones of the infinite plate.

When both the symmetric thickness-shear and thickness-stretch deformations are taken into account, important properties of the frequency spectrum contained in the exact theory are reproduced in the resulting approximate equations of the second order, whereas they do not appear in the previous approximation of the first order [1]. These properties of the exact frequency spectrum include the imaginary loop discovered by Aggarwal and Shaw [10]; the anomalous behavior of the second and third branches

---

[1] This investigation was supported by the Office of Naval Research and the U. S. Army Signal Engineering Laboratories.

[2] Formerly Research Assistant, Department of Civil Engineering and Engineering Mechanics, Columbia University, New York, N. Y.

[3] Numbers in brackets designate References at end of paper.

Presented at the Summer Conference of the Applied Mechanics Division, Troy, N. Y., June 18–20, 1959, of The American Society of Mechanical Engineers.

Discussion of this paper should be addressed to the Secretary, ASME, 29 West 39th Street, New York, N. Y., and will be accepted until January 10, 1960, for publication at a later date. Discussion received after the closing date will be returned.

Note: Statements and opinions advanced in papers are to be understood as individual expressions of their authors and not those of the Society. Manuscript received by ASME Applied Mechanics Division, June 25, 1958. Paper No. 59—APM-4.

with variation of Poisson's ratio [5]; the frequency minimum of the second branch [5] with its associated zero group velocity at a nonzero wave number and phase and group velocities of opposite sign at smaller wave numbers, as described by Tolstoy and Usdin [11]; and, finally, a pair of complex branches which account for edge vibrations observed in experiments.

Expansion in Infinite Series

We refer the plate to rectangular co-ordinates x_i $(i = 1, 2, 3)$ with x_1 and x_3 in the middle plane and the faces at $x_2 = \pm b$. The components of displacement u_j $(j = 1, 2, 3)$ are expressed as

$$u_j = \sum_{n=0}^{\infty} P_n(\eta) u_j^{(n)}(x_1, x_3, t), \qquad (1)$$

where $\eta = x_2/b$, the $P_n(\eta)$ are the Legendre polynomials

$$P_0(\eta) = 1, \quad P_1(\eta) = \eta, \quad P_2(\eta) = (3\eta^2 - 1)/2, \ldots$$

$$\ldots P_n(\eta) = \frac{1}{2^n n!} \frac{d^n(\eta^2 - 1)^n}{d\eta^n}, \ldots$$

and the $u_j^{(n)}$, it is to be noted, are functions of the co-ordinates x_1, x_3, and the time t, only. The $u_j^{(n)}$ are the amplitudes of polynomial distributions of displacements across the thickness of the plate. For convenience, however, they will be referred to as displacements of order n or, simply, as displacements.

Stress Equations of Motion. The series expression for the displacement is substituted in the equation

$$\int_V (\tau_{ij,i} - \rho \ddot{u}_j)\delta u_j dV = 0, \qquad (2)$$

which is obtained from the variational equation of motion.[4] In Equation (2) the integration is over the volume V of the plate; the τ_{ij} are the components of stress; and the summation convention for repeated indexes is employed, as are the comma notation for differentiation with respect to the co-ordinates x_i and the dot notation for differentiation with respect to time.

When the integration with respect to η, from -1 to $+1$, is performed in Equation (2), the result is

$$\int_A \sum_{n=0}^{\infty} (b\tau_{ij,i}^{(n)} - \sum_{m=1,3\ldots}^{n} D_{mn}\tau_{2j}^{(n-m)} + F_j^{(n)} - \rho b C_n \ddot{u}_j^{(n)})\delta u_j^{(n)} dA = 0, \qquad (3)$$

where A is the area of the plate and

$$\tau_{ij}^{(n)} = \int_{-1}^{1} P_n(\eta)\tau_{ij} d\eta, \quad F_j^{(n)} = [P_n(\eta)\tau_{2j}]_{-1}^{1},$$
$$D_{mn} = 2(n - m) + 1, \quad C_n = 2/(2n + 1). \qquad (4)$$

The $\tau_{ij}^{(n)}$ and $F_j^{(n)}$ are defined as the nth-order components of stress and face traction, respectively; while the constants D_{mn} and C_n arise from the operations

$$\frac{dP_n}{d\eta} = \sum_{m=1,3\ldots}^{n} D_{mn}P_{n-m}, \quad \int_{-1}^{1} P_m P_n d\eta = \begin{cases} 0, & m \neq n, \\ C_n, & m = n. \end{cases} \qquad (5)$$

The appearance of $F_j^{(n)}$ and $\tau_{2j}^{(n-m)}$, in Equation (3), follows from an integration, by parts, of the terms in Equation (2) that contain $\partial/\partial x_2$.

Since Equation (3) must hold for all A and arbitrary $\delta u_j^{(n)}$, the quantity in parentheses must vanish and we arrive at the stress equations of motion of order n:

$$b\tau_{ij,i}^{(n)} - \sum_{m=1,3\ldots}^{n} D_{mn}\tau_{2j}^{(n-m)} + F_j^{(n)} = \rho b C_n \ddot{u}_j^{(n)}. \qquad (6)$$

In the analogous equations of motion obtained from an expansion in power series,[5] an infinite series appears on the right-hand side and no series appears on the left-hand side. However, the series in Equation (6) contributes more than one term only for $n > 2$.

Strain. In the three-dimensional theory, the components of strain ϵ_{ij} are expressed in terms of the components of displacement by

$$2\epsilon_{ij} = u_{i,j} + u_{j,i}. \qquad (7)$$

Inserting the series expansion from Equation (1) and using the formula for the derivative from Equation (5), we find

$$2\epsilon_{ij} = \sum_{n=0}^{\infty} \sum_{m=1,3\ldots}^{n} [(u_{i,j}^{(n)} + u_{j,i}^{(n)})P_n + (\delta_{2j}u_i^{(n)} + \delta_{2i}u_j^{(n)})b^{-1}D_{mn}P_{n-m}], \qquad (8)$$

where δ_{2j} is the Kronecker symbol δ_{ij} with $i = 2$.

In order to define components of strain of order n, the summand in Equation (8) must be expressed as the product of P_n and a function independent of x_2. Considering the double sum as a triangular array and interchanging the order of summation of columns and rows, we find

$$\epsilon_{ij} = \sum_{n=0}^{\infty} P_n \epsilon_{ij}^{(n)} \qquad (9)$$

where the $\epsilon_{ij}^{(n)}$, defined as the components of strain of order n, are given by

$$2\epsilon_{ij}^{(n)} = u_{i,j}^{(n)} + u_{j,i}^{(n)} + b^{-1}(2n + 1) \sum_{m=1,3\ldots}^{\infty} (\delta_{2j}u_i^{(m+n)} + \delta_{2i}u_j^{(m+n)}). \qquad (10)$$

In the analogous expression obtained by a power-series expansion[6] there is no sum over m. However, the additional terms in Equation (10) do not appear until $m = 3$.

Stress-Strain Relations. The relations between the $\tau_{ij}^{(n)}$ and the $\epsilon_{ij}^{(n)}$ may be obtained by inserting the three-dimensional expressions

$$\tau_{ij} = c_{ijkl}\epsilon_{kl} = c_{ijkl} \sum_{n=0}^{\infty} P_n \epsilon_{kl}^{(n)} \qquad (11)$$

in the first of Equations (4). After performing the integration, we find

$$\tau_{ij}^{(n)} = C_n c_{ijkl} \epsilon_{kl}^{(n)}. \qquad (12)$$

For later use, it is convenient to have Equation (12) expressed in the reduced indicial notation, in which double indexes ranging from 1 to 3 are replaced by single indexes, ranging from 1 to 6, as follows:

$$\tau_1 = \tau_{11}, \quad \tau_4 = \tau_{23}, \quad \epsilon_1 = \epsilon_{11}, \quad \epsilon_4 = 2\epsilon_{23},$$
$$\tau_2 = \tau_{22}, \quad \tau_5 = \tau_{31}, \quad \epsilon_2 = \epsilon_{22}, \quad \epsilon_5 = 2\epsilon_{31},$$
$$\tau_3 = \tau_{33}, \quad \tau_6 = \tau_{12}; \quad \epsilon_3 = \epsilon_{33}, \quad \epsilon_6 = 2\epsilon_{12}.$$

Then Equation (12) becomes

$$\tau_p^{(n)} = C_n c_{pq} \epsilon_q^{(n)}; \quad p, q = 1, 2, \ldots 6, \; c_{pq} = c_{qp}. \qquad (13)$$

[4] Reference [12], p. 167.

[5] Reference [5], p. 3.04.

[6] Ibid., p. 3.08.

Energy Densities. Using the strain energy density U given in the three-dimensional theory by

$$2U = c_{ijkl}\epsilon_{ij}\epsilon_{kl} = c_{pq}\epsilon_p\epsilon_q, \tag{14}$$

we define a plate-strain energy density

$$\bar{U} = \int_{-1}^{1} U d\eta \tag{15}$$

and find, with the aid of Equations (14), (9), and the second of (5),

$$2\bar{U} = c_{pq}\sum_{n=0}^{\infty} C_n\epsilon_p^{(n)}\epsilon_q^{(n)}. \tag{16}$$

We also note that
$$2\bar{U} = \sum_{n=0}^{\infty} \tau_p^{(n)}\epsilon_p^{(n)}, \tag{17}$$

$$\tau_p^{(n)} = \partial\bar{U}/\partial\epsilon_p^{(n)}. \tag{18}$$

Similarly, using the kinetic energy density K as given in the three-dimensional theory by

$$2K = \rho\dot{u}_j\dot{u}_j, \tag{19}$$

we define a kinetic energy density of the plate by

$$2\bar{K} = 2\int_{-1}^{1} K d\eta = \rho\sum_{n=0}^{\infty} C_n\dot{u}_j^{(n)}\dot{u}_j^{(n)}. \tag{20}$$

Extensional Vibrations of Isotropic Plates

When the plate is isotropic, motions symmetric (extensional) and antisymmetric (flexural), with respect to the middle plane, may be considered separately. In the case of the former, only those components of displacement $u_j^{(n)}$ are retained for which $j + n$ is odd. As a result, only those components of stress $\tau_{ij}^{(n)}$ and strain $\epsilon_{ij}^{(n)}$ appear for which $i + j + n$ is even.

The stress-equations of motion (6) then are

$$\begin{aligned}
&\frac{\partial\tau_1^{(0)}}{\partial x_1} + \frac{\partial\tau_5^{(0)}}{\partial x_3} + \frac{F_1^{(0)}}{b} = 2\rho\frac{\partial^2 u_1^{(0)}}{\partial t^2}\\
&\frac{\partial\tau_5^{(0)}}{\partial x_1} + \frac{\partial\tau_3^{(0)}}{\partial x_3} + \frac{F_3^{(0)}}{b} = 2\rho\frac{\partial^2 u_3^{(0)}}{\partial t^2}\\
&\frac{\partial\tau_6^{(1)}}{\partial x_1} + \frac{\partial\tau_4^{(1)}}{\partial x_3} - \frac{\tau_2^{(0)}}{b} + \frac{F_2^{(1)}}{b} = \frac{2\rho}{3}\frac{\partial^2 u_2^{(1)}}{\partial t^2}\\
&\frac{\partial\tau_1^{(2)}}{\partial x_1} + \frac{\partial\tau_5^{(2)}}{\partial x_3} - \frac{3\tau_6^{(1)}}{b} + \frac{F_1^{(2)}}{b} = \frac{2\rho}{5}\frac{\partial^2 u_1^{(2)}}{\partial t^2}\\
&\frac{\partial\tau_5^{(2)}}{\partial x_1} + \frac{\partial\tau_3^{(2)}}{\partial x_3} - \frac{3\tau_4^{(1)}}{b} + \frac{F_3^{(2)}}{b} = \frac{2\rho}{5}\frac{\partial^2 u_3^{(2)}}{\partial t^2}\\
&\frac{\partial\tau_6^{(3)}}{\partial x_1} + \frac{\partial\tau_4^{(3)}}{\partial x_3} - \frac{5\tau_2^{(2)}}{b} - \frac{\tau_2^{(0)}}{b} + \frac{F_2^{(3)}}{b} = \frac{2\rho}{7}\frac{\partial^2 u_2^{(3)}}{\partial t^2}\\
&\cdot \qquad \cdot\\
&\cdot \qquad \cdot\\
&\cdot \qquad \cdot
\end{aligned} \tag{21}$$

The components of strain that remain, in Equations (10), are

$$\begin{aligned}
\epsilon_1^{(0)} &= \partial u_1^{(0)}/\partial x_1\\
\epsilon_2^{(0)} &= (u_2^{(1)} + u_2^{(3)} + \ldots)/b\\
\epsilon_3^{(0)} &= \partial u_3^{(0)}/\partial x_3\\
\epsilon_5^{(0)} &= \partial u_3^{(0)}/\partial x_1 + \partial u_1^{(0)}/\partial x_3\\
\epsilon_4^{(1)} &= \partial u_2^{(1)}/\partial x_3 + 3(u_3^{(2)} + u_3^{(4)} + \ldots)/b\\
\epsilon_6^{(1)} &= \partial u_2^{(1)}/\partial x_1 + 3(u_1^{(2)} + u_1^{(4)} + \ldots)/b\\
\epsilon_1^{(2)} &= \partial u_1^{(2)}/\partial x_1\\
\epsilon_2^{(2)} &= 5(u_2^{(3)} + u_2^{(5)} + \ldots)/b\\
\epsilon_3^{(2)} &= \partial u_3^{(2)}/\partial x_3\\
\epsilon_5^{(2)} &= \partial u_3^{(2)}/\partial x_1 + \partial u_1^{(2)}/\partial x_3\\
\epsilon_4^{(3)} &= \partial u_2^{(3)}/\partial x_3 + 7(u_3^{(4)} + u_3^{(6)} + \ldots)/b\\
\epsilon_6^{(3)} &= \partial u_2^{(3)}/\partial x_1 + 7(u_1^{(4)} + u_1^{(6)} + \ldots)/b\\
&\cdot \qquad \cdot\\
&\cdot \qquad \cdot\\
&\cdot \qquad \cdot
\end{aligned} \tag{22}$$

The stress-strain relations (13) reduce to

$$\begin{aligned}
\tau_1^{(0)} &= 2[(\lambda + 2\mu)\epsilon_1^{(0)} + \lambda(\epsilon_2^{(0)} + \epsilon_3^{(0)})]\\
\tau_2^{(0)} &= 2[(\lambda + 2\mu)\epsilon_2^{(0)} + \lambda(\epsilon_3^{(0)} + \epsilon_1^{(0)})]\\
\tau_3^{(0)} &= 2[(\lambda + 2\mu)\epsilon_3^{(0)} + \lambda(\epsilon_1^{(0)} + \epsilon_2^{(0)})]\\
\tau_5^{(0)} &= 2\mu\epsilon_5^{(0)}\\
\tau_4^{(1)} &= 2\mu\epsilon_4^{(1)}/3\\
\tau_6^{(1)} &= 2\mu\epsilon_6^{(1)}/3\\
\tau_1^{(2)} &= 2[(\lambda + 2\mu)\epsilon_1^{(2)} + \lambda(\epsilon_2^{(2)} + \epsilon_3^{(2)})]/5\\
\tau_2^{(2)} &= 2[(\lambda + 2\mu)\epsilon_2^{(2)} + \lambda(\epsilon_3^{(2)} + \epsilon_1^{(2)})]/5\\
\tau_3^{(2)} &= 2[(\lambda + 2\mu)\epsilon_3^{(2)} + \lambda(\epsilon_1^{(2)} + \epsilon_2^{(2)})]/5\\
\tau_5^{(2)} &= 2\mu\epsilon_5^{(2)}/5\\
\tau_4^{(3)} &= 2\mu\epsilon_4^{(3)}/7\\
\tau_6^{(3)} &= 2\mu\epsilon_6^{(3)}/7\\
&\cdot \qquad \cdot\\
&\cdot \qquad \cdot\\
&\cdot \qquad \cdot
\end{aligned} \tag{23}$$

where λ and μ are Lamé's constants.

The strain energy density, in the form given in Equation (17), becomes

$$\begin{aligned}
2\bar{U} = {} & \tau_1^{(0)}\epsilon_1^{(0)} + \tau_3^{(0)}\epsilon_3^{(0)} + \tau_5^{(0)}\epsilon_5^{(0)}\\
&+ \tau_2^{(0)}\epsilon_2^{(0)} + \tau_4^{(1)}\epsilon_4^{(1)} + \tau_6^{(1)}\epsilon_6^{(1)}\\
&+ \tau_1^{(2)}\epsilon_1^{(2)} + \tau_3^{(2)}\epsilon_3^{(2)} + \tau_5^{(2)}\epsilon_5^{(2)}\\
&+ \tau_2^{(2)}\epsilon_2^{(2)} + \tau_4^{(3)}\epsilon_4^{(3)} + \tau_6^{(3)}\epsilon_6^{(3)}\\
&+ \ldots
\end{aligned} \tag{24}$$

and, finally, the kinetic energy density (20) is

$$\begin{aligned}
\bar{K} = {} & \rho(\dot{u}_1^{(0)}\dot{u}_1^{(0)} + \dot{u}_3^{(0)}\dot{u}_3^{(0)} + {}^1/_3\dot{u}_2^{(1)}\dot{u}_2^{(1)}\\
&+ {}^1/_5\dot{u}_1^{(2)}\dot{u}_1^{(2)} + {}^1/_5\dot{u}_3^{(2)}\dot{u}_3^{(2)} + {}^1/_7\dot{u}_2^{(3)}\dot{u}_2^{(3)} + \ldots).
\end{aligned} \tag{25}$$

Truncation of Series

We begin by setting

$$\begin{aligned}
u_1^{(n)} = 0, \quad u_3^{(n)} = 0, \quad n > 2,\\
u_2^{(n)} = 0, \quad n > 3.
\end{aligned} \tag{26}$$

This leaves only the components $u_1^{(0)}$, $u_3^{(0)}$, $u_2^{(1)}$, $u_1^{(2)}$, $u_3^{(2)}$, and $u_2^{(3)}$, as illustrated in Fig. 1. The terms $u_1^{(0)}$ and $u_3^{(0)}$ are the amplitudes of uniform distributions that represent the thickness distributions of displacements which occur in low-frequency extensional and shear motions in the plane of the plate; $u_2^{(1)}$ is the amplitude of a linear distribution of displacement which is

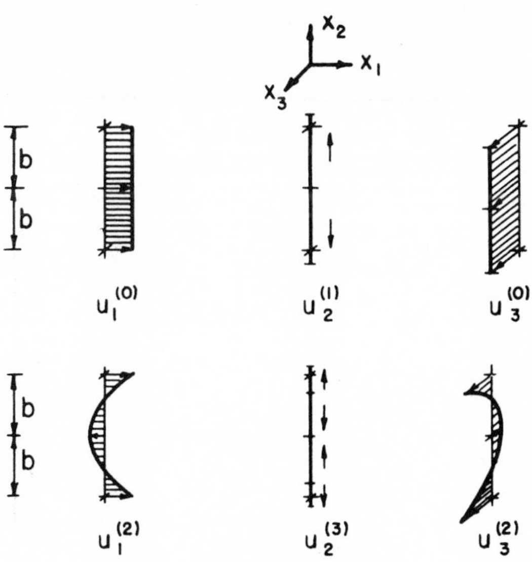

Fig. 1 Components of displacement

an approximation to the exact sinusoidal distribution in the lowest, symmetric, thickness-stretch mode; $u_1^{(2)}$ and $u_3^{(2)}$ are the amplitudes of quadratic distributions of displacements which are approximations to the sinusoidal distributions in the lowest, symmetric, thickness-shear mode, and the face-shear mode of the same order.

The last "displacement" retained ($u_2^{(3)}$) is the amplitude of a cubic distribution which is an approximation to the sinusoidal distribution of displacement in the second, symmetric, thickness-stretch mode. This is a mode of higher order than is to be included in the approximate equations of the second order which we seek. However, $u_2^{(3)}$ produces the second-order strain $\epsilon_2^{(2)}$ [see the eighth of Equations (22)] which, in turn, appears in the second-order stress-strain relations [see the seventh, eighth, and ninth of Equations (23)]. In order to permit the alternating expansion and contraction, through the thickness, which should accompany the stresses $\tau_1^{(2)}$ and $\tau_3^{(2)}$ (by coupling through Poisson's ratio) and at the same time avoid coupling with the undesired higher mode, the thickness stress $\tau_2^{(2)}$ and the velocity $\dot{u}_2^{(3)}$ are set equal to zero. The eighth of Equations (23) is then used to express $\epsilon_2^{(2)}$ in terms of $\epsilon_1^{(2)}$ and $\epsilon_3^{(2)}$:

$$\epsilon_2^{(2)} = -\lambda(\epsilon_3^{(2)} + \epsilon_1^{(2)})/(\lambda + 2\mu) \qquad (27)$$

and this result is substituted in the expressions for $\tau_1^{(2)}$ and $\tau_3^{(2)}$ to obtain

$$\begin{aligned} \tau_1^{(2)} &= 2E'(\epsilon_1^{(2)} + \nu\epsilon_3^{(2)})/5, \\ \tau_3^{(2)} &= 2E'(\epsilon_3^{(2)} + \nu\epsilon_1^{(2)})/5, \end{aligned} \qquad (28)$$

where $\nu = \lambda/2(\lambda + \mu)$ is Poisson's ratio and

$$E' = 4\mu(\lambda + \mu)/(\lambda + 2\mu) = E/(1 - \nu^2), \qquad (29)$$

in which E is Young's modulus. In addition, the contributions of the stresses $\tau_4^{(3)}$ and $\tau_6^{(3)}$ to the strain energy [see Equation (24)] are neglected in order to destroy coupling with the unwanted higher mode when the displacements vary with x_1 and x_3.

At this stage, the stress equations of motion end with the fifth of Equations (21); the strain-displacement relations end with $\epsilon_6^{(2)}$ in Equations (22); the stress-strain relations end with $\tau_5^{(2)}$ in Equations (23), with the expressions for $\tau_1^{(2)}$, $\tau_2^{(2)}$, and $\tau_3^{(2)}$ replaced by Equations (27) and (28); the strain energy density ends with the third line in Equation (24); and the kinetic energy density ends with the fifth term in Equation (25).

Up to this point, the process of truncating the series expressions and adjusting the remaining terms is similar to the one employed by Poisson [2] to obtain the zero-order equations of extensional motions of isotropic plates,[7] the main difference being that here the process is carried on at a level two orders higher. In Poisson's equations only the zero-order displacements $u_1^{(0)}$ and $u_3^{(0)}$ survive. These are the amplitudes of uniform distributions, across the thickness of the plate, which are good approximations to the nearly uniform distributions found in the lowest extensional mode of the exact theory (at long wave lengths) and are exact for the lowest face-shear mode at all wave lengths. The additional terms $u_2^{(1)}$, $u_1^{(2)}$, and $u_3^{(2)}$, which are now included, are the amplitudes of first and second-degree polynomials and these are not good approximations to the distributions in the thickness-stretch and thickness-shear modes of the exact theory. For example, at infinite wave length along the plate, the exact distributions are sinusoidal across the thickness. It is advisable, therefore, to introduce additional adjustments to compensate, as well as possible, for the omission of the polynomials of higher degrees.

The incorrect distributions of displacements affect the frequencies mainly through the thickness strains and velocities. Accordingly, we make the substitutions

$$\kappa_1\epsilon_2^{(0)} \text{ for } \epsilon_2^{(0)}, \qquad \kappa_3\dot{u}_2^{(1)} \text{ for } \dot{u}_2^{(1)},$$

$$\kappa_2\epsilon_4^{(1)} \text{ for } \epsilon_4^{(1)}, \qquad \kappa_4\dot{u}_1^{(2)} \text{ for } \dot{u}_1^{(2)},$$

$$\kappa_2\epsilon_6^{(1)} \text{ for } \epsilon_6^{(1)}, \qquad \kappa_4\dot{u}_3^{(2)} \text{ for } \dot{u}_3^{(2)}$$

in the strain energy and kinetic energy densities, so that the coefficients κ_r (r = 1, 2, 3, 4) will be available for appropriate adjustments of the equations.

Finally, as a matter of expediency, we omit the term $u_2^{(3)}$ from the strain $\epsilon_2^{(0)}$. This term complicates the equations and may be shown to have little influence on the long wave-length end of the spectrum.

Second-Order Approximation

As a result of the truncations and adjustments described in the preceding section, the variables and equations of the second-order approximation are

Kinetic Energy Density:

$$\bar{K}^{(2)} = \rho\,[\dot{u}_1^{(0)}\dot{u}_1^{(0)} + \dot{u}_3^{(0)}\dot{u}_3^{(0)} + {}^1/_3\,\kappa_3^2\dot{u}_2^{(1)}\dot{u}_2^{(1)} + {}^1/_5\,\kappa_4^2(\dot{u}_1^{(2)}\dot{u}_1^{(2)} + \dot{u}_3^{(2)}\dot{u}_3^{(2)})] \qquad (30)$$

Strain Energy Density:

$$\begin{aligned} \bar{U}^{(2)} = {} & (\lambda + 2\mu)(\epsilon_1^{(0)}\epsilon_1^{(0)} + \kappa_1^2\epsilon_2^{(0)}\epsilon_2^{(0)} + \epsilon_3^{(0)}\epsilon_3^{(0)}) \\ & + 2\lambda(\kappa_1\epsilon_2^{(0)}\epsilon_3^{(0)} + \epsilon_3^{(0)}\epsilon_1^{(0)} + \kappa_1\epsilon_1^{(0)}\epsilon_2^{(0)}) \\ & + \mu\epsilon_5^{(0)}\epsilon_5^{(0)} + \mu\kappa_2^2(\epsilon_4^{(1)}\epsilon_4^{(1)} + \epsilon_6^{(1)}\epsilon_6^{(1)})/3 \\ & + E'(\epsilon_1^{(2)}\epsilon_1^{(2)} + \epsilon_3^{(2)}\epsilon_3^{(2)} + 2\nu\epsilon_1^{(2)}\epsilon_3^{(2)})/5 + \mu\epsilon_5^{(2)}\epsilon_5^{(2)}/5 \end{aligned} \qquad (31)$$

Stress-Strain Relations ($\tau_p^{(n)} = \partial\bar{U}^{(2)}/\partial\epsilon_p^{(n)}$):

$$\tau_1^{(0)} = 2[(\lambda + 2\mu)\epsilon_1^{(0)} + \lambda(\kappa_2\epsilon_2^{(0)} + \epsilon_3^{(0)})]$$

$$\tau_2^{(0)} = 2[\kappa_1^2(\lambda + 2\mu)\epsilon_2^{(0)} + \lambda\kappa_1(\epsilon_3^{(0)} + \epsilon_1^{(0)})]$$

$$\tau_3^{(0)} = 2[(\lambda + 2\mu)\epsilon_3^{(0)} + \lambda(\epsilon_1^{(0)} + \kappa_1\epsilon_2^{(0)})]$$

---

[7] Reference [12], p. 497.

$$
\begin{aligned}
\tau_5^{(0)} &= 2\mu\epsilon_5^{(0)} \\
\tau_4^{(1)} &= 2\mu\kappa_2^2\epsilon_4^{(1)}/3 \\
\tau_6^{(1)} &= 2\mu\kappa_2^2\epsilon_6^{(1)}/3 \\
\tau_1^{(2)} &= 2E'(\epsilon_1^{(2)} + \nu\epsilon_3^{(2)})/5 \\
\tau_3^{(2)} &= 2E'(\epsilon_3^{(2)} + \nu\epsilon_1^{(2)})/5 \\
\tau_5^{(2)} &= 2\mu\epsilon_5^{(2)}/5
\end{aligned}
\tag{32}
$$

(The components of stress are illustrated in Fig 2.)

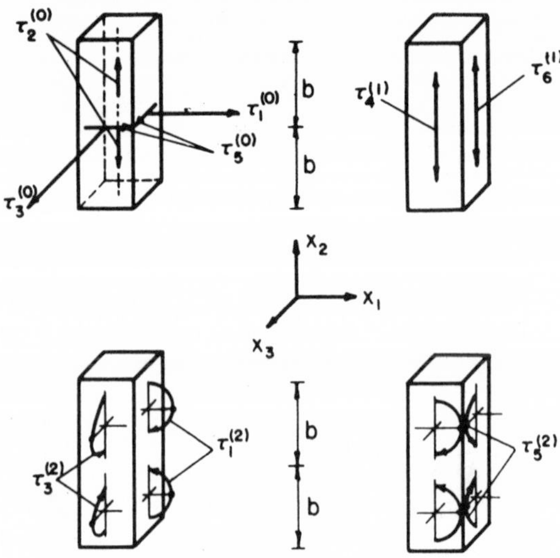

Fig. 2 Components of stress

Strain-Displacement Relations:

$$
\begin{aligned}
\epsilon_1^{(0)} &= \partial u_1^{(0)}/\partial x_1 \\
\epsilon_2^{(0)} &= u_2^{(1)}/b \\
\epsilon_3^{(0)} &= \partial u_3^{(0)}/\partial x_3 \\
\epsilon_5^{(0)} &= \partial u_3^{(0)}/\partial x_1 + \partial u_1^{(0)}/\partial x_3 \\
\epsilon_4^{(1)} &= 3u_3^{(2)}/b + \partial u_2^{(1)}/\partial x_3 \\
\epsilon_6^{(1)} &= \partial u_2^{(1)}/\partial x_1 + 3u_1^{(2)}/b \\
\epsilon_1^{(2)} &= \partial u_1^{(2)}/\partial x_1 \\
\epsilon_3^{(2)} &= \partial u_3^{(2)}/\partial x_3 \\
\epsilon_5^{(2)} &= \partial u_3^{(2)}/\partial x_1 + \partial u_1^{(2)}/\partial x_3
\end{aligned}
\tag{33}
$$

(The components of strain are illustrated in Fig. 3.)

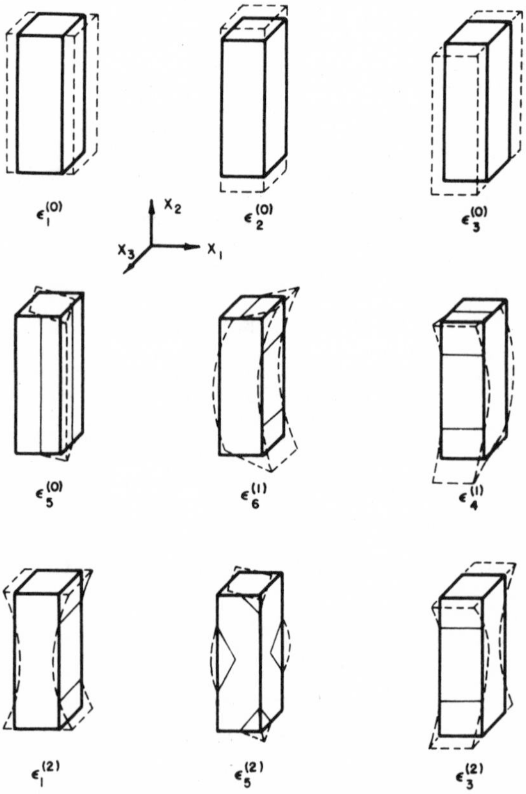

Fig. 3 Components of strain

Stress Equations of Motion:

$$
\begin{aligned}
&\frac{\partial\tau_1^{(0)}}{\partial x_1} + \frac{\partial\tau_5^{(0)}}{\partial x_3} + \frac{F_1^{(0)}}{b} = 2\rho\,\frac{\partial^2 u_1^{(0)}}{\partial t^2} \\
&\frac{\partial\tau_5^{(0)}}{\partial x_1} + \frac{\partial\tau_3^{(0)}}{\partial x_3} + \frac{F_3^{(0)}}{b} = 2\rho\,\frac{\partial^2 u_3^{(0)}}{\partial t^2} \\
&\frac{\partial\tau_6^{(1)}}{\partial x_1} + \frac{\partial\tau_4^{(1)}}{\partial x_3} - \frac{\tau_2^{(0)}}{b} + \frac{F_2^{(1)}}{b} = \frac{2\rho\kappa_3^2}{3}\,\frac{\partial^2 u_2^{(1)}}{\partial t^2} \\
&\frac{\partial\tau_1^{(2)}}{\partial x_1} + \frac{\partial\tau_5^{(2)}}{\partial x_3} - \frac{3\tau_6^{(1)}}{b} + \frac{F_1^{(2)}}{b} = \frac{2\rho\kappa_4^2}{5}\,\frac{\partial^2 u_1^{(2)}}{\partial t^2} \\
&\frac{\partial\tau_5^{(2)}}{\partial x_1} + \frac{\partial\tau_3^{(2)}}{\partial x_3} - \frac{3\tau_4^{(1)}}{b} + \frac{F_3^{(2)}}{b} = \frac{2\rho\kappa_4^2}{5}\,\frac{\partial^2 u_3^{(2)}}{\partial t^2}
\end{aligned}
\tag{34}
$$

Displacement Equations of Motion:

[obtained by substituting (33) in (32) and the result in (34)]:

$$
\begin{aligned}
&\mu\nabla^2 u_1^{(0)} + (\lambda+\mu)\frac{\partial e_0}{\partial x_1} + \frac{\lambda\kappa_1}{b}\frac{\partial u_2^{(1)}}{\partial x_1} + \frac{F_1^{(0)}}{2b} = \rho\,\frac{\partial^2 u_1^{(0)}}{\partial t^2} \\
&\mu\nabla^2 u_3^{(0)} + (\lambda+\mu)\frac{\partial e_0}{\partial x_3} + \frac{\lambda\kappa_1}{b}\frac{\partial u_2^{(1)}}{\partial x_3} + \frac{F_3^{(0)}}{2b} = \rho\,\frac{\partial^2 u_3^{(0)}}{\partial t^2} \\
&\mu\kappa_2^2\nabla^2 u_2^{(1)} - \frac{3\lambda\kappa_1 e_0}{b} - \frac{3\kappa_1^2(\lambda+2\mu)u_2^{(1)}}{b^2} \\
&\qquad + \frac{3\mu\kappa_2^2 e_2}{b} + \frac{3F_2^{(1)}}{2b} = \rho\kappa_3^2\,\frac{\partial^2 u_2^{(1)}}{\partial t^2} \\
&\frac{E'}{2}\left[(1-\nu)\nabla^2 u_1^{(2)} + (1+\nu)\frac{\partial e_2}{\partial x_1}\right] \\
&\qquad - \frac{5\mu\kappa_2^2}{b}\left(\frac{\partial u_2^{(1)}}{\partial x_1} + \frac{3u_1^{(2)}}{b}\right) + \frac{5F_1^{(2)}}{2b} = \rho\kappa_4^2\,\frac{\partial^2 u_1^{(2)}}{\partial t^2} \\
&\frac{E'}{2}\left[(1-\nu)\nabla^2 u_3^{(2)} + (1+\nu)\frac{\partial e_2}{\partial x_3}\right] \\
&\qquad - \frac{5\mu\kappa_2^2}{b}\left(\frac{\partial u_2^{(1)}}{\partial x_3} + \frac{3u_3^{(2)}}{b}\right) + \frac{5F_3^{(2)}}{2b} = \rho\kappa_4^2\,\frac{\partial^2 u_3^{(2)}}{\partial t^2}
\end{aligned}
\tag{35}
$$

where

$$
\nabla^2 = \frac{\partial^2}{\partial x_1^2} + \frac{\partial^2}{\partial x_3^2},\quad e_0 = \frac{\partial u_1^{(0)}}{\partial x_1} + \frac{\partial u_3^{(0)}}{\partial x_3},\quad e_2 = \frac{\partial u_1^{(2)}}{\partial x_1} + \frac{\partial u_3^{(2)}}{\partial x}.
$$

The Coefficients κ_i

The coefficients κ_i are determined from a comparison of properties of the solutions of Equations (35), for the case of straight-crested waves in an infinite plate, with the corresponding properties of the analogous solution of the three-dimensional equations. Taking the wave normal in the direction of x_1, we find that Equations (35) separate into a group of three coupled equations on $u_1^{(0)}$, $u_2^{(1)}$, and $u_1^{(2)}$, and two independent equations: one on $u_3^{(0)}$ and one on $u_3^{(2)}$. The coupled equations govern three "compressional modes" and the remaining two equations govern "face-shear" modes. To consider, first, the group of three, we set $F_j^{(n)} = 0$ and

$$\begin{aligned} u_1^{(0)} &= A \sin \xi x_1 \, e^{i\omega t}, \quad u_3^{(0)} = 0, \\ u_2^{(1)} &= B \cos \xi x_1 \, e^{i\omega t}, \\ u_1^{(2)} &= C \sin \xi x_1 \, e^{i\omega t}, \quad u_3^{(2)} = 0 \end{aligned} \tag{36}$$

in Equations (35) and obtain the secular equation

$$\begin{vmatrix} a_{11} & a_{12} & 0 \\ a_{12} & a_{22} & a_{23} \\ 0 & a_{23} & a_{33} \end{vmatrix} = 0 \tag{37}$$

where

$$\begin{aligned} a_{11} &= k^2 z^2 - \Omega^2 \\ a_{22} &= \kappa_2^2 z^2/3 + 4\kappa_1^2 k^2/\pi^2 - \kappa_3^2 \Omega^2/3 \\ a_{33} &= E' z^2/5\mu + 12\kappa_2^2/\pi^2 - \kappa_4^2\Omega^2/5 \\ a_{12} &= 2\kappa_1(k^2 - 2)z/\pi \\ a_{23} &= -2\kappa_2^2 z/\pi \end{aligned}$$

and

$$\begin{aligned} z &= 2\xi b/\pi, \quad \Omega = \omega/\omega_s, \quad \omega_s = \pi(\mu/\rho)^{1/2}/2b, \\ k^2 &= (\lambda + 2\mu)/\mu = 2(1 - \nu)/(1 - 2\nu). \end{aligned} \tag{38}$$

We also obtain the three sets of amplitude ratios

$$A_i : B_i : C_i = 1 : a_i/z_i : a_i/b_i, \quad i = 1, 2, 3 \tag{39}$$

where

$$\begin{aligned} a_i &= \pi(\Omega^2 - k^2 z_i^2)/2\kappa_1(k^2 - 2) \\ b_i &= \frac{3}{2\pi}\left(\frac{\pi^2\kappa_4^2\Omega^2}{15\kappa_2^2} - 4\right) - \frac{2\pi(k^2 - 1)z_i^2}{5\kappa_2^2 k^2} \end{aligned}$$

and the z_i are the roots of Equation (37).

It may be seen that z, which is proportional to the wave number ξ, is the ratio of the thickness of the plate to the real half wave length along the plate; k is the ratio of the velocity of dilatational waves to the velocity of equivoluminal waves in an infinite medium; and Ω is the ratio of the frequency to the frequency of the lowest antisymmetric thickness-shear mode.

The relation between Ω and z, in Equation (37), should match, as closely as possible, the corresponding relation obtained from the three-dimensional equations. The match is improved, within the framework of the present approximation, by choosing appropriate values for the coefficients κ_i; but, since the κ_i are constants, a perfect match can be made only at one value of z for each of them. Now, large enough z corresponds to frequencies high enough to enter the range of modes that have not been included in the approximate equations. In a plate vibrating at such high frequencies under practical (i.e., not mixed) edge conditions, the high modes would, in general, couple with the ones of lower order. Thus the applicability of the approximate equations to bounded plates is limited to frequencies below the lowest frequency of the lowest neglected mode. There is, then, little advantage to be gained in matching the approximate and exact solutions at short wave lengths (large z) at the expense of a good match at long wave lengths. In fact, we go to the extreme and do all of the matching in the neighborhood of $z = 0$ primarily because of the intricate behavior of the exact solution at long wave lengths and also because this choice results in a reasonably good match out to as short wave lengths as the frequency limitation permits.

When $z = 0$, Equation (37) has the three roots

$$\Omega^2 = 0, \quad 12\kappa_1^2 k^2/\pi^2\kappa_3^2, \quad 60\kappa_2^2/\pi^2\kappa_4^2 \tag{40}$$

corresponding to the amplitude ratios

$$A:B:C = 1:0:0, \quad 0:1:0, \quad 0:0:1$$

respectively. It may be seen that, at infinite wave length, the three compressional modes uncouple to form an extensional, a thickness-stretch, and a thickness-shear mode. The three limiting frequencies are to be compared with the exact values

$$\Omega^2 = 0, \quad k^2, \quad 4 \tag{41}$$

of the cut-off frequencies of the extensional, first thickness-stretch and first symmetric thickness-shear modes in Rayleigh's exact solution [7]. Hence, we set

$$\kappa_1/\kappa_3 = \pi/\sqrt{12}, \quad \kappa_2/\kappa_4 = \pi/\sqrt{15} \tag{42}$$

and these relations between pairs of κ_i are used in what follows.

We proceed to examine the roots of Equation (37) in the neighborhood of the roots given in Equations (40) or (41) (which are now the same).

First, for $z \ll 1$, $\Omega \ll 1$, we find

$$\Omega^2 = 4z^2(k^2 - 1)/k^2, \tag{43}$$

i.e., phase velocity $[E/\rho(1 - \nu^2)]^{1/2}$. This is exactly the result obtained from the three-dimensional theory for the lowest mode at long wave lengths.

Next, for $z \ll 1$, $\Omega = k + \epsilon$, $\epsilon \ll 1$, Equation (37) gives

$$\frac{d\Omega}{dz} = \frac{z}{k^2 - 4}\left(\frac{\pi^2 k\kappa_2^2}{12\kappa_1^2} + \frac{(k^2 - 2)^2(k^2 - 4)}{k^3}\right) \tag{44a}$$

and in the limit, as $k \to 2$ and $z \to 0$,

$$\frac{d\Omega}{dz} = \pm\frac{\pi\kappa_2}{4\kappa_1\sqrt{3}}. \tag{45a}$$

Finally, for $z \ll 1$, $\Omega = 2 + \epsilon$, $\epsilon \ll 1$, we find

$$\frac{d\Omega}{dz} = \frac{\pi^2 z}{6(k^2 - 4)}\left(\frac{4(k^2 - 1)(k^2 - 4)}{5k^2\kappa_2^2} - \frac{\kappa_2^2}{\kappa_1^2}\right) \tag{46a}$$

and in the limit, as $k \to 2$ and $z \to 0$,

$$\frac{d\Omega}{dz} = \mp\frac{\pi\kappa_2}{4\kappa_1\sqrt{3}}. \tag{47a}$$

Equations (44a) to (47a) are to be compared with the exact values[8] given in the following correspondingly numbered equations:

$$\frac{d\Omega}{dz} = \frac{4kz}{\pi}\left(\frac{\pi}{4} + \frac{4}{k^3}\cot\frac{\pi k}{2}\right) \tag{44b}$$

$$\frac{d\Omega}{dz} = \pm\frac{2}{\pi} \tag{45b}$$

$$\frac{d\Omega}{dz} = \frac{z}{\pi}\left(\frac{\pi}{2} - \frac{4}{k}\tan\frac{\pi}{k}\right) \tag{46b}$$

[8] Reference [5], p. 2.43.

$$\frac{d\Omega}{dz} = \mp\frac{2}{\pi} \qquad (47b)$$

Equating (44a) to (44b) and (46a) to (46b), we find

$$\frac{\kappa_2^2}{\kappa_1^2} = \frac{48(k^2-4)}{\pi^3 k^4}\left[\pi(k^2-1) + 4k\cot\frac{\pi k}{2}\right],$$

$$\frac{1}{\kappa_1^2} = \frac{5k^2\kappa_2^2}{4\kappa_1^2(k^2-1)(k^2-4)} \left[\frac{\kappa_2^2}{\kappa_1^2} + \frac{6(k^2-4)}{\pi^3}\left(\frac{\pi}{2} - \frac{4}{k}\tan\frac{\pi}{k}\right)\right]. \qquad (48)$$

We also note, from Equations (48), that

$$\lim_{k\to 2}\frac{\kappa_2^2}{\kappa_1^2} = \frac{192}{\pi^4},$$

$$\lim_{k\to 2}\frac{1}{\kappa_1^2} = \frac{3840(\pi^2-6)}{\pi^8}. \qquad (49)$$

Equations (42) and (48) give the values of the four coefficients κ_i in terms of Poisson's ratio. As a result of these relations, all of the ordinates, slopes, and curvatures of the curves Ω versus z, characterizing the frequency spectrum of the compressional modes of the approximate theory, are identical with those of the exact solution at $z = 0$. Equations (48) and, in fact, all of the equations of this second-order approximation should be used only for Poisson's ratios less than about $^7/_{16}$. Above that value, the frequency of the thickness-stretch mode is so high in the spectrum that, in the exact theory, coupling with the second symmetric thickness-shear mode becomes important and that mode is not included in the present approximation.

Turning, now, to the two face-shear modes, we set $F_j^{(n)} = 0$,

$$u_1^{(0)} = u_2^{(1)} = u_1^{(2)} = 0,$$

$$u_3^{(0)} = A_1 \cos \xi x_1\, e^{i\omega t}, \qquad (50)$$

$$u_3^{(2)} = A_2 \cos \xi x_1\, e^{i\omega t},$$

in Equations (35), and find that $u_3^{(0)}$ and $u_3^{(2)}$ are not coupled and have frequencies given by

$$\Omega^2 = z^2, \quad 4 + z^2/\kappa_4^2 \qquad (51)$$

respectively. These are to be compared with the frequencies

$$\Omega^2 = z^2, \quad 4 + z^2 \qquad (52)$$

of the face-shear modes of orders zero and two obtained from the exact equations. It may be seen that the zero-order face-shear mode is reproduced exactly, in the approximate equations, for all wave lengths. The second order face-shear mode has the correct ordinate (Ω) and slope ($d\Omega/dz$) at $z = 0$ but it does not have the correct curvature because of the presence of the coefficient κ_4. By choosing $\kappa_4 = 1$, this discrepancy could be eliminated, but only at the expense of incorrect behavior of one of the coupled modes.

Frequency Spectrum

The spectrum of frequencies of the five possible modes of vibration of an infinite plate, which the approximate theory contains, is given by the five independent roots of Equations (37) and (51). We consider, first, the three compressional modes.

Equation (37) relates Ω^2 and z^2 but only real, positive values of Ω have physical significance. For real, positive Ω, the roots z^2 may be three real positive or two real positive and one real negative or one real positive and two conjugate complex. Hence the roots

$$z = x + iy$$

may be three real or two real and one imaginary or one real and two conjugate complex. In addition, the character of the spectrum is different according as Poisson's ratio is less than equal to, or greater than $^1/_3$.

We consider the case $\nu < {}^1/_3$ (i.e., $k < 2$) first. Then, when $\Omega > 2$, the three roots of Equation (37) are real ($y = 0$). They are illustrated in Fig. 4(a) (for $\nu = {}^1/_4$, i.e., $k^2 = 3$) by the three curves (full lines) marked ϕ_1, ϕ_2, ϕ_3 in the plane $y = 0$ and the region $\Omega > 2$. As Ω drops below $\Omega = 2$, the largest root ϕ_1 approaches zero with asymptotic behavior given by Equation (43); the intermediate root ϕ_2 approaches a minimum Ω^*, below which there are no real roots other than ϕ_1; the smallest root ϕ_3 approaches a minimum in the real plane $y = 0$ at $\Omega = 2$, $x = 0$ with asymptotic behavior given by Equation (46a) or (46b) which are the same in view of Equations (48). Continuing with the root ϕ_3, as Ω drops below $\Omega = 2$, the root is imaginary ($x = 0$), again with asymptotic behavior given by Equation (46a) or (46b) in the neighborhood of $\Omega = 2$. As Ω approaches k, from above, the root ϕ_3 forms a loop in the imaginary plane ($x = 0$) and approaches $\Omega = k$, $y = 0$ with asymptotic behavior given by Equation (44a) or (44b). As Ω continues to drop below $\Omega = k$, the root ϕ_3 becomes real, with behavior in the neighborhood of $\Omega = k$, $x = 0$ again given by (44a) but now x is negative. Upon further diminution of Ω, the root ϕ_3 approaches a minimum at $\Omega = \Omega^*$ and negative x. This portion of ϕ_3 (i.e., between $\Omega = k$ and $\Omega = \Omega^*$) is identified in Fig. 4(a) by [ϕ_3], where the brackets indicate that it is the reflection in the plane $x = 0$ that is shown. Finally, when $\Omega < \Omega^*$, the roots ϕ_2 and ϕ_3 are conjugate complex ($z = \pm x + iy$). One of them is shown, in Fig. 4(a), as the curve marked ϕ_2; this curve is also [ϕ_3], i.e., the reflection of the conjugate root ϕ_3 in the plane $x = 0$.

As Poisson's ratio approaches $^1/_3$ from below, the frequency of the thickness-stretch mode increases, approaching the frequency of the thickness-shear mode; i.e., at $z = 0$ the intercept $\Omega = k$ approaches $\Omega = 2$. Conjointly, the curvature of ϕ_3 at $\Omega = k$, $z = 0$ approaches negative infinity in the plane $y = 0$ and positive infinity in the plane $x = 0$ while both slopes remain zero; all in accordance with Equation (44a) or (44b). At the same time, the curvature of ϕ_3 at $\Omega = 2$, $z = 0$ approaches positive infinity in the plane $y = 0$ and negative infinity in the plane $x = 0$ while the slopes remain zero: all in accordance with Equation (46a) or (46b). Meanwhile, the imaginary loop shrinks toward the point $\Omega = 2$, $y = 0$.

At $\nu = {}^1/_3$, the thickness-stretch and thickness-shear modes have the same frequency and the slopes of the two branches of ϕ_3 become $2/\pi$ in accordance with Equations (45a) and (47a). This situation is illustrated in Fig. 4(b). Here, again, the portions marked [ϕ_3] are the reflections, in the plane $x = 0$ of the actual branches.

When $\nu > {}^1/_3$ the thickness-stretch mode has a frequency higher than that of the thickness-shear mode. The spectrum [illustrated in Fig. 4(c) for $\nu = {}^2/_5$] has now undergone an important change in that the imaginary branch loops up, from $\Omega = 2$, instead of down.

Since only powers of z^2 occur in Equation (35), there is another set of physically significant roots given by the reflections of the curves of Fig. 4 in the planes $x = 0$ and $y = 0$.

Turning, now, to the face-shear modes, the first root in Equation (51) yields the straight line marked H_0 in Fig. 4. The second face-shear mode gives the roots marked H_2 in the figures; these roots are real for $\Omega > 2$ and imaginary for $\Omega < 2$. As before, there is an additional set of roots given by the reflections of the curves H_0 and H_2 in the planes $x = 0$ and $y = 0$.

The spectra of the corresponding five modes, as computed from

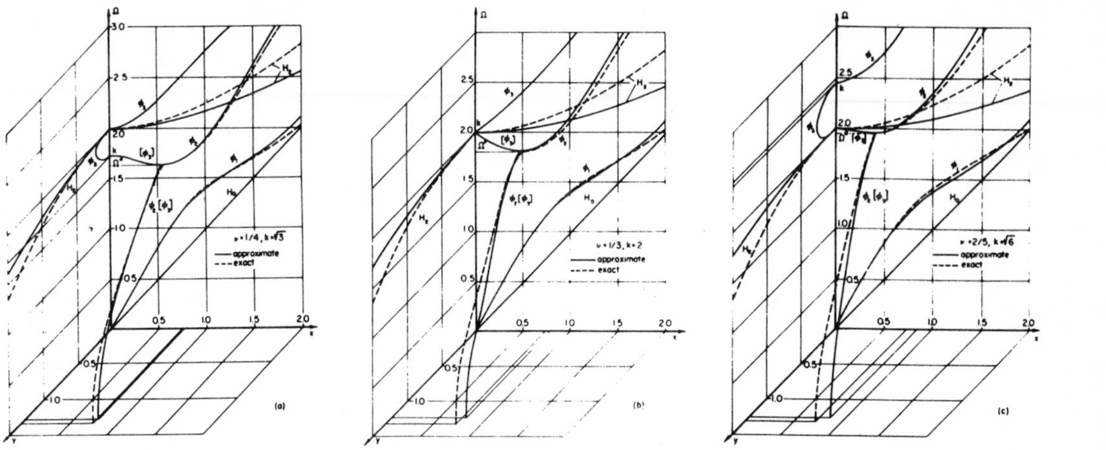

Fig. 4 Frequency spectrum of an infinite plate. Comparison of approximate and exact branches for three values of Poisson's ratio.

Rayleigh's solution of the exact equations, are also shown[9] in Figs. 4(*a*), (*b*), and (*c*) (as dashed lines). The importance of the introduction of the coefficients κ_i and their definitions, in terms of Poisson's ratio, is apparent. Without these coefficients, the extraordinarily complicated behavior of the branches of the exact frequency spectrum at long wave lengths would not be reproduced in the approximate equations. In the whole range of frequencies and wave numbers depicted, the approximate spectrum is reasonably close to the exact one; the poorest representation occurring in portions of the complex branches and the spectrum of the second face-shear mode. On the whole, fair results may be expected from solutions of the approximate equations in the case of finite plates.

The shapes of modes in the various ranges of frequency can be anticipated from the real, imaginary, or complex character of the roots. In rectangular co-ordinates, for example, real roots correspond to trigonometric mode-shapes; imaginary roots to exponential or hyperbolic mode-shapes; and complex roots correspond to modes whose shapes are given by products of trigonometric and exponential or hyperbolic functions. A striking example of the latter is to be found in the experiments with circular disks by Shaw [13].

The phase velocities (v) and group velocities (v_g) can be visualized readily from Fig. 4 inasmuch as

$$v = \frac{\omega}{\xi} = \frac{\Omega}{x}\left(\frac{\mu}{\rho}\right)^{1/2}, \qquad v_g = \frac{d\omega}{d\xi} = \frac{d\Omega}{dx}\left(\frac{\mu}{\rho}\right)^{1/2}. \tag{53}$$

For example, the phenomenon of phase and group velocities of opposite sign, noticed by Tolstoy and Usdin [11] in Rayleigh's solution, is represented by the branch $[\phi_3]$ in the real plane. Also, the minimum at Ω^* represents zero group velocity and nonzero phase velocity. As the wave length approaches zero ($x \to \infty$) the frequencies rise beyond the range of applicability of the equations and the asymptotic behavior of the velocities of the three compressional modes are

$$v = v_g = \frac{\kappa_2}{\kappa_3}\left(\frac{\mu}{\rho}\right)^{1/2}, \quad \left(\frac{\lambda + 2\mu}{\rho}\right)^{1/2}, \quad \frac{1}{\kappa_4}\left[\frac{E}{\rho(1-\nu^2)}\right]^{1/2}. \tag{54}$$

According to the exact theory, the first of these should be the velocity of Rayleigh surface waves and the second and third should be the velocity of equivoluminal waves.

[9] The complex roots were kindly supplied by Dr. Morio Onoe.

Additional Results

(a) The equations of compatibility may be obtained by eliminating the displacements from Equations (33), with the result:

$$\begin{aligned}
&\frac{\partial^2 \epsilon_1^{(0)}}{\partial x_3^2} + \frac{\partial^2 \epsilon_3^{(0)}}{\partial x_1^2} = \frac{\partial^2 \epsilon_5^{(0)}}{\partial x_1 \partial x_3} \\
&\frac{\partial^2 \epsilon_2^{(0)}}{\partial x_1^2} + \frac{3\epsilon_1^{(2)}}{b^2} = \frac{1}{b}\frac{\partial \epsilon_6^{(1)}}{\partial x_1} \\
&\frac{3\epsilon_3^{(2)}}{b^2} + \frac{\partial^2 \epsilon_2^{(0)}}{\partial x_3^2} = \frac{1}{b}\frac{\partial \epsilon_4^{(1)}}{\partial x_3} \\
&\frac{\partial^2 \epsilon_1^{(2)}}{\partial x_3^2} + \frac{\partial^2 \epsilon_3^{(2)}}{\partial x_1^2} = \frac{\partial^2 \epsilon_5^{(2)}}{\partial x_1 \partial x_3} \\
&\frac{3}{b}\frac{\partial \epsilon_1^{(2)}}{\partial x_3} = \frac{1}{2}\frac{\partial}{\partial x_1}\left(-\frac{\partial \epsilon_4^{(1)}}{\partial x_1} + \frac{3\epsilon_5^{(2)}}{b} + \frac{\partial \epsilon_6^{(1)}}{\partial x_3}\right) \\
&\frac{\partial^2 \epsilon_2^{(0)}}{\partial x_1 \partial x_3} = \frac{1}{2b}\left(-\frac{3\epsilon_5^{(2)}}{b} + \frac{\partial \epsilon_6^{(1)}}{\partial x_3} + \frac{\partial \epsilon_4^{(1)}}{\partial x_1}\right) \\
&\frac{3}{b}\frac{\partial \epsilon_3^{(2)}}{\partial x_1} = \frac{1}{2}\frac{\partial}{\partial x_3}\left(-\frac{\partial \epsilon_6^{(1)}}{\partial x_3} + \frac{\partial \epsilon_4^{(1)}}{\partial x_1} + \frac{3\epsilon_5^{(2)}}{b}\right)
\end{aligned} \tag{55}$$

The first of (55) is the usual compatibility equation of generalized plane stress and the remaining six equations correspond to the ordinary six compatibility equations; the main differences being that here the components $\epsilon_p^{(2)}$ have a factor 3 and the operator $\partial/\partial x_2$ is replaced by $1/b$.

(b) It also may be shown that there are nine dislocations when the nine components of strain and their derivatives are continuous. Three of them are the two translational and one rotational dislocations of generalized plane stress. Of the remaining six, three are translational (in the displacements $u_1^{(2)}$, $u_2^{(1)}$, $u_3^{(2)}$) and three are rotational, in the component of rotation

$$\frac{1}{2}\left(\frac{3u_3^{(2)}}{b} - \frac{\partial u_2^{(1)}}{\partial x_3}\right), \quad \frac{1}{2}\left(\frac{\partial u_3^{(2)}}{\partial x_1} - \frac{\partial u_1^{(2)}}{\partial x_3}\right), \quad \frac{1}{2}\left(\frac{\partial u_2^{(1)}}{\partial x_1} - \frac{3u_1^{(2)}}{b}\right). \tag{56}$$

(c) A theorem of uniqueness of solutions of the approximate equations of motion may be established along the lines of Neumann's theorem.[10] This leads to the following conditions sufficient for a unique solution (in the absence of discontinuities and singularities):

(i) Initial values of $u_1^{(0)}$, $u_3^{(0)}$, $u_2^{(1)}$, $u_1^{(2)}$, and $u_3^{(2)}$ throughout the plate.

(ii) Initial values of $\dot{u}_1^{(0)}$, $\dot{u}_3^{(0)}$, $\dot{u}_2^{(1)}$, $\dot{u}_1^{(2)}$, and $\dot{u}_3^{(2)}$ throughout the plate.

(iii) One member of each of the five products $F_1^{(0)}u_1^{(0)}$, $F_3^{(0)}u_3^{(0)}$, $F_2^{(1)}u_2^{(1)}$, $F_1^{(2)}u_1^{(2)}$, $F_3^{(2)}u_3^{(2)}$ at each point of the plate; i.e., five surface conditions.

(iv) At each point on the edge of the plate (normal n tangent s) one member of each of the five products $\tau_{nn}^{(0)}u_n^{(0)}$, $\tau_{ns}^{(0)}u_s^{(0)}$, $\tau_{n2}^{(1)}u_2^{(1)}$, $\tau_{nn}^{(2)}u_n^{(2)}$, $\tau_{ns}^{(2)}u_s^{(2)}$; i.e., five edge conditions.

The requirement that the strain energy-density $\bar{U}^{(2)}$, Equation (31), be positive definite is satisfied by the addition of the requirements $\kappa_1 > 0$, $\kappa_2 > 0$ to the usual requirements $3\lambda + 2\mu > 0$, $\mu > 0$.

(d) In the case of steady vibrations, the displacements may be expressed conveniently in terms of potentials that satisfy Helmholtz equations. Omitting a factor $e^{i\omega t}$ the results are

$$u_1^{(0)} = \frac{\partial\phi_1}{\partial x_1} + \frac{\partial\phi_2}{\partial x_1} + \frac{\partial\phi_3}{\partial x_1} - \frac{\partial H_0}{\partial x_3}$$

$$u_3^{(0)} = \frac{\partial\phi_1}{\partial x_3} + \frac{\partial\phi_2}{\partial x_3} + \frac{\partial\phi_3}{\partial x_3} + \frac{\partial H_0}{\partial x_1}$$

$$u_2^{(1)} = \alpha_1\phi_1 + \alpha_2\phi_2 + \alpha_3\phi_3 \quad (57)$$

$$u_1^{(2)} = \beta_1\frac{\partial\phi_1}{\partial x_1} + \beta_2\frac{\partial\phi_2}{\partial x_1} + \beta_3\frac{\partial\phi_3}{\partial x_1} - \frac{\partial H_2}{\partial x_3}$$

$$u_3^{(2)} = \beta_1\frac{\partial\phi_1}{\partial x_3} + \beta_2\frac{\partial\phi_2}{\partial x_3} + \beta_3\frac{\partial\phi_3}{\partial x_3} + \frac{\partial H_2}{\partial x_1}$$

$$\mu\nabla^2 H_0 + \rho\omega^2 H_0 = 0$$

$$\mu\nabla^2 H_2 + (\kappa_4{}^2\rho\omega^2 - 15b^{-2}\kappa_2{}^2\mu)H_2 = 0 \quad (58)$$

$$\nabla^2\phi_i + \xi_i{}^2\phi_i = 0, \quad i = 1, 2, 3$$

where

$$\alpha_i = b[(\lambda + 2\mu)\xi_i{}^2 - \rho\omega^2]/\kappa_1\lambda$$

$$\beta_i = \alpha_i b\Big/\left[\frac{3}{4}(\Omega^2 - 4) - \frac{E'b^2\xi_i{}^2}{5\mu\kappa_2{}^2}\right] \quad (59)$$

and the $\xi_i{}^2$ are, again, the roots of Equation (37). It may be seen that H_0 and H_2 are the potentials of the two face-shear modes and the ϕ_i are the potentials of the three compressional modes.

(e) The tensor, and hence invariant, characters of the quantities that appear in the second-order equations are, for the most part, apparent. For example, in the displacement equations of motion (35), if the first and second equations are regarded as the rectangular components of a vector and the fourth and fifth equations the rectangular components of another vector, the only differential operators that appear are the gradient, divergence, Laplacian, and $\partial^2/\partial t^2$: all invariants. The dependent variables are the scalar $u_2^{(1)}$ and the two vectors

$$\mathbf{u}^{(0)} = u_1^{(0)}\mathbf{k}_1 + u_3^{(0)}\mathbf{k}_3 = u_\alpha^{(0)}\mathbf{k}_\alpha + u_\gamma^{(0)}\mathbf{k}_\gamma,$$
$$\mathbf{u}^{(2)} = u_1^{(2)}\mathbf{k}_1 + u_3^{(2)}\mathbf{k}_3 = u_\alpha^{(2)}\mathbf{k}_\alpha + u_\gamma^{(2)}\mathbf{k}_\gamma, \quad (60)$$

[10] References [11], p. 176, and [9].

while the gradient operator is

$$\nabla = \mathbf{k}_1\frac{\partial}{\partial x_1} + \mathbf{k}_3\frac{\partial}{\partial x_3} = \mathbf{k}_\alpha\frac{\partial}{\partial s_\alpha} + \mathbf{k}_\gamma\frac{\partial}{\partial s_\gamma} \quad (61)$$

where $\mathbf{k}_1$, $\mathbf{k}_3$, and $\mathbf{k}_\alpha$, $\mathbf{k}_\gamma$ are unit vectors in the rectangular directions 1, 3, and the orthogonal curvilinear directions α, γ, respectively: all in the plane of the plate.

The appropriate strain tensors and their expression in terms of vector displacements are not quite as apparent. We define another vector displacement

$$\mathbf{u}' = \mathbf{u}^{(2)} + \mathbf{k}_2 u_2^{(1)} \quad (62)$$

and a gradient operator

$$\nabla' = \nabla + \mathbf{k}_2/b \quad (63)$$

where $\mathbf{k}_2$ is a unit vector normal to the plane of the plate. Then the two tensors

$$\boldsymbol{\varepsilon}^{(0)} = {}^1\!/_2(\nabla\mathbf{u}^{(0)} + \mathbf{u}^{(0)}\nabla),$$
$$\boldsymbol{\varepsilon}' = {}^1\!/_2(\nabla'\mathbf{u}' + \mathbf{u}'\nabla') \quad (64)$$

constitute the strain. The seven equations of compatibility, for example, become

$$\nabla \times \boldsymbol{\varepsilon}^{(0)} \times \nabla = 0,$$
$$\nabla' \times \boldsymbol{\varepsilon}' \times \nabla' = 0. \quad (65)$$

References

1 T. R. Kane and R. D. Mindlin, "High-Frequency Extensional Vibrations of Plates," Journal of Applied Mechanics, vol. 23, Trans. ASME, vol. 78, 1956, pp. A-277–283.

2 S. D. Poisson, "Mémoire sur l'Équilibre et le Mouvement des Corps Élastiques," *Mémoires de l'Académie des Sciences*, series 2, vol. 8, Paris, 1829, pp. 357–570.

3 A. L. Cauchy, "Sur l'Équilibre et le Mouvement d'une Plaque Solide," *Exercices de Mathématique*, vol. 3, 1828, pp. 328–355.

4 G. Kirchhoff, "Über das Gleichgewicht und die Bewegung einer Elastichen Scheibe," *Crelles Journal*, vol. 40, 1850, pp. 51–88.

5 R. D. Mindlin, "An Introduction to the Mathematical Theory of Vibrations of Elastic Plates," Signal Corps Engineering Laboratories, Fort Monmouth, N. J., 1955.

6 E. Reissner, "On the Calculation of Three-Dimensional Corrections for the Two-Dimensional Theory of Plane Stress," *Proceedings of the Fifteenth Semi-Annual Eastern Photoelasticity Conference*, Addison-Wesley Press, Cambridge, Mass., 1942.

7 Lord Rayleigh, "On the Free Vibrations of an Infinite Plate of Homogeneous Isotropic Elastic Matter," *Proceedings of The London Mathematical Society*, vol. 20, 1889, pp. 225–234.

8 S. Timoshenko, "On the Correction for Shear of the Differential Equation for Transverse Vibrations of Prismatic Bars," *Philosophical Magazine*, series 6, vol. 41, 1921, pp. 744–746.

9 R. D. Mindlin, "Influence of Rotatory Inertia and Shear on Flexural Vibrations of Isotropic, Elastic Plates," Journal of Applied Mechanics, vol. 18, Trans. ASME, vol. 73, 1951, pp. 31–38.

10 R. R. Aggarwal and E. A. G. Shaw, "Axially Symmetrical Vibrations of a Finite Isotropic Disk. IV," *Journal of the Acoustical Society of America*, vol. 26, 1954, pp. 341–342.

11 I. Tolstoy and E. Usdin, "Wave Propagation in Elastic Plates: Low and High Mode Dispersion," *Journal of the Acoustical Society of America*, vol. 29, 1957, pp. 37–42.

12 A. E. H. Love, "A Treatise on the Mathematical Theory of Elasticity," Cambridge University Press, London, 1927.

13 E. A. G. Shaw, "On the Resonant Vibrations of Thick Barium Titanate Disks," *Journal of the Acoustical Society of America*, vol. 28, 1956, pp. 38–50.

13

Reprinted from J. Appl. Mech., **27**, 681–689 (1960)

Flexural Stress Waves in an Infinite Elastic Plate Due to a Suddenly Applied Concentrated Transverse Load[1]

JULIUS MIKLOWITZ[2]

Associate Professor of Applied Mechanics, California Institute of Technology, Pasadena, Calif. Mem. ASME

The problem solved is that of an infinite plate subjected to a suddenly applied concentrated transverse shear load. The solution is derived from a plate theory that incorporates, in addition to bending, the effect of shear force and rotatory inertia on the deflection. These added effects give the present theory true wave character along with greater accuracy in the waves predicted. Numerical evaluation of the solution brings out the effects of dispersion and distortion on the moment and shear-force response of the plate. A criterion is developed for judging the accuracy of this response. It is based on a comparison, employing the stationary phase method, of the present approximate and exact (three-dimensional) theories.

THE problem treated in the paper is that of a thin elastic plate, infinite in extent, subjected to a step concentrated transverse shear load at its center, Fig. 1. The classical Lagrange theory could be employed for this problem. However, it is known that this theory does not give very accurate response information for sharp inputs; i.e., in having an infinite velocity associated with its leading disturbance, it is basically nonwave in nature. A more accurate displacement equation of motion was contributed by Uflyand [1][3] whose plate theory is the analog of the Timoshenko beam theory and which, in a similar manner, includes the effects of shear force and rotatory inertia.

Both of these theories have a bounding wave velocity, and wave information from the beam theory has been shown to compare favorably with experimental results involving reasonably sharp inputs. Unfortunately, Uflyand did not contribute the correct boundary conditions for his theory; however, these were supplied in a later and more complete analysis by Mindlin [2]. Mindlin also showed how much more favorably the plate, incorporating shear force and rotatory-inertia physics, compared with the three-dimensional theory of elasticity on the basis of phase velocities.

Uflyand [1] was the first to derive a solution for the present problem in accord with his higher order bending theory. He employed the Laplace transform and a contour integration-inversion method. The afore-mentioned erroneous boundary conditions, however, led to an invalid result. A correct mathematical solution for the problem was contributed by Lubkin [3] who worked with the nonhomogeneous equations of the present approximate theory. Lubkin's input was created by defining a limiting unit, disk loading in the neighborhood of $r = 0$, having impulsive time character (delta function). The response to the step is then obtained through a convolution.

As will be shown later, a somewhat simpler analysis results if the homogeneous equations and boundary conditions (at $r = 0$) of the theory are used for the problem. Further, the method could be used for other plate problems involving boundaries. On the other hand, Lubkin's analysis offers possibilities for solutions to plate problems that involve distributed loadings. It will be seen that the mathematical solution arrived at here is in agreement with Lubkin's. It is evaluated numerically and discussed for the first time here.

Statement of Problem

The method used by the author [4, 5] to derive traveling-wave solutions from the Timoshenko beam theory will also yield analogous solutions from this related plate theory and, in particular, the solution to the problem of interest here, Fig. 1.

Using a transformation analogous to that used in [4]

$$w = w_b + w_s, \quad \psi_x = -\frac{\partial w_b}{\partial x}, \quad \psi_y = -\frac{\partial w_b}{\partial y} \tag{1}$$

applied to Mindlin's equations [equation (36) and the third of equations (16) in reference 2], the governing equations for the plate can be written in a form similar to the Timoshenko beam equations in [4]; i.e., for the homogeneous case (plate normal pressure, $q = 0$)

$$\left. \begin{aligned} \left(-D\nabla^2\cdot\nabla^2 + \frac{\rho h^3}{12}\frac{\partial}{\partial t^2}\cdot\nabla^2\right) w_b - \kappa^2 G h \nabla^2 w_s = 0 \\ \kappa^2 G h \nabla^2 w_s - \rho h \frac{\partial^2}{\partial t^2}(w_b + w_s) = 0 \end{aligned} \right\} \tag{2}$$

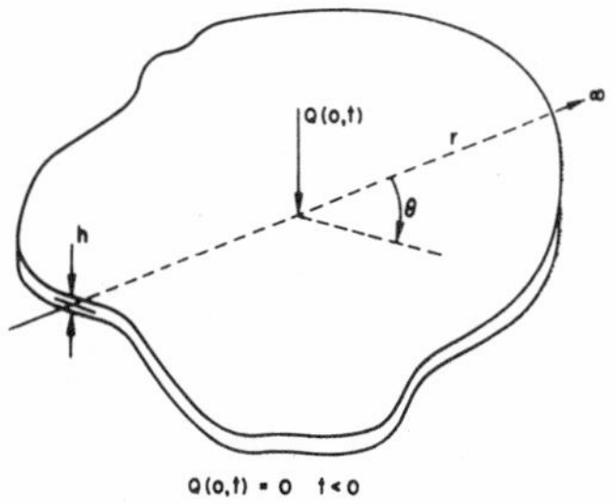

Fig. 1 Problem of infinite plate subjected to a step-concentrated, transverse-shear load

[1] This paper has been abstracted from Space Technology Laboratories Report TR-59-0000-00701, EM 9-10, June 5, 1959.

[2] Also, Consultant, Space Technology Laboratories, Inc.

[3] Numbers in brackets designate References at end of paper.

Presented at the West Coast Conference of the Applied Mechanics Division, Pasadena, Calif., June 27–29, 1960, of THE AMERICAN SOCIETY OF MECHANICAL ENGINEERS.

Discussion of this paper should be addressed to the Secretary, ASME, 29 West 39th Street, New York, N. Y., and will be accepted until January 10, 1961, for publication at a later date. Discussion received after the closing date will be returned.

Manuscript received by ASME Applied Mechanics Division, January 15, 1960. Paper No. 60—APMW-11.

where w_b and w_s are, respectively, the bending and shear components of the total displacement w, and ∇^2 is the two-dimensional Laplacian for the axially symmetric case; D is given by $Eh^3/12(1-\nu^2)$; κ^2, a correction factor relating to Rayleigh surface waves, is given by $0.76 + 0.3\nu$; G is the shear modulus equal to $E/2(1+\nu)$; ν is Poisson's ratio, ρ is the material density, and h the plate thickness. It should be emphasized that the second two equations of (1) are valid transformations in Mindlin's theory only in the special case where the plate rotation (in plane of plate) is zero; i.e., the rotation potential H of equations (52) in [2] is zero. Hence (1) and (2) apply only in the one-dimensional (Timoshenko beam) and axially symmetric cases.

For the present problem, it can be shown that the appropriate initial conditions are

$$w_b(r, 0) = w_s(r, 0) = \frac{\partial w_b(r, 0)}{\partial t} = \frac{\partial w_s(r, 0)}{\partial t} = 0, \quad r > 0 \qquad (3)$$

and the boundary conditions at $r = 0$

$$\lim_{r\to 0}\left[r\frac{\partial w_s(r, t)}{\partial r}\right] = \lim_{r\to 0}\left[r\frac{Q_r(r, t)}{\kappa^2 Gh}\right] = \frac{Q(0, t)}{2\pi\kappa^2 Gh} \qquad (4)$$

$$\lim_{r\to 0}\left[\frac{\partial^2 w_b(r, t)}{\partial t\partial r}\right] = 0, \qquad t > 0 \qquad (5)$$

where $Q(0, t)$ is the step function

$$Q(0, t) = 0 \qquad t < 0$$
$$= Q_0 \qquad t > 0$$

The conditions at infinity are

$$\lim_{r\to\infty}[w_b(r, t)] = 0, \qquad \lim_{r\to\infty}[w_s(r, t)] = 0 \qquad (6)$$

Equations (3), (4), (5), and (6) are consistent with the requirements of the theory as stated in [2].[4] Hence information in the present shear problem lies in the solution of the boundary-value problem given by (2), (3), (4), (5), and (6).

Solution of Problem

Following the earlier method [4] based on the Laplace transform, the governing equations (2), subject to the initial conditions (3), transform to

$$\left.\begin{aligned}\left(-D\nabla^2\cdot\nabla^2 + \frac{\rho h^3}{12}p^2\nabla^2\right)\bar{w}_b - \kappa^2 Gh\nabla^2\bar{w}_s &= 0\\ \kappa^2 Gh\nabla^2\bar{w}_s - \rho hp^2(\bar{w}_b + \bar{w}_s) &= 0\end{aligned}\right\} \qquad (7)$$

where the bar indicates the transformed variable and p the transformation parameter. A possible solution of this set of ordinary differential equations is given by

$$\left.\begin{aligned}\bar{w}_b(r, p) &= A(p)K_0[n(p)r]\\ \bar{w}_s(r, p) &= B(p)K_0[n(p)r]\end{aligned}\right\} \qquad (8)$$

where K_0 is the modified Bessel function of the second kind of order zero. Substitution of (8) into (7) yields the characteristic equation of the system, the roots of which are given by

$$n_j = \pm M\sqrt{p[p \mp N(p^2 - a^2)^{1/2}]^{1/2}}, \qquad (j = 1, 2, 3, 4) \qquad (9)$$

where

$$M = \frac{\sqrt{2}}{2}\left(\frac{v_1^2 + v_2^2}{v_1^2 v_2^2}\right)^{1/2}, \qquad N = \frac{v_1^2 - v_2^2}{v_1^2 + v_2^2}, \qquad a = \frac{2\sqrt{V}v_1^2 v_2^2}{v_1^2 - v_2^2},$$

$$V = \frac{12}{h^2 v_1^2} \qquad v_1^2 = \frac{E}{\rho(1-\nu^2)}, \qquad v_2^2 = \frac{\kappa^2 G}{\rho}, \qquad v_1 > v_2$$

v_1 is the so-called "plate velocity," v_2 the Rayleigh surface wave velocity [respectively, c_p and κc_s in [2], where c_s is the shear-wave velocity $(G/\rho)^{1/2}$].

Hence the general solution to (7) is given by

$$\left.\begin{aligned}\bar{w}_b(r, p) &= \sum_{j=1}^{4} A_j K_0(n_j r)\\ \bar{w}_s(r, p) &= \sum_{j=1}^{4} \phi_j A_j K_0(n_j r)\end{aligned}\right\} \qquad (10)$$

where

$$\phi_j = \left(\frac{p^2}{v_1^2} - n_j^2\right)\Big/ v_2^2 V$$

The conditions (6) transform to

$$\lim_{r\to\infty}[\bar{w}_b(r, p)] = 0, \qquad \lim_{r\to\infty}[\bar{w}_s(r, p)] = 0 \qquad (11)$$

Applying this to (10), taking into account the behavior of $K_0(n_j r)$ for large argument (see Erdelyi [6]),

$$K_0(n_j r) = \left(\frac{\pi}{2n_j r}\right)^{1/2} e^{-n_j r} \qquad (12)$$

it is seen that the general solutions in the present infinite medium problem reduce to

$$\left.\begin{aligned}\bar{w}_b(r, p) &= A_1 K_0(n_1 r) + A_2 K_0(n_2 r)\\ \bar{w}_s(r, p) &= \phi_1 A_1 K_0(n_1 r) + \phi_2 A_2 K_0(n_2 r)\end{aligned}\right\} \qquad (13)$$

where n_1 and n_2 are the positive roots in (9), agreeing with

$$Re\, n_j > 0 \qquad (14)$$

which is imposed by (11) and (12). A_1 and A_2 are evaluated by transforming and applying the boundary conditions (4) and (5) to (13); i.e.,

$$A_1 = -A_2 = -\frac{Q_0}{2\pi Dp(n_2^2 - n_1^2)} \qquad (15)$$

With prime interest here in the stresses, solutions for the moments and shear force were needed. In the present theory, the expressions for M_θ and M_r are given by

$$\left.\begin{aligned}M_\theta &= -D\left[\nu\frac{\partial^2 w_b}{\partial r^2} + \frac{1}{r}\frac{\partial w_b}{\partial r}\right]\\ M_r &= -D\left[\frac{\partial^2 w_b}{\partial r^2} + \frac{\nu}{r}\frac{\partial w_b}{\partial r}\right]\end{aligned}\right\} \qquad (16)$$

and that for Q_r by

$$Q_r = \kappa^2 Gh\frac{\partial w_s}{\partial r} \qquad (17)$$

where M_θ and M_r are moments per unit length, acting on radial and circumferential sections, respectively, and Q_r is shear force per unit length acting on a circumferential section. Maximum stresses differ from M_θ, M_r, and Q_r only by a constant as in the elementary bending theory.

[4] See the section on initial and boundary conditions, and note that the third boundary condition in the present case is automatically satisfied by the axial symmetry involved.

Substitution of (13) into (16) and (17) give the transformed solutions for the moments and shear force. They are

$$\bar{M}_\theta(r, p) = -D\left\{A_1\left[\nu n_1^2K_0(n_1r) - \frac{(1-\nu)}{r}n_1K_1(n_1r)\right] + A_2\left[\nu n_2^2K_0(n_2r) - \frac{(1-\nu)}{r}n_2K_1(n_2r)\right]\right\} \qquad (18)$$

$$\bar{M}_r(r, p) = -D\left\{A_1\left[n_1^2K_0(n_1r) + \frac{(1-\nu)}{r}n_1K_1(n_1r)\right] + A_2\left[n_2^2K_0(n_2r) + \frac{(1-\nu)}{r}n_2K_1(n_2r)\right]\right\} \qquad (19)$$

and

$$\bar{Q}_r(r, p) = -\kappa^2Gh[\phi_1n_1A_1K_1(n_1r) + \phi_2n_2A_2K_1(n_2r)] \qquad (20)$$

The solutions for M_θ, M_r, and Q_r are obtained by inverting (18), (19), and (20). Since the Laplace transform pairs for these complicated functions do not exist in tables, appeal must be made to the complex inversion integral and contour integration. As Uflyand [1] pointed out, the analysis is considerably reduced, however, by noting that the characteristic equation of the present transformed plate equations is, except for constants, the same as that associated with the Timoshenko beam equations. In comparison the reader will note that if v_1, the plate velocity, were replaced by the bar velocity $c_1 = (E/\rho)^{1/2}$, and the constant V by the analogous $C = A/Ic_1^2$, where A and I are the area and moment of inertia of the beam section, respectively, the roots given by (9) would be those associated with the Timoshenko beam theory. Velocity c_2 is also taken equal to v_2.

This equivalence in characteristic equations means that the contour and related analysis, used to obtain the solutions for the Timoshenko beam, also may be used to invert (18), (19), and (20). In this work, this was done by drawing upon the contour analysis and results in the author's earlier work on beams [5]. That analysis in turn was shown to be in agreement with Uflyand's work.

On the basis of the existing work [1, 5] much of the detail of the inversion method employed here can be omitted. Briefly, then, use of the Mellin inversion theorem leads to the solution statements

$$M_\theta(r, t) = M_{\theta 1}(r, t) + M_{\theta 2}(r, t) = \frac{1}{2\pi i}\int_{\mathrm{Br}_1}[\bar{M}_{\theta 1}(r, p) + \bar{M}_{\theta 2}(r, p)]e^{pt}dp \qquad (21)$$

$$M_r(r, t) = M_{r1}(r, t) + M_{r2}(r, t) = \frac{1}{2\pi i}\int_{\mathrm{Br}_1}[\bar{M}_{r1}(r, p) + \bar{M}_{r2}(r, p)]e^{pt}dp \qquad (22)$$

$$Q_r(r, t) = Q_{r1}(r, t) + Q_{r2}(r, t) = \frac{1}{2\pi i}\int_{\mathrm{Br}_1}[\bar{Q}_{r1}(r, p) + \bar{Q}_{r2}(r, p)]e^{pt}dp \qquad (23)$$

where Br_1 is the Bromwich contour in the right half of the p-plane (see McLachlan [7]) and where, from (18), (19), and (20) in which $j = 1, 2$,

$$\bar{M}_{\theta j}(r, p) = -DA_j\left[\nu n_j^2K_0(n_jr) - \frac{(1-\nu)}{r}n_jK_1(n_jr)\right]$$

$$\bar{M}_{rj}(r, p) = -DA_j\left[n_j^2K_0(n_jr) + \frac{(1-\nu)}{r}n_jK_1(n_jr)\right]$$

$$\bar{Q}_{rj}(r, p) = -\kappa^2Gh\phi_jA_jn_jK_1(n_jr)$$

Completing the contour Br_1 to the right, and employing Cauchy's theorem and Jordan's lemma, it is readily shown that

$$\left.\begin{aligned} M_{\theta j}(r, t) &= 0 \\ M_{rj}(r, t) &= 0 \\ Q_{rj}(r, t) &= 0 \end{aligned}\right\} \quad \text{for} \quad t < \frac{r}{v_j} \qquad (24)$$

stemming from the fact that the singularities of the integrals appearing in (21), (22), and (23) lie to the left of Br_1 and that

$$\lim_{|p|\to\infty} n_j = \frac{p}{v_j} \qquad (25)^5$$

Equations (24) establish the wave character of the solutions (21), (22), and (23).

For $t > r/v_j$, inversion of the integrals in (21), (22), and (23) was accomplished by integrating along a contour completed to the left of Br_1. Introducing Uflyand's transformation $p = aq$, and noting the common structure of terms composing $\bar{M}_{\theta j}$ and $\bar{M}_{rj}$ in (21) and (22), the solutions for M_θ, M_r, and Q_r in terms of the dimensionless q-plane become

$$\frac{M_\theta(r, t)}{Q_0} = \left.\begin{cases} 0 & 0 < t < \dfrac{r}{v_1} \\ -D\left[\nu\alpha_1 - \dfrac{(1-\nu)}{r}\beta_1\right] & \dfrac{r}{v_1} < t < \dfrac{r}{v_2} \\ -D\left[\nu(\alpha_1 + \alpha_2) - \dfrac{(1-\nu)}{r}(\beta_1 + \beta_2)\right] & t > \dfrac{r}{v_2} \end{cases}\right\} \qquad (26)$$

$$\frac{M_r(r, t)}{Q_0} = \left.\begin{cases} 0 & 0 < t < \dfrac{r}{v_1} \\ -D\left[\alpha_1 + \dfrac{(1-\nu)}{r}\beta_1\right] & \dfrac{r}{v_1} < t < \dfrac{r}{v_2} \\ -D\left[\alpha_1 + \alpha_2 + \dfrac{(1-\nu)}{r}(\beta_1 + \beta_2)\right] & t > \dfrac{r}{v_2} \end{cases}\right\} \qquad (27)$$

$$\frac{hQ_r(r, t)}{Q_0} = \left.\begin{cases} 0 & 0 < t < \dfrac{r}{v_1} \\ -\kappa^2Gh\gamma_1 & \dfrac{r}{v_1} < t < \dfrac{r}{v_2} \\ -\kappa^2Gh(\gamma_1 + \gamma_2) & t > \dfrac{r}{v_2} \end{cases}\right\} \qquad (28)$$

where use has been made of (24), and where

$$\alpha_j(r, t) = \frac{(-1)^jH}{i}\int_{\mathrm{Br}_1}\frac{z_jK_0(n_jr)e^{aqt}dq}{q(q^2-1)^{1/2}}$$

$$\beta_j(r, t) = \frac{(-1)^jJ}{i}\int_{\mathrm{Br}_1}\frac{\sqrt{z_j}\,K_1(n_jr)e^{aqt}dq}{q^{3/2}(q^2-1)^{1/2}}$$

$$\gamma_j(r, t) = \frac{(-1)^jL}{i}\int_{\mathrm{Br}_1}\frac{[q - v_1^2M^2z_j]\sqrt{z_j}\,K_1(n_jr)e^{aqt}dq}{\sqrt{q}(q^2-1)^{1/2}}$$

and where

$$n_j = P\sqrt{q}\sqrt{z_j}, \qquad z_j = q + (-1)^jN(q^2-1)^{1/2}$$

---

[5] The behavior of $K_0(n_jr)$ and $K_1(n_jr)$, in the right half plane for large $|p|$, is given by the right-hand side of (12).

$$P = Ma, \quad H = \frac{1}{8\pi^2 DN'}, \quad J = \frac{H}{P'}, \quad L = \frac{Hha}{Mv_1^2 v_2^2 V}$$

The integrand singularities of α_j, β_j, and γ, are the branch points

$$q = 0, \quad q = \pm 1, \quad \text{and} \quad q = \pm i\alpha \tag{29}$$

where

$$\alpha = \frac{N}{(1 - N^2)^{1/2}}$$

The first of these is a branch point of $\sqrt{q}$, hence of $q^{3/2}$, n_j, $K_0(n_j r)$, and $K_1(n_j r)$. The second are branch points of $(q^2 - 1)^{1/2}$, hence of z_j, n_j, $K_0(n_j r)$ and $K_1(n_j r)$. The last are branch points of $\sqrt{z_1}$, and hence only of n_1, $K_0(n_1 r)$, and $K_1(n_1 r)$. Further consideration must be given to the multivaluedness of the K-functions in their own right, due to their logarithmic behavior (see the series definition in [6]). Since their multivaluedness depends directly on that of

$$\log n_j r = \log |n_j r| + i \arg (n_j r) \tag{30}$$

hence on that of n_j, it can be seen that single valuedness is obtained for the K-functions through the single valuedness of the n_j-functions.

The cuts and contours used for the Timoshenko beam solution, shown in Fig. 2, are sufficient to render the integrals defining α_j, β_j, and γ_j single valued and led to their evaluation. In order to use these contours to get a convergent inversion for α_j and β_j in the present case, however, they first had to be written in terms of the convolution theorem; a necessity imposed by the contribution from the small circle about $q = 0$. According to the theorem the inversion integral can be written as

$$\frac{1}{2\pi i} \int_{\text{Br}_1} \phi_1(aq)\phi_2(aq)e^{aqt}d(aq)$$

which has its inverse in

$$\int_0^t f_1(t - \tau)f_2(\tau)d\tau$$

where ϕ_1 and f_1, ϕ_2 and f_2 are transform pairs. In the case of α_j and β_j, ϕ_1 was taken as $1/aq$ and ϕ_2 as the rest of the integrand. Since the inverse $f_1(t - \tau)$ is the unit function, α_j and β_j may be written as

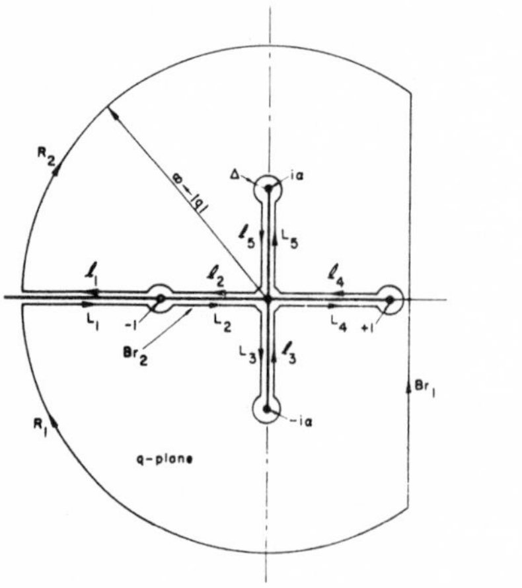

Fig. 2 Integration contours used in evaluating inverse transforms

$$\alpha_j(r, t) = \frac{(-1)^j Ha}{i} \int_{\frac{r}{v_j}}^{t} \left[\int_{\text{Br}_1} \frac{z_j K_0(n_j r)e^{aq\tau}}{(q^2 - 1)^{1/2}} dq \right] d\tau$$

$$= \frac{(-1)^j Ha}{i} \int_{\frac{r}{v_j}}^{t} \alpha_j^*(r, \tau)d\tau \tag{31}$$

and

$$\beta_j(r, t) = \frac{(-1)^j Ja}{i} \int_{\frac{r}{v_j}}^{t} \left[\int_{\text{Br}_1} \frac{\sqrt{z_j}K_1(n_j r)e^{aq\tau}}{\sqrt{q}(q^2 - 1)^{1/2}} dq \right] d\tau$$

$$= \frac{(-1)^j Ja}{i} \int_{\frac{r}{v_j}}^{t} \beta_j^*(r, \tau)d\tau \tag{32}$$

It remains, then, to carry out the contour integration of the inner integral in (31) and (32) (i.e., evaluate α_j^* and β_j^*) over the paths shown in Fig. 2 and subsequently, the simple outer τ-integration. γ_j in (28) is evaluated directly by contour integration.

Making use of (12) and (25) it may be shown that the paths R_1 and R_2 do not contribute to the solutions for α_j, β_j, and γ_j, since on these paths as $|q| \to \infty$ the integrands of α_j^*, β_j^*, and γ_j behave as

$$\frac{\exp\left[aq\left(\tau - \frac{r}{v_j}\right)\right]}{\sqrt{q}}, \quad \frac{\exp\left[aq\left(\tau - \frac{r}{v_j}\right)\right]}{q^{3/2}},$$

$$\frac{\exp\left[aq\left(\tau - \frac{r}{v_j}\right)\right]}{\sqrt{q}}$$

respectively, (see [7], page 77). The replacement of contour Br_1 with Br_2 in Fig. 2 (composed of the linear paths L_1, L_2, L_3, l_3, L_4, l_4, L_5, l_5, l_2, and l_1, and the small circles about the branch points) follows from Cauchy's theorem.

The integrals on these component paths are written by employing the same technique as used in [1, 5]. Briefly then, particular branches of the multivalued integrands of α_j^*, β_j^*, and γ_j are selected to agree with $Re\, q > 0$, and the condition $Re\, n_j > 0$ (14). These branches are then continued analytically to, and on Br_2. The use of basic geometrical variables (complex vectors emanating from the branch points) to write the integrands, which are composed of these variables, and the use of the binomial theorem to obtain representations in the vicinity of the branch points, facilitates the continuation. The table in [5], giving the resultant values of these integrand elements on each of the lineal paths comprising Br_2, has been used to construct on these paths the integrals of α_j^*, β_j^*, and γ_j. A limiting process, similar to that in the Appendix of [5] which used for the integrand elements on the small circles, the limiting end point values of the adjacent lineal paths, showed that the small circles contributed nothing to α_j^* and β_j^*. This was also true of γ_j, with the exception of a contribution from the small circle around $q = 0$, reflecting the boundary condition of the step input in shear. Hence it remains to consider the possible contributions from just the lineal paths of Br_2.

Let us consider α_1^* on l_1. The table in [5] shows this is

$$[\alpha_1^*]_{l_1} = \frac{Ha}{i} \int_1^\infty \frac{\gamma^2 K_0(Pr\sqrt{\rho}\gamma e^{\pi i})e^{-a\rho\tau}\, d\rho}{(\rho^2 - 1)^{1/2}} \tag{33a}$$

ρ being the real variable $|q|$, and $\gamma = [\rho - N(\rho^2 - 1)^{1/2}]^{1/2}$ representing $|\sqrt{z_1}|$ on the path. Similarly on L_1

$$[\alpha_1^*]_{L_1} = \frac{Ha}{i} \int_\infty^1 \frac{\gamma^2 K_0(Pr\sqrt{\rho}\gamma e^{-\pi i})e^{-a\rho\tau}\, d\rho}{(\rho^2 - 1)^{1/2}} \tag{33b}$$

These integrals combine since we have the relation (given in [6], page 80),

$$K_n(ze^{im\pi}) = e^{-imn\pi}K_n(z) - i\pi m(-1)^{n(m+1)}I_n(z) \quad (34)$$

where $I_n(z)$ is the modified Bessel function of the first kind of order n, m and n are integers, and z is a complex variable. Equation (34) applied to (33a) and (33b) combines these integrals into the single real definite integral

$$[\alpha_1^*]_{l_1+L_1} = -2Ha\pi \int_1^\infty \frac{\gamma^2 I_0(Pr\sqrt{\rho}\gamma)e^{-a\rho\tau}\,d\rho}{(\rho^2-1)^{1/2}} \quad (35)$$

Similarly, contributions of this type from the paths l_1 and L_1 are obtained for α_2^*, β_j^*, and γ_j.[6]

Consider next α_1^* on l_2. The table in [5] shows that this integral can be written as

$$[\alpha_1^*]_{l_2} = Ha\int_0^1 \frac{(\rho+i\delta)K_0[Pr\sqrt{\rho}e^{i\pi}(\lambda+i\mu)]e^{-a\rho\tau}\,d\rho}{(1-\rho^2)^{1/2}} \quad (36)$$

where $(\lambda+i\mu)e^{i\pi/2}$ represents $\sqrt{z_1}$ on the path in terms of the rectangular real variables

$$\lambda = \left(\frac{\Delta+\rho}{2}\right)^{1/2}, \qquad \mu = \left(\frac{\Delta-\rho}{2}\right)^{1/2},$$

$$\Delta = (1-N^2)^{1/2}(\rho^2+\alpha^2)^{1/2}$$

and $-(\rho+i\delta)$ represents z_1 on the path, where $\delta = 2\lambda\mu$. Similarly, α_1^* on L_2 is given by

$$[\alpha_1^*]_{L_2} = Ha\int_1^0 \frac{(-\rho+i\delta)K_0[Pr\sqrt{\rho}e^{-i\pi}(\lambda-i\mu)]e^{-a\rho\tau}\,d\rho}{(1-\rho^2)^{1/2}} \quad (37)$$

where $(\lambda-i\mu)e^{-i\pi/2}$ represents $\sqrt{z_1}$, and $-\rho+i\delta$ represents z_1 on the path.

Since $K_0(\bar{z}) = \overline{K_0(z)}$, after an interchange of the limits on this last integral it may be seen that its integrand is the complex conjugate of that in (36). Hence using (34) again and the complex conjugate character of the integrands in (36) and (37), they combine to give

$$[\alpha_1^*]_{l_2+L_2} = 2Ha\int_0^1 Re\{z_a[K_0(z_b) - i\pi I_0(z_b)]\}\frac{e^{-a\rho\tau}\,d\rho}{(1-\rho^2)^{1/2}} \quad (38)$$

where $z_a = \rho + i\delta$ and $z_b = Pr\sqrt{\rho}(\lambda+i\mu)$.

Similar considerations for the paths l_4 and L_4 produce the result

$$[\alpha_1^*]_{l_4+L_4} = -2Ha\int_0^1 Re[z_aK_0(z_b)]\frac{e^{a\rho\tau}\,d\rho}{(1-\rho^2)^{1/2}} \quad (39)$$

so that adding there results the real definite integral

$$[\alpha_1^*]_{l_2+L_2+l_4+L_4} = -2Ha\int_0^1 \{2\sinh a\rho\tau Re[z_aK_0(z_b)] + \pi e^{-a\rho\tau}Re[iz_aI_0(z_b)]\}\frac{d\rho}{(1-\rho^2)^{1/2}} \quad (40)$$

The analysis for α_2^*, β_j^*, and γ_j is quite similar, producing contributions like (40) for these integrals.

Consider finally the α_1^* on l_5. The table in [5] gives

$$[\alpha_1^*]_{l_5} = Ha\int_\alpha^0 \frac{\omega^2K_0(Pr\sqrt{\rho}\omega e^{i\pi})e^{ia\rho\tau}\,d\rho}{(\rho^2+1)^{1/2}} \quad (41)$$

[6] In agreement with Lubkin's findings. Uflyand erroneously concluded that these paths contributed nothing.

where $\omega = [N(\rho^2+1)^{1/2} - \rho]^{1/2}$ represents $|\sqrt{z_1}|$ on the path. Likewise the integral α_1^* on L_5 is given by

$$[\alpha_1^*]_{L_5} = Ha\int_0^\alpha \frac{\omega^2K_0(Pr\sqrt{\rho\omega})e^{ia\rho\tau}\,d\rho}{(\rho^2+1)^{1/2}} \quad (42)$$

Using (34) again, (41) and (42) combine to give

$$[\alpha_1^*]_{l_5+L_5} = \pi Ha\int_0^\alpha \frac{\omega^2I_0(Pr\sqrt{\rho\omega})ie^{ia\rho\tau}\,d\rho}{(\rho^2+1)^{1/2}} \quad (43)$$

Considering the paths l_3 and L_3 in the same way then, results in

$$[\alpha_1^*]_{l_3+L_3} = -\pi Ha\int_0^\alpha \frac{\omega^2I_0(Pr\sqrt{\rho\omega})ie^{-ia\rho\tau}\,d\rho}{(\rho^2+1)^{1/2}} \quad (44)$$

Equations (43) and (44) combine to give the real definite integral

$$[\alpha_1^*]_{l_3+L_3+l_5+L_5} = -2\pi Ha\int_0^\alpha \frac{\omega^2\sin a\rho\tau I_0(Pr\sqrt{\rho\omega})\,d\rho}{(\rho^2+1)^{1/2}} \quad (45)$$

Here, analogous results are similarly produced for β_1^* and γ_1, but α_2^*, β_2^*, and γ_2 give a zero contribution on these paths since $q = \pm i\alpha$ are branch points only of $\sqrt{z_1}$ (and not of $\sqrt{z_2}$), as noted earlier.

Adding the results in (35), (40), and (45), and performing the simple outer integration indicated in (31), we have

$$\alpha_1(r,t) = 2\pi H\left\langle \int_1^\infty \frac{X_1\gamma^2I_0(Pr\sqrt{\rho}\gamma)\,d\rho}{\rho(\rho^2-1)^{1/2}} - \int_0^1\left\{\frac{2}{\pi}Y_1Re[z_aK_0(z_b)] - X_1Re[iz_aI_0(z_b)]\right\}\frac{d\rho}{\rho(1-\rho^2)^{1/2}} + \int_0^\alpha \frac{\omega^2Z_1I_0(Pr\sqrt{\rho\omega})\,d\rho}{\rho(\rho^2+1)^{1/2}}\right\rangle \quad (46)$$

and similarly, in light of the previously discussed development for α_2^*, β_j^*, and γ_j, we have

$$\alpha_2(r,t) = -2\pi H\left\langle \int_1^\infty \frac{X_2\beta^2I_0(Pr\sqrt{\rho}\beta)\,d\rho}{\rho(\rho^2-1)^{1/2}} - \int_0^1\left\{\frac{2}{\pi}Y_2Re[z_aK_0(z_b)] + X_2Re[iz_aI_0(z_b)]\right\}\frac{d\rho}{\rho(1-\rho^2)^{1/2}}\right\rangle \quad (47)$$

$$\beta_1(r,t) = -2\pi J\left\langle \int_1^\infty \frac{X_1\gamma I_1(Pr\sqrt{\rho}\gamma)\,d\rho}{\rho^{1/2}(\rho^2-1)^{1/2}} + \int_0^1\left\{\frac{2}{\pi}Y_1Re[z_cK_1(z_b)] + X_1Re[iz_cI_1(z_b)]\right\}\frac{d\rho}{\rho^{1/2}(1-\rho^2)^{1/2}} + \int_0^\alpha \frac{\omega Z_1I_1(Pr\sqrt{\rho\omega})\,d\rho}{\rho^{1/2}(\rho^2+1)^{1/2}}\right\rangle \quad (48)$$

$$\beta_2(r,t) = 2\pi J\left\langle \int_1^\infty \frac{X_2\beta I_1(Pr\sqrt{\rho}\beta)\,d\rho}{\rho^{1/2}(\rho^2-1)^{1/2}} + \int_0^1\left\{\frac{2}{\pi}Y_2Re[z_cK_1(z_b)] - X_2Re[iz_cI_1(z_b)]\right\}\frac{d\rho}{\rho^{1/2}(1-\rho^2)^{1/2}}\right\rangle \quad (49)$$

$$\gamma_1(r,t) = -2\pi L\left\langle \int_1^\infty \frac{\Gamma\gamma I_1(Pr\sqrt{\rho\gamma})e^{-a\rho t}\,d\rho}{\rho^{1/2}(\rho^2-1)^{1/2}} + \int_0^1\left\{\frac{2}{\pi}\cosh a\rho t Re[z_g z_c K_1(z_b)]\right.\right.$$

$$+ e^{-apt}Re[iz_g z_e I_1(z_b)]\Big\} \frac{d\rho}{\rho^{1/2}(1-\rho^2)^{1/2}} - \int_0^\alpha \frac{\Omega\omega \cos(apt) I_1(Pr\sqrt{\rho\omega})\,d\rho}{\rho^{1/2}(\rho^2+1)^{1/2}}\Big\rangle - \frac{1}{4\pi\kappa^2 Gr} \qquad (50)$$

$$\gamma_2(r, t) = 2\pi L\Big\langle \int_1^\infty \frac{B\beta I_1(Pr\sqrt{\rho\beta})e^{-apt}d\rho}{\rho^{1/2}(\rho^2-1)^{1/2}} + \int_0^1 \Big\{\frac{2}{\pi}\cosh apt Re[z_g z_e K_1(z_b)] - e^{-apt}Re[iz_g z_e I_1(z_b)]\Big\} \frac{d\rho}{\rho^{1/2}(1-\rho^2)^{1/2}}\Big\rangle - \frac{1}{4\pi\kappa^2 Gr} \qquad (51)$$

where

$$X_1 = e^{-apt} - e^{-apr/v_1}, \qquad X_2 = e^{-apt} - e^{-apr/v_2}$$

$$Y_1 = \cosh apt - \cosh ap\frac{r}{v_1}, \qquad Y_2 = \cosh apt - \cosh ap\frac{r}{v_2}$$

$$Z_1 = \cos apt - \cos ap\frac{r}{v_1}$$

$$\beta = [\rho + N(\rho^2 - 1)^{1/2}]^{1/2}, \qquad \Gamma = (1 - v_1^2M^2)\rho + v_1^2M^2N(\rho^2 - 1)^{1/2}$$

$$\Omega = \rho + v_1^2M^2\omega^2, \qquad B = (1 - v_1^2M^2)\rho - v_1^2M^2N(\rho^2 - 1)^{1/2}$$

$$z_e = \frac{z_b}{Pr\sqrt{\rho}} = \lambda + i\mu, \qquad z_g = (1 - v_1^2M^2)\rho - iv_1^2M^2\delta$$

Equations (26) to (28) together with (46) to (51) give the complete mathematical solutions for the dimensionless moments and shear in the present problem.

Numerical Evaluation and Discussion of Solution

Figs. 3 to 5 present the results that were derived from (26) to (28) and (46) to (51) by numerical integration.[7] They represent the response of the infinite plate to a (*a*) step (solid and dashed line) and (*b*) rectangular pulse (solid line) input of concentrated shear load at $r = 0$.

The latter was obtained from the former by simple graphical addition according to

$$[M_r(r, t)]_{RP} = [M_r(r, t)]_S - [M_r(r, t - k)]_S \qquad (52)$$

for the moment M_r, for instance, where RP stands for the response to the rectangular pulse and S the step input.

It will be noted that three dimensionless stations r/h (ratio of station r to plate thickness) are represented in Figs. 3 to 5. Consistent generality is contained in the abscissa variable in these curves, $(1/h)(t - r/v_1)$. This is a modified "time after arrival of first wave front" which yields an absolute time for a particular plate thickness. The ordinate variables are all dimensionless. The material constants represented in these figures are $E = 28 \times 10^6$ psi $\nu = 0.3$, and density, $\rho = 7.41$ lb-sec^2/in^4, reflecting a high-strength stainless steel.

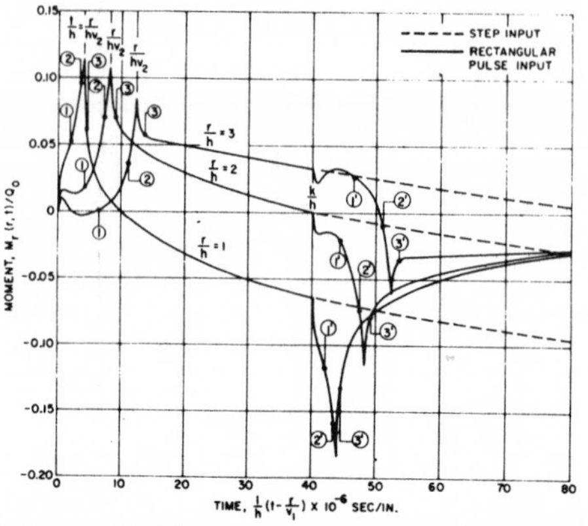

Fig. 3 Moment $M_r(r, t)/Q_0$ response of plate at various stations r/h due to step and rectangular pulse inputs of concentrated shear load at $r = 0$

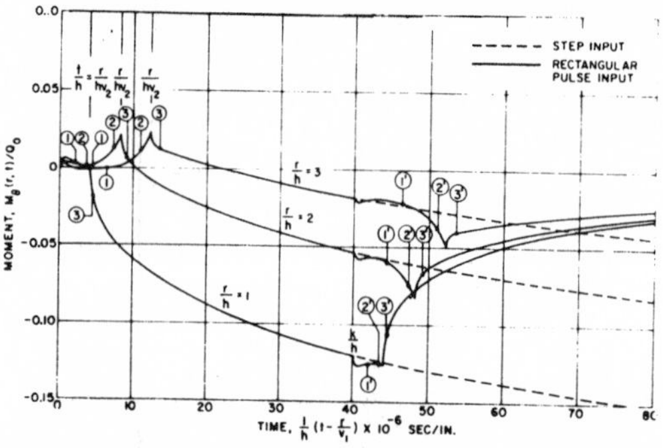

Fig. 4 Moment $M_\theta(r, t)/Q_0$ response of plate at various stations r/h due to step and rectangular pulse inputs of concentrated shear load at $r = 0$

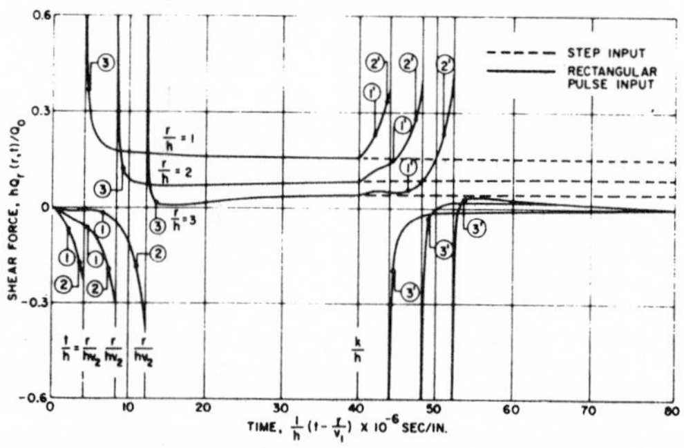

Fig. 5 Shear force $hQ_r(r, t)/Q_0$ response of plate at various stations r/h due to step and rectangular pulse inputs of concentrated shear load at $r = 0$

[7] The numerical integration of the integrals appearing in (46) to (51) was carried out on an 1103A Univac scientific computer. The integration scheme employed a high-order polynomial and a sufficient number of points to represent the integrand in all cases. The modified Bessel functions appearing in these equations were represented in series form for both real and complex arguments. In the case of the latter this meant a series composed of both real and imaginary parts, enabling an evaluation of the $\int_0^1$ integrals in (46) to (51) in the form shown rather than reduce them further. Convergence of an integration to an acceptable numerical value was based on a sufficiency criterion in the number of points used to represent the integrand in the full range of integration; that is, when the number of points used reached a value such that a significant increase in this number did not make a significant change in the numerical integration result, this evaluation was accepted. An even stronger check on results is given through wave-front expansion terms, which are established independently of the solution given in (26) to (28) and (46) to (51). These expansions are derived from the inversion integral-solution statements (21) to (23) for $t > r/v_j$ by letting $|p| \rightarrow \infty$ in these integrals and then imposing $t \rightarrow r/v_j$ on the resultant inversion which can be found in transform tables. The terms in the present case may be found in the report from which this paper was abstracted [8] along with numerical comparisons with the present integral solution which show good agreement.

Focusing attention first on the curves in Figs. 3 to 5, representing the plate response to the step excitation, it may be seen that consistent with the solution statements, (26) to (28), two resultant waves are involved, the fronts of these waves propagating with the velocities v_1 and v_2, as in a Timoshenko beam. It may be noted in the present case that the concentrated step shear excitation causes no discontinuities in the moment-response curves, although there are discontinuities at the wave fronts in the moment-time derivatives. In the case of the shear force there is no discontinuity associated with the first wave front but an infinite discontinuity occurs at the second wave front (shear front). Hence the step concentrated shear load at $r = 0$ becomes in propagation a more severe discontinuity. It should be emphasized, however, that these wave-front characteristics, although a property of the solution, are not to be considered as accurate information since they represent the very short waves and very high frequencies in the present theory. As Mindlin [9] points out, this theory is good only up to moderately short waves and moderately high frequencies as a comparison with the three-dimensional theory shows (see Fig. 5.071 in this reference).

Considering next the long-time behavior of the moment solutions shown in Figs. 3 and 4 for the step input, it may be seen that the response of both moments at a particular station r/h increases indefinitely with time. In the limit, this long-time behavior should be equivalent to that for the static problem of an infinitely large circular plate subjected to a concentrated load at $r = 0$. It may be shown that the solution for a problem of this type according to the present theory is the same as in the elementary theory. Timoshenko [10] gives the solution for the case of built-in outer edge as

$$\left.\begin{aligned} \frac{M_r}{Q_0} &= \frac{1}{4\pi}\left[(1+\nu)\log\frac{a}{r} - 1\right] \\ \frac{M_\theta}{Q_0} &= \frac{1}{4\pi}\left[(1+\nu)\log\frac{a}{r} - \nu\right] \end{aligned}\right\} \qquad (53)$$

where a is the plate radius. Since the stations close to $r = 0$ will be independent of the type of boundaries at large r, this solution suffices to show that for fixed r the moments in an infinitely large plate ($a \to \infty$) are infinite, in agreement with the long-time character exhibited in Figs. 3 and 4. In turn, the long-time shear-load solution should be just $Q_r = Q_0/2\pi r$ or

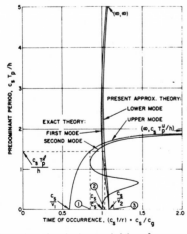

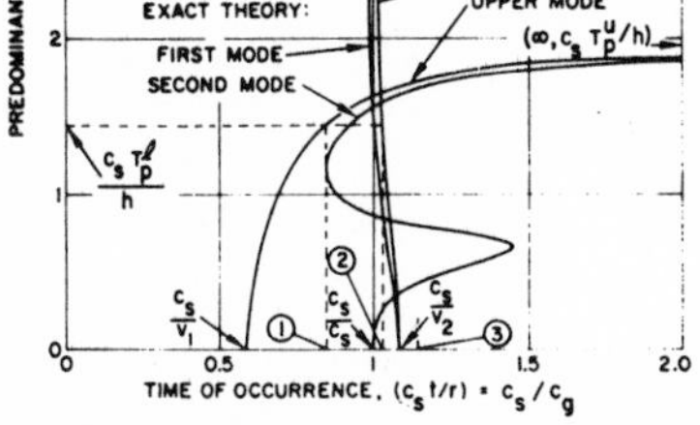

Fig. 6 Comparison of predominant period-time of occurrence curves in present and exact theories (ν = 0.3)

$$\frac{hQ_r}{Q_0} = \frac{h}{2\pi r} \qquad (54)$$

Letting $r/h = 1$ there results $hQ_r/Q_0 = 1/2\pi = 0.159$, in agreement with the long-time limit of the relative curve in Fig. 5. It may be noted that (54) results from just the two nonintegral terms in (50) and (51), added according to the third of (28).

Consider next the curves in Figs. 3 to 5 corresponding to the rectangular pulse input (of duration k/h). In the limit, as t becomes large, the long-time response of the plate (at a station r/h) to the rectangular pulse input should be equivalent to that for the static problem of an unloaded infinitely large circular plate; i.e., zero moments and shear force. Figs. 3 to 5 exhibit this character which follows from (52) and the related equations for M_θ and Q_r.

Of importance in the present problem is the influence of the spacial variable r on the response amplitudes. Before this kind of information can be deduced from the curves in Figs. 3 to 5, however, it is necessary to have some criterion for judging their accuracy. This is necessary since as pointed out earlier the solution given here is derived from a theory that is a good approximation only when limited to moderately short waves and moderately high frequencies. Therefore what is needed is a guide that will establish what parts of the curves in Figs. 3 to 5 are composed predominantly of the longer and lower frequency waves in the theory. Such a guide is contained in a comparison of the predominant period T_p versus time of occurrence t-relations from the exact and the present approximate theories that result from applying Kelvin's stationary phase method to the problem of an infinite plate, subjected initially to a shear load that is infinite at $r = 0$ and zero elsewhere. In the method this initial distribution of shear load is expressed as a Fourier integral. Physically the input might be thought of as the superposition of an infinite number of sinusoidal wave trains, agreeing in phase at the origin at time $t = 0$, but interfering to give zero shear load elsewhere. At time t later, the result at a station r is obtained by summing the contributions due to all the waves, which have traveled ct (where c is the phase velocity). The main effect at station r is due to a small group of waves that have nearly equal phase velocities, periods, and wave lengths, and that are in phase at the station at time t.

The $T_p - t$ relation can be derived from the phase velocity-wave number relation of the theory, related group velocity c_g-wave-number relation, and the stationary phase condition $r - c_g t = 0$. It is convenient to introduce dimensionless variables similar to those used by Davies [11] who first used the present scheme in an analysis of the elastic rod. In the present case the dimensionless predominant period and time of occurrence are

$$\frac{c_s T_p}{h} = \frac{\Lambda c_s}{hc} \quad \text{and} \quad \frac{c_s t}{r} = \frac{c_s r}{r c_g} = \frac{c_s}{c_g} \qquad (55)$$

where Λ is wave length.

The phase velocity-wave number relations from the exact (three-dimensional theory) and present approximate theories for the free elastic plate are given in [2]. Use of the well-known $c_g - c$ relation yields the group velocity-wave number relations for the two theories. A particular dimensionless wave number h/Λ, then, yields the dimensionless variables given by (55). The numerical results are given in Fig. 6.[8]

First it should be noted that Fig. 6 reflects the important features of the group velocity-wave number relations of the two theories and their comparison. It shows that as the wave length decreases (accompanying frequency gets higher and period shorter) the

[8] For the present purposes, only this figure is necessary. The phase and group velocity-wave number relations for the first two modes from the exact theory are given in a paper by Tolstoy and Usdin [12].

group velocity of the second mode of the exact theory goes through a maximum and a minimum before reaching its limiting short-wave velocity c_s. On the other hand, the group velocity from the corresponding mode of the approximate theory increases monotonically to its limiting short-wave velocity v_1 which for $\nu = 0.3$ is about 70 percent greater than c_s.

The first mode in the present approximate theory is seen to have excellent agreement with the exact theory for all periods. This, of course, was shown in the phase velocity-wave number comparison in [2]. Fig. 6 shows that there is also excellent agreement of the two theories for the second mode of wave transmission from the upper bound on periods $c_sT_p{}^u/h$ (an inverse cut-off frequency) down to moderately short periods. The lower bound on second-mode periods $c_sT_p{}^l/h$ is taken as that predominant period of the approximate theory that occurs at the same time as the related period from the exact theory. The dashed lines in the figure indicate this. The corresponding initial time of interest in Figs. 3 to 5 is based on this criterion then; i.e., the initial dimensionless time will be taken as that marked ① in Fig. 6. This criterion says that for dimensionless times earlier than ① there will be no representation in the second mode of the approximate theory that corresponds to the second mode of the exact theory.[9] For all times later than ① then, as far as their long period (and accompanying low frequency and long wave-length) character is concerned, the solutions (in Figs. 3 to 5 for the step-input case) should be good approximations; i.e., Fig. 6 shows in this region of time good agreement between the present approximate and exact theories for long period waves. Note that the initial time ① has been indicated in Fig. 3 to 5 for each station r/h. It is arrived at easily by noting that the time of occurrence in Fig. 6 can be written as

$$\left(\frac{h}{h}\right)\frac{c_s t}{r} = \frac{c_s}{c_g} \quad \text{hence} \quad \frac{t}{h} = \frac{(r/h)}{c_s}\left(\frac{c_s}{c_g}\right)$$

pointing out the linear dependence the time of occurrence has on the station position.

One further thing must be taken into account in our criterion for judging the accuracy of the curves in Figs. 3 to 5. This is a means for dealing with the discontinuity that occurs in the shear force at $t = r/v_2$, and its coupled effect on the moments occurring at the same time. The literature contains enough conclusive evidence that in bounded elastic media, such as beams and plates, no propagating discontinuities are present. It follows that no such character would be expected from the exact theory. Hence aside from being of interest in pointing out the limitations of the present approximate theory, information from the vicinity of the wave front at $t = r/v_2$ in the present problem will be excluded.[10] The stationary phase solution in Fig. 6 can give no fixed means of deciding on a region about $t = r/v_2$ to exclude in the solutions given by Figs. 3 to 5; i.e., it offers no means of judging amplitudes. It does, however, offer a seemingly logical approximation in this direction.

The scheme is as follows: The lower bound on upper-mode periods $c_sT_p{}^l/h$ is also taken as the lower bound on the lower-mode periods. The time of occurrence of this period is marked in Fig. 6 as ②, arrived at by extending the dashed horizontal line through $c_sT_p{}^l/h$ to the right until it intersects the lower-mode curve of the approximate theory. Then if it is assumed that the discontinuity affects the same region before c_s/v_2 as after, the excluded region is taken as that between ② and ③; i.e.,

$$c_s/v_2 + c_s/v_2 - ② = ③$$

Hence the scheme amounts to imposing the same limitations on both modes of the theory; i.e., periods less than $c_sT_p{}^l/h$ are not recognized. It should be noted that this lower-bound period limitation is consistent with the upper bound on frequencies argued in [9].

The times ①, ②, and ③ indicated in Figs. 3 to 5 refer to the step-input curves. On each of the curves this criterion means that the disturbance in advance of ① and between ② and ③ is not recognized. This, however, is not a severe limitation on the solution since it may be seen that most of the response curve is acceptable. The same remark applies to the rectangular pulse-input curves where this criterion only adds another set of the same limitations delayed by k/h, indicated by the primed times ①, ②, and ③. It should be remarked that this stationary phase criterion is best suited to large time analysis. Hence use of it in judging basically short-time results, like those in Figs. 3 to 5 (i.e., r/h stations are in close), may be subject to criticism. However, the author has found, in some earlier work with waves in an elastic rod employing an approximate theory quite similar to that used here, that results from the stationary phase and integral-solution analysis agree very well even for stations close in [14].

The character of the spatial influence in the present solutions is amply demonstrated by the results in Figs. 3 to 5, in spite of the limitations on numerical evaluation at larger r/h stations. Consider the curves representing the response to the rectangular pulse input. The mean of the moment M_r, Fig. 3, in the "valid" region between ① and ② decays very rapidly with r/h, going from $M_r/Q_0 \cong 0.075$ at $r/h = 1$ to ~ 0.018 at $r/h = 3$. In the region between ③ and k/h the three curves exhibit nearly equal values for the mean moment, although the maximum value in this region is the lowest for the $r/h = 3$ station. The later "valid" regions between primed ① and ② and after primed ③, both show sharp decays in the mean moments with the r/h increase. It may be noted that the tail of the disturbance exhibits rather sizable moment values that seem to decay rather slowly with time. It must be pointed out, however, that the extent of absolute time represented in Figs. 3 to 5 is rather short, an $h = 1/4$-in. plate, for instance, is represented by just $\sim 18 \times 10^{-6}$ sec in these figures.

The mean of the moment M_θ, Fig. 4, decays rapidly with r/h in the important region between ③ and k/h and in the later regions of validity between primed ① and ② and after primed ③. The mean moment in the early region between ① and ② is relatively unimportant from a magnitude point of view and exhibits no particular trend in its dependence on r/h.

The mean of the shear force Q_r, Fig. 5, decays rapidly with r/h in the region between ③ and k/h. The early region between ① and ② exhibits a slow decay in mean shear force. The later region of validity between primed ① and ② and after primed ③ also exhibit rapid decays in the mean shear force.

It is of interest to make a calculation representative of the outer fiber (maximum) tensile stresses in the present problem. Since the present approximation has greater validity with larger distance from the origin, a moment value representing the $r/h = 3$ station will be selected. The value $M_r/Q_0 = 0.05$ will be used being representative of the region between ③ and k/h, Fig. 3. It is also representative of the M_θ/Q_0 response at $r/h = 3$ for the

---

[9] It should be emphasized that ① signifies the earliest arriving disturbance, governed by the exact theory, only when the lowest two modes are considered. Actually when higher modes of transmission are taken into account they govern, as Tolstoy and Usdin [13] have shown, earlier arrival times (larger c_g) than the second mode. These higher modes become more important as the impact becomes sharper and of shorter duration.

[10] It should be noted that this is still consistent with the results in Fig. 6. Even though the predominant periods are in agreement at c_s/v_2 it does not necessarily follow that amplitudes there have to have like agreement. In particular an approximate theory, based on "plane sections remain plane," such as the present one, would not be expected to give reliable information based on its very short-wave character.

later region after primed ③, Fig. 4. The outer fiber stresses then are given by

$$\sigma_{\theta 0} = \sigma_{r0} = \frac{0.3Q_0}{h^2} \tag{56}$$

With say a load Q_0 of 5000 lb and a plate thickness h of $^1/_4$ in. (56) gives stresses of 24,000 psi.

Figs. 3 to 5 bring out another interesting point worth mentioning. They indicate that, if the pulse time k is made smaller, the magnitudes of the moment response becomes less. Figs. 3 to 5 bring this out with a shift of the k/h point to the left which cuts off the long-time step response earlier. Hence with a shorter pulse it appears that the stresses are reduced. Said in other words, the longer pulse with its relatively longer wave influence excites the higher stresses. It appears that the shear force Q_r is not as sensitive to this effect.

Acknowledgments

Acknowledgment is due Mr. Francis Welsh, Jr., of the Computing Center at Space Technology Laboratories, Inc., who contributed the numerical evaluation work contained herein.

References

1 Ya. S. Uflyand, "The Propagation of Waves in the Transverse Vibrations of Bars and Plates," Akad. Nauk, *USSR Prikladnaya Matematika i Mekhanika*, vol. 12, 1948, pp. 287–300.

2 R. D. Mindlin, "Influence of Rotary Inertia and Shear on Flexural Motions of Isotropic, Elastic Plates," Journal of Applied Mechanics, vol. 18, Trans. ASME, vol. 73, 1951, pp. 31–38.

3 J. L. Lubkin, "Propagation of Elastic Impact Stresses," ONR Progress Report No. 5, Contract No. Nonr-704(00), April, 1954.

4 J. Miklowitz, "Flexural Wave Solutions of Coupled Equations Representing the More Exact Theory of Bending," Journal of Applied Mechanics, vol. 20, Trans. ASME, vol. 75, 1953, pp. 511–514.

5 J. Miklowitz, "Flexural Waves in Beams According to the More Exact Theory of Bending," NAVORD Report No. 2049, NOTS, China Lake, Calif., September, 1953.

6 A. Erdelyi, W. Magnus, F. Oberhettinger, and F. G. Tricomi, "Higher Transcendental Functions," Bateman Project, McGraw-Hill Book Company, Inc., New York, N. Y.,1953, p. 86.

7 N. W. McLachlan, "Complex Variable Theory and Transform Calculus," Cambridge University Press, Cambridge, England, second edition, 1953, p. 65.

8 J. Miklowitz, "Flexural Stress Waves Generated by the Sudden Punching Process in a Stretched Elastic Plate," Space Technology Laboratories, Inc., Report No. TR-59-0000-00701, June 5. 1959, pp. 28–29.

9 R. D. Mindlin, "An Introduction to the Mathematical Theory of Vibrations of Elastic Plates," U. S. Army Signal Corps Engineering Laboratories Monograph, Fort Monmouth, N. J., 1955, pp. 5.25–5.29.

10 S. Timoshenko, "Theory of Plates and Shells," McGraw-Hill Book Company, Inc., New York, N. Y., 1940, p. 75.

11 R. M. Davies, "A Critical Study of the Hopkinson Pressure Bar," *Philosophical Transactions of The Royal Society of London*, series A, vol. 240, 1948, pp. 375–457.

12 I. Tolstoy and E. Usdin, "Dispersive Properties of Stratified Elastic and Liquid Media: A Ray Theory," *Geophysics*, vol. 18, 1953, pp. 844–870.

13 I. Tolstoy and E. Usdin, "Wave Propagation in Elastic Plates: Low and High Mode Dispersion," *Journal of the Acoustical Society of America*, vol. 29, 1957, pp. 37–42.

14 J. Miklowitz, "The Propagation of Compressional Waves in a Dispersive Elastic Rod," Journal of Applied Mechanics, vol. 24, Trans. ASME, vol. 79, 1957, pp. 231–239.

14

Reprinted from *J. Appl. Mech.*, **35**, 467–475 (1968)

C.-T. SUN
Research Engineer.

J. D. ACHENBACH
Associate Professor. Mem. ASME

GEORGE HERRMANN
Professor. Mem. ASME

Department of Civil Engineering,
The Technological Institute,
Northwestern University,
Evanston, Ill.

Continuum Theory for a Laminated Medium[1]

A system of displacement equations of motion is presented, pertaining to a continuum theory to describe the dynamic behavior of a laminated composite. In deriving the equations, the displacements of the reinforcing layers and the matrix layers are expressed as two-term expansions about the mid-planes of the layers. Dynamic interaction of the layers is included through continuity relations at the interfaces. By means of a smoothing operation, representative kinetic and strain energy densities for the laminated medium are obtained. Subsequent application of Hamilton's principle, where the continuity relations are included through the use of Lagrangian multipliers, yields the displacement equations of motion. The distinctive traits of the system of equations are uncovered by considering the propagation of plane harmonic waves. Dispersion curves for harmonic waves propagating parallel to and normal to the layering are presented, and compared with exact curves. The limiting phase velocities at vanishing wave numbers agree with the exact limits. The lowest antisymmetric mode for waves propagating in the direction of the layering shows the strongest dispersion, which is very well described by the approximate theory over a substantial range of wave numbers.

1 Introduction

The customary approach in constructing a theory to describe the mechanical behavior of a laminated composite consists of replacing the composite by a homogeneous but usually anisotropic medium whose material constants are determined in terms of the geometry and in terms of the material properties of the constituents of the composite. Theories of this type are termed "effective modulus theories." For a laminated medium, the effective elastic constants have been computed by Riznichenko [1],[2] Postma [2], White and Angona [3], and Rytov [4] on the basis of both static and dynamic considerations.

It appears that the effective modulus theory yields good results for static analysis. In considering dynamic problems, however, this theory has to be viewed with some suspicion. Indeed, for a laminated or a fiber reinforced composite one would expect that the heterogeneous structure of the medium should entail dispersion of free harmonic waves. This expectation has been confirmed for layered composites by solving appropriate boundary value problems of the theory of elasticity as discussed in [5] and in the Appendix of this paper. By analyzing a directionally reinforced composite as a homogeneous medium, however, the phase velocities are obtained as constants, and dispersion, as well as higher modes of motion, is ignored.

Recently, Herrmann and Achenbach [6] have proposed a conceptually different approach to constructing continuum models for the dynamic analysis of directionally reinforced composites. Instead of introducing a representative homogeneous medium by means of "effective moduli," representative elastic moduli are used for the matrix, and the elastic and geometric properties of the reinforcing elements are combined into effective stiffnesses. With the aid of certain assumptions regarding the deformation of the reinforcing elements and by employing a smoothing operation, approximate kinetic and strain energy densities for the composite material are obtained. Subsequent application of Hamilton's principle yields the displacement equations of motion. The equations of motion as derived in [6] represent a simple version of what was termed the "effective stiffness" theory, which is distinguished from the well-known effective modulus theory primarily because bending, shear, and extensional stiffnesses of the reinforcing elements enter the strain energy density of the laminated medium. In the simple version of the effective stiffness theory of [6], only restricted interaction between the reinforcing elements and the matrix is allowed.

It is found in [6] that for a layered medium the approximate dispersion curve of the lowest transverse mode shows good agreement over a considerable range of dimensionless wave numbers with an "exact" curve obtained by employing the equations of elasticity for all layers [5]. The agreement is particularly good for the large ratios of the shear moduli that are predominant in actual laminates. An undesirable feature of the theory of [5] is that the representative elastic moduli in the matrix are not the actual elastic constants of the matrix material but certain fictitious moduli to be computed from limiting phase velocities at vanishing wave numbers.

In this paper a more accurate system of displacement equations of motion is presented for a laminated medium. For both the reinforcing layers and the matrix layers the displacements are expressed as linear expansions about the midplanes of the layers. In this manner effective stiffnesses are introduced for both the matrix layers and the reinforcing layers. The dynamic interaction between matrix and reinforcing layers is better described by simulating continuity of the displacements at the interfaces. In this theory only the actual material constants of the reinforcing layers and the matrix layers enter the analysis.

The displacement equations of motion are used to study the propagation of plane harmonic waves in the directions parallel to and normal to the layering. The limiting phase velocities at vanishing wave numbers agree with the constant phase velocities according to the effective modulus theory, and also with the limiting phase velocities obtained from the exact treatment; see [5] and the Appendix of this paper. For the lowest modes the dispersion curves according to the effective stiffness theory agree over a significant range of wave numbers with the exact dispersion curves.

2 Kinematics and Smoothing Operation

A stratified medium consisting of a large number of alternating plane, parallel layers of two homogeneous, isotropic materials is considered. The elastic constants and the thicknesses of the stiff reinforcing layers and the soft matrix layers are denoted by λ_f, μ_f, d_f and λ_m, μ_m, d_m, respectively. We focus the attention on the kth pair of reinforcing and matrix layers whose midplane

---

[1] This work was supported by the Office of Naval Research under Contract ONR Nonr.1228(34) with Northwestern University.

[2] Numbers in brackets designate References at end of paper.

Contributed by the Applied Mechanics Division and presented at the Applied Mechanics Conference, Providence, R. I., June 12–14, 1968, of The American Society of Mechanical Engineers.

Discussion of this paper should be addressed to the Editorial Department, ASME, United Engineering Center, 345 East 47th Street, New York, N. Y. 10017, and will be accepted until October 20, 1968. Discussion received after the closing date will be returned. Manuscript received by the ASME Applied Mechanics Division, November 20, 1967; revised draft, January 30, 1968. Paper No. 68—APM-19.

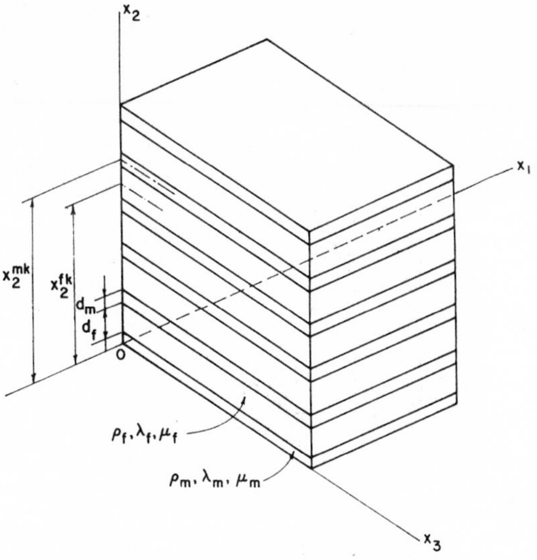

Fig. 1 Laminated medium

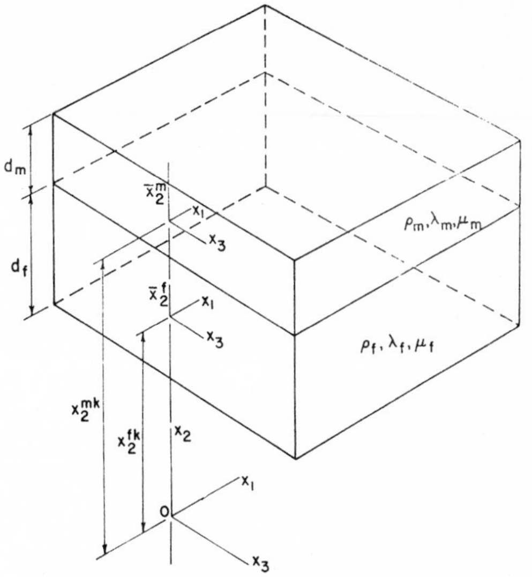

Fig. 2 Pair of reinforcing and matrix layers

positions are defined by x_2^{fk} and x_2^{mk}, respectively, see Fig. 1. The displacements at the midplanes are denoted by u_{0i}^{fk} (x_1, x_2^{fk}, x_3, t) and u_{0i}^{mk} (x_1, x_2^{mk}, x_3, t) for reinforcing and matrix layers, respectively. Two local coordinate systems $(x_1, \bar{x}_2^f, x_3)$ and $(x_1, \bar{x}_2^m, x_3)$ are defined with axes parallel to x_i and with origins in the midplanes of the reinforcing layer and the matrix layer, respectively, see Fig. 2.

Analogously to a procedure that was introduced earlier for approximate plate theories [7], the displacements in the kth reinforcing layer, u_i^{fk}, and the displacement in the kth matrix layer, u_i^{mk}, are expanded in an infinite series of Legendre polynomials. For the kth reinforcing layer, we write

$$u_i^{fk} = \sum_{n=0}^{\infty} P_n(\bar{x}_2^f/d_f) u_{ni}^{fk}(x_1, x_2^{fk}, x_3, t), \tag{1}$$

where $P_n(\bar{x}_2^f/d_f)$ are the Legendre polynomials,

$$\begin{aligned} P_0(\bar{x}_2^f/d_f) &= 1; \qquad P_1(\bar{x}_2^f/d_f) = \bar{x}_2^f/d_f; \\ P_2(\bar{x}_2^f/d_f) &= [3(\bar{x}_2^f/d_f)^2 - 1]/2, \text{ etc.} \end{aligned} \tag{2}$$

A similar expansion can be written for $u_i^{mk}(x_1, x_2^{mk}, \bar{x}_2^m, x_3, t)$. Conceptually, any degree of accuracy of the displacement approximation can be achieved by retaining a sufficient number of terms in the series. With many terms the manipulation becomes, however, rather cumbersome and the approximation does not result in the desired simplification. In this paper it is, therefore, stipulated that the thicknesses of the layers are sufficiently small as compared with any characteristic length of the deformation such that only linear terms in equations (1, 2) need be retained,

$$u_i^{fk} = u_{0i}^{fk}(x_1, x_2^{fk}, x_3, t) + \bar{x}_2^f \psi_{2i}^{fk}(x_1, x_2^{fk}, x_3, t) \tag{3}$$

$$u_i^{mk} = u_{0i}^{mk}(x_1, x_2^{mk}, x_3, t) + \bar{x}_2^m \psi_{2i}^{mk}(x_1, x_2^{mk}, x_3, t). \tag{4}$$

In equation (3), u_{0i}^{fk} is the displacement at the median plane of the kth reinforcing layer, ψ_{21}^{fk} and ψ_{23}^{fk} represent antisymmetric thickness shear deformations, and ψ_{22}^{fk} represents symmetric thickness stretch deformation. Since the displacements must be continuous at the interface of the kth reinforcing and the kth matrix layers, we write

$$\begin{aligned} &u_{0i}^{mk}(x_1, x_2^{mk}, x_3, t) - u_{0i}^{fk}(x_1, x_2^{fk}, x_3, t) \\ &= \tfrac{1}{2} d_f \psi_{2i}^{fk}(x_1, x_2^{fk}, x_3, t) + \tfrac{1}{2} d_m \psi_{2i}^{mk}(x_1, x_2^{mk}, x_3, t). \end{aligned} \tag{5}$$

With a view toward constructing representative potential and kinetic energy densities for the laminated medium, we first compute approximate expressions for the strain energy stored in elements of unit surface area of the kth pair of reinforcing and matrix layers.

In an isotropic, linearly elastic body, the strain energy density can be written as

$$W = \tfrac{1}{2}\lambda \epsilon_{ii}\epsilon_{jj} + \mu \epsilon_{ij}\epsilon_{ij}, \tag{6}$$

where λ and μ are Lamé's constants, and the strain tensor ϵ_{ij} is defined as

$$\epsilon_{ij} = \tfrac{1}{2}(\partial_j u_i + \partial_i u_j), \tag{7}$$

in which

$$\partial_j u_i = \partial u_i / \partial x_j. \tag{8}$$

For the kth reinforcing layer, approximations for the components of the strain tensor are obtained by substituting the assumed displacement components, equation (3), into equation (7), where, however, the differentiation in x_2-direction should be, with respect to the local coordinate, $\bar{x}_2^f$. We find

$$\epsilon_{11}^{fk} = \partial_1 u_{01}^{fk} + \bar{x}_2^f \partial_1 \psi_{21}^{fk} \tag{9}$$

$$\epsilon_{22}^{fk} = \psi_{22}^{fk} \tag{10}$$

$$\epsilon_{33}^{fk} = \partial_3 u_{03}^{fk} + \bar{x}_2^f \partial_3 \psi_{23}^{fk} \tag{11}$$

$$\epsilon_{12}^{fk} = \epsilon_{21}^{fk} = \tfrac{1}{2}(\partial_1 u_{02}^{fk} + \psi_{21}^{fk} + \bar{x}_2^f \partial_1 \psi_{22}^{fk}) \tag{12}$$

$$\epsilon_{32}^{fk} = \epsilon_{23}^{fk} = \tfrac{1}{2}(\partial_3 u_{02}^{fk} + \psi_{23}^{fk} + \bar{x}_2^f \partial_3 \psi_{22}^{fk}) \tag{13}$$

$$\epsilon_{13}^{fk} = \epsilon_{31}^{fk} = \tfrac{1}{2}[\partial_3 u_{01}^{fk} + \partial_1 u_{03}^{fk} + \bar{x}_2^f(\partial_3 \psi_{21}^{fk} + \partial_1 \psi_{23}^{fk})]. \tag{14}$$

By replacing superscripts f by superscripts m in equations (9–14), the corresponding expressions for the strain components in the kth matrix layer are obtained.

Substitution of the expressions (9)–(14) into equation (6) and integration over the thickness d_f yield the strain energy stored per unit surface area of the kth reinforcing layer as

$$\begin{aligned} W^{fk} = {} & \tfrac{1}{2} d_f(\lambda_f + 2\mu_f)[(\partial_1 u_{01}^{fk})^2 + (\partial_3 u_{03}^{fk})^2 + (\psi_{22}^{fk})^2] \\ & + d_f \lambda_f(\partial_1 u_{01}^{fk} \partial_3 u_{03}^{fk} + \partial_1 u_{01}^{fk} \psi_{22}^{fk} + \partial_3 u_{03}^{fk} \psi_{22}^{fk}) \\ & + \tfrac{1}{2} d_f \mu_f[(\partial_1 u_{02}^{fk} + \psi_{21}^{fk})^2 + (\partial_3 u_{02}^{fk} + \psi_{23}^{fk})^2 \\ & + (\partial_3 u_{01}^{fk} + \partial_1 u_{03}^{fk})^2] \end{aligned}$$

$$+ \frac{1}{24} d_f^3(\lambda_f + 2\mu_f)[(\partial_1\psi_{21}^{fk})^2 + (\partial_3\psi_{23}^{fk})^2]$$

$$+ \frac{1}{12} d_f^3\lambda_f\partial_1\psi_{21}^{fk}\partial_3\psi_{23}^{fk}$$

$$+ \frac{1}{24} d_f^3\mu_f[(\partial_1\psi_{22}^{fk})^2 + (\partial_3\psi_{22}^{fk})^2 + (\partial_3\psi_{21}^{fk} + \partial_1\psi_{23}^{fk})^2]. \quad (15)$$

A similar computation yields an expression for the strain energy stored in an element of unit surface area of the kth matrix layer. The actual expression for W^{mk} can be written by replacing in equation (15) subscripts and superscripts f by subscripts and superscripts m.

The kinetic energy density of a continuum is defined by

$$T = \sum_{i=1}^{3} \tfrac{1}{2}\rho(\dot{u}_i)^2, \quad (16)$$

where ρ is the mass density. By the use of the displacement expansion, equation (3), the kinetic energy per unit surface area of the kth reinforcing layer is obtained as

$$T^{fk} = \sum_{i=1}^{3} \{\tfrac{1}{2}(d_f + d_m)\eta\rho_f(\dot{u}_{0i}^{fk})^2 + \tfrac{1}{2}(d_f + d_m)I_f(\dot{\psi}_{2i}^{fk})^2\}, \quad (17)$$

in which ρ_f is the mass density of the reinforcing material, and

$$\eta = d_f/(d_f + d_m) \quad (18)$$

$$I_f = \frac{1}{12} d_f^2\eta\rho_f. \quad (19)$$

The kinetic energy per unit surface area of the kth matrix layer is obtained by replacing in equation (17) superscripts and subscripts f by m, and defining

$$I_m = \frac{1}{12} d_m^2(1 - \eta)\rho_m. \quad (20)$$

Suppose there are n reinforcing layers and n matrix layers within a certain length l in the x_2-direction; the total strain and kinetic energies stored in a rectangular parallelepiped of sides l, unity and unity, of the laminated medium are then given by

$$W_l = \sum^{n} (W^{fk} + W^{mk}) \quad (21)$$

and

$$T_l = \sum^{n} (T^{fk} + T^{mk}), \quad (22)$$

respectively. The basic premise of the effective stiffness theory is that the summations over $2n$ discrete points x_2^{fk} and x_2^{mk} may be approximated by a weighted integration over x_2. Thus

$$W_l = \sum^{n} (W^{fk} + W^{mk}) \simeq \int_l \frac{1}{d_f + d_m} (W^f + W^m)dx_2 \quad (23)$$

$$T_l = \sum^{n} (T^{fk} + T^{mk}) \simeq \int_l \frac{1}{d_f + d_m} (T^f + T^m)dx_2. \quad (24)$$

The superscript k has been removed on the right-hand sides of equations (23, 24) to indicate that W^f, W^m, T^f, and T^m are now defined for all x_2. By means of this "smoothing operation," we have, in fact, replaced the layered medium by a homogeneous continuum whose strain and kinetic energy densities are functions of x_2 as well as of x_1, x_3, and t:

$$W = [W^f(x_i, t) + W^m(x_i, t)]/(d_f + d_m) \quad (25)$$

$$T = [T^f(x_i, t) + T^m(x_i, t)]/(d_f + d_m). \quad (26)$$

The field variables u_{0i}^{fk}, u_{0i}^{mk}, ψ_{2i}^{fk}, ψ_{2i}^{mk}, which were thus far defined only on certain discrete parallel planes, $x_2 = x_2^{fk}$ and $x_2 = x_2^{mk}$, have now become functions of x_2, and henceforth we delete the superscript k.

The state of deformation in the laminated medium is now described by twelve field variables u_{0i}^f, u_{0i}^m, ψ_{2i}^f, and ψ_{2i}^m, which are functions of x_1, x_2, x_3, and t. The number of twelve field variables is reduced to nine by observing that u_{0i}^f and u_{0i}^m should be considered as representing the same quantity, namely, the "gross displacement," henceforth referred to as u_i. The same cannot be said of the field variables ψ_{2i}^f and ψ_{2i}^m, which describe "local deformations" in the reinforcing layers and the matrix layers, respectively. The local deformations and the gradients of the gross displacements are related, however, by conditions of continuity at the interfaces of the layers. In view of equation (5), we may write in terms of the gross displacements and the local deformations

$$u_i(x_1, x_2^{mk}, x_3, t) - u_i(x_1, x_2^{fk}, x_3, t)$$
$$= \tfrac{1}{2}d_f\psi_{2i}^f(x_1, x_2^{fk}, x_3, t) + \tfrac{1}{2}d_m\psi_{2i}^m(x_1, x_2^{mk}, x_3, t). \quad (27)$$

Noting that

$$x_2^{mk} = x_2^{fk} + \tfrac{1}{2}(d_m + d_f), \quad (28)$$

we assume that the thicknesses of the layers are sufficiently small that the difference relation (27) can be replaced by a differential relation between the local deformations and the gradient of the gross displacement:

$$(d_f + d_m)\partial_2u_i(x_1, x_2, x_3, t)$$
$$= d_f\psi_{2i}^f(x_1, x_2, x_3, t) + d_m\psi_{2i}^m(x_1, x_2, x_3, t). \quad (29)$$

The continuity condition, equation (5), has thus been generalized to hold at any point in the continuum. Since originally the local deformations ψ_{2i}^f and ψ_{2i}^m were only defined in the domains of the reinforcing layers and the matrix layers, respectively, we have neglected the differences between the local deformations at x_2^{fk} and x_2^{mk} in deriving equation (29).

In view of equations (15, 25), the approximate strain energy function of the laminated medium may now be written in terms of the derivatives of the gross displacement, and of the local deformations and their derivatives,

$$W = \tfrac{1}{2}a_1[(\partial_1u_1)^2 + (\partial_3u_3)^2] + \tfrac{1}{2}a_2(\partial_3u_1 + \partial_1u_3)^2$$
$$+ a_3(\partial_1u_1)(\partial_3u_3) + a_4(\partial_1u_1 + \partial_3u_3)\psi_{22}^f$$
$$+ \tfrac{1}{2}a_5(\psi_{22}^f)^2 + \tfrac{1}{2}a_6[(\partial_1\psi_{21}^f)^2 + (\partial_3\psi_{23}^f)^2]$$
$$+ a_7(\partial_1\psi_{21}^f)(\partial_3\psi_{23}^f) + \tfrac{1}{2}a_8[(\partial_1u_2 + \psi_{21}^f)^2 + (\partial_3u_2 + \psi_{23}^f)^2]$$
$$+ \tfrac{1}{2}a_9[(\partial_1\psi_{22}^f)^2 + (\partial_3\psi_{22}^f)^2 + (\partial_3\psi_{21}^f + \partial_1\psi_{23}^f)^2]$$
$$+ a_{10}(\partial_1u_1 + \partial_3u_3)\psi_{22}^m + \tfrac{1}{2}a_{11}(\psi_{22}^m)^2$$
$$+ \tfrac{1}{2}a_{12}[(\partial_1\psi_{21}^m)^2 + (\partial_3\psi_{23}^m)^2] + a_{13}(\partial_1\psi_{21}^m)(\partial_3\psi_{23}^m)$$
$$+ \tfrac{1}{2}a_{14}[(\partial_1u_2 + \psi_{21}^m)^2 + (\partial_3u_2 + \psi_{23}^m)^2]$$
$$+ \tfrac{1}{2}a_{15}[(\partial_1\psi_{22}^m)^2 + (\partial_3\psi_{22}^m)^2 + (\partial_3\psi_{21}^m + \partial_1\psi_{23}^m)^2], \quad (30)$$

where

$$a_1 = \eta(\lambda_f + 2\mu_f) + (1 - \eta)(\lambda_m + 2\mu_m) \quad (31)$$

$$a_2 = \eta\mu_f + (1 - \eta)\mu_m; \quad a_3 = \eta\lambda_f + (1 - \eta)\lambda_m \quad (32)$$

$$a_4 = \eta\lambda_f; \quad a_5 = \eta(\lambda_f + 2\mu_f) \quad (33)$$

$$a_6 = d_f^2\eta(\lambda_f + 2\mu_f)/12; \quad a_7 = d_f^2\eta\lambda_f/12 \quad (34)$$

$$a_8 = \eta\mu_f; \quad a_9 = d_f^2\eta\mu_f/12 \quad (35)$$

$$a_{10} = (1 - \eta)\lambda_m; \quad a_{11} = (1 - \eta)(\lambda_m + 2\mu_m) \quad (36)$$

$$a_{12} = d_m^2(1 - \eta)(\lambda_m + 2\mu_m)/12; \quad a_{13} = d_m^2(1 - \eta)\lambda_m/12 \quad (37)$$

$$a_{14} = (1 - \eta)\mu_m; \quad a_{15} = d_m^2(1 - \eta)\mu_m/12. \quad (38)$$

It is noted that some of the "material constants" of the laminated continuum are combinations of the elastic moduli of the original reinforcing layers and the matrix layers while others represent stiffnesses of the layers. The appearance of stiffnesses in the representative strain energy density has motivated the terminology "effective stiffness theory."

The approximate kinetic energy density is obtained from equations (17, 26) as

$$T = \sum_{i=1}^{3} [\tfrac{1}{2}\rho_c(\dot{u}_i)^2 + \tfrac{1}{2}I_f(\dot{\psi}_{2i}{}^f)^2 + \tfrac{1}{2}I_m(\dot{\psi}_{2i}{}^m)^2], \qquad (39)$$

in which

$$\rho_c = \eta\rho_f + (1 - \eta)\rho_m, \qquad (40)$$

and I_f and I_m are defined by equations (19, 20).

3 Displacement Equations of Motion

Let V be a fixed regular region of the laminated medium. Hamilton's principle for independent variations of the dependent field quantities in V at times t_0 and t_1 may be written as

$$\delta \int_{t_0}^{t_1} (T_V - W_V)dt + \int_{t_0}^{t_1} \delta W_1 dt = 0, \qquad (41)$$

where δW_1 is the variation of the work done by external forces, and T_V and W_V are the total kinetic and strain energies,

$$T_V = \int_V T dV \qquad (42)$$

$$W_V = \int_V W dV. \qquad (43)$$

In equations (42, 43), dV denotes the scalar volume element.

In this section we are interested only in the displacement equations of motion, and we restrict the admissible variations to vanish identically on the bounding surface of V. In the absence of body forces, the variational problem then reduces to finding the Euler equations for

$$\delta \int_{t_0}^{t_1} \int_V (T - W)dt dV = 0. \qquad (44)$$

In view of the continuity condition (29), six field variables may be chosen as independent; for example, u_i and $\psi_{2i}{}^f$. One way of finding the six corresponding field equations would be to eliminate $\psi_{2i}{}^m$ by means of equation (29) from the strain and kinetic energy densities, and then determine the system of Euler equations for equation (44). This procedure was followed in [10]. A perhaps more elegant method, however, is to introduce the continuity conditions as subsidiary conditions through Lagrangian multipliers by whose use the variational problem may be redefined as [8],

$$\delta \int_{t_0}^{t_1} \int_V (T - W - \lambda_1 S_1 - \lambda_2 S_2 - \lambda_3 S_3) dt dV = 0, \qquad (45)$$

where the Lagrangian multipliers λ_1, λ_2, and λ_3 are functions of x_i and t, and

$$S_i = (d_f + d_m)\partial_2 u_i - d_f\psi_{2i}{}^f - d_m\psi_{2i}{}^m. \qquad (46)$$

Since $(T - W - \lambda_1 S_1 - \lambda_2 S_2 - \lambda_3 S_3)$ depends only on the dependent field variables and their first order derivatives, the system of Euler equations may be written as [9]

$$\sum_{r=1}^{4} \frac{\partial}{\partial q_r}\left[\frac{\partial(T - W - \lambda_1 S_1 - \lambda_2 S_2 - \lambda_3 S_3)}{\partial(\partial f_s/\partial q_r)}\right] - \frac{\partial(T - W - \lambda_1 S_1 - \lambda_2 S_2 - \lambda_3 S_3)}{\partial f_s} = 0. \qquad (47)$$

In equation (47), f_s represent the twelve dependent variables u_i, $\psi_{2i}{}^f$, $\psi_{2i}{}^m$ and λ_i, and q_r are the spatial variables x_i and the time t.

Substitution of the strain energy density (30) and the kinetic energy density (39) in equation (47) yields a system of twelve equations:

$$a_1\partial_{11}u_1 + a_2\partial_{33}u_1 + (a_2 + a_3)\partial_{13}u_3 + a_4\partial_1\psi_{22}{}^f + a_{10}\partial_1\psi_{22}{}^m + \partial_2\lambda_1 = \rho_c\ddot{u}_1 \qquad (48)$$

$$a_2\partial_{11}u_2 + a_2\partial_{33}u_2 + a_8\partial_1\psi_{21}{}^f + a_8\partial_3\psi_{23}{}^f + a_{14}\partial_1\psi_{21}{}^m + a_{14}\partial_3\psi_{23}{}^m + \partial_2\lambda_2 = \rho_c\ddot{u}_2 \qquad (49)$$

$$a_1\partial_{33}u_3 + a_2\partial_{11}u_3 + (a_2 + a_3)\partial_{13}u_1 + a_4\partial_3\psi_{22}{}^f + a_{10}\partial_3\psi_{22}{}^m + \partial_2\lambda_3 = \rho_c\ddot{u}_3 \qquad (50)$$

$$a_6\partial_{11}\psi_{21}{}^f + (a_7 + a_9)\partial_{13}\psi_{23}{}^f + a_9\partial_{33}\psi_{21}{}^f - a_8\partial_1 u_2 - a_8\psi_{21}{}^f + \eta\lambda_1 = I_f\ddot{\psi}_{21}{}^f \qquad (51)$$

$$a_{12}\partial_{11}\psi_{21}{}^m + (a_{13} + a_{15})\partial_{13}\psi_{23}{}^m + a_{15}\partial_{33}\psi_{21}{}^m - a_{14}\partial_1 u_2 - a_{14}\psi_{21}{}^m + (1 - \eta)\lambda_1 = I_m\ddot{\psi}_{21}{}^m \qquad (52)$$

$$a_6\partial_{33}\psi_{23}{}^f + (a_7 + a_9)\partial_{13}\psi_{21}{}^f + a_9\partial_{11}\psi_{23}{}^f - a_8\partial_3 u_2 - a_8\psi_{23}{}^f + \eta\lambda_3 = I_f\ddot{\psi}_{23}{}^f \qquad (53)$$

$$a_{12}\partial_{33}\psi_{23}{}^m + (a_{13} + a_{15})\partial_{13}\psi_{21}{}^m + a_{15}\partial_{11}\psi_{23}{}^m - a_{14}\partial_3 u_2 - a_{14}\psi_{23}{}^m + (1 - \eta)\lambda_3 = I_m\ddot{\psi}_{23}{}^m \qquad (54)$$

$$a_9\partial_{11}\psi_{22}{}^f + a_9\partial_{33}\psi_{22}{}^f - a_4\partial_1 u_1 - a_4\partial_3 u_3 - a_5\psi_{22}{}^f + \eta\lambda_2 = I_f\ddot{\psi}_{22}{}^f \qquad (55)$$

$$a_{15}\partial_{11}\psi_{22}{}^m + a_{15}\partial_{33}\psi_{22}{}^m - a_{10}\partial_1 u_1 - a_{10}\partial_3 u_3 - a_{11}\psi_{22}{}^m + (1 - \eta)\lambda_2 = I_m\ddot{\psi}_{22}{}^m \qquad (56)$$

$$\partial_2 u_1 - \eta\psi_{21}{}^f - (1 - \eta)\psi_{21}{}^m = 0 \qquad (57)$$

$$\partial_2 u_2 - \eta\psi_{22}{}^f - (1 - \eta)\psi_{22}{}^m = 0 \qquad (58)$$

$$\partial_2 u_3 - \eta\psi_{23}{}^f - (1 - \eta)\psi_{23}{}^m = 0. \qquad (59)$$

In equations (48–59) we have used the notation

$$\partial_{ij}u_k = \partial^2 u_k/\partial x_i \partial x_j. \qquad (60)$$

4 Propagation of Plane Harmonic Waves

The displacement equations of motion (48–59) may be used to study the propagation of plane harmonic waves in an arbitrary direction, i.e., we might consider solutions of the form

$$A \exp[ik(n_j x_j - ct)]. \qquad (61)$$

In equation (61), A is a constant amplitude, k is the wave number, c is the phase velocity, and n_j are the components of the unit vector defining the direction of propagation. Since exact solutions for the propagation of plane waves normal to or parallel to the direction of the layering are available (see [5] and the Appendix), we restrict ourselves here to the corresponding approximate solutions obtained from equations (48–59).

Case 1: Propagation of longitudinal waves in the direction of the layering ($n_1 = 1$, $n_2 = n_3 = 0$):

For waves of this type the field variables are of the form

$$(u_1, \psi_{22}{}^f, \psi_{22}{}^m, \lambda_2) = (A_1, A_{22}{}^f, A_{22}{}^m, B_2) \exp[ik(x_1 - ct)]. \qquad (62)$$

All other field variables vanish identically. The substitution of equation (62) in the displacement equations of motion yields a system of four homogeneous equations for A_1, $A_{22}{}^f$, $A_{22}{}^m$, and B_2. The dispersion equation is obtained by requiring that the determinant of the coefficients vanishes. In nondimensional form the dispersion equation can be written as

$$[(1-\eta)+\eta\theta]^2\xi^2\beta^4 - \{[(1-\eta)+\eta\theta][(1-\eta)+\eta\gamma+\eta\gamma\delta_f + (1-\eta)\delta_m]\xi^2 + 12\eta[(1-\eta)+\eta\theta][\gamma\delta_f+\eta\delta_m/(1-\eta)]\}\beta^2 + \{[\eta\gamma\delta_f+(1-\eta)\delta_m][(1-\eta)+\eta\gamma]\xi^2 + 12\eta[\gamma\delta_f+\eta\delta_m/(1-\eta)][\eta\gamma\delta_f+(1-\eta)\delta_m] - 12\eta^2(\gamma\epsilon_f-\epsilon_m)^2\} = 0. \quad (63)$$

In equation (63) we have introduced the following dimensionless quantities:

Dimensionless phase velocity

$$\beta = c/(\mu_m/\rho_m)^{1/2}, \quad (64)$$

Dimensionless wave number

$$\xi = kd_f, \quad (65)$$

Ratio of mass densities

$$\theta = \rho_f/\rho_m, \quad (66)$$

Ratio of shear moduli

$$\gamma = \mu_f/\mu_m, \quad (67)$$

and constants δ_f, δ_m, ϵ_f, and ϵ_m, which depend on Poisson's ratios of the reinforcing layers and the matrix layers:

$$\delta_f = 2(1-\nu_f)/(1-2\nu_f); \quad \epsilon_f = 2\nu_f/(1-2\nu_f) \quad (68)$$

$$\delta_m = 2(1-\nu_m)/(1-2\nu_m); \quad \epsilon_m = 2\nu_m/(1-2\nu_m). \quad (69)$$

For vanishing wave number ξ the dimensionless phase velocity reduces to

$$\beta_{11} = \left\{\frac{[\gamma\delta_f+\eta\delta_m/(1-\eta)][\eta\gamma\delta_f+(1-\eta)\delta_m] - \eta(\gamma\epsilon_f-\epsilon_m)^2}{[(1-\eta)+\eta\theta][\gamma\delta_f+\eta\delta_m/(1-\eta)]}\right\}^{1/2} \quad (70)$$

The phase velocity (70) agrees with the limiting phase velocity at vanishing wave number obtained from an exact analysis which was discussed in detail in [5].

Case 2: Propagation of *SV* waves in the direction of the layering ($n_1 = 1, n_2 = n_3 = 0$):

We now consider a shear wave for which the nonvanishing field variables are u_2, ψ_{21}^f, ψ_{21}^m and λ_1. We write

$$(u_2, \psi_{21}^f, \psi_{21}^m, \lambda_1) = (A_2, A_{21}^f, A_{21}^m, B_1)\exp[ik(x_1-ct)]. \quad (71)$$

Following the previously described procedure, we obtain the dispersion equation corresponding to the system of solutions (71) in nondimensional form as

$$[(1-\eta)+\eta\theta]^2\xi^2\beta^4 - \{[(1-\eta)+\eta\theta][(1-\eta)+\eta\gamma]\xi^2 + [(1-\eta)+\eta\theta][\eta\gamma\delta_f+(1-\eta)\delta_m]\xi^2 + 12[(1-\eta)+\eta\theta][(1-\eta)\gamma+\eta]\eta/(1-\eta)\}\beta^2 + \{[(1-\eta)+\eta\gamma][\eta\gamma\delta_f+(1-\eta)\delta_m]\xi^2 + 12\eta\gamma/(1-\eta)\} = 0. \quad (72)$$

For vanishing dimensionless wave number, we obtain

$$\beta_{12} = \left\{\frac{\gamma}{[(1-\eta)+\eta\theta][(1-\eta)\gamma+\eta]}\right\}^{1/2}. \quad 73)$$

Equation (73) agrees with the limiting phase velocity for the transverse wave according to the exact solution as derived in [5].

Case 3: Propagation of *SH* waves in the direction of the layering ($n_1 = 1$; $n_2 = n_3 = 0$):

In this case the displacements are parallel to the x_3-direction while the waves propagate in the x_1-direction. The nonvanishing field variables are u_3, ψ_{23}^f, ψ_{23}^m, and λ_3. Closer inspection reveals that for this type of wave motion the system of equations separates into two uncoupled systems, governing symmetric and antisymmetric motion, respectively. The field quantity representing the symmetric motion is

$$u_3 = A_3 \exp[ik(x_1-ct)] \quad (74)$$

for which the constant phase velocity is obtained as

$$\beta_{13} = \left\{\frac{\eta\gamma+(1-\eta)}{\eta\theta+(1-\eta)}\right\}^{1/2}. \quad (75)$$

The phase velocity (75) agrees with the limiting phase velocity for symmetric SH waves propagating in the x_1-direction according to an exact analysis, as discussed in the Appendix.

The antisymmetric system is represented by the following solutions:

$$(\psi_{23}^f, \psi_{23}^m, \lambda_3) = (A_{23}^f, A_{23}^m, B_3)\exp[ik(x_1-ct)]. \quad (76)$$

The phase velocity of this motion is obtained as

$$\beta = \left\{\frac{[(1-\eta)+\eta\gamma]\xi^2 + 12\eta[\gamma+\eta/(1-\eta)]}{[(1-\eta)+\eta\theta]\xi^2}\right\}^{1/2}. \quad (77)$$

It is observed that β increases beyond bound as ξ decreases. By introducing the dimensionless frequency as

$$\Omega = \beta\xi, \quad (78)$$

equation (77) may, however, be rewritten as

$$\Omega = \left\{\frac{[(1-\eta)+\eta\gamma]\xi^2 + 12\eta[\gamma+\eta/(1-\eta)]}{[(1-\eta)+\eta\theta]}\right\}^{1/2}. \quad (79)$$

As $\xi \to 0$, the dimensionless frequency becomes

$$\Omega_{13} = \left\{\frac{12\eta[\gamma+\eta/(1-\eta)]}{[(1-\eta)+\eta\theta]}\right\}^{1/2}. \quad (80)$$

It is noted that the cut-off frequency given by equation (80) agrees with the limiting dimensionless frequency for the same type of wave motion according to an exact analysis, as carried out in the Appendix.

Case 4: Propagation of longitudinal waves normal to the layering ($n_2 = 1, n_1 = n_3 = 0$):

The solutions

$$(u_2, \psi_{22}^f, \psi_{22}^m, \lambda_2) = (A_2, A_{22}^f, A_{22}^m, B_2)\exp[ik(x_2-ct)] \quad (81)$$

represent a train of longitudinal waves traveling in the x_2-direction. The dispersion equation is obtained in the usual manner as

$$\{\theta\xi^4 + 12[(1-\eta)+\eta\theta]^2\xi^2\eta/(1-\eta)\}\beta^4 - \{12\xi^2[\gamma\delta_f+\eta^2\theta\delta_m/(1-\eta)^2] + 144\eta^2[(1-\eta)+\eta\theta][(1-\eta)\delta_f\gamma + \eta\delta_m]/(1-\eta)^2\}\beta^2 + 144\eta^2\gamma\delta_m\delta_f/(1-\eta)^2 = 0. \quad (82)$$

The limiting dimensionless phase velocity at vanishing wave number is

$$\beta_{22} = \left\{\frac{\gamma\delta_m\delta_f}{[(1-\eta)+\eta\theta][(1-\eta)\gamma\delta_f+\eta\delta_m]}\right\}^{1/2} \quad (83)$$

Equation (83) is exactly the same expression as obtained by means of an exact analysis; see Appendix.

Case 5: Propagation of transverse waves normal to the layering ($n_2 = 1, n_1 = n_3 = 0$):

Since the laminated medium has cylindrical symmetry with respect to the x_2-axis, we may consider, for convenience, a displacement parallel to the x_1-direction. Thus we write

$$(u_1, \psi_{21}^f, \psi_{21}^m, \lambda_1) = (A_1, A_{21}^f, A_{21}^m, B_1) \exp [ik(x_2 - ct)]. \quad (84)$$

In nondimensional form the corresponding dispersion equation is expressed as

$$\{\theta\xi^4 + 12\eta[(1 - \eta) + \eta\theta]^2\xi^2/(1 - \eta)\}\beta^4$$
$$- \{12[\gamma + \theta\eta^2/(1 - \eta)^2]\xi^2 + 144\eta^2[(1 - \eta) + \eta\theta][(1 - \eta)\gamma$$
$$+ \eta]/(1 - \eta)^2\}\beta^2 + 144\eta^2\gamma/(1 - \eta)^2 = 0. \quad (85)$$

As $\xi \to 0$, we have

$$\beta_{21} = \left\{\frac{\gamma}{[(1 - \eta) + \eta\theta][(1 - \eta)\gamma + \eta]}\right\}^{1/2} \quad (86)$$

which agrees with the phase velocity at vanishing wave number obtained from an exact solution; see Appendix.

5 Discussion of Dispersion Curves

In the previous section it was shown that for waves propagating normal to or parallel to the layering, the displacement equations of motion proposed in this paper yield the correct limiting values at vanishing wave number for the phase velocities and the cut-off frequencies of the lowest modes. It is of interest to point out that for longitudinal, transverse, and horizontally polarized waves propagating in both the x_1- and x_2-directions, the limiting phase velocities equations (70, 73, 75, 83, and 86) are identical to the values computed according to the effective modulus theory, see [2–4]. The antisymmetric SH mode, whose cut-off frequency is given by equation (80), cannot be described by the effective modulus theory. In this section, we compare the approximate dispersion curves with the exact dispersion curves and with the constant phase velocities according to the effective modulus theory for waves of finite length, thus exhibiting the merits of the effective stiffness theory over the effective modulus theory. The numerical computations were carried out for three values of γ, namely, $\gamma = 100$, 50, and 10, and for $\eta = 0.8$, $\theta = 3$, $\nu_f = 0.3$, and $\nu_m = 0.35$.

For transverse waves propagating in the x_1-direction, which are governed by equation (72), the dispersion curves are shown in Fig. 3. It is noted that the approximate curves show a very marked dispersive behavior, even at small values of ξ, and especially for large values of γ. This behavior is also predicted by the elasticity solution [5], and Fig. 3 shows good agreement up to values of kd_f as large as $kd_f = 4$, which corresponds to a wavelength of $1.57d_f$. The approximate and exact dispersion curves for longitudinal waves propagating in the x_1-direction, equation (63), are shown in Fig. 4. For the same set of values for the parameters θ, η, ν_f, and ν_m, the exact dispersion curves do not show such pronounced dispersion behavior at very long wave lengths as for transverse motion. At increasing wave numbers, however, the phase velocity decreases steeply, and the approximate curves then depart substantially from the exact curves. Considering the curve in Fig. 4 for $\gamma = \mu_f/\mu_m = 100$, it is seen that acceptable agreement is achieved up to, say $kd_f = 1$. In the range $0 < kd_f < 1$, the phase velocity is virtually constant, and equal to the phase velocity according to the effective modulus theory. For longitudinal motion the present version of the effective stiffness theory appears, therefore, to be of less interest, at least for the values of θ, η, ν_f and ν_m that are considered here. Returning to transverse motion, Fig. 3, we note, however, that for $\mu_f/\mu_m = 100$ and at $kd_f = 1$, the approximate and exact solutions agree, and the phase velocity is more than double the value according to the effective modulus theory. It is thus seen that, although longitudinal motion restricts the theory to rather large wavelength, the very pronounced dispersion of transverse motion at such wavelengths is very well described by the effective stiffness theory.

In a previous study of time-harmonic waves in a stratified medium propagating in the direction of the layering [5], it was surmised that for large γ the rapid decrease of the phase velocity of the lowest symmetric (longitudinal) mode was due to a shift of the participation in the motion of both reinforcing and matrix layers to the matrix layers only, where the latter are in symmetric motion and behave essentially as clamped layers. The lowest modes of symmetric motion in an elastic layer include both thickness

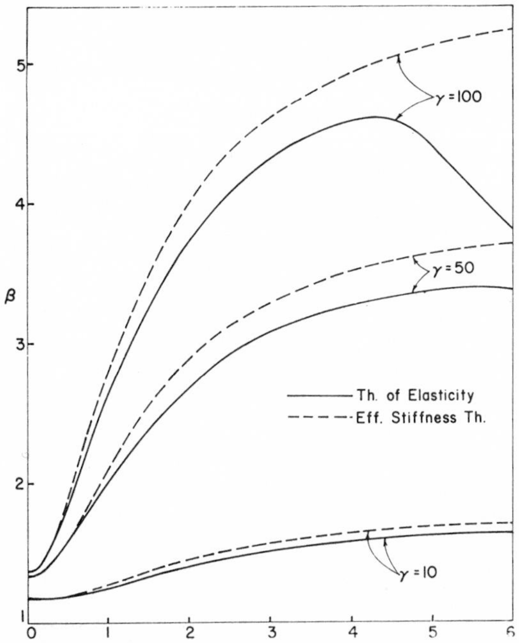

Fig. 3 Phase velocity vs. wave number for the lowest antisymmetric mode propagating in the direction of the layering

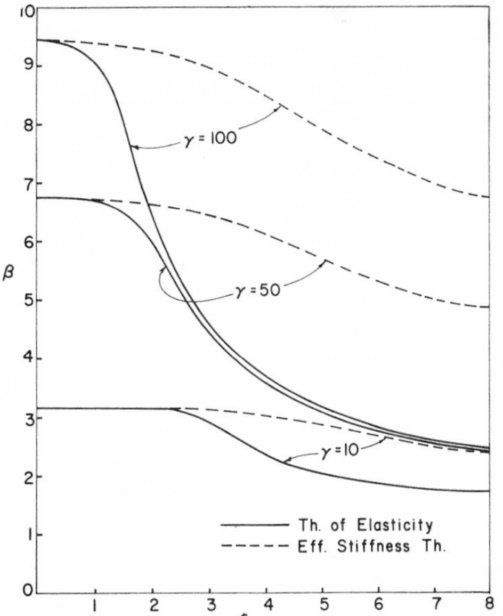

Fig. 4 Lowest symmetric mode propagating in the direction of the layering

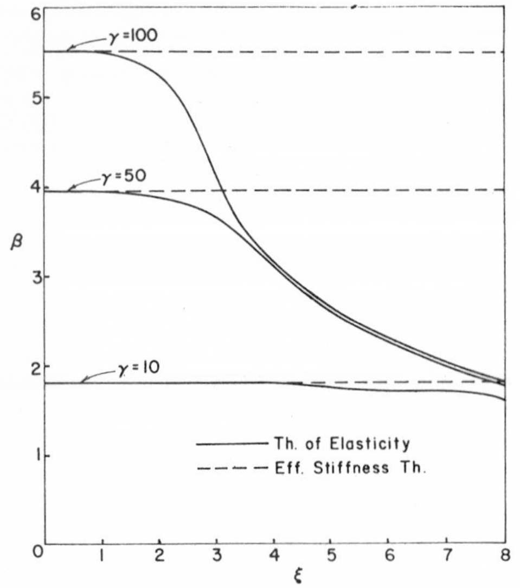

Fig. 5 Lowest symmetric *SH* mode propagating in the direction of the layering

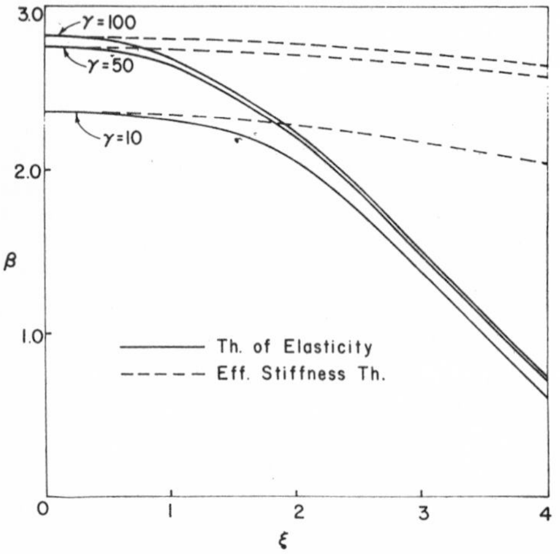

Fig. 6 Lowest longitudinal mode propagating normal to the layering

stretch and thickness shear motions. The first-order approximation, however, can account only for thickness stretch motion, as is evident from equations (3, 4). In a free plate the cut-off frequency of the lowest symmetric thickness shear mode is less than the cut-off frequency of the lowest symmetric thickness stretch mode if $\nu < 1/3$. In that case, no useful purpose is served by including in an approximate plate theory a representation of the lowest thickness stretch mode and ignoring the lowest symmetric thickness shear mode. A similar situation exists for the layered medium, except that not only Poisson's ratio, but also the parameters γ, θ, and η, determine the magnitudes of the cut-off frequencies; see [10, 11] for details.

The dispersion curves for symmetric horizontally polarized waves propagating in the x_1-direction, equation (75), are shown in Fig. 5. It is noted that no improvement over the effective modulus theory is obtained. Dispersion curves for longitudinal and transverse waves propagating in the x_2-direction are shown in Figs. 6 and 7. The improvements over the effective modulus theory are negligible. The phase velocities for anti-

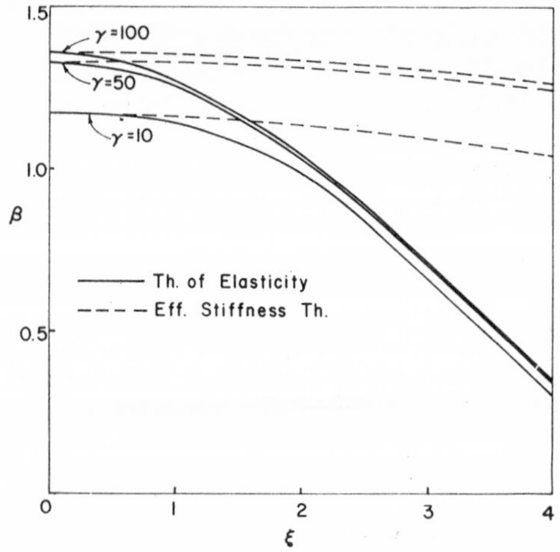

Fig. 7 Lowest transverse mode propagating normal to the layering

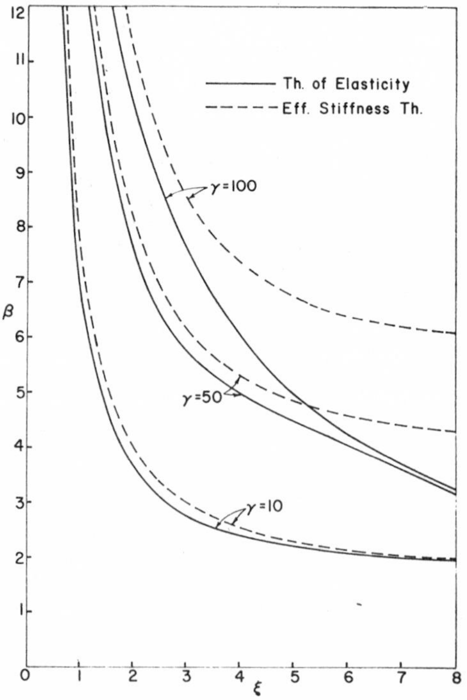

Fig. 8 Lowest antisymmetric *SH* mode propagating in the direction of the layering

symmetric horizontally polarized shear waves propagating in the x_1-direction are shown in Fig. 8. Agreement between exact and approximate dispersion curves is good for a substantial range of wave numbers, and it is noted that this type of motion cannot be described by the effective modulus theory.

It was noted at the beginning of section 4 that the displacement equations of motion (48–59) may be used to study plane harmonic waves propagating in an arbitrary direction. The authors do not have available, however, exact dispersion relations for waves propagating in directions other than normal or parallel to the layering, assuming that such waves are admitted by the exact equations. The question whether the waves that, according to the approximate equations, propagate under an angle with the layering agree with an exact solution can thus not be answered at this time.

6 Concluding Remarks

A system of displacement equations of motion for a laminated medium was derived. Dispersion relations for harmonic waves propagating parallel to and normal to the direction of the layering were presented, and the approximate dispersion curves were compared with exact curves. The limiting phase velocities at vanishing wave numbers agree with the exact limits. The lowest antisymmetric mode for waves propagating in plane strain in the direction of the layering shows the strongest dispersion, which is very adequately described by the approximate theory over a substantial range of wave numbers. In this range, both the approximate and the exact theory show negligible dispersion in the other modes of wave propagation that are considered here, except for the antisymmetric horizontally polarized shear wave propagating in the x_1-direction which cannot be described by the effective modulus theory, but which is adequately approximated by the effective stiffness theory. Further ingredients of a complete theory, such as explicit constitutive equations, boundary conditions, etc., will be discussed in a forthcoming sequel to this paper.

The approximate displacement distributions, equations (3, 4), may, for shorter wavelengths, lead to serious inaccuracies in the strain distributions, which, in turn, may affect the phase velocities. In plate theories, see [7], such inaccuracies can be compensated for by introducing appropriate correction factors. In [10] and [11] such correction factors were used to adjust the strain distributions in the layers of the laminated medium. If one is interested in wave propagation in one direction, for example the direction of the layering, somewhat better agreement between approximate and exact dispersion curves can then be achieved.

To obtain a radical improvement of the present theory, higher modes must be represented in the approximate displacement distributions. In particular, the first symmetric shear mode should be included. An effective stiffness theory which includes a representative of the latter mode was discussed in [10] and [11].

It is worth noting that the various dynamical continuum theories for laminated composites of the effective stiffness type advanced here and in the earlier paper [6] bear close resemblance to theories of linear elasticity with microstructure, which were discussed by Mindlin [12]. In interpreting these theories as pertinent to a laminated medium, their material constants can be written as functions of the geometric layout and of the classical constants of the constituent materials. For details, the reader is referred to [13].

References

1 Riznichenko, Yu. V., "Seismic Quasi-Anisotropy," *Izvest. Akad. Nauk. SSSR, Ser. Geofiz. i. Geogr.*, Vol. 13, 1949, pp. 518–544.

2 Postma, G. W., "Wave Propagation in a Stratified Medium," *Geophysics*, Vol. 20, 1955, pp. 780–806.

3 White, J. E., and Angona, F. A., "Elastic Wave Velocities in Laminated Media," *Journal of the Acoustical Society of America*, Vol. 27, 1955, pp. 311–317.

4 Rytov, S. M., "Acoustical Properties of a Thinly Laminated Medium," *Soviet Phys. Acoustics*, Vol. 2, 1956, pp. 68–80.

5 Sun, C. T., Achenbach, J. D., and Herrmann, G., "Time-Harmonic Waves in a Stratified Medium Propagating in the Direction of the Layering," JOURNAL OF APPLIED MECHANICS, Vol. 35, TRANS. ASME, Vol. 90, Series E, June 1968, pp. 408–411.

6 Herrmann, G., and Achenbach, J. D., "On Dynamic Theories of Fiber-Reinforced Composites," *Proceedings of the AIAA/ASME Eighth Structures, Structural Dynamics and Materials Conference*, March 29–31, 1967, Palm Springs, Calif., pp. 112–118; also Northwestern University Structural Mechanics Laboratory Technical Report No. 67-2, Jan. 1967.

7 Mindlin, R. D., and Medick, M. A., "Extensional Vibrations of Elastic Plates," JOURNAL OF APPLIED MECHANICS, Vol. 26, Trans. ASME, Vol. 81, Series E, 1959, pp. 561–569.

8 Courant, R., and Hilbert, D., *Methods of Mathematical Physics*, Vol. I, Interscience Publishers, New York, 1953, p. 219.

9 Hildebrand, F. B., *Methods of Applied Mathematics*, Prentice-Hall, New York, 1952, p. 136.

10 Sun, C.-T., Achenbach, J. D., and Herrmann, G., "Effective Stiffness Theory for Laminated Media," Northwestern University Structural Mechanics Laboratory Technical Report No. 67-4, July 1967.

11 Achenbach, J. D., and Herrmann, G., "Effective Stiffness Theory for a Laminated Medium," *Proceedings of the Tenth Midwestern Mechanics Conference*, forthcoming.

12 Mindlin, R. D., "Micro-structure in Linear Elasticity," *Arch. Rat. Mech. Anal.*, Vol. 16, 1964, pp. 51–78.

13 Herrmann, G., and Achenbach, J. D., "Applications of Theories of Generalized Cosserat Continua to the Dynamics of Composite Materials," *Proceedings IUTAM Symposium on the Generalized Cosserat Continuum Theory*, Springer-Verlag, Berlin, Göttingen, Heidelberg, 1968; also Northwestern University Structural Mechanics Laboratory Technical Report No. 67-7, October 1967.

APPENDIX

Exact Dispersion Relations for Harmonic Waves in a Layered Medium

In a homogeneous isotropic elastic material, the scalar and vector displacement potentials satisfy the following wave equations:

$$\nabla^2\varphi = \ddot{\varphi}/c_L^2, \qquad \nabla^2\mathbf{H} = \ddot{\mathbf{H}}/c_T^2, \tag{87}$$

where

$$c_L^2 = (\lambda + 2\mu)/\rho, \qquad c_T^2 = \mu/\rho. \tag{88}$$

The displacement vector may be written as

$$\mathbf{u} = \nabla\varphi + \nabla \times \mathbf{H} \tag{89}$$

Cases 1 and 2: Longitudinal and vertically polarized transverse waves propagating in the direction of the layering were considered in [5].

Case 3: For horizontally polarized transverse waves propagating in the direction of the layering, the nontrivial displacement components are u_3^f and u_3^m. For the reinforcing layers we have solutions of the form

$$u_3^f = [A_f \sin (ks_f\bar{x}_2^f) + B_f \cos (ks_f\bar{x}_2^f)] \exp [ik(x_1 - ct)]. \tag{90}$$

The corresponding shear stress is obtained as

$$\sigma_{23}^f = \mu_f ks_f[A_f \cos (ks_f\bar{x}_2^f) - B_f \sin (ks_f\bar{x}_2^f)] \exp [ik(x_1 - ct)]. \tag{91}$$

In equations (90) and (91),

$$s_f = (\rho_f c^2/\mu_f - 1)^{1/2}. \tag{92}$$

The corresponding expressions for u_3^m and σ_{23}^m are obtained by changing the subscripts and superscripts in equations (90) and (91). It is observed that the solutions can be separated into a symmetric system ($A_f = A_m \equiv 0$) and an antisymmetric system ($B_f = B_m \equiv 0$). Referring to Fig. 2, we write in view of continuity of the displacement and the stress at the interface of the matrix layer and the reinforcing layer,

$$u_3^f|_{\bar{x}_2^f = d_f/2} = u_3^m|_{\bar{x}_2^m = -d_m/2};$$
$$\sigma_{23}^f|_{\bar{x}_2^f = d_f/2} = \sigma_{23}^m|_{\bar{x}_2^m = -d_m/2}. \tag{93}$$

Because symmetric and antisymmetric motions with respect to the mid-planes of the layers may be considered separately, the displacements and stresses at $\bar{x}_2^f = -d_f/2$ are automatically matched with the corresponding displacements and stresses at $x_2^m = d_m/2$.

The continuity conditions (93) yield the dispersion relations in the usual manner. For symmetric SH waves we obtain

$$s_f\gamma \tan (\tfrac{1}{2}ks_fd_f) + s_m \tan (\tfrac{1}{2}ks_md_m) = 0. \tag{94}$$

For antisymmetric SH waves we find

$$s_f\gamma \tan (\tfrac{1}{2}ks_md_m) + s_m \tan (\tfrac{1}{2}ks_fd_f) = 0. \tag{95}$$

For the lowest modes, the solutions of equations (94) and (95) are plotted in Figs. 5 and 8.

Case 4: To study longitudinal waves propagating in the direction normal to the layering, we shift, for convenience, the origin of the coordinate system to the interface plane of one pair of layers.

For the reinforcing layers, solutions for the scalar potential are sought of the form

$$\varphi^f = \Phi_f(x_2) \exp [ik(x_2 - ct)]. \tag{96}$$

Solving the wave equation we find

$$\Phi_f = A_f \exp [-ik(1 + c/c_L^f)x_2] + B_f \exp [-ik(1 - c/c_L^f)x_2]. \tag{97}$$

Analogous expressions are obtained for the matrix layers. The displacement components u_2^f and u_2^m are obtained from equations (89, 96) and the nontrivial stresses follow from

$$\sigma_{22}^f = (\lambda_f + 2\mu_f)\partial u_2^f/\partial x_2, \qquad \sigma_{22}^m = (\lambda_m + 2\mu_m)\partial u_2^m/\partial x_2. \tag{98}$$

In view of the periodicity of the laminated medium, we can conclude from Floquet's theorem that the functions Φ_f and Φ_m are periodic with a period $d_f + d_m$. The condition of continuity at $x_2 = 0$ yields two equations for A_f, B_f, A_m, and B_m. Continuity at $x_2 = d_m$, and invoking the periodicity of Φ_f and Φ_m, yield two additional equations for the constants. In the usual manner the dispersion equation is then obtained as

$$\cos [k(d_f + d_m)] = \cos (kd_m c/c_L^m) \cos (kd_f c/c_L^f) - [(1 + p^2)/2p] \sin (kd_m c/c_L^m) \sin (kd_f c/c_L^f), \tag{99}$$

wherein

$$p = (\lambda_f + 2\mu_f)c_L^m/(\lambda_m + 2\mu_m)c_L^f. \tag{100}$$

The dispersion curves obtained from equation (99) are plotted in Fig. 6.

Case 5: The analysis for transverse waves propagating in the direction normal to the layering is very similar to the analysis for Case 4. It may be checked that the dispersion curve may be obtained by replacing in equations (99) and (100) $\lambda_m + 2\mu_m$ and $\lambda_f + 2\mu_f$ by μ_m and μ_f, respectively. The dispersion curves are plotted in Fig. 7.

15

Reprinted from *J. Acoust. Soc. Amer.*, **40**(6), 1489–1494 (1966)

Pressure Radiated by a Point-Excited Elastic Plate

David Feit

Cambridge Acoustical Associates, Inc., Cambridge, Massachusetts 02138

The acoustic pressure radiated by an infinite elastic plate excited by a time-harmonic point force or moment is obtained. The expression is valid for frequencies above and below the coincidence frequency. Below coincidence the pressure pattern is relatively nondirective and, for all practical purposes, insensitive to structural damping. Above coincidence, the pressure pattern becomes directive, and the peak pressures can be significantly reduced by structural damping.

LIST OF SYMBOLS

c	sound velocity of the fluid
c_1	plate wave velocity $[\{Eh/m(1-\sigma^2)\}^{\frac{1}{2}}]$
c_2	modified shear wave velocity $[\kappa\{Eh/2m(1+\sigma)\}^{\frac{1}{2}}]$
D	bending stiffness of plate $[Eh^3/\{12(1-\sigma^2)\}]$
E	Young's modulus
F_0	amplitude of point force
h	thickness of plate
I	rotatory inertia coefficient $[h^2/12]$
i	$(-1)^{\frac{1}{2}}$
k	Hankel-transform parameter
k_0	acoustic wavenumber $[\omega/c]$
l	separation distance between two forces
M_0	amplitude of point moment
m	mass of plate per unit area
P	acoustic power
p	acoustic pressure
(R,φ,θ)	spherical coordinates
(r,θ,z)	polar coordinates
S	shear deformation coefficient $[D/k^2Gh]$
t	time
v	velocity of the plate
$\bar{v}$	Hankel transform of plate velocity
η	loss factor
κ^2	shear correction factor
ρ_0	density of fluid medium
σ	Poisson's ratio
Φ	fluid-velocity potential
ω	angular frequency ($e^{-i\omega t}$ assumed throughout)
ω_c	coincidence frequency (Timoshenko–Mindlin plate theory) $[\omega_0\{(1-c^2/c_1^2)(1-c^2/c_2^2)\}^{-\frac{1}{2}}]$
ω_0	coincidence frequency (classical plate theory) $[c^2(m/D)^{\frac{1}{2}}]$

INTRODUCTION

The acoustic power radiated by a thin, isotropic or orthotropic plate of infinite extent excited by a time-harmonic point force has been investigated by Heckl[1–3] and by Maidanik and Kerwin.[4] The case in which the excitation is in the form of a point moment has been studied by Thompson and Rattayya.[5] For certain applications, it might be useful to have the directivity of the radiated field as well as the power radiated. An expression for the radiated sound pressure has been given by Morse and Ingard,[6] but there are errors in their expression and these have already been pointed out by Yarmush.[7]

The present study derives an expression for the sound pressure radiated by an isotropic elastic plate in response to either a force or a moment acting at a

[1] M. Heckl, "Schallabstrahlung von Platten bei punktförmiger Anregung," Acustica **9**, 371–380 (1959).

[2] M. Heckl, "Untersuchungen an orthotropen Platten," Acustica **10**, 109–115 (1960).

[3] M. Heckl, "Abstrahlung von einer punktförmig angeregten unendlich grossen Platte unter Wasser," Acustica **13**, 182 (1963).

[4] G. Maidanik and E. M. Kerwin, Jr., *Acoustic Radiation from Ribbed Plates Including Fluid-Loading Effects*, Bolt Beranek and Newman Inc. Rept. No. 1024 (Oct. 1963).

[5] W. Thompson, Jr., and J. V. Rattayya, "Acoustic Power Radiated by an Infinite Plate Excited by a Concentrated Moment," J. Acoust. Soc. Am. **36**, 1488–1490 (1964).

[6] P. M. Morse and K. U. Ingard, in *Encyclopedia of Physics* S. Flugge, Ed. (S. Hirzel Verlag, Berlin, 1961), Vol. 11/1, p. 113.

[7] D. Yarmush, *Theoretical Study of Resonant Stiffening Devices*, TRG, Inc., Rept. No. TRG-142-TN-64-10, pp. 183–184 (1964).

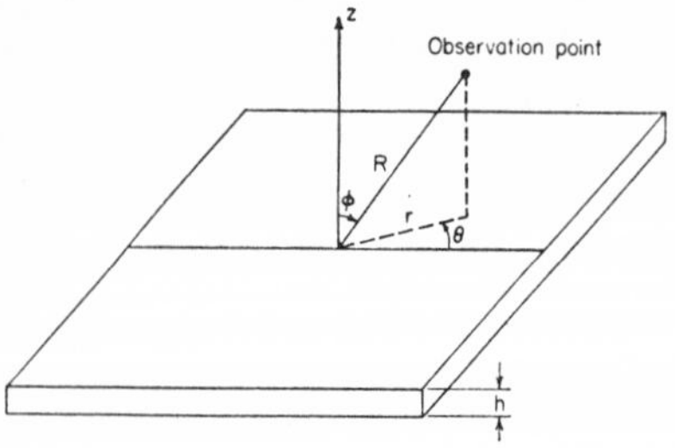

FIG. 1. Geometry of the plate and coordinate defining observation point.

point. Gutin[8] gives an expression, valid only at low frequencies, for the farfield pressure radiated by a thin elastic plate. Because of the high sound velocity of water as compared to the atmosphere, the coincidence frequency, i.e., the frequency above which the flexural wavelength in the plate exceeds the acoustic wavelength, is relatively high, even for steel plates. For this reason, it is meaningless to use thin-plate theory when analyzing sound radiation by plates driven above their coincidence frequency. In order to provide a better description of the phenomenon at higher frequencies, the Timoshenko–Mindlin theory of plate vibrations is used.[9] The Timoshenko–Mindlin theory, by including rotatory-inertia and transverse-shear terms, accounts for the coupling of flexural vibrations with thickness-shear vibrations, which becomes important at high frequencies. This theory has been used by Lubliner in studying the effect of an elastic plate on the acoustic field of a point source.[10]

The solution for the point-force excitation is derived first; then the solution for the moment excitation is found by taking the limit of the response due to two forces, equal in amplitude but opposite in phase, as the separation distance between the two forces vanishes. In order to assess the effects of structural damping, a complex Young's modulus is introduced and computations performed both with and without damping.

I. MATHEMATICAL ANALYSIS, FORCE EXCITATION

A. Formulation

Consider the half-space $z>0$ occupied by an acoustic fluid of density ρ_0 and sound velocity c, bounded by a thin elastic plate of infinite extent and thickness h lying in the plane $z=0$ (as shown in Fig. 1). The region lying below the plate is assumed to be a vacuum. The plate is excited by a time-harmonic point force $F_0e^{-i\omega t}\delta(r)/2\pi r$, where F_0 is the magnitude of the force, $\delta(r)$ is the Dirac delta function, r is the radial coordinate measured in the plane of the plate, and ω is the frequency of excitation in radians/second (time dependence of the form $e^{-i\omega t}$ is assumed throughout). Using polar coordinates (r,θ,z), the fluid velocity potential Φ (independent of θ, the circumferential angle measured in the plane of the plate) satisfies the reduced wave equation

$$\frac{\partial^2\Phi}{\partial r^2}+\frac{1}{r}\frac{\partial\Phi}{\partial r}+\frac{\partial^2\Phi}{\partial z^2}+k_0^2\Phi=0, \tag{1}$$

where $k_0=\omega/c$, the acoustic wavenumber. The pressure in terms of the velocity potential Φ is given by

$$p=i\rho_0\omega\Phi. \tag{2}$$

The Timoshenko–Mindlin plate equation of motion[9] is given by

$$\left[\left(\nabla_1^2+\frac{m\omega^2}{D}S\right)(D\nabla_1^2+m\omega^2I)-m\omega^2\right]v(r)$$
$$=-i\omega\left(1-S\nabla_1^2-\frac{m\omega^2}{D}IS\right)\frac{F_0\delta(r)}{2\pi r}$$
$$-\rho_0\omega^2\left(1-S\nabla_1^2-\frac{m\omega^2}{D}IS\right)\Phi(r,0), \tag{3}$$

where the operator $\nabla_1^2=\partial^2/\partial r^2+(1/r)\partial/\partial r$, v is the transverse velocity of the plate, m is the mass density of the plate material per unit area, $D=Eh^3/[12(1-\sigma^2)]$ is the bending stiffness of the plate, E is Young's modulus, and in the cases where structural damping is included is given by $E(1-i\eta)$, where η is the loss factor, $\sigma=$ Poisson's ratio, $I=h^2/12$, $S=D/(\kappa^2Gh)$, and $\kappa^2=\pi^2/12$, the shear correction factor. (For a discussion of this correction factor see Ref. 9.) The constants I and S represent, respectively, the effects of rotatory inertia and shear deformation. If the constants I and S are set equal to zero the classical equation of plate vibrations is obtained.

Introducing the condition of continuity of normal velocity at the plate–fluid interface

$$\frac{\partial\Phi}{\partial z}(r,0)=v(r), \tag{4}$$

and then transforming Eq. 3 by use of a Hankel transform, the boundary condition on the Hankel transform of the potential Φ is obtained:

$$f(k^2,\omega^2)\frac{d\tilde{\Phi}}{dz}(k,0)+\rho_0\omega^2g(k^2,\omega^2)\tilde{\Phi}(k,0)=-i\omega g(k^2,\omega^2)\frac{F_0}{2\pi}, \tag{5}$$

where

$$f(k^2,\omega^2)=D\left[k^4-\frac{m\omega^2}{D}(I+S)k^2+\left(\frac{m\omega}{D}\right)^2IS-\frac{m\omega^2}{D}\right], \tag{6}$$

[8] L. Y. Gutin, "Sound Radiation from an Infinite Plate Excited by a Normal Point Force," Soviet Phys.—Acoust. **10**, 369–371 (1964).

[9] R. D. Mindlin, "Influence of Rotatory Inertia and Shear on Flexural Motions of Isotropic, Elastic Plates," J. Appl. Mech. **18**, 31–38 (1951).

[10] J. Lubliner, "Effect of an Elastic Plate on the Acoustic Field of a Point Source," Columbia Univ. Tech. Rept. No. 32, ONR-226-(86) (July 1963).

$$g(k^2,\omega^2)=1+Sk^2-\frac{m\omega^2}{D}IS, \tag{7}$$

and f can be factored into the form

$$f(k^2,\omega^2)=D(k^2-\delta_1^2)(k^2-\delta_2^2), \tag{8}$$

where

$$\delta_{1,2}^2=\frac{1}{2}\frac{m\omega^2}{D}\{I+S\pm[(I-S)^2+4D/m\omega^2]^{\frac{1}{2}}\}. \tag{9}$$

The Hankel-transform pair is defined by

$$\begin{aligned}\tilde{\Phi}(k,z)&=\int_0^\infty \Phi(r,z)J_0(kr)rdr,\\ \Phi(r,z)&=\int_0^\infty \tilde{\Phi}(z,k)J_0(kr)kdk,\end{aligned} \tag{10}$$

where $J_0(kr)$ is the zero-order Bessel function.

Applying the Hankel transform to Eq. 1, it can be shown that $\tilde{\Phi}$ must satisfy the equation

$$\frac{d^2\tilde{\Phi}}{dz^2}+(k_0^2-k^2)\tilde{\Phi}=0. \tag{11}$$

The solution of this equation is given by

$$\tilde{\Phi}(k,z)=Ae^{+i(k_0^2-k^2)^{\frac{1}{2}}z}. \tag{12}$$

In order that $\tilde{\Phi}$ be bounded as z approaches infinity, it is necessary that

$$\mathrm{Im}[(k_0^2-k^2)^{\frac{1}{2}}]>0. \tag{13}$$

Introducing this solution into Eq. 5, solving the resulting equation for A and then using Eq. 2 to solve for the pressure, one obtains

$$p(r,z)=\frac{\rho_0\omega^2F_0}{2\pi D}\int_0^\infty \frac{g(k^2,\omega^2)}{\Delta(k^2,\omega^2)}J_0(kr)e^{+iz(k_0^2-k^2)^{\frac{1}{2}}}kdk, \tag{14}$$

where

$$\Delta(k^2,\omega^2)=i(k_0^2-k^2)^{\frac{1}{2}}(k^2-\delta_1^2)(k^2-\delta_2^2)+\frac{\rho_0\omega^2}{D}g(k^2,\omega^2). \tag{15}$$

An expression for the transverse velocity of the plate is obtained by using Eqs. 2, 4, and 14;

$$v(r)=\frac{\omega F_0}{2\pi D}\int_0^\infty (k_0^2-k^2)^{\frac{1}{2}}\frac{g(k^2,\omega^2)}{\Delta(k^2,\omega^2)}J_0(kr)kdk. \tag{16}$$

The integrals appearing in Eqs. 14 and 16 are transformed into contour integrals in the complex k plane.[11]

[11] A. Sommerfeld, *Partial Differential Equations* (Academic Press Inc., New York, 1949), p. 251.

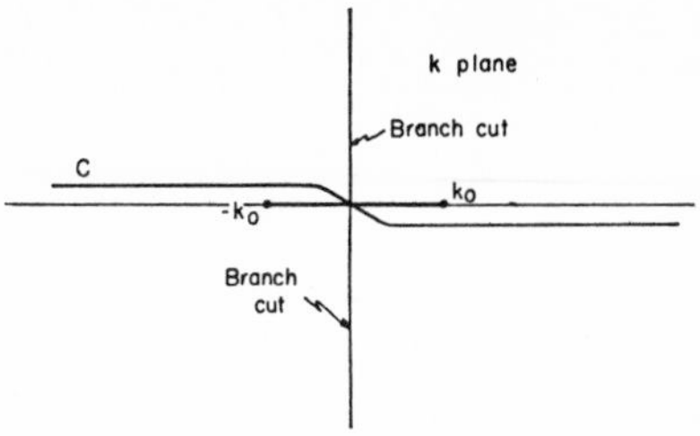

FIG. 2. Integration path in k plane for integrals defined in Eqs. 17 and 18.

Equations 14 and 16 become, respectively,

$$p(r,z)=\frac{\rho_0\omega^2F_0}{4\pi D}\int_C \frac{g(k^2,\omega^2)}{\Delta(k^2,\omega^2)}H_0^{(1)}(kr)e^{+iz(k_0^2-k^2)^{\frac{1}{2}}}kdk, \tag{17}$$

$$v(r)=\frac{\omega F_0}{4\pi D}\int_C (k_0^2-k^2)^{\frac{1}{2}}\frac{g(k^2,\omega^2)}{\Delta(k^2,\omega^2)}H_0^{(1)}(kr)kdk, \tag{18}$$

where $H_0^{(1)}(kr)$ is the zero-order Hankel function of the first kind, and the contour C goes from $-\infty$ to ∞ passing above the negative and below the positive real axis. The points $k=\pm k_0$ are branch points in the complex k plane and Eq. 13 defines the Riemann sheet on which the integration path C must be taken in order that the integrals defined in Eq. 17 and Eq. 18 be bounded for $z>0$. The path C, together with the branch cuts defined by

$$\mathrm{Im}[(k_0^2-k^2)^{\frac{1}{2}}]=0, \tag{19}$$

is shown in Fig. 2.

B. Evaluation of the Farfield Pressure

Using Cauchy's theorem, the integrals along the contour C can be transformed into the sum of a branch-line integral plus residue contributions. In order to obtain an expression for the farfield sound pressure, it is convenient to introduce the new complex variable w via

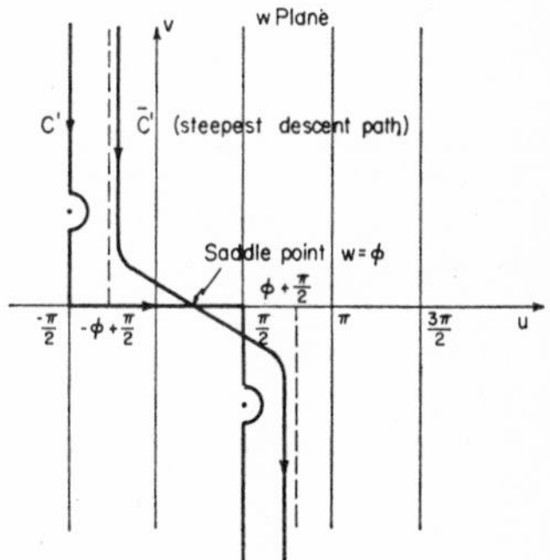

FIG. 3. Integration path and steepest-descent path in w plane for evaluation of integral in Eq. 21.

the transformation

$$k=k_0 \sin w. \tag{20}$$

This transformation maps the entire k plane onto a new complex plane $w=u+iv$. The path C in Fig. 2 is mapped into the path C' shown in Fig. 3. Under this transformation and introducing spherical coordinates, by letting $z=R\cos\varphi$, $r=R\sin\varphi$, where R is the distance of the observation point to the origin, and φ is the polar angle shown in Fig. 1, the integral in Eq. 17 becomes

$$p(R,\varphi,\theta)=\frac{F_0k_0^2}{4\pi}\int_{C'} H_0^{(1)}(k_0R\sin\varphi\sin w)\times\exp(ik_0R\cos\varphi\cos w)V(w)\sin w dw, \tag{21}$$

where $V(w)$ is given by

$$V(w)=\frac{\left(1+Sk_0^2\sin^2 w-\frac{m\omega^2}{D}IS\right)\cos w}{\left(1+Sk_0^2\sin^2 w-\frac{m\omega^2}{D}IS\right)+i\frac{Dk_0^5}{\rho_0\omega^2}\cos w\left(\sin^2 w-\frac{\delta_1^2}{k_0^2}\right)\left(\sin^2 w-\frac{\delta_2^2}{k_0^2}\right)}, \tag{22}$$

and θ is the circumferential angle measured in the plane of the plate (Fig. 1).

The integral in Eq. 21 is then evaluated asymptotically, i.e., at distances large as compared to the wavelength in the fluid, $k_0R\gg1$ by a saddle-point integration. In order to perform this integration, the path C' is deformed into the steepest-descent path $\bar{C}'$, which passes through the saddle point at $w=\varphi$ as shown in Fig. 3. Brekhovskikh[12] evaluates an integral of exactly the same form, and his results are used yielding

$$p(R,\varphi,\theta)\sim-\frac{iF_0k_0}{2\pi}\frac{e^{ik_0R}}{R}V(\varphi). \tag{23}$$

Using the definition of V given by Eq. 22 and the definitions of I and S in terms of $c_1^2=Eh/[m(1-\sigma^2)]$, $c_2^2=\kappa^2Eh/[2m(1+\sigma)]$, where c_1 is the plate-wave velocity, and c_2 is a modified shear-wave velocity, Eq. 23 can be written as

$$p(R,\varphi,\theta)\sim-\frac{iF_0k_0}{2\pi}\frac{e^{ik_0R}}{R}\cos\varphi\left\{\left[1+\frac{\omega^2h^2}{12c^2}\frac{c_1^2}{c_2^2}\left(\sin^2\varphi-\frac{c^2}{c_1^2}\right)\right]\Big/\left[1+\frac{\omega^2h^2}{12c^2}\frac{c_1^2}{c_2^2}\left(\sin^2\varphi-\frac{c^2}{c_1^2}\right)-\frac{i\omega m}{\rho_0c}\cos\varphi\left[1-\frac{\omega^2h^2c_1^2}{12c^4}\left(\sin^2\varphi-\frac{c^2}{c_1^2}\right)\left(\sin^2\varphi-\frac{c^2}{c_2^2}\right)\right]\right]\right\}. \tag{24}$$

In the low-frequency limit, $\omega h/c_1$, $\omega h/c_2$, $\omega h/c\ll1$ and assuming that $c/c_1\ll\sin\varphi$ and $c/c_2\ll\sin\varphi$, the above expression reduces to

$$p(R,\varphi,\theta)=-\frac{ik_0F_0}{2\pi}\frac{e^{ik_0R}}{R}\times\cos\varphi\Big/\left[1-\frac{i\omega m}{\rho_0c}\cos\varphi\left(1-\frac{\omega^2}{\omega_0^2}\sin^4\varphi\right)\right], \tag{25}$$

where ω_0, the coincidence frequency (the frequency as predicted by classical plate theory above which the flexural wavelength in the plate exceeds the acoustic wavelength) is given by

$$\omega_0^2=12c^4/h^2c_1^2. \tag{26}$$

Equation 25 is exactly the same as the expression obtained by Gutin.[8] This expression can be simplified even further if one assumes that the excitation frequency is well below coincidence; i.e., $\omega/\omega_0\ll1$. In this range, if $\omega m/\rho_0c\ll1$, the pressure in the fluid is the same as if the force were acting on the fluid alone, and is given by

$$p(r,\varphi,\theta)=-ik_0F_0(e^{ik_0R}/2\pi R)\cos\varphi, \tag{27}$$

which represents a dipole-type radiation.

If $\omega m/\rho_0c\gg1$, the pressure can be approximated by

$$p(r,\varphi,\theta)=\rho_0F_0e^{ik_0R}/2\pi mR, \tag{28}$$

which represents a monopole-type radiation. The case $\omega m/\rho_0c\ll1$ is a case of heavy fluid loading whereas the case $\omega m/\rho_0c\gg1$ is a case of light fluid loading.

The directivity pattern of the farfield pressure (using Eq. 24) normalized to the high-frequency pressure on the z axis $\varphi=0$ is shown in Fig. 4, for frequencies above and below coincidence. The calculations in this Figure as well as the following Figure were made for a steel plate in water. As can be seen in this Figure, the frequencies are given in terms of a ratio ω/ω_c, where ω_c is the frequency at which the phase speed of the propagating flexural wave of the Timo-

[12] L. M. Brekhovskikh, *Waves in Layered Media* (Academic Press Inc., New York, 1960), pp. 242–258.

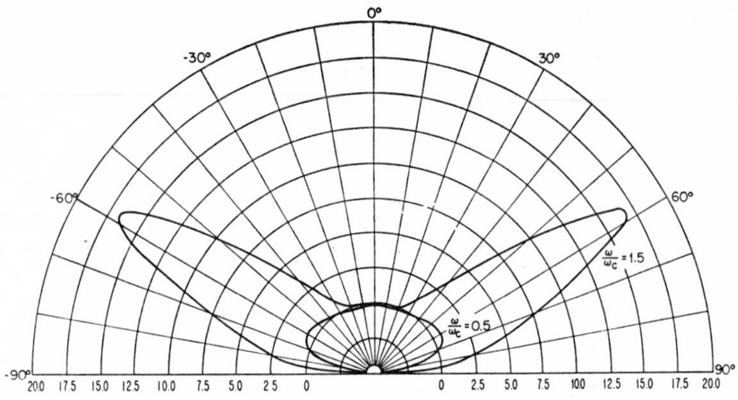

FIG. 4. Directivity function in dB normalized to high-frequency directivity at $\varphi=0°$.

shenko–Mindlin plate theory becomes equal to the sound speed of the water. It is of interest to note that this coincidence frequency is higher than the coincidence frequency predicted by classical plate theory and is given by

$$\omega_c/\omega_0=[(1-c^2/c_1^2)(1-c^2/c_2^2)]^{-\frac{1}{2}}. \tag{29}$$

For the case of a steel plate in water, ω_c is about 23% higher than ω_0.

Below coincidence, the propagating flexural wave does not radiate energy away from the plate. Therefore, the radiated sound pressure can be associated with the nonpropagating nearfield deflection of the plate, which acts like a small piston and therefore is relatively nondirective as shown in Fig. 4, and, like the nearfield response of the structure, insensitive to structural damping.

Above coincidence the situation is quite different. In this range, the phase speed of the propagating flexural wave exceeds that of the sound wave, and therefore the propagating wave can itself radiate energy away from the plate in a definite direction as shown in Fig. 4. Above coincidence, the peak pressure can be reduced by structural damping, and this is shown in Fig. 5. A loss factor of 10% is admittedly quite large but might be possible when a special damping treatment is used. Below coincidence, even such a large loss factor has essentially no effect on the radiated pressure pattern.

It might be of interest to note that, at a sufficiently high driving frequency, the so-called nearfield deflection mentioned earlier can itself become a propagating wave and at this frequency and above the directivity pattern will show yet another lobe. This would never be predicted by an analysis using classical plate theory.

C. Nearfield Pressure

To obtain the nearfield pressure, one must add to the steepest-descent result contributions due to singularities in the w plane that might be encountered in deforming the original path C' into the steepest-descent path $\bar{C}'$. For the present case in particular, the only singularities present would be poles of the function $V(w)$. If such a pole is crossed, then the residue at the pole must be added. The positions of the poles are determined by the zeros of the denominator of $V(w)$, which can be written as

$$\left(1+Sk_0^2\sin^2 w-\frac{m\omega^2}{D}IS\right)$$
$$+i\frac{Dk_0}{\rho_0\omega^2}\cos w\,(k_0^2\sin^2 w-\delta_1^2)(k_0^2\sin^2 w-\delta_2^2)=0. \tag{30}$$

Making the substitution $z=-i\cos w$, a fifth-degree equation with real coefficients[13] is obtained:

$$z^5+z^3\left(2-\frac{\delta_1^2}{k_0^2}-\frac{\delta_2^2}{k_0^2}\right)-z^2\frac{S}{D}\frac{\rho_0\omega^2}{k_0^3}+z\left(1-\frac{\delta_1^2}{k_0^2}\right)\left(1-\frac{\delta_2^2}{k_0^2}\right)$$
$$-\frac{\rho_0\omega^2}{Dk_0^5}\left(1+Sk_0^2-\frac{m\omega^2}{D}IS\right)=0. \tag{31}$$

It can be shown by using a zero counting theorem[14] that Eq. 31 has one real positive root, and four complex roots.

To each of the five zeros of Eq. 31, there correspond two values of w in the strip $-\pi/2\leq \mathrm{Re}w\leq 3\pi/2$ and, hence, there are 10 poles in the strip $-\pi/2<\mathrm{Re}w<3\pi/2$. Two of the poles fall on the contour of integration (those corresponding to the positive real root of Eq. 31) and must be avoided by identation as shown in Fig. 3. The pole contributions arise only for a range of φ greater than some minimum value φ_c, where φ_c is the smallest value of φ in the range $0<\varphi<90°$ for which the steepest descent path just crosses a pole. The pole waves can be of either the surface-wave type (contribution decays exponentially in a direction perpendicular

[13] This only holds for $\eta=0$ (no structural damping).

[14] E. C. Titchmarsh, *The Theory of Functions* (Oxford University Press, London, 1939), 2nd ed., p. 116.

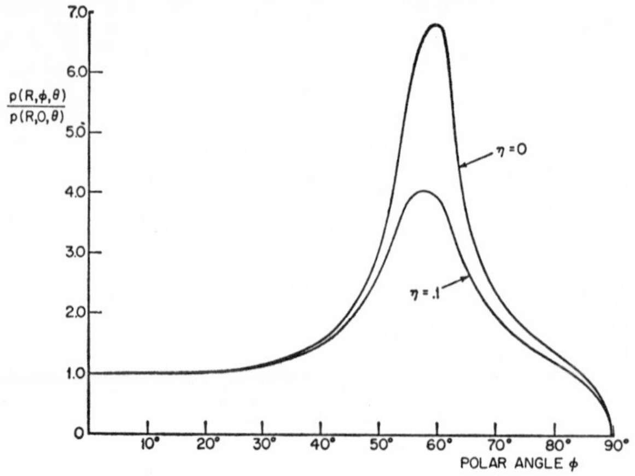

FIG. 5. Farfield pressure for $\omega/\omega_c = 1.5$, with and without structural damping.

to the surface) or the leaky-wave type (waves that grow exponentially with z, but decay with r, the radial distance away from the source in the plane of the plate). The pole contributions can be written as[15]

$$p(R,\varphi,\theta) = i\frac{F_0 k_0^2 \rho_0 \omega^2}{2D} \sum_{n=1}^{j} H_0^{(1)}(k_0 R \sin\varphi \sin w_n)$$

$$\times \exp(ik_0 R \cos\varphi \cos w_n) \sin w_n \left. \frac{g(k^2 \sin^2 w, \omega^2)}{\frac{\partial\Delta}{\partial w}(k_0^2 \sin^2 w, \omega^2)} \right|_{w=w_n},$$

$$\text{with } j \leq 4, \quad (32)$$

where j is the number of poles crossed in deforming C' into the steepest-descent $\bar{C}'$. For field points located on the plate, $\varphi = 90°$ and $j = 4$, the pressure on the plate is predominantly associated with the pole contributions given by Eq. 32, rather than with contributions from the branch-line integral.

II. MOMENT EXCITATION

As described in the Introduction, the farfield pressure due to moment excitation is obtained by utilizing the solution of the point-force excitation. Equation 23 gives the pressure at a point in the farfield owing to a force acting at the origin. By adding to this, the pressure due to a second force of equal magnitude but of opposite sign acting at a distance l at some arbitrary angle θ_0 in the plane of the plate, one obtains

$$p(R,\varphi,\theta) = -\frac{iF_0 k_0}{2\pi}\left(\frac{e^{ik_0 R}}{R} - \frac{e^{ik_0 R'}}{R'}\right) V(\varphi), \quad (33)$$

where

$$R' = R\left[1 - \frac{2l}{R}\cos(\theta - \theta_0)\sin\varphi + \frac{l^2}{R^2}\right]^{\frac{1}{2}}. \quad (34)$$

[15] R. V. Churchill, *Complex Variables and Applications* (McGraw–Hill Book Co., Inc., New York, 1960), 2nd ed., p. 160.

Now, taking the limit of the above expression as $l \to 0$ and $F_0 \to \infty$ in such a way that $F_0 l \to M_0$, a constant, one obtains

$$p(r,\varphi,\theta) = \frac{M_0 k_0^2}{2\pi} \frac{e^{ik_0 R}}{R} \cos(\theta - \theta_0) \sin\varphi V(\varphi). \quad (35)$$

In the low-frequency limit $\omega h/c_1$, $\omega h/c_2$, $\omega h/c \ll 1$, $V(\varphi)$ can be simplified considerably and Eq. 35 reduces to

$$p(r,\varphi,\theta) = \frac{M_0 k_0^2}{2\pi} \frac{e^{ik_0 R}}{R} \frac{\cos(\theta-\theta_0)\sin\varphi\cos\varphi}{1 - \frac{i\omega m}{\rho_0 c}\cos\varphi\left(1 - \frac{\omega^2}{\omega_0^2}\sin^4\varphi\right)}. \quad (36)$$

This formula for the case of moment excitation is analogous to Eq. 25 for the case of force excitation. The radiated acoustic power P can be shown to be

$$P = \frac{M_0^2 k_0^2}{8\pi\rho c} \int_0^{\pi/2} \left\{ \sin^3\varphi \cos^2\varphi d\varphi \middle/ \left[1 + \frac{\omega^2 m^2}{\rho_0^2 c^2}\cos^2\varphi\left(1 - \frac{\omega^2}{\omega_0^2}\sin^4\theta\right)^2\right]\right\}, \quad (37)$$

which agrees exactly with the expression for the radiated power obtained by Thompson and Rattayya[5] before simplifications necessary for the evaluation of the integral are made.

ACKNOWLEDGMENTS

This work was supported by the U. S. Department of the Navy, Bureau of Ships, Code 345.

The author is grateful to Dr. M. C. Junger and W. Thompson, Jr., of Cambridge Acoustical Associates, Inc., for helpful discussions during the course of this work.

Part III

FREE VIBRATION OF SHELLS

Editor's Comments on Papers 16 Through 25

The common mathematical models for thin shells admit a strain energy expression that consists of two parts: extensional (or stretching, or membrane) and flexural (or bending). In general, the extensional energy is produced by the extensional and shear strains in the middle surface, and it is proportional to thickness. Flexural energy is produced by the changes in curvature and torsion of the middle surface, and it is proportional to the cube of thickness.

For beams and plates, the extensional and flexural deformations are governed by separate systems of differential equations, and one type of deformation can occur without the other. As soon as curvature is introduced (i.e., the object is a shell), the two systems of governing equations are coupled, and pure extensional or flexural modes cannot occur.

The question of whether the flexural or extensional energy terms are the dominant ones in vibration of thin shells produced a rather lively debate at the end of the nineteenth century. Lord Rayleigh (*The Theory of Sound,* vol. 1, 2nd ed., London and Basingstoke: Macmillan, pp. 395–432, 1926) proposed the hypothesis that the low-frequency modes must be flexural and that for a sufficiently thin shell the contribution of extensional energy to the total strain energy must be negligible. An excerpt of his book expressing this view is included as Paper 16.

Following his hypothesis, Lord Rayleigh assumed that during vibration the displacement field is such that the two extensional strains and the shearing strain of the middle surface are zero. From these three differential equations, he obtained the three displacement components which he called *inextensional* displacements. (A concise derivation of the differential equations for the inextensional displacements for an arbitrary shell of revolution is given in the book by W. Flügge, *Stresses in Shells,* Berlin: Springer-Verlag, pp. 89–96, 1962. For general geometries, the inextensional displacements can be found by numerical integration of the three differential equations.)

Lord Rayleigh used the inextensional displacements to calculate approximate natural frequencies from the identification of the strain and kinetic energies (Rayleigh's Principle). Since the kinetic energy is proportional to thickness and to the square of frequency, and the flexural energy is proportional to the cube of thickness, Lord Rayleigh concluded that the natural frequency of the inextensional modes is proportional to thickness. He failed to emphasize the restrictions of this conclusion.

Following Lord Rayleigh's first article on this subject [*Proc. London Math. Soc.,* **13,** 4–16 (1881)], A. E. H. Love published a paper [*Philo. Trans. Roy. Soc. London,* Ser. A, **179,** 491–546 (1888)] in which he laid down the foundations of the defomation of thin shells. The summary of this paper is reproduced as Paper 17. Love criticized Rayleigh's hypothesis of the inextensible middle surface and pointed out that the inextensional displacements cannot satisfy specified boundary conditions; therefore, he reasoned, some stretching of the middle surface of a shell must occur. The essence of Love's argument is that for a sufficiently thin shell, the flexural energy must become smaller than the extensional energy, because the former is proportional to a higher power of thickness, and therefore Lord Rayleigh's hypothesis cannot be correct.

The crucial test in this debate is whether or not there exist modes of free vibration for which the natural frequency vanishes as the thickness is decreased indefinitely. The vanishing of the thickness can be associated with a shell for which the bending stiffness has diminished indefinitely (it is a membrane). The vanishing of the frequency means that certain displacements can be superimposed on the membrane without any resistance. It is clear that they must be inextensional displacements. But are there any conditions under which they cannot be superimposed? The answer to this question was provided by E. W. Ross, Jr., in Paper 18. He found that if the boundary conditions involve any one of the displacements that are tangential to the middle surface (Case V), then the lowest natural frequency is independent of thickness. For all other boundary conditions (Cases I–IV), the lowest natural frequency depends on thickness raised to various powers. For a free edge (Case I), it is proportional to thickness, just as Lord Rayleigh predicted. With Ross' help, the debate can be regarded as settled. It is all a matter of boundary conditions. If the tangential displacements are restricted on the edge, flexural energy will become negligible in comparison to extensional energy for sufficiently small values of the thickness, and Love is then correct. For a free edge, extensional energy is negligible and Lord Rayleigh is correct. For the other boundary conditions, the decision is not clear-cut.

One final point in this matter must be explained. When using inextensional displacements in Rayleigh's Principle, the predicted frequencies are ensured upper bounds only when no displacement boundary conditions are violated. This means that Rayleigh's Principle will ensure an upper bound for a free edge only (Ross' Case I), and only for this case can the prediction be regarded as accurate. For the other boundary conditions (Cases II–IV), the natural frequencies approach zero at the different rates [$\omega = 0(h^{3/4})$ for Case II, and $\omega = 0(h^{1/4})$ for Cases III and IV] and the predictions by Rayleigh's Principle may be much too low for a thin shell.

The desire to simplify vibration analyses of shells by considering the extensional and flexural deformations separately paralleled the same trend in static analyses that followed the derivation of the general shell equations at the end of the nineteenth century. Even though the general shell equations can be regarded as known since 1888, and the specific procedure for obtaining the complete solution of the free vibration problem of a finite cylindrical shell was already contained in Love's work (*A Treatise on the Mathematical Theory of Elasticity*, 4th ed., Cambridge, Eng.: Cambridge University Press, p. 545, 1927), numerical solutions of the frequency equations of the complete free-vibration problem began to appear only in the 1930s. W. Flügge [*Zeitschr. f. angew. Math. u. Mech.*, **13,** 425–427 (1933)] did it for cylindrical shells and K. Federhofer [*Sitzungsber. Akard. Wiss. Wien, Math.–Naturw. K. 2A*, **146,** 57–69, 505–514 (1937)] for spherical shells. With the appearance of these publications, it was recognized that the approximate treatment of shell vibrations, by considering either only the extensional or only the flexural strain energy, could provide useful information in some special cases, but that it covered only a small part of the possible vibration modes. For this reason, the remaining papers in this chapter deal exclusively with the complete problem.

The complete boundary-value problem of free vibration of a finite cylindrical shell was solved in detail by R. N. Arnold and G. B. Warburton in what is here Paper 19. The theoretical predictions of this paper are matched against experimental results (Figures 4 and 5) and excellent agreement is noted, which supports the soundness of the mathematical model and the analysis. A significant contribution of this paper is the strain energy calculation which shows (Figures 6 and 7) the fractions of stretching and bending energies that are contained in the total energy, for various types of modes. These figures show that, except for some transition regions, one or the other kind of energy tends to predominate, suggesting that, on the basis of their energy content, the free vibration modes of shells could be generally classified as bending or stretching modes. Arnold and Warburton limited their results to a certain type of simply-supported edges, with no axial constraint. The effect of the edge conditions on the natural frequency of free vibration of a finite cylindrical shell is examined by K. Forsberg in Paper 20. As shown clearly by Forsberg, the effect can be considerable. Especially, when constraints on the tangential displacements are imposed, the natural frequencies are raised considerably.

The approach that had been used for the free-vibration analysis of cylindrical shells in Papers 19 and 20 can be applied in principle equally well to other shells. However, the only other geometry for which such an approach has been practical is the spherical shell. This is because, for spherical shells, the governing equations are separable and their solutions involve Bessel and Legendre functions, which are well discussed in the literature. For other shell geometries, the determination of comparable solutions is simply too difficult.

A complete analysis of the transverse vibrations of a shallow spherical shell was done by E. Reissner, and is included in Paper 21. Reissner had shown in an earlier paper [*Quar. Appl. Math.*, **13,** 169–176 (1955)] that for transverse vibration the longitudinal inertia terms can be neglected. Starting with this assumption, Reissner was able to formulate the problem in a simpler way and obtained exact solutions. In Paper 22 the same problem was reconsidered by Kalnins with all the inertia terms included. He solved the complete frequency equation and found an additional set of modes that was not considered by Reissner. After the examination of their mode shapes, the additional modes were shown to have tangential displacements larger than the transverse displacement. The results of this paper show that the complete frequency spectrum of a shallow spherical shell consists of two infinite sets of modes with separate increasing node counts in the displacements. They show also that when curvature is diminished indefinitely these two sets of modes reduce to the purely longitudinal and purely transverse modes of a circular plate. The longitudinal modes contain mostly extensional strain energy, while the transverse modes contain mostly flexural strain energy.

In Paper 23, Kalnins further explores the distinction between the stretching and bending modes. His results show that for a nonshallow spherical bowl two distinct types of modes are again encountered. Energy calculations show that for the first type the bending energy is predominant while for the second the stretching energy is predominant. It can therefore be concluded that the idea of having separate groups of extensional and flexural modes can be carried over to shells of general shape as well. The significance of this classification lies in the variation of the natural frequencies with thickness, as shown in Figures 4 through 6. The extensional (or membrane) modes show a very small variation with thickness, while the bending modes show an approximately parabolic variation with thickness.

In reference to the discussion of Papers 16–18 with regard to extensional vs. inextensional modes, Figures 4 through 6 of Paper 23 make it clear that two different kinds of "extensional" modes are involved as the thickness is diminished indefinitely. One kind consists of the true extensional modes that are identified by *M* in the figures. The other kind are not extensional modes at all, even though their strain energy is all extensional. They have been designated as degenerate bending modes for a zero-thickness shell.

Figure 6 shows the pitfalls in relying on membrane theory for the determination of the natural frequencies of a closed, thin shell, which has been advocated by Lamb [*Proc. London Math. Soc.,* **14,** 50–56 (1882)] and endorsed by many others. This is explained at the end of Paper 23. The brief note by J. P. Wilkinson, Paper 24, illustrates this point even better. Referring to Wilkinson's Figure 1, it is seen that membrane theory models the extensional modes (Branch II) very accurately, while it is unable to predict the flexural Branch I beyond a few lower modes. Wilkinson's Figure 1 shows another general feature of shell vibrations: the modes represented by Branch I are flexural; by Branch II, extensional; and by Branch III, thickness-shear modes. The last branch is the result of the inclusion of transverse shear strain in Wilkinson's analysis. Unfortunately, Wilkinson does not plot the frequencies for other thicknesses. Had he done so, the results would have shown that, as the thickness is diminished, the natural frequencies of the flexural modes would go down (Classical I and Improved I both approach Membrane I), those of extensional modes would stay fixed, and those of thickness shear modes would go up (Improved III moves higher).

The free-vibration problems discussed up to this point are intended to reveal general behavior of a vibrating shell. When it comes to practical applications, the methods used in these papers do not go very far. They can be applied to cylindrical and spherical shells, with constant thickness, but that is the practical limit of their application. The technology of this day presents demands for solutions to shells with much more complicated geometries. The use of the electronic computer has made it possible to satisfy some of the demands.

For the free-vibration analysis of arbitrary shells, one has to rely on the finite-element method. No other method can surpass it in versatility as far as shell geometry is concerned. For certain classes of shell geometries, however, specialized methods are available. One such method is given in the next paper.

In Paper 25 Kalnins uses a computer-oriented method that can be applied to the free-vibration problem of an arbitrary shell of revolution, i.e., a shell with an arbitrary contour that is revolved about one straight axis of symmetry. The method used in this paper is applicable to general boundary-value problems governed by any number of ordinary first-order differential equations. Although an axisymmetric shell is two dimensional, its governing equations can be brought to a one-dimensional form by the separation of variables. The natural frequencies of the shell are evaluated in this paper by a version of the determinant-plotting technique. A discussion of the difficulties inherent in this technique, when applied to shell analysis problems, is given by R. E. Blum and R. E. Fulton [*AIAA J.*, **4,** 2231–2232 (1966)]. Other techniques for the determination of the eigenvalues, that can be coupled with the same numerical integration method, have been given by W. H. Wittrick and F. W. Williams [*Quar. Jour. Mech. Appl. Math.*, **24,** 263–284 (1971)], and by G. Cohen [*AIAA J.*, **3,** 2305–2312 (1965)].

16

Reprinted from *The Theory of Sound,* vol. 1, 2nd ed., Lord Rayleigh, Macmillan, London and Basingstoke, 1926, pp. 395–397

CURVED PLATES OR SHELLS

Lord Rayleigh

235 *b*. IN the last chapter (§§ 232, 233) we have considered the comparatively simple problem of the vibration in two dimensions of a cylindrical shell, so far at least as relates to vibrations of a *flexural* character. The shell is supposed to be thin, to be composed of isotropic material, and to be bounded by infinite coaxal cylindrical surfaces. It is proposed in the present chapter to treat the problem of the cylindrical shell more generally, and further to give the theory of the flexural vibrations of spherical shells.

In considering the deformation of a thin shell the most important question which presents itself is whether the middle surface, viz. the surface which lies midway between the boundaries, does, or does not, undergo extension. In the former case the deformation may be called *extensional*, and its potential energy is proportional to the thickness of the shell, which will be denoted by $2h$. Since the inertia of the shell, and therefore the kinetic energy of a given motion, is also proportional to h, the frequencies of vibration are in this case *independent* of h, § 44. On the other hand, when no line traced upon the middle surface undergoes extension, the potential energy of a deformation is of a higher order in the small quantity h. If the shell be conceived to be divided into laminæ, the extension in any lamina is proportional to its distance from the middle surface, and the contribution to the potential energy is proportional to the square of that distance. When the integration over the thickness is carried out, the whole potential energy is found to be proportional to h^3. Vibrations of this kind may be called inextensional, or flexural, and (§ 44) their frequencies are proportional to h, so that the sounds become graver without limit as the thickness is reduced.

Vibrations of the one class may thus be considered to depend upon the term of order h, and vibrations of the other class upon the term of order h^3, in the expression for the potential energy. In general both terms occur; and it is only in the limit that the separation into two classes becomes absolute. This is a question which has sometimes presented difficulty. That in the case of extensional vibrations the term in h^3 should be negligible in comparison with the term in h seems reasonable enough. But is it permissible in dealing with the other class of vibrations to omit the term in h while retaining the term in h^3?

The question may be illustrated by consideration of a statical problem. It is a general mechanical principle (§ 74) that, if given displacements (not sufficient by themselves to determine the configuration) be produced in a system originally in equilibrium by forces of corresponding types, the resulting deformation is determined by the condition that the potential energy shall be as small as possible. Apply this principle to the case of an elastic shell, the given displacements being such as not of themselves to involve a stretching of the middle surface. The resulting deformation will, in general, include both stretching and bending, and any expression for the energy will be of the form

$$Ah\,(\text{extension})^2 + Bh^3\,(\text{bending})^2 \quad \ldots\ldots\ldots\ldots\ldots(1).$$

This energy is to be as small as possible. Hence, when the thickness is diminished without limit, the actual displacement will be one of pure bending, if such there be, consistent with the given conditions.

At first sight it may well appear strange that of the two terms the one proportional to the cube of the thickness is to be retained, while that proportional to the first power may be neglected. The fact, however, is that the large potential energy that would accompany any stretching of the middle surface is the very reason why such stretching does not occur. The comparative largeness of the coefficient (proportional to h) is more than neutralized by the smallness of the stretching itself, to the square of which the energy is proportional.

An example may be taken from the case of a rod, clamped at one end *A*, and deflected by a lateral force; it is required to trace the effect of constantly increasing stiffness of the part included between *A* and a neighbouring point *B*. In the limit we may regard the rod as clamped at *B*, and neglect the energy of the part *AB*, in spite of, or rather in consequence of, its infinite stiffness.

It would thus be a mistake to regard the omission of the term in h as especially mysterious. In any case of a constraint which is supposed to be gradually introduced (§ 92 *a*), the vibrations tend to arrange themselves into two classes, in one of which the constraint is observed, while in the other, in which the constraint is violated, the frequencies increase without limit. The analogy with the shell of gradually diminishing thickness is complete if we suppose that at the same time the elastic constants are increased in such a manner that the resistance to *bending* remains unchanged. The resistance to extension then becomes infinite, and in the limit one class of vibrations is purely inextensional, or flexural.

In the investigation which we are about to give of the vibrations of a cylindrical shell, the extensional and the inextensional classes will be considered separately. It would apparently be more direct to establish in the first instance a general expression for the potential energy complete as far as the term in h^3, from which the whole theory might be deduced. Such an expression would involve the extensions and the curvatures of the middle surface. It appears, however, that this method is difficult of application, inasmuch as the potential energy (correct to h^3) does not depend only upon the above-mentioned quantities, but also upon the manner of application of the normal forces, which are in general implied in the existence of middle surface extensions[1].

[*Editor's Note:* Material has been omitted at this point.]

[1] On the Uniform Deformation in Two Dimensions of a Cylindrical Shell, with Application to the General Theory of Deformation of Thin Shells. *Proc. Math. Soc.*, vol. xx. p. 372, 1889.

17

Reprinted from *Phil. Trans. Roy. Soc. London,* Ser. A, **179,** 543–546 (1888)

ON THE SMALL FREE VIBRATIONS AND DEFORMATION OF A THIN ELASTIC SHELL

A. E. H. Love

§ 9. *Summary.**

This paper is really an attempt to construct a theory of the vibrations of bells. In any actual bell complications will arise, which have been omitted in this discussion, partly from variations of the thickness in different parts, and partly from the want of isotropy in the material. We can hardly expect a metal which has been subjected to the process of bell-manufacture to be other than very æolotropic, while it is notorious that bells are usually thickest at the rim. The difficulty of the problem in its general form seems to make it advisable to begin with the limiting case of an indefinitely thin perfectly isotropic shell, whose thickness is everywhere constant, and so small compared with its linear dimensions, that powers of it above the first may be neglected in mathematical expressions, which contain the first and higher powers multiplied by quantities of the same order of magnitude.

Of previous theoretical work we have examples in Lord RAYLEIGH'S 'Theory of Sound,' and in his paper on the "Bending of Surfaces of Revolution," in ARON'S and MATHIEU'S memoirs, and in IBBETSON'S treatise on the Mathematical Theory of Elasticity. In the 'Theory of Sound' Lord RAYLEIGH treats the vibrations of a thin ring or infinite cylinder of matter, supposed to be deformed in such a way that the motion is in one plane and the elements remain unextended, and remarks that at the time of publication this was the nearest approximation to a theoretical treatment of bells. He afterwards applies his theory of the bending of surfaces to obtain a more exact analytical method of treating the problem, but his disregard of the boundary-conditions which hold at a free edge appears to vitiate this theory. ARON can hardly be said to have attained a theory of bells, and the interest of his memoir is mainly mathematical; his inaccuracies have been already referred to. I have also previously referred to the objection which lies against MATHIEU'S method of treatment; this and the complexity and difficulty of some of his analysis seem to render a new method desirable. I shall have to refer to IBBETSON later.

The theory here put forward rests on the form of the function expressing the potential-energy of deformation per unit area of the middle-surface of the shell. Supposing that the surface is stretched and has its curvature changed, we find that the energy consists of two terms. One of these contains only the functions defining the stretching, while the other contains also those defining the bending of the middle-surface. The modulus of stretching is proportional to the thickness, while the modulus of bending is proportional to its cube. Unless, therefore, the functions expressing the stretching, viz., the extensions and shear of rectangular line-elements of the middle-surface, are of a higher order of small quantities than those defining the bending, viz., the changes of the principal curvatures and of the directions of the principal planes, the vibrations depend on the term which involves the stretching, and not on that which involves the bending. Now, it seems to have been universally

* Partly rewritten, July, 1888.

assumed by English writers that the reverse of this is the case, viz., that the vibrations take place in such a way that no line on the middle-surface is altered in length. This will be borne out by a reference to Lord RAYLEIGH and IBBETSON. The theory of the present paper rests on the fact that the functions expressing the stretching and those expressing the changes in magnitude and direction of curvature are of the same order of small quantities. This is proved in the following way:—The potential energy consists of two parts; one, Q_2, proportional to the thickness h; and the other, Q_1, proportional to h^3. The first is expressed in terms of the stretching, and the second in terms of the bending of the middle-surface. Some previous theories have proceeded as if Q_1 only occurred. If this were the case, we ought to get an approximation by supposing that $Q_2/h = 0$. This is equivalent to assuming that there is no stretching of the middle-surface. We should therefore get an approximation by supposing the surface inextensible to the first order. The stretching and the bending are expressed, to the first order, by linear functions of certain differential coefficients of the displacements. Our supposed method of getting an approximation is then to make the functions expressing the stretching vanish. Now, I have shown that the functions expressing the displacement are thus, to a certain extent, determined, and *that* in such a way that the boundary-conditions cannot be satisfied. The boundary-conditions referred to are the exact conditions found by retaining the complete expression for the potential energy. It is inferred that the functions expressing the stretching cannot be taken equal to zero for an approximation; or, in other words, small compared with those expressing the bending; and, thus, Q_1/h^3 and Q_2/h, are of the same order of magnitude. The conclusion that Q_1 is small compared with Q_2 seems inevitable.

The argument breaks down for a plane plate through the vanishing of the curvatures; Q_1 is then alone of importance. In the case of an open shell or bowl whose linear dimension is small compared with its radius of curvature, and large compared with its thickness, both terms are important. When this is so, we get a class of cases for which the linear dimensions concerned are of three different orders of magnitude, and this case will not come under the method of the present paper. It may be compared with the problem of the watch-spring mentioned in THOMSON and TAIT'S 'Natural Philosophy,' Part 2, p. 264, which stands between a bar and a plate. The very open shell or bowl stands in the same way between a plate and what I have called a shell.

The theory of this paper proceeds as if Q_2 alone occurred. It is to be regarded as the limiting form for indefinitely thin shells. A complete theory of bells, even when regarded as uniformly thick and isotropic, could only be obtained by using the exact equations formed by retaining both terms of the potential energy.

Again, English writers have assumed that the potential energy, which they suppose to depend only on the bending, will be the same quadratic function of the changes of principal curvature as it is for a plane plate. The same authorities as before may be quoted, and we may also refer to a question set in the Mathematical Tripos,

January 18th, morn., 1878, question η. To test this assumption involved the investigation of Artt. 7, 8, and the result is that it is only in the case of a sphere supposed unstretched that the potential energy has this form. This is the case treated by Lord RAYLEIGH, but his method still fails, for a complete sphere cannot be bent without stretching, while, if the sphere be incomplete, the conditions which hold at a free edge cannot be satisfied ; this is explicitly proved in Art. 14.

A general result is derived from the consideration of the functions expressing the kinetic and potential energies, Q_2 only being retained. Both these functions are proportional to the thickness of the shell, and thus the periods of vibration are independent of the thickness. That this result holds for a complete thin spherical shell vibrating in any manner has been demonstrated by LAMB ('London Math. Soc. Proc.,' vol. 14, 1882, p. 52). His equations (7) and (9) when reduced are independent of the thickness.

Two general results are obtained without solution from the equations of motion. The first is, that vibrations involving displacement along the normal only are impossible except in the cases of the plane, complete sphere, and infinitely long circular cylinder. IBBETSON's treatment of the problem appears to assume (1) inextensibility, (2) the incorrect formula for the energy, (3) normal displacements. The other result is that any surface of revolution can execute purely tangential vibrations which are symmetrical with respect to the axis of revolution, and in which the motion is purely torsional, or perpendicular to the planes through the axis. These must not be confounded with the familiar vibrations of finger-bowls, which are most probably a type with two nodal meridians.*

The theory of the vibrations of a thin spherical shell bounded by a small circle is an interesting example of the general theory of vibrations of an elastic solid. In an infinite solid there are two types of vibratory motion, the longitudinal and the distortional, both of which are propagated as waves. In a bounded solid this state of things is modified by reflexions at the bounding-surfaces, so that the purely longitudinal and purely tangential waves do not in general exist separately. Again, in all cases of displacement in one direction only, as in the vibrations of strings, bars, and plates, there may be displacements in different directions which are independent of each other, with their corresponding nodal lines or points. This also is modified in the general solid. The types of vibration, for example, of a portion of a spherical shell bounded by a small circle are partially made out in this essay. One immediate result is that there are in general no nodal lines, properly so called. In any type the displacement along the parallels vanishes at one set of meridians ; the other displacements vanish together at another set of meridians. These sets are ranged at equal intervals round the sphere. There appears to be good reason to suppose generally that the corresponding proposition will not obtain with reference to nodal parallels. The establishment of the fact would require a solution of the general frequency equa-

* RAYLEIGH, 'Sound,' vol. 1, Art. 234.

tion, and this I have not been able to effect. One case, however, is readily solved, and that is where the displacement is symmetrical with respect to the pole of the sphere. It appears here that the vibrations divide themselves into two types, one purely tangential with displacement along the parallels, the other partly radial and partly consisting of displacements along the meridians. There are no nodal meridians. In the purely tangential vibrations there exists a series of nodal parallels, whose number corresponds to the type of vibration. The intervals for the various tones are each of them nearly a fifth. In the partly radial vibrations the radial displacement vanishes at one set of small circles, and the tangential displacement at another set. The number and position of the nodal circles for the purely tangential vibration coincide exactly with the number and position of the circles along which the tangential displacement vanishes in the corresponding partly radial mode. The vibrations of the two types belong to different normal modes of vibration, and have different frequencies. If we like to extend the meaning of "nodal lines," so as to include the small circles just referred to, then we may state another result in the form that for partly radial vibrations there are two periods and modes of vibration which have the same set of "nodal lines." The tones of one of these sets are all very near together; those of the other set are separated by intervals nearly the same as for a harmonic scale.

A discussion of the vibrations of an elastic shell in the form of a circular cylinder closed at one end by a rigid disc perpendicular to its axis leads to similar conclusions as to types of vibration and their definition by nodal lines.

It is unfortunate that solutions of the frequency equation for the case of two "nodal" meridians dividing the shell into four equal portions could not be obtained, as these probably include the gravest mode of vibration of which the shell is capable. The tones of the symmetrical vibrations discussed are very high, and the theory in its present state cannot easily be tested by experiment. There is, however, one result which would seem to admit of practical verification, viz., it is found that, for similar thin shells, the frequency is independent of the thickness, and varies inversely as the linear dimension.

18

Reprinted from *J. Appl. Mech.*, **35**, 516–523 (1968)

On Inextensional Vibrations of Thin Shells

E. W. ROSS, Jr.
Mathematician,
U. S. Army Natick Laboratories,
Natick, Mass.

In this paper the nonsymmetric, free, elastic vibrations of thin domes of revolution are studied. It is assumed that the frequency is low. The asymptotic approximations previously given by the writer are used to estimate the general solution to the shell vibration equations at low frequencies. Approximations for the low natural frequencies and modes are derived systematically under a variety of edge conditions. Low natural frequencies are found only when the edge conditions impose no forces tangent to the shell surface. When the edge is free (and only then) Rayleigh's inextensional frequencies are recovered. For certain other edge conditions, new natural frequencies are found that are above Rayleigh's frequencies but still low compared, e.g., with the lowest membrane frequency. The displacement modes associated with these new frequencies are mostly of inextensional type. The general results are applied to estimate these new frequencies for spherical domes.

Introduction

The inextensional vibrations of thin shells were first studied by Lord Rayleigh [1],[1] and since that time his procedure has often been used to estimate natural frequencies for various shell shapes. The frequencies obtained by this procedure are much lower (for a thin shell) than those predicted by any other method and are therefore of great practical interest. For example, a recent paper by Goodier and McIvor [2] shows the important part played by the inextensional modes in the transient response of an elastic cylindrical shell.

Despite the importance of Rayleigh's procedure, there is good reason for skepticism about its generality. For example, Love [4] has shown that the modes in general satisfy neither the motion equations nor (with a few exceptions) the edge conditions. Arnold and Warburton [5] observed that Rayleigh's procedure gave good agreement with experiments in some cases, but that the agreement was very sensitive to the edge conditions. In the same vein, Forsberg [3] has emphasized the care that is needed in choosing the boundary conditions for an approximate analysis.

The present investigation is an attempt to clarify this situation by studying how the inextensional modes may be derived from the general shell theory. We do not assume (as is usually done) that the mode is inextensional. Rather, we merely assume that the frequency is low (in a sense that will later be made more precise) and derive inextensional modes from the general shell theory. This change of procedure is important for two reasons. First, we find that inextensional modes can be derived only for certain edge conditions, and this sheds light on the questions raised in [3–5]. Second, the new procedure leaves the way open to find all low frequencies, whereas Rayleigh's procedure is limited to frequencies for which the modal bending energy greatly exceeds the modal stretching energy. For certain edge conditions, we shall find inextensional modes with frequencies different from those obtained by Rayleigh.

To demonstrate the procedure in a context general enough to be convincing but simple enough to avoid unessential manipulations, we consider a general dome of revolution executing small, nonsymmetric vibrations. We shall use the approximations obtained by the author [6] to write down an approximate general solution of the differential equation system when the frequency is low. This solution is substituted into the boundary conditions, the resulting frequency determinant is solved, and the ratios of the arbitrary constants are found. This entire process is carried through for four different edge conditions, starting with a free edge and proceeding at each stage to the "freest" of the remaining edge conditions. The frequency increases with each new edge condition until we exhaust all edge conditions for which low frequencies can be found.

For the two freest edge conditions this procedure gives complete estimates of the mode but only an order-of-magnitude estimate of the frequency. To find frequency estimates we use Rayleigh's principle for these cases. In the other two cases, explicit estimates are found for the inextensional frequencies.

[1] Numbers in brackets designate References at end of paper.

Contributed by the Applied Mechanics Division for publication (without presentation) in the Journal of Applied Mechanics.
Discussion of this paper should be addressed to the Editorial Department, ASME, United Engineering Center, 345 East 47th Street, New York, N. Y. 10017, and will be accepted until October 20, 1968. Discussion received after the closing date will be returned. Manuscript received by ASME Applied Mechanics Division, September 20, 1967; final draft, February 16, 1968. Paper No. 68—APM-R.

Nomenclature

For most of the quantities defined below, we list the equations in which they are defined or first occur.

$A_1, A_2, \ldots, A_8$ = arbitrary constants in general solution, (14) and (16)
a, a^* = angle function in asymptotic approximations, (16)
B_1, B_2, B_5, B_7 = modified arbitrary constants in general solution
b_s, b_θ = meridional and circumferential rotations, (2)
$c(a), c(a^*)$ = abbreviations for cos (a), cos (a^*), (16)
D = $dw/d\sigma$, (12)
E = Young's modulus
E_K, E_S, E_B = kinetic, stretching, and bending energies, (13)
$f(\sigma)$ = cot ϕ/r_θ, (10)
G_n, G_s, G_v = coefficient functions in asymptotic approximations, (16)
H = coefficient function in asymptotic approximation, (16)
h = shell thickness, assumed constant
K = E_K/Ω^2, (13)
$k_{ss}, k_{\theta\theta}, k_{s\theta}$ = dimensionless meridional, circumferential, and torsional curvature changes, (3)
L_5, L_Ω = combinations of direct stresses arising in certain modes, (34) and (35)
$M(\sigma)$ = $m$csc ϕ/r_θ, (10)
m = circumferential wave number
$m_{ss}, m_{\theta\theta}, m_{s\theta}$ = dimensionless meridional, circumferential, and twisting moments, (5)

(Continued on next page)

The general formulas are applied to a spherical dome, and numerical results are obtained for the previously unknown inextensional frequencies.

Fundamental Equations and Solutions

We shall adopt as our starting point the equations of thin-shell theory propounded by Sanders [7] and modified by the inclusion of translational (but not rotational) inertia. We may write the system in dimensionless form as in [6].

$$\gamma_{ss} = u' + wr_s^{-1}, \qquad \gamma_{\theta\theta} = uf + vM + wr_\theta^{-1} \qquad \gamma_{s\theta} = (1/2)(v' - uM - vf) \tag{1}$$

$$b_s = -w' + ur_s^{-1}, \qquad b_\theta = vr_\theta^{-1} + wM \tag{2}$$

$$k_{ss} = b_s', \qquad k_{\theta\theta} = fb_s + Mb_\theta \qquad k_{s\theta} = (1/2)\{b_\theta' - Mb_s - fb_\theta + (1/2)(r_\theta^{-1} - r_s^{-1})(v' + vf + uM)\} \tag{3}$$

$$n_{ss} = \gamma_{ss} + \nu\gamma_{\theta\theta}, \qquad n_{\theta\theta} = \gamma_{\theta\theta} + \nu\gamma_{ss} \qquad n_{s\theta} = (1 - \nu)\gamma_{s\theta} \tag{4}$$

$$m_{ss} = k_{ss} + \nu k_{\theta\theta}, \qquad m_{\theta\theta} = k_{\theta\theta} + \nu k_{ss} \qquad m_{s\theta} = (1 - \nu)k_{s\theta} \tag{5}$$

$$q_s = m_{ss}' + f(m_{ss} - m_{\theta\theta}) + Mm_{s\theta} \qquad q_\theta = m_{s\theta}' + 2fm_{s\theta} - Mm_{\theta\theta} \tag{6}$$

The motion equations are

$$(1 - \nu^2)^{-1}\{n_{ss}' + f(n_{ss} - n_{\theta\theta}) + Mn_{s\theta}\} + \Omega^2 u + \epsilon^2\{q_s r_s^{-1} + (1/2)Mm_{s\theta}(r_s^{-1} - r_\theta^{-1})\} = 0 \tag{7}$$

$$(1 - \nu^2)^{-1}\{n_{s\theta}' + 2fn_{s\theta} - Mn_{\theta\theta}\} + \Omega^2 v + \epsilon^2\{q_\theta r_\theta^{-1} + (1/2)[(r_\theta^{-1} - r_s^{-1})M_{s\theta}]'\} = 0 \tag{8}$$

$$(1 - \nu^2)^{-1}\{n_{ss}r_s^{-1} + n_{\theta\theta}r_\theta^{-1}\} - \Omega^2 w - \epsilon^2\{q_s' + fq_s + Mq_\theta\} = 0 \tag{9}$$

where the various definitions are given in the Nomenclature, and

$$\Omega = \omega R(\rho/E)^{1/2} \qquad \epsilon^2 = h^2/[12R^2(1 - \nu^2)] \ll 1 \qquad f(\sigma) = r_\theta^{-1}\cot\phi, \qquad M(\sigma) = mr_\theta^{-1}\csc\phi \tag{10}$$

The boundary conditions at an edge have been given by Sanders [7] and consist of prescribing

$$n_{ss} \text{ or } u \qquad N_{s\theta} \text{ or } v \qquad Q_s \text{ or } w \qquad m_{ss} \text{ or } D \tag{11}$$

where

$$N_{s\theta} = n_{s\theta} + \epsilon^2(1/2)(1 - \nu^2)(3r_\theta^{-1} - r_s^{-1})m_{s\theta} \qquad Q_s = q_s + Mm_{s\theta} \qquad D = w' = -b_s + ur_s^{-1} \tag{12}$$

The principle of conservation of energy for the vibrating shell states that

$$E_K - E_S - E_B = 0$$

where

$$E_K = \Omega^2\int(u^2 + v^2 + w^2)r_\theta \sin\phi\, d\sigma = \Omega^2 K$$
$$E_S = (1 - \nu^2)^{-1}\int(n_{ss}\gamma_{ss} + n_{\theta\theta}\gamma_{\theta\theta} + 2n_{s\theta}\gamma_{s\theta})r_\theta \sin\phi\, d\sigma \tag{13}$$
$$E_B = \epsilon^2\int(m_{ss}k_{ss} + m_{\theta\theta}k_{\theta\theta} + 2m_{s\theta}k_{s\theta})r_\theta \sin\phi\, d\sigma$$

and the integrals are extended over a meridian.

Nomenclature

$N_{s\theta}$ = modified dimensionless shear stress arising in boundary conditions, (11) and (12)

$n_{ss}, n_{\theta\theta}, n_{s\theta}$ = dimensionless meridional, circumferential, and shearing direct stresses, (4)

$n_{ss}^{(\epsilon)}, n_{\theta\theta}^{(\epsilon)}, n_{s\theta}^{(\epsilon)}$ = contribution of bending terms to direct stresses of inextensional solution

$n_{ss}^{(\Omega)}, n_{\theta\theta}^{(\Omega)}, n_{s\theta}^{(\Omega)}$ = contribution of inertia terms to direct stresses of inextensional solution

$n_{ss}^{(\lambda)}, n_{s\theta}^{(\lambda)}$ = $n_{ss}^{(\epsilon)}/r_\theta^2$, $n_{s\theta}^{(\epsilon)}/r_\theta^2$, respectively

Q_s = modified meridional shear force arising in boundary conditions, (11) and (12)

q_s, q_θ = dimensionless meridional and circumferential shear forces, (11) and (12)

R = characteristic radius of curvature, (10)

r_s, r_θ = dimensionless principal radii of curvature, (1)

$s(a), s(a^*)$ = abbreviations for sin a, sin a^*, (16)

u, v, w = dimensionless meridional, circumferential, and normal (outward) displacements, (1)

$X_{12}, X_{22}, X_{32}, X_{13}, X_{23}, X_{33}, X_{14}, X_{24}, X_{34}$ = elements of frequency determinant

$x(\sigma)$ = rapidly changing function arising in asymptotic approximations (15) and (17)

z = $n_{ss}^{(\Omega)} + n_{s\theta}^{(\Omega)}$, (45)

$\alpha_{I}, \alpha_{II}, \alpha_{III}, \alpha_{IV}$ = factors in expressions for inextensional frequencies, (27), (33), and (35)

β_5 = constant in determination of mode, (28)

γ = $x/2^{1/2}$, (16)

$\gamma_{ss}, \gamma_{\theta\theta}, \gamma_{s\theta}$ = meridional, circumferential, and shear strains, (1)

$\Delta(\sigma)$ = function arising in evaluation of E_s

ϵ = small dimensionless parameter, $(h/R)[12(1 - \nu^2)]^{-1/2}$

ζ = rapidly varying function, (31)

η_{11}, η_{21} = functions arising from higher-order terms in asymptotic approximations

θ = circumferential angle

$\Lambda(\sigma)$ = $dx/d\sigma$, (18)

λ = $\epsilon^{-1/2}$, large parameter, (15)

ν = Poisson ratio, (4) and (5)

ρ = mass density of shell material, (10)

σ = dimensionless meridional arc length

σ_0 = σ at shell edge

$\phi(\sigma)$ = angle between axis of revolution and shell normal

ϕ_0 = $\phi(\sigma_0)$

$\chi(\sigma)$ = functions occurring in modes, (30), (32), (34) and (36)

Ω = dimensionless circular frequency, (10)

ω = circular frequency

$(\ )'$ = $d(\ )/d\sigma$

The classical shell theory embodied in (1)–(9) neglects effects of transverse shear, thickness change, and rotational inertia and therefore can treat accurately only wavelengths much greater than the shell thickness, which implies

$$m\epsilon \ll 1.$$

This differential equation system is linear, of eighth order, and has singularities where $\sin\phi = 0$. We assume that $\sin\phi = 0$ at, and only at, the axis, and that the apex of the dome is of second degree.

The analysis depends on approximations to the eight solutions of this system which are described in [6] and will be summarized here. The essential step is an asymptotic analysis very similar to the one given earlier by the writer for the axisymmetric case [8] but now valid only when

$$m^2\epsilon \ll 1,$$

which we henceforth assume to be true.[2] It is found that four of the solutions usually vary rapidly with σ (i.e., along a meridian) and are called bending solutions, and four vary much more slowly and are called membrane or inextensional solutions. Two solutions of each type are singular at the dome apex, where $\sin\phi = 0$.

We shall now list the approximations to the four bending solutions, first near $\sin\phi = 0$, then for $\sin\phi \neq 0$. The latter are linear combinations of the approximations obtained in [6] for the case $\Omega r_\theta < 1$.

For $\sin\phi \simeq 0$:

$$w \simeq A_1\,\mathrm{ber}_m(x) + A_2\,\mathrm{bei}_m(x) + A_3\,\mathrm{ker}_m(x) + A_4\,\mathrm{kei}_m(x) \qquad (14)$$

$$x = (1-\Omega^2)^{1/4}\lambda\phi, \qquad \lambda = \epsilon^{-1/2} \gg 1 \qquad (15)$$

For $\sin\phi \neq 0$:

$$\begin{bmatrix} w \\ n_{\theta\theta} \\ b_\theta \end{bmatrix} \simeq H \begin{bmatrix} 1 \\ G_n \\ M \end{bmatrix} \{A_1e^{\gamma}c(a) + A_2e^{\gamma}s(a) + A_3\pi e^{-\gamma}c(a^*) - A_4\pi e^{-\gamma}s(a^*)\}$$

$$\begin{bmatrix} u \\ n_{ss} \\ n_{s\theta} \\ q_s \end{bmatrix} \simeq \Lambda^{-1}H \begin{bmatrix} -G_u \\ fG_n \\ MG_n \\ \Lambda^4 \end{bmatrix} \{A_1e^{\gamma}s(a^*) - A_2e^{\gamma}c(a^*) + A_3\pi e^{-\gamma}s(a) + A_4\pi e^{-\gamma}c(a)\} \qquad (16)$$

$$\begin{bmatrix} b_s \\ m_{s\theta} \end{bmatrix} \simeq \Lambda H \begin{bmatrix} -1 \\ (1-\nu)M \end{bmatrix} \{A_1e^{\gamma}c(a^*) + A_2e^{\gamma}s(a^*) - A_3\pi e^{-\gamma}c(a) + A_4\pi e^{-\gamma}s(a)\}$$

$$\begin{bmatrix} v \\ m_{ss} \\ m_{\theta\theta} \\ q_\theta \end{bmatrix} \simeq \Lambda^2 H \begin{bmatrix} -\Lambda^{-4}MG_v \\ -1 \\ -\nu \\ M \end{bmatrix} \{-A_1e^{\gamma}s(a) + A_2e^{\gamma}c(a) + A_3\pi e^{-\gamma}s(a^*) + A_4\pi e^{-\gamma}c(a^*)\}$$

where the A's are arbitrary constants and

$$x \equiv \lambda \int_{\sigma'=0}^{\sigma'=\sigma} (r_\theta^{-2} - \Omega^2)^{1/4}\,d\sigma' \qquad (17)$$

$$\Lambda \equiv dx/d\sigma = \lambda(r_\theta^{-2} - \Omega^2)^{1/4} \gg 1 \qquad (18)$$

$$H \equiv [(1-\Omega^2)\epsilon]^{1/4}(2\pi r_\theta \sin\phi)^{-1/2}(r_\theta^{-2} - \Omega^2)^{-3/8}$$

$$\gamma \equiv 2^{-1/2}x, \qquad a \equiv \gamma - (\pi/8) + (1/2)m\pi,$$

$$a^* \equiv a + (1/4)\pi$$

$$c(a) \equiv \cos a, \qquad s(a) \equiv \sin a$$

$$c(a^*) \equiv \cos a^* \qquad s(a^*) \equiv \sin a^*$$

$$G_u \equiv r_s^{-1} + \nu r_\theta^{-1}, \qquad G_n \equiv (1-\nu^2)r_\theta^{-1}$$

$$G_v \equiv (2+\nu)r_\theta^{-1} - r_s^{-1}$$

[2] Note that this is a more stringent inequality on m than the preceding one.

In deriving these formulas we have assumed that σ is measured from the apex of the dome. Also, R is chosen as the common value of the principal radii of curvature at the apex. Then $r_s(0) = r_\theta(0) = 1$ and the definition of x for $\sin\phi \simeq 0$ is a continuation of that for $\sin\phi \neq 0$.

In general, we shall neglect terms that are $0(\lambda^{-1})$ compared with terms that are $0(1)$. The asymptotic analysis shows that the errors in the approximations (14) and (16) are all $0(\lambda^{-1})$, and we shall neglect them henceforth. Moreover, if we expand these approximations in powers of Ω^2, we see that the static asymptotic approximations (the "edge-effect" approximations) are obtained when

$$\Omega^2 \leq 0(\lambda^{-2}).$$

The approximations for the remaining four solutions are also described in [6]. These are all membrane solutions when Ω^2 is not small but, as $\Omega^2 \to 0$, they must separate into two pairs of solutions, one pair (labeled 5 and 6) that approach the two static membrane solutions, and another pair (labeled 7 and 8) that approach the two approximately inextensional, static solutions. To be more precise, we now assume that the frequency is low, by which we mean

$$\Omega^2 \leq 0(\lambda^{-1}). \qquad (19)$$

The solutions 5 and 6, the membrane solutions, are found in general by solving the system (1)–(9) with ϵ^2 set to zero in (7)–(9). When $\Omega^2 \leq 0(1)$, all the dimensionless variables associated with these two solutions are $0(1)$; i.e., the variables do not increase in size under differentiation. Hence, when $\Omega^2 \leq 0(\lambda^{-1})$, the inertia terms in (7)–(9) are negligible according to our criterion, and the membrane solutions are approximately just the static ones.

In estimating the inextensional solutions when $\Omega^2 \leq 0(\lambda^{-1})$ we are guided by the static case (see Love [4]). The displacements are obtained by solving (1) with $\gamma_{ss} = \gamma_{\theta\theta} = \gamma_{s\theta} = 0$, and all the other variables except n_{ss}, $n_{\theta\theta}$, $n_{s\theta}$ are then found from (2), (3), (5), and (6). These variables are all of the same order, say, $0(1)$, and are exactly the functions obtained in the static case. To estimate the small direct stresses we solve the motion equations, (7)–(9), as a system of three linear, nonhomogeneous equations for n_{ss}, $n_{\theta\theta}$, $n_{s\theta}$, taking the bending and inertia terms as already known. This gives the same results as in the static case except for additional terms arising from the inertia terms in (7)–(9), i.e., the two inextensional solutions for the direct stresses may be written

$$\begin{bmatrix} n_{ss} \\ n_{\theta\theta} \\ n_{s\theta} \end{bmatrix} \simeq A_7 \begin{bmatrix} \Omega^2 n_{ss7}^{(\Omega)} + \epsilon^2 n_{ss7}^{(\epsilon)} \\ \Omega^2 n_{\theta\theta 7}^{(\Omega)} + \epsilon^2 n_{\theta\theta 7}^{(\epsilon)} \\ \Omega^2 n_{s\theta 7}^{(\Omega)} + \epsilon^2 n_{s\theta 7}^{(\epsilon)} \end{bmatrix} + A_8 \begin{bmatrix} \Omega^2 n_{ss8}^{(\Omega)} + \epsilon^2 n_{ss8}^{(\epsilon)} \\ \Omega^2 n_{\theta\theta 8}^{(\Omega)} + \epsilon^2 n_{\theta\theta 8}^{(\epsilon)} \\ \Omega^2 n_{s\theta 8}^{(\Omega)} + \epsilon^2 n_{s\theta 8}^{(\epsilon)} \end{bmatrix} \qquad (20)$$

where the functions multiplying ϵ^2 give the static bending contribution and the functions multiplying Ω^2 give the contribution of the inertia terms. The functions $n_{ss7}^{(\Omega)}$, $n_{ss7}^{(\epsilon)}$, $n_{ss8}^{(\Omega)}\ldots$ are all $0(1)$ in general. It will eventually be seen that these estimates of the inertia contributions to the direct stresses are very important to our development.

Of the eight solutions, four are singular at $\phi = 0$ and must be discarded for a dome. We may take two of these as the membrane solution numbered 6 and the inextensional solution numbered 8, and we see from (14) that the remaining two are the bending solutions numbered 3 and 4. Thus we must take

$$A_3 = A_4 = A_6 = A_8 = 0. \qquad (21)$$

We may now without confusion set

$$n_{ss7}^{(\Omega)} \equiv n_{ss}^{(\Omega)}, \qquad n_{ss7}^{(\epsilon)} \equiv n_{ss}^{(\epsilon)}$$

and, similarly, for $n_{\theta\theta}$ and $n_{s\theta}$. The approximate general solution may finally be written

$$w \simeq He^{\gamma}\{A_1c(a) + A_2s(a)\} + A_5w^{(5)} + A_7w^{(7)}$$

$$u \simeq \Lambda^{-1}G_uHe^{\gamma}\{-A_1s(a^*) + A_2c(a^*)\} + A_5u^{(5)} + A_7u^{(7)}$$

$$v \simeq \Lambda^{-2}MG_vHe^{\gamma}\{A_1s(a) - A_2c(a)\} + A_5v^{(5)} + A_7v^{(7)}$$

$$b_s \simeq \Lambda He^{\gamma}\{-A_1c(a^*) - A_2s(a^*)\} + A_5b_s^{(5)} + A_7b_s^{(7)}$$

$$b_\theta \simeq M\dot{H}e^{\gamma}\{A_1c(a) + A_2s(a)\} + A_5b_\theta^{(5)} + A_7b_\theta^{(7)}$$

$$m_{ss} \simeq \Lambda^2He^{\gamma}\{A_1s(a) - A_2c(a)\} + A_5m_{ss}^{(5)} + A_7m_{ss}^{(7)}$$

$$m_{\theta\theta} \simeq \Lambda^2\nu He^{\gamma}\{A_1s(a) - A_2c(a)\} + A_5m_{\theta\theta}^{(5)} + A_7m_{\theta\theta}^{(7)}$$

$$m_{s\theta} \simeq \Lambda M(1-\nu)He^{\gamma}\{A_1c(a^*) + A_2s(a^*)\} + A_5m_{c\theta}^{(5)} + A_7m_{s\theta}^{(7)} \quad (22)$$

$$q_s \simeq \Lambda^3He^{\gamma}\{A_1s(a^*) - A_2c(a^*)\} + A_5q_s^{(5)} + A_7q_s^{(7)}$$

$$q_\theta \simeq \Lambda^2MHe^{\gamma}\{-A_1s(a) + A_2c(a)\} + A_5q_\theta^{(5)} + A_7q_\theta^{(7)}$$

$$n_{ss} \simeq \Lambda^{-1}fG_nHe^{\gamma}\{A_1s(a^*) - A_2c(a^*)\} + A_5n_{ss}^{(5)} + A_7\{\Omega^2n_{ss}^{(\Omega)} + \lambda^{-4}n_{ss}^{(\epsilon)}\}$$

$$n_{\theta\theta} \simeq G_nHe^{\gamma}\{A_1c(a) + A_2s(a)\} + A_5n_{\theta\theta}^{(5)} + A_7\{\Omega^2n_{\theta\theta}^{(\Omega)} + \lambda^{-4}n_{\theta\theta}^{(\epsilon)}\}$$

$$n_{s\theta} \simeq \Lambda^{-1}MG_nHe^{\gamma}\{A_1s(a^*) - A_2c(a^*)\} + A_5n_{s\theta}^{(5)} + A_7\{\Omega^2n_{s\theta}^{(\Omega)} + \lambda^{-4}n_{s\theta}^{(\epsilon)}\}$$

These approximations are accurate when $\Omega^2 \leq 0(\lambda^{-1})$, $m \geq 2$ and $\sin\phi \neq 0$. When $\Omega^2 \leq 0(\lambda^{-2})$, all the quantities are approximately static (independent of Ω) except the direct stresses of the inextensional solution. These formulas form the basis for our analysis of low frequencies and modes.

Calculation of Low Frequencies and Modes

In this section we shall put the general solution (22) into various sets of edge conditions and calculate the natural modes and frequencies. The derivation will be carried out in detail for two cases but results will be given for the rest. We shall begin with the case of a completely free edge and then consider successively "tighter" sets of edge conditions until the frequency is increased above the range, $\Omega^2 \leq 0(\lambda^{-1})$, in which (22) is applicable.

We assume the edge is at $\sigma = \sigma_0$, and $\sin\phi(\sigma_0) \neq 0$. A new set of constants, B_j, $j = 1, 2, 5, 7$ is introduced, defined by

$$B_1 = A_1H(\sigma_0)e^{\gamma(\sigma_0)}, \qquad B_2 = A_2H(\sigma_0)e^{\gamma(\sigma_0)}$$

$$B_5 = A_5, \qquad B_7 = A_7,$$

and we also set

$$\lambda^{-4}n_{ss}^{(\epsilon)} = \Lambda^{-4}r_\theta^{-2}n_{ss}^{(\epsilon)} \equiv \Lambda^{-4}n_{ss}^{(\lambda)}$$

$$\lambda^{-4}n_{s\theta}^{(\epsilon)} = \Lambda^{-4}r_\theta^{-2}n_{s\theta}^{(\epsilon)} \equiv \Lambda^{-4}n_{s\theta}^{(\lambda)}$$

In deducing the natural frequency and evaluating the constants it is to be understood that all quantities are evaluated at $\sigma = \sigma_0$.

Case (I): Free Edge. $n_{ss} = N_{s\theta} = Q_s = m_{ss} = 0$ **at** $\sigma = \sigma_0$. The four conditions are (keeping the leading terms only)

$$n_{ss} = B_1\Lambda^{-1}fG_ns(a^*) - B_2\Lambda^{-1}fG_nc(a^*) + B_5n_{ss}^{(5)} + B_7\{\Omega^2n_{ss}^{(\Omega)} + \Lambda^{-4}n_{ss}^{(\lambda)}\} = 0 \quad (23)$$

$$N_{s\theta} = B_1\Lambda^{-1}MG_ns(a^*) - B_2\Lambda^{-1}MG_nc(a^*) + B_5n_{s\theta}^{(5)} + B_7\{\Omega^2n_{s\theta}^{(\Omega)} + \Lambda^{-4}[n_{s\theta} + gm_{s\theta}^{(7)}]\} = 0 \quad (24)$$

$$Q_s = B_1\Lambda^3s(a^*) - B_2\Lambda^3c(a^*) + B_5[q_s^{(5)} + Mm_{s\theta}^{(5)}] + B_7\{q_s^{(7)} + Mm_{s\theta}^{(7)}\} = 0 \quad (25)$$

$$m_{ss} = B_1\Lambda^2s(a) - B_2\Lambda^2c(a) + B_5m_{ss}^{(5)} + B_7m_{ss}^{(7)} = 0 \quad (26)$$

where

$$g = (1/2)r_\theta^{-2}(1-\nu^2)(3r_\theta^{-1} - r_s^{-1})$$

If we eliminate B_1 using the condition (23), we obtain a three-square system that can be written in matrix form

$$\begin{bmatrix} \eta_{11} & X_{12} & (\Omega^2X_{13} + \Lambda^{-4}X_{14}) \\ \eta_{21} & X_{22} & (\Omega^2X_{23} + \Lambda^{-4}X_{24}) \\ 2^{-1/2} & X_{32}\Lambda^{-1} & (\Omega^2\Lambda^{-1}X_{33} + \Lambda^{-4}X_{34}) \end{bmatrix} \begin{bmatrix} B_2\Lambda^{-2} \\ B_5 \\ B_7 \end{bmatrix} = [0]$$

where

$$X_{12} = n_{s\theta}^{(5)} - n_{ss}^{(5)}(M/f), \qquad X_{22} = -n_{ss}^{(5)}/(fG_n), \qquad X_{32} = -X_{22}s(a)$$

$$X_{13} = n_{s\theta}^{(\Omega)} - n_{ss}^{(\Omega)}(M/f), \qquad X_{23} = -n_{ss}^{(\Omega)}/(fG_n), \qquad X_{33} = -X_{23}s(a)$$

$$X_{14} = n_{s\theta}^{(\lambda)} + gm_{s\theta}^{(7)} - n_{ss}^{(7)}(M/f)$$

$$X_{24} = -\{n_{ss}^{(\lambda)}/fG_n\} + q_s^{(7)} + Mm_{s\theta}^{(7)}$$

$$X_{34} = -m_{ss}^{(7)}s(a^*)$$

When B_1 is eliminated, the leading terms in the coefficients of B_2 cancel in equations (24) and (25). The dominant terms in these coefficients then arise from later terms in the asymptotic expansions of n_{ss}, $n_{s\theta}$, and q_s for the bending solutions. These are not known explicitly, but we know their orders of magnitude and designate the unknown functions η_{11} and η_{21}, both of which are $0(1)$.

The frequency is found by annulling the determinant of this system, with the result

$$\Omega^2\Lambda^4 = \frac{X_{14}X_{22} - X_{12}X_{24} + 2^{1/2}(\eta_{21}X_{12} - \eta_{11}X_{22})X_{34}}{X_{12}X_{23} - X_{13}X_{22}} = \alpha_1^2 \quad (27)$$

The ratios of the coefficients are found to be

$$B_1/B_7 = \Lambda^{-2}2^{1/2}m_{ss}^{(7)}c(a^*)$$

$$B_2/B_7 = \Lambda^{-2}2^{1/2}m_{ss}^{(7)}s(a^*) \quad (28)$$

$$B_5/B_7 = \Lambda^{-4}\beta_5$$

where $\beta_5 = 0(1)$ is a constant.

The denominator in the frequency condition (27) is

$$X_{12}X_{23} - X_{13}X_{22} = n_{ss}^{(5)}n_{s\theta}^{(\Omega)} - n_{s\theta}^{(5)}n_{ss}^{(\Omega)}$$

and cannot vanish because the direct stresses associated with the membrane and inextensional solutions must be linearly independent. Thus, in the range $\Omega^2 \leq 0(\lambda^{-1})$, the frequency condition can be satisfied only when

$$\Omega \simeq \alpha_1\Lambda^{-2}$$

Since $\Lambda \simeq \lambda r_\theta^{-1/2}$ whenever $\Omega^2 \leq 0(\lambda^{-1})$, we find that there is only one natural frequency for each $m \geq 2$ in the range $\Omega^2 \leq 0(\epsilon^{1/2})$, and it is given by

$$\Omega \simeq \alpha_1\epsilon r_\theta(\sigma_0). \quad (29)$$

We cannot determine α_1 and β_5 because we do not know η_{11} and η_{21}, which are found from the second terms in the asymptotic expansions of n_{ss}, $n_{s\theta}$, and q_s. Hence (29) is not of much practical value in calculating the frequency.

However, a first approximation to the mode is completely determined by the coefficients obtained in (28), even though we do not know β_5 precisely.

$$\left.\begin{aligned} w &\simeq w^{(7)}(\sigma), \qquad v \simeq v^{(7)}(\sigma), \qquad u \simeq u^{(7)}(\sigma)\\ b_s &\simeq b_s^{(7)}(\sigma), \qquad b_\theta \simeq b_\theta^{(7)}(\sigma)\\ \begin{bmatrix} m_{ss}\\ m_{\theta\theta}\end{bmatrix} &\simeq \chi\begin{bmatrix}1\\ \nu\end{bmatrix}\sin\{\zeta-(\pi/4)\} + \begin{bmatrix} m_{ss}^{(7)}(\sigma)\\ m_{\theta\theta}^{(7)}(\sigma)\end{bmatrix}\\ m_{s\theta} &\simeq m_{s\theta}^{(7)}(\sigma)\\ q_s &\simeq \Lambda(\sigma)\,\chi\sin\zeta\\ q_\theta &\simeq -\chi M(\sigma)\sin\{\zeta-(\pi/4)\} + q_\theta^{(7)}(\sigma)\\ \begin{bmatrix} n_{ss}\\ n_{s\theta}\end{bmatrix} &\simeq \Lambda^{-3}(\sigma)\chi G_n(\sigma)\begin{bmatrix} f(\sigma)\\ M(\sigma)\end{bmatrix}\sin\zeta\\ n_{\theta\theta} &\simeq \Lambda^{-2}(\sigma)\chi G_n(\sigma)\cos\{\zeta-(\pi/4)\}\end{aligned}\right\}\qquad(30)$$

$$\begin{aligned}\chi &= 2^{1/2}\,\frac{H(\sigma)}{H(\sigma_0)}\left\{\frac{\Lambda(\sigma)}{\Lambda(\sigma_0)}\right\}^2 e^{\zeta} m_{ss}^{(7)}(\sigma_0)\\ &= 2^{1/2}\left\{\frac{r_\theta(\sigma_0)}{r_\theta(\sigma)}\right\}^{3/4}\left\{\frac{\sin\phi(\sigma_0)}{\sin\phi(\sigma)}\right\}^{1/2} e^{\zeta} m_{ss}^{(7)}(\sigma_0)\end{aligned}\qquad(31)$$

$$\zeta = 2^{-1/2}(x-x_0) = -2^{-1/2}\lambda\int_\sigma^{\sigma_0}\{r_\theta(\sigma')\}^{-1/2}d\sigma'$$

It is noteworthy that all the quantities occurring in these formulas for the mode are static and relatively easy to evaluate. The displacements and rotations are dominated by the inextensional solution, the membrane solution is entirely negligible, and the bending (edge-effect) solutions have a strong influence on the stresslike quantities, making possible the satisfaction of all boundary conditions.

Although this procedure has yielded only an order-of-magnitude estimate for the frequency, it has delivered an estimate of the mode that is both more general and more complete than any previously known.

Case (II): $n_{ss} = N_{s\theta} = Q_s = D = 0$ **at** $\sigma = \sigma_0$. Among the possible edge conditions this is the freest except for the free edge of Case (I). The analysis strongly resembles that of Case (I), and we shall merely record the results. Only one frequency is found in the range $\Omega \leq 0(\lambda^{-1})$ for each $m \geq 2$, namely,

$$\Omega \simeq \alpha_{II}\Lambda^{-3/2} \simeq \alpha_{II}\{\epsilon r_\theta(\sigma_0)\}^{3/4}$$

$\alpha_{II} = 0(1)$ cannot be found explicitly because of cancellation of the leading terms, as was true of α_I. A complete first approximation for the mode is found

$$\left.\begin{aligned} w &\simeq w^{(7)}(\sigma), \qquad u \simeq u^{(7)}(\sigma) \qquad v \simeq v^{(7)}(\sigma)\\ b_\theta &\simeq b_\theta^{(7)}(\sigma)\\ b_s &\simeq \chi\cos\zeta + b_s^{(7)}(\sigma)\\ \begin{bmatrix} m_{ss}\\ m_{\theta\theta}\end{bmatrix} &\simeq -\Lambda(\sigma)\chi\begin{bmatrix}1\\ \nu\end{bmatrix}\sin\{\zeta-(\pi/4)\}\\ m_{s\theta} &\simeq -(1-\nu)M(\sigma)\chi\cos\zeta + m_{s\theta}^{(7)}(\sigma)\\ q_s &\simeq -\Lambda^2(\sigma)\chi\sin\zeta\\ q_\theta &\simeq \Lambda(\sigma)M(\sigma)\chi\sin\{\zeta-(\pi/4)\}\\ \begin{bmatrix} n_{ss}\\ n_{s\theta}\end{bmatrix} &\simeq -\Lambda^{-2}(\sigma)G_n(\sigma)\chi\begin{bmatrix} f\\ M\end{bmatrix}\sin\zeta\\ n_{\theta\theta} &\simeq -\Lambda^{-1}(\sigma)G_n(\sigma)\chi\cos\{\zeta-(\pi/4)\}\end{aligned}\right\}\qquad(32)$$

where ζ is defined as in (31), and

$$\chi = \frac{H(\sigma)}{H(\sigma_0)}\,\frac{\Lambda(\sigma)}{\Lambda(\sigma_0)}\,D^{(7)}(\sigma_0)e^{\zeta} = \frac{r_\theta^{1/4}(\sigma_0)}{r_\theta^{1/4}(\sigma)}\,\frac{\sin^{1/2}\phi(\sigma_0)}{\sin^{1/2}\phi(\sigma)}\,D^{(7)}(\sigma_0)e^{\zeta}$$

The frequency is somewhat higher than in Case (I). The modal displacements are wholly inextensional, and the stresslike quantities are almost entirely derived from the bending solutions.

We see that Cases (I) and (II) are quite similar. In neither case can we calculate the frequency directly but, in both cases, we have very good knowledge of the mode. However, if we use Rayleigh's principle, we can translate accurate information about the mode into accurate information about the frequency. This is exactly what Rayleigh did for spherical domes and cylinders, and we shall derive general formulas for domes with the edge conditions of these two cases in the section, "Applications of Rayleigh's Principle."

Case (III): $n_{ss} = N_{s\theta} = w = m_{ss} = 0$ **at** $\sigma = \sigma_0$. This is the freest of the remaining boundary conditions. The analysis proceeds as in Case (I) except that (25) is replaced by

$$w = B_1c(a) + B_2s(a) + B_5w^{(5)} + B_7w^{(7)} = 0.$$

After eliminating B_1, the matrix equation of the system is

$$\begin{bmatrix}\eta_{11} & X_{12} & (\Omega^2X_{13}+\Lambda^{-4}X_{14})\\ 2^{-1/2}\Lambda^2 & \Lambda X_{22} & (\Omega^2\Lambda X_{23}+X_{24})\\ 2^{-1/2} & \Lambda^{-1}X_{32} & (\Omega^2\Lambda^{-1}X_{33}+\Lambda^{-}X_{34})\end{bmatrix}\begin{bmatrix}\Lambda^{-2}B_2\\ B_5\\ B_7\end{bmatrix} = [0]$$

where

$$X_{22} = -\frac{n_{ss}^{(5)}c(a)}{fG_n},\qquad X_{23} = -\frac{n_{ss}^{(\Omega)}c(a)}{fG_n},\qquad X_{24} = w^{(7)}s(a^*)$$

$$X_{32} = \frac{n_{ss}^{(5)}s(a)}{fG_n},\qquad X_{33} = \frac{n_{ss}^{(\Omega)}s(a)}{fG_n}$$

and the remaining X's are defined as in Case (I). We find for the frequency

$$\Omega^2\Lambda \simeq \frac{X_{12}X_{24}}{X_{12}(X_{33}-X_{23}) + X_{13}(X_{22}-X_{32})} = \alpha_{III}^2$$

and after some reduction

$$\alpha_{III}^2 = 2^{-1/2}\left[\cdot w^{(7)}G_n\left\{\frac{fn_{s\theta}^{(5)} - Mn_{ss}^{(5)}}{n_{ss}^{(\Omega)}n_{s\theta}^{(5)} - n_{s\theta}^{(\Omega)}n_{ss}^{(5)}}\right\}\right]_{\sigma=\sigma_0}\qquad(33)$$

$$\Omega \simeq \alpha_{III}\Lambda^{-1/2} \simeq \alpha_{III}\{\epsilon r_\theta(\sigma_0)\}^{1/4}$$

The frequency is now higher by a factor of roughly $\epsilon^{-1/2}$ than in Case (II). The mode is

$$\left.\begin{aligned} w &\simeq w^{(7)}(\sigma) - \chi\cos\zeta, \qquad u \simeq u^{(7)}(\sigma),\\ v &= v^{(7)}(\sigma)\\ b_s &\simeq \Lambda(\sigma)\chi\cos\{\zeta+(\pi/4)\},\\ b_\theta &= b_\theta^{(7)}(\sigma) - M\chi\cos\zeta\\ \begin{bmatrix} m_{ss}\\ m_{\theta\theta}\end{bmatrix} &\simeq -\Lambda^2(\sigma)\chi\begin{bmatrix}1\\ \nu\end{bmatrix}\sin\zeta,\\ m_{s\theta} &\simeq -\Lambda(\sigma)M(1-\nu)\chi\cos\{\zeta+(\pi/4)\}\\ q_s &\simeq -\Lambda^3(\sigma)\chi\sin\{\zeta+(\pi/4)\},\\ q_\theta &\simeq \Lambda^2(\sigma)M\chi\sin\zeta\\ \begin{bmatrix} n_{ss}\\ n_{s\theta}\end{bmatrix} &\simeq -\Lambda^{-1}(\sigma)G_n(\sigma)\chi\begin{bmatrix} f\\ M\end{bmatrix}\sin\{\zeta+(\pi/4)\}\\ &\quad + \Lambda^{-1}(\sigma_0)w^{(7)}(\sigma_0)G_n(\sigma_0)2^{-1/2}\\ &\quad \times\left\{L_5\begin{bmatrix} n_{ss}^{(5)}(\sigma)\\ n_{s\theta}^{(5)}(\sigma)\end{bmatrix} + L_\Omega\begin{bmatrix} n_{ss}^{(\Omega)}(\sigma)\\ n_{s\theta}^{(\Omega)}(\sigma)\end{bmatrix}\right\}\\ n_{\theta\theta} &\simeq -G_n\chi\cos\zeta\end{aligned}\right\}\qquad(34)$$

where

$$\chi = \{H(\sigma)/H(\sigma_0)\}e^{\zeta}w^{(7)}(\sigma_0)$$

$$L_5 = \left[\frac{Mn_{ss}^{(\Omega)} - fn_{s\theta}^{(\Omega)}}{n_{ss}^{(\Omega)}n_{s\theta}^{(5)} - n_{s\theta}^{(\Omega)}n_{ss}^{(5)}}\right]_{\sigma=\sigma_0}$$

$$L_\Omega = \left[\frac{fn_{s\theta}^{(5)} - Mn_{ss}^{(5)}}{n_{ss}^{(\Omega)}n_{s\theta}^{(5)} - n_{s\theta}^{(\Omega)}n_{ss}^{(5)}}\right]_{\sigma=\sigma_0}$$

This mode differs from those in the two preceding cases in two important ways. First, the displacements are no longer completely inextensional, for the bending solutions make a contribution to w near the edge. Second, the effect of the membrane solution is not now completely negligible but is felt in the formulas for the direct stresses.

Case (IV): $n_{ss} = N_{s\theta} = w = D = 0$ **at** $\sigma = \sigma_0$. The analysis is the same in this case as in Case (III) with an obvious change in the last boundary condition. The natural frequency is found to be

$$\Omega \simeq \alpha_{IV}\Lambda^{-1/2}$$

$$\alpha_{IV}^2 = 2^{1/2}\left[w^{(7)}G_n\left\{\frac{fn_{s\theta}^{(5)} - Mn_{ss}^{(5)}}{n_{ss}^{(\Omega)}n_{s\theta}^{(5)} - n_{s\theta}^{(\Omega)}n_{ss}^{(5)}}\right\}\right]_{\sigma=\sigma_0} \tag{35}$$

and the mode is

$$\left.\begin{aligned}
&w \simeq w^{(7)}(\sigma) + \chi \sin\{\zeta - (\pi/4)\},\\
&\qquad u \simeq u^{(7)}(\sigma), \qquad v \simeq v^{(7)}(\sigma)\\
&b_s \simeq -\Lambda(\sigma)\chi \sin \zeta,\\
&\qquad b_\theta \simeq b_\theta^{(7)}(\sigma) + \chi \sin\{\zeta - (\pi/4)\}\\
&\begin{bmatrix} m_{ss} \\ m_{\theta\theta}\end{bmatrix} \simeq -\Lambda^2(\sigma)\chi\begin{bmatrix} 1 \\ \nu\end{bmatrix}\cos\{\zeta - (\pi/4)\},\\
&\qquad m_{s\theta} \simeq \Lambda(\sigma)M\chi(1-\nu)\sin\zeta\\
&q_s \simeq -\Lambda^3(\sigma)\chi\cos\zeta,\\
&\qquad q_\theta \simeq \Lambda^2(\sigma)M\chi\cos\{\zeta - (\pi/4)\}\\
&\begin{bmatrix} n_{ss} \\ n_{s\theta}\end{bmatrix} \simeq -\Lambda^{-1}(\sigma)G_n(\sigma)\begin{bmatrix} f \\ M\end{bmatrix}\chi\cos\zeta\\
&\quad + \Lambda^{-1}(\sigma_0)w^{(7)}(\sigma_0)G_n(\sigma_0)2^{-1/2}\\
&\quad \times\left\{L_5\begin{bmatrix} n_{ss}^{(5)}(\sigma) \\ n_{s\theta}^{(5)}(\sigma)\end{bmatrix} + L_\Omega\begin{bmatrix} n_{ss}^{(\Omega)} \\ n_{s\theta}^{(\Omega)}\end{bmatrix}\right\}\\
&n_{\theta\theta} \simeq G_n(\sigma)\chi\sin\{\zeta - (\pi/4)\}
\end{aligned}\right\} \tag{36}$$

where

$$\chi = 2^{1/2}\{H(\sigma)/H(\sigma_0)\}e^{\zeta}w^{(7)}(\sigma_0)$$

This frequency and mode are qualitatively much like those in Case (III). We see from (35) and (33) that the frequency estimate in the present case is larger than in Case (III) by a simple factor $2^{1/2}$.

The edge conditions considered in Cases (I)–(IV) all have $n_{ss} = N_{s\theta} = 0$ at $\sigma = \sigma_0$; i.e., the edges of the shell have been free to move in directions tangent to the middle surface. We have now exhausted all the cases with this property. If we work out similar analyses for Case (V): $n_{ss} = v = Q_s = m_{ss} = 0$ at $\sigma = \sigma_0$, and Case (VI): $u = N_{s\theta} = Q_s = m_{ss} = 0$ at $\sigma = \sigma_0$, we see that natural frequencies in the range $\Omega^2 \leq 0(\lambda^{-1})$ cannot occur; i.e., for these edge conditions all the natural frequencies obey

$$\Omega^2 \geq 0(1).$$

But all the remaining edge conditions are obtained from (V) or (VI) by "tightening" some of the conditions. Hence, in all the remaining cases, the natural frequencies are at least as high as in Cases (V) or (VI). We conclude that only for Cases (I)–(IV) can we find natural frequencies in the range

$$\Omega^2 \leq 0(\lambda^{-1}).$$

Now, it is easy to see from the motion equations that inextensional solutions, i.e., solutions having the property that

$$n_{ss}, \quad n_{\theta\theta}, \quad n_{s\theta} \leq 0(\lambda^{-1})$$

and all other quantities are 0(1), cannot occur when $\Omega^2 \geq 0(1)$. Hence, for a dome, inextensional modes and frequencies can occur only in Cases (I)–(IV); i.e., only when the edge is free to move tangentially.

Applications of Rayleigh's Principle

In this section we shall see how estimates of the inextensional frequencies can be obtained for the Cases (I) and (II) by using Rayleigh's principle.

Rayleigh's principle states that for any field of displacements satisfying the edge conditions on displacements,

$$\Omega^2 \leq \Omega_E^2 = (E_s + E_B)/K \tag{37}$$

where E_s, E_B, and K are to be calculated from the given field of displacements by means of (13). The accuracy of the estimated frequency, Ω_E, depends on (and is usually much better than) the accuracy of the assumed displacement field. We must emphasize (because it is occasionally overlooked) the effect of the edge conditions on the displacements. If these edge conditions are not satisfied by the chosen displacement field, Ω_E may differ wildly from Ω and need not even be the larger of the two.

In applying Rayleigh's principle to Cases (I) and (II), we shall take as the trial displacements the approximate modes given for these cases by our previous analysis. From (13), (4), and (5), we have

$$E_S = (1-\nu^2)^{-2}\int_{\sigma=0}^{\sigma_0}\{n_{ss}^2 + n_{\theta\theta}^2 - 2\nu n_{ss}n_{\theta\theta} + 2(1+\nu)n_{s\theta}^2\}r_\theta \sin\phi\, d\sigma \tag{38}$$

$$E_B = \lambda^{-4}(1-\nu^2)^{-1}\int_{\sigma=0}^{\sigma_0}\{m_{ss}^2 + m_{\theta\theta}^2 - 2\nu m_{ss}m_{\theta\theta} + 2(1+\nu)m_{s\theta}^2\}r_\theta \sin\phi\, d\sigma \tag{39}$$

$$K = \int_{\sigma=0}^{\sigma_0}(u^2 + v^2 + w^2)r_\theta \sin\phi\, d\sigma \tag{40}$$

Referring to the formulas (30) and (32), for the modes in the two cases, we see that two kinds of terms occur; namely, terms of inextensional and edge-effect types. The integrals of the inextensional terms are of the same order as the terms themselves and cannot be evaluated explicitly until the shell shape is specified. The integrals of the edge-effect terms are smaller by an order of magnitude than the terms themselves and can be evaluated explicitly (though approximately) by use of the Laplace method for asymptotic approximation of definite integrals.

For example, in Case (II), (38) and (32) lead to

$$\begin{aligned}
E_s &\simeq (1-\nu^2)^{-2}\int_0^{\sigma_0} n_{\theta\theta}^2 r_\theta \sin\phi\, d\sigma\\
&\simeq (1-\nu^2)^{-2}\int_0^{\sigma_0}[\Lambda^{-2}(\sigma)\chi^2(\sigma)r_\theta(\sigma)\sin\phi(\sigma)G_n^2(\sigma)]\\
&\qquad \times \cos^2[\zeta - (\pi/4)]d\sigma\\
&\simeq \int_0^{\sigma_0}\{\Lambda^{-2}(\sigma)r_\theta^{-1}(\sigma)\sin\phi(\sigma)\Delta^2(\sigma)[D^{(7)}(\sigma)]^2\}\\
&\qquad \times e^{2\zeta}\cos^2[\zeta - (\pi/4)]d\sigma
\end{aligned}$$

where

$$\Delta(\sigma) = H(\sigma)\Lambda(\sigma)/[H(\sigma_0)\Lambda(\sigma_0)]$$

The function $e^{2\zeta}$ has the value unity for $\sigma = \sigma_0$ and decreases rapidly to zero as σ decreases from σ_0. Hence this integral is of Laplace type; i.e., only the region near $\sigma = \sigma_0$ contributes appreciably to the integral. We may therefore approximate it by

$$E_s \simeq \{\Lambda^{-2}r_\theta^{-1}[D^{(7)}]^2\sin\phi\}_{\sigma_0}(1/2)\int_0^{\sigma_0}e^{2\zeta}(1 + \sin 2\zeta)d\sigma$$

The integral in this expression can be evaluated approximately with the aid of (31) to give

$$E_s \simeq \{\Lambda^{-3} r_\theta^{-1} [D^{(7)}]^2 \sin\phi\}|_{\sigma_0} (2^{1/2}/8)$$

We know that $\Omega^2 = 0(\lambda^{-3})$, hence

$$\Lambda(\sigma_0) \simeq \lambda r_\theta^{-1/2}(\sigma_0).$$

Thus, finally, we find

$$E_s \simeq (1/8)\lambda^{-3}[2r_\theta(\sigma_0)]^{1/2}[D^{(7)}(\sigma_0)]^2 \sin\phi\ (\sigma_0)$$

In a similar manner E_B may be estimated,

$$E_B \simeq (3/8)\lambda^{-3}[2r_\theta(\sigma_0)]^{1/2}[D^{(7)}(\sigma_0)]^2 \sin\phi\ (\sigma_0)$$

Hence for Case (II)

$$\Omega_E^2 \simeq \frac{\lambda^{-3}[r_\theta(\sigma_0)/2]^{1/2}[D^{(7)}(\sigma_0)]^2 \sin\phi\ (\sigma_0)}{K^{(7)}} \qquad (41)$$

where

$$K^{(7)} = \int_{\sigma=0}^{\sigma_0} \{[u^{(7)}(\sigma)]^2 + [v^{(7)}(\sigma)]^2 + [w^{(7)}(\sigma)]^2\} r_\theta \sin\phi d\sigma$$

Applying the same analysis in Case (I), we find

$$E_s = 0(\lambda^{-5})$$

$$E_B = E_B^{(7)} + 0\ (\lambda^{-5})$$

$$E_B^{(7)} = \lambda^{-4} \int_{\sigma=0}^{\sigma_0} \{[m_{ss}^{(7)}]^2 + [m_{\theta\theta}^{(7)}]^2 - 2\nu m_{ss}^{(7)} m_{\theta\theta}^{(7)} + 2(1+\nu)[m_{s\theta}^{(7)}]^2\} r_\theta \sin\phi d\sigma$$

$$K = K^{(7)}$$

Hence we obtain the estimate

$$\Omega_E^2 \simeq E_B^{(7)}/K^{(7)} = 0(\lambda^{-4}) \qquad (42)$$

We see that in this case the estimate given by Rayleigh's principle can be derived solely from the inextensional displacements. This is not true of the estimate just obtained in Case (II), nor is it true of the Rayleigh estimates that are obtained for the inextensional frequencies in Cases (III) and (IV). Equation (42) is of course just the estimate that Rayleigh used to find the inextensional frequencies for a spherical dome. However, neither Rayleigh nor any subsequent investigator seems to have been sure of the conditions under which the estimate is accurate. We now see that it is accurate only when the edge of the dome is free.

Inextensional Frequencies for a Spherical Dome

In this section we carry out the calculation of the two lowest inextensional frequencies for a spherical dome under the edge conditions of Cases (II) and (III), using formulas (41) and (33), respectively.

For a spherical dome the inextensional solution that is finite at $\phi = 0$ has

$$u^{(7)} = v^{(7)} = \sin\phi \tan^m(\phi/2), \qquad m \geq 2$$

$$w^{(7)} = -(m + \cos\phi)\tan^m(\phi/2) \qquad (43)$$

$$D^{(7)} = \{\sin\phi - m(m+\cos\phi)\csc\phi\} \tan^m(\phi/2)$$

and the kinetic energy is given by (recall that ϕ_0 is the edge angle)

$$K^{(7)} = K(\phi_0, m) = \int_0^{\phi_0} \tan^{2m}(\phi/2)\{2\sin^2\phi + (m+\cos\phi)^2\} \sin\phi d\phi.$$

In Case (II) equation (41) reduces to

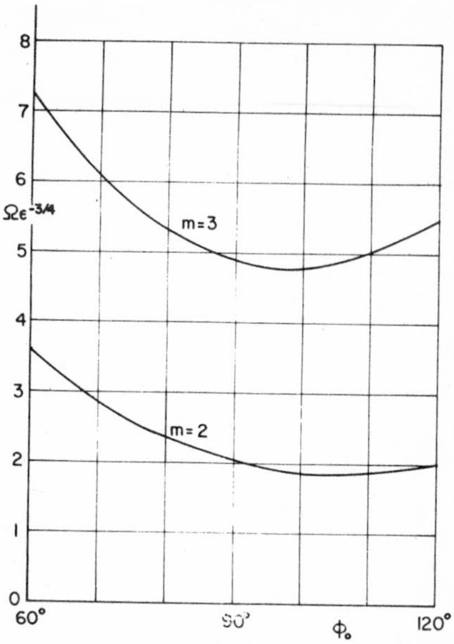

Fig. 1 Inextensional frequencies as functions of edge angle, ϕ_0, for the edge condition $n_{ss} = N_{s\theta} = Q_s = D = 0$ from equation (44)

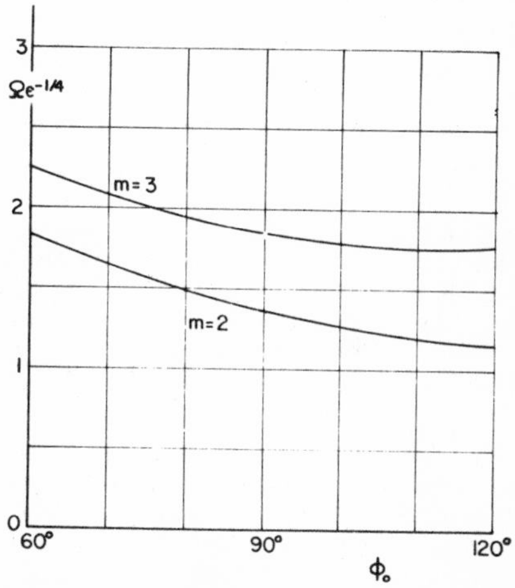

Fig. 2 Inextensional frequencies as functions of edge angle, ϕ_0, for the edge condition $n_{ss} = N_{s\theta} = w = m_{ss} = 0$ from equation (46)

$$\Omega_E^2 = \frac{2^{-1/2}\epsilon^{3/2}}{K(\phi_0, m)} \sin\phi_0 \tan^{2m}(\phi_0/2)\{\sin\phi_0 - m(m+\cos\phi_0)\csc\phi_0\}^2 \qquad (44)$$

Fig. 1 shows graphs of the relations between Ω_E and ϕ_0 for $m = 2$ and 3, obtained from (44).

In Case (III) the frequency is given by (33). To evaluate this for a sphere, we observe first that the membrane solution finite at $\phi = 0$ has (see Love [4])

$$n_{s\theta}^{(5)} = -n_{ss}^{(5)},$$

and (33) reduces to

$$\alpha_{\rm III}^2 = \frac{-2^{-1/2}(1-\nu^2)(m+\cos\phi_0)^2 \csc\phi_0 \tan^m(\phi_0/2)}{\{n_{ss}^{(\Omega)} + n_{s\theta}^{(\Omega)}\}|_{\sigma_0}} \qquad (45)$$

To find

$$z = n_{ss}^{(\Omega)} + n_{s\theta}^{(\Omega)}$$

we must solve the system of equations obtained from the motion equations, (7)–(9), by setting $\epsilon = 0$ and taking for u, v, and w the inextensional displacements (43). The governing equation is

$$\frac{dz}{d\phi} + (m + 2 \cos \phi) \csc \phi \, z$$
$$= -(1 - \nu^2)\{u^{(7)} + v^{(7)} - w^{(7)}(m + \cos \phi) \csc \phi\}$$

and a particular solution (which is all we need) is

$$z(\phi_0) = -(1 - \nu^2)K(\phi_0, m) \sin^{-2} \phi_0 \tan^{-m} (\phi_0/2)$$

Combining this with (45) and (33), we find

$$\Omega^2 \simeq \frac{\epsilon^{1/2}(m + \cos \phi_0)^2 \sin \phi_0 \tan^{2m} (\phi_0/2)}{2^{1/2}K(\phi_0, m)} \quad (46)$$

The frequencies predicted by (46) when $m = 2$ and 3 are shown in Fig. 2

The inextensional frequencies in Case (IV) are $2^{1/2}$ times those of Case (III).

Discussion

We may make the following comments about the results of the preceding sections.

(i) Although it has been customary to assume that inextensional modes of vibration could, in general, be derived from a classical shell theory (see Kalnins [9], for example), it seems not to have been done before. Our derivation is of course only approximate, but the results are reassuring.

(ii) The observation that inextensional modes can be found for a dome only when the edge is free to move tangentially is similar to a conclusion reached by Arnold and Warburton [5] for a cylinder.

(iii) The analysis shows that Rayleigh's procedure (i.e., using the static inextensional solution as trial displacements in Rayleigh's variational principle) works for a dome only when the edge is free. For cases (II)–(IV) the inextensional displacements do not satisfy the edge conditions required by the variational principle. Alternatively, using the inextensional displacements alone can work only when the modal energy of bending greatly exceeds that of stretching, a condition which the true modes do not satisfy in cases (II)–(IV).

(iv) Recently, Hwang [10] described difficulties in calculating the inextensional frequencies from the general solution for a free-edged hemisphere. The difficulty took the form of severe accuracy loss in computing the frequency determinant. The writer has commented elsewhere on this difficulty [11], but here we add that it may be partly related to the cancellation of the leading bending terms for n_{ss} and $n_{s\theta}$ in Cases (I) and (II). A sufficiently accurate solution would be free of this difficulty and, evidently, the calculations described by Kalnins [9] were successful in this case.

(v) The inextensional frequencies are acutely sensitive to the edge conditions. For example, if $h/R = 1/30$, the lowest inextensional frequency in Case (IV) may be larger than in Case (I) by a factor like 20–40. Conceivably this sensitivity could be useful for experimentally determining what the edge conditions actually are.

(vi) We have not considered boundary conditions of elastic constraint at the edge. In general, we may expect that these will produce frequencies lying between those associated with the two "pure" edge conditions that are combined to give the elastic condition. For example, the lowest natural frequency associated with the boundary condition

$$n_{ss} = N_{s\theta} = w = \xi m_{ss} + (1 - \xi)D = 0,$$

where $0 \leq \xi \leq 1$, should satisfy

$$\alpha_{III}\Lambda^{-1/2} \leq \Omega \leq \alpha_{IV}\Lambda^{-1/2}.$$

Although we have chosen to demonstrate this procedure for domes, it ought to work equally well for shells with two edges. However, it remains always subject to the condition that $m^2\epsilon \ll 1$.

References

1 Lord Rayleigh, *The Theory of Sound*, 2nd ed., Vol. I, Dover Publications, New York, 1944.

2 Goodier, J. N., and McIvor, I. K., "The Elastic Cylindrical Shell Under Nearly Uniform Radial Impulse," Journal of Applied Mechanics, Vol. 31, No. 2, Trans. ASME, Vol. 86, Series E, June 1964, pp. 259–266.

3 Forsberg, K., "Influence of Boundary Conditions on the Modal Characteristics of Thin Cylindrical Shells," *American Institute of Aeronautics and Astronautics Journal*, Vol. 2, 1964, pp. 2150–2157.

4 Love, A. E. H., "A Treatise on the Mathematical Theory of Elasticity," 4th ed., Dover Publications, New York, 1944.

5 Arnold, R. N., and Warburton, G. B., "Flexural Vibrations of the Walls of Thin Cylindrical Shells Having Freely Supported Ends," *Proceedings of the Royal Society*, London, Series A, Vol. 197, 1949, pp. 238–256.

6 Ross, E. W., Jr., "Approximations in Nonsymmetric Shell Vibrations," U. S. Army Materials Research Agency Report No. TR 67-09, Apr. 1967, submitted to *Journal of the Acoustical Society of America*.

7 Sanders, J. L., "An Improved First-Approximation Theory for Thin Shells," NASA Report 24, 1959.

8 Ross, E. W., Jr., "Asymptotic Analysis of the Axisymmetric Vibrations of Shells," Journal of Applied Mechanics, Vol. 33, No. 1, Trans. ASME, Vol. 88, Series E, Mar. 1966, pp. 85–92.

9 Kalnins, A., Discussion of "Some Experiments on the Vibration of a Hemispherical Shell," Journal of Applied Mechanics, Vol. 34, No. 3, Trans. ASME, Vol. 89, Series E, Sept. 1967, pp. 792–793.

10 Hwang, C., "Some Experiments on the Vibrations of a Hemispherical Shell," Journal of Applied Mechanics, Vol. 33, No. 4, Trans. ASME, Vol. 88, Series E, Dec. 1966, pp. 817–824

11 Ross, E. W., Jr., "On Membrane Frequencies for Spherical Shell Vibrations," *American Institute of Aeronautics and Astronautics Journal*, Vol. 6, No. 5, May 1968, pp. 803–808

19

Reprinted from *Proc. Roy. Soc. London*, Ser. A, **197**, 238–256 (1949)

Flexural vibrations of the walls of thin cylindrical shells having freely supported ends

By R. N. Arnold, D.Sc. and G. B. Warburton
Department of Engineering, University of Edinburgh

(*Communicated by M. Born, F.R.S.—Received* 25 *October* 1948—
Revised 14 *December* 1948)

The paper deals with the general equations for the vibration of thin cylinders and a theoretical and experimental investigation is made of the type of vibration usually associated with bells. The cylinders are supported in such a manner that the ends remain circular without directional restraint being imposed. It is found that the complexity of the mode of vibration bears little relation to the natural frequency; for example, cylinders of very small thickness-diameter ratio, with length about equal to or less than the diameter, may have many of their higher frequencies associated with the simpler modes of vibration. The frequency equation which is derived by the energy method is based on strain relations given by Timoshenko. In this approach, displacement equations are evolved which are comparable to those of Love and Flügge, though differences are evident due to the strain expressions used by each author. Results are given for cylinders of various lengths, each with the same thickness-diameter ratio, and also for a very thin cylinder in which the simpler modes of vibration occur in the higher frequency range. It is shown that there are three possible natural frequencies for a particular nodal pattern, two of these normally occurring beyond the aural range.

Introduction

A thin cylindrical shell may vibrate in a variety of ways depending on the particular straining actions involved. The present paper is confined to that type of vibration in which bending and stretching of the shell predominate, as, for example, that associated with the tones emitted by church bells. Some of the vibration forms, which are possible under such conditions, are illustrated in figure 1. For sections perpendicular to the axis of the cylinder, the vibration consists of both radial and tangential movement, a number of stationary waves being formed around the circumference. In the simplest case ($n = 2$)* there are four positions at which the radial movement is zero. These will be referred to, for convenience, as circumferential nodes even although some tangential motion is known to exist. The form of vibration increases in complexity with the number of nodes, but theoretically there is no limit to the number which may be present. For a cylinder with free ends, the fundamental vibration form for a given circumferential arrangement exists when all cross-sections of the cylinder describe precisely the same motion. In such a case, each axial strip of the shell remains straight during vibration. Further complication, however, may result due to the superposition of waves in the axial direction. Some examples of this are given in figure 1*b* for the case in which the ends of the cylinder are maintained circular and no directional restraint is imposed, a condition which will be termed 'freely supported'.

* A list of symbols is given in appendix 1, p. 251.

It will be observed that certain cross-sections of the cylinder remain at rest, depending on the number of axial waves which are present. The positions of these define the axial nodes. Thus an infinite number of axial vibration forms are theoretically possible, each of which may be combined with a corresponding number of circumferential forms. To define any particular nodal arrangement, it is only necessary to specify the number of circumferential waves n and the number of axial half-waves m; figure 1*c*, for example, illustrates the vibration form $n = 3$, $m = 4$. It will be shown later that three distinct natural frequencies are associated with each nodal pattern, depending on the relative amplitudes of the three component vibrations executed by an element of the cylinder.

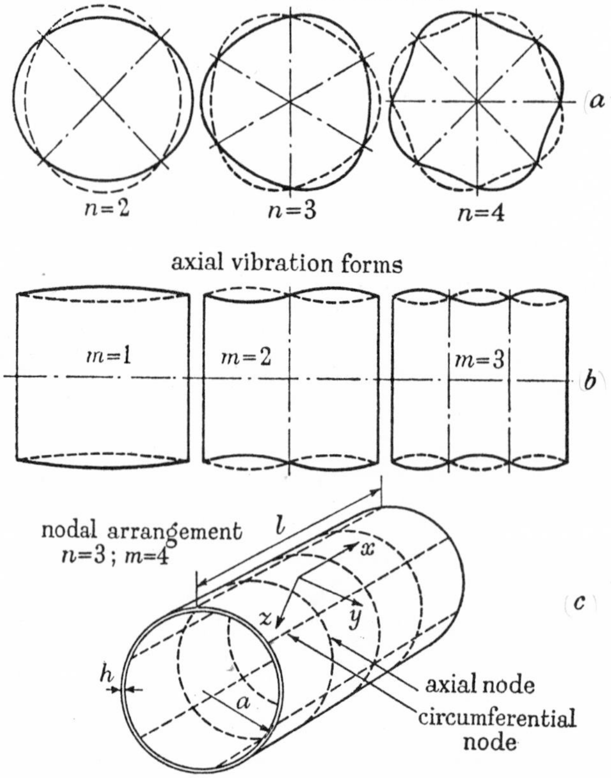

FIGURE 1. Forms of vibration of thin cylinders with freely supported ends.

THEORY*

Rayleigh (1894, chapter x) derived an expression for the natural frequencies of thin cylindrical shells with free ends vibrating with circumferential nodes. For this, he assumed the deformations to be inextensional; this means that the length of any line drawn on the middle surface of the shell (an imaginary surface situated at mid thickness) remains unaltered during vibrations of small amplitude. Bending was thus accepted as the predominant straining action, and stretching of the shell was ignored. While applicable to the particular problem which Rayleigh investigated,

* The mathematical analysis is given in the appendices.

this concept has limitations. For example, while it may be legitimate to consider a line inscribed circumferentially on the middle surface to remain unextended during vibration, it is incompatible to assume in general that the middle surface as a whole is not subject to extension. Rayleigh did not discuss this aspect but merely confined himself to those vibrations in which all cross-sections throughout the entire length of the cylinder performed precisely the same motion. By equating kinetic energy at the mean position to strain energy at the maximum displacement, he obtained an expression for frequency in the form

$$f = \frac{1}{2\pi}\sqrt{\left\{\left[\frac{Eh^2}{12\rho a^4(1-\sigma^2)}\right]\left[\frac{n^2(n^2-1)^2}{n^2+1}\right]\right\}},$$

in which the second term expresses the nodal configuration.

For modes of vibration in which both circumferential and axial nodes exist, the assumption of inextension of the middle surface is inadmissible and the resulting analysis more complicated. In such cases, both bending and stretching of the shell wall must be considered.

The initial procedure involves the derivation of equations connecting the three component displacements u, v and w in the directions x, y and z (figure 1), of a point in the middle surface of the shell. This may be achieved in two ways:

(*a*) The equilibrium of a small element may be considered and the forces and moments acting on it expressed first in terms of strain and change of curvature of the middle surface, and finally as functions of the displacements. The resulting equations express u, v and w and their derivatives in terms of the elasticity and density of the material and the dimensions of the cylinder.

(*b*) A vibration form compatible with the end conditions may be assumed, the strain energy expressed in terms of the displacements and the kinetic energy in terms of the rates of change of the displacement. As both strain and kinetic energy are functions of the three independent variables u, v and w, the Lagrange dynamical equations may be used to form three equations connecting the displacements and their derivatives (see appendix 2).

It may be argued that the latter does not provide a direct attack, but it has the merit of brevity in the present case, when the form of the vibration is readily determined. Moreover, in the former method, a vibration form which will satisfy the derived equations must eventually be obtained. The equations resulting from either method will, however, only be as accurate as the original expressions adopted for the strain-displacement relations.

Procedure (*a*) has been followed by Love (1927), Flügge (1934) and Timoshenko (1940), but in view of the assumptions and approximations made by each during the analysis, the results are not identical. It appears advisable, therefore, before proceeding further to give a brief outline of the origin of the discrepancies. The three main differences appear to be as follows:

(i) Love and Flügge both make allowance, when considering an element, for the trapezoidal shape of the two faces, which lie in the planes perpendicular to the cylinder axis (figure 9, p. 252). Timoshenko assumes the inner and outer edges of these faces to be of equal length.

(ii) Each author makes use of different approximations in expressing the strains at a point distant z from the middle surface in terms of the extensional strain ϵ and change of curvature κ of that surface. Terms of the following types are obtained:

Love	Flügge	Timoshenko
ϵ	ϵ; $\frac{z}{a}\epsilon$; $\frac{z^2}{a^2}\epsilon$	ϵ
$z\kappa$; $\frac{z^2}{a}\kappa$	$z\kappa$; $\frac{z^2}{a}\kappa$	$z\kappa$

In this, Love assumes that ϵ is small in comparison with $z\kappa$ and thus neglects terms of the type $\frac{z}{a}\epsilon$ and $\frac{z^2}{a^2}\epsilon$, but considers $\frac{z^2}{a}\kappa$. As the maximum value of z is $\frac{1}{2}h$ and the ratio of thickness to radius is small for a thin cylinder, terms $\frac{z^2}{a}\kappa$ and $\frac{z}{a}\epsilon$ are always small compared with $z\kappa$ and ϵ respectively. On this basis, Timoshenko elects to ignore all terms involving z/a.

(iii) Love considers the existence of a direct stress acting radially in the cylinder walls, satisfying the condition of zero stress at the inner and outer boundaries.

It seems relevant at this stage to indicate the objective of each of these analyses. That due to Timoshenko is only concerned with the straining action of thin cylinders, and terms involving inertia forces are absent. No attempt is therefore made to apply the result to problems of vibration. Love proceeds only so far as to indicate the type of functions which will satisfy the vibration equations, but does not investigate the implications arising from the use of such expressions. Only Flügge produces a direct calculation of frequency by obtaining the solution for a cylinder with freely supported ends. A strange fact, which emerges from his analysis, is the existence of three natural frequencies for any particular nodal pattern. Investigation reveals that each is associated with a unique arrangement of the ratios of the maximum vibration amplitudes occurring in the three component directions. While Flügge's equations allow for the presence of axial nodes, no attempt is made to investigate their existence.

Working independently the present authors derived expressions for frequency using the strain relations of Timoshenko in conjunction with the energy method described in paragraph (*b*), p. 240. They assumed that the axial, circumferential and radial vibration displacements were of the form

$$u = A\cos\frac{m\pi x}{l}\cos n\phi\cos\omega t,$$

$$v = B\sin\frac{m\pi x}{l}\sin n\phi\cos\omega t,$$

and
$$w = C\sin\frac{m\pi x}{l}\cos n\phi\cos\omega t,$$

which satisfies the conditions for freely supported ends and allows for the existence of axial nodes. It was later found that Flügge had made precisely similar assumptions.

In the above, A, B and C are constants, $\omega/2\pi$ is the frequency, and n and m are integers defining the circumferential and axial nodes; the number of nodes is respectively $2n$ and $(m+1)$ for freely supported ends.

From the three displacement equations, it is possible to eliminate A, B and C to form the cubic

$$\Delta^3 - K_2\Delta^2 + K_1\Delta - K_0 = 0,$$

where K_0, K_1 and K_2 are as specified in appendix 2, equation (15), and frequency

$$f = \frac{1}{2\pi a}\sqrt{\left\{\left[\frac{Eg}{\rho(1-\sigma^2)}\right]\Delta\right\}}.$$

It is found that the above equation supplies three real positive values for Δ, and consequently three frequencies are obtained for any given nodal configuration of the cylinder. The essential difference between these vibrations is the relative amplitudes of the motions u, v and w.

The above mathematical result does not offer any direct interpretation of the manner in which the frequency varies with the cylinder dimensions or nodal pattern. Direct calculation for specific cases was therefore undertaken. Four non-dimensional factors were chosen, namely,

$$\sqrt{\Delta} = \text{frequency} \times \sqrt{\left\{\frac{4\pi^2 a^2 \rho(1-\sigma^2)}{Eg}\right\}}.$$

$$\lambda = \frac{\text{mean circumference}}{\text{axial wave-length}} = \frac{2\pi a}{2l/m} = \frac{m\pi a}{l},$$

$$n = \text{number of circumferential waves},$$

$$\alpha = \frac{\text{thickness}}{\text{mean radius}} = \frac{h}{a}.$$

Of these, the first is proportional to frequency, the second to the number of axial waves and the fourth to the cross-sectional proportions.

Frequency calculations were performed for a series of thickness ratios α over a range of values of λ, and typical results are shown in figure 2a to d. In these, curves of $\sqrt{\Delta}$ are plotted to a base of λ for various nodal arrangements n, the value of α being constant for each diagram. For an infinitely thin cylinder (figure 2a) it will be seen that the natural frequency decreases as n increases, a result which one would not readily have predicted. That this frequency arrangement is not confined to hypothetical cylinders is shown by the remaining diagrams, where α has respectively the values 0·002, 0·01 and 0·0525. The last value, similar to that of a cylinder used in the experimental investigation, shows the same tendency at the higher values of λ, which are related to small axial wave-length. These theoretical results were so unexpected that a comprehensive experimental investigation was considered advisable.

Experimental investigations

Steel cylinders were used for the experimental work in order that the natural frequencies could be stimulated by magnetic means. Difficulty was experienced in reproducing end conditions in accordance with the theoretical assumptions. The

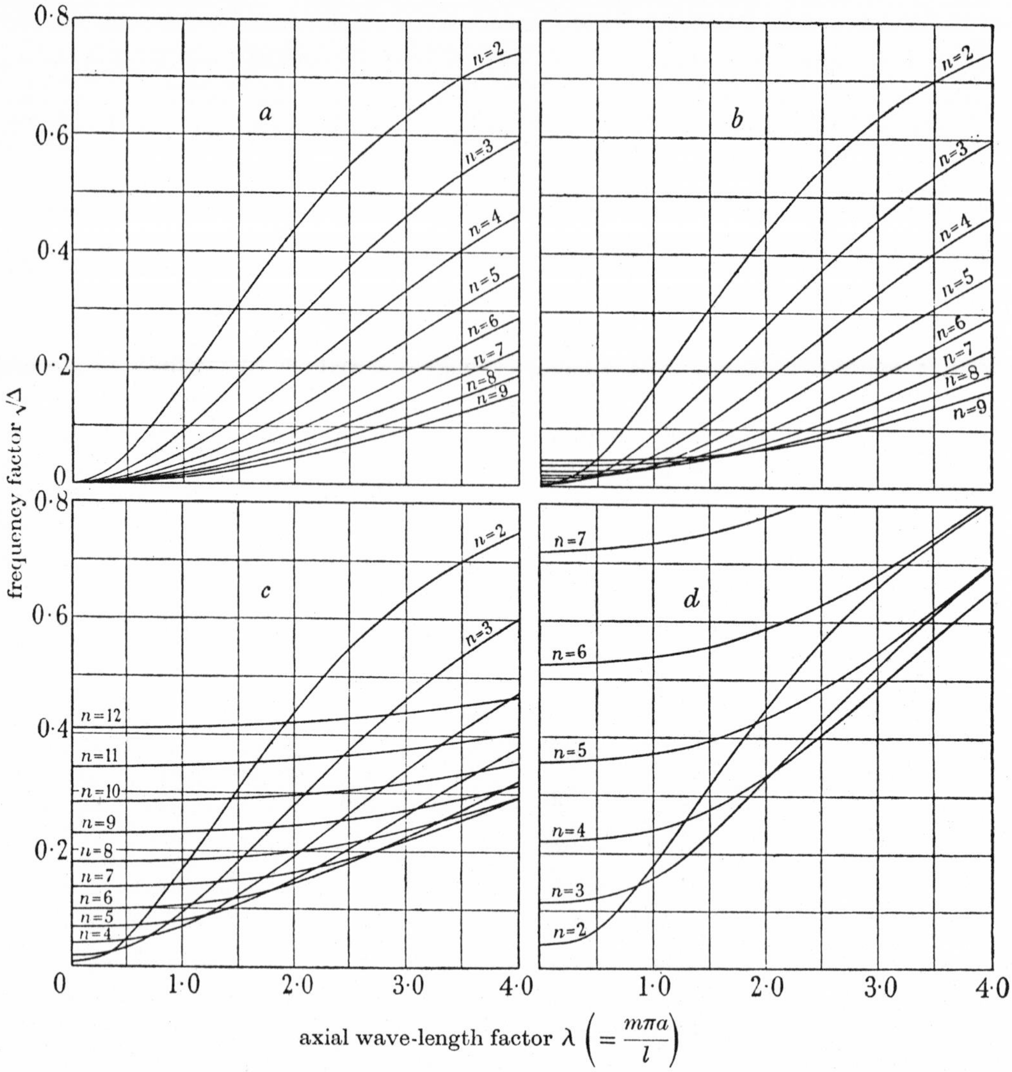

FIGURE 2. Theoretical frequency curves for cylinders with freely supported ends. α = thickness/mean radius. Values of α in $a = 0$, in $b = 0{\cdot}002$, in $c = 0{\cdot}01$ and in $d = 0{\cdot}0525$.

type of end-piece finally adopted is illustrated in figure 3. It consisted of a circular steel plate accurately machined to fit the bore of the cylinder and shaped to provide as near as possible line contact with the inner surface. Slackness of fit in the end-pieces was found to produce inconsistent results. The cylinder, with end-pieces in place, was supported between centres so that it could be freely rotated and was set vibrating by a small electromagnet which was excited by an audio-frequency oscillator. The magnet was fixed to a support which could be moved to any axial position along the cylinder, and provision was made for adjusting the gap between magnet and cylinder.

When investigating a particular cylinder the magnet was placed midway along the axis and the frequency increased slowly by means of the oscillator. In passing through a natural frequency the cylinder was heard to respond and the oscillator was then adjusted in frequency so that the note emitted was of maximum intensity. On slowly turning the cylinder it was found that the sound intensity increased and decreased periodically, the number of fluctuations being a measure of the number of circumferential nodes. The nodes rotate with the cylinder due to the presence of minute imperfection, for if the cylinder were mathematically perfect, no preferential location of the circumferential nodes could exist. As a result of imperfection, two natural frequencies (Rayleigh 1894, p. 389) are present in association with two preferential nodal directions. For an accurately made cylinder, these frequencies are very close to one another, and this may lead to difficulty of interpretation if the resonance of each is not clearly defined. Only in a few cases could the results be determined by the unaided ear, and in general a medical stethoscope was used to

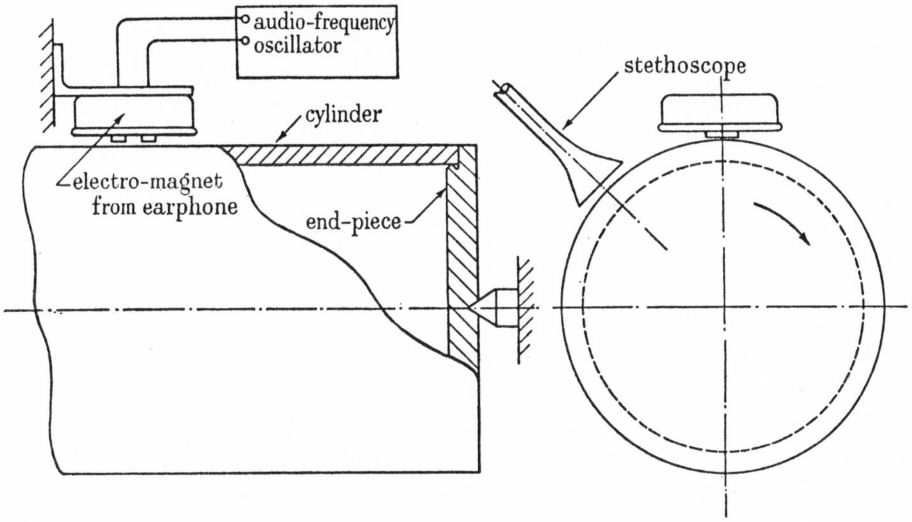

FIGURE 3. Experimental arrangement.

listen to the response. For the axial nodes it was only necessary to move the stethoscope along the length of the cylinder and count the number of fluctuations of sound intensity. For doubtful responses, an electrical pick-up of condenser type was available; this recorded the vibration amplitude of the cylinder on a cathode-ray oscilloscope, but it was rarely necessary to make use of this instrument.

After progressing through the aural range of frequencies and recording the responses of the cylinder, the magnet was moved axially to another location and the procedure repeated to discover natural frequencies which had not been elicited at the mid-position. It was not possible for the magnet when situated centrally to elicit vibrations having a whole number of axial waves.

Some vibration configurations were found to be more readily stimulated than others, notably those having 6 or 8 circumferential nodes. The 4-nodal pattern at times presented difficulty, especially when its frequency was higher than those of the more complicated forms. It should be mentioned that the frequency values were slightly influenced by the axial pressure exerted on the supporting centres. Theoretically, a uniform axial compressive stress should not affect the frequency, but the pressure in this case resulted from the ends of the cylinder coming into contact with the end-pieces, and thus a small fixing moment was also introduced. This effect, however, was unimportant, as no appreciable error was introduced by the small axial forces used in the experiments.

Experimental work was conducted on five different cylinders. One of these, of thickness ratio $\alpha = 0{\cdot}01$, was specially designed to demonstrate the peculiar effects evident in figure 2*c*. The remaining cylinders ($\alpha = 0{\cdot}0525$) were identical in all dimensions except that of length. They were machined to give relative length ratios of 1, $\frac{1}{2}$, $\frac{1}{3}$ and $\frac{1}{4}$ and thus allow a comparison to be made when each vibrated with the same axial wave-length.

Experimental results

The collected results for four experimental steel cylinders are plotted in figure 4. These cylinders were of outside diameter 3·950 in., internal diameter 3·748 in. and had the following effective lengths when freely supported at the ends:

cylinder	length (in.)
1	15·63
2	7·77
3	5·15
4	3·84

The full curves of figure 4 represent the calculated variation of frequency with λ for $n = 2$ (4 nodes) to $n = 6$ (12 nodes) and are similar to those of figure 2*d*. The calculations were based on the following constants: $E = 29{\cdot}6 \times 10^6$ lb./in.^{2}; $\sigma = 0{\cdot}29$; $\rho = 0{\cdot}283$ lb./in.^{3}. The experimental frequencies obtained from all the cylinders have been superposed on these curves, the points for cylinders 2 to 4 being numbered while those for cylinder 1 have been left unmarked. In the latter, it was possible to identify 41 natural frequencies, a list of which is given in table 1.

Table 1. Experimental frequencies (c./sec.) cylinder no. 1

number of axial half-waves	number of circumferential nodes ($2n$)				
	4	6	8	10	12
1	960	2,130	3,985	6,400	9,270
2	2,070	2,420	4,130	6,500	9,370
3	3,725	3,130	4,430	6,700	9,570
4	5,270	4,180	4,950	7,030	9,850
5	6,880	5,380	5,690	7,520	10,230
6	8,270	6,670	6,630	8,180	10,730
7	—	7,940	7,580	8,900	11,340
8	—	9,160	8,660	9,770	—
9	—	10,330	9,720	10,650	—
10	—	—	10,870	—	—

The experimental results are found to be in fair agreement with those predicted by theory, the greatest divergence being when $n = 2$. In this case, the experimental curve has been drawn dashed for comparison. These results are a striking demonstration of the strange phenomenon, to which reference has been made, namely, that certain vibration patterns may have frequencies higher than those related to vibrations which are much more complicated. It will be observed, for example, that at $\lambda \doteqdot 2{\cdot}3$ (wave-length 5·15 in.) the frequency for $n = 2$ is higher than those for $n = 3$, 4 or 5, and that the frequency for $n = 3$ has just overtaken that for $n = 4$.

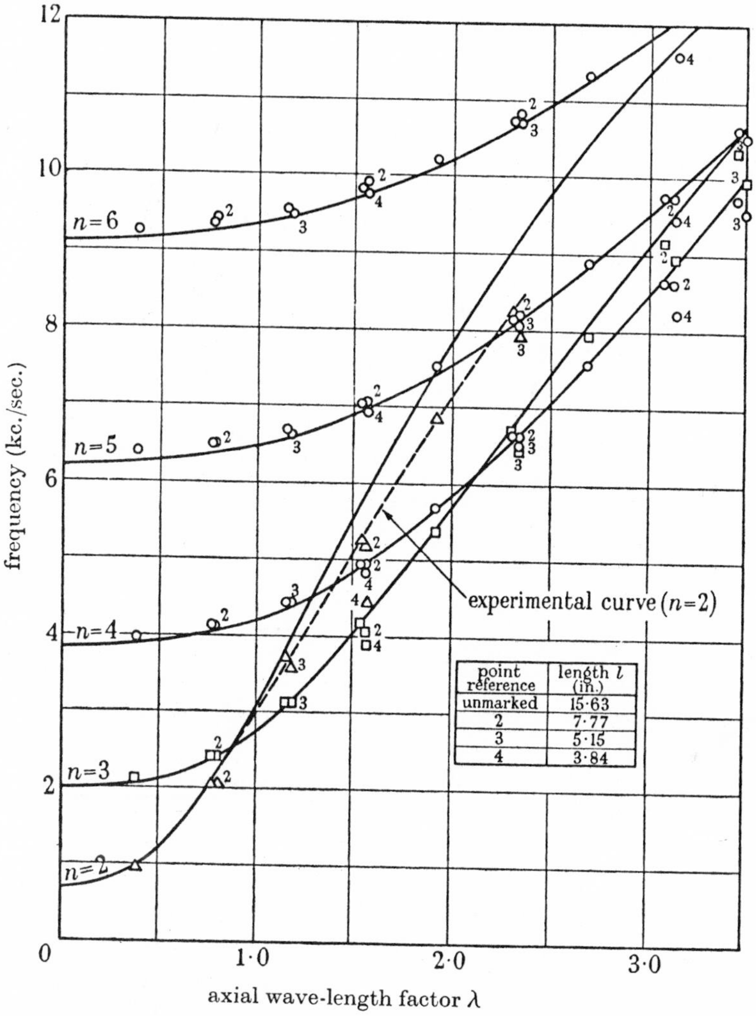

FIGURE 4. Natural frequencies of experimental cylinders ($\alpha = 0{\cdot}0525$; $h = 0{\cdot}101$ in., $a = 1{\cdot}925$ in.).

There is a tendency as λ is increased for each curve to continue to cross that immediately above it, though the reverse curvature evident on the curve for $n = 2$ shows that limitations must exist when λ is large.

The experimental results obtained from a thin cylinder 0·025 in. thick, 5·02 in. mean diameter and 2·065 in. long are shown in figure 5, where frequency is plotted to a base of circumferential nodes for both free and freely supported end-conditions. It was found possible in both cases to identify, within the aural range, all vibration patterns up to 32 nodes when no additional nodes were present in the axial direction.

For freely supported ends, it will be seen that the frequency initially decreases as the circumferential nodes increase, until a turning value is reached at approximately 14 nodes. Thereafter, the frequency increases with the number of nodes. The theoretical curve has been plotted for this condition, and while it is by no means coincident with that obtained experimentally, it is sufficiently close to corroborate the theory. It may be mentioned that the end-conditions were found to be extremely

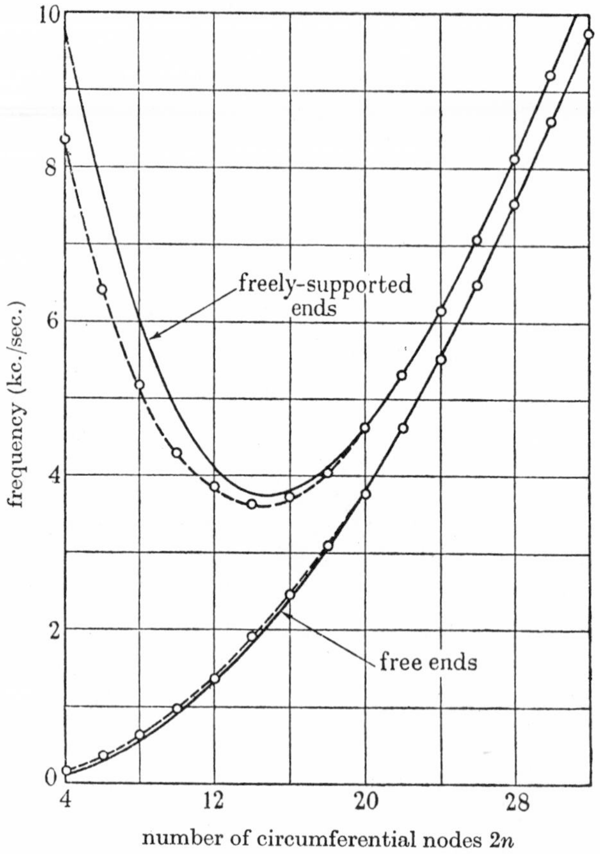

FIGURE 5. Experimental and theoretical frequency curves ($\alpha = 0{\cdot}01$).

critical in this particular cylinder, the initial experiments being entirely unsuccessful owing to a small degree of slackness at the supports. It was only after special end-pieces had been accurately machined that the above results were obtained. The theoretical and experimental curves for the same cylinder, when the ends are free, are almost coincident. For such a case, the assumption of inextension is valid and the calculated results were obtained from the equation due to Rayleigh (1894, p. 386). In comparing the results for both end-conditions, it will be seen that beyond 20 nodes the frequencies are approximately the same. Below this, however, the curves rapidly diverge, giving, at 4 nodes, frequencies of 150 and 8350 c./sec. This is rather striking, for it shows that by the introduction of conditions which maintain circularity at the ends, the frequency is increased 55 times. This immense change is a clear demonstration of the marked difference between extensional and inextensional vibration. It is due, as will be shown later, to the large strain energy involved when the cylinder walls are compelled to stretch.

Discussion

It has been shown that the higher frequencies may be associated with relatively simple modes of vibration, particularly in very thin cylinders, when the axial wave-length is small. In an attempt to understand the physical significance of this phenomenon, calculations were made to estimate the proportion of strain energy due to bending and stretching for various nodal arrangements. The calculations were based on equations (19) and (20) of appendix 3, and the results are shown plotted in figures 6 and 7. A comparison is made in figure 6 a and b between $n = 2$ and $n = 4$

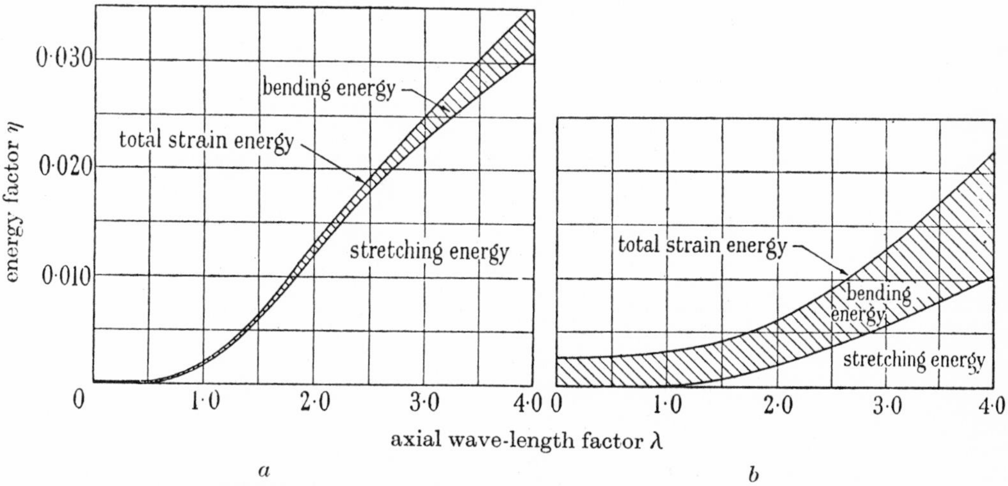

Figure 6. Strain energy due to bending and stretching ($\alpha = 0{\cdot}0525$. a, $n = 2$; b, $n = 4$).

over a range of λ for $\alpha = 0{\cdot}0525$, corresponding to the experimental cylinders for which frequencies are plotted in figure 4. It will be seen from figure 6a that with four circumferential nodes the strain energy is extremely small for values of λ up to 0·5, and that it is almost entirely composed of bending. At higher values, however, the stretching energy increases rapidly and becomes predominant, as may be observed from the shaded area representing the contribution of bending. With eight circumferential nodes (figure 6b), the situation is very different. In this case the bending effect is prominent throughout the whole range and never contributes less than 50 % of the total energy. Comparison shows that the total energy is approximately the same in each diagram when $\lambda = 1{\cdot}2$, even although the bending and stretching components are entirely different. This is in approximate agreement with figure 4, where it will be seen that the curves for $n = 2$ and $n = 4$ cross at $\lambda = 1{\cdot}3$. That these are not identical is due to the fact that the frequency is influenced by the kinetic energy function which is dependent on the component amplitudes A, B and C.

Calculations of strain energy were also made for the thin experimental cylinder ($\alpha = 0{\cdot}01$), the results being plotted in figure 7. In this case, the stretching energy decreases rapidly with the increase of circumferential nodes, while the bending energy varies in the reverse manner. This results in a curve for total strain energy of somewhat parabolic form having a minimum value at $n = 7$. It will be observed that this curve corresponds in shape to the frequency curve given in figure 5 for freely supported ends. The curve for bending energy (figure 7) is also useful for com-

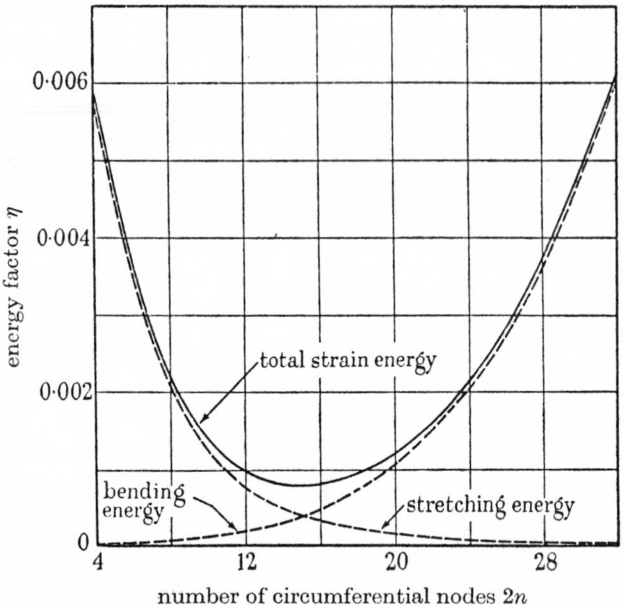

FIGURE 7. Strain energy due to bending and stretching ($\alpha = 0{\cdot}01$; $h = 0{\cdot}025$ in.; $a = 2{\cdot}51$ in.; $l = 2{\cdot}065$ in.).

parison with the curve for free ends given in figure 5. In the latter case, Rayleigh's theory of inextension is applicable. With freely supported ends, however, the energy due to stretching is extremely small when n is greater than 10, and in this region the vibration approximates to one of inextension. The frequencies for both end-conditions are thus practically equal when n is large, as may be seen from figure 5.

As previously stated the cubic equation resulting from the mathematical analysis has, in general, three real positive roots, and thus for a given nodal configuration of the cylinder three frequencies are possible. In figure 8, calculated values of such frequencies are plotted for the experimental cylinder ($\alpha = 0{\cdot}0525$) for the case of eight circumferential nodes. The two higher frequencies, f_2 and f_3, lie considerably beyond the aural range, whatever value of n be considered. For $n = 4$, it will be seen that f_2 has a minimum value of 41,500 c./sec. Each curve of the diagram represents a particular type of vibration, but the nodal arrangements for each are identical at the same value of λ. The particular form of these vibrations may be appreciated by considering the relative magnitudes of the maximum component amplitudes in the axial, circumferential and radial directions. These may be calculated from equations (17) and (18) of appendix 3, and a series of values are given in table 2 for the case considered in figure 8.

TABLE 2. COMPONENT AMPLITUDE RATIOS ($\alpha = 0{\cdot}0525$; $n = 4$)

	axial wave-length factor λ							
	0·3867		1·547		2·707		3·867	
frequency	A/C	B/C	A/C	B/C	A/C	B/C	A/C	B/C
f_1	0·024	0·252	0·072	0·258	0·073	0·246	0·052	0·220
f_2	27·5	1·82	6·75	2·14	4·53	3·02	4·10	3·66
f_3	0·36	4·00	1·60	4·32	3·28	5·05	5·79	5·96

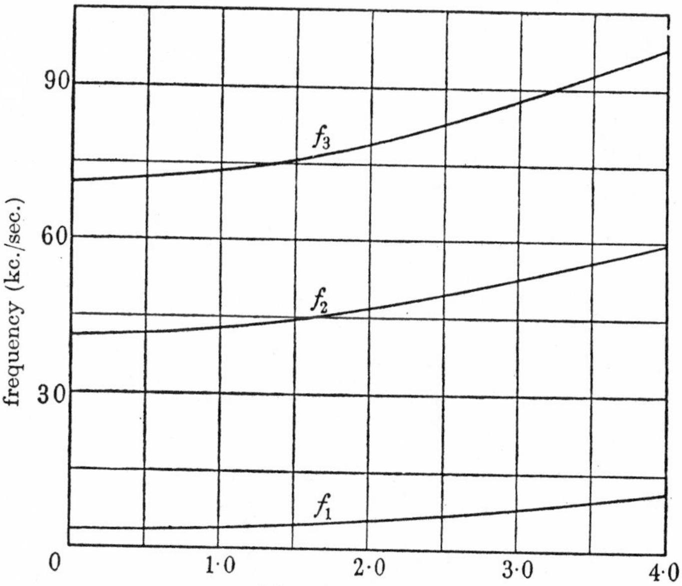

FIGURE 8. Calculated frequencies for similar nodal arrangements ($h = 0{\cdot}101$ in.; $a = 1{\cdot}925$ in.; $n = 4$).

It will be seen that for the lowest frequency f_1, which has been studied at length in this paper, the predominant amplitude is C; this means that the vibratory motion is mainly radial. The circumferential amplitude B is approximately $C/4$ for all values of λ, and the axial amplitude A has a maximum value of $C/14$. The former value corresponds to the case of free ends in which the ratio of circumferential to radial movement is given as $1/n$ (Rayleigh 1894, p. 386).

The vibration associated with f_2 is comprised mainly of axial motion when λ is small, but for larger values of λ the axial and circumferential components become almost equal. There is little bending of the cylinder walls in this case.

The highest frequency f_3 is shown to have a vibration which is predominantly circumferential, except at high values of λ when the axial motion becomes comparable.

The above results indicate that f_2 and f_3 are associated almost entirely with extensional vibration. In such cases, the strain energy is large, and this results in extremely high frequencies. It is this large energy contribution of stretching which is responsible for the strange variation of f_1 obtained from cylinders with freely supported ends. The higher frequencies have yet to be observed experimentally, but as this involves rather elaborate apparatus, such work must be left to the future.

Appendix 1

List of symbols

a	mean radius of cylinder.
A, B, C	maximum amplitudes of component vibrations.
e_x, e_y, e_z	direct strains.
e_{xy}, e_{yz}, e_{zx}	shear strains.
E	Young's modulus.
$f = \omega/2\pi$	frequency.
f_1, f_2, f_3	frequencies of similar nodal patterns.
g	acceleration due to gravity.
h	thickness of cylinder.
K_0, K_1, K_2	coefficients in frequency equation.
l	length of cylinder.
m	number of axial half-waves.
n	number of circumferential waves.
p_x, p_y, p_z	direct stresses.
p_{xy}, p_{yz}, p_{zx}	shear stresses.
S	total strain energy at any displacement.
S_b	strain energy due to bending.
S_s	strain energy due to stretching.
S_m	total strain energy at maximum displacement.
t	time.
T	total kinetic energy at any displacement.
u, v, w	component displacements of a point on the middle surface.
$U = A\cos\omega t$.	
$V = B\cos\omega t$.	
$W = C\cos\omega t$.	
x, y, z	co-ordinate distances in axial, circumferential and radial directions.
$\alpha = h/a$.	
$\beta = h^2/12a^2$.	
γ	shear strain in middle surface.
$\Delta = \dfrac{\rho a^2(1-\sigma^2)\,4\pi^2 f^2}{Eg}$	frequency factor.
ϵ_1, ϵ_2	direct strain in middle surface in directions of x and y.
η_b	energy factor (bending).
η_s	energy factor (stretching).
κ_1, κ_2	change of curvature of middle surface in directions of x and y.
$\lambda = \dfrac{\pi m a}{l}$	axial wave-length factor.
ρ	density.
σ	Poisson's ratio.
ϕ	angular co-ordinate.
τ	twist of middle surface.
$\omega = 2\pi f$	circular frequency.

Appendix 2

Derivation of frequency equation for thin cylindrical shells with freely supported ends

Consider a cylindrical shell of length l, thickness h and mean radius a, an element of which is shown in figure 9.

The element is bounded by two parallel planes which are perpendicular to the axis and distance δx apart, and by two radial planes which intersect at angle $\delta\phi$. The direct stresses acting on the element parallel to the X, Y and Z axes will be denoted respectively by p_x, p_y and p_z, and the shear stress acting on the face perpendicular to the X-axis in direction Y by p_{xy}. Shear stresses $p_{yz}, p_{zx}, p_{yx}, \ldots$ have similar definitions. The direct strains corresponding to the above will be denoted by e_x, e_y and e_z and the shear strains by $e_{xy}, e_{yz}, e_{zx}, \ldots$.

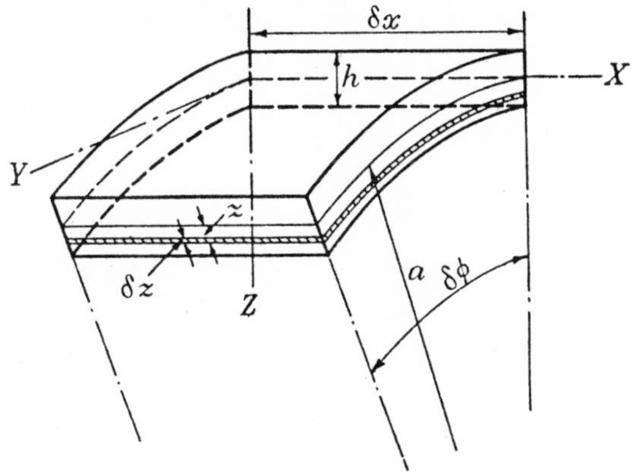

FIGURE 9. Element of thin cylinder.

To the first approximation the direct stress p_z and the shear strains e_{yz} and e_{zx} are zero (Love 1927, p. 529). The total strain energy for the deformed cylinder may thus be written, neglecting the trapezoidal form of the faces perpendicular to the X-axis, as

$$S = \int_0^{2\pi}\int_0^l\int_{-\frac{1}{2}h}^{+\frac{1}{2}h} \tfrac{1}{2}[p_x e_x + p_y e_y + p_{xy} e_{xy}]\, a\, d\phi\, dx\, dz. \tag{1}$$

Now from Hooke's law

$$p_x = \frac{E}{1-\sigma^2}(e_x+\sigma e_y), \quad p_y = \frac{E}{1-\sigma^2}(e_y+\sigma e_x), \quad p_{xy} = \frac{E}{2(1+\sigma)} e_{xy}, \tag{2}$$

which leads to

$$S = \frac{E}{2(1-\sigma^2)}\int_0^{2\pi}\int_0^l\int_{-\frac{1}{2}h}^{+\frac{1}{2}h}\left[e_x^2+e_y^2+2\sigma e_x e_y + \left(\frac{1-\sigma}{2}\right)e_{xy}^2\right] a\, d\phi\, dx\, dz. \tag{3}$$

Let the following symbols refer to the middle surface:

$\epsilon_1\,\epsilon_2$ strains in directions x and y.

$\kappa_1\,\kappa_2$ changes of curvature in directions x and y.

γ shear strain.

τ twist.

It can be shown that the strains at a distance z from the middle surface of a deformed shell can be expressed approximately (Love 1927, p. 529) as

$$e_x = \epsilon_1 - z\kappa_1, \quad e_y = \epsilon_2 - z\kappa_2, \quad e_{xy} = \gamma - 2z\tau. \tag{4}$$

If u, v and w are the displacements in the directions X, Y and Z of a point on the middle surface at any instant, then the strains and changes of curvature are given (Love 1927, p. 543) in terms of the displacements and their derivatives by

$$\left.\begin{aligned} \epsilon_1 &= \frac{\partial u}{\partial x}, \quad \epsilon_2 = \frac{1}{a}\frac{\partial v}{\partial \phi} - \frac{w}{a}, \quad \gamma = \frac{\partial v}{\partial x} + \frac{1}{a}\frac{\partial u}{\partial \phi}, \\ \kappa_1 &= \frac{\partial^2 w}{\partial x^2}, \quad \kappa_2 = \frac{1}{a^2}\frac{\partial^2 w}{\partial \phi^2} + \frac{1}{a^2}\frac{\partial v}{\partial \phi}, \quad \tau = \frac{1}{a}\frac{\partial^2 w}{\partial x \partial \phi} + \frac{1}{a}\frac{\partial v}{\partial x}, \end{aligned}\right\} \tag{5}$$

where ϕ defines the angular position of the point considered. Let

$$u = U\cos n\phi \cos\frac{m\pi x}{l}, \quad v = V\sin n\phi \sin\frac{m\pi x}{l}, \quad w \doteq W\cos n\phi \sin\frac{m\pi x}{l}, \tag{6}$$

where n and $\frac{1}{2}m$ are respectively the number of circumferential and axial wave-lengths, and U, V and W are functions of time only. The above expressions satisfy the condition that the ends of the cylinder remain circular during vibration.

By expressing the strains in terms of U, V and W and substituting in (3), the integral may be evaluated to give

$$S = \frac{\pi E h l}{4a(1-\sigma^2)}\Big[U^2\lambda^2 + (nV - W)^2 + 2\sigma\lambda U(W - nV) + \left(\frac{1-\sigma}{2}\right)(\lambda V - nU)^2 + \frac{h^2}{12a^2}\{\lambda^4 W^2 + (nV - n^2 W)^2 + 2\sigma\lambda^2 W(n^2 W - nV) + 2(1-\sigma)(\lambda V - \lambda n W)^2\}\Big], \tag{7}$$

where $\lambda = \dfrac{m\pi a}{l}$.

The kinetic energy at any instant is given by

$$T = \frac{\rho}{2g}\int_0^{2\pi}\int_0^l\int_{-\frac{1}{2}h}^{+\frac{1}{2}h}\left[\left(\frac{\partial u}{\partial t}\right)^2 + \left(\frac{\partial v}{\partial t}\right)^2 + \left(\frac{\partial w}{\partial t}\right)^2\right] a\,d\phi\,dx\,dz$$

or

$$T = \frac{\pi\rho h l a}{4g}[\dot{U}^2 + \dot{V}^2 + \dot{W}^2]. \tag{8}$$

Since U, V and W are independent variables, the Lagrange equation is applicable. This gives

$$\frac{d}{dt}\left(\frac{\partial T}{\partial \dot{U}}\right) - \frac{\partial T}{\partial U} = -\frac{\partial S}{\partial U}, \tag{9}$$

and two similar equations in V and W.

From (7), (8) and (9)

$$\frac{\pi\rho h l a}{2g}\ddot{U} - 0 = -\frac{\pi E h l}{2a(1-\sigma^2)}\left[U\lambda^2 + \sigma\lambda(W - nV) - n\left(\frac{1-\sigma}{2}\right)(\lambda V - nU)\right]. \tag{10}$$

U, V and W are periodic with respect to time and may be written

$$U = A\cos\omega t, \quad V = B\cos\omega t, \quad W = C\cos\omega t, \tag{11}$$

where A, B and C are constants and $\omega/2\pi$ is the natural frequency of the vibration.

Hence
$$\left[\lambda^2+\left(\frac{1-\sigma}{2}\right)n^2-\Delta\right]A-\left(\frac{1+\sigma}{2}\right)\lambda nB+\sigma\lambda C = 0, \tag{12}$$

where
$$\Delta = \frac{\rho a^2(1-\sigma^2)\,\omega^2}{Eg}.$$

Similarly, by substituting in the Lagrange equations for V and W

$$-\left(\frac{1+\sigma}{2}\right)\lambda nA+\left[\left(\frac{1-\sigma}{2}\right)\lambda^2+n^2-\Delta+\beta\{n^2+2(1-\sigma)\,\lambda^2\}\right]B$$
$$-[n+\beta\{n^3+(2-\sigma)\,\lambda^2 n\}]\,C = 0 \tag{13}$$

and
$$\sigma\lambda A-[n+\beta\{n^3+(2-\sigma)\,\lambda^2 n\}]\,B+[1-\Delta+\beta(\lambda^2+n^2)^2]\,C = 0, \tag{14}$$

where $\beta = h^2/12a^2$.

Eliminating A, B and C from equations (12) to (14) leads to a cubic equation in Δ, the roots of which define the natural frequencies of the vibration. The equation is

$$\Delta^3-K_2\Delta^2+K_1\Delta-K_0 = 0, \tag{15}$$

where

$$K_0 = \tfrac{1}{2}(1-\sigma)^2(1+\sigma)\,\lambda^4+\tfrac{1}{2}(1-\sigma)$$
$$\times\beta[(\lambda^2+n^2)^4-2(4-\sigma^2)\,\lambda^4 n^2-8\lambda^2 n^4-2n^6+4(1-\sigma^2)\,\lambda^4+4\lambda^2 n^2+n^4],$$

$$K_1 = \tfrac{1}{2}(1-\sigma)\,(\lambda^2+n^2)^2+\tfrac{1}{2}(3-\sigma-2\sigma^2)\,\lambda^2+\tfrac{1}{2}(1-\sigma)\,n^2$$
$$+\beta[\tfrac{1}{2}(3-\sigma)\,(\lambda^2+n^2)^3+2(1-\sigma)\,\lambda^4-(2-\sigma^2)\,\lambda^2 n^2-\tfrac{1}{2}(3+\sigma)\,n^4+2(1-\sigma)\,\lambda^2+n^2],$$

and

$$K_2 = 1+\tfrac{1}{2}(3-\sigma)\,(\lambda^2+n^2)+\beta[(\lambda^2+n^2)^2+2(1-\sigma)\,\lambda^2+n^2].$$

Appendix 3

Strain energy due to bending and stretching of a cylindrical shell

By substituting the values for U, V and W of (11) in the expression for strain energy (7), the maximum strain energy may be written in terms of the component amplitudes

$$S_m = \frac{\pi Ehl}{4a(1-\sigma^2)}\left[A^2\lambda^2+(C-nB)^2+2\sigma\lambda A(C-nB)+\left(\frac{1-\sigma}{2}\right)(\lambda B-nA)^2\right.$$
$$\left.+\beta\{\lambda^4C^2+(n^2C-nB)^2+2\sigma\lambda^2C(n^2C-nB)+2(1-\sigma)\,(\lambda B-\lambda nC)^2\}\right]. \tag{16}$$

In this, the first four terms give the strain energy due to stretching and the remainder (dependent on ratio h/a) that due to bending. To determine what each effect contributes to the total energy, the ratio $A:B:C$ is required. By elimination between (12), (13) and (14)

$$\frac{A}{C}=\frac{\begin{array}{r}(1-\sigma)^2\lambda n^2-(1+\sigma)^2\lambda n^2\Delta-4\sigma^2\lambda\Delta+2\sigma^2(1-\sigma)\lambda^3+\beta[(1+\sigma)^2\lambda n^2(\lambda^2+n^2)^2\\ -4\sigma(1+\sigma)\lambda n^4-4\sigma(2-\sigma)(1+\sigma)\lambda^3n^2+4\sigma^2\lambda n^2+8\sigma^2(1-\sigma)\lambda^3]\end{array}}{\begin{array}{r}[2\lambda^2-2\Delta+(1-\sigma)n^2][(1-\sigma)n^2+2\sigma\Delta-\sigma(1-\sigma)\lambda^2\\ +\beta\{(1+\sigma)n^4+(2-\sigma)(1+\sigma)\lambda^2n^2-2\sigma n^2-4\sigma(1-\sigma)\lambda^2\}]\end{array}} \tag{17}$$

and

$$\frac{B}{C}=\frac{(1-\sigma)n-(1+\sigma)\Delta n+\beta[(1+\sigma)n(\lambda^2+n^2)^2-2\sigma(2-\sigma)\lambda^2n-2\sigma n^3]}{\begin{array}{r}(1-\sigma)n^2+2\sigma\Delta-\sigma(1-\sigma)\lambda^2+\beta[(1+\sigma)n^4\\ +(2-\sigma)(1+\sigma)\lambda^2n^2-2\sigma n^2-4\sigma(1-\sigma)\lambda^2]\end{array}}. \tag{18}$$

These results may be used to evaluate the stretching energy

$$S_s=\left[\frac{\pi EhlC^2}{4a(1-\sigma^2)}\right]\left[\lambda^2\left(\frac{A}{C}\right)^2+\left(1-n\frac{B}{C}\right)^2+2\sigma\lambda\left(\frac{A}{C}\right)\left(1-n\frac{B}{C}\right)+\left(\frac{1-\sigma}{2}\right)\left(\lambda\frac{B}{C}-n\frac{A}{C}\right)^2\right], \tag{19}$$

and the bending energy

$$S_b=\beta\left[\frac{\pi EhlC^2}{4a(1-\sigma^2)}\right]\left[(\lambda^2+n^2)^2+\{n^2+2(1-\sigma)\lambda^2\}\left(\frac{B}{C}\right)^2-2\{(2-\sigma)\lambda^2n+n^3\}\left(\frac{B}{C}\right)\right], \tag{20}$$

for any particular case in which Δ (proportional to square of frequency) has previously been obtained from (15). Equations (19) and (20) can be rewritten in the form

$$S_s=\eta_s\left[\frac{E\pi}{4(1-\sigma^2)}\right]lC^2 \quad\text{and}\quad S_b=\eta_b\left[\frac{E\pi}{4(1-\sigma^2)}\right]lC^2, \tag{21}$$

where η_s and η_b are non-dimensional energy factors for stretching and bending respectively, and the term in the bracket is a constant for the material.

Appendix 4

Note regarding the influence of approximations in the expressions for strain

The frequency equation derived by considering equilibrium is found to differ slightly from that obtained by the energy method depending on the accuracy of the expressions for strain. For example, if inertia terms are introduced into the displacement equations of Timoshenko (1940, p. 440) and expressions (6) are substituted for u, v and w, three equations are obtained corresponding to (12), (13) and (14). Comparison shows, however, that while two pairs are identical, that corresponding to (13) differs in the coefficients of two small terms. This divergence does not occur if the strain expressions given by Flügge (1934) are used, presumably because he does not approximate in deriving the strain relations. In general, any strain approximation results in divergences between the equations derived by the equilibrium and energy methods.

If the coefficients K_0, K_1 and K_2 of (15) be examined, it will be seen that each consists of two parts, one dependent on β, the other independent of β. In each case, the former contains one term of high power, e.g. $(\lambda^2+n^2)^4$ in the expression for K_0. In deriving K_0, K_1 and K_2 by the equilibrium method, using the strain expressions derived by Timoshenko (1940, p. 354), it is found that some of the terms of lower order dependent on β differ from those of (15), but the term of highest power and those independent of β are identical. In general, whatever strain expressions are used, the answers derived by the two methods differ only in the coefficients of some of the smaller terms. Moreover, these coefficients are only slightly different numerically, when a practical value for σ is introduced.

For the field covered by the present paper, namely, $h/a = 0$ to $0{\cdot}1$, $\lambda = 0$ to 4 and $n = 2$ to 7, the calculated frequencies are found to be practically independent of the strain approximations and method of derivation of the frequency equation. The greatest divergence in frequency recorded by using different strain expressions is only 0·7 %, and for most modes of vibration the difference is considerably less.

References

Flügge, W. 1934 *Statik und Dynamik der Schalen.* Berlin: Springer.
Love, A. E. H. 1927 *Mathematical theory of elasticity*, 4th ed. Cambridge.
Rayleigh, Lord 1894 *Theory of sound*, 2nd ed. London: Macmillan.
Timoshenko, S. 1940 *Theory of plates and shells*, 1st ed. New York: McGraw-Hill.

20

Reprinted from *AIAA Jour.*, **2**(12), 2150–2157 (1964)

Influence of Boundary Conditions on the Modal Characteristics of Thin Cylindrical Shells

KEVIN FORSBERG*
Lockheed Missiles and Space Company, Palo Alto, Calif.

The modal characteristics of thin cylindrical shells have previously been determined only for three sets of boundary conditions. In the present analysis, all sixteen sets of homogeneous boundary conditions have been examined at each end of the shell (each set contains four conditions). The equations of motion developed by Flügge for thin, circular cylindrical shells are used. The general solution to these equations can easily be written down. One can select a circumferential nodal pattern, eight boundary conditions, and a length of shell, and then iterate numerically to find the frequency of vibration that will meet these conditions. The advantage of this approach is that one can obtain a solution to the basic equations for any boundary conditions desired. Results indicate that, contrary to the rather common assumption, the condition placed on the longitudinal displacement u in many cases is more influential than restrictions on the slope $\partial w/\partial x$ or moment M_x. It has been found that even for long cylinders (length to radius ratio of 40 or more) the minimum natural frequency may differ by more than 50% depending upon whether $u = 0$ or the longitudinal stress resultant $N_x = 0$ at both ends.

Nomenclature

a = radius of cylinder
h = thickness of cylinder wall
k = $h^2/12a^2$
l = length of shell
m = number of axial half-waves
n = number of circumferential waves
u = longitudinal displacement
v = tangential displacement
w = radial displacement
x = dimensionless axial coordinate, ξ/a
E = Young's modulus of elasticity
$M_x, M_{x\varphi}, N_x, N_{x\varphi}, Q_x$ = stress resultants, see Fig. 1
S_x = $Q_x + \partial M_{x\varphi}/a\partial\varphi$
T_x = $N_x - M_{x\varphi}/a$
γ^2 = $\rho a^2(1 - \nu^2)/E$
ν = Poisson's ratio
ξ = axial coordinate
ρ = mass density of shell material
φ = circumferential coordinate
ω = circular frequency
ω_0 = lowest extensional frequency of a ring in plane strain, = $1/\gamma$
$(\ )'$ = $\partial(\ldots)/\partial x$
$(\ )^{\cdot}$ = $\partial(\ldots)/\partial\varphi$

Introduction

THE determination of the modal characteristics for vibrations of thin cylindrical shells is a problem of great technical interest. However, in spite of the great number of papers devoted to this topic,[1] only three sets of boundary conditions (i.e., one set for a force-free end and two other sets loosely called "freely supported" end and "clamped" end) have been considered in the literature. Additionally there appears to be no discussion of how the modal characteristics are altered when longitudinal displacement is prevented at the ends.

The primary contribution of this paper is the presentation of results obtained from an exact solution of the basic differential equations of motion. The method, which was outlined by Flügge in 1934, requires numerical evaluation of an eighth-order determinant to find its eigenvalues, and this is certainly the reason this approach was not feasible before the advent of a high-speed digital computer. Although the method requires numerical computation, the results are exact in the same sense that the numerical solution to the transcendental frequency equation for a beam yields an exact solution.

This study is identical in approach with the recently published work of Sobel[2] on the closely related area of stability of cylindrical shells. The results of these two independent studies lead to the same conclusions regarding the importance of the various boundary conditions.

The major purpose of the present report is to determine the ranges of the length-to-radius ratio l/a and radius-to-thickness ratio a/h for which a change in the boundary conditions will appreciably alter the modal characteristics. The modal behavior of a simply supported shell without axial constraint is used as the reference for comparison. This investigation will be directed into three areas: 1) the effect of the various boundary conditions on the frequency envelope (10 cases are considered here); 2) a study of a portion of the over-all frequency pattern for one value of a/h and for two sets of boundary conditions; and 3) a brief study of the modal stress resultants for several cases.

This study will be presented in two parts. The second part, which is in preparation, will consider the cases for which $n = 0$ and $n = 1$. Also in the second part a brief investigation will be made of shells having one or both ends entirely force free (all values of n will be considered). A comparison with experimental data will be made for a cylinder having force-free ends.

The Flügge equations used in the present analysis are based on the usual assumptions of linear thin shell theory, i.e., that the shell is thin (usually considered to mean $a/h > 10$), of constant wall thickness, and of a linear, homogeneous, isotropic material. The results apply only for small deflections and, since the effects of shear distortion and rotatory inertia of the shell wall have been neglected, the results apply only when the half-wave length of the mode shape is

Presented as Preprint 64-77 at the AIAA Aerospace Sciences Meeting, New York, January 20–22, 1964; revision received July 27, 1964. This work was partially sponsored by the Lockheed Independent Research Program. The author would like to express his appreciation to C. W. Coale for the helpful advice given during the writing of this paper.

* Research Specialist, Aerospace Sciences Laboratory. Member AIAA.

more than ten times the shell wall thickness ($l/ma > 10\ h/a$, $\pi/n > 10\ h/a$), where m and $2n$ are the number of axial and circumferential half-waves.

Method of Solution

The present analysis is based on the following equations of motion developed by Flügge[3] for free vibrations of thin, circular cylindrical shells (see Fig. 1):

$$\left.\begin{aligned}
&u'' + \frac{1-\nu}{2}(1+k)u^{\cdot\cdot} + \frac{1+\nu}{2}v'^{\cdot} - kw''' + \frac{1-\nu}{2}kw'^{\cdot\cdot} + \nu w' - \gamma^2\frac{\partial^2 u}{\partial t^2} = 0\\
&\frac{1+\nu}{2}u'^{\cdot} + v^{\cdot\cdot} + \frac{1-\nu}{2}(1+3k)v'' - \frac{3-\nu}{2}kw''^{\cdot} + w^{\cdot} - \gamma^2\frac{\partial^2 v}{\partial t^2} = 0\\
&-ku''' + \frac{1-\nu}{2}ku'^{\cdot\cdot} + \nu u' - \frac{3-\nu}{2}kv''^{\cdot} + v^{\cdot} + w + k[w^{IV} + 2w''^{\cdot\cdot} + w^{\cdot\cdot\cdot\cdot} + 2w^{\cdot\cdot} + w] + \gamma^2\frac{\partial^2 w}{\partial t^2} = 0
\end{aligned}\right\}\quad (1)$$

where

$$(\ )' = \partial(\ )/\partial x \qquad (\ )^{\cdot} = \partial(\ )/\partial\varphi$$

$$\gamma^2 = \rho a^2(1-\nu^2)/E \qquad k = h^2/12a^2$$

For a complete cylinder, the general solution for modal vibration can easily be written in the following form:

$$\begin{aligned}
u &= \left(\sum_{n=0}^{\infty}\sum_{s=1}^{8}\alpha_{sn}A_{sn}e^{\lambda_{sn}x}\cos n\varphi\right)e^{i\omega t}\\
v &= \left(\sum_{n=0}^{\infty}\sum_{s=1}^{8}\beta_{sn}A_{sn}e^{\lambda_{sn}x}\sin n\varphi\right)e^{i\omega t}\qquad (2)\\
w &= \left(\sum_{n=0}^{\infty}\sum_{s=1}^{8}A_{sn}e^{\lambda_{sn}x}\cos n\varphi\right)e^{i\omega t}
\end{aligned}$$

The arbitrary constants A_{sn} will be evaluated by considering the boundary conditions at each end of the shell. Any of the boundary conditions can be a general function of the circumferential coordinate φ and can be expressed by a Fourier series in φ. Thus, in the general case of nonuniform boundary conditions, the frequency of modal vibration ω will depend upon all of the harmonics n. For the special cases in which the boundary conditions depend on only one harmonic, or when the boundary conditions are homogeneous, the modal frequency will be a function of a single value of n, and the summation over n in Eq. (2) can be discarded. Only this latter case will be considered in the present paper. Thus the index n will be dropped.

Substitution of these expressions into the homogeneous differential equations leads to an eighth-order algebraic equation for λ_s:

$$\lambda_s^8 + g_{s6}\lambda_s^6 + g_{s4}\lambda_s^4 + g_{s2}\lambda_s^2 + g_{s0} = 0 \qquad (3)$$

where

$$g_{sk} = g_{sk}\,(h/a, \nu, n, \omega)$$

The preceding solution is readily obtainable from Ref. 3. In contrast to the associated statics problem for which the roots of Eq. (3) are, in general, all complex, it can be shown that the solution for the vibration problem will usually have the form

$$\lambda = \pm a,\ \pm ib,\ \pm(c \pm id)$$

where a, b, c, d are real quantities. For a finite shell, there will always be at least two roots of the form $(\pm ib)$. This leads to a solution of the form

$$\begin{aligned}
w = \{&C_1e^{ax} + C_2e^{-ax} + C_3\cos bx + C_4\sin bx +\\
&e^{cx}(C_5\cos dx + C_6\sin dx) + e^{-cx}(C_7\cos dx +\\
&C_8\sin dx)\}\cos n\varphi e^{i\omega t}\qquad (4)
\end{aligned}$$

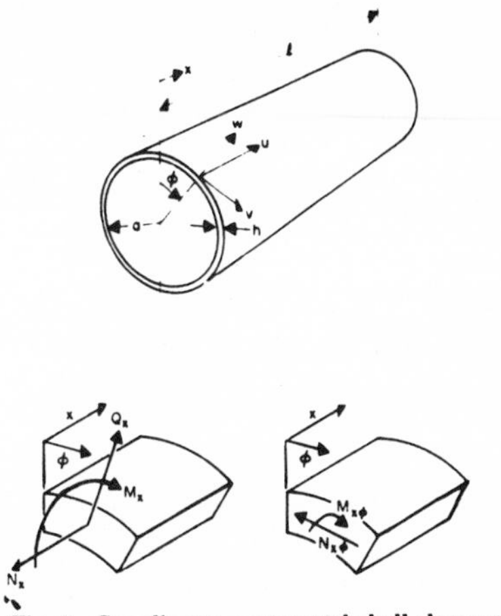

Fig. 1 Coordinate system and shell element.

with similar expressions for u and v. Note that Eq. (4) has been rewritten so that the complex constants A_s have been replaced by real constants C_s. The expressions for u and v involve combinations of the constants C_s and the real and imaginary parts of α_s or β_s. It should be noted that the parameters α_s and β_s depend on λ_s, h/a, ν, n, and ω. When the solutions to Eq. (3) are obtained, α_s and β_s can be evaluated.

Once the boundary conditions are specified (four at each end of the shell), the problem is entirely determined. The detailed statement of these conditions leads directly to eight equations for the eight unknown constants C_s. These equations involve the four quantities a, b, c, and d. Since the boundary conditions are homogeneous, the determinant D of these equations must be zero for a nontrivial solution.

It does not appear feasible to seek analytic expressions for the quantities a, b, c, and d. Thus, at this point in the analysis, a numerical evaluation of the solution is introduced. We now select a given shell (i.e., fix a/h, l/a, ν), an assumed number of circumferential waves n, and a specific set of boundary conditions at each end. Starting from some initial estimate for the frequency ω, we can iterate to find the values of ω which will make the determinant D go to zero. We can cover the entire range of problems of interest by varying the initial input to the determinant, i.e., by varying a/h, l/a, ν, n, or the boundary conditions.

No assumptions or simplifications beyond those underlying Eqs. (1) have been introduced in the numerical evaluation, and the solution can be obtained with any desired degree of accuracy. In the present instance, the frequency was determined to six significant figures. Such accuracy, required only for intermediate computations, is necessary in order to obtain accurate values for the mode shapes and corresponding modal stress resultants. The final result is only meaningful for three or four significant figures of course. The number of iterations required for convergence is greatly reduced if good initial estimates are available. The solutions developed by Arnold and Warburton[4,5] are excellent for this purpose. It should be noted that, for any given modal pattern (fixed number of axial half-waves m and cir-

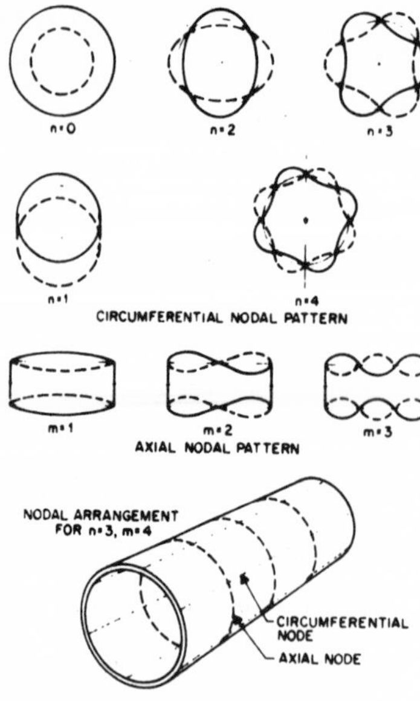

Fig. 2 Nodal patterns.

cumferential waves n), there are three frequencies, corresponding to different amplitude ratios in u, v, and w.[4] If longitudinal and tangential inertia terms are omitted, there will be only one value of ω for each pair of m and n. For $n \geq 1$, this eigenvalue is an approximation to the lowest of the three frequencies just mentioned. In the present approach, the number of axial half-waves m cannot be specified conveniently in advance, and thus one must take care to determine the mode shape as well as the frequency in each instance, since there are an infinite number of eigenvalues for any fixed set of values of n, a/h, l/a, and ν. Although it is possible for two mode shapes to have identical natural frequencies, this can occur only if they have different values of n. For a given number of circumferential waves, the frequency increases monotonically as the number of axial waves increases. However, for long shells the frequencies for $m = 1, 2, 3, \ldots$, are very closely spaced; the longer the shell the closer are the frequencies of adjacent axial modes.

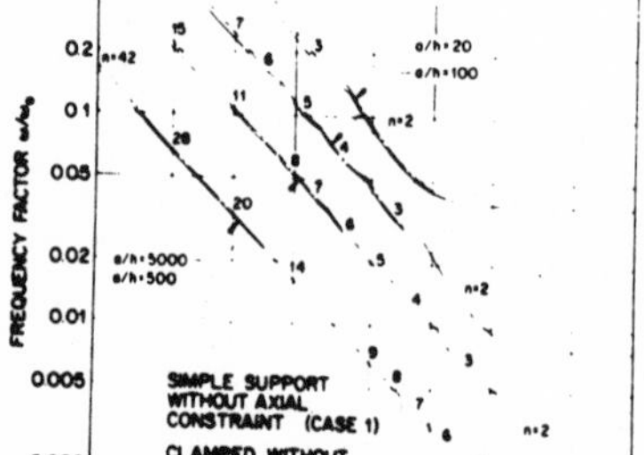

Fig. 3 Frequency envelope, cases 1 and 6.

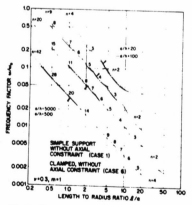

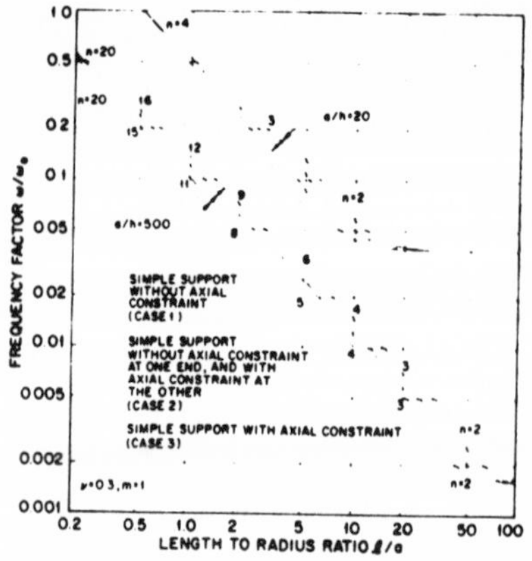

Fig. 4 Frequency envelope, cases 1–3.

Finally it should be pointed out that, for mode shapes in which $n \geq 2$, if one starts from a zero frequency, the minimum nonzero eigenvalue obtained will be that of an infinite cylinder (or a ring in a state of plane strain). This value is independent of the axial wavelength (and hence independent of the boundary conditions) and represents the asymptotic value for a given n, a/h, and ν as $l/a \rightarrow \infty$. It does not, in general, represent a solution for a finite shell. The numerical computation was done on an IBM 7094 computer.

Boundary Conditions

There are 16 possible sets of homogeneous boundary conditions which can be specified independently at each end of the shell. These consist of all combinations of the following:

$$w = 0 \text{ or } S_x = Q_x + (1/a)\,\partial M_{x\varphi}/\partial\varphi = 0 \tag{5a}$$

$$\partial w/\partial x = 0 \text{ or } M_x = 0 \tag{5b}$$

$$u = 0 \text{ or } N_x = 0 \tag{5c}$$

$$v = 0 \text{ or } T_x = N_{x\varphi} - (1/a)M_{x\varphi} = 0 \tag{5d}$$

(For a discussion of the origin of the force boundary conditions indicated in Eqs. (5a) and (5d), see Ref. 3, p. 233.) For completeness, all 16 sets of conditions were studied in the course of the present investigation for shells having the same boundary conditions at both ends. Free ends (cases with $S_x = 0$) will be discussed in a subsequent paper. Additionally, selected combinations of one set at one end and a different set at the other were examined.

The significant findings of this study can be summarized by discussing the ten cases indicated in Table 1. Certain symbols used therein are defined in Fig. 1. The exact solution for case 1 has been discussed in detail by Arnold and Warburton.[4] It is included here for comparison. Arnold and Warburton also presented an approximate solution[5] for case 7. A comparison of their results with the present analysis is given below.

Discussion of Results

Frequency Envelope for $n \geq 2$

The general character of the various mode shapes is indicated in Fig. 2. Modes for which $n < 2$ have a somewhat different behavior from those having $n \geq 2$ and will be treated at a later date.

For any fixed values of n and m, there are three natural frequencies corresponding to three mode shapes (having different u, v, w amplitude ratios). The asymptotic values of these frequencies for long axial wavelengths are:

Flexural Vibrations of a Ring ($w_{max} = nv_{max}$, $u = 0$)

$$(\omega_1/\omega_0)^2 = (h^2/12a^2)n^2(n^2 - 1)^2/(n^2 + 1)$$

Axial Shear Vibration ($w = v = 0$)

$$(\omega_2/\omega_0)^2 = (1 - \nu)n^2/2$$

Extensional Vibrations of Ring ($v_{max} = nw_{max}$, $u = 0$)

$$(\omega_3/\omega_0)^2 = n^2 + 1$$

In general, two of these three frequencies are several orders of magnitude higher than the minimum value and hence are not usually of immediate interest.

These higher frequencies will arise only if all three inertia terms are retained in Eqs. (1). Although these higher frequencies are not to be studied here, all three inertia terms have been retained in developing the results presented herein. One approach for determining the region of influence of the various boundary conditions on the modal behavior of cylindrical shells consists of looking at the minimum natural frequency (which is the envelope of the frequency curves drawn for constant values of n). The value of n varies from point to point as indicated on the curves shown in Figs. 3–7. Moreover, for a given shell of fixed length, changing the boundary conditions may alter the value of n associated with the minimum frequency. Since the minimum frequency always occurs for a mode having one axial half-wave ($m = 1$), Figs. 3–7 are all drawn for $m = 1$.

Table 1 List of boundary conditions used in present analysis

Case no.	Description	Boundary conditions $x = 0$	$x = l/a$
1	Simple support without axial constraint (called "freely supported" in Ref. 3)	$w = 0$ $v = 0$ $M_x = 0$ $N_x = 0$	$w = 0$ $v = 0$ $M_x = 0$ $N_x = 0$
2	Simple support without axial constraint at one end, with axial constraint at other	$w = 0$ $v = 0$ $M_x = 0$ $N_x = 0$	$w = 0$ $v = 0$ $M_x = 0$ $u = 0$
3	Simple support with axial constraint	$w = 0$ $u = 0$ $v = 0$ $M_x = 0$	$w = 0$ $u = 0$ $v = 0$ $M_x = 0$
4	Simple support, no tangential constraint (similar to case 1, with $v \neq 0$)	$w = 0$ $M_x = 0$ $N_x = 0$ $T_x = 0$	$w = 0$ $M_x = 0$ $N_x = 0$ $T_x = 0$
5	Simple support, axial constraint but no tangential constraint (similar to case 3, with $v \neq 0$)	$w = 0$ $u = 0$ $M_x = 0$ $T_x = 0$	$w = 0$ $u = 0$ $M_x = 0$ $T_x = 0$
6	Clamped end, without axial constraint	$w = 0$ $w' = 0$ $v = 0$ $N_x = 0$	$w = 0$ $w' = 0$ $v = 0$ $N_x = 0$
7	Clamped end, with axial constraint (called "fixed end" in Ref. 4)	$w = 0$ $w' = 0$ $u = 0$ $v = 0$	$w = 0$ $w' = 0$ $u = 0$ $v = 0$
8	Clamped end, no tangential constraint (similar to case 6, but with $v \neq 0$)	$w = 0$ $w' = 0$ $N_x = 0$ $T_x = 0$	$w = 0$ $w' = 0$ $N_x = 0$ $T_x = 0$
9	Clamped end, with axial constraint but no tangential constraint (similar to case 7, but with $v \neq 0$)	$w = 0$ $w' = 0$ $u = 0$ $T_x = 0$	$w = 0$ $w' = 0$ $u = 0$ $T_x = 0$
10	Simple support without axial constraint at one end, clamped with axial constraint at the other end	$w = 0$ $v = 0$ $M_x = 0$ $N_x = 0$	$w = 0$ $v = 0$ $w' = 0$ $u = 0$

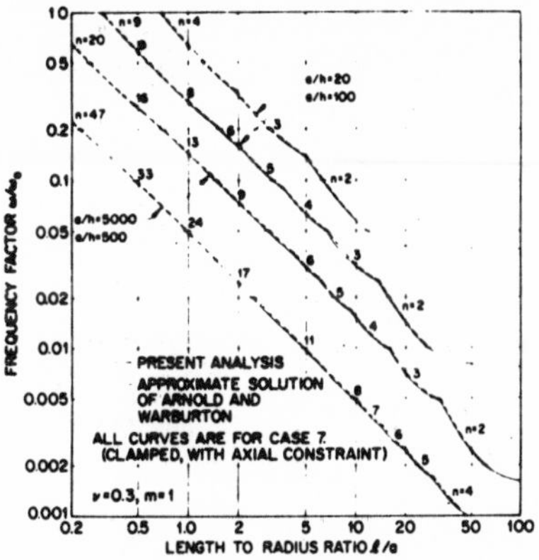

Fig. 5 Frequency envelope, case 7 and Arnold and Warburton's approximate solution.

One should keep in mind the fact that, for long shells, the minimum frequency will occur for $n = 1$. The values of l/a for which this change takes place depend upon a/h. For $a/h = 20$, the crossover occurs for $l/a = 12$ to 18; for $a/h = 5000$, the change occurs for $l/a > 100$.

The effect of clamping ($\partial w/\partial x = 0$) is illustrated in Fig. 3, in which the minimum frequency for a simply supported shell without axial constraint is compared with that of a clamped shell without axial constraint. Clearly, the effect of clamping rapidly diminishes as the length increases. The magnitude of the increase in minimum frequency due to clamping is quite small for all but very short shells ($l/a < 1$).

The influence of axial constraint ($u = 0$) is illustrated in Fig. 4. Here the minimum frequency for a simply supported shell without axial constraint is compared with that of a simply supported shell with axial constraint at one or both ends. In direct contrast to the previous case, the effect of axial constraint is significant even for very long shells and for all values of a/h. Note that the minimum frequency for case 3 is about 40 to 60% higher than in that of case 1 through-

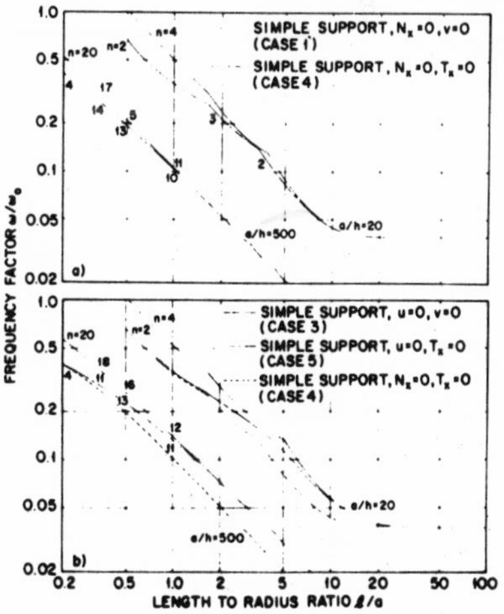

Fig. 6 Frequency envelope, cases 1 and 3–5.

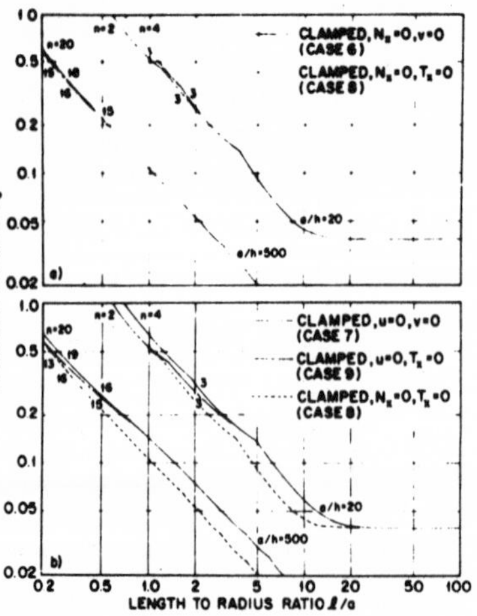

Fig. 7 Frequency envelope, cases 6–9.

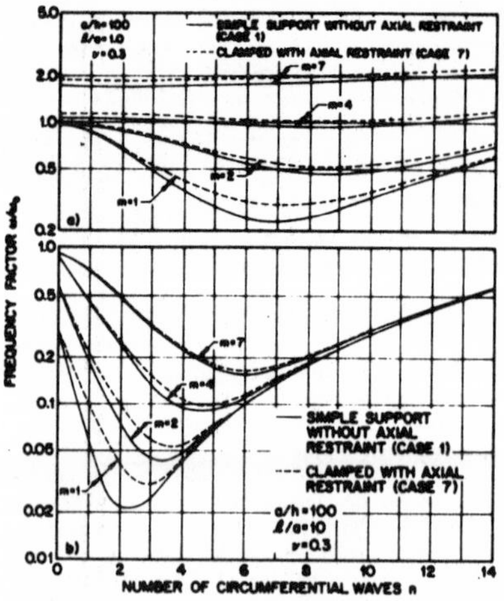

Fig. 9 Frequency distribution for l/a = 1 and l/a = 10.

out most of the region of interest. This difference may be even greater for certain modes (see Fig. 8). The physical reason for the difference in the influence of $u = 0$ as compared with $\partial w/\partial x = 0$ perhaps can best be understood by examining the modal stress resultants. This will be done in another section. Although the curves on Fig. 4 have been drawn for $m = 1$, the frequencies for modes having one axial node ($m = 2$), for a simply supported shell with axial constraint (case 3), can be determined directly from the intermediate curve (case 2) on Fig. 4. It can be shown that a shell of length $2l/a$, boundary conditions of case 3, and with $m = 2$, has the same modal behavior as a similar shell of length l/a, with $m = 1$ and boundary conditions of case 2. Further study indicates that, no matter what homogeneous boundary conditions are enforced at the ends of the shell, as m increases, the modal characteristics gradually approach those of a simply supported shell without axial constraint (case 1). This trend can be seen (for case 7) in Figs. 8–10.

Arnold and Warburton[5] used the Rayleigh-Ritz procedure to obtain an approximate solution for what they called a shell with "fixed ends." This is a shell clamped, with axial constraint, at both ends (case 7). Although they took only a one-term approximation to the mode shape, their solution for the frequency agrees very closely with the results of the present analysis (see Fig. 5), theirs being a maximum of 5% higher. For most cases, the results agree within 2%. As is to be expected, however, the modal stress resultants as predicted by their solutions are quite seriously in error. Arnold and Warburton's paper is misleading, however, in so far as it implies that the primary change from the "freely supported" case (studied earlier by them[4]) was the addition of clamping. The effect of elastic moment restraint which they considered must actually be quite small.

In all of the preceding cases, we have assumed that the tangential displacement v is zero. If this requirement is relaxed, with all other parameters remaining the same, the frequency will be lower (see Figs. 6 and 7). The greatest change in the frequency due to relaxing the condition $v = 0$ occurs for $n = 1$. It is important to note at this point that for this boundary condition the minimum frequency may occur for $n = 1$ even for short shells.

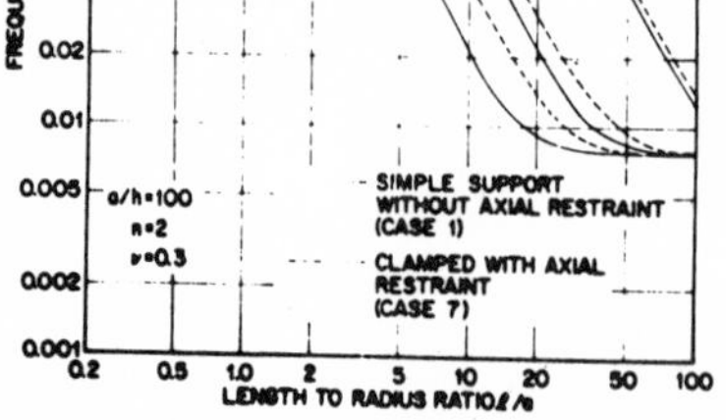

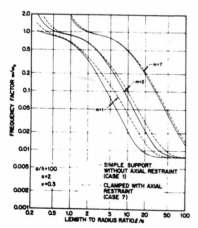

Fig. 8 Frequency distribution for $n = 2$, $m \geq 1$.

Over-All Frequency Pattern

It is not practical to try to compare on a two-dimensional plot the influence of different boundary conditions on the over-all frequency distribution. This was the reason for discussing only the frequency envelope in the preceding figures. In order to understand how these figures relate to the over-all pattern, one case will be studied in more depth. Figure 11 is a three-dimensional plot of the frequency as a function of n and l/a, for a specific shell ($a/h = 100$, $\nu = 0.3$), and for one axial half-wave. Two different sets of boundary conditions are shown: simple support without axial constraint (case 1) and clamped with axial constraint (case 7). Note that the scale for l/a is reversed from the sense used in preceding figures. The minimum curves shown in Fig. 11 are the type of curves plotted in Figs. 3–7.

The difference between the surface for case 1 and the surface for case 7, for $l/a < 1$, is primarily due to the effect of moment restraint. For $l/a > 1$, the difference is due to the effect of axial restraint. The frequencies associated with the various values of n for a fixed l/a are very close in magnitude for short shells, as evidenced by the nearly horizontal grid lines in the n direction. For long shells, the influence of the boundary conditions extends over a much narrower band of circumferential waves.

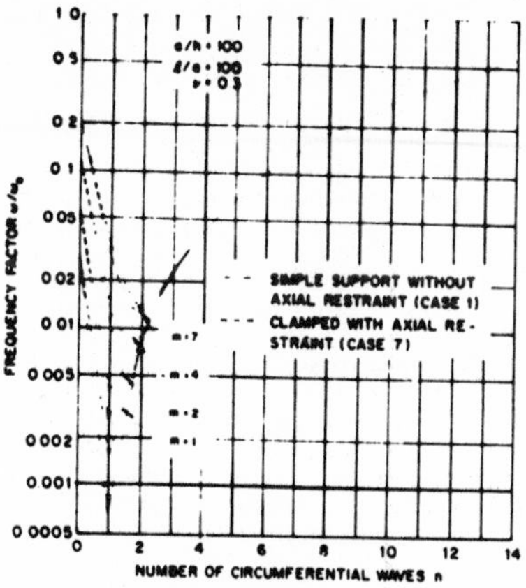

Fig. 10 Frequency distribution for l/a = 100.

To further illustrate these points, a cross section of Fig. 11 has been drawn for three different values of l/a (i.e., 1, 10, 100). Figure 9a is the cross section for $l/a = 1$. (Only the $m = 1$ curve appears in Fig. 11.) Note that the value of n for which the minimum frequency occurs depends upon the half-wavelength l/ma. Note also that, for $m = 1$, there are nine values of n which have frequencies less than the minimum value for $m = 2$. For $l/a = 10$ (Fig. 9b), there are three values. Figure 10 shows a change in character of the ordered frequencies. The minimum frequency occurs for $n = 1$, and there are no other values of n which have frequencies less than that for $n = 1$ and $m = 3$. For $n = 2$, there are 10 values of m which have corresponding frequencies that are less than the minimum value for $n = 3$.

The detailed behavior of the higher modes for the entire range of l/a is illustrated in Fig. 8 for $n = 2$. These curves

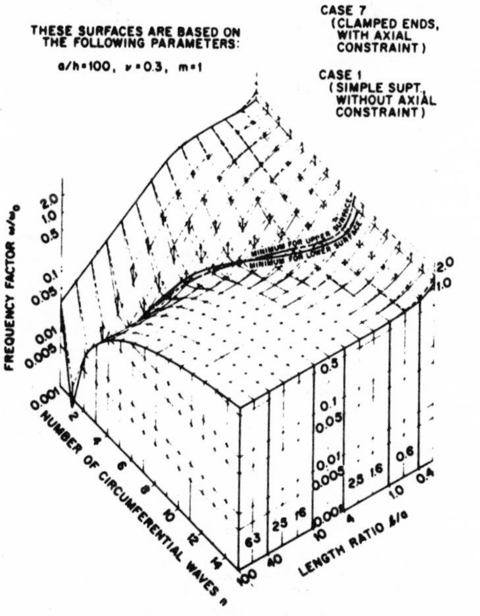

Fig. 11 Over-all frequency pattern for a/h = 100.

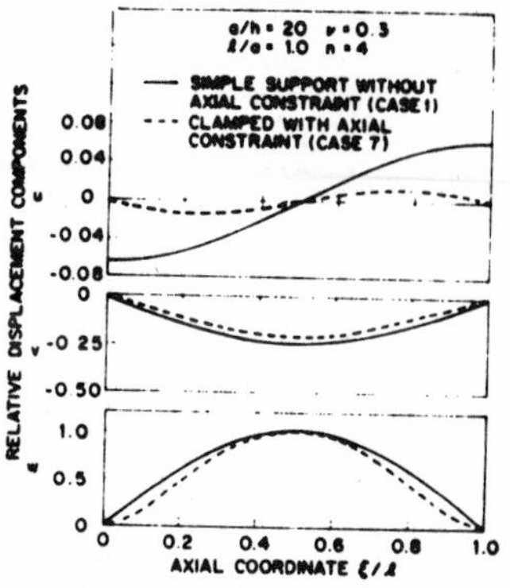

Fig. 12 Mode shape for a/h = 20, l/a = 1, n = 4.

are typical of those obtained for any value of n. The diminishing influence of the boundary conditions on higher values of n is clearly seen here. Note also that, for $m = 1$ and $5 < l/a < 15$, the frequency for case 7 is almost 100% higher than for case 1. In this same region, the difference in minimum frequencies is about 50%.

Modal Stress Resultants

A study of selected stress resultants that arise during modal vibration is essential for full understanding of the influence of the various boundary conditions on the modal behavior of thin shells. The results shown in Figs. 12-16 are all for $a/h = 20$. For thinner shells, the moment distribution is not influenced nearly as much by the various boundary conditions and, in fact, is very close to that for a simply supported shell in all cases. The axial force distribution, on the other hand, is essentially independent of a/h.

There is no unique basis for comparing the magnitude of the various stress resultants for different conditions. For our purpose here, the maximum radial deflection has been taken as the normalizing factor. The important items to compare from case to case are not so much the magnitudes of the stress resultants, but rather their distributions.

In Figs. 14 and 15, distributions of both the axial force N_x and the axial moment M_x are compared for two sets of bound-

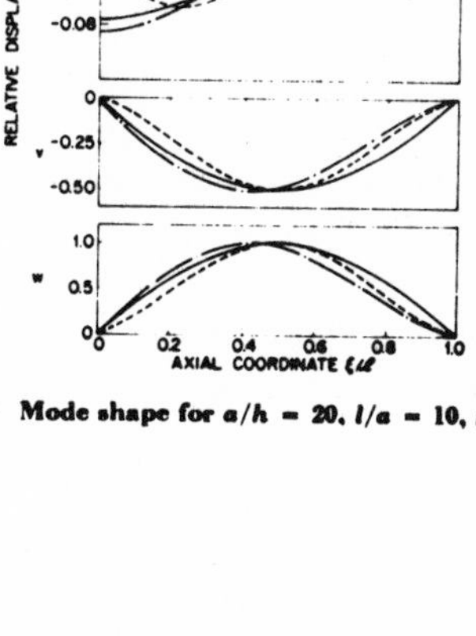

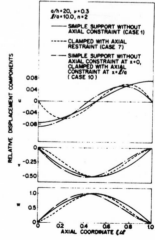

Fig. 13 Mode shape for a/h = 20, l/a = 10, n = 2.

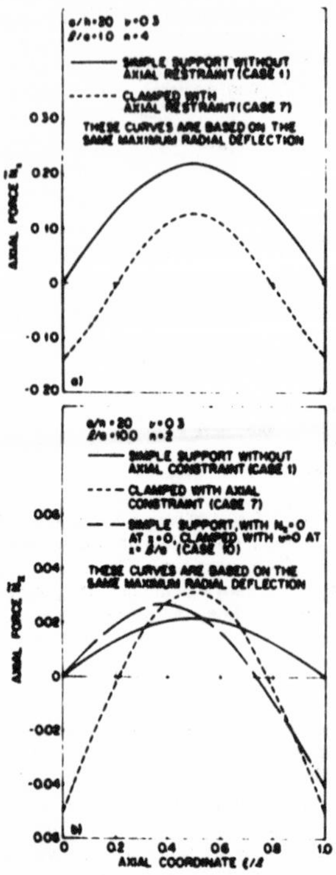

Fig. 14 Distribution of axial force during modal vibration.

ary conditions, cases 1 and 7. From these figures, we can see that the region in which the moment is affected decreases as the shell gets longer. Furthermore, a study of other values of a/h shows that, as the shell becomes thinner, the boundary conditions have less and less influence on the moment distribution. The influence of the condition $\partial w/\partial x = 0$ is generally localized, and this underlies the frequency distribution shown in Fig. 3. The shape of the axial force distribution shown in Fig. 14, on the other hand, could apply to any shell. It is apparent that the boundary condition $u = 0$ always has a strong influence on the axial force distribution, and this contributes to an understanding of the significant increase in the frequency shown in Fig. 4.

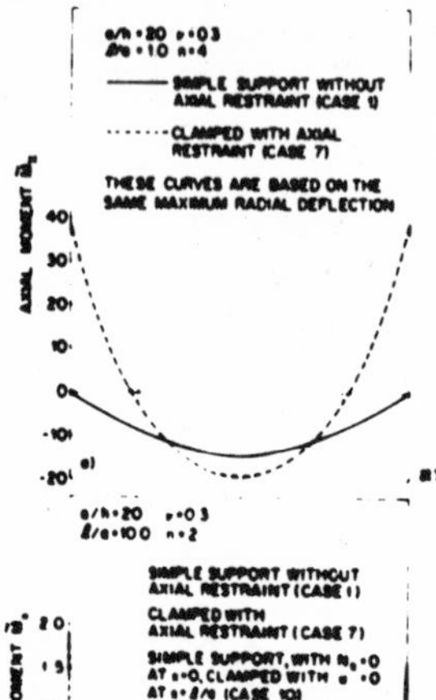

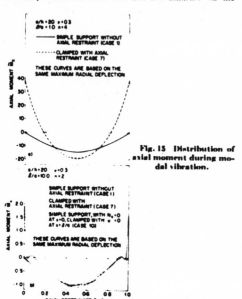

Fig. 15 Distribution of axial moment during modal vibration.

Figure 16 rather clearly demonstrates the increasing localization of the effects of the boundary conditions as the number of axial waves increases. This is in accord with the trend already noted in Figs. 8–10.

The difference in behavior of N_x and M_x for the various boundary conditions is similar to the static response of a thin cylindrical shell to various types of edge loading. It has been shown[6] that, for cylindrical shells, certain types of edge loading produce significant stresses only in a very localized boundary zone, whereas other types of loading will propagate far into the interior of the shell.

Conclusions

The present approach provides a powerful tool for examining a wide variety of boundary conditions and their influence on the modal behavior of cylindrical shells. Since no approximations have been introduced beyond those underlying Flügge's equations, the mode shapes, modal accelerations, and modal stress resultants can be computed quite accurately. The results of this study clearly indicate that care must be taken in any approximate analysis to use appropriate boundary conditions.

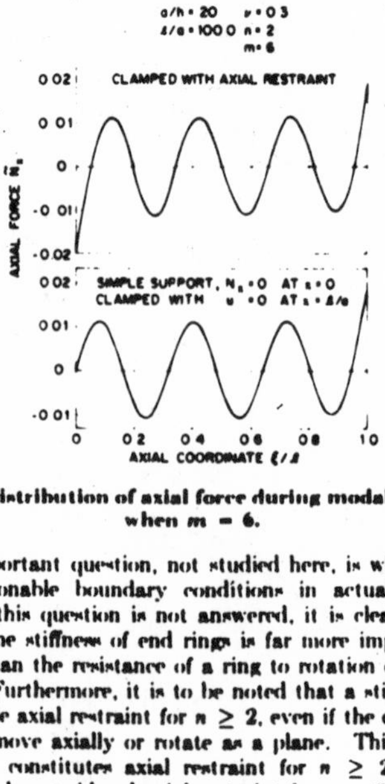

Fig. 16 Distribution of axial force during modal vibration when $m = 6$.

One important question, not studied here, is what constitutes reasonable boundary conditions in actual practice. Although this question is not answered, it is clear that the out of plane stiffness of end rings is far more important, in general, than the resistance of a ring to rotation of its cross section. Furthermore, it is to be noted that a stiff end ring will provide axial restraint for $n \geq 2$, even if the ends of the shell can move axially or rotate as a plane. This indicates that what constitutes axial restraint for $n \geq 2$ may not necessarily be considered axial restraint for $n = 0$ or $n = 1$.

References

[1] Gros, C. G. and Forsberg, K., "Vibrations of thin shells: a partially annotated bibliography," Lockheed Missiles and Space Co., SB 63-43 (1963).

[2] Sobel, L., "Effects of boundary conditions on the stability of cylinders subject to lateral and axial pressures," AIAA J. 2, 1437–1440 (1964).

[3] Flügge, W., *Stresses in Shells* (Springer-Verlag, Berlin, 1960), Chap. 5, pp. 219 and 233.
[4] Arnold, R. N. and Warburton, G. B., "Flexural vibrations of the walls of thin cylindrical shells having freely supported ends," Proc. Roy. Soc. (London) **A197**, 238–256 (1949).
[5] Arnold, R. N. and Warburton, G. B., "The flexural vibrations of thin cylinders," Inst. Mech. Engrs. (London), Proc. Automobile Div. **167**, 62–80 (1953).
[6] Steele, C. R., "Shells with edge loads of rapid variation," Lockheed Missiles and Space Co., TR 6-90-63-84 (1963).

21

Reprinted from *Quar. Appl. Math.*, **13**(3), 279–290 (1955)

ON AXI-SYMMETRICAL VIBRATIONS OF SHALLOW SPHERICAL SHELLS*

BY

ERIC REISSNER

Massachusetts Institute of Technology

1. Introduction. The present note may be considered as a sequel to an earlier paper on the same subject [4]. In this earlier paper the solution of the differential equations for axi-symmetrical vibrations of shallow spherical shells was given in terms of certain Bessel functions. The problem of the frequency determination for a shell segment with clamped edge was considered as an example of application of this solution. It led to the vanishing of a third-order determinant each element of which involved Bessel functions and the solution of a certain cubic (Eqs. 32 and 35 of Ref. 4). Similar results, by a somewhat different method, had earlier been obtained by Federhofer [2]. The Bessel-function frequency determinant being difficult to evaluate, no numerical results have yet been obtained by its use. Instead, an approximate solution for the lowest frequency was obtained by means of the procedure of Rayleigh and Ritz (Equation 40 of Ref. 4 and a similar result in Ref. 2).

The Bessel-function solution of Ref. 4 was obtained on the basis of assumptions which had previously been made for the problem of static deformations of shallow spherical shells [3]. It was not observed at that time that an additional approximation would be appropriate for the problem of transverse vibrations. This additional approximation is based on the fact that for transverse vibrations of shallow shells the magnitude of the longitudinal inertia terms is negligibly small compared with the magnitude of the transverse inertia terms [5]. Upon omission of longitudinal inertia terms it becomes possible to reduce the differential equations of dynamics to the same form as the equations of statics except that the transverse load function must include the d'Alembert term $-\rho h w_{tt}$.

The present paper contains applications of this result to three specific problems of axi-symmetrical vibrations of spherical shells.

(1) *Determination of the lowest frequency of free vibrations for a shell segment with clamped edge.* The numerical results are compared with the corresponding results from the earlier Rayleigh-Ritz formula.

(2) *Determination of the frequencies of free vibrations of a shell segment with free edge.* The relation between the present frequency equation and the corresponding frequency equation for a flat plate is of such nature that known numerical results for the flat plate can be translated with little difficulty so as to furnish the corresponding results for the shallow spherical shell.

(3) *Forced vibrations due to point load at apex of the shell.* We determine first the point-force singularity which occurs regardless of the form of the boundary conditions at the edge of the shell. We then consider further the case for which the boundary is sufficiently far removed to assume it at infinity and use the condition that sufficiently far from the point of load application the solution must represent outward travelling waves.

---

*Received August 10, 1954. A report on work supported by Office of Naval Research under Contract N5-ori-07834 with Massachusetts Institute of Technology.

Our results generalize corresponding results of H. Cremer and L. Cremer [1] for the unlimited flat plate.

2. Differential equations for transverse vibrations of shallow spherical shells. Let the equation of the shell surface be given by

$$z = \frac{a^2}{2R} - \frac{x^2 + y^2}{2R} = \frac{a^2 - r^2}{2R}. \tag{1}$$

In the absence of longitudinal loads we have as differential equations for the displacement w in the direction of z and for an Airy stress function F,

$$D\nabla^2\nabla^2 w + \frac{1}{R}\nabla^2 F = -\rho h \frac{\partial^2 w}{\partial t^2} + p(r, \theta, t) \tag{2}$$

$$\nabla^2\nabla^2 F - \frac{hE}{R}\nabla^2 w = 0, \tag{3}$$

where ρ = density of shell material, h = wall thickness (assumed constant), E = modulus of elasticity, $D = Eh^3/12(1 - \nu^2)$ and $\nabla^2 = \partial^2/\partial r^2 + r^{-1}\partial/\partial r + r^{-2}\partial^2/\partial\theta^2$ (Fig. 1).

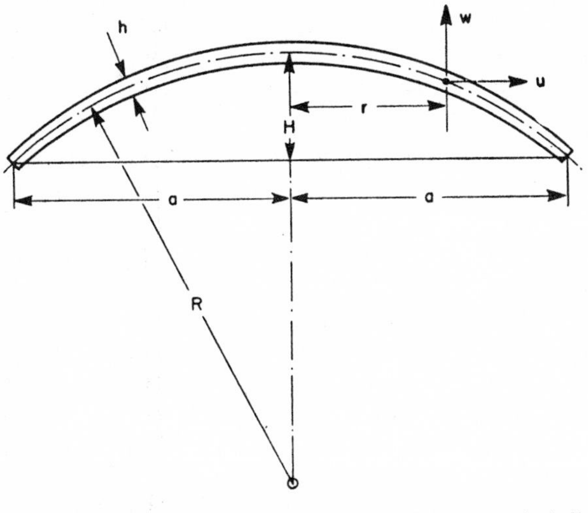

FIG. 1. Spherical shell segment, showing notations for geometrical dimensions.

We shall further need the relations

$$\epsilon_\theta = \frac{N_\theta - \nu N_r}{Eh}, \qquad N_\theta = \frac{\partial^2 F}{\partial r^2}, \qquad N_r = \frac{1}{r}\frac{\partial F}{\partial r} + \frac{1}{r^2}\frac{\partial^2 F}{\partial \theta^2}, \tag{4}$$

$$M_r = -D\left(\frac{\partial^2 w}{\partial r^2} + \frac{\nu}{r}\frac{\partial w}{\partial r} + \frac{\nu}{r^2}\frac{\partial^2 w}{\partial\theta^2}\right), \tag{5}$$

$$Q_r = -D\frac{\partial \nabla^2 w}{\partial r}, \qquad V_r = Q_r + N_r\frac{\partial z}{\partial r} + N_{r\theta}\frac{1}{r}\frac{\partial z}{\partial \theta}, \tag{6}$$

which are the same as in problems of statics.

3. Axi-symmetrical solutions. We set

$$w = e^{i\omega t}W(r), \qquad F = e^{i\omega t}f(r), \qquad p = e^{i\omega t}q(r). \tag{7}$$

Restricting attention to the case $q = 0$ it is found that solutions of the homogeneous equations (2) and (3) are of the form

$$W = C_1J_0(\lambda r) + C_2Y_0(\lambda r) + C_3I_0(\lambda r) + C_4K_0(\lambda r) + C_5 + C_6 \log r, \tag{8}$$

$$f = B_1J_0(\lambda r) + B_2Y_0(\lambda r) + B_3I_0(\lambda r) + B_4K_0(\lambda r) + B_5 r^2 + B_6 r^2 \log r + C_7 + C_8 \log r. \tag{9}$$

The constants C_n in (8) and (9) are arbitrary, the usual notation for Bessel functions and modified Bessel functions is employed, and furthermore

$$\lambda^4 = \frac{\rho h}{D}\omega^2 - \frac{hE}{R^2D}. \tag{10}$$

Taking account of the fact that

$$\nabla^2(J_0, Y_0) = -\lambda^2(J_0, Y_0), \qquad \nabla^2(I_0, K_0) = \lambda^2(I_0, K_0), \tag{11}$$

there results for the constants B_n

$$\left.\begin{aligned} B_{1,2} &= -\frac{hE}{R\lambda^2}C_{1,2} \\ B_{3,4} &= \frac{hE}{R\lambda^2}C_{3,4} \\ B_5 &= \frac{1}{4}\rho h\omega^2R(C_5 - C_6) \\ B_6 &= \frac{1}{4}\rho h\omega^2RC_6 \end{aligned}\right\}. \tag{12}$$

Of the eight constants C_n only six are physically significant. The constant C_7 has no effect on stresses and displacements. The condition of vanishing circumferential displacements as in problems of statics [3], is of the form $(hE/R)W = \nabla^2 f + \text{const}$. From this $(hE/R)C_6 = 4B_6$. With B_6 from (12) this implies $\lambda^4C_6 = 0$. Since $\lambda \neq 0$ we have

$$B_6 = C_6 = 0 \tag{12'}$$

and therewith altogether

$$W = C_1J_0(\lambda r) + C_2Y_0(\lambda r) + C_3I_0(\lambda r) + C_4K_0(\lambda r) + C_5, \tag{13}$$

$$f = -\frac{hE}{R\lambda^2}[C_1J_0(\lambda r) + C_2Y_0(\lambda r) - C_3I_0(\lambda r) - C_4K_0(\lambda r)] + \frac{1}{4}\rho h\omega^2RC_5 r^2 + C_8 \log r. \tag{14}$$

The solution (13) and (14) will be applied to the three problems indicated earlier.

4. Frequency equation for shell segment with clamped edge. Let $r = a$ be the edge of the shell segment. Since W and f must be regular for $r = 0$ we have

$$C_2 = C_4 = C_8 = 0. \tag{15}$$

The conditions of vanishing edge displacement W and edge slope W' are

$$C_1 J_0(\mu) + C_3 I_0(\mu) + C_5 = 0, \tag{16}$$

$$C_1 J_0'(\mu) + C_3 I_0'(\mu) = 0, \tag{17}$$

where

$$\mu = \lambda a. \tag{18}$$

We assume that the third edge condition stipulates vanishing horizontal displacement or, equivalently, vanishing circumferential strain ϵ_θ. With the help of (4) this relation becomes $(F'' - \nu r^{-1}F')_a \equiv [\nabla^2 F - (1 + \nu)r^{-1}F']_a = 0$. Introduction of (14) gives as third relation for the constants C_1, C_3, C_5,

$$\frac{hE}{R}\left\{C_1 J_0(\mu) + C_3 I_0(\mu) - \frac{1+\nu}{\mu}\left[-C_1 J_0'(\mu) + C_3 I_0'(\mu)\right]\right\} + \frac{1}{2}\,\rho h\omega^2 R(1-\nu)C_5 = 0. \tag{19}$$

The frequency equation of the problem is the condition of vanishing of the determinant of the system (16), (17) and (19). Setting $J_0' = -J_1$, $I_0' = I_1$, this frequency equation may be brought into the following form

$$J_0(\mu)I_1(\mu) + J_1(\mu)I_0(\mu) + \frac{4\kappa^4 J_1(\mu)I_1(\mu)}{\mu(\mu^4 - \kappa^4)} = 0. \tag{20}$$

The parameter κ^4 is

$$\kappa^4 = \frac{1+\nu}{1-\nu}\,\frac{hEa^4}{R^2 D} = 48(1+\nu)^2\,\frac{H^2}{h^2} \tag{21}$$

and ω is given in terms of μ and κ, as follows

$$\rho\omega^2\,\frac{ha^4}{D} = \frac{1-\nu}{1+\nu}\,\kappa^4 + \mu^4. \tag{22}$$

The quantity H in (21) is the rise of the shell. It is related to the radii R and a through the formula,

$$H = \frac{a^2}{2R}. \tag{23}$$

The frequency equation (20) furnishes μ as a function of κ. Approximate values of μ for the lowest frequency may be found in Table 1.* Having μ as function of κ we obtain further from (22)

$$\omega = \left(\frac{E}{\rho}\right)^{1/2}\frac{h}{a^2}\left[\frac{\mu^4 + \kappa^4(1-\nu)/(1+\nu)}{12(1-\nu^2)}\right]^{1/2}. \tag{24}$$

Let ω_0 be the value of ω corresponding to $\kappa = 0$ and $\nu = 0$, that is the value of the frequency ω for a flat plate and for vanishing Poisson's ratio. We have

$$\omega_0 = \left(\frac{E}{\rho}\right)^{1/2}\frac{h}{a^2}\left(\frac{\mu_0^4}{12}\right)^{1/2} = 2.948\left(\frac{E}{\rho}\right)^{1/2}\frac{h}{a^2}, \tag{25}$$

*For these and all other computations in this paper the author is indebted to Millard W. Johnson.

TABLE 1. Numerically smallest solution of Eq. (20)

κ^4	μ
0	3.196
10	3.235
20	3.273
50	3.380
100	3.537
215.56	3.832
300	4.000
400	4.180
500	4.328
600	4.460
700	4.575
800	4.680
900	4.770
1000	4.855
1200	4.990
1400	5.112
1600	5.212
1800	5.285
2000	5.360
2500	5.482
3000	5.570
4000	5.665
5000	5.723
10000	5.828
15900	5.865
21150	5.875
∞	5.910

where $\mu_0 = 3.195$ and $\mu_0^4 = 104.27$. It is convenient to express the actual values of ω in units of ω_0, as follows,

$$\frac{\omega}{\omega_0} = \left[\frac{1}{1-\nu^2}\left(\frac{\mu}{\mu_0}\right)^4 + \frac{48}{\mu_0^4}\left(\frac{H}{h}\right)^2\right]^{1/2}. \tag{26}$$

Numerical values of ω/ω_0 as a function of ν and H/h may be found in Table II and in Fig. 2.

When H/h is large enough so that the second term under the square root in (26) dominates the first term, practically when H/h is larger than about 25, then Eq. (26) may be replaced by the approximation

$$\omega = 2\left(\frac{E}{\rho}\right)^{1/2}\frac{H}{a^2}. \tag{26*}$$

It is interesting to note that the thickness h which occurs in the flat-plate frequency formula (25) has been replaced by the shell rise H and the frequency has become independent of wall thickness. It should be emphasized, however, that the simple formula (26*) depends on two limitations. On the one hand we must have that $H/h \gtrsim 25$ and on the other hand we must have $H/a \lesssim \frac{1}{4}$ in order that the theory of shallow shells is

TABLE 2. Frequency of clamped-edge shell in units of corresponding flat-plate frequency for $\nu = 0$, as function of ν and H/h

$\frac{H}{h}$	ω/ω_0		
	$\nu = 0$	$\nu = .3$	$\nu = .5$
0	1.000	1.0483	1.1547
.5	1.08	1.149	1.28
1.0	1.31	1.40	1.59
1.5	1.61	1.75	2.00
2.0	1.94	2.16	2.42
2.5	2.31	2.57	2.87
3.0	2.67	2.99	3.32
3.5	3.06	3.40	3.73
4.0	3.43	3.78	4.12
4.5	3.81	4.16	4.49
5	4.18	4.51	4.89
6	4.90	5.17	5.45
7	5.59	5.80	6.01
8	6.22	6.41	6.60
9	6.86	7.01	7.20
10	7.50	7.62	7.82
11	8.13	8.25	8.44
12	8.76	8.88	9.05
14	10.05	10.15	10.28
16	11.35	11.42	11.54
18	12.67	12.73	12.82
20	14.00	14.1	14.2

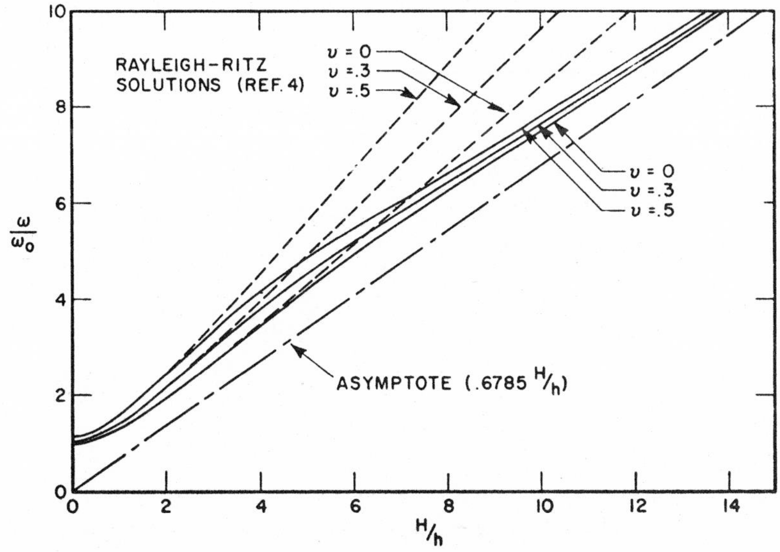

FIG. 2. Lowest frequency for shell segment with clamped edge, in units of corresponding frequency for flat plate with zero Poisson's ratio.

applicable. We may note that it is this latter restriction which insures that the transverse vibrations of the shell take place at frequencies which are low compared with the frequencies of longitudinal vibrations of flat plates, these latter being of order $(E/\rho)^{1/2}(1/a)$.

Comparison with Rayleigh-Ritz formula. Equation (40) of Ref. 4 may be written in the following form, which is equivalent to (26)

$$\left[\frac{\omega}{\omega_0}\right]_{RR} = \left[\frac{1}{1-\nu^2} + \frac{.7 + .5\nu - .2\nu^2}{1-\nu^2}\left(\frac{H}{h}\right)^2\right]^{1/2}. \tag{27}$$

Figure 2 contains values of $(\omega/\omega_0)_{RR}$ as dotted lines. It is seen that the Rayleigh-Ritz solution agrees remarkably well with the differential equation solution for sufficiently small values of H/h, practically up to values of H/h of about three. As H/h increases the error becomes larger, the percentage error approaching a finite limiting value of from 60 per cent to 25 per cent as ν decreases from 0.5 to 0.

5. Frequency equation for shell segment with free edge. The boundary conditions for a free edge are

$$M_r = 0, \qquad V_r = 0, \qquad N_r = 0, \tag{28}$$

where M_r, V_r and N_r are given in (4) to (6). To this are again added regularity conditions for $r = 0$ which again mean that Eq. (15) must hold. We observe further that (28) contains two relations involving W', W'' and W''' but neither W itself nor f and its derivatives. This means that the frequency equation for the shell with free edge does not involve the constant C_5 in Eq. (13) for W. In other words the admissible values of μ are the same as those for a flat plate with free edge.

The frequency equation for axially symmetric vibrations of a flat plate with free edge, according to Kirchhoff, is

$$\frac{I_0(\mu)}{I_1(\mu)} + \frac{J_0(\mu)}{J_1(\mu)} = \frac{2(1-\nu)}{\mu}. \tag{29}$$

Examples of the numerically smallest values of μ are $\nu = 0$, $\mu = 2.87$; $\nu = .3$, $\mu = 3.00$; $\nu = .5$, $\mu = 3.07$. Having μ we obtain ω by means of (10), (18) and (23) in the form

$$\omega = \left(\frac{E}{\rho}\right)^{1/2}\left[\frac{\mu^4}{12(1-\nu^2)}\frac{h^2}{a^4} + 4\frac{H^2}{a^4}\right]^{1/2}. \tag{30}$$

Let μ_0 and ω_0 be appropriate values for the flat plate and for $\nu = 0$. We may then write, in analogy to the result (26) for the shell with clamped edge,

$$\frac{\omega}{\omega_0} = \left[\frac{1}{1-\nu^2}\left(\frac{\mu}{\mu_0}\right)^4 + \frac{48}{\mu_0^4}\left(\frac{H}{h}\right)^2\right]^{1/2}. \tag{31}$$

Equation (31) is the same as Eq. (26) except that for the shell with free edge the quantity μ is independent of the values of H/h while for the shell with clamped edge μ was found to be a function of H/h. Values of ω/ω_0 for the numerically smallest solution of (29) as function of H/h and ν may be found in Table 3 and Fig. 3.

We note that when H/h is greater than about 10, Eq. (30) may be simplified to $\omega = 2(E/\rho)^{1/2}(H/a^2)$ which is the same limiting expression which was previously obtained for the shell segment with clamped edge.

We further note that very probably for this problem of a shell with free edge the

TABLE 3. Frequency of free-edge shell in units of corresponding flat-plate frequency for $\nu = 0$, as function of ν and H/h

$\frac{H}{h}$	ω/ω_0		
	$\nu = 0$	$\nu = .3$	$\nu = .5$
0	1	1.148	1.315
.5	1.084	1.224	1.380
1.0	1.305	1.423	1.562
1.5	1.610	1.705	1.821
2.0	1.96	2.035	2.135
2.5	2.33	2.40	2.48
3.0	2.71	2.77	2.85
3.5	3.11	3.16	3.22
4.0	3.51	3.55	3.62
4.5	3.91	3.96	4.01
5	4.32	4.36	4.41
6	5.15	5.17	5.21
7	5.98	6.00	6.03
8	6.81	6.82	6.86
9	7.64	7.66	7.70
10	8.47	8.49	8.53
11	9.31	9.32	9.35
12	10.15	10.15	10.18
14	11.82	11.82	11.82
16	13.50	13.50	13.50
18	15.20	15.20	15.20
20	16.86	16.86	16.86

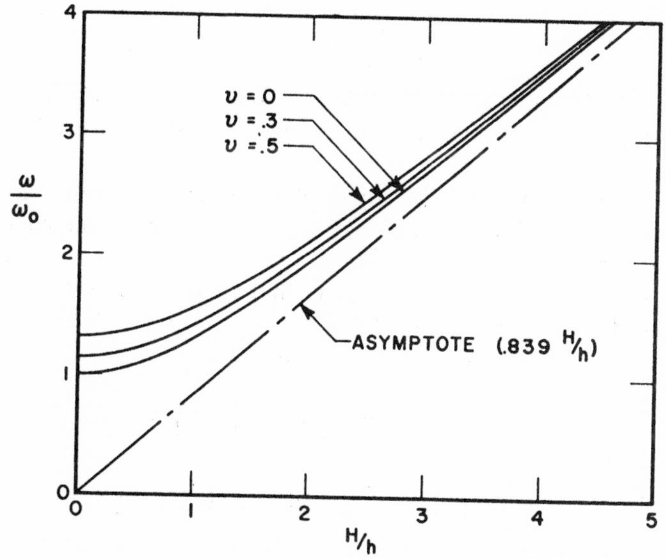

FIG. 3. Lowest frequency of axi-symmetrical vibrations for shell segment with free edge, in units of corresponding frequency for flat plate with zero Poisson's ratio.

lowest frequency of axi-symmetrical free vibrations is higher than the frequencies of certain non-symmetrical vibrations. This fact is concluded from corresponding known results for the special case of the flat plate.

6. Oscillating point load at apex of shell. The following conditions must be satisfied for an oscillating point load P exp $(i\omega t)$ at the apex of the shell

$$\lim_{r \to 0} (2\pi r V_r) = Pe^{i\omega t}, \tag{32}$$

$$w(0, t), \qquad N_r(0, t), \qquad N_\theta(0, t) \qquad \text{finite.} \tag{33}$$

In view of (4), (5), (6), (7) these conditions assume the following form for the solution functions W and f as given by (13) and (14)

$$\lim_{r \to 0} \left(r \frac{d\nabla^2 w}{dr} \right) = -\frac{P}{2\pi D}, \tag{32'}$$

$$r = 0; \qquad W, \nabla^2 f, \qquad \frac{1}{r}\frac{df}{dr} \qquad \text{finite.} \tag{33'}$$

We begin by omitting terms from the solutions (13) and (14) which automatically satisfy the conditions (32′) and (33′) and retain the singular portion

$$W_s = C_2 Y_0(\lambda r) + C_4 K_0(\lambda r), \tag{34}$$

$$f_s = -\frac{hE}{R\lambda^2} [C_2 Y_0(\lambda r) - C_4 K_0(\lambda r)] + C_8 \log r. \tag{35}$$

Equations (34) and (35) are valid, with real values of the constants C_2 and C_4, as long as λ is real. According to (10) this means as long as

$$\frac{1}{R} \left(\frac{E}{\rho} \right)^{1/2} < \omega. \tag{36a}$$

When

$$\omega < \frac{1}{R} \left(\frac{E}{\rho} \right)^{1/2} = \frac{2H}{a^2} \left(\frac{E}{\rho} \right)^{1/2} \tag{36b}$$

the representations (34) and (35) are no longer convenient. It will, however, be shown that the results obtained in the range of frequencies (36a) are readily transferred to the range (36b).

We note that the frequency which divides the ranges (36a) and (36b) is exactly that lowest frequency of *free* vibrations which occurs in the range of sufficiently large values of H/h [See Eq. (26*)].

In order to determine the constants C_2, C_4 and C_8 we observe the following relations in which use has been made of (11),

$$\nabla^2 W_s = \lambda^2[-C_2 Y_0 + C_4 K_0], \tag{37a}$$

$$\nabla^2 f_s = \frac{hE}{R} [C_2 Y_0 + C_4 K_0], \tag{37b}$$

$$\frac{1}{r}\frac{df_s}{dr} = -\frac{hE}{R}\frac{1}{\lambda r} [C_2 Y_0' - C_4 K_0'] + \frac{C_8}{r^2}, \tag{37c}$$

$$r \frac{d\nabla^2 W_s}{dr} = \lambda^2(\lambda r)[-C_2 Y_0' + C_4 K_0'], \tag{37d}$$

$$\left.\begin{aligned} Y_0(\lambda r) &\approx \frac{2}{\pi}\left[\log\frac{\lambda r}{2} + .577 \cdots\right], & Y_0'(\lambda r) &\approx \frac{2}{\pi}\frac{1}{\lambda r} \\ K_0(\lambda r) &\approx -\left[\log\frac{\lambda r}{2} + .577 \cdots\right], & K_0'(\lambda r) &\approx -\frac{1}{\lambda r} \end{aligned}\right\}, \tag{38}$$

Eqs. (38) being valid when $\lambda r \ll 1$.

Introduction of (34), (35), (37) and (38) into the finiteness conditions (33′) shows that these conditions are satisfied provided

$$\frac{2}{\pi} C_2 - C_4 = 0, \tag{39a}$$

$$\frac{2}{\pi} C_2 + C_4 = \frac{\lambda^2 R}{Eh} C_8 . \tag{39b}$$

The load condition (32′) leads to the further relation

$$\frac{2}{\pi} C_2 + C_4 = \frac{P}{2\pi D\lambda^2}. \tag{39c}$$

Introduction of C_2, C_4 and C_8 from (39) into (34) and (35) leads to the following expressions for the singular solutions W_s and f_s,

$$W_s = \frac{P}{4\pi D\lambda^2}\left[\frac{\pi}{2} Y_0(\lambda r) + K_0(\lambda r)\right], \tag{40}$$

$$f_s = -\frac{Eh}{R\lambda^2}\frac{P}{4\pi D\lambda^2}\left[\frac{\pi}{2} Y_0(\lambda r) - K_0(\lambda r) - 2\log r\right]. \tag{41}$$

It will now be shown that (40) and (41) remain valid, but are conveniently written in different form, in the range of small ω given by (36b). For values of ω in the range (36b) we write

$$\lambda = i^{1/2}\left[\frac{hE}{R^2 D} - \frac{\rho h\omega^2}{D}\right]^{1/4} \equiv i^{1/2}\gamma \tag{42}$$

and we introduce Kelvin functions through the following known relations

$$Y_0(i^{1/2}\gamma r) = \frac{2}{\pi}\,[-\mathrm{ker}\,\gamma r + i\,\mathrm{kei}\,\gamma r], \tag{43a}$$

$$K_0(i^{1/2}\gamma r) = \mathrm{ker}\,\gamma r + i\,\mathrm{kei}\,\gamma r. \tag{43b}$$

If (42) and (43) are introduced into (40) and (41) there follow as expressions for W_s and f_s which are valid when $\omega < R^{-1}(E/\rho)^{1/2}$,

$$W_s = \frac{P}{2\pi D\gamma^2}\,\mathrm{kei}\,\gamma r, \tag{44}$$

$$f_s = -\frac{Eh}{R\gamma^2}\frac{P}{2\pi D\gamma^2}\,[\mathrm{ker}\,\gamma r + \log r]. \tag{45}$$

Equations (44) and (45) contain as special case, when $\omega = 0$, the previously given formulas [3, Eq. (48)] for the corresponding problem of static deflection. Writing

$\gamma = (hE/R^2D)^{1/4} = [12(1-\nu^2)]^{-1/4}(Rh)^{-1/2}$ Eqs. (44) and (45) become for $\omega = 0$

$$W_s = \frac{[12(1-\nu^2)]^{1/2}}{2\pi}\frac{PR}{Eh}\operatorname{kei}\frac{r}{[12(1-\nu^2)]^{1/4}(Rh)^{1/2}}, \tag{44'}$$

$$f_s = -\frac{PR}{2\pi}\left[\operatorname{ker}\frac{r}{[12(1-\nu^2)]^{1/4}(Rh)^{1/2}} + \log r\right]. \tag{45'}$$

We finally note the following important fact. As long as the solution (44) applies, that is as long as $\omega < (2H/a^2)(E/\rho)^{1/2}$, we have that the deflection amplitude W decreases exponentially at large distances r from the point of load application. When ω is larger than $(2H/a^2)(E/\rho)^{1/2}$ so that the solution (40) applies then the deflection amplitude W decreases as $r^{-1/2}$, that is, much more slowly than exponentially. In this latter case an important distinction can be made between standing wave and travelling wave solutions.

7. Travelling wave solution for effectively unlimited shell segment. We inquire for a solution in which the deflection $w = \exp(i\omega t)W(r)$ behaves for sufficiently large values of r as $r^{-1/2}\exp[i(\omega t - \lambda r)]$. For a solution with this type of behavior we have that the wave produced by the pulsating point load travels *outward*, with energy being dissipated through radiation.

In view of the asymptotic behavior of the Hankel function of the second kind

$$H_0^{(2)}(\lambda r) = J_0(\lambda r) - iY_0(\lambda r) \sim \left(\frac{2}{\pi\lambda r}\right)^{1/2}\exp[-i(\lambda r - \pi/4)], \tag{46}$$

an appropriate result is obtained by adding in (40) and (41) a suitable multiple of the non-singular solution $J_0(\lambda r)$ as follows,

$$W_{ST} = \frac{P}{4\pi D\lambda^2}\left\{\frac{\pi}{2}[Y_0(\lambda r) + iJ_0(\lambda r)] + K_0(\lambda r)\right\}, \tag{47}$$

$$f_{ST} = -\frac{Eh}{R\lambda^2}\frac{P}{4\pi D\lambda^2}\left\{\frac{\pi}{2}[Y_0(\lambda r) + iJ_0(\lambda r)] - K_0(\lambda r) - 2\log r\right\}. \tag{48}$$

We note that the solution (47) and (48) contains as a special case the corresponding result for a flat plate, as given by H. Cremer and L. Cremer [1]. The case of the flat plate follows if we set $R = \infty$ in the expression for λ, making $\lambda = (\rho h\omega^2/D)^{1/4}$. We note further that while for the spherical shell the solution (47) and (48) is valid, subject to the restriction that $\omega > (E/\rho)^{1/2}R$, this relation ceases to be a restriction for the case of the flat plate.

In order that the travelling wave solution (47) have a meaning it is necessary that the boundary $r = a$ be sufficiently far removed from the point of load application. Let us assume that sufficiently far means

$$a\lambda > 5, \qquad \text{say}. \tag{49}$$

Equation (49) may be written as a restriction on frequencies as follows

$$\omega > \frac{2H}{a^2}\left(\frac{E}{\rho}\right)^{1/2}\left[1 + \frac{625}{48(1-\nu^2)}\frac{h^2}{H^2}\right]^{1/2}. \tag{49'}$$

8. Two formulas of acoustical significance. We consider the ratio of velocity at the point of load application to the force causing this velocity, $(\partial w/\partial t)_{r=0}/P\exp(i\omega t)$.

According to Eq. (47)

$$\frac{(\partial w/\partial t)_{r=0}}{Pe^{i\omega t}} = \frac{i\omega}{4\pi D\lambda^2}\lim_{r\to 0}\left\{\frac{\pi}{2}\left[Y_0(\lambda r) + iJ_0(\lambda r)\right] + K_0(\lambda r)\right\}.$$

Since $J_0(0) = 1$ and, according to (38), $\lim_{r\to 0}\left[\frac{1}{2}\pi Y_0(\lambda r) + K_0(\lambda r)\right] = 0$ there follows,

$$\frac{(\partial w/\partial t)_{r=0}}{Pe^{i\omega t}} = \frac{-\omega}{8D\lambda^2} = \frac{-\omega}{8\left[\rho h D\left(\omega^2 - 4\dfrac{E}{\rho}\dfrac{H^2}{a^4}\right)\right]^{1/2}}, \qquad \frac{2H}{a^2}\left(\frac{E}{\rho}\right)^{1/2} < \omega. \tag{50a}$$

For $H = 0$, Eq. (50a) reduces to a result in Ref. 1. We note that a ratio which is independent of frequency for the flat plate is a function of frequency for the shallow spherical shell. In the range of applicability of (50a), which is given by (49′), this ratio decreases with increasing frequency towards a limiting value which is the constant flat plate value.

When $\omega < (2H/a^2)(E/\rho)^{1/2}$ so that Eq. (44) applies we have instead of (50a)

$$\frac{(\partial w/\partial t)_{r=0}}{Pe^{i\omega t}} = \frac{i\omega\,\mathrm{kei}(0)}{2\pi D\gamma^2}.$$

In view of the fact that $\mathrm{kei}(0) = -\frac{1}{4}\pi$ and with γ defined in (42) this may be written in the following form

$$\frac{(\partial w/\partial t)_{r=0}}{Pe^{i\omega t}} = \frac{-i\omega}{8\left[\rho h D\left(4\dfrac{E}{\rho}\dfrac{H^2}{a^4} - \omega^2\right)\right]^{1/2}}, \qquad \frac{2H}{a^2}\left(\frac{E}{\rho}\right)^{1/2}. \tag{50b}$$

We finally observe, on the basis of (50a) and (50b) that the following result holds for the work of the force $P\exp(i\omega t)$ per cycle

$$\frac{\text{Work}}{\text{Cycle}} = \begin{cases} 0, & \omega < \dfrac{2H}{a^2}\left(\dfrac{E}{\rho}\right)^{1/2}, \\[2ex] \dfrac{P^2}{4\left[\rho h D\left(\omega^2 - 4\dfrac{E}{\rho}\dfrac{H^2}{a^4}\right)\right]^{1/2}}, & \dfrac{2H}{a^2}\left(\dfrac{E}{\rho}\right)^{1/2} < \omega. \end{cases} \tag{51}$$

It is recalled that Eq. (51) is derived without consideration of damping sources other than radiation damping. In addition to this, while formally the second part of Eq. (51) holds for all ω greater than $(2H/a^2)(E/\rho)^{1/2}$, for the solution to have physical meaning a stronger restriction such as (49′) actually applies.

References

1. H. and L. Cremer, *Theorie der Entstehung des Klopfschalls*, Z. f. Schwingungs-und Schwachstromtechnik **2**, 61–72 (1948)
2. K. Federhofer, *Zur Berechnung der Eigenschwingungen der Kugelschale*, Sitzber. Akad. Wiss. Wien **146**, 57-69 (1937)
3. E. Reissner, *Stresses and small displacements of shallow spherical shells*, J. Math. and Phys. **25**, 80-85, 279-300 (1945); **27**, 240 (1948)
4. E. Reissner, *On vibrations of shallow spherical shells*, J. Appl. Phys. **17**, 1038-1042 (1946)
5. E. Reissner, *On transverse vibrations of thin shallow elastic shells*, Q. Appl. Math. **13**, 169-176 (1955).

22

Reprinted from *Proc. 4th U.S. Natl. Cong. Appl. Mech.*, 225–233 (1963)

Free nonsymmetric vibrations of shallow spherical shells

A. Kalnins [1]
Yale University

Introduction

In order to obtain the complete frequency spectrum of free vibrations of shallow spherical shells as predicted by the classical bending theory of shallow shells, the longitudinal as well as the transverse inertia terms must be included in the analysis[2]. Although the presence of the longitudinal inertia complicates the analysis, its inclusion brings out an infinite number of normal modes of vibration which are not predicted when the longitudinal inertia is omitted. The advantage of the theory of *transverse* vibrations [3], which neglects the longitudinal inertia, lies in its simpler equations, but it is capable of predicting only some of the normal modes of vibration of shallow shells.

Frequency equations valid for *transverse* vibrations of shallow spherical shells have been derived and solved for axisymmetric deformation by Reissner [4] and for non-symmetric deformation by Johnson and Reissner [5]. Treatments of free vibrations of shallow spherical shells with the inclusion of the longitudinal inertia terms have been heretofore limited to only axisymmetric deformation for which frequency equations of a spherical cap with various edge conditions have been deduced and studied by Federhofer [6], Reissner [7], and by the present author and Naghdi [8]. In [8], the equations governing the axisymmetric vibration of shallow spherical shells were derived from the general results of [1], and the lowest natural frequencies for free, simply supported, and clamped edges were compared to those given by the theory of *transverse* vibrations. For all edge conditions the effect of longitudinal inertia on the lowest natural frequencies was found to be negligible. More recently, several natural frequencies of axisymmetric vibration were calculated by Hoppmann [9] from the frequency equations of a shallow spherical cap with simply supported and clamped edges.

The main object of the present paper is the investigation of the natural frequencies and mode shapes of free nonsymmetric vibrations of shallow spherical shells and the examination of the character of all the modes predicted by the classical theory of shallow shells within a certain frequency band. For completeness, a study of the axisymmetric modes, which extends beyond the results given in [6, 7, 8, 9], is also included.

After the presentation of the governing system of equations, separable solutions for the three displacement components of the middle surface are deduced explicitly in terms of Bessel functions. From these solutions the frequency equation of free vibration of a shallow spherical shell with a clamped edge is derived as a function of the geometric parameters, Poisson's ratio, and the circumferential wave number n. Then the natural frequencies are calculated numerically for various wave numbers, and the variation of the natural frequencies with the thickness and the curvature of the shell is studied. The displacement amplitudes corresponding to the first

[1] Assistant Professor of Civil Engineering.

[2] An analysis of the importance of the longitudinal inertia terms is included in [1]. Employing the linearized equations of [1], it was recently shown by Nordgren [2] that in order to predict longitudinal type vibrations of pretwisted plates, equations which include the longitudinal inertia terms are required.

six natural frequencies are determined for $n = 0, 1$, and it is found that the modes within the frequency band considered can be dividedinto two distinct groups distinguished by a predominant transverse displacement or by a predominant longitudinal displacement. A convenient definition of these groups of modes is given with reference to the transverse and longitudinal modes of vibration of a plate.

Since the analysis of the theory of *transverse* vibrations is simpler than that of the classical theory of shallow shells, it is of interest to determine those natural frequencies of the complete frequency spectrum which can be predicted and to isolate those modes which cannot be predicted by the theory of *transverse* vibration. For this purpose, the corresponding frequency equation of the theory of *transverse* vibrations is derived, the natural frequencies are calculated, and the obtained frequency spectrum is compared to that of the classical theory. This comparison shows that the infinite number of modes given by the theory of *transverse* vibrations are interspersed throughout the spectrum and their frequencies differ little from a set of corresponding modes of the complete spectrum which are designated transverse.

While this investigation is concerned with the free vibration of a shallow spherical shell with a clamped edge, the main features of the corresponding analysis for a shallow spherical shell subjected to any other natural boundary condition are expected to be similar to those presented here. Frequency equations for different boundary conditions can be derived from the displacement expressions given in this paper with the use of the stress-displacement relations of shallow shells in a straightforward manner.

System of Equations and Solution

A convenient system of uncoupled equations which governs free nonsymmetric vibration of shallow spherical shells according to the classical theory can be obtained from the results of [10] by neglecting the effects of the transverse shear deformation and the rotatory inertia and by setting the surface load terms equal to zero. With reference to polar coordinates (r, θ) this system of equations can be written as

$$\nabla^6 w + r_1 \nabla^4 w + r_2 \nabla^2 w + r_3 w = 0 \tag{1a}$$

$$\nabla^2 \psi = 2(1+\nu)\, k \frac{\Omega^2}{\omega^2 a^2} \frac{\partial^2 \psi}{\partial t^2} \tag{1b}$$

$$r_1 = -\frac{1-\nu^2}{a^2} k \frac{\Omega^2}{\omega^2}\frac{\partial^2}{\partial t^2}; \quad r_2 = \frac{12(1-\nu^2)}{a^2 h^2}\left[\left(\frac{a}{R}\right)^2 + \frac{\Omega^2}{\omega^2}\frac{\partial^2}{\partial t^2}\right]$$

$$r_3 = -\frac{12(1-\nu^2)^2 k}{a^4 h^2}\frac{\Omega^2}{\omega^2}\left[\frac{2}{1-\nu}\left(\frac{a}{R}\right)^2 \frac{\partial^2}{\partial t^2} + \frac{\Omega^2}{\omega^2}\frac{\partial^4}{\partial t^4}\right]$$

where

$$u_r = \frac{\partial U}{\partial r} - r\psi \tag{2a}$$

$$u_\theta = \frac{1}{r}\frac{\partial U}{\partial \theta} \tag{2b}$$

and U is determined from

$$\frac{\partial^2 U}{\partial t^2} = -\frac{a^2 R}{(1+\nu)(1-\nu^2)}\frac{\omega^2}{\Omega^2}\left\{\frac{h^2}{12}\nabla^4 w + \frac{1-\nu^2}{a^2}\left[\left(\frac{a}{R}\right)^2 + \frac{\Omega^2}{\omega^2}\frac{\partial^2}{\partial t^2}\right] w - \frac{1-\nu^2}{2R}\, r\, \frac{\partial \psi}{\partial r}\right\} \tag{3}$$

In the above equations w, u_r, u_θ are the transverse, meridional, and circumferential displacement components respectively, h denotes the thickness and R the radius of curvature of the shell; ν is Poisson's ratio, E is Young's modulus, ρ is mass density, ω is circular frequency, a denotes a characteristic length of the shell, t is time, and

$$\nabla^2 \equiv \frac{\partial^2}{\partial r^2} + \frac{1}{r}\frac{\partial}{\partial r} + \frac{1}{r^2}\frac{\partial^2}{\partial \theta^2}$$

$$\Omega^2 = \frac{\rho \omega^2 a^2}{E}$$

The tracer k identifies the contribution of the longitudinal inertia; if $k = 0$, then longitudinal inertia is neglected, otherwise $k = 1$.

Separable solutions for the displacement components are obtained from (2) with the aid of (3) and the solutions of (1) in the form

$$w = \sum_{\alpha=1}^{3} A_\alpha J_n\left(\mu_\alpha \frac{r}{a}\right)\cos n\theta \cos \omega t \tag{4a}$$

$$u_r = KR\frac{1}{r}\left\{\sum_{\alpha=1}^{3} A_\alpha \delta_\alpha\left[\mu_\alpha \frac{r}{a} J_{n-1}\left(\mu_\alpha \frac{r}{a}\right) - n J_n\left(\mu_\alpha \frac{r}{a}\right)\right] + A_4 n^2 J_n\left(\mu_4 \frac{r}{a}\right)\right\}\cos n\theta \cos \omega t \tag{4b}$$

$$u_\theta = -KR\frac{n}{r}\left\{\sum_{\alpha=1}^{3} A_\alpha \delta_\alpha J_n\left(\mu_\alpha \frac{r}{a}\right) + A_4\left[\mu_4 \frac{r}{a} J_{n-1}\left(\mu_4 \frac{r}{a}\right) - n J_n\left(\mu_4 \frac{r}{a}\right)\right]\right\}\sin n\theta \cos \omega t \tag{4c}$$

where A_α, A_4 are arbitrary constants, $J_n(r)$ denotes the Bessel function of the first kind, the parameters μ_1, μ_2, μ_3 are the roots of

$$\left(\frac{\mu}{a}\right)^6 - r_1\left(\frac{\mu}{a}\right)^4 + r_2\left(\frac{\mu}{a}\right)^2 - r_3 = 0 \tag{5a}$$

where in r_1, r_2, r_3 each second time derivative is replaced by $(-\omega^2)$, and

$$\delta_\alpha = \frac{1}{12}\left(\frac{h}{a}\right)^2 \mu_\alpha^4 + (1-\nu^2)\left[\left(\frac{a}{R}\right)^2 - \Omega^2\right] \tag{5b}$$

$$\mu_4 = \Omega\sqrt{2(1+\nu)} \tag{5c}$$

$$K = \frac{1}{(1+\nu)(1-\nu^2)\,\Omega^2} \tag{5d}$$

Since only shells that include the apex $(r=0)$ are considered here, the Bessel function of the second kind has been omitted from (4).

With the use of (4), stress resultants can easily be found from the stress-displacement relations of the theory of shallow spherical shells. Since only displacement boundary conditions will be considered in this paper, no stress resultants need be calculated here.

Frequency Equation – Classical Theory

The main part of this paper consists of the investigation of free vibrations of a shallow spherical shell with a clamped edge. The appropriate frequency equation is deduced with the aid of (4) from the four boundary conditions at the clamped edge $r = a$

$$w(a,\theta,t) = \left.\frac{\partial w}{\partial r}\right|_{r=a} = 0$$
$$u_r(a,\theta,t) = u_\theta(a,\theta,t) = 0 \tag{6}$$

Vanishing of the determinant of the coefficient matrix of (6) gives the frequency equation

$$[\mu_4 J_{n-1}(\mu_4) - nJ_n(\mu_4)]\,D_0 + n\mu_4 J_{n-1}(\mu_4)\,D_1 = 0 \tag{7}$$

where D_0 and D_1 are the determinants

$$D_0 = \begin{vmatrix} J_n(\mu_1) & J_n(\mu_2) & J_n(\mu_3) \\ \mu_1 J_{n-1}(\mu_1) & \mu_2 J_{n-1}(\mu_2) & \mu_3 J_{n-1}(\mu_3) \\ \delta_1\mu_1 J_{n-1}(\mu_1) & \delta_2\mu_2 J_{n-1}(\mu_2) & \delta_3\mu_3 J_{n-1}(\mu_3) \end{vmatrix} \tag{8a}$$

$$D_1 = \begin{vmatrix} J_n(\mu_1) & J_n(\mu_2) & J_n(\mu_3) \\ \mu_1 J_{n-1}(\mu_1) & \mu_2 J_{n-1}(\mu_2) & \mu_3 J_{n-1}(\mu_3) \\ \delta_1 J_n(\mu_1) & \delta_2 J_n(\mu_2) & \delta_3 J_n(\mu_3) \end{vmatrix} \tag{8b}$$

For the axisymmetric case $(n = 0)$ (7) reduces to

$$\mu_4 J_1(\mu_4)\,D_0 = 0 \tag{9}$$

where $D_0 = 0$, previously employed in [7, 8, 9], gives the natural frequencies for axisymmetric torsionless vibrations, and the zeros of $J_1(\mu_4)$ represent the natural frequencies of the axisymmetric torsional vibrations. It is seen from (7) that only for $n = 0$ the torsional modes are uncoupled.

The amplitudes of the displacements corresponding to a particular natural frequency are given by (4), where the arbitrary constants are calculated from

$$\frac{A_2}{A_1} = \frac{1}{D}[\mu_3 J_{n-1}(\mu_3)J_n(\mu_1) - \mu_1 J_{n-1}(\mu_1)J_n(\mu_3)] \tag{10a}$$

$$\frac{A_3}{A_1} = \frac{1}{D}[\mu_1 J_{n-1}(\mu_1)J_n(\mu_2) - \mu_2 J_{n-1}(\mu_2)J_n(\mu_1)] \tag{10b}$$

$$\frac{A_4}{A_1} = -\frac{1}{\mu_4 J_{n-1}(\mu_4)}\left[\delta_1\mu_1 J_{n-1}(\mu_1) + \frac{A_2}{A_1}\delta_2\mu_2 J_{n-1}(\mu_2) + \frac{A_3}{A_1}\delta_3\mu_3 J_{n-1}(\mu_3)\right] \tag{10c}$$

where

$$D = \mu_2 J_{n-1}(\mu_2)J_n(\mu_3) - \mu_3 J_n(\mu_2)J_{n-1}(\mu_3)$$

Frequency Equation – Transverse Vibrations

The theory of *transverse* vibrations of shallow shells permits the utilization of an Airy stress function [3] for the reduction of the primitive equations to a system of two equations involving the transverse displacement w and the stress function. For nonsymmetric vibrations the stress function approach is convenient for cases where the boundary conditions are imposed on the stress resultants and no explicit expressions of the longitudinal displacements u_r, u_θ are needed. Such cases were treated in [5].

For problems with displacement boundary conditions, u_r and u_θ for *transverse* vibrations may be determined by the scheme employed in [10]. The final system of equations consists of (1) when k is set equal to zero. Furthermore, for transverse vibrations with $n \geq 1$ (3) must be replaced by

$$\nabla^2 U = -\frac{1+\nu}{R}w + \frac{1+\nu}{2}r\frac{\partial\psi}{\partial r} + 2\psi \tag{11a}$$

and the following relation must be satisfied

$$\left[\nabla^4 - \left(\frac{\mu}{a}\right)^4\right]w = \frac{6(1-\nu^2)}{h^2 R}\,r\frac{\partial\psi}{\partial r} \tag{11b}$$

where

$$\mu = \left\{12(1-\nu^2)\left(\frac{a}{h}\right)^2\left[\Omega^2 - \left(\frac{a}{R}\right)^2\right]\right\}^{\frac{1}{4}} \tag{11c}$$

The separable solutions of the displacement components

are obtained from (2) with the aid of (1) (with $k = 0$) and (11) in the form

$$w = \left\{ \sum_{\alpha=1}^{2} A_\alpha J_n \left(\mu_\alpha \frac{r}{a}\right) + A_3 r^n \right\} \cos n\theta \cos \omega t \tag{12a}$$

$$u_r = \left\{ (1+\nu) \frac{a}{R} \sum_{\alpha=1}^{2} \frac{A_\alpha}{\mu_\alpha^2} \left[\mu_\alpha J_{n-1} \left(\mu_\alpha \frac{r}{a}\right) - n \frac{a}{r} J_n \left(\mu_\alpha \frac{r}{a}\right) \right] \right.$$

$$- \frac{n+2}{4(n+1)} \frac{r^{n+1}}{R} A_3 \left(\frac{R}{a}\right)^2 \left[\left(1+\nu-\frac{4}{n+2}\right) \Omega^2 + \frac{4}{n+2} \left(\frac{a}{R}\right)^2 \right]$$

$$\left. + A_4 n r^{n-1} \right\} \cos n\theta \cos \omega t \tag{12b}$$

$$u_\theta = -n \frac{a}{r} \left\{ (1+\nu) \frac{a}{R} \sum_{\alpha=1}^{2} \frac{A_\alpha}{\mu_\alpha^2} J_n \left(\mu_\alpha \frac{r}{a}\right) \right.$$

$$- \frac{r^{n+2}}{4(n+1) Ra} A_3 \left(\frac{R}{a}\right)^2 \left[\left(1+\nu+\frac{4}{n}\right) \Omega^2 - \frac{4}{n} \left(\frac{a}{R}\right)^2 \right]$$

$$\left. + A_4 \frac{r^n}{a} \right\} \sin n\theta \cos \omega t \tag{12c}$$

where $n \geq 1$ and $\mu_1 = \mu$, $\mu_2 = i\mu$, $i = \sqrt{-1}$.
Boundary conditions (6) give the following frequency equation for *transverse* vibration of a shallow spherical shell with a clamped edge when $n \geq 1$

$$\left[N_n - 4n \left(\frac{a}{R}\right)^2 \right] I_n(\mu) J_{n-1}(\mu) - \left[N_n + 4n \left(\frac{a}{R}\right)^2 \right] J_n(\mu) I_{n-1}(\mu)$$

$$+ \left(\frac{a}{R}\right)^2 \left[\frac{8n^2}{\mu} J_n(\mu) I_n(\mu) + 2\mu J_{n-1}(\mu) I_{n-1}(\mu) \right] = 0 \tag{13a}$$

where $I_n(\mu)$ denotes the modified Bessel function of the first kind and

$$N_n = \frac{\mu^2}{2(1+\nu)(n+1)} \left[(3-\nu) \Omega^2 - 4 \left(\frac{a}{R}\right)^2 \right] \tag{13b}$$

The corresponding frequency equation for $n = 0$, which does not follow from (13) by setting $n = 0$, was previously derived and investigated in detail in [4].

Transverse and Longitudinal Modes for Shallow Shells

In order to distinguish between the transverse and the longitudinal modes of free vibration of a shallow spherical shell, it is of interest to examine the limiting expressions of (7) and (13) as $R \to \infty$.

When $R = \infty$, (5a) takes the form

$$[\mu^2 - (1-\nu^2)\Omega^2] \left[\mu^4 - 12(1-\nu^2) \left(\frac{a}{h}\right)^2 \Omega^2 \right] = 0 \tag{14}$$

from which it follows that

$$\mu_1 = \mu_0 \ ; \ \mu_2 = i\mu_0 \ ; \ \mu_3 = (1-\nu^2)\Omega^2 \tag{15a}$$

where

$$\mu_0 = \left[12(1-\nu^2) \left(\frac{a}{h}\right)^2 \Omega^2 \right]^{\frac{1}{4}}$$

and (5b) gives

$$\delta_1 = \delta_2 = 0 \ ; \ \delta_3 = -(1-\nu^2)\Omega^2 \tag{15b}$$

With the aid of (15), the determinants of (8) can be written as

$$D_0 = \mu_3 \delta_3 J_{n-1}(\mu_3) D_2 \tag{16a}$$

$$D_1 = n \delta_3 J_n(\mu_3) D_2 \tag{16b}$$

where

$$D_2 = \mu_2 J_n(\mu_1) J_{n-1}(\mu_2) - \mu_1 J_n(\mu_2) J_{n-1}(\mu_1) \tag{16c}$$

Thus, in the limit as $R \to \infty$, (7) reduces to

$$\delta_3 D_2 D_3 = 0 \tag{17a}$$

where

$$D_3 = \mu_3 J_{n-1}(\mu_3)[\mu_4 J_{n-1}(\mu_4) - n J_n(\mu_4)] + n \mu_4 J_n(\mu_4) \tag{17b}$$

and $\mu_1, \mu_2, \mu_3, \delta_3$ have the special values given by (15).

The corresponding limit of (13), which were deduced from the theory of *transverse* vibrations, as $R \to \infty$, is

$$\frac{(3-\nu)\mu_0 \Omega^2}{2(1+\nu)(n+1) i^n} D_2 = 0 \tag{18}$$

It should be noted that the natural frequencies given by (17) are determined from the vanishing of two uncoupled expressions. The first, $D_2 = 0$, depends on the parameter μ_0 and is the well-known frequency equation of transverse vibrations of a circular plate with a clamped edge.[3] The second, $D_3 = 0$, contains only μ_3 and μ_4, and it gives the natural frequencies of the longitudinal modes of a circular plate. An examination of (18) reveals that the frequency equation (13) in the limit as $R \to \infty$ can only predict the transverse modes of a circular plate. Since a continuous transition from a spherical shell to a plate should give a continuous natural frequency, it is clear that the frequency equation for a shallow spherical shell derived from the theory of *transverse* vibrations cannot predict that part of the complete frequency spectrum which reduces to the longitudinal modes of a plate when $R \to \infty$.

From the above analysis the longitudinal modes of a shallow shell can be defined as those modes which, when $R \to \infty$, coincide with the longitudinal modes of a plate. Similarly, the transverse modes of a shallow shell are those modes which, when $R \to \infty$, coincide with the transverse modes of a plate[4]. Thus, by definition, the

[3] The general theory of transverse vibrations of a circular plate was obtained by Kirchhoff [11]; convenient expressions for the calculation of the natural frequencies were given by Airey [12]. Longitudinal vibrations of plates were studied earlier by Poisson, and are included in [13].

[4] For a shallow shell which is not spherical, R for this definition should be regarded as any principal radius of curvature.

TABLE I

NATURAL FREQUENCIES Ω_m FOR $\nu = 0.3$, $a/h = 10$

			Classical				Transverse
	m	Type	$a/R = 0$	0.1	0.3	0.5	0.5
	1	T	0.3091	0.3386	0.5205	0.7515	0.7580
	2	T	1.204	1.205	1.247	1.331	1.339
	3	L_t	2.376	2.376	2.376	2.376	–
$n = 0$	4	T	2.696	2.698	2.711	2.737	2.749
	5	L_r	4.017	4.018	4.032	4.060	–
	6	L_t	4.351	4.351	4.351	4.351	–
	7	T	4.787	4.789	4.796	4.814	4.815
	1	T	0.6434	0.6520	0.7140	0.8195	0.8412
	2	T	1.841	1.841	1.844	1.852	1.913
	3	L	2.052	2.054	2.087	2.151	–
$n = 1$	4	L	3.332	3.332	3.332	3.334	–
	5	T	3.632	3.637	3.648	3.667	3.670
	6	L	5.250	5.251	5.253	5.255	–
	7	L	5.628	5.629	5.634	5.650	–
	1	T	1.056	1.060	1.099	1.168	1.182
	2	T	2.560	2.562	2.571	2.589	2.611
$n = 2$	3	L	3.195	3.196	3.208	3.230	–
	4	L	4.283	4.283	4.284	4.288	–
	5	T	4.655	4.657	4.664	4.683	4.683
	1	T	1.548	1.550	1.572	1.622	1.633
	2	T	3.360	3.361	3.369	3.383	3.399
$n = 3$	3	L	4.154	4.155	4.160	4.173	–
	4	L	5.266	5.267	5.269	5.273	–
	5	T	5.759	5.760	5.766	5.780	5.783

TABLE II

NATURAL FREQUENCIES Ω_m FOR $\nu = 0.3$, CLASSICAL THEORY

			$a/h = 20$			$a/h = 5$	
	m	Type	$a/R = 0$	0.5	Type	$a/R = 0$	0.5
	1	T	0.1545	0.6561	T	0.6182	0.9291
	2	T	0.6020	0.8675	L_t	2.376	2.376
	3	T	1.348	1.449	T	2.408	2.457
$n = 0$	4	L_t	2.376	2.376	L_r	4.017	4.058
	5	T	2.394	2.444	L_t	4.351	4.351
	6	T	3.738	3.757	T	5.392	5.410
	7	L_r	4.017	4.077	L_t	6.309	6.309
	1	T	0.3217	0.6161	T	1.287	1.346
	2	T	0.9205	1.047	L	2.052	2.110
	3	T	1.816	1.857	L	3.332	3.333
$n = 1$	4	L	2.052	2.135	T	3.682	3.716
	5	T	3.012	3.056	L	5.250	5.254
	6	L	3.332	3.336	L	5.628	5.655
	7	T	4.506	4.529	L	7.261	7.264
	1	T	0.5280	0.7361	T	2.112	2.144
	2	T	1.280	1.368	L	3.195	3.222
$n = 2$	3	T	2.327	2.373	L	4.283	4.288
	4	L	3.195	3.221	T	5.120	5.137
	5	T	3.672	3.713	L	6.142	6.144
	1	T	0.7740	0.9293	T	3.096	3.103
	2	T	1.680	1.751	L	4.154	4.172
$n = 3$	3	T	2.879	2.919	L	5.266	5.275
	4	L	4.154	4.160	T	6.720	6.724
	5	T	4.375	4.410	L	7.011	7.019

modes predicted by the classical theory of shallow shells are only either longitudinal or transverse. It follows from the results of this paper that the predominant character of these modes for a shallow spherical shell is indeed either longitudinal or transverse. This is expected to be true only if the shell is shallow; non-shallow shells, as for example hemispherical shells which were treated in [14], are expected to possess modes with a character other than just transverse or longitudinal.

Numerical Results

The actual values of the natural frequencies of a shallow spherical shell with a clamped edge can be found from (7) by evaluating numerically the left-hand side of (7) as a function of the nondimensional frequency parameter Ω; the natural frequencies Ω_m $(m = 1, 2, \ldots)$ are those values of Ω which satisfy (7). The displacement amplitudes corresponding to a particular value of Ω_m are then calculated from (4) with the aid of (10). In the present paper, calculations of Ω_m are carried out for $\nu = 0.3$ and various values of h/R, a/h, and n. For each set of parameters the natural frequencies given by (7) within the frequency band bounded by $\Omega = 0$ and approximately $\Omega = 6$ are determined and shown in Tables I and II. Similarly, all the natural frequencies predicted by the theory of *transverse* vibrations in the same frequency band are calculated from (13) and also included in Table I. Only the case $a/R = 0.5$ is shown since it represents the maximum deviation from the exact frequency. The frequencies in the Tables corresponding to $a/R = 0$ are the natural frequencies of both $D_2 = 0$ and $D_3 = 0$. Those of $D_2 = 0$ (transverse) are identified by the letter T and those given by $D_3 = 0$ (longitudinal) are marked L. Moreover, for $n = 0$, the longitudinal modes further uncouple into torsional and radial modes which are identified as L_t and L_r, respectively. While most of the numbers given in the tables are accurate to four significant figures, an error analysis has shown that the error is not more than one unit in the fourth significant figure.

Some important features of the natural frequencies of transverse and longitudinal modes can be best seen from Figs. 1–4 where the variation of the frequencies with the thickness of the shell is shown for $\nu = 0.3$, $a/R = 0.5$, and $n = 0, 1, 2, 3$. The slanted curves in each figure represent the transverse modes, while the nearly horizontal curves are the longitudinal modes. The dotted lines are the corresponding natural frequencies of transverse modes of a plate, and the vertical distance between a dotted line and the nearest full curve represents the rise of the natural frequency due to curvature.

In order to demonstrate that the transverse displace-

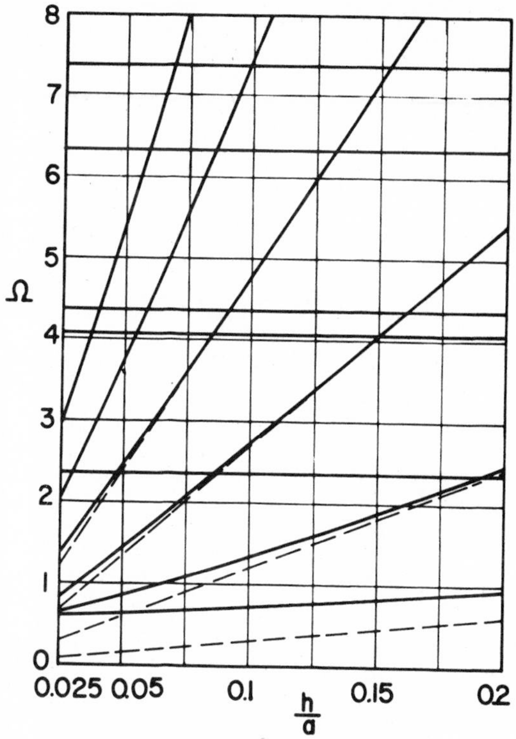

FIGURE 1. NATURAL FREQUENCIES Ω_m VS. RATIO OF THICKNESS TO BOUNDING RADIUS FOR $\nu = 0.3$, $a/R = 0.5$, $n = 0$.

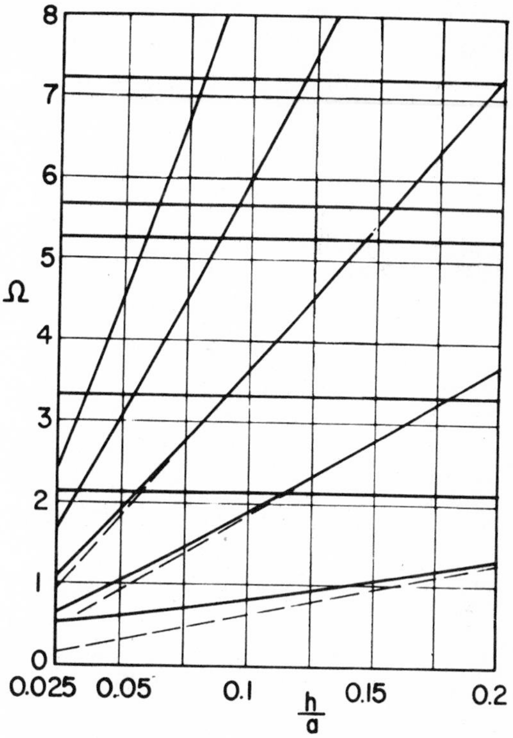

FIGURE 2. NATURAL FREQUENCIES Ω_m VS. RATIO OF THICKNESS TO BOUNDING RADIUS FOR $\nu = 0.3$, $a/R = 0.5$, $n = 1$.

ment is predominant in transverse modes and the longitudinal displacements are predominant in the longitudinal modes, the r-dependent parts of the displacement amplitudes are calculated for $\nu = 0.3$, $a/R = 0.5$, $a/h = 10$, $n = 0$, 1 and shown in Figs. 5–6. The ordinates are not marked because the curves depict relative amplitudes and the absolute values of the ordinates are meaningless. It should be noted that in Fig. 6 u_r, u_θ, and the slope of w do not vanish at $r = 0$. It can be readily seen from (4) that this is only true for $n = 1$, since Bessel functions near zero argument behave as

$$\lim_{r \to 0} J_n\left(\frac{\mu r}{a}\right) = \lim_{r \to 0} \left(\frac{\mu r}{a}\right)^n$$

For the purpose of studying the effect of the curvature on the natural frequencies of the shell, Figs. 7 and 8 show for $a/h = 10$, $\nu = 0.3$, $n = 0$, 1 the natural frequency curves as functions of a/R. The case $a/R = 0.5$ represents the minimum allowable curvature for a shallow shell which was suggested by Reissner [4]. The corresponding curves for $n = 2$, 3 are similar to those of Fig. 8.

Conclusions

The natural frequencies given in Tables I and II indicate a considerable effect of curvature on the lowest frequency for all wave numbers considered. For higher modes the effect of curvature is decreasing with increasing m as shown in Figs. 7 and 8. The variation of the natural frequencies with the thickness of the shell, given in Figs. 1–4, shows that the transverse modes of a plate, indicated by dotted lines going through the origin, are asymptotes to the corresponding transverse modes of the shallow spherical shell. The increasing effect of curvature on the frequencies when the thickness is decreasing should be noted.

The relative displacement amplitudes shown in Fig. 5 include also the purely torsional axisymmetric modes ($m = 3$, 6) whose natural frequencies are given by (9).

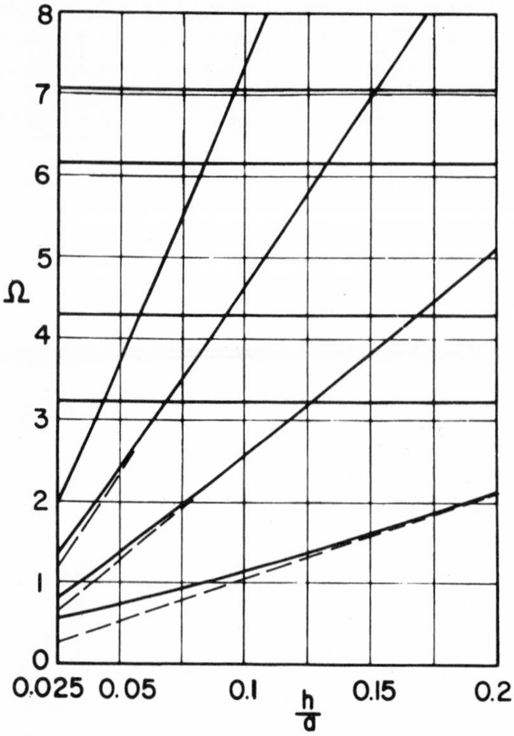

FIGURE 3. NATURAL FREQUENCIES Ω_m VS. RATIO OF THICKNESS TO BOUNDING RADIUS FOR $\nu = 0.3$, $a/R = 0.5$, $n = 2$.

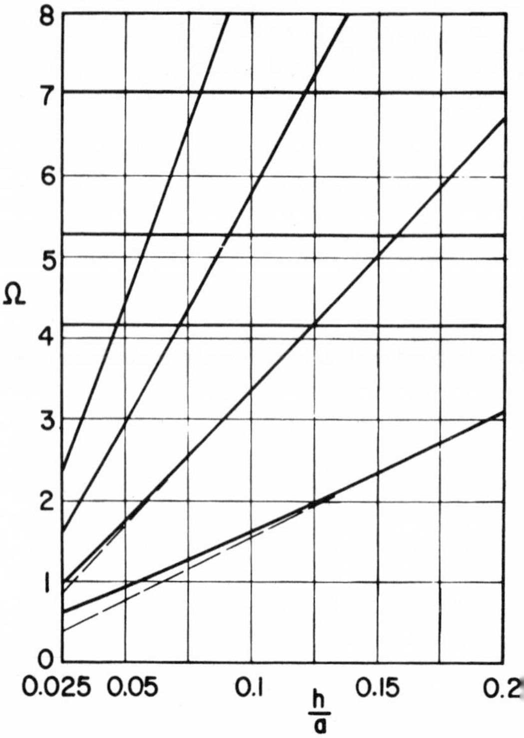

FIGURE 4. NATURAL FREQUENCIES Ω_m VS. RATIO OF THICKNESS TO BOUNDING RADIUS FOR $\nu = 0.3$, $a/R = 0.5$, $n = 3$.

Examination of the mode shapes of Fig. 6 reveals that in the absence of axial symmetry there are no modes for which the torsional motion u_θ is dominant. It is interesting to note that in Figs. 5 and 6 the ratios of the maximum values of the transverse to the longitudinal displacements for the transverse modes are larger than the inverse ratios for the longitudinal modes.

The results of this paper show that a part of the complete natural frequency spectrum (i.e., the transverse modes) of a shallow spherical shell can be calculated with small error from the theory of *transverse* vibrations, and that the predicted frequencies are always higher than those of the classical theory. The error is larger for the nonsymmetric ($n = 1$, 3) than for the symmetric ($n = 0,2$) modes and the cases considered show a maximum error of 3.3 per cent in the second mode for $n = 1$. However, when the theory of *transverse* vibrations is used, a part of the frequency spectrum (i.e., the longitudinal modes) cannot be obtained.

Acknowledgment

The author wants to express his appreciation to the Department of Civil Engineering of Yale University and to the Yale Computer Center for funds made available to carry out the numerical computations of this paper. Support of the National Science Foundation, Grant NSF # G – 23922, is also gratefully acknowledged.

References

1. Naghdi, P. M., "On the General Problem of Elastokinetics in the Theory of Shallow Shells," *Proceedings of the IUTAM Symposium on the Theory of Thin Elastic Shells*, North Holland Publishing Co., 1960, pp. 301-330.

2. Nordgren, R. P., "On Vibrations of Pretwisted Rectangular Plates," *Journal of Applied Mechanics*, Vol. 29, 1962, pp. 30-32.

3. Reissner, E., "On Transverse Vibrations of Thin Shallow Elastic Shells," *Quarterly of Applied Mathematics*, Vol. 13, 1955, pp. 169-176.

4. Reissner, E., "On Axi-symmetrical Vibrations of Shallow Spherical Shells," *Quarterly of Applied Mathematics*, Vol. 13, 1955, pp. 279-290.

5. Johnson, M. W. and Reissner, E., "On Transverse

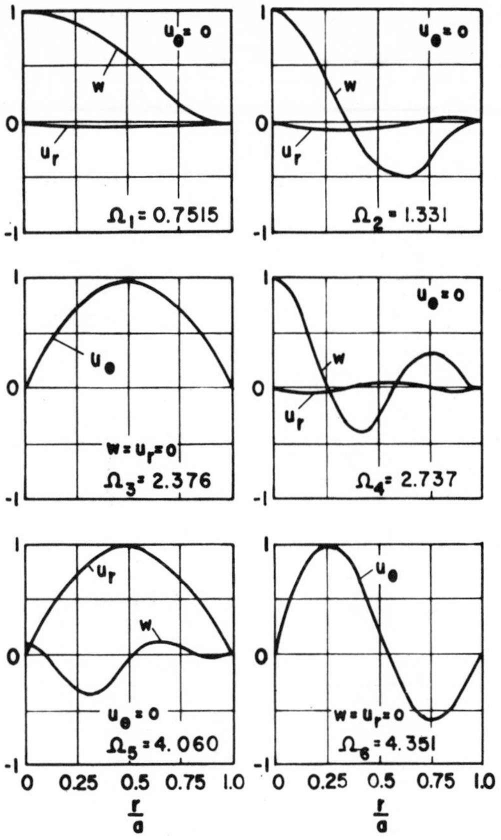

FIGURE 5. RELATIVE AMPLITUDES OF DISPLACEMENTS VS. RADIAL DISTANCE FROM APEX FOR $\nu = 0.3$, $a/R = 0.5$, $a/h = 10$, $n = 0$.

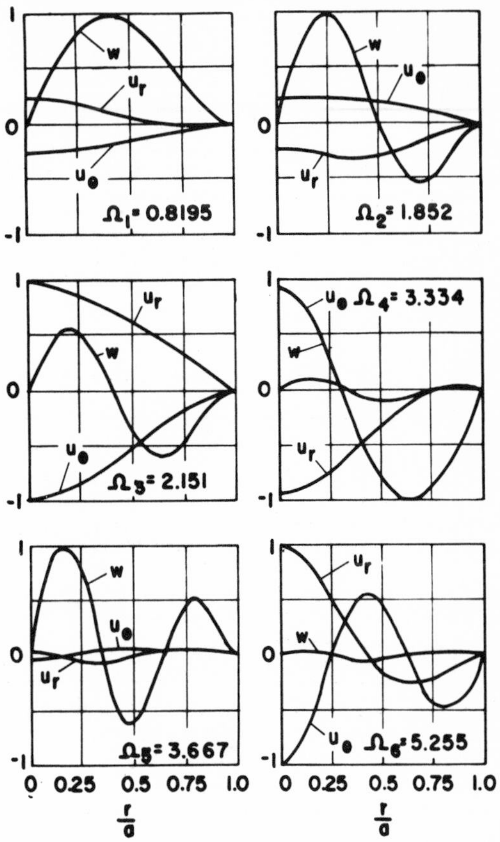

FIGURE 6. RELATIVE AMPLITUDES OF DISPLACEMENTS VS. RADIAL DISTANCE FROM APEX FOR $\nu = 0.3$, $a/R = 0.5$, $a/h = 10$, $n = 1$.

Vibrations of Shallow Spherical Shells," *Quarterly of Applied Mathematics*, Vol. 15, 1958, pp. 367-380.

6. Federhofer, K., "Zur Berechnung der Eigenschwingungen der Kugelschale," *Akademie der Wissenschaften in Wien, Sitzungsberichte, Math.-naturw. Klasse*, Vol. 146:2A, 1937, pp. 57-69.

7. Reissner, E., "On Vibrations of Shallow Spherical Shells," *Journal of Applied Physics*, Vol. 17, 1946, pp. 1038-1042.

8. Kalnins, A. and Naghdi, P. M., "Axisymmetric Vibrations of Shallow Elastic Spherical Shells," *Journal of the Acoustical Society of America*, Vol. 32, 1960, pp. 342-347.

9. Hoppmann II, W. H., "Frequencies of Vibration of Shallow Spherical Shells," *Journal of Applied Mechanics*, Vol. 28, 1961, pp. 305-307.

10. Kalnins, A., "On Vibrations of Shallow Spherical Shells," *Journal of the Acoustical Society of America*, Vol. 33, 1961, pp. 1102-1107.

11. Kirchhoff, G, "Über das Gleichgewicht und die Bewegung einer elastischen Scheibe," *Journal für die reine und angewandte Mathematik*, Vol. 40, 1850, pp. 51-88.

12. Airey, J. R., "The Vibrations of Circular Plates and their Relation to Bessel Functions," *Proceedings of the Physical Society of London*, Vol. 23, 1911, pp. 225-232.

13. Love, A. E. H., *A Treatise on the Mathematical Theory of Elasticity*, Fourth Edition, Dover Publications, 1944, p. 498.

14. Naghdi, P. M. and Kalnins, A., "On Vibrations of Elastic Spherical Shells," to appear in the Journal of Applied Mechanics, Vol. 29, 1962, pp. 65-72.

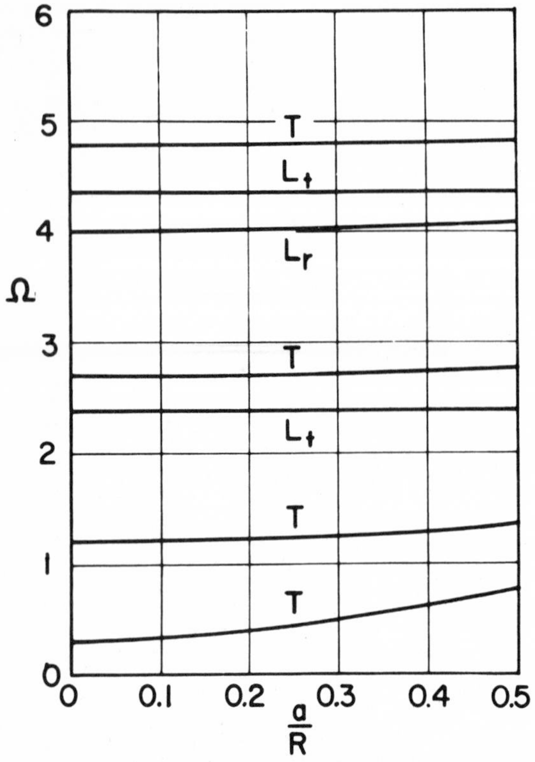

FIGURE 7. NATURAL FREQUENCIES Ω_m VS. RATIO OF BOUNDING RADIUS TO RADIUS OF CURVATURE FOR $\nu = 0.3$, $a/h = 10$, $n = 0$.

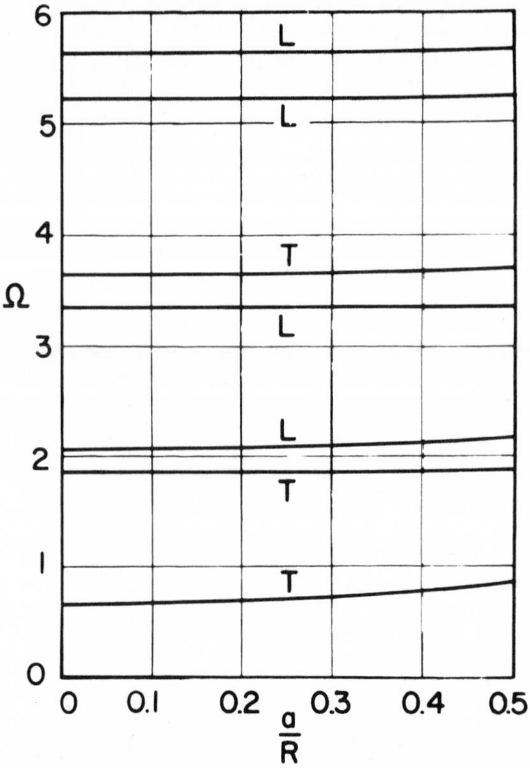

FIGURE 8. NATURAL FREQUENCIES Ω_m VS. RATIO OF BOUNDING RADIUS TO RADIUS OF CURVATURE FOR $\nu = 0.3$, $a/h = 10$, $n = 1$.

23

Reprinted from *J. Acoust. Soc. Amer.*, **36**(1), 74–81 (1964)

Effect of Bending on Vibrations of Spherical Shells

A. KALNINS

Department of Engineering and Applied Science, Yale University, New Haven, Connecticut

(Received 11 July 1963)

This paper is concerned with the vibration analysis of spherical shells, closed at one pole and open at the other, by means of the linear classical bending theory of shells. Frequency equations are derived in terms of Legendre functions with complex indices, and for axisymmetric vibration the natural frequencies and mode shapes are deduced for opening angles ranging from a shallow to a closed spherical shell. It is found that for all opening angles the frequency spectrum consists of two coupled infinite sets of modes that can be labeled as bending (or flexural) and membrane modes. This distinction is made on the basis of the comparison of the strain energies due to bending and stretching of each mode. It is also found that the membrane modes are practically independent of thickness, whereas the bending modes vary with thickness. Previous analyses with the use of membrane theory have shown that one of two infinite sets of modes is spaced within a finite interval of the frequency spectrum. It is shown in this paper that this set of modes is a degenerate case of bending modes, and, if deduced by means of membrane theory, it is applicable only when the thickness of the shell is zero. When the bending theory is employed, then the frequency interval for this set of modes extends to infinity for every value of thickness that is greater than zero.

INTRODUCTION

ALTHOUGH the first papers on the vibration of thin, elastic, spherical shells precede the general formulation of the classical bending theory of shells, detailed investigations on this subject are still limited to some special configurations. In particular, only the vibration problems of closed and shallow spherical shells have received considerable previous attention. The closed shell was first examined by Lamb[1] by means of the membrane theory, and then by Federhofer,[2] who employed the classical bending theory of shells. More-detailed treatments of vibration of a closed spherical shell were given by Silbiger[3,4] and by Baker,[5] who based their analyses on the results of Ref. 1. Investigations of axisymmetric, as well as nonsymmetric, vibrations of shallow spherical shells within the scope of the classical bending theory are contained in numerous papers. A complete analysis of the natural frequencies and the mode shapes of free vibration of shallow spherical shells, together with references to other contributions, is given in a paper by the present author.[6] In the case of nonshallow spherical shells, a system of equations of bending theory is derived for axisymmetric vibration by Naghdi and Kalnins.[7] In Ref. 7, the lowest natural frequency of a hemispherical shell is calculated as a function of thickness, and calculations of natural frequencies by means of the membrane theory are also included. In addition, some results on the vibration of a hemispherical shell may be found in papers by Love[8] and by Zwingli.[9]

In all the references cited above, where the membrane theory is employed, the following conclusions are reached: (a) the natural frequency spectrum consists of two infinite sets of modes; (b) one set of an infinite number of modes is spaced within a finite frequency interval. The second conclusion means that the intervals between the natural frequencies may become infinitely small, and such a situation appears to be physically meaningless.

It is the main object of this paper to present a com-

[1] H. Lamb, Proc. London Math. Soc. **14**, 50–56 (1882).
[2] K. Federhofer, Sitzber. Akad. Wiss. Wien, Math.-Naturw. Kl. **2A**, **146**, 57–69 (1937).
[3] A. Silbiger, J. Acoust. Soc. Am. **34**, 862 (1962).
[4] A. Silbiger, "Free and Forced Vibrations of a Spherical Shell," CAA, Inc., Rept. U-106-48 (Dec. 1960).
[5] W. E. Baker, J. Acoust. Soc. Am. **33**, 1749–1758 (1961).
[6] A. Kalnins, Proc. Natl. Congr. Appl. Mech. 4th U. S., 225–233 (1963).
[7] P. M. Naghdi and A. Kalnins, J. Appl. Mech. **29**, 65–72 (1962).
[8] A. E. H. Love, Phil. Trans. Roy. Soc. (London) **A179**, 491–546 (1888).
[9] H. Zwingli, "Elastische Schwingungen von Kugelschalen," dissertation, Tech. Hochschule in Zürich (1930).

plete analysis of the natural frequencies and the mode shapes of free axisymmetric vibration of a nonshallow spherical shell having an axisymmetric configuration, and to reexamine the above conclusions with the use of the classical bending theory of shells. It is shown in the present paper that if the effects of bending are included in the analysis then, for spherical shells: (a) the natural frequency spectrum can be divided into two infinite sets of modes, which are interspersed and distributed over the infinite frequency range; (b) the intervals between the natural frequencies of these sets do not approach zero.

The analysis of this paper is concerned with the vibration of a thin spherical shell that has a circular opening, axisymmetrically located, around one of its poles. The conical angle ϕ_0 of the opening may be in the interval defined by $0° < \phi_0 \leq 180°$. Explicit solutions in terms of Legendre functions are obtained from the general equations derived in Refs. 2 and 7, and the frequency equations are deduced for various natural boundary conditions imposed on the edge of the opening for an arbitrary opening angle. By setting the bending stiffness in the solutions equal to zero, the corresponding expressions given by membrane theory are obtained. Then, from the derived frequency equations, the natural frequencies and their corresponding mode shapes are calculated as functions of the opening angle and the thickness of the shell. In addition, the strain energy due to the bending and stretching associated with each mode is calculated and a bending-energy coefficient is obtained, which represents the ratio of the bending energy contained in the total energy of each mode.

The results show that, in analogy to the results predicted by the membrane theory, the modes of free vibration can again be classified into two groups, which are here designated as the bending and the membrane modes. This distinction is made on the basis of the value of the bending-strain-energy coefficient. The distinct difference in the character of the bending and membrane modes is substantiated by the analysis of the corresponding mode shapes of each mode. Moreover, if in the results obtained by the bending theory the thickness of the shell is set equal to zero, then it is established that the bending modes degenerate to an infinite set of modes within a finite frequency interval, and that they are identical to the lower set of modes predicted by the membrane theory. It is concluded that, while the membrane theory of shells is capable of predicting accurately the membrane modes, the bending modes may be considerably in error if the bending effects are omitted.

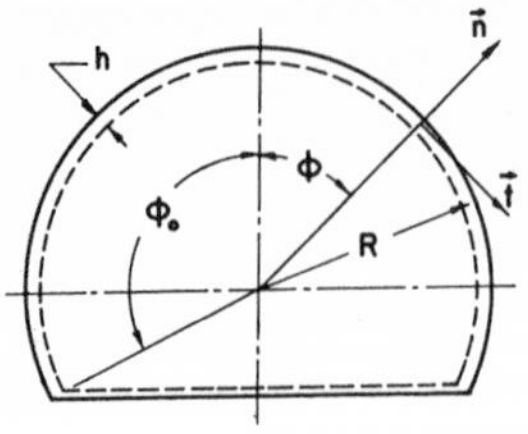

FIG. 1. Geometry of a spherical shell.

SOLUTIONS BY BENDING AND MEMBRANE THEORIES

The separable homogeneous solutions to the axisymmetric vibration problem of thin elastic spherical shells with the inclusion of the effects of bending can be obtained from the equations given in Refs. 2 and 7. If y denotes any relevant dependent variable, the solutions can be written in the following form:

$$y(\phi,t) = \sum_{i=1}^{3} y_i(\phi)\cos\omega t, \tag{1}$$

where ω is the circular frequency, t is time, and ϕ denotes the angle measured from the z axis, as shown in Fig. 1. The actual ϕ-dependent coefficients of the variables are given by

$$w_i = A_i P_{n_i}(x) + B_i Q_{n_i}(x), \tag{2a}$$

$$u_i = -(1+\nu)C_i[A_i P_{n_i}'(x) + B_i Q_{n_i}'(x)], \tag{2b}$$

$$N_{\phi i} = \frac{Eh}{(1-\nu)R}\{(1+C_i\lambda_i)[A_i P_{n_i}(x) + B_i Q_{n_i}(x)] + (1-\nu)\cot\phi\, C_i[A_i P_{n_i}'(x) + B_i Q_{n_i}'(x)]\}, \tag{2c}$$

$$N_{\theta i} = \frac{Eh}{(1-\nu)R}\{(1+\nu C_i\lambda_i)[A_i P_{n_i}(x) + B_i Q_{n_i}(x)] - (1-\nu)\cot\phi\, C_i[A_i P_{n_i}'(x) + B_i Q_{n_i}'(x)]\}, \tag{2d}$$

$$M_{\phi i} = \frac{D}{R^2}[1+(1+\nu)C_i]\{\lambda_i[A_i P_{n_i}(x) + B_i Q_{n_i}(x)] + (1-\nu)\cot\phi[A_i P_{n_i}'(x) + B_i Q_{n_i}'(x)]\}, \tag{2e}$$

$$M_{\theta i} = \frac{D}{R^2}[1+(1+\nu)C_i]\{\nu\lambda_i[A_i P_{n_i}(x) + B_i Q_{n_i}(x)] - (1-\nu)\cot\phi[A_i P_{n_i}'(x) + B_i Q_{n_i}'(x)]\}, \tag{2f}$$

$$Q_i = \frac{D}{R^3}[1+(1+\nu)C_i](\nu+\lambda_i-1) \times [A_i P_{n_i}'(x) + B_i Q_{n_i}'(x)], \tag{2g}$$

where

$$C_i = \frac{1+(\lambda_i-2)/[(1+\nu)(1+\xi)]}{1-\nu-\lambda_i+\xi(1-\nu^2)\Omega^2/(1+\xi)}, \tag{3a}$$

$$n_i = -0.5 + (0.25+\lambda_i)^{\frac{1}{2}}, \tag{3b}$$

$$\xi = 12R^2/h^2, \tag{3c}$$

$$x = \cos\phi, \tag{3d}$$

and the parameters λ_i are the three roots of the cubic

$$\lambda^3-[4+(1-\nu^2)\Omega^2]\lambda^2$$
$$+[4+(1-\nu)(1-\nu^2)\Omega^2+(1+\xi)(1-\nu^2)(1-\Omega^2)]\lambda$$
$$+(1-\nu)(1-\nu^2)\left[\Omega^2-\frac{2}{1-\nu}\right]$$
$$\times\left[1+\xi(1+\nu)\left(\Omega^2+\frac{1}{1+\nu}\right)\right]=0. \quad (4)$$

The nondimensional frequency parameter is defined as

$$\Omega^2=\rho\omega^2R^2/E.$$

In the above equations, w, u denote the displacements of the middle surface in the direction of n and t (see Fig. 1); N_ϕ, N_θ are the membrane stress resultants; M_ϕ, M_θ are the moment resultants; Q is the transverse shear resultant; ν is Poisson's ratio; E is Young's modulus; h is the thickness of the shell; R denotes the radius of curvature of the middle surface; ρ is the mass density of the shell; and D is the bending stiffness defined by

$$D=Eh^3/[12(1-\nu^2)].$$

The symbols $P_n(x)$, $Q_n(x)$ denote the Legendre functions of the first and second kind, respectively, as defined in Kratzer and Franz,[10] and a prime denotes derivatives with respect to ϕ; A_i and B_i are arbitrary constants.

If in the above equations the bending stiffness is set equal to zero, the corresponding, separable, homogeneous solutions according to the membrane theory of spherical shells are obtained in the form

$$w=[AP_n(x)+BQ_n(x)]\cos\omega t, \quad (5a)$$

$$u=C[AP_n'(x)+BQ_n'(x)]\cos\omega t, \quad (5b)$$

$$N_\phi=-\frac{EhC}{(1+\nu)R}\{\cot\phi[AP_n'(x)+BQ_n'(x)]$$
$$+[1+(1+\nu)\Omega^2][AP_n(x)+BQ_n(x)]\}\cos\omega t, \quad (5c)$$

$$N_\theta=\frac{EhC}{(1+\nu)R}\left\{\cot\phi[AP_n'(x)+BQ_n'(x)]\right.$$
$$\left.+\frac{1+(1+2\nu)\Omega^2+\nu(1+\nu)\Omega^4}{1-\Omega^2}[AP_n(x)+BQ_n(x)]\right\}$$
$$\times\cos\omega t, \quad (5d)$$

where A and B are arbitrary constants and

$$C=(1-\Omega^2)/[1+(1+\nu)\Omega^2], \quad (6a)$$

$$n=-0.5+\{2.25+(1+\nu)\Omega^2$$
$$\times[3-(1-\nu)\Omega^2]/(1-\Omega^2)\}^{\frac{1}{2}}. \quad (6b)$$

[10] A. Kratzer and W. Franz, *Transzendente Funktionen* (Geest and Portig, Leipzig, 1960).

It should be added that some of the roots of Eq. (4) and the corresponding Legendre functions are, in general, complex. However, since one root of the cubic is always real, and if the other two roots are complex then they are complex conjugates, it is possible to write all physical variables in terms of real functions, which in this case are the real and imaginary parts of the Legendre functions. For this purpose, the arbitrary constants A_i and B_i are expressed as complex conjugates in terms of other real arbitrary constants.

The solutions of Eqs. (2) and (5) can be applied to the study of free vibration of an elastic spherical shell bounded, in general, by any two concentric openings. In this paper, the opening around $\phi=0$ is regarded as being degenerated to a point. Since the Legendre function of the second kind is singular at $\phi=0$, then the arbitrary constants B in Eqs. (2) and (5) must be set equal to zero. For this reason, in the remainder of the paper the terms involving $Q_n(x)$ are omitted.

FREQUENCY EQUATIONS

The solutions that include bending, derived above, are now applied to the study of free vibration of a spherical shell closed at one pole and open at the other. The frequency equation consists of the determinantal equation

$$|D_{ji}|=0 \quad (j, i=1,2,3), \quad (7)$$

where the elements D_{ji} depend on the boundary conditions prescribed at the open edge $\phi=\phi_0$. After the natural frequency, which satisfies Eq. (7), is determined, the corresponding mode shapes for all dependent variables are given by Eqs. (2), where the arbitrary constants A_2 and A_3 are obtained in terms of a single multiplier A_1 in the form

$$A_2=A_1(D_{21}D_{33}-D_{31}D_{23})/D_0, \quad (8a)$$

$$A_3=A_1(D_{31}D_{22}-D_{21}D_{32})/D_0, \quad (8b)$$

where

$$D_0=D_{32}D_{23}-D_{22}D_{33}. \quad (9)$$

The elements D_{ji} for various conditions are:

(a) **Clamped edge:** $w(x_0)=u(x_0)=w'(x_0)=0.$

$$D_{1i}=P_{n_i}(x_0), \quad (10a)$$

$$D_{2i}=C_iP_{n_i}'(x_0), \quad (10b)$$

$$D_{3i}=P_{n_i}'(x_0). \quad (10c)$$

(b) **Free edge:** $Q(x_0)=N_\phi(x_0)=M_\phi(x_0)=0.$

$$D_{1i}=[1+(1+\nu)C_i][\nu+\lambda_i-1]P_{n_i}'(x_0), \quad (11a)$$

$$D_{2i}=(1-\nu)\cot\phi_0C_iP_{n_i}'(x_0)$$
$$+(1+C_i\lambda_i)P_{n_i}(x_0), \quad (11b)$$

$$D_{3i}=[1+(1+\nu)C_i]\lambda_iP_{n_i}(x_0)$$
$$+(1-\nu)\cot\phi_0P_{n_i}'(x_0). \quad (11c)$$

(c) **Roller-hinged edge**: $w(x_0)=N_\phi(x_0)=M_\phi(x_0)=0$.

$$D_{1i}=P_{n_i}(x_0), \quad (12a)$$

$$D_{2i}=(1-\nu)\cot\phi_0 C_i P_{n_i}'(x_0)+(1+C_i\lambda_i)P_{n_i}(x_0), \quad (12b)$$

$$D_{3i}=[1+(1+\nu)C_i]\lambda_i P_{n_i}(x_0) + (1-\nu)\cot\phi_0 P_{n_i}'(x_0). \quad (12c)$$

(d) **Fixed-hinged edge**: $w(x_0)=u(x_0)=M_\phi(x_0)=0$.

$$D_{1i}=P_{n_i}(x_0), \quad (13a)$$

$$D_{2i}=C_i P_{n_i}'(x_0), \quad (13b)$$

$$D_{3i}=[1+(1+\nu)C_i]\lambda_i P_{n_i}(x_0) + (1-\nu)\cot\phi_0 P_{n_i}'(x_0). \quad (13c)$$

In the above equations, $x_0=\cos\phi_0$, and C_i and n_i are given by Eqs. (3a) and (3b), respectively.

When the problem of free vibration of a spherical shell is solved by means of the membrane theory, the solutions given by Eqs. (5) show that not all of the four boundary conditions given above can be satisfied. Clearly, since the membrane theory assumes that $Q=M_\phi=0$, then the only other variable that may be prescribed at $\phi=\phi_0$ is either u or N_ϕ. The case of $N_\phi(x_0)=0$ in membrane theory corresponds to a free edge, and the natural frequencies for this case are obtained from

$$\cot\phi_0 P_n'(x_0)+[1+(1+\nu)\Omega^2]P_n(x_0)=0. \quad (14)$$

The remainder of the paper is concerned with the modes of free vibration predicted by the frequency equations given above.

DETERMINATION OF MODES OF VIBRATION

The character of the solution given by Eqs. (2) is strongly dependent on the character of the three indices n_1, n_2, n_3 given by Eq. (3b). For the purpose of illustration of the various combinations of complex and real values that the indices may assume, Fig. 2 shows a plot of n_i $(i=1, 2, 3)$ vs Ω for a given constant value of ν and h/R. The character of n_i varies little with the latter two parameters.

Since one index is always real and positive, it may be arbitrarily denoted by n_1. The indices n_2 and n_3 are seen to be of a certain type within three distinct zones of Ω as shown in Fig. 2. The first of the critical frequencies, which are the limit values of these zones, is denoted by Ω_m, and it varies only slightly with ν and h/R. Within the range of the applicability of the classical theory of shells (i.e., for $0<h/R\leq 0.05$), this critical frequency lies in the interval $1<\Omega_m\leq 1.12$. The case $\Omega_m=1$ corresponds to $h=0$ and $\Omega_m=1.12$ is the value for $h/R=0.05$, where $\nu=0.3$. The second critical frequency, denoted by Ω_r, also varies only slightly with ν and h/R, and $\Omega_r=1.64975$ for $\nu=0.3$, $h/R=0.02$. The frequency for which $n_2=0$ is independent of h/R and is given by

$$\Omega=[2/(1-\nu)]^{\frac{1}{2}}.$$

It should be noted that this frequency represents the natural frequency of free radial vibration of a thin spherical shell.[11]

Thus, the variation of the character of n_2 and n_3 can be summarized as follows:

$$\text{Zone I} \quad 0<\Omega<\Omega_m \quad \begin{cases} n_2=b_2+ib_3 \\ n_3=b_2-ib_3 \end{cases};$$

$$\text{Zone II} \quad \Omega_m<\Omega<\Omega_r \quad \begin{cases} n_2=-\frac{1}{2}+ib_2 \\ n_3=-\frac{1}{2}+ib_3 \end{cases};$$

$$\text{Zone III} \quad \Omega_r<\Omega<\infty \quad \begin{cases} n_2 \text{ is real} \\ n_3=-\frac{1}{2}+ib_3 \end{cases}$$

Here, b_2 and b_3 are real numbers.

It is clear that in Zone I it is necessary to make use of the identity

$$P_{a\pm ib}(x)=\text{Re}[P_{a+ib}(x)]\pm i\,\text{Im}[P_{a+ib}(x)], \quad (15)$$

which indicates that in Zone I the Legendre functions occurring in Eqs. (2) with indices n_2 and n_3 are complex conjugates. However, since C_2, C_3 and A_2, A_3 are then also complex conjugates, the form of the solution in Zone I remains the same, if the real and imaginary parts of the products of these complex conjugates are interpreted as the terms occurring in Eqs. (2) with the subscript $i=2$ and $i=3$, respectively. In this way, the solution in Zone I can be expressed in terms of real functions.

It should be recalled that in the theory of Legendre functions the index $n=-0.5+ib$ plays a special role and by some authors[12] the function $P_{-0.5+ib}(x)$ is called the conical function. Since the conical function is real and also the corresponding values of λ_2, λ_3 are real, the solution given in the form of Eqs. (2) is directly applicable in Zones II and III.

Figure 2 also gives a clear illustration of the character of the solution that is obtained by means of the membrane theory of shells, using the solution given by Eqs. (5) and (6). Equation (6b) gives only one value for the index n, which is plotted in Fig. 2 versus Ω and shown by the dashed curve. The index n is real in Zones I and III and it follows closely the corresponding indices n_1 and n_2 in these zones. It should be noted that now the first critical frequency $\Omega_m=1$. However, in the proximity of $\Omega=1$, n departs from n_1 and becomes infinite at $\Omega=1$. This fact was noted in Ref. 7, and it is of great significance in the evaluation of the natural frequencies by means of membrane theory, which are discussed in the following sections. In Zone II, the index n has the form $n=-0.5+ib$, where b is real, and since the conical function possesses no zeroes with

[11] A. E. H. Love, *A Treatise on the Mathematical Theory of Elasticity* (Dover Publications, Inc., New York, 1944), p. 287.

[12] W. Magnus and F. Oberhettinger, *Formulas and Theorems for the Functions of Mathematical Physics* (Chelsea Publishing Co., New York, 1954), p. 74.

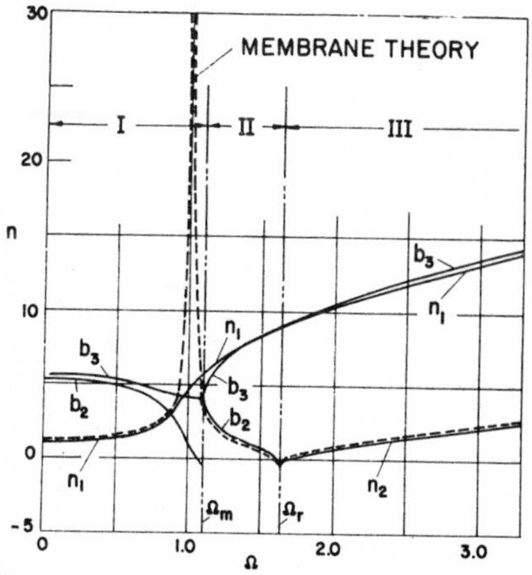

FIG. 2. Indices of Legendre functions vs frequency.

respect to b, then it follows that the membrane theory of shells can predict no natural frequencies within Zone II, i.e., within $1<\Omega<\Omega_r$.

With this interpretation of the character of the indices of the Legendre functions, the actual calculation of the natural frequencies is carried out for given values of the opening angle ϕ_0, Poisson's ratio ν, thickness ratio h/R, and the prescribed boundary conditions at $\phi=\phi_0$. The natural frequencies Ω_i (where i represents the mode number) for the bending theory are obtained as the infinite number of roots of Eq. (7)[13] with respect to Ω. This is done by evaluating the determinant given by Eq. (7) for successively incremented values of Ω. If the determinant changes sign, the natural frequency Ω_i is determined by inverse interpolation. The accuracy of the natural frequency can be as great as desired, depending on the size of the frequency increment. Once Ω_i is known, the actual mode shapes are determined for all relevant variables from Eqs. (8) and (2). The same procedure is followed for the membrane solution, where the zeroes of Eq. (14) determine Ω_i, and the mode shapes are directly given by Eqs. (5).

CLASSIFICATION OF MODES OF VIBRATION

When the consecutive natural frequencies Ω_i (say, for $i=1, 2, \ldots, 9$) and their corresponding mode shapes are evaluated in the manner described in the preceding section for a given set of ν, h/R, ϕ_0, and boundary conditions, it is observed that some of the modes are of a different character than others. This is reminiscent of the previously observed phenomenon in Ref. 6 that for shallow spherical shells the modes of vibration fall into two distinct groups, which were designated in Ref. 6 as the transverse and longitudinal modes. It may be of interest to find out whether or not such a classification of modes is also possible if the conditions of shallowness do not apply. Instead of using the ratio w/u as the criterion for the classification of the modes as it was done in Ref. 6, it turns out that for nonshallow shells the comparison of the strain energy due to the bending and membrane strains is more applicable.

Denoting the strain energy due to the stretching of the middle surface of the shell by V_s and due to the bending of the shell by V_b, the strain-energy expressions are

$$V_s=\frac{Eh\pi R^2}{1-\nu^2}\int_0^{\phi_0}(\epsilon_\phi^2+2\nu\epsilon_\phi\epsilon_\theta+\epsilon_\theta^2)\sin\phi d\phi, \quad (16a)$$

$$V_b=\frac{Eh^3\pi R^2}{12(1-\nu^2)}\int_0^{\phi_0}(k_\phi^2+2\nu k_\phi k_\theta+k_\theta^2)\sin\phi d\phi, \quad (16b)$$

where the middle surface strains are expressed as

$$\epsilon_\phi=(1/R)(u'+w), \quad (17a)$$

$$\epsilon_\theta=(1/R)(\cot\phi u+w), \quad (17b)$$

and the bending strains as

$$k_\phi=(1/R^2)(u'-w''), \quad (18a)$$

$$k_\theta=(1/R^2)(u-w')\cot\phi, \quad (18b)$$

in terms of the displacements that are given by Eqs. (2a, b). A convenient parameter η is introduced through the ratio

$$\eta=V_b/(V_b+V_s) \quad (19)$$

for the estimate of relative bending in each mode of vibration.

It is apparent from Table I that the parameter η serves its purpose. When the first successive modes of vibration of a spherical shell subjected to four different boundary conditions are calculated and the corresponding η for each mode obtained, it is discovered that, in general, the ratios η tend to increase with the mode number in a smooth fashion. However, some modes are detected in all cases for which η is an order of magnitude smaller than the numbers forming the smooth curve. For this reason, the modes that have a large percentage of energy due to the bending of the shell are called bending modes and those having a small η are called membrane modes.

TABLE I. Natural frequencies Ω_i for $h/R=0.05$, $\phi=60°$, $\nu=0.3$.

i	Free edge Ω	Free edge η	Fixed-hinged Ω	Fixed-hinged η	Roller-hinged Ω	Roller-hinged η	Clamped Ω	Clamped η
1	0.931	0.053	0.962	0.318	0.995	0.233	1.006	0.041
2	1.088	0.520	1.334	0.448	1.381	0.656	1.391	0.699
3	1.533	0.745	2.128	0.802	2.110	0.875	2.375	0.755
4	2.348	0.900	3.176	0.933	2.546	0.005	3.486	0.850
5	2.544	0.0006	3.988	0.054	3.183	0.953	3.991	0.049
6	3.497	0.954	4.575	0.935	4.563	0.985	4.974	0.950
7	4.951	0.974	6.231	0.968	5.530	0.001	6.690	0.955
8	5.230	0.0009	...	...	6.235	0.990	7.113	0.034
9	6.693	0.987	...	...	...	...	...	...

[13] All calculations carried out in this paper were performed on the IBM-709 computer at the Yale Computer Center. Since the elements of Eq. (7) and the solutions given by Eqs. (2) involve Legendre functions, the author has prepared for this purpose a subroutine that can calculate Legendre functions $P_n(\cos\phi)$ with arbitrary complex indices valid in the interval $0\leq\phi<180°$.

For example, with reference to Table I, among the first nine modes for the free edge, the fifth and the eighth are membrane modes, while the others are bending modes. Similar conclusions are easily reached for the other boundary conditions.

It is also of interest to examine the corresponding mode shapes of the bending and the membrane modes. This is done for the eight modes shown in Table I for the case of a roller-hinged edge, and the displacements w and u are plotted in Fig. 3. Again, it is seen that the two membrane modes have a different character. If the number of nodes of w and u are examined, it is seen that this number increases by one for each successive mode, as it should, except for the membrane modes, which appear to have a separate count of nodes.

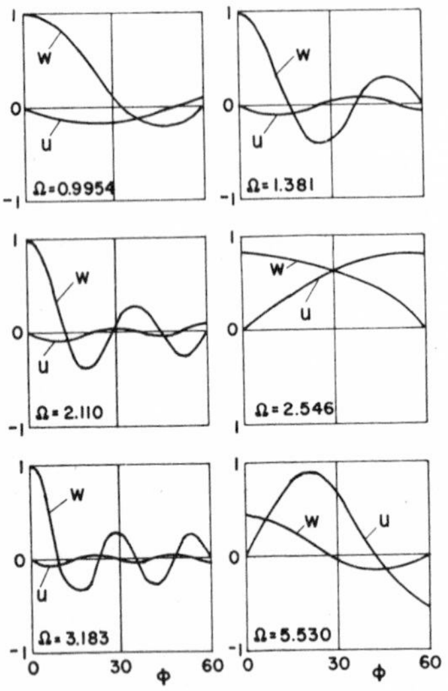

FIG. 3. Relative displacements of modes of vibration vs conical angle for $h/R=0.05$, $\nu=0.3$, $\phi_0=60°$, and a roller-hinged edge.

The distinction between the bending and the membrane modes can be made for all values of opening angle and boundary conditions. The easiest test for the determination of the character of each mode is based on the strain-energy coefficient η. This classification of the modes is of great importance in the study of vibration characteristics of nonshallow spherical shells, and it may be applicable to other shells of revolution.

DISCUSSION OF RESULTS

The different characters of the modes of vibration according to the classification of the preceding section can be further studied from the examination of the variation of the natural frequencies with the thickness ratio h/R. Such plots are shown in Figs. 4–6. In Figs. 4 and 5, the first three bending modes and the first membrane mode are shown for the case of a free edge and $\phi_0=60°$ and $\phi_0=90°$, respectively. Figure 6 shows the first ten bending modes and the first two membrane modes for a closed shell. For all figures, the Poisson's ratio $\nu=0.3$.

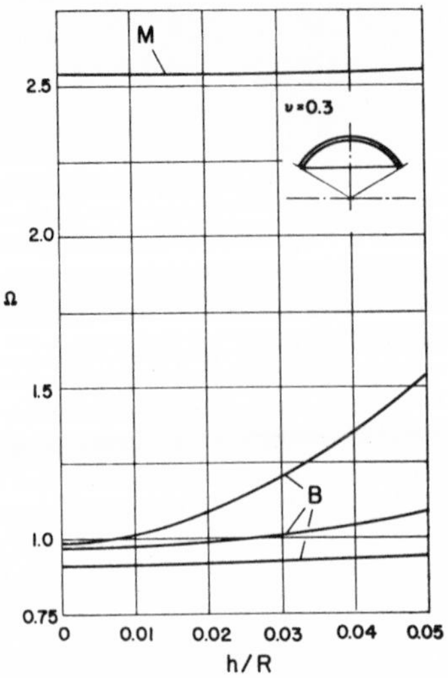

FIG. 4. Natural frequency vs h/R for free edge at $\phi_0=60°$, bending theory.

It should be noted that the variation of the natural frequency of the bending modes increases with the thickness and with the mode number. Also, for smaller opening angles, the variation is more pronounced than for larger values of ϕ_0. Since the membrane modes occur at relatively high values of Ω in comparison to the first bending mode, the variation of only a few of the membrane modes with h/R was investigated. The conclusion was reached that the membrane modes have very small variations with h/R. The results of Figs. 4–6 are now compared to the corresponding results obtained by means of membrane theory.

Since all natural frequencies in the absence of the effects of bending are independent of the thickness, a plot of Ω_i predicted by membrane theory vs the only other relevant parameter ϕ_0 is shown in Fig. 7 for the

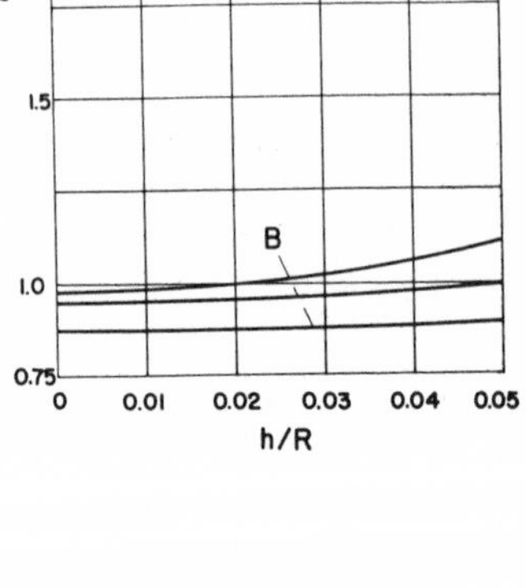

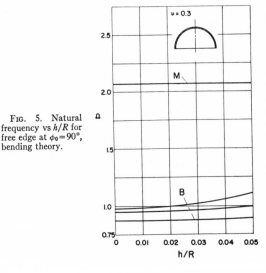

FIG. 5. Natural frequency vs h/R for free edge at $\phi_0=90°$, bending theory.

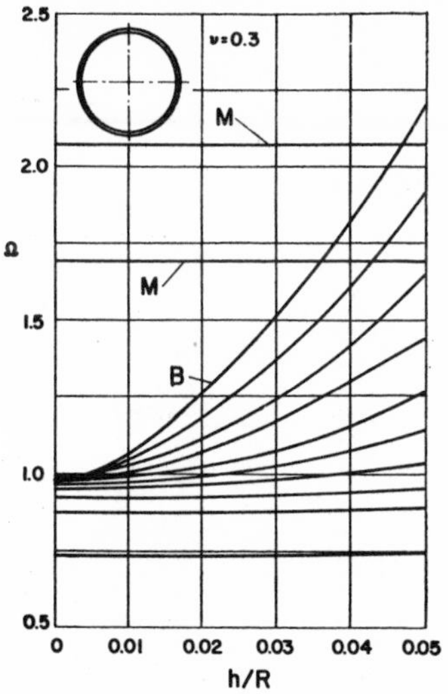

FIG. 6. Natural frequency vs h/R for a closed spherical shell, bending theory.

case of a free edge. It is seen that for all opening angles a set of an infinite number of modes (called the lower set) is spaced within the dotted area shown in Fig. 7, while another infinite set of modes (the upper set) is found to lie above the line $\Omega=1$. The two distinct sets of modes obtained by means of membrane theory were first discovered by Lamb[1] for a closed spherical shell, and their existence was subsequently confirmed for the axisymmetric vibration of a hemispherical shell by Love[8] and for nonsymmetric vibration by Naghdi and Kalnins.[7]

While the upper set of modes seems to be physically acceptable, the lower set contains an infinite number of modes within a finite frequency interval, $0<\Omega<1$. Moreover, the intervals between the natural frequencies of the lower set tend to zero when the critical frequency $\Omega=1$ is approached. A complete explanation of this seemingly anomalous vibration behavior predicted by membrane theory is found in the study of the effects of bending on the lower set of modes of vibration.

Examination of the bending modes shown in Figs. 4–6 reveals that if $h=0$ the natural frequencies are identical to those of the lower set predicted by membrane theory, and for $h=0$ they all fall below the limit value $\Omega=1$. As soon as the shell is allowed to have a thickness greater than zero, however small, this limiting frequency disappears and the range of the natural frequencies of the bending modes extends to infinity.

Thus, the conclusion is reached that the lower set of modes predicted by membrane theory is really a degenerate set of bending modes that is valid only for $h=0$. Since the bending energy of these modes is predominant, it seems reasonable to expect that they cannot be accurately predicted by membrane theory, where the bending effects are completely neglected. On the other hand, the membrane modes shown in Figs. 4–6 agree well with the upper set of modes predicted by membrane theory. The value of the natural frequencies of the membrane modes at $h=0$ is almost the same as that at $h/R=0.05$. Again, this should be expected since in these modes the membrane strain energy is predominant, and, consequently, the membrane theory should give good results.

COMPARISON TO SHALLOW-SHELL THEORY

In order to obtain some bounds on the validity of the theory of shallow spherical shells, it is of interest to compare the actual natural frequencies obtained in this paper with those that are predicted by the theory of shallow shells and given in Ref. 6. For this purpose, the natural frequencies of the lowest mode of vibration are shown in Fig. 8 plotted vs a/R (where $a=R\sin\phi_0$), which is a measure of shallowness. The dashed curve shown in Fig. 8 represents the results obtained in Ref. 6 by means of the theory of shallow shells. At $a/R=0.5$, which corresponds to $\phi_0=30°$, the percentage error in the lowest natural frequency of shallow-shell theory is 2.5%. If this error is regarded as the maximum allowable error, then it may be concluded that $a/R=0.5$ is the limit of the theory of shallow spherical shells. Such a limit is in agreement with that which was previously suggested by Reissner[14] (i.e., $H/a\leq 0.25$, where H represents the maximum shell rise), since for shallow spherical shells $H\approx a^2/2R$.

When the modes for the case $\phi_0=30°$ are compared to the corresponding results of Ref. 6, it follows that the transverse modes are bending modes and the longitudinal modes are membrane modes. However, the ratio of the maximum values of transverse and longi-

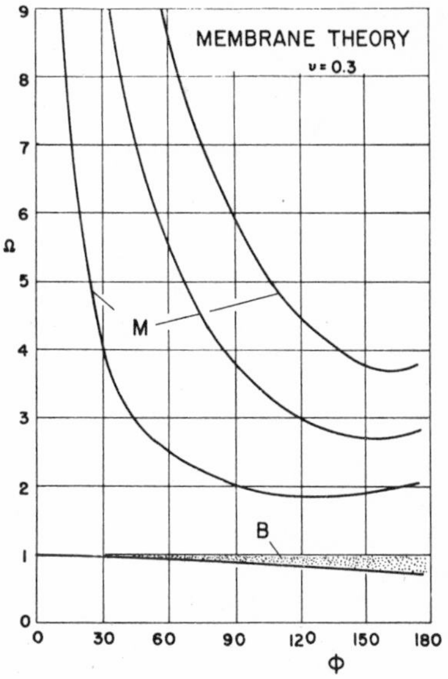

FIG. 7. Natural frequency vs opening angle, membrane theory.

[14] E. Reissner, Quart. Appl. Math. 13, 279–290 (1955).

tudinal displacements of the bending and membrane modes does not appear to be a valid criterion for the classification of the modes if the shell cannot be considered shallow (i.e., for $a/R>0.5$). This can be seen from Fig. 9, where the mode shapes of the lowest mode of vibration,[15] which is a bending mode, are shown for increasing values of the opening angle ϕ_0. Clearly, the ratio u/w is small for $\phi_0=30°$, but it increases rapidly with ϕ_0. An important conclusion drawn from the results shown in Fig. 9 is that the longitudinal inertia is not negligible in bending modes for nonshallow shells. Another interesting point can be made with respect to the transverse vibrations of shallow spherical shells. It was established in Ref. 6 that if the longitudinal-inertia terms in the theory of shallow shells are neglected, then the longitudinal modes cannot be obtained. In this paper, it is shown that the longitudinal modes for shallow shells are membrane modes, and that only these modes can be accurately predicted by membrane theory. It must be concluded that membrane theory cannot be used to predict transverse modes of vibration. If it is used, then the result was shown in Ref. 16 to be a single, degenerate, natural frequency at $\Omega=1$, which, with reference to Fig. 7, represents the limiting line of all degenerate bending modes predicted by membrane theory.

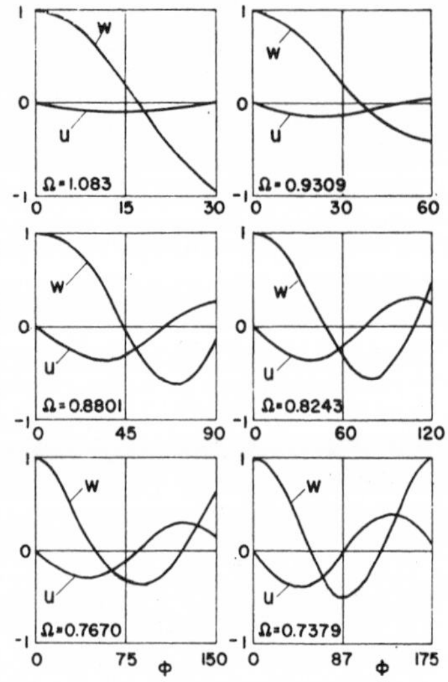

FIG. 9. Relative displacements of modes of vibration vs conical angle for lowest mode, free edge, $h/R=0.05$, $\nu=0.3$, and various values of opening angle.

CONCLUSION

From the results presented in this paper, it is concluded that the complete vibration spectrum of a nonshallow spherical shell consists of two infinite, interspersed sets of modes distinguished by a predominant bending or membrane strain-energy component. These modes are defined as bending and membrane modes. It is shown that the transverse modes of shallow spherical shells are bending modes and the longitudinal modes are membrane modes. Moreover, the distinction of the modes according to the ratio of the maximum values of the displacements is not meaningful for nonshallow spherical shells, and it follows that the neglect of the tangential-inertia term for nonshallow shells is not justified.

It is further concluded that the membrane theory of shells can determine accurately the membrane modes, but that it predicts bending modes that are valid only at $h=0$. The accuracy of membrane theory is still good for small values of h and for low mode numbers.

Finally, it is of interest to remark that a theorem in the theory of surfaces was proved by Jellett[17] in 1855. It states: "*The most general displacement, which a closed, oval, inextensible surface admits of, is that of a rigid body.*" The precise interpretation of this theorem with regard to the theory of shells is that a closed shell cannot undergo purely inextensional deformation. However, the theorem should not be interpreted to mean that the membrane (or extensional) theory will always give accurate results if it is applied to a closed shell. The results of this paper support this statement. For example, with reference to Fig. 6, the natural frequency of the ninth mode of a closed spherical shell for $h/R=0.05$ is about double that predicted by membrane theory. On the other hand, Jellett's theorem, if interpreted correctly, is also supported by the results of this paper, since no modes are found for which the extension of the middle surface is identically zero.

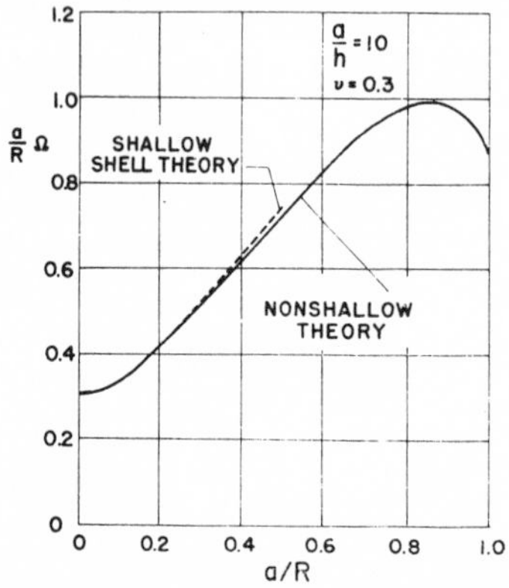

FIG. 8. Lowest natural frequency vs a/R for a clamped edge, bending theory.

ACKNOWLEDGMENT

The results reported in this paper were obtained in the course of research supported by the National Science Foundation through grant G-23922 to Yale University.

[15] The mode shapes of the lowest mode of vibration of a shallow spherical shell with a free edge are included in A. Kalnins and P. M. Naghdi, J. Acoust. Soc. Am. **32**, 342–347 (1960). In comparing the graph of this reference with the $\phi_0=30°$ mode of Fig. 9, it should be recalled that in this reference the displacement components that are parallel and perpendicular to the axis of symmetry were employed.

[16] M. W. Johnson and E. Reissner, Quart. Appl. Math. **15**, 367–380 (1958).

[17] J. H. Jellett, Trans. Roy. Irish Acad. **22**, 375 (1855).

24

Reprinted from *J. Acoust. Soc. Amer.*, **38**(2), 367–368 (1965)

Natural Frequencies of Closed Spherical Shells

J. P. WILKINSON

North American Aviation, Inc., Downey, California

Equations are presented for the natural frequencies of free vibration of closed elastic spherical shells. They are derived from a theory of shells that includes the effects of transverse, shear, and rotatory inertia. It is shown that five branches appear in the frequency spectrum, whereas only three are known to be predicted by the classical bending theory of shells.

List of Symbols.

w radial displacement
u tangential displacement
n integer
r $n(n+1)$
$P_n(\cos\phi)$, $P_n^1(\cos\phi)$ Legendre polynomials
ϕ meridional coordinate
ρ mass density
E Young's modulus
ν Poisson's ratio
λ $[\rho a^2 p^2(1-\nu^2)/E]^{\frac{1}{2}}$, a nondimensional frequency parameter
p circular frequency
t time
h shell thickness
a shell radius
k_s shear coefficient, usually 6/5
ξ $12a^2/h^2$
k_1 $1+1/\xi$
k_r $1+1.8/\xi$
c_1 $2/\xi$
c_r 2

IN A RECENT STUDY OF THE VIBRATIONS OF ELASTIC SPHERICAL shells according to the improved theory of shells, Wilkinson and Kalnins[1] included the effects of transverse shear and rotatory inertia. The natural frequencies of closed spherical shells may be obtained from their analysis.

It may be shown that the torsionless and torsional axisymmetric modes of vibration are uncoupled. The radial displacement w of the torsionless modes is given by

$$w=P_n(\cos\phi)\sin pt. \quad (1)$$

The tangential displacement u of the torsional modes is given by

$$u=P_n^1(\cos\phi)\sin pt. \quad (2)$$

A relation between the natural frequency λ^2 and the integer n may be found from their paper. According to the improved theory of shells, in the torsionless case it is a cubic equation in λ^2:

$$2\lambda^6 k_s k_1(k_r k_1-c_r c_1)/(1-\nu)-\lambda^4\{(k_r k_1-c_r c_1)[r+4k_s(1+\nu)/(1-\nu)] +k_1[\xi(k_1+c_1)+c_r+k_r+2k_s(k_1+k_r)(r/(1-\nu)-1)]\} +\lambda^2\{(\xi c_1+c_r)(1+\nu)(2-r)+k_r[r(r-3-\nu)+2(1+\nu) \times((r-2)k_s+1)]+k_1[2k_s r(r+4\nu)/(1-\nu)+r(r+\xi+\nu) +(1+3\nu)(\xi-2k_s)-(1-\nu)]\}-(r-2)[r(r-2)+2k_s(1+\nu) \times(r-1+\nu)+(1-\nu^2)(\xi+1)]=0. \quad (3)$$

In the torsional case, it is a quadratic equation:

$$4\lambda^4 k_s(k_r k_1-c_r c_1)/(1-\nu) -2\lambda^2[\xi(k_1+c_1)+c_r+k_r+k_s(k_1+k_r)(r-2)] +(1-\nu)(r-2)[\xi+1+(r-2)k_s]=0. \quad (4)$$

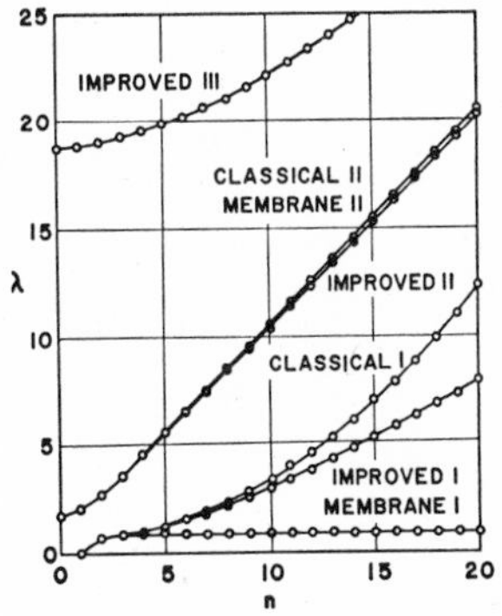

FIG. 1. Natural frequencies of torsionless vibrations of a closed spherical shell.

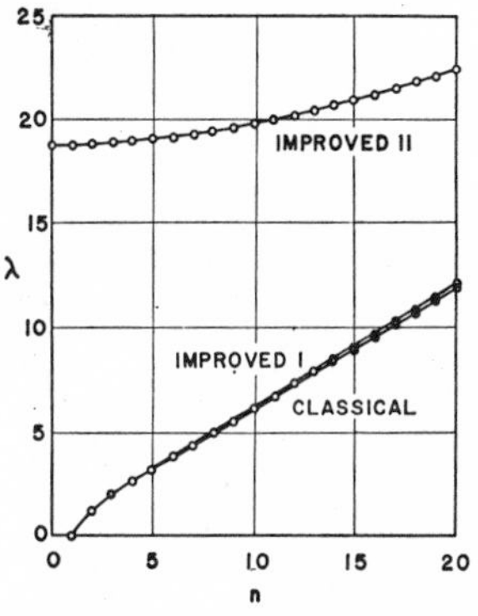

FIG. 2. Natural frequencies of torsional vibrations of a closed spherical shell.

The values of the natural frequencies corresponding to each integral value of n are plotted in Figs. 1 and 2 for a shell thickness/radius ratio h/a of $\frac{1}{10}$ and Poisson's ratio $\nu=0.3$. Significantly, the improved theory predicts that there are five frequencies associated with each value of n. Thus, a total of five branches appears in the frequency spectrum. Three of them govern the torsionless modes while the other two govern the torsional modes. The modes of vibration of the highest branches occur at very high frequencies and correspond to the first thickness-shear modes of an infinite plate. As h/a decreases, the frequencies associated with these two highest branches become rapidly higher.

As a comparison, the frequency branches predicted by the classical bending theory of shells[2] (for which $k_1=1$, $k_s=k_r=c_r=c_1=0$) and by the membrane theory of shells[3] (for which $k_1=1$, $k_s=k_r=c_r=c_1=0$, and $h/a=0$) are shown on the same Figures. These theories predict only the existence of the two lowest torsionless branches and of the lowest torsional branch.

[1] J. P. Wilkinson and A. Kalnins, "On Nonsymmetric Dynamic Problems of Elastic Spherical Shells," J. Appl. Mech. Paper No. 65-APM-9 (1965).
[2] A. Kalnins, "Effect of Bending on Vibrations of Spherical Shells," J. Acoust. Soc. Am. 36, 74–81 (1964).
[3] H. Lamb, "On the Vibrations of a Spherical Shell," Proc. London Math. Soc. 14, 50–56 (1882); W. E. Baker, "Axisymmetric Modes of Vibration of Thin Spherical Shell," J. Acoust. Soc. Am. 33, 1749–1758 (1961).

Reprinted from *J. Acoust. Soc. Amer.*, **36**(7), 1355–1365 (1964)

Free Vibration of Rotationally Symmetric Shells

A. Kalnins

Department of Engineering and Applied Science, Yale University, New Haven, Connecticut

(Received 16 December 1963)

This paper is concerned with a theoretical investigation of the free vibration of arbitrary shells of revolution by means of the classical bending theory of shells. A method is developed that is applicable to rotationally symmetric shells with meridional variations (including discontinuities) in Young's modulus, Poisson's ratio, radii of curvature, and thickness. By means of the method of this paper, the natural frequencies and the corresponding mode shapes of axisymmetric or nonsymmetric free vibration of rotationally symmetric shells can be obtained without a limitation on the length of the meridian of the shell. To illustrate the application of the method given in this paper to particular shells, some results of free vibration of spherical and conical shells obtained earlier by means of the bending theory are reproduced by the general method of this paper, and a detailed comparison is made. In addition, paraboloidal shells and a sphere–cone shell combination are considered, which have been previously analyzed by means of the inextensional theory of shells, and natural frequencies and mode shapes predicted by the bending theory are given.

SYMBOLS

$A(x)$	(m,m) matrix, denotes coefficients of differential equations
D	$=Eh^3/12(1-\nu^2)$
E	Young's modulus
G	$=1/(1+D\sin^2\phi/Kr^2)$
H	$=1/R_\phi-\sin\phi/r$
I	unit matrix
J	$=1/R_\phi+\sin\phi/r$
K	$=Eh/(1-\nu^2)$
L	some characteristic length of shell
M	number of segments
M_ϕ, M_θ, $M_{\theta\phi}$	moment resultants
N	effective tangential shear resultant
N_ϕ, N_θ, $N_{\theta\phi}$	membrane-stress resultants
P	$=1/R_\phi+\nu\sin\phi/r$
Q	effective transverse-shear resultant
Q_ϕ, Q_θ	transverse-shear resultants
R_ϕ	radius of curvature of meridian
$T(x)$	(m,m) transformation matrix
$U(x)$	(m,m) matrix, related to $Y(x)$
$Y(x)$	(m,m) matrix, denotes homogeneous solutions
$Y_i^j(x)$	i identifies segment S_i, j denotes quadrant of partitioned matrix $Y(x)$
a, b	end points of shell
c	$=(E/\rho)^{\frac{1}{2}}$—speed of sound
h	thickness of shell
m	number of first-order differential equations
n	integer, denotes Fóurier component
r	distance from axis of symmetry
s	distance along meridian in positive direction of ϕ
t	time
$u(x)$	$(m,1)$ matrix, related to $y(x)$
w, u_ϕ, u_θ	displacements of middle surface in normal, meridional, circumferential direction
x	independent variable, can be ϕ or s
$y(x)$	$(m,1)$ matrix, denotes fundamental variables
$y_j^i(x)$	$j=1$, 2 denotes upper and lower half of partitioned $y(x)$; i denotes ith element of each submatrix
β_ϕ	angle of rotation of normal in meridional direction
ϕ	angle between normal and axis of symmetry
ν	Poisson's ratio
ω	circular frequency, rad/sec
ρ	mass density
Ω	$\omega L/c$, nondimensional frequency parameter

INTRODUCTION

IN the analysis of free vibration of a shell of revolution, the natural frequencies of the shell are the roots of a determinant whose elements are related to certain solutions of the homogeneous field equations. These solutions, from which the frequency equation is constructed, are characteristic of a particular shell, and for simple shell configurations, such as cylindrical and spherical shells, they are known hypergeometric functions. However, for more-complicated shells such solutions are not available.

The main purpose of the present paper is to develop a method for the determination of the homogeneous solutions, natural frequencies, and mode shapes of free vibration for arbitrary rotationally symmetric shells by means of the linear-bending theory of shells. The method given here is applicable to any eigenvalue problem that is governed in an interval by a system of m linear homogeneous first-order ordinary differential equations and $m/2$ homogeneous boundary conditions at each end of the interval.

It was shown by the present author[1] that the basic equations of the classical static theory of shells of revolution can be reduced to a system of 8 first-order ordinary differential equations involving 8 unknowns in such a way that no derivatives of the shell properties appear in the coefficients of these equations. For the special case of vibration of a conical shell, a similar system of equations was previously derived by Goldberg and Bogdanoff.[2] The essential point in these derivations is the definition of the 8 variables as exactly those quantities that enter into the appropriate boundary conditions on a rotationally symmetric edge of a shell of revolution.

To obtain a similar system of differential equations for the analysis of free vibration of arbitrary rotationally symmetric shells, the reduction scheme given in Ref. 1 is started from the homogeneous equations of the classical dynamic theory of shells. Again, a system of 8 first-order differential equations results, which together with the boundary conditions constitutes an eigenvalue problem. The solution to this problem is obtained in this paper by means of a multisegment, direct, numerical integration approach, which is an extension to eigenvalue problems of the method used successfully by the author[1] for the analysis of static deformation of shells of revolution.

A part of the method given here is similar to the one used in the analysis of axisymmetric modes of free vibration of relatively short conical shells by Goldberg, Bogdanoff, and Marcus.[3] It turns out that, if the method of Ref. 3 is applied to sufficiently long shells of revolution, then a complete loss of accuracy invariably results in the process of calculation of the natural frequencies and mode shapes. The loss of accuracy is attributed to the subtraction of almost equal numbers, and its cause is explained in detail in this paper. It is also shown in this paper that the loss of accuracy can be avoided and that the natural frequencies and mode shapes of long shells can be obtained if the shell is divided into a number of sufficiently short segments and the numerical integration is carried out over each segment separately. By requiring the continuity of all relevant variables at the end points of the segments, a linear homogeneous system of matrix equations is obtained that possesses a nontrivial solution if the determinant of a certain $(m/2,m/2)$ matrix vanishes. Thus, the free-vibration problem of an arbitrary rotationally symmetric shell is reduced to the determination of the homogeneous solutions for a particular frequency and to the calculation of the value of an $(m/2,m/2)$ determinant.

For the purpose of illustration of the use of the method given in this paper, the natural frequencies and mode shapes of free vibration are investigated for some special cases of rotationally symmetric shells. To demonstrate the accuracy of the method, results are obtained for free vibration of a conical and a spherical shell and compared to those obtained previously by means of the bending theory of shells. Then, the lowest natural frequencies and mode shapes of a paraboloidal shell and a sphere–cone shell combination are calculated and compared to previous results, which have been obtained earlier by means of the inextensional theory of shells.

I. FUNDAMENTAL SYSTEM OF EQUATIONS FOR VIBRATION OF SHELLS OF REVOLUTION

In Ref. 1, the governing system of static equations for shells of revolution is reduced to a system of 8 differential equations that contain 8 unknowns. This system of equations and the unknowns are called the fundamental system and the fundamental variables, respectively, because they are necessary and sufficient for a complete statement of the problem.

In this paper, we are concerned with the fundamental system of equations that is reduced from the dynamic theory of shells of revolution. In the absence of any external loads, the fundamental equations can be written in matrix form:

$$dy(x)/dx = A(x)y(x), \qquad (1)$$

[1] A. Kalnins, "Analysis of Shells of Revolution Subjected to Symmetrical and Nonsymmetrical Loads," Paper No. 64-APM-33, J. Appl. Mech. (to be published).

[2] J. E. Goldberg, and J. L. Bogdanoff, Proc. Symp. Ballistic Missile Aerospace Technol., 6th, 1, 219–238 (1961).

[3] J. E. Goldberg, J. L. Bogdanoff, and L. Marcus, J. Acoust. Soc. Am. 32, 738–742 (1960).

where x is an independent variable, $y(x)$ is a column matrix whose elements represent the m fundamental variables, and $A(x)$ is an (m,m) coefficient matrix whose elements are piecewise continuous known functions of x defined in an interval of x denoted by (a,b). The system of Eqs. (1), together with $m/2$ homogeneous boundary conditions at each end point of the interval ($x=a$ and $x=b$), forms an eigenvalue problem, for which a method of solution is given in the following sections.

It should be noted that the method given in Ref. 1 for static analysis, as well as that in this paper for free vibration, can be conveniently used with any version of the consistent linear shell theories available in the literature, including the anisotropic and the improved theories in which the effects of transverse-shear deformation are accounted for. The differences in these theories are reflected only in the elements of $A(x)$ and the number of the fundamental variables. For example, for nonsymmetric deformation of a shell of revolution for improved theory, $m=10$; for classical theory, $m=8$, while, for axisymmetric deformation for both theories, $m=6$. For brevity, in this paper we consider only the symmetric and nonsymmetric free vibration by means of the classical theory of isotropic rotationally symmetric shells. Extensions are straightforward.

The variables of the classical theory of shells used in this paper are assumed to be separable in the form

$$\begin{Bmatrix} w \\ u_\phi \\ \beta_\phi \\ Q \\ Q_\phi \\ N_\phi \\ N_\theta \\ M_\phi \\ M_\theta \end{Bmatrix} = \begin{Bmatrix} w_n \\ u_{\phi n} \\ \beta_{\phi n} \\ Q_n \\ Q_{\phi n} \\ N_{\phi n} \\ N_{\theta n} \\ M_{\phi n} \\ M_{\theta n} \end{Bmatrix} \cos n\theta \cos\omega t,$$

$$\begin{Bmatrix} u_\theta \\ N \\ N_{\theta\phi} \\ Q_\theta \\ M_{\theta\phi} \end{Bmatrix} = \begin{Bmatrix} u_{\theta n} \\ N_n \\ N_{\theta\phi n} \\ Q_{\theta n} \\ M_{\theta\phi n} \end{Bmatrix} \sin n\theta \cos\omega t.$$

The fundamental variables, in terms of which the problem is stated, are taken as those quantities that appear in the appropriate boundary conditions of the classical theory on an axisymmetric circular edge. Thus, The independent variable x can be regarded as either the angle ϕ between the normal and the axis of symmetry of the shell or the distance s measured along the

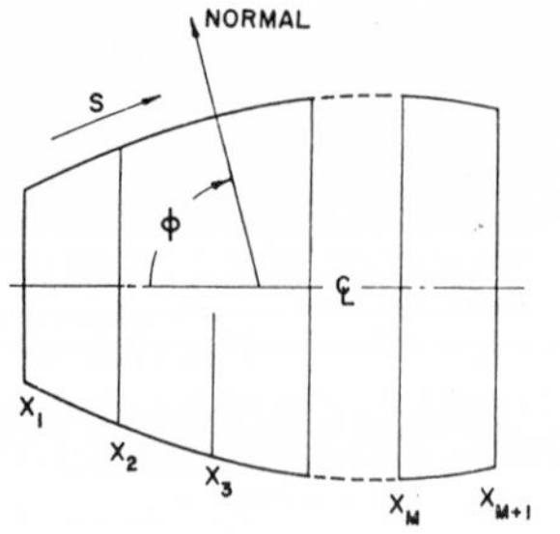

FIG. 1. Coordinates of the shell.

we define the (8,1) matrix $y(x)$ by

$$y(x) = \begin{Bmatrix} w_n(x) \\ u_{\phi n}(x) \\ u_{\theta n}(x) \\ \beta_{\phi n}(x) \\ Q_n(x) \\ N_{\phi n}(x) \\ N_n(x) \\ M_{\phi n}(x) \end{Bmatrix}$$

meridian, as shown in Fig. 1. For some shells (e.g., paraboloidal), ϕ is more convenient, while for shells with straight generators, when ϕ is a constant (e.g., conical), s should be employed. However, the distinction between ϕ and s is necessary only in the derivatives where the transformation $d/ds=(1/R_\phi)d/d\phi$ holds.

The fundamental equations for vibration of shells of revolution in the form of Eq. (1) can be derived by following the steps given in Ref. 1. Based on the linear classical theory of shells derived by Reissner,[4] with the inclusion of the effects of translatory and rotatory inertia, the nonzero elements A_{ij} of the coefficient matrix $A(x)$ can be written in the form

$$A_{12}=1/R_\phi,\quad A_{14}=-1,$$

$$A_{21}=-P,\quad A_{22}=-\nu\cos\phi/r,\quad A_{23}=-\nu n/r,\quad A_{26}=1/K,$$

$$A_{31}=nGD\sin 2\phi/Kr^3,\quad A_{32}=n(1-GDJ\sin\phi/Kr)/r,$$

$$A_{33}=(1-GDH\sin\phi/Kr)\cos\phi/r,\quad A_{34}=2nGD\sin\phi/Kr^2,$$

$$A_{37}=2(1-GD\sin^2\phi/Kr^2)/(1-\nu)K,$$

$$A_{41}=-\nu n^2/r^2,\qquad A_{43}=-\nu n\sin\phi/r^2,$$

$$A_{44}=-\nu\cos\phi/r,\quad A_{48}=1/D,$$

$$A_{51}=2n^2GD(1-\nu)\cos^2\phi/r^4+n^4D(1-\nu^2)/r^4$$
$$+K(1-\nu^2)\sin^2\phi/r^2-\Omega^2Eh(1+h^2n^2/12r^2)/L^2,$$

$$A_{52}=K(1-\nu^2)\sin 2\phi/2r^2-n^2GDJ(1-\nu)\cos\phi/r^3,$$

[4] E. Reissner, Am. J. Math. **63**, 177–184 (1941).

$$A_{53}=n(1-\nu^2)(K+n^2D/r^2)\sin\phi/r^2$$
$$-nGDH(1-\nu)\cos^2\phi/r^3-n\Omega^2Eh^3\sin\phi/12r^2L^2,$$
$$A_{54}=n^2D(1-\nu)(1+\nu+2G)\cos\phi/r^3,$$
$$A_{55}=-\cos\phi/r,\quad A_{56}=P,\quad A_{57}=-A_{31},\quad A_{58}=-A_{41},$$
$$A_{61}=K(1-\nu^2)\sin2\phi/2r^2-n^2GDJ(1-\nu)\cos\phi/r^3,$$
$$A_{62}=K(1-\nu^2)\cos^2\phi/r^2+n^2GDJ^2(1-\nu)/2r^2-\Omega^2Eh/L^2,$$
$$A_{63}=nK(1-\nu^2)\cos\phi/r^2+nGDJH(1-\nu)\cos\phi/2r^2,$$
$$A_{64}=-n^2GDJ(1-\nu)/r^2,\quad A_{65}=-1/R_\phi,$$
$$A_{66}=-(1-\nu)\cos\phi/r,\quad A_{67}=-n(1-GDJ\sin\phi/Kr),$$
$$A_{71}=A_{53},\quad A_{72}=A_{63},$$
$$A_{73}=n^2(K+D\sin^2\phi/r^2)(1-\nu^2)/r^2$$
$$+GDH^2(1-\nu)\cos^2\phi/2r^2$$
$$-\Omega^2Eh(1+h^2\sin^2\phi/12r^2)/L^2,$$
$$A_{74}=-nD(1-\nu)[GH-(1+\nu)\sin\phi/r]\cos\phi/r^2,$$
$$A_{76}=-A_{23},\quad A_{77}=-(2-GDH\sin\phi/Kr)\cos\phi/r,$$
$$A_{78}=-A_{43},$$
$$A_{81}=A_{54},\quad A_{82}=A_{64},\quad A_{83}=A_{74},$$
$$A_{84}=D(1-\nu)[(1+\nu)\cos^2\phi+2n^2G]/r^2-\Omega^2Eh^3/12L^2,$$
$$A_{85}=1,\quad A_{87}=-A_{34},\quad A_{88}=-(1-\nu)\cos\phi/r.$$

The given coefficients A_{ij} are those which arise when x in Eq. (1) is identified with s. If the independent variable is taken as ϕ, then every element A_{ij} given above must be multiplied by R_ϕ.

The eigenvalue problem as defined above determines the fundamental variables $y(x)$. Other variables are given in terms of $y(x)$ by

$$N_{\theta n}=\nu N_{\phi n}+(1-\nu^2)K(w_n\sin\phi+u_{\phi n}\cos\phi+nu_{\theta n})/r,$$
$$M_{\theta n}=\nu M_{\phi n}+(1-\nu^2)$$
$$\times D(n^2w_n/r+\beta_{\phi n}\cos\phi+n\sin\phi u_{\theta n}/r)/r,$$
$$M_{\theta\phi n}=D(1-\nu)[-2n\cos\phi w_n/r+nJu_{\phi n}$$
$$+H\cos\phi u_{\theta n}-2n\beta_{\phi n}]/2r+\sin\phi DN_n/Kr,$$
$$N_{\theta\phi n}=N_n-\sin\phi M_{\theta\phi n}/r,$$
$$Q_{\phi n}=Q_n-nM_{\theta\phi n}/r,$$
$$Q_{\theta n}=-nM_{\theta n}/r+(dM_{\theta\phi n}/ds)$$
$$+2\cos\phi M_{\theta\phi n}/r+\Omega^2Eh^3(nw_n+u_{\theta n}\sin\phi)/12rL^2.$$

The method given for static problems in Ref. 1 is equally applicable for finding the forced dynamic response of rotationally symmetric shells to harmonic surface and edge (or ring) loads if the coefficients $A(x)$ derived in Ref. 1 are replaced by those given in this paper. The main purpose of this paper is the development of a method for the analysis of free vibration of rotationally symmetric shells, which is given in the following sections.

II. METHOD OF SOLUTION FOR A ONE-SEGMENT SHELL

In order to find the solution of the system of Eqs. (1), it is desirable to reduce the system of differential equations to one equation and one unknown. This has been accomplished, for example, for cylindrical and spherical shells with constant thickness and material properties by eliminating all other unknowns but the normal displacement w. Regardless of whether the solution is obtained from a single equation involving a single unknown or from the simultaneous system of Eqs. (1), the solution for the fundamental variables of a shell of revolution in the absence of external loads can be written in the form

$$y(x)=W(x)C, \tag{2}$$

where $W(x)$ is an (m,m) matrix whose columns represent m linearly independent solutions of the homogeneous governing equations, and C denotes a column matrix of m arbitrary constants.

It should be noted that for cylindrical shells the columns of $W(x)$ consist of independent trigonometric functions, while for shallow spherical shells they are Bessel functions, and for nonshallow spherical shells they are Legendre functions, all of which can be obtained either from their hypergeometric series or by means of numerical integration from the corresponding differential equations. For more-complicated shell configurations, especially in the presence of meridionally variable properties, the reduction of Eqs. (1) to a single equation with one unknown and a series solution may not be possible. For such shells, the method of direct numerical integration of the system of Eqs. (1) for obtaining the solutions $W(x)$ leads to a powerful method for the analysis of arbitrary rotationally symmetric shells.

The solution is obtained by defining the columns of $W(x)$ as the solutions of m initial-value problems in the interval (a,b) governed by the system of Eqs. (1) and subjected to arbitrary linearly independent initial conditions at $x=a$. If the independence requirement is met at $x=a$, then the solutions will be independent at any other value of x in the interval (a,b).

Since the only requirement of the columns of $W(x)$ is that they be linearly independent solutions of the system of Eqs. (1), in place of $W(x)$ we may employ in the interval (a,b) a matrix of linear combinations of the solutions of Eqs. (1), which at $x=a$ reduces to a unit matrix I. This is done by evaluating Eq. (2) at $x=a$

$$y(a)=W(a)C, \tag{3}$$

solving for C

$$C=W^{-1}(a)y(a), \tag{4}$$

and replacing C in Eq. (2) by Eq. (4) to give

$$y(x)=W(x)W^{-1}(a)y(a).$$

Now, we define

$$Y(x)=W(x)W^{-1}(a), \tag{5}$$

and obtain the expression for the solution in the form

$$y(x)=Y(x)y(a). \tag{6}$$

Thus, the columns of $Y(x)$ are solutions of the m initial-value problems given by

$$dY(x)/dx=A(x)Y(x), \tag{7a}$$

$$Y(a)=I. \tag{7b}$$

The elements of the rows of $Y(x)$ represent those fundamental variables that are contained in the corresponding rows of $y(x)$ in Eq. (1).

It is important to note that the solutions $Y(x)$ depend only on the geometric and material properties of the shell [i.e., on the coefficient matrix $A(x)$] but not on the boundary conditions. The same solutions $Y(x)$ can be used for any appropriate boundary conditions imposed at the edges of a given shell. For this reason, the solution of the free-vibration problem of a rotationally symmetric shell is completely determined by $Y(x)$.

At this point, it should be recalled that at each edge of the shell, i.e., at $x=a$ and $x=b$, $m/2$ quantities must be prescribed. If these quantities are the fundamental variables, then $m/2$ elements of $y(a)$ and $m/2$ elements of $y(b)$ are known, and the solution for the $m/2$ unknown elements of $y(a)$ is directly obtained from Eq. (6) when evaluated at $x=b$ in the form

$$y(b)=Y(b)y(a). \tag{8}$$

It should be clear that for free-vibration problems, when the prescribed variables vanish, Eq. (8) constitutes a linear homogeneous system of $m/2$ equations with $m/2$ unknown elements of $y(a)$. Requiring the vanishing of the determinant of the coefficient matrix of this system of equations gives the frequency equation and natural frequencies of the system and a solution for $y(a)$. Once all elements of $y(a)$ are known, the mode shapes corresponding to a particular frequency can be obtained at any desired value of x from Eq. (6).

The above analysis is easily extended to such cases when the boundary conditions at $x=a$ and $x=b$ are imposed on new variables $u(x)$, which are linear combinations of the fundamental variables and are related to $y(x)$ by

$$u(x)=T(x)y(x), \tag{9}$$

where $T(x)$ is a given (m,m) nonsingular matrix. In this case, we obtain an analogous equation to Eq. (6) in terms of $u(x)$ by substituting $y(x)$ from Eq. (9) into (6) and solving for $u(x)$ in the form

$$u(x)=T(x)Y(x)T^{-1}(a)u(a). \tag{10}$$

If we define

$$U(x)=T(x)Y(x)T^{-1}(a), \tag{11}$$

we obtain

$$u(x)=U(x)u(a), \tag{12}$$

which is of the same form as Eq. (6). Thus, the frequency equation and the solution for $u(x)$ can be obtained in the same manner as those for $y(x)$ by using the solutions $Y(x)$ defined by Eqs. (7) and simply transforming them as shown by Eq. (11).

In Sec. III, it becomes necessary to rearrange the elements of $u(x)$ at some values of x in such a way that the known and unknown variables of $u(a)$ and $u(b)$ are separated into two partitioned matrices. The transformation matrix $T(x)$ can easily contain such a rearrangement.

In the remainder of this section, it is regarded that the solution of the problem is given by Eq. (12), and that $u(a)$, $u(b)$ contain those quantities that are prescribed at $x=a$ and $x=b$, respectively. Moreover, for added simplicity of the analysis, it is regarded that the first $m/2$ elements of $u(a)$, denoted by $u_1(a)$, and the last $m/2$ elements of $u(b)$, denoted by $u_2(b)$, are the prescribed variables. All the necessary transformations for going from $y(x)$ to $u(x)$ are given by Eqs. (9) and (11), and it should be clear that no loss of generality is involved in such a transformation.

With this transformation, Eq. (12) evaluated at $x=b$ can be written as a partitioned matrix product in the form

$$\begin{bmatrix} u_1(b) \\ \hline u_2(b) \end{bmatrix}=\left[\begin{array}{c|c} U^1(b) & U^2(b) \\ \hline U^3(b) & U^4(b) \end{array}\right]\begin{bmatrix} u_1(a) \\ \hline u_2(a) \end{bmatrix}, \tag{13}$$

where the $(m/2,m/2)$ matrices $U^i(b)$ are the partitioned matrices of $U(b)$. If we assume that for free vibration $u_1(a)=u_2(b)=0$, then the unknowns $u_2(a)$ are directly obtained from

$$U^4(b)u_2(a)=0. \tag{14}$$

Since a nontrivial solution for $u_2(a)$ is possible if the matrix $U^4(b)$ is of rank $m/2-1$, the frequency equation of the system is given by

$$|U^4(b)|=0. \tag{15}$$

Once a frequency is found that satisfies Eq. (15), the corresponding solution for $u_2(a)$ is obtained from

$$u_2^i(a)=d(-1)^{i+1}|M_i|, \tag{16}$$

where $u_2^i(a)$ denotes the ith element of $u_2(a)$, d is an arbitrary constant, and $|M_i|$ is the determinant obtained from any $(m/2-1, m/2)$ submatrix of rank

$m/2-1$ contained in $U^4(b)$ by deleting the ith column. After $u_2(a)$ is calculated from Eq. (16), the corresponding mode shapes for a particular natural frequency are found from Eq. (12).

For the determination of the mode shapes from Eq. (12) in terms of the quantities that are prescribed at $x=b$, values of $U(x)$ are necessary at any x at which the output is desired. For this purpose, during the integration process for the determination of $Y(b)$, the intermediate values of $Y(x)$ should be stored from which the corresponding $U(x)$ are obtained by Eq. (11). If the mode shapes are desired in terms of the fundamental variables, then, after determination of $u(a)$ by means of Eq. (16), the corresponding $y(a)$ can be obtained from Eq. (9), and then the mode shapes are calculated from Eq. (6) at those points at which $Y(x)$ is stored.

Thus, the free-vibration problem of an arbitrary rotationally symmetric shell is reduced to (1) integration for a given Ω of Eqs. (1) m times from $x=a$ to $x=b$ to determine the fundamental solutions $Y(x)$, starting with $Y(a)=I$; (2) transformation of $Y(x)$ at certain values of x to the boundary-condition variables by Eq. (11); (3) calculation of the value of the determinant of $U^4(b)$; (4) repetition of this process for the evaluation of the natural frequency Ω_i at which Eq. (15) is satisfied; and (5) calculation of the mode shapes from Eq. (16) and either Eq. (12) or Eq. (6) at a particular natural frequency.

The method thus far is essentially a generalization of the one employed in the analysis of axisymmetric vibration of a conical shell in Ref. 3. It works very well for a shell with a relatively short interval (a,b). However, when the length of the meridian of the shell is increased, the elements of $U(b)$ increase rapidly in magnitude while the value of the frequency determinant does not, and, consequently, an increasing number of significant digits is subtracted out in the process of calculation of the determinant of $U^4(b)$. Similarly, because of the large values of the elements of $U(x)$ at large values of x, accuracy is invariably lost when the mode shapes are obtained from Eqs. (6) or (12). This phenomenon was pointed out previously in Ref. 1, and also by Galletly[5] *et al.* and Sepetoski *et al.*[6] in connection with shells of revolution, and by Fox[7] with regard to some boundary-value problems of second-order differential equations.

To make the direct integration technique applicable also to long shells, a method was developed in Ref. 1 for the static analysis of shells of revolution in which the initial-value problems for the determination of $Y(b)$ are defined over suitably selected short segments. This method is extended in Sec. III to apply to eigenvalue problems.

III. EXTENSION TO A MULTISEGMENT SHELL

The method of the analysis of shells of revolution proposed in Ref. 1 is based on the idea that if the m initial-value problems for the determination of $Y(x)$ are defined over sufficiently short segments of the shell, then the elements of $Y(x)$ are not large in magnitude and no loss of accuracy occurs due to subtraction of large almost equal numbers. This idea is also applicable in the analysis of free vibration of shells of revolution.

As it was done in Ref. 1, let the shell be divided into M segments, which are denoted by S_i in which $x_i \leq x \leq x_{i+1}$, where $i=1, 2, \cdots, M$. The ends of the shell, denoted in Sec. II by $x=a$ and $x=b$, are now at $x=x_1$ and $x=x_{M+1}$, as shown in Fig. 1. The solution in each segment S_i is given by Eq. (6) in the form

$$y(x)=Y_i(x)y(x_i), \tag{17}$$

where $Y_i(x)$ are obtained from the initial-value problems defined in S_i by

$$dY_i(x)/dx=A(x)Y_i(x), \quad Y_i(x_i)=I. \tag{18}$$

Continuity requirements on all fundamental variables[8] at the end points of the segments (i.e., at $x=x_i$, $i=2, 3, \cdots, M+1$) lead from Eq. (17) to

$$y(x_{i+1})=Y_i(x_{i+1})y(x_i), \tag{19}$$

where $i=1, 2, \cdots, M$.

The analysis and the prescription of general boundary conditions in terms of linear combinations of the fundamental variables are considerably simplified if the $T(x)$ transformation introduced in Sec. II is employed to obtain the quantities prescribed at the end points ($x=x_1$ and $x=x_{M+1}$) of the shell in the form

$$u(x_1)=T(x_1)y(x_1), \tag{20a}$$

$$u(x_{M+1})=T(x_{M+1})y(x_{M+1}). \tag{20b}$$

According to Eq. (11), we must then transform the solutions $Y_1(x_2)$ and $Y_M(x_{M+1})$ by

$$U_1(x_2)=Y_1(x_2)T^{-1}(x_1), \tag{21a}$$

$$U_M(x_{M+1})=T(x_{M+1})Y_M(x_{M+1}). \tag{21b}$$

The matrix $T(x_1)$ is such that the first $m/2$ elements of $u(x_1)$ are the prescribed quantities at x_1, while $T(x_{M+1})$ is such that the last $m/2$ elements of $u(x_{M+1})$ are the prescribed quantities at $x=x_{M+1}$. The variables $y(x)$ at interior points of the shell (i.e., at x_i with

[5] G. D. Galletly, W. T. Kyner, and C. E. Moller, J. Soc. Ind. Appl. Math. **9**, 489–513 (1961).

[6] W. K. Sepetoski, C. E. Pearson, I. W. Dingwell, and A. W. Adkins, J. Appl. Mech. **29**, 655–661 (1962).

[7] L. Fox, *Numerical Solution of Ordinary and Partial Differential Equations* (Addison-Wesley Publishing Co., Reading, Mass., 1962), p. 61.

[8] It should be mentioned here that the continuity requirement of the fundamental variables ensures continuity of the stress resultants but not necessarily of the stresses, which is consistent with the formulation of the theory of shells in terms of resultants.

$i=2, 3, \cdots, M$) need not be transformed, and they can be left in terms of the fundamental variables in the order in which the appear in Eq. (3).

In the following, it is assumed that the transformations given by Eqs. (20) and (21) are carried out, but that for simplicity of notation the symbols $y(x_1)$, $y(x_{M+1})$, $Y_1(x)$, $Y_M(x)$ are used instead of $u(x_1)$, $u(x_{M+1})$, $U_1(x)$, $U_M(x)$. In this way, the continuity Eqs. (19) can be employed without change when it is understood that the transformations of Eqs. (20) and (21) have been performed.

Using the partitioned matrix product as given by Eq. (13), Eq. (19) can be written as

$$y_1(x_{i+1}) = Y_i^1(x_{i+1})y_1(x_i) + Y_i^2(x_{i+1})y_2(x_i), \quad (22a)$$

$$y_2(x_{i+1}) = Y_i^3(x_{i+1})y_1(x_i) + Y_i^4(x_{i+1})y_2(x_i), \quad (22b)$$

where $i=1, 2, \cdots, M$, and the superscripts of Y and subscripts of y have the same meaning as in Eq. (13).

Equations (22) constitute a system of $2M$ linear homogeneous matrix equations with $2M$ unknowns: $y_1(x_i)$, $i=2, 3, \cdots, M+1$, and $y_2(x_i)$, $i=1, 2, \cdots, M$. For free-vibration problems, we assume that the prescribed quantities $y_1(x_1)=y_2(x_{M+1})=0$, and then Eqs. (22) can be arranged in matrix form as

$$\left[\begin{array}{cccc|cc} Y_1^2 & -1 & 0 & 0 & 0 & 0 \\ Y_1^4 & 0 & -I & 0 & 0 & 0 \\ 0 & Y_2^1 & Y_2^2 & -I & 0 & 0 \\ 0 & Y_2^3 & Y_2^4 & 0 & -I & 0 \\ \hline 0 & 0 & 0 & Y_M^1 & Y_M^2 & -I \\ 0 & 0 & 0 & Y_M^2 & Y_M^4 & 0 \end{array}\right] \left[\begin{array}{c} y_2(x_1) \\ y_1(x_2) \\ y_2(x_2) \\ y_1(x_M) \\ \hline y_2(x_M) \\ y_1(x_{M+1}) \end{array}\right] = 0, \quad (23)$$

where the dashes indicate a pair of rows and/or columns of Eqs. (22) with $i=3, 4, \cdots, M-1$, and for brevity we have written Y_i^j in place of $Y_i^j(x_{i+1})$.

As it was stated in Refs. 1 and 6, if the system of Eqs. (23) is solved by means of Gaussian elimination, then the loss of accuracy due to subtraction is avoided. For this purpose, we triangularize the coefficient matrix of Eqs. (23) to the form

$$\left[\begin{array}{cccc|cc} E_1 & -I & 0 & 0 & 0 & 0 \\ 0 & C_1 & -I & 0 & 0 & 0 \\ 0 & 0 & E_2 & -I & 0 & 0 \\ 0 & 0 & 0 & C_2 & -I & 0 \\ \hline 0 & 0 & 0 & 0 & E_M & -I \\ 0 & 0 & 0 & 0 & 0 & C_M \end{array}\right] \left[\begin{array}{c} y_2(x_1) \\ y_1(x_2) \\ y_2(x_2) \\ y_1(x_M) \\ \hline y_2(x_M) \\ y_1(x_{M+1}) \end{array}\right] = 0. \quad (24)$$

The $(m/2,m/2)$ matrices E_i and C_i ($i=1, 2, \cdots, M$), obtained by the standard Gaussian elimination procedure from Eqs. (23), are evaluated successively from

$$E_1 = Y_1^2, \quad (25a)$$

$$C_1 = Y_1^4 E_1^{-1}, \quad (25b)$$

$$E_i = Y_i^2 + Y_i^1 C_{i-1}^{-1}, \quad (25c)$$

$$C_i = (Y_i^4 + Y_i^3 C_{i-1}^{-1}) E_i^{-1}, \quad (25d)$$

where $i=2, 3, \cdots, M$.

A nontrivial solution of the system of Eqs. (23) is possible if the $(m/2,m/2)$ matrix C_M is of rank $m/2-1$, and then the solution for the elements of $y_1(x_{M+1})$, denoted by $y_1^i(x_{M+1})$, where $i=1, 2, \cdots, m/2$, is given by

$$y_1^i(x_{M+1}) = d(-1)^{i+1}|M_i|, \quad (26)$$

where again d is an arbitrary constant and the determinant $|M_i|$ is obtained from any $(m/2-1, m/2)$ submatrix of rank $m/2-1$ contained in C_M by deleting the ith column. Once $y_1(x_{M+1})$ is known, the remaining unknowns in the column matrix of Eq. (24) are determined in successive order directly from Eq. (24). In this process, we utilize the inverses of E_i and C_i, which are needed in Eqs. (25).

The above free-vibration analysis is applicable to any rotationally symmetric shell with two circular edges, at each of which $m/2$ boundary conditions in terms of linear combinations of the fundamental variables are specified. Situations may arise when the shell has no open edges, and we may then regard that the end points $x=x_1$ and $x=x_{M+1}$ are joined together. If this joint occurs on the axis of symmetry (e.g., a closed spherical shell) where $r=0$, then by introducing exact solutions in the vicinity of the axis of symmetry, this shell can be treated in the same way as a shell with two open edges.

If, however, the joint of $x=x_1$ and $x=x_{M+1}$ does not occur on the axis of symmetry (e.g., a complete torus), then, in general, integration must be carried out over the closed curve of the cross section of the shell, which, of course, should not cross or touch the axis of symmetry. The above analysis can be extended to such closed shells by requiring in Eqs. (22) that $y_1(x_1) = y_1(x_{M+1})$ and $y_2(x_1)=y_2(x_{M+1})$, i.e., that the fundamental variables at the starting edge are equal to those at the final edge. The resulting matrix is then triangularized to the form of Eq. (24), and the frequency equation and mode shapes are obtained in the same manner as given above. Since no boundary conditions are given, the transformations to $u(x)$ in this case are not necessary.

IV. CALCULATION OF NATURAL FREQUENCIES AND MODE SHAPES

On the basis of the coefficients $A(x)$ and the multisegment analysis given above, the author has prepared a digital-computer program[9] that is applicable to any rotationally symmetric shell for the analysis of free vibration. The input of this program requires (1) properties of the shell: $R_\phi(x)$, $r(x)$, $h(x)$, $c(x)$, $\nu(x)$, $E(x)$ at every x; (2) boundary conditions [transformation matrix $T(x)$] at $x=x_1$ and $x=x_{M+1}$; (3) the frequency interval that is to be investigated for natural frequencies; (4) the number of segments in which the shell is to be divided. The program outputs each natural frequency Ω_i and the corresponding mode shapes of all fundamental variables.

The numerical integration for the determination of the solutions $Y_i(x)$ in each segment S_i is carried out by means of a predictor–corrector method in which the step size is automatically selected according to a prescribed accuracy of the solution. Since the shell can have discontinuities in its properties, it is regarded that such an automatic selection of step size is absolutely necessary for a controlled accuracy and optimum efficiency of the solution.

In order to get the precise effect of the discontinuities in the shell properties, the segments should be selected in such a way that the end point of one segment coincides with the location of a discontinuity. Otherwise, the segments may have arbitrary lengths, so that within each segment $\beta \leq 3-5$, where $\beta = l[3(1-\nu^2)]^{\frac{1}{4}}/(Rh)^{\frac{1}{2}}$, l is the meridional length of the segment, and R is a radius of curvature. The program outputs the mode shapes at the ends of the segments twice—once from Eqs. (22) and then from Eq. (17)—and their comparison offers a direct check of the satisfaction of the continuity requirements. If these repeated values do not match a sufficient number of significant digits, then accuracy has been lost by subtraction and the number of segments must be increased.

From the results of Sec. III, it follows that for the multisegment analysis the natural frequencies Ω_i are the roots of the determinantal equation

$$|C_M| = 0, \tag{27}$$

which are found by calculating the determinant of C_M for incremented values of Ω until a change in sign occurs. Then, by inverse interpolation the particular $\Omega_i (i=1, 2, \cdots)$ are obtained at which $|C_M| = 0$.

After applying the program to the analysis of free vibration of many shell configurations, it was observed that at some values of Ω, denoted by Ω_F, a change of sign of any one of the determinants of the matrices $E_i (i=1, 2, \cdots, M)$ or $C_i (i=1.. 2, \cdots, M-1)$ may occur. From Eqs. (25), it follows that at these frequencies $|C_M|$ becomes infinite, which is accompanied by a sign change of $|C_M|$ and all the corresponding values of $y_j(x_i)$ obtained from Eqs. (26) and (24). Since $|C_M|$ does not actually vanish, no nontrivial solution exists at Ω_F, and they are not the natural frequencies of the system. However, since an automatic search of the roots of Eq. (27) is based on the changes of the sign of $|C_M|$, then at $\Omega = \Omega_F$ false natural frequencies are indicated at which the inverse-interpolation technique fails.

We can remove the false sign changes from the frequency equation by using the following procedure: Recalling that at Ω_F both $|C_M|$ and $y_j(x_i)$ change sign, we should seek the roots of $|C_M|$ divided by some nonzero element of $y_j(x_i)$. It is convenient to choose for this purpose the first element of $y_2(x_1)$, denoted by $y_2^1(x_1)$. Thus, the modified frequency equation is

$$Z = |C_M| / y_2^1(x_1), \tag{28}$$

which gives the same roots as Eq. (27), but is free from the sign changes at $\Omega = \Omega_F$.

We should also note that if in Eq. (22) $y_2(x_{M+1})$ is not set equal to zero, then $y_2(x_{M+1})$ appears in the last row on the right-hand side of Eq. (23). After triangularization by means of Eqs. (25), the last row of Eq. (24) then reads

$$y_2(x_{M+1}) = C_M y_1(x_{M+1}). \tag{29}$$

The use of the frequency Eq. (28) involves the determination of all the $y_j(x_i)$'s by means of Eqs. (26) and (24) at every trial frequency Ω, so that $y_2^1(x_1)$ can be found. According to Eq. (29), the values of $y_j(x_i)$, obtained from Eqs. (26) and (24) for any Ω (except Ω_i), represent the forced response to the nonzero boundary values $y_2(x_{M+1})$, which are given by Eq. (29). Clearly, for a natural-frequency Ω_i, it follows from Eq. (29) that $y_2(x_{M+1}) = 0$.

While the computer time for this additional calculation is found to be negligible, the zeros of the modified frequency Eq. (28) are obtained much more easily than those of Eq. (27). This is clearly shown in Fig. 2, where

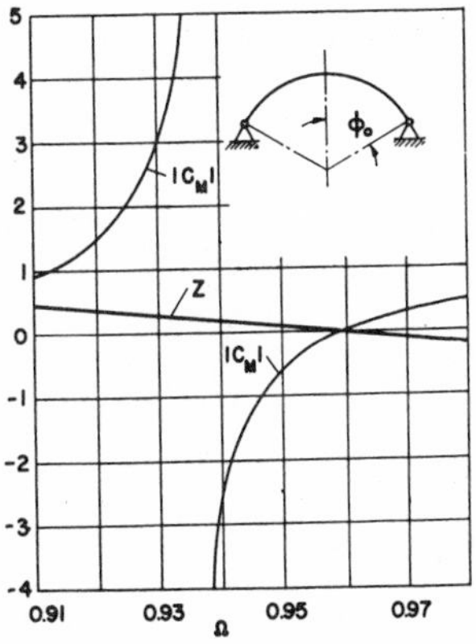

FIG. 2. Two types of frequency equations vs Ω.

[9] The program was prepared and all calculations were carried out on the IBM-709 computer at the Yale Computer Center.

TABLE I. Natural frequencies Ω_i of axisymmetric vibration of truncated conical shell.

	Ref. 3	This paper
Ω_1	0.00238	0.00238
Ω_2	0.00292	0.00291
Ω_3	0.00358	0.00358

TABLE II. Natural frequencies Ω_i of axisymmetric vibration of spherical shell with fixed–hinged edge at $\phi=60°$.

	Ref. 10	This paper
Ω_1	0.962	0.959
Ω_2	1.334	1.328
Ω_3	2.128	2.114

both $|C_M|$ and Z are plotted versus Ω near the lowest mode of a spherical shell, which is considered in Sec. V. While a sign change of $|C_M|$ occurs at $\Omega=0.936$, the F curve is free of such a disturbance. Of course, both curves predict correctly the first natural frequency at $\Omega=0.959$.

Once a natural frequency is obtained, the corresponding mode shapes at x_{M+1} are found from Eq. (26) and then at the remaining x_i from Eq. (24) starting with $i=M$ and ending with $i=1$. If the fundamental variables of the mode shapes are desired at more points than at the ends of the segments, then they are directly obtained from Eq. (17), provided that the values of $Y_i(x)$ at these points are stored during the integration from $x=x_i$ to $x=x_{i+1}$ in each segment. Before using Eq. (17), we should recall that $y(x_1)$ and $y(x_{M+1})$ are really the transformed variables $u(x_1)$ and $u(x_{M+1})$, and in order to use $Y(x)$ in Eq. (17) we must transform them back to the fundamental variables by means of Eqs. (20).

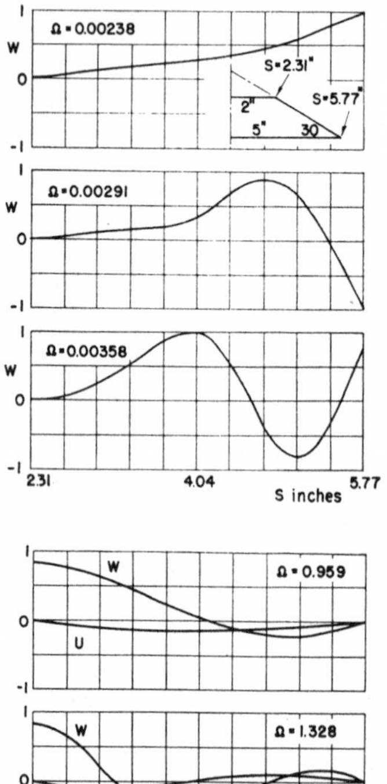

FIG. 3. Mode shapes of normal displacement w for truncated conical shell.

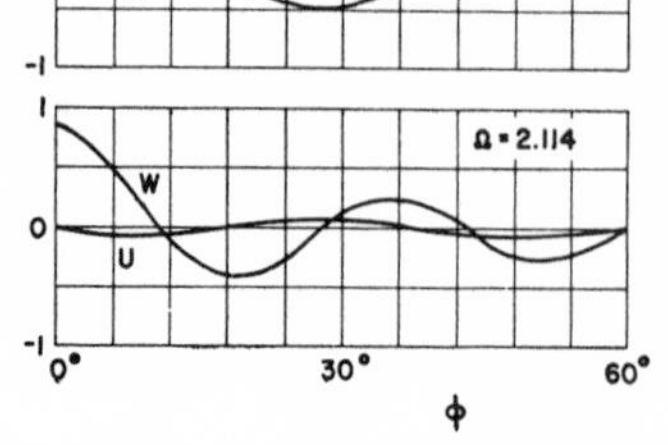

FIG. 4. Mode shapes of displacements for spherical shell.

V. ILLUSTRATIVE EXAMPLES

A. Conical Shell

In Ref. 3, the natural frequencies and mode shapes of free axisymmetric vibration are determined for a truncated conical shell in the shape of a loudspeaker cone. To show the accuracy of the present method, exactly the same problem as in Ref. 3 is considered by means of the general program. The comparison of the first three natural frequencies is given in Table I. Letting $L=h$, the frequency parameter is $\Omega=\omega h/c$, where the speed of sound in Ref. 3 is $c=7.071\times10^4$ in./sec, $h=0.025$ in., and $\nu=0.25$. The corresponding mode shapes for the first three modes obtained by the general program, together with other dimensions of the shell, are shown in Fig. 3. The agreement with the results of Ref. 3 is excellent.

B. Spherical Shell

The first three modes of axisymmetric vibration of a spherical shell with a fixed–hinged edge at $\phi=60°$ have been calculated by means of the hypergeometric series solutions of bending theory and are included in a recent paper.[10] The natural frequencies $\Omega=\omega R/c$ for $\nu=0.3$, $h/R=0.05$, are also calculated by the general program and are given in Table II, while the corresponding mode shapes are shown in Fig. 4.

The lowest natural frequency of a nonsymmetric mode ($n=2$) as predicted by the classical bending theory of shallow shells is given in Ref. 11: $\Omega_1=2.34$. For $h/R=0.05$, $\nu=0.3$, and a clamped edge at $\phi=30°$, the corresponding result obtained by the general program is $\Omega_1=2.15$.

While the agreement with the results of Ref. 10 is within a few percent, the nonsymmetric frequency pre-

[10] A. Kalnins, "Effect of Bending on Vibrations of Spherical Shells," J. Acoust. Soc. Am. **36**, 74–81 (1964).

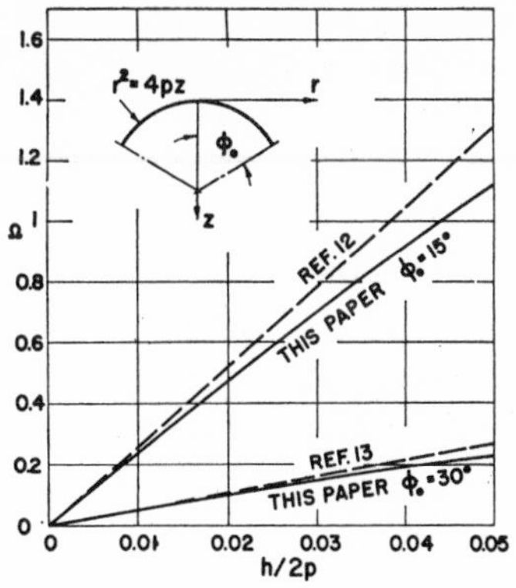

FIG. 5. Natural frequencies of inextensional modes of paraboloidal shell vs $h/2p$ for $n=2$ and $\nu=0.3$.

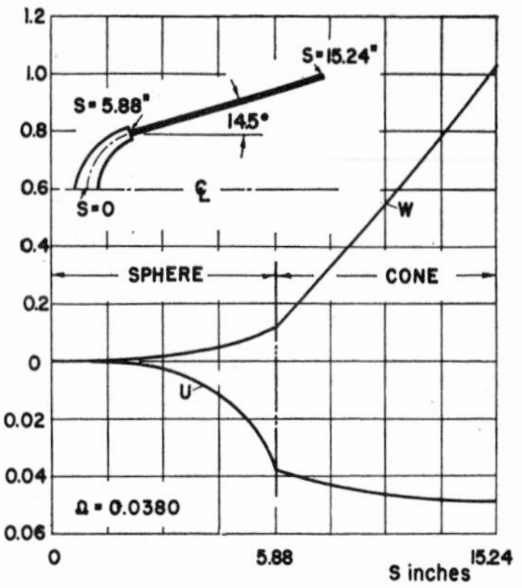

FIG. 6. Transverse and meridional displacements for inextensional mode ($n=3$) of sphere–cone shell combination.

dicted by the theory of shallow shells in Ref. 11 is higher by about 9%. The reason for this difference may be that the admissible maximum limiting angle in the theory of shallow shells is smaller for $n>0$ than for $n=0$.

The inextensional theory of deformation of shells as formulated by Lord Rayleigh can be used in the analysis of the inextensional modes of shells of revolution. Since the inextensional modes must be included in the general bending theory of shells, it may be of interest to compare specific results obtained by the general program of this paper and the inextensional theory. The comparison is made here with the results given by Johnson and Reissner,[12] by Lin and Lee[13] for paraboloidal shells of revolution, and by Saunders and Paslay[14] for a sphere–cone combination. Since in the inextensional theory complete boundary conditions are not prescribed, we compare here the results of Refs. 12–14 with those of this paper where the respective shells are assumed to be closed at one end while the other edge is free.

C. Paraboloidal Shell of Revolution

We consider a paraboloidal shell whose properties are given by

$$R_\phi = 2p/\cos^3\phi; \quad r = 2p\tan\phi, \tag{30}$$

where p denotes the focal distance. Let $L=2p$. The natural frequencies for various ratios of $h/2p$ and two values of the limiting angle ϕ_0 are calculated for $n=2$ and $\nu=0.3$ by the general program and shown in Fig. 5.

In Ref. 12, a formula for the frequency of a shallow first-degree paraboloidal shell is found in the form

$$\Omega_n = \omega_n 2p/c = (h/2p)n[(n^2-1)/3(1+\nu)]^{\frac{1}{2}}/\sin^2\phi_0, \tag{31}$$

where ϕ_0 should be sufficiently small so that the condition of shallowness is satisfied. The corresponding relation for $\phi_0=30°$ and $n=2$ given in Ref. 13 is

$$\Omega_2 = [4\Delta/3(1-\nu^2)]^{\frac{1}{2}} h/2p, \tag{32}$$

where $\Delta=18$. The natural frequencies given by these two formulas are also included in Fig. 5.

D. Sphere–Cone Shell Combination

An interesting application of the general program of this paper can be made to the free-vibration problem of a sphere–cone shell combination considered by means of the inextensional theory in Ref. 14. The properties of the spherical part are given by $h=h_1\cos\phi$, $R_\phi=R$, $r=R\sin\phi$, while those of the conical part are $h=h_2$, $R_\phi=\infty$, $\phi=75.5°$, $r=b\cos\phi$, where the distance b is measured from the projected apex of the conical shell.

It should be noted that at the shell joint (at $\phi=75.5°$ of the spherical shell) the thickness h and R_ϕ are discontinuous. It can be seen from Eq. (1), together with the coefficients $A(x)$, that if the fundamental variables on an edge $\phi=$const are continuous, then their derivatives may not be continuous at points where the shell has discontinuities in its properties. For example, the first equation of Eq. (1) reads $dw_n/ds = -\beta_{\phi n} + u_{\phi n}/R_\phi$. Obviously, if the rotation of the normal $\beta_{\phi n}$ and the tangential displacement $u_{\phi n}$ are continuous but R_ϕ is discontinuous, then the derivative of w_n cannot be continuous. This conclusion in no way contradicts physical reasoning, and it follows directly from the mathematical formulation of the theory of shells.

---

[11] A. Kalnins, Proc. U. S. Natl. Congr. Appl. Mech., 4th, pp. 225–233 (1963).
[12] M. W. Johnson and E. Reissner, J. Math. Phys. 34, 335–346 (1955).
[13] Y. K. Lin and F. A. Lee, J. Appl. Mech. 27, 743–744 (1960).
[14] H. Saunders and P. R. Paslay, J. Acoust. Soc. Am. 31, 579–583 (1959).

For the particular sphere–cone combination considered in Ref. 14, we let $h_1=0.5$ in., $h_2=0.045$ in., $\nu=0.3$, $R=4.46$ in., and calculate the lowest natural frequency for $n=3$. The result obtained from the general program of this paper is $\Omega_1=\omega_1 R/c=0.0381$. The corresponding value, given in Ref. 14 and obtained by means of the inextensional theory, is $\Omega_1=0.042$. The discrepancy in the predicted frequency may be due to the fact that, as pointed out in Ref. 14, in the inextensional theory not all fundamental variables can be made continuous at the shell joint. To illustrate the type of the mode of free vibration, which is possible at this frequency, the mode shapes obtained by the present method are given in Fig. 6. The absence of nodes indicates that this is indeed the inextensional mode.

ACKNOWLEDGMENT

Results reported in this paper were obtained in a course of research supported by the National Science Foundation, grant No. 23922.

Part IV

DYNAMIC RESPONSE OF SHELLS

Editor's Comments on Papers 26 Through 31

This chapter is concerned with the response of a shell to time-dependent loads. An overview of response problems reveals different methods of analysis for time distributions of loads that are periodic, have been acting for all times, and for which a steady-state response is sought, and for time distributions in the form of a pulse for which a transient response is sought.

The simplest response problems arise for finite shells when the time distribution of the loads is periodic and the steady-state response is desired. This means that the loads are assumed to have been acting for a long time, many reflections have occurred, and any transient motions that were produced by the starting of the excitation (progressing wave fronts) have died out. The spatial distribution of the loads is arbitrary.

The solutions to such problems are obtained by first representing the periodic time distributions by Fourier series and then obtaining the response to each harmonic Fourier component separately. For finite systems, such problems are solved with the same methods as those used for static loads. The only difference is that the inertia terms with harmonic time variations are added. Because the mathematical problem is the same as that in the static case, no specific papers that illustrate the analysis for finite shells are included in this volume. Most of the automated computer programs that are available for static shell analysis can also be used to calculate the harmonic response.

The response problems for periodic loads can be extended further to those cases when the loads are applied to a small portion of the shell surface or to one edge, and when other edges are sufficiently far away. Then, before any reflections have occurred, the motion of the shell resembles continuous wave motion, with the nodal lines moving away from the region of load application. The periodic time distributions of the loads are again represented by their Fourier series, and the solution is obtained first for each harmonic, and then the harmonic waves are summed. The interesting feature of these problems is that, at a given frequency, only definite types of waves can be propagated; and, depending on the space distribution of the loads, it is possible that in some low-frequency interval no wave motion at all can be propagated. These problems are usually formulated for systems that can be imagined to have infinite extent, such as beams, plates, and shells of revolution. Solutions of these problems are valid at times well after the initial wave front has passed and the transients caused by the starting of the excitation have died out. Examples of this type of problem are provided by Paper 8 for an infinite plate, by Section 7 of Paper 21 for an unlimited shallow spherical shell (which is represented mathematically by an unlimited parabolic shell of revolution), and by the following two papers for cylindrical shells.

In Papers 26a and 26b, D. C. Gazis considers continuous harmonic waves in an infinite cylindrical shell that are propagated in the axial direction. The waves can be imagined generated by sources that are distributed in one section of the shell (say, $z = 0$), and they impart some disturbance to the shell that is harmonic in time, with a frequency ω. Such excitation produces continuous harmonic waves that travel along the infinite shell in both directions from $z = 0$. They are superpositions of spherical waves which emanate from the points within the section $z = 0$, where the sources are applied, and are then reflected from the two cylindrical bounding surfaces of the shell; they establish standing waves with fixed nodes in the radial and circumferential directions, but produce wave motions in the axial direction.

The object of this type of paper is to predict the axial wavelengths and velocities (phase velocities), and the radial and circumferential standing-wave patterns, that are possible to be propagated from given sources at $z = 0$. Such information is revealed by the dispersion curves shown in Figures 3 through 10. For example, referring to Figure 8, no waves with a circumferential wave number $n = 2$ can be propagated at a frequency $\omega = \frac{1}{2}\omega_s$. What happens is that, at that frequency, the harmonic motion has an axial distribution that decays exponentially when going away from $z = 0$. (This motion is sometimes called an "evanescent wave.") At $\omega = 2\omega_s$, three types of waves can be propagated. Their wavelengths can be read off Figure 8, and their radial standing-wave patterns are described by the code numbers 1, 2, and 3 (attached to the curves) and are given by the r-dependent functions of the solution. It must be emphasized that only these waves can be propagated at each of the given frequencies, regardless of the radial distribution of the sources. For a general radial distribution of the sources, the other expected components are simply nonpropagating (evanescent).

Gazis' results are based on the three-dimensional theory of elasticity, and they include an infinite number of radial standing-wave patterns. At a given frequency, the patterns having a specific number of cylindrical nodal surfaces are propagating and the higher patterns are not. These conclusions provide a logical basis for a mathematical model of a "shell" theory, which would be applicable only for frequencies that are below a limiting frequency. This means that only standing-wave patterns with a low number of radial nodes would be propagated. Therefore, if the limitation to low frequencies is acceptable, it would make sense to limit the theory at the outset to those radial standing-wave patterns that are propagated and leave out all the nonpropagating patterns with a higher number of radial nodes. The usual mathematical models for shells assume displacement components with at most one node across the thickness of the shell. Then, for a thin shell, at most six modes can be propagated. Starting from such a shell model, further approximations can be made. If the normal displacement is restricted to zero radial nodes, which eliminates the lowest thickness-stretch mode, then at most five modes are propagated. If the model is simplified further by assuming that the transverse shear strains are zero (classical shell theory), which eliminates the lowest thickness-shear modes, then at most three modes can be propagated. This information can be obtained from a number of papers that consider the same problem considered by Gazis, except that they use a mathematical model of shell theory (for references, see those listed in Gazis' papers). Such results as those in Figure 6 of Gazis' paper can be used to estimate the ranges of valid frequencies for each of the shell models.

Once it is recognized that the mathematical models of beam, plate, and shell theories are valid only within a certain low-frequency interval, then an important conclusion can be made with regard to the class of time distributions that can be modeled meaningfully by these models. The conclusion is that, as far as the predictions of the behavior of the physical systems are concerned, differences in the responses have no meaning when they are produced by two time distributions of loads that differ only in those Fourier components that have frequencies outside the low-frequency interval of validity.

It should be remembered that Gazis' results apply only after the initial wave front has passed and a steady-state motion is reached. When reading Gazis' paper, it is important to distinguish between the constant wave velocity with which the spherical waves emanate from the source points into a three-dimensional medium, and the phase velocity of the summed disturbance, consisting of the reflections from the cylindrical surfaces, that is described by Gazis. As seen in Figure 11, the phase velocity can range from infinity to zero, while the velocity of spherical waves has a fixed value. Excellent discussions of the meaning of phase velocity are given by M. Redwood (*Mechanical Waveguides*, Elmsford, N.Y.: Pergamon Press, pp. 29–30, 1960) and by J. D. Achenbach (*Wave Propagation in Elastic Solids*, Amsterdam: North-Holland, pp. 30–31, 1973). In order to give a physical interpretation of phase velocity, Redwood (page 30) draws the analogy of a continuous harmonic wave in water, striking a wall at an angle. The point of intersection of one crest of the wave with the wall travels along the wall at the phase velocity, which is different from the velocity at which the straight crests of the waves themselves propagate. The water wave is analogous to the spherical wave produced by a source point and the wall is analogous to the two cylindrical surfaces.

It is clear from the analogy that the phase velocity can approach infinity when the wave approaches normal incidence with respect to the wall. Also, the energy of a single harmonic wave is propagated with the so-called group velocity (see Achenbach, p. 211), which is zero when the phase velocity is infinite. So, an infinite phase velocity does not imply any physical contradiction.

When the time distribution of the loads is not periodic, the Fourier series representation must be replaced by the Fourier integral (see Achenbach, p. 39). Then the harmonic solutions can be regarded as solutions in the Fourier transform plane, and the response can be obtained by using them in the integral of the inverse Fourier transform. In the evaluation of this integral, it must again be remembered that in the high-frequency range the predictions of the mathematical model are unreliable.

The very process of inverting the Fourier transform can be a formidable task if it must be carried out numerically. There are four alternate methods that are commonly used: (1) modal method; (2) numerical integration in time; (3) Laplace transform in time; and (4) the method of characteristics. It should be understood that all four methods can be said to apply to all time-dependent problems, and, theoretically, they all will give identical results. However, in order to achieve that, one would have to use an infinite number of modes, an infinitesimal time step, invert the Laplace transform exactly, or trace and sum all individual reflecting waves. Because of such limitations, each of the methods is better suited for certain classes of problems. In general, it can be said that the first two methods are intended for finite systems. The Laplace transform method is intended for semi-infinite or infinite systems, and the method or characteristics is intended for the study of an individual wavefront as it enters a region of quiescence.

In order to discuss the limitations of the modal method, its main features will be indicated here. The loads will be assumed in the form $p(x,t) = P(x)f(t)$, where $p(x,t)$ and $P(x)$ can be regarded as vectors, and x and t denote position in space and time, respectively. In the modal method, $P(x)$ is represented by the series

$$P(x) = \sum_i c_i \rho h u_i(x) \tag{1}$$

where

$$c_i = \frac{\int_s P(x) \cdot u_i(x) dS}{\int_s \rho h u_i(x) \cdot u_i(x) dS} \tag{2}$$

and $u_i(x)$ is the displacement vector of the ith mode of free vibration, x stands for any number of space coordinates, S is the reference surface of the shell, ρ is mass density, and h is thickness. The solution is given by

$$V(x,t) = \sum_i V_i(x) c_i D_i(t) \tag{3}$$

where $V(x,t)$ is any variable in the solution; $V_i(x)$ is its value in the solution of the ith mode of free vibration; and, in the absence of damping and with zero initial conditions,

$$D_i(t) = \frac{1}{\omega_i} \int_0^t f(\tau) \sin \omega_i (t - \tau) d\tau \tag{4}$$

where ω_i is the natural frequency of free vibration in the ith mode.

In the modal method, each term in the load series [equation (1)] gives the corresponding term in the solution series [equation (3)] as an exact solution of the governing equations. The only error, then, comes from having an inexact space distribution of the load. The difficulty of representing the load distribution by the modal series is the same as that in any other Fourier series. It is clear that sharp, localized load distributions, which are zero over a part of the shell surface, will be modeled particularly inaccurately; and the zero values of the loads will never be achieved. Therefore, whatever the actual load distribution, the modal method will model it as if it were covering the whole shell surface. For the same reason, the solution will also be unable to predict accurately any discontinuities or regions of quiescence.

It can be seen from this discussion that the modal method is best suited for space distributions of loads that are smooth and cover the whole of the shell surface. When sharp variations in the space distributions occur, the number of modes required for the series will increase drastically; and, if a smaller number is used, the sharp variations will be simply modeled inaccurately.

It must be pointed out here, however, that this inaccuracy of the modal method refers to the solution of the mathematical model. Since the sole purpose of the mathematical model is to represent the behavior of a physical system, such inaccuracies must be viewed within the scope of limitations of the model itself. It makes no sense to strive toward additional accuracy in solutions which the model is unable to transfer to the physical system. Fortunately, this argument removes much of the significance of the inability of the modal method to represent sharp space variations of the loads, because a shell model cannot recognize any statically equivalent redistributions of loads over areas that can be enclosed within circles with minimum radii that are comparable to, or less than, the thickness. This means that a sinusoidal space distribution of a load, with wavelength comparable to the thickness, or smaller, can be replaced by a zero load without affecting meaningfully the prediction of the behavior of the shell. Since the wavelengths of the displacement distributions, $u_i(x)$, in the modal series become shorter as i increases, it is clear that a shell theory can digest beneficially only a finite number of terms of the modal series and, consequently, that it is worth including only such load distributions that can be made up of these finite number of terms. Addition of further terms simply does not filter through the model to the system.

A good example of this argument is provided by a point load on a shell. First of all, the shell model will be incapable of predicting the difference in the behavior of the shell when subjected to a point load or a statically equivalent load that is distributed over an area with a diameter of the order of magnitude of thickness, or less. When such a distributed load is represented by the modal series, it will not be meaningful to differentiate between a series that is truncated up to the terms with wavelengths of order of thickness or a series that is truncated to higher-order terms. Clearly, such truncated series will no longer give an exact "region of quiescence" at the moment when the load is applied; but, as argued in the foregoing, the differences in the predictions between short-wavelength and zero-load distributions are physically meaningless.

It should be added that it is impossible to say how many terms in the modal series are relevant for a given problem. It can only be said that theoretical predictions for load distributions that can be represented by terms with longer wavelengths will agree better with the behavior of the physical system, and that the classical shell theory will be limited to fewer terms than the higher-order shell theories which account for transverse shear and normal strains.

Another comment is in order here. From Figure 1 of Paper 24, by Wilkinson, it can be seen that the node count (represented by n in the Legendre function and related to wavelength) for each of the three infinite sequences of modes (bending, stretching, and shearing) begins with zero. This shows that the modal series should be written separately for the bending, stretching, and shearing modes of a shell, because the limitation of the ability to represent loads is applied to each of the mode types separately. Whether or not the sequence of a mode type is relevant in an application will depend on the character of the loads, which is revealed by the numerator of c_i in equation (2). For axisymmetric deformation of a spherical shell, the integrand of the numerator is given by

$$pw + p_\phi u_\phi + m_\phi \beta_\phi$$

where p, w and p_ϕ, u_ϕ are the normal and meridional (tangential) surface load and displacement components, respectively; m_ϕ is a distributed surface couple, and β_ϕ is the rotation of the normal in the meridional direction. In general it can be said that w determines the wavelengths in bending modes, u_ϕ in stretching modes, and β_ϕ in shearing modes. It follows, then, that in the presence of normal surface loads (e.g., pressure), the bending mode sequence is most relevant; for tangential loads, the stretching modes are most relevant; and for surface couples, the shear modes are most relevant. The last situation is very seldom actually encountered, which explains why the classical shell theory, which is incapable of predicting the shearing modes, is sufficient in most applications.

As far as the decision on the relevant terms that are to be included in the modal series is concerned, it should be viewed at first as a matter of the load representation in space, as given by the series of equation

(1). It is important to realize that the best load representation may not be achieved by blindly summing consecutive modes of the series, starting from the lowest, because of certain properties of the load distribution. For example, the loads may have certain symmetries that rule out some modes; and, for a thin shell, normal surface loads will rule out most of the stretching modes and all shearing modes. After a selected list of modes for the load distribution has been arrived at, the decision on the number of modes from this list that is actually used must be made on the decreasing magnitudes of the terms in the solution series, as given by equation (3). Once the point has been reached where the change in the solution by the addition of a further term is within an acceptable error, then the series can be declared converged.

Examination of the solution series reveals that its convergence may differ from that of the load series because of the additional factor $D_i(t)$, which means that the time distribution of the loads and the natural frequencies of the system will play their parts in the decision on the number of modes to be included. In order to see their role, it is useful to consider the integral $D_i(t)$ for some time inputs $f(t)$. Expressions of this integral for various $f(t)$ can be found in the book by S. P. Timoshenko et al. (*Vibration Problems in Engineering,* 4th ed., New York: John Wiley, pp. 99–101, 1974). For example, when $f(t)$ is a single rectangular pulse with height Q and width t_1 then

$$D_i(t) = \frac{Q}{\omega_i^2}(1 - \cos \omega_i t) \qquad \text{for } 0 \leq t \leq t_1 \tag{5}$$

and

$$D_i(t) = A \sin \omega_i\left(t - \frac{t_1}{2}\right) \qquad \text{for } t_1 \leq t \tag{6}$$

with the amplitude

$$A = \frac{2Q}{\omega_i^2} \sin \pi \frac{t_1}{T_i} \tag{7}$$

and

$$T_i = \frac{2\pi}{\omega_i}$$

where T_i is the natural period of free vibration.

It can be seen that, for a given t, the modal series converges in general as $1/\omega_i^2$, but that the convergence can be affected also by the duration of the pulse t_1. Equation (7) shows that for a pulse that is much shorter than the lowest natural period, the amplitude of $D_i(t)$ may not have the largest value in the first term of the modal expansion. Depending on the increasing values of the natural frequencies, the amplitude A may not reach a maximum until a number of modes have been added to the response. This illustration provides the reason why the convergence can be slower for some time distributions than for others, even though the space distributions of the loads are identical.

When using the modal method, it is important to recognize the two load properties that affect the convergence of the modal series. The space distribution affects the decay of the numerator of c_i in equation (2) and the time distribution affects the decay of $D_i(t)$. As long as the sources of convergence of the modal series are understood, the user of this method should be able to make sound decisions with regard to the terms that must be considered.

The numerical integration method in time replaces the time derivatives in the inertia terms by finite differences and expresses the governing equations at discrete points in time. The problem at each time step is solved in the same way as in the static case.

The numerical integration method shifts the difficulty of load representation to the static solution at each time step. Since a static solution is felt instantaneously everywhere in the shell, the method is also incapable of predicting discontinuities at wave fronts and zones of quiescence. For this reason, like the modal method, it is suited best for load distributions that are smooth and cover the whole shell. Its main difficulty lies in selecting the time steps that give an acceptable solution, and in knowing when the solution is acceptable. This difficulty corresponds to that in the modal method in deciding where to truncate the modal series. However, in the modal method a definite criterion is found in the comparison of each successive term with the sum of the preceding terms, while the only accuracy assessment in numerical integration is provided by the repetition of the integration with a smaller time step. The modal method requires some judgement on the part of the analyst in examining the loads and the mode shapes and in deciding what types of modes to include in the series. The bulk of its computational work lies in the determination of the modal solutions throughout the system. Usually it is a simple matter to calculate the coefficients c_i, as given by equation (2), together with the modal solutions. Once that is done, it only remains to evaluate the integral for $D_i(t)$ for a given time input, as required by equation (3). Each of the terms can be subjected to close scrutiny and its relevancy to the solution ascertained. The modal method possesses the exciting possibility that for some special space and time distributions of the loads the analyst can construct an acceptable solution with only a few terms, or even one term, of the series.

The numerical integration method bypasses all that and relies on the computer to produce answers. While having acquired more automation in its execution, it has obscured the make-up of the solution from which its accuracy could be assessed. It has also given up the capacity of producing simpler solutions to special problems because it obtains all solutions in the same way, by stepping away in time. Since the errors at progressive time steps are cumulative, the numerical integration method is most effective in obtaining a short-time response. Also, the numerical integration method can be readily extended to nonlinear problems, while this is not the case with the modal method.

The next three papers are examples of the modal and numerical integration methods when applied to shells. In Paper 27 Kraus and Kalnins use the straightforward modal approach. They do not start with equation (1) but arrive at its equivalent by their equation (6). Starting with the modal expansion of the load [equation (1)] would bring out in a clearer fashion the fact that each load term gives the corresponding solution term as an exact solution.

In Paper 28, J. P. Wilkinson uses a variant of the modal method, which makes use of a known static solution to the same problem obtained by omitting the time variation from all the load terms that are applied either on the shell surface or on the edges. This variant is often called Williams' method. Wilkinson calls it the "mode acceleration" method, because in its corresponding Duhamel integral [equation (4)] the second derivative of $f(t)$ (acceleration) appears. However, after deriving the integral, Wilkinson has integrated it by parts twice and obtained his equation (13).

The idea in Williams' method is that the modal expansion is added to the static solution. As shown by W. Ramberg [*Natl. Bur Stds. Res.*, **42,** 437–447 (1949)], for some problems it can lead to faster convergence than the ordinary modal method. J. Sheng [*AIAA J.*, **3,** 701–709 (1965)] claims that Williams' method will converge faster during the time when the loads are applied. A great advantage of Williams' method is that it can be applied to find the response to time-dependent edge loads while the usual modal method cannot. For general shells, such edge loads were considered by Kalnins [*Nucl. Eng. Design,* **20,** 131–147 (1972)]. A disadvantage is that a static solution must also be calculated, which is not needed in the ordinary modal method.

Paper 29, by D. E. Johnson and R. Greif, is a classic that was motivated by the many new applications arising in the U.S. space program in the early 1960s. It provided the groundwork for many further papers on this topic.

The remaining two methods for obtaining a dynamic response of a shell, those of Laplace transform and characteristics, have not as yet reached the state of automation that has been enjoyed by the modal and numerical integration methods for some time. Nevertheless, they offer powerful techniques for the extraction of specific information of the response.

The applicability of integral transform methods to response problems is discussed by L. Meirovitch (*Analytical Methods in Vibrations,* New York: Macmillan, pp. 328–387, 1967). Leaving out a few simple shell geometries, the inversion of the transformed variables for shell response problems involves the numerical integration of an integral. For Laplace transforms, the complex inversion integral is solved by contour integration. An example of such an inversion process of the Laplace transform can be found in Paper 13, by Miklowitz, which is an application to an infinite plate. There are no fundamental differences in the extension of this process to infinite shells, such as cylindrical, conical, or paraboloidal shells, except that the calculation is bound to be more complicated. When applied to a finite system, the inversion of the transform results in an infinite series which is identical to the modal expansion (see Meirovitch's book, p. 348, equation 8.83). This is so because the poles of the kernel of the inversion integral correspond to the natural frequencies of free vibration. For this reason, integral transforms are best suited for semi-infinite or infinite systems for which modal and numerical integration methods do not apply.

Theoretically, the Laplace transform method can be used to calculate the response of a point at any value of time. The difficulty of such a calculation lies in evaluation of the inversion integrals. The complexity of this procedure is evident from Paper 13. Some specific information about the response can be obtained from approximate evaluations of the inversion integral. For large values of time, the methods of saddle point, stationary phase, and steepest descent are applicable (see Achenbach's book, pp. 271–283). This approach has been used by R. Skalak for the analysis of the longitudinal impact problem of a semi-infinite, circular, elastic bar [*J. Appl. Mech.,* **24,** 59–64 (1957)]. Another approximate inversion of Laplace transform is based on asymptotic expansions with respect to a large transform parameter. This procedure is used by M. P. Mortell in Paper 30 to study waves on a spherical shell. Mortell uses asymptotic expansions for the transformed variables, of which the first term represents the behavior at the wavefront. The location of the wavefront is established with the method of characteristics which, for a spherical shell, predicts three constant wave speeds. Success of Mortell's method, however, depends on the possibility of finding the inverse of the variables at the wavefront [equation (48)]. He does achieve that for a spherical shell through equations (49) and (50). For other shell geometries this inversion may be much more complicated. In spite of this possible limitation of the applicability of Mortell's approach, his paper provides a valuable technique for the study of wave-front behavior which can offer competition to the method of characteristics.

As stated earlier, the method of characteristics is intended for the study of individual waves. Such waves arise, for example, when the loads are either localized surface loads or applied at some edge, and the shell before, say, $t = 0$, is completely at rest. Since disturbances propagate at finite wave speeds, photographs taken at consecutive small steps in time would show the disturbed region of the shell spreading into a region of quiescence, where the shell is still at rest. If any variable were plotted over the shell surface at a fixed time, it would be represented by a function that is nonzero over the disturbed region, but identically zero within the region of quiet. Such a function cannot be analytic; discontinuities in it or in some of its derivatives must arise at the borderline between the nonzero and zero regions. The borderlines that propagate the discontinuities into the undisturbed shell are called wave fronts. Usually, information is desired about the discontinuities on the borderline of the region of quiescence. However, such discontinuities in loads can be also introduced at times after the loads have already had some nonzero values. In those cases, the discontinuities are propagated on the borderline of two disturbed regions.

The method of characteristics is applicable to such wave propagation problems. They must be governed by systems of partial differential equations of hyperbolic type (see C. Chester, *Techniques in Partial Differential Equations,* New York: McGraw-Hill, pp. 266–273, 1971). Hyperbolic equations possess the property that they can admit discontinuities in some variables of solution, or their derivatives, on certain families of curves (or surfaces) called characteristics. For one-dimensional problems in space, such as axisymmetric waves in shells of revolution, characteristics are curves in the (x,t) plane (x is the space coordinate). For general shell problems, characteristics are surfaces in the (x_1,x_2,t) space. Only the illustration of a one-dimensional space problem will be given here, simply because solutions to two-dimensional problems for shells are not available. Paper 31, by W. R. Spillers, considers such a one-dimensional space problem with respect to the propagation of axisymmetric waves along a thin cylindrical shell.

Spillers uses the mathematical model of thin-shell theory, and he shows clearly that the method of characteristics is capable of predicting in a very simple manner the speeds of discontinuities of certain variables and the variation of the strengths of the discontinuities with time. This information is obtained from equations (17) and (21)–(27) of Spillers' paper. It is also capable of giving the solution for the entire (x,t)

plane. Such solutions are usually obtained by numerical integration. Spillers integrates the system of equations (28)–(34) numerically along the characteristics and displays the results in Figures 4 through 12. However, his explanation of the numerical integration is too brief. The reader is advised to study Paper 3, by Leonard and Budiansky, for a much more detailed description of this procedure.

26A

Reprinted from *J. Acoust. Soc. Amer.*, **31**(5), 568–573 (1959)

Three-Dimensional Investigation of the Propagation of Waves in Hollow Circular Cylinders. I. Analytical Foundation

Denos C. Gazis
Research Laboratories, General Motors Corporation, Detroit 2, Michigan

(Received November 28, 1958)

The propagation of free harmonic waves along a hollow circular cylinder of infinite extent is discussed within the framework of the linear theory of elasticity. A characteristic equation appropriate to the circular hollow cylinder is obtained by use of the Helmholtz potentials for arbitrary values of the physical parameters involved. Axially symmetric waves, the limiting modes of infinite wavelength, and a special family of equivoluminal modes are derived and discussed as degenerate cases of the general equations.

INTRODUCTION

THE propagation of free harmonic waves in an infinitely long cylindrical rod has been discussed on the basis of the linear theory of elasticity by Pochhammer[1] and Chree.[2] Similar waves in a hollow circular cylinder have been investigated, under the restriction of axial symmetry of motion, by McFadden,[3] Ghosh,[4] and Herrmann and Mirsky.[5] A three-dimensional solution of the more general problem of wave propagation in a hollow cylinder without the stipulation of axial symmetry is desirable, not only as a further contribution to the elastic theory, but also as a means of estimating the range of applicability of various shell theories.

This paper contains the analytical foundation for the investigation of the most general type of harmonic waves in a hollow circular cylinder of infinite extent. A characteristic equation has been obtained, in the framework of the linear theory of elasticity, for the eigenmodes of an isotropic continuum bounded by two concentric cylindrical surfaces. This equation appears, in general, rather intractable, but its evaluation can always be achieved numerically by the use of a modern high-speed electronic computer. In this manner the frequency spectrum has been completely determined for a wide range of the physical parameters that are involved.

The paper is divided into two parts. In Part I the frequency equation is obtained for an arbitrary number of waves around the circumference n, Poisson's ratio ν, ratio of wall thickness to internal radius h/a, and longitudinal wave number ξ (Fig. 1). The Helmholtz displacement potentials are used for the sake of clarity of presentation. An added advantage of this presentation is that it demonstrates the mechanism of coupling of dilatational and equivoluminal motion; it also facilitates the derivation and discussion of some simple degenerate cases, namely the cases of axial symmetry and/or infinite wavelength, and the case of the Lamé-type equivoluminal modes. A derivation and discussion of these degenerate cases is included in Part I.

Part II contains the numerical results obtained by the use of an IBM 704 digital computer, and a comparison with the corresponding results of a shell theory.

FREQUENCY EQUATION

The equations of motion for an isotropic elastic medium are, in invariant form,

$$\mu\nabla^2\mathbf{u}+(\lambda+\mu)\nabla\nabla\cdot\mathbf{u}=\rho(\partial^2\mathbf{u}/\partial t^2), \tag{1}$$

where $\mathbf{u}$ is the displacement vector, ρ is the density, λ and μ are Lamé's constants, and ∇^2 is the three-dimensional Laplace operator.

The vector $\mathbf{u}$ is expressed in terms of a dilatational scalar potential ϕ and an equivoluminal vector potential $\mathbf{H}$ according to

$$\mathbf{u}=\nabla\phi+\nabla\times\mathbf{H} \tag{2}$$

with

$$\nabla\cdot\mathbf{H}=F(\mathbf{r},t). \tag{3}$$

In Eq. (3) F is a function of the coordinate vector $\mathbf{r}$ and the time, which can be chosen arbitrarily due to the *gauge invariance*[6] of the field transformation described by Eq. (2). The displacement equations of motion are satisfied if the potentials ϕ and $\mathbf{H}$ satisfy the

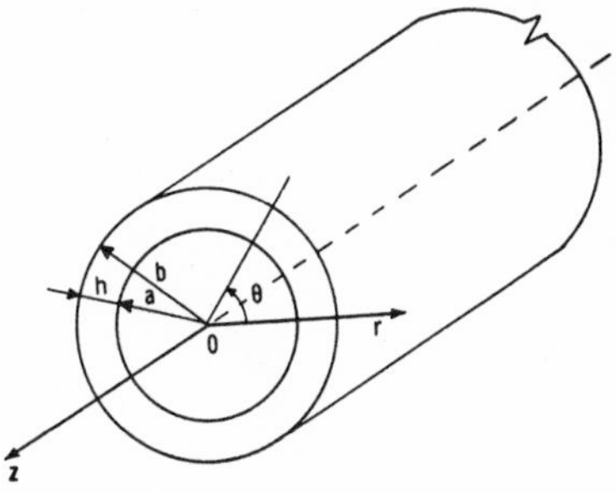

Fig. 1. Reference coordinates and dimensions.

[1] L. Pochhammer, J. für Math. (Crelle) **81**, 324–336 (1876).
[2] C. Chree, Quart. J. Math. **21**, 287–298 (1886).
[3] J. A. McFadden, J. Acoust. Soc. Am. **26**, 714–715 (1954).
[4] J. Ghosh, Bull. Calcutta Math. Soc. **14**, 1, 31–40 (1923–1924).
[5] G. Herrmann and I. Mirsky, Trans. Am. Soc. Mech. Engrs. **78**, 563–568 (1956).

[6] See, for example, P. M. Morse and H. Feshbach, *Methods of Theoretical Physics* (McGraw-Hill Book Company, Inc., New York, 1953), Part 1, p. 297.

wave equations

$$v_1^2\nabla^2\phi=\partial^2\phi/\partial t^2$$
$$v_2^2\nabla^2\mathbf{H}=\partial^2\mathbf{H}/\partial t^2, \quad (4)$$

where

$$v_1^2=(\lambda+2\mu)/\rho$$
$$v_2^2=\mu/\rho. \quad (5)$$

Let

$$\begin{aligned}\phi&=f(r)\cos n\theta\cos(\omega t+\xi z)\\ H_r&=g_r(r)\sin n\theta\sin(\omega t+\xi z)\\ H_\theta&=g_\theta(r)\cos n\theta\sin(\omega t+\xi z)\\ H_z&=g_3(r)\sin n\theta\cos(\omega t+\xi z);\end{aligned} \quad (6)$$

then Eqs. (4) yield

$$\begin{aligned}(\nabla^2+\omega^2/v_1^2)\phi&=0\\ (\nabla^2+\omega^2/v_2^2)H_z&=0\\ (\nabla^2-1/r^2+\omega^2/v_2^2)H_r-(2/r^2)(\partial H_\theta/\partial\theta)&=0\\ (\nabla^2-1/r^2+\omega^2/v_2^2)H_\theta+(2/r^2)(\partial H_r/\partial\theta)&=0.\end{aligned} \quad (7)$$

Furthermore, using the differential operator notation

$$\mathcal{B}_{n,x}=\left[\frac{\partial^2}{\partial x^2}+\frac{1}{x}\frac{\partial}{\partial x}-\left(\frac{n^2}{x^2}-1\right)\right]$$

one obtains, from Eqs. (6) and (7),

$$\begin{aligned}\mathcal{B}_{n,\alpha r}[f]&=0\\ \mathcal{B}_{n,\beta r}[g_3]&=0\\ \mathcal{B}_{n+1,\beta r}[g_r-g_\theta]&=0\\ \mathcal{B}_{n-1,\beta r}[g_r+g_\theta]&=0,\end{aligned} \quad (8)$$

where

$$\alpha^2=\omega^2/v_1^2-\xi^2,\quad \beta^2=\omega^2/v_2^2-\xi^2. \quad (9)$$

The general solution of Eqs. (9) is given in terms of the Bessel functions[7] J and Y, or the modified Bessel functions I and K of the arguments $\alpha_1 r=|\alpha r|$ and $\beta_1 r=|\beta r|$, depending on whether α and β, as determined by Eqs. (9), are real or imaginary. The proper selection of Bessel functions to be used is shown in Table I.

TABLE I. Bessel functions used at different intervals of the frequency ω.

Interval	Functions used
$v_1\xi<\omega$:	$J(\alpha r)$, $Y(\alpha r)$, $J(\beta r)$, $Y(\beta r)$
$v_2\xi<\omega<v_1\xi$:	$I(\alpha_1 r)$, $K(\alpha_1 r)$, $J(\beta r)$, $Y(\beta r)$
$\omega<v_2\xi$:	$I(\alpha_1 r)$, $K(\alpha_1 r)$, $I(\beta_1 r)$, $K(\beta_1 r)$.

[7] For the notation used see, for example, G. N. Watson, *Theory of Bessel Functions* (Cambridge University Press, London, 1952).

Then the general solution of Eqs. (9) is

$$\begin{aligned}f&=AZ_n(\alpha_1 r)+BW_n(\alpha_1 r)\\ g_3&=A_3Z_n(\beta_1 r)+B_3W_n(\beta_1 r)\\ 2g_1&=(g_r-g_\theta)=2A_1Z_{n+1}(\beta_1 r)+2B_1W_{n+1}(\beta_1 r)\\ 2g_2&=(g_r+g_\theta)=2A_2Z_{n-1}(\beta_1 r)+2B_2W_{n-1}(\beta_1 r),\end{aligned} \quad (10)$$

where, for brevity, Z denotes a J or I function, and W denotes a Y or K function, according to Table I.

The property of the *gauge invariance* can now be utilized in order to eliminate two of the integration constants entering Eqs. (10). It may be shown that any one of the three potentials g_i, $(i=1, 2, \text{ or } 3)$ can be set equal to zero, without loss of the generality of solution. Physically, this implies that the displacement field corresponding to an equivoluminal potential g_i of Eq. (10) can also be derived by a combination of the other two equivoluminal potentials. This is seen to be true for the potential functions (6).

Setting $g_2=0$ one obtains

$$g_r=-g_\theta=g_1 \quad (11)$$

and hence the displacement field

$$\begin{aligned}u_r&=[f'+(n/r)g_3+\xi g_1]\cos n\theta\cos(\omega t+\xi z)\\ u_\theta&=[-(n/r)f+\xi g_1-g_3']\sin n\theta\cos(\omega t+\xi z)\\ u_z&=[-\xi f-g_1'-(n+1)(g_1/r)]\cos n\theta\sin(\omega t+\xi z),\end{aligned} \quad (12)$$

where primes denote differentiation with respect to r.

The strain-displacement relations

$$\begin{aligned}\epsilon_{rr}&=\partial u_r/\partial r\\ \epsilon_{rz}&=(1/2)[\partial u_r/\partial z+\partial u_z/\partial r]\\ \epsilon_{r\theta}&=(1/2)\left[r\frac{\partial}{\partial r}\left(\frac{u_\theta}{r}\right)+\frac{1}{r}\frac{\partial u_r}{\partial\theta}\right]\end{aligned} \quad (13)$$

and the stress-strain relations

$$\begin{aligned}\sigma_{rr}&=\lambda\Delta+2\mu\epsilon_{rr}\\ \sigma_{rz}&=2\mu\epsilon_{rz}\\ \sigma_{r\theta}&=2\mu\epsilon_{r\theta}\end{aligned} \quad (14)$$

are now utilized in order to express the boundary stresses in terms of the potentials f, g_1, and g_3. In the first of Eqs. (14), Δ is the dilatation given by

$$\Delta=\nabla^2\phi=-(\alpha^2+\xi^2)f\cos n\theta\cos(\omega t+\xi z). \quad (15)$$

The boundary conditions, for free motion, are

$$\sigma_{rr}=\sigma_{rz}=\sigma_{r\theta}=0 \quad \text{at} \quad r=a \quad \text{and} \quad r=b, \quad (16)$$

and the stress components entering Eqs. (16) are, in terms of the displacement potentials and their derivatives,

$$\sigma_{rr}=\left\{-\lambda(\alpha^2+\xi^2)f+2\mu\left[f''+\frac{n}{r}\left(g_3'-\frac{g_3}{r}\right)+\xi g_1'\right]\right\}\times\cos n\theta\cos(\omega t+\xi z)$$

$$\sigma_{r\theta}=\mu\left\{-\frac{2n}{r}\left(f'-\frac{f}{r}\right)-(2g_3''-\beta^2 g_3)-\xi\left(\frac{n+1}{r}g_1-g_1'\right)\right\}\sin n\theta\cos(\omega t+\xi z)\quad(17)$$

$$\sigma_{rz}=\mu\left\{-2\xi f'-\frac{n}{r}\left[g_1'+\left(\frac{n+1}{r}-\beta^2+\xi^2\right)g_1\right]-\frac{n\xi}{r}g_3\right\}\times\cos n\theta\sin(\omega t+\xi z).$$

Substitution of Eqs. (17) into Eqs. (16) yields the characteristic equation, formed by the determinant of the coefficients of the amplitudes A, B, A_1, B_1, A_3, and B_3, as follows:

$$|c_{ij}|=0,\quad(i,j=1\text{ to }6),\quad(18)$$

where i identifies the row and j the column of the determinant. The first three rows of this determinant are given by

$$\begin{aligned}
c_{11}&=[2n(n-1)-(\beta^2-\xi^2)a^2]Z_n(\alpha_1 a)+2\lambda_1\alpha_1 aZ_{n+1}(\alpha_1 a)\\
c_{12}&=2\xi\beta_1 a^2Z_n(\beta_1 a)-2\xi a(n+1)Z_{n+1}(\beta_1 a)\\
c_{13}&=-2n(n-1)Z_n(\beta_1 a)+2\lambda_2 n\beta_1 aZ_{n+1}(\beta_1 a)\\
c_{14}&=[2n(n-1)-(\beta^2-\xi^2)a^2]W_n(\alpha_1 a)+2\alpha_1 aW_{n+1}(\alpha_1 a)\\
c_{15}&=2\lambda_2\xi\beta_1 a^2W_n(\beta_1 a)-2(n+1)\xi aW_{n+1}(\beta_1 a)\\
c_{16}&=-2n(n-1)W_n(\beta_1 a)+2n\beta_1 aW_{n+1}(\beta_1 a)\\
c_{21}&=2n(n-1)Z_n(\alpha_1 a)-2\lambda_1 n\alpha_1 aZ_{n+1}(\alpha_1 a)\\
c_{22}&=-\xi\beta_1 a^2Z_n(\beta_1 a)+2\xi a(n+1)Z_{n+1}(\beta_1 a)\\
c_{23}&=-[2n(n-1)-\beta^2a^2]Z_n(\beta_1 a)-2\lambda_2\beta_1 aZ_{n+1}(\beta_1 a)\\
c_{24}&=2n(n-1)W_n(\alpha_1 a)-2n\alpha_1 aW_{n+1}(\alpha_1 a)\\
c_{25}&=-\lambda_2\xi\beta_1 a^2W_n(\beta_1 a)+2\xi a(n+1)W_{n+1}(\beta_1 a)\\
c_{26}&=-[2n(n-1)-\beta^2a^2]W_n(\beta_1 a)-2\beta_1 aW_{n+1}(\beta_1 a)\\
c_{31}&=2n\xi\alpha_1 Z_n(\alpha_1 a)-2\lambda_1\xi\alpha_1 a^2Z_{n+1}(\alpha_1 a)\\
c_{32}&=n\beta_1 aZ_n(\beta_1 a)-(\beta^2-\xi^2)a^2Z_{n+1}(\beta_1 a)\\
c_{33}&=-n\xi aZ_n(\beta_1 a)\\
c_{34}&=2n\xi aW_n(\alpha_1 a)-2\xi\alpha_1 a^2W_{n+1}(\alpha_1 a)\\
c_{35}&=\lambda_2 n\beta_1 aW_n(\beta_1 a)-(\beta^2-\xi^2)a^2W_{n+1}(\beta_1 a)\\
c_{36}&=-n\xi aW_n(\beta_1 a).
\end{aligned}\quad(19)$$

The remaining three rows are obtained from the first three by substitution of b for a. In the foregoing Eqs. (19) λ_1 and λ_2 are parameters which are introduced in order to account for the differences in the recursion and differentiation formulas between the different kinds of Bessel functions. The value of these parameters is 1 when J and Y functions are used, and -1 when I and K functions are used.

By reference to Table I it is seen that λ_i vary as follows:

$$\begin{aligned}
v_1\xi<\omega:&\quad\lambda_1=1,\ \lambda_2=1\\
v_2\xi<\omega<v_1\xi:&\quad\lambda_1=-1,\ \lambda_2=1\\
\omega<v_2\xi:&\quad\lambda_1=-1,\ \lambda_2=-1.
\end{aligned}\quad(20)$$

For given dimensions and elastic constants of the cylinder, Eq. (18) constitutes an implicit transcendental function of h/L and ω/ω_s, where L is the wavelength and

$$\omega_s=\pi v_2/h\quad(21)$$

is the lowest simple thickness-shear frequency of an infinite plate of thickness h. The roots ω/ω_s may be computed for a fixed h/L, or vice versa, by an iteration procedure; this is done numerically in Part II, for a wide range of the physical parameters involved.

The characteristic equation (18) degenerates to simpler forms when the wave number ξ and/or the number of waves around the circumference n are taken equal to zero. Another degenerate case is that of pure equivoluminal modes analogous to the modes first discussed by Lamé[8] for an isotropic plate. These degenerate cases are discussed in the following.

MOTION INDEPENDENT OF z

When the wave number ξ is taken equal to zero, Eq. (18) breaks into the product of subdeterminants

$$D_1D_2=0,\quad(22)$$

where

$$D_1=\begin{vmatrix}c_{11}&c_{13}&c_{14}&c_{16}\\c_{21}&c_{23}&c_{24}&c_{26}\\c_{41}&c_{43}&c_{44}&c_{46}\\c_{51}&c_{53}&c_{54}&c_{56}\end{vmatrix}\quad\text{and}\quad D_2=\begin{vmatrix}c_{32}&c_{35}\\c_{62}&c_{65}\end{vmatrix}\quad(23)$$

and the terms c_{ij} are given by Eqs. (19) with $\xi=0$. It should be remarked that for $\xi=0$ both α^2 and β^2 are positive as seen from Eq. (9), and hence the J and Y Bessel functions enter the solution of the wave equations according to Table I.

Equation (22) is satisfied if either D_1 or D_2 is equal to zero. The case of $D_1=0$ corresponds to plane-strain vibrations which have been discussed in a previous paper.[9] The case of $D_2=0$ corresponds to *longitudinal shear* vibrations, i.e., motion involving only longitudinal displacements u_z. It is seen that these two types of vibration are uncoupled when the motion is independent of the longitudinal coordinate z. They are coupled for a nonzero wave number ξ, as may be ascertained from Eq. (18). The frequencies of the uncoupled plane-strain

---

[8] M. G. Lamé, *Leçons sur la théorie mathématique de l' élasticité des corps solides* (Gauthier-Villars, Paris, 1866), second edition, p. 170. See also H. Lamb, Proc. Roy. Soc. (London) **93**, 122 (1917), footnote.

[9] D. C. Gazis, J. Acoust. Soc. Am. **30**, 786–794 (1958).

and longitudinal shear modes are the cutoff frequencies of waves with the same number of circumferencial waves n in a diagram of the frequency as a function of the wave number.

For the longitional shear vibratións, it may be ascertained that the displacement field is derived from a potential g_1 alone and is given by

$$u_r = u_\theta = 0$$
$$u_z = [A_1\beta J_n(\beta r) + B_1\beta Y_n(\beta r)] \cos n\theta \sin\omega t. \quad (24)$$

The frequency equation is

$$J_n'(\beta a) Y_n'(\beta b) - J_n'(\beta b) Y_n'(\beta a) = 0 \quad (25)$$

and the amplitude ratio

$$A_1/B_1 = -Y_1'(\beta a)/J_1'(\beta a). \quad (25a)$$

The lowest mode of this type corresponds essentially to a shearing of the cylinder as a whole across its diameter; the displacement u_z is zero along $2n$ radial planes corresponding to the zeros of the function $\cos n\theta$. The second and all higher longitudinal shear modes involve, in addition, a number of concentric nodal cylindrical surfaces, and correspond essentially to a shearing of the cylinder across its thickness.

Equation (25) may be used for the determination of the frequency ratio

$$\beta h = \pi\omega/\omega_s$$

for any arbitrary ratio h/a. A numerical computation of the longitudinal shear cutoff frequencies is included in Part II. A brief investigation of Eq. (25) for limiting values of h/a is given in the following.

For thin cylindrical shells, that is $h/a \ll 1$, and under the assumption of nonzero βh it is seen that $\beta a \gg 1$ and $\beta b \gg 1$. Accordingly, using the Hankel-Kirchhoff asymptotic approximations[10] for the Bessel functions, one obtains

$$\sin\beta h - [(4n^2+3)\beta h/8\beta^2 ab] \cos\beta h \approx 0 \quad (26)$$

and finally

$$\omega \approx \frac{q\pi}{h} v_2 \left[1 + \frac{4n^2+3}{8(q\pi)^2}\left(\frac{h}{a}\right)^2\right], \quad q = 1, 2, 3 \cdots. \quad (27)$$

The preceding approximations are valid for all the longitudinal shear modes except the lowest one for which $\beta h \to 0$ as $h/a \to 0$. As is to be expected, for $h/a \to 0$ the frequencies of the second and higher modes tend to the frequencies of the simple thickness-shear modes of a plate of thickness h.

When $a/h \to 0$, that is for an almost solid cylinder, Eq. (25) tends asymptotically tо the corresponding frequency equation for a solid cylinder of radius h,

$$nJ_n(\beta h) - \beta h J_{n+1}(\beta h) = 0. \quad (28)$$

[10] See, for example, reference 7, p. 194.

MOTION INDEPENDENT OF θ

For motion independent of the angular coordinate θ, ($n=0$), the determinantal Eq. (18) breaks into the product of subdeterminants

$$D_3 D_4 = 0, \quad (29)$$

where

$$D_3 = \begin{vmatrix} c_{11} & c_{12} & c_{14} & c_{15} \\ c_{31} & c_{32} & c_{34} & c_{35} \\ c_{41} & c_{42} & c_{44} & c_{45} \\ c_{61} & c_{62} & c_{64} & c_{65} \end{vmatrix} \quad \text{and} \quad D_4 = \begin{vmatrix} c_{23} & c_{26} \\ c_{53} & c_{56} \end{vmatrix}, \quad (30)$$

and the terms c_{ij} are given by Eqs. (19) with $n=0$.

Longitudinal Waves

The frequency equation

$$D_3 = 0 \quad (31)$$

corresponds to longitudinal waves, i.e., waves involving displacements u_r and u_z which are independent of θ. A frequency equation equivalent to Eq. (31) has been given by J. Ghosh,[4] who also derived a simplified equation for thin cylindrical shells, and the corresponding frequency equation for a cylinder which is rigidly clamped along one of its boundaries and free along the other.

The displacement field is derived from a dilatational potential f and an equivoluminal potential g_1. As may be seen from Eqs. (17) the dilatational and equivoluminal parts of the solution of the wave equations are in general coupled through the boundary conditions. However, some pure equivoluminal modes may exist uncoupled and are discussed in the following.

Equivoluminal Lamé-Type Modes

A particular type of equivoluminal waves, analogous to the ones first discussed by Lamé[8] for the infinite isotropic plate, may be obtained in the following manner.

For $n=0$, the boundary conditions (17) are satisfied if

$$\beta^2 = \xi^2 > 0$$
$$f = g_3 = 0 \quad (32)$$
$$g_1'(\beta a) = g_1'(\beta b) = 0,$$

where

$$g_1(\beta r) = A_1 J_1(\beta r) + B_1 Y_1(\beta r). \quad (33)$$

Hence, the frequency equation

$$J_1'(\beta a) Y_1'(\beta b) - J_1'(\beta b) Y_1'(\beta a) = 0 \quad (34)$$

and the amplitude ratio

$$A_1/B_1 = -Y_1'(\beta a)/J_1'(\beta a). \quad (35)$$

The complete solution is

$$u_r=\xi[A_1J_1(\beta r)+B_1Y_1(\beta r)]\cos(\omega t+\xi z)$$
$$u_z=-\xi[A_1J_0(\beta r)+B_1Y_0(\beta r)]\sin(\omega t+\xi z) \quad (36)$$
$$\sigma_{rr}=-2\mu\xi[A_1J_1'(\beta r)+B_1Y'(\beta r)]\cos(\omega t+\xi z)$$
$$u_\theta=\sigma_{r\theta}=\sigma_{rz}=0.$$

A superposition of two waves of this type, of the same amplitude and traveling in opposite directions yields standing equivoluminal waves with traction-free planes $z=\text{constant}$ at intervals $2\pi/\beta$. A number of cylindrical surfaces $r=c_j$ may also be traction free, the c_j being determined by

$$J_1'(\beta c_j)Y_1'(\beta a)-J_1'(\beta a)Y_1'(\beta c_j)=0. \quad (37)$$

Equation (34) is of the same form as the frequency equation (25) of the longitudinal shear vibrations, for $n=1$. The results of the investigation of the latter are directly applicable to the case of the Lamé-type equivoluminal modes, insofar as the determination of the frequency ratio βh is concerned. Accordingly, for the limiting values of the ratio h/a one obtains the following:

For thin cylindrical shells, i.e., $h/a\ll 1$

$$\sin\beta h-(7\beta h/8\beta^2 ab)\cos\beta h\approx 0 \quad (38)$$

and, finally, by virtue of the first of Eqs. (32)

$$\omega\approx\sqrt{2}\frac{q\pi}{h}v_2\left[1+\frac{7}{8(q\pi)^2}\left(\frac{h}{a}\right)^2\right],\quad q=1,2,3\cdots \quad (39)$$

For $h/a=0$, Eq. (39) yields the frequencies of the straight-crested Lamé modes[8] of a plate of thickness h.

When $a/h\to 0$, Eq. (34) tends asymptotically to the frequency equation for the Lamé-type modes of a solid cylinder of radius h

$$J_1'(\beta h)=0. \quad (40)$$

One last remark on the frequency equation (34): for some specific values of the ratio of the inner to the outer diameter of the hollow cylinder, a/b, it is possible to obtain a β such as to make $J_1'(\beta a)$ and $J_1'(\beta b)$ vanish simultaneously. As a consequence the amplitude B_1 [Eq. (33)] vanishes and one obtains the equivoluminal modes obtained by Goodman[11] for sections of an infinite plate. The appropriate ratios a/b are the ratios of any two roots of Eq. (40). Similarly, ratios of any two roots of the equation

$$dY_1(x)/dx=0, \quad (41)$$

if set equal to a/b, make it possible to obtain solutions with zero amplitude A_1 [Eq. (33)] and hence another family of free contour surfaces analogous to those obtained by Goodman. It may be seen that the ratios a/b corresponding to the preceding modes are independent of the elastic constants of the material. The first five roots of Eqs. (40) and (41) are given in Table II.

TABLE II.

$dJ_1(x)/dx=0$	$dY_1(x)/dx=0$
1.841185	3.683025
5.331445	6.941504
8.536320	10.123409
11.706009	13.285762
14.863590	16.440059

[11] L. E. Goodman, "Circular-crested vibrations of an elastic solid bounded by two parallel planes," Proceedings of the 1st U. S. National Congress of Applied Mechanics, p. 70 (1951).

Torsional Waves

For $f=g_1=0$ and

$$D_4=0, \quad (42)$$

one obtains motion involving displacements u_θ only, i.e., torsional modes. Equation (42) may be reduced to

$$J_2(\beta a)Y_2(\beta b)-J_2(Bb)Y_2(\beta a)=0, \quad (43)$$

where β is given by (9). It may be ascertained that no roots of Eq. (42) exist for $\beta^2<0$; hence, it follows from Eq. (9) that the phase velocity of the torsional waves is always greater than or equal to v_2.

Equation (43) is identical in form to the frequency equation of axially symmetric shear vibrations in plane strain, which has been investigated by the author in reference 9. It follows that the frequencies of these plane-strain shear modes are the cutoff frequencies ($\xi=0$) of the torsional waves. The frequency corresponding to a wave number $\xi\neq 0$ is then given, for the jth mode, by

$$\omega_j=[(\omega_0)_j^2+\xi^2 v_2^2]^{\frac{1}{2}}, \quad (44)$$

where $(\omega_0)_j$ is the jth cutoff frequency.

As in the case for the cylindrical rod,[12] the phase velocity for all modes decreases monotonically from infinity to v_2 as ξ increases from 0 to ∞. At the same time the group velocity increases monotonically from zero to v_2.

In analogy to the case of the cylindrical rod, the lowest torsional mode, corresponding to

$$\beta^2=\omega^2/v_2^2-\xi^2=0, \quad (45)$$

is not adequately described by the general expressions for the displacements. However, it is seen that a displacement field

$$u_r=u_z=0,\quad u_\theta=Ar\sin(\omega t+\xi z) \quad (46)$$

satisfies the equations of motion (1) and the boundary conditions (17), if Eq. (45) holds true. The displacement field (46) corresponds to a rotation of each transverse section of the cylinder as a whole about its center.

[12] See, for example, H. Kolsky, *Stress Waves in Solids* (Clarendon Press, Oxford, 1953), p. 67.

There is no dispersion for waves of this type, both the phase velocity and the group velocity being equal to v_2.

SUMMARY

The displacement field is derived from a dilatational potential f and two equivoluminal ones g_3 and g_1, all three of them periodic in θ and sinusoidal in z. When the number of waves around the circumference n and the longitudinal wave number ξ are both zero, i.e., for axially symmetric motion and infinite wavelength, the three potentials generate three uncoupled families of modes identified as plane-strain extentional, plane-strain shear, and longitudinal shear modes, respectively.

For $\xi=0$ and $n\neq 0$ the potentials f and g_3 are coupled through the boundary conditions and generate the nonaxially symmetric plane-strain modes. An uncoupled family of longitudinal shear modes is again derived from g_1 alone.

For $\xi\neq 0$ and $n=0$ the potentials f and g_1 are similarly coupled yielding the axially symmetric longitudinal modes, which are uncoupled from the g_3-generated torsional modes. A special family of the longitudinal modes are the Lamé-type equivoluminal modes which are derived from g_1 alone for some discrete values of the frequency and wavelength given by Eqs. (37) and (38).

Finally, for $\xi\neq 0$ and $n\neq 0$ all three potentials are coupled through the boundary conditions. Thus nonaxially symmetric waves may be considered as the result as coupling of motion in the plane r, θ, analogous to the motion in plane-strain and longitudinal shear motion. Alternatively, they may be considered as the result of coupling of modes analogous to the longitudinal and torsional modes but periodic in θ. This may provide an approach for obtaining the coupled frequency spectrum with the aid of the spectra of two uncoupled families of modes. A more direct approach is a numerical computation of the frequency spectrum such as given in Part II of this paper.

26B

Reprinted from *J. Acoust. Soc. Amer.*, **31**(5), 573–578 (1959)

Three-Dimensional Investigation of the Propagation of Waves in Hollow Circular Cylinders. II. Numerical Results

DENOS C. GAZIS
Research Laboratories, General Motors Corporation, Detroit 2, Michigan
(Received November 28, 1958)

The results are given of a numerical evaluation of a characteristic equation derived in Part I, appropriate to free harmonic waves propagated along a hollow cylinder of infinite extent. This equation is evaluated for some representative cylinders covering the entire range from thin shells to solid cylinders, and the results are compared with the corresponding results of a shell theory. Observations are made regarding the variation of the frequency spectrum with the physical parameters, as well as the range of applicability of shell theories.

INTRODUCTION

THE frequency equation for the propagation of free harmonic waves along a hollow cylinder of infinite extent is given in Part I of this paper, together with a discussion of some degenerate cases of simple motion.

Part II contains the results of a numerical evaluation of the complete frequency spectrum which was obtained, for various sets of the physical parameters involved, by means of an IBM 704 digital computer. It also includes a comparison with the corresponding results obtained by Mirsky and Herrmann[1] on the basis of their Timoshenko-type shell theory.

The reference coordinates and dimensions of the cylinder are shown in Fig. 9, and the covered range of parameters is as follows:

Poisson's ratio, $\nu=0.30$;
number of circumferential waves, $n=1$ and 2;
ratio of thickness to mean radius, $m=h/R=1/30$, 1/4, 1 and 2;
ratio of thickness to wavelength, $0\leqslant h/L\leqslant 1.0$.

[1] I. Mirsky and G. Herrmann, J. Acoust. Soc. Am. 29, 1116–1123 (1957).

The case of $n=1$ includes the ordinary flexural mode of a bar, while $n=2$ is a typical case of nonaxially symmetric motion with $n>1$, insofar as the character of the frequency spectrum is concerned. The values of m which have been used cover the range from the case of a thin shell ($m=1/30$) to the Pochhammer case of a cylindrical rod ($m=2$). The first two values of m, namely $m=1/30$ and $m=1/4$, were used also for the sake of comparison with the Mirsky-Herrmann[1] results. This comparison is given in the last section of this paper, while the first two sections contain a description of the computation procedure and a discussion of the numerical results.

NUMERICAL COMPUTATION

As mentioned in Part I, for given dimensions and elastic constants of the cylinder, the frequency equation (18) of Part I constitutes an implicit transcendental function of $\delta=h/L$ and $\Omega=\omega/\omega_s$. The frequency spectrum ω/ω_s *versus* h/L was computed by an "interval halving" iteration technique, as follows.

For a fixed value of δ the determinant (18), Part I,

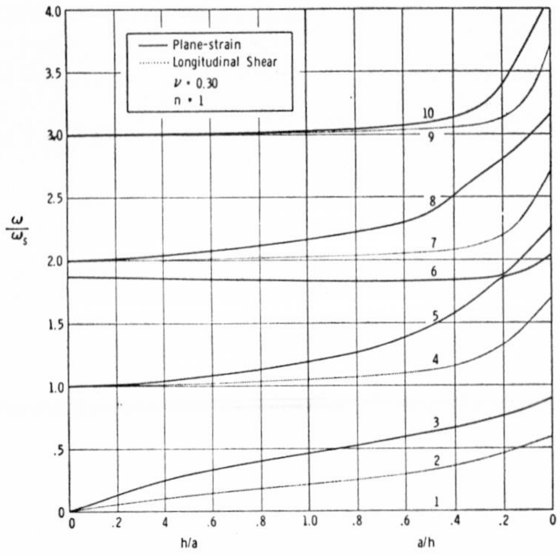

FIG. 1. Cutoff frequencies for $n=1$ and $\nu=0.3$. Variation of plane-strain and longitudinal shear frequencies with the ratio of wall thickness to internal radius, h/a.

is a function of Ω alone. The value of this determinant is evaluated at a prescribed starting point Ω_1 and at intervals of a specified $\Delta\Omega$ thereafter, up to and including a prescribed end point Ω_2. A change of sign of the determinant across a $\Delta\Omega$ interval indicates a root in that interval. The interval is then halved and the direction of scanning is reversed; the process is repeated until a root is obtained within a prescribed accuracy. Additional roots are obtained by re-entering successively into the iteration subroutine with a new starting point just beyond the last obtained root, until the entire region $\Omega_1 \leqslant \Omega \leqslant \Omega_2$ is covered, or until a prescribed number of roots is obtained.[2] The present computations yielded a maximum of ten roots in the interval $0 \leqslant \Omega \leqslant 4.0$. The allowable "relative error" (i.e., the error divided by the root) was 10^{-4}.

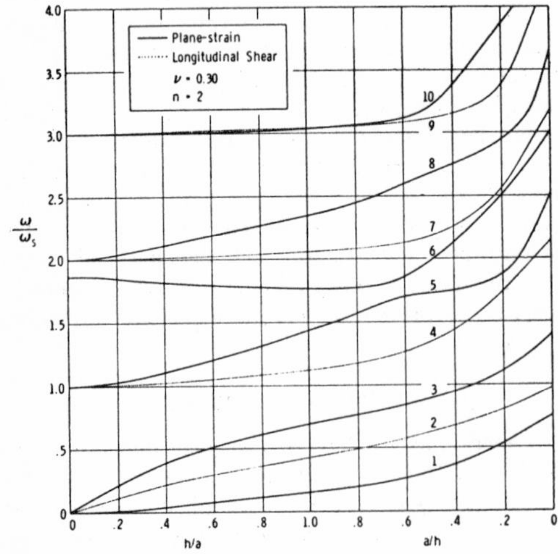

FIG. 2. Cutoff frequencies for $n=2$ and $\nu=0.3$. Variation of plane-strain and longitudinal shear frequencies with the ratio of wall thickness to internal radius, h/a.

It should be remarked that the switch from the Bessel functions J and Y to the modified Bessel functions I and K across the lines

$$\omega/v_1=\xi \quad \text{and} \quad \omega/v_2=\xi \tag{1}$$

according to the Table I, Part I, gave rise to spurious roots along these lines. This simply meant that an undesirable change of sign was registered across these lines during the iteration. The spurious roots were eliminated by changing the sign of the computed determinant in the interval

$$\Omega/e<2\delta<\Omega, \tag{2}$$

where

$$e=v_1/v_2=[(\lambda+2\mu)/\mu]^{\frac{1}{2}}. \tag{3}$$

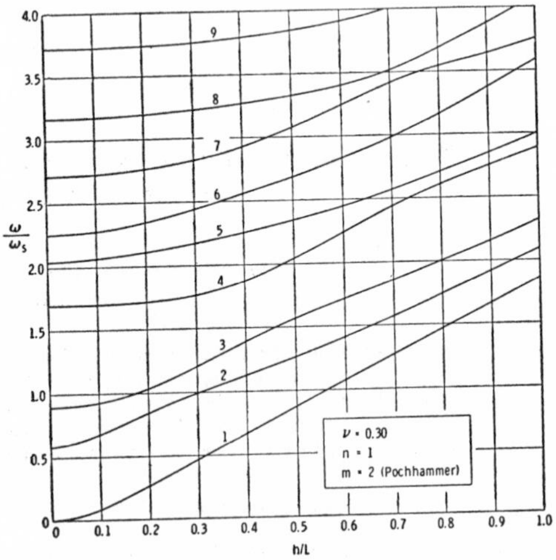

FIG. 3. Frequency *versus* wavelength for $n=1$ and $m=h/R=2$ (Pochhammer modes).

The variation of the mode shapes with the wavelength was investigated by computing the displacement field for a few points of the frequency spectrum. Referring to Part I, for a given root of the frequency equation (18) the five amplitude ratios B/A, A_1/A, B_1/A, A_3/A, and B_3/A were determined by using five of the six compatible boundary conditions. These ratios were inserted in Eqs. (12) and the displacement components u_r, u_θ, and u_z were computed for twenty points across the thickness of the cylinder, assuming $A=1$. The displacements were then normalized with reference to the maximum displacement value appearing therein.

DISCUSSION OF THE RESULTS

The numerical results are shown in Figs. 1 to 11. Figures 1 and 2 show the cutoff frequencies (infinite

[2] The role of Ω and δ may be reversed in the coded program if one wants to obtain roots δ for a constant Ω.

wavelength) for the entire range of the ratio of wall thickness to internal radius h/a from zero to infinity, and for $n=1$ and $n=2$, respectively. The lowest mode for $n=1$ is the flexural mode which degenerates to a simple translation and hence zero frequency as the wavelength increases to infinity, for any value of h/a. This is not true for $n>1$, as is exemplified by the case $n=2$ for which the frequency of the lowest (ring flexural) mode rises monotonically, with increasing h/a, from zero for $h/a=0$ to a finite Pochhammer frequency for $a/h=0$. A more detailed discussion of the plane-strain vibrations may be found in a previous paper,[3] and an investigation of the longitudinal shear modes is included in Part I of this paper.

Figures 3 to 10 contain the frequency spectrum in the interval $0\leqslant\delta\leqslant 1.0$ and $0\leqslant\Omega\leqslant 4.0$, for $n=1$ and $n=2$ and for the four assumed values of m. It is remarked that the frequency spectrum for $m=1/30$ (Fig. 6) is almost identical for either $n=1$ or $n=2$, except for the lowest three modes near the origin $\delta=\Omega=0$. A detail of these modes near the origin is given in Fig. 7. Attention is called to the fact that for $n=1$ the lowest (flexural) mode tends to zero frequency as the wavelength increases to infinity (i.e., as $\delta\to 0$), whereas for $n=2$ the lowest mode tends to a nonzero cutoff frequency, as seen also in Fig. 2. The phase velocity also goes to zero, for $n=1$ and $\delta\to 0$, but it tends to infinity for $n=2$ and $\delta\to 0$, as shown in Fig. 11.

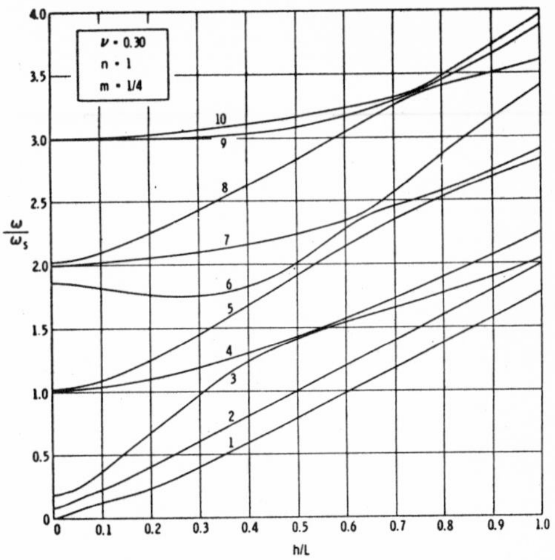

FIG. 5. Frequency *versus* wavelength for $n=1$ and $m=h/R=1/4$.

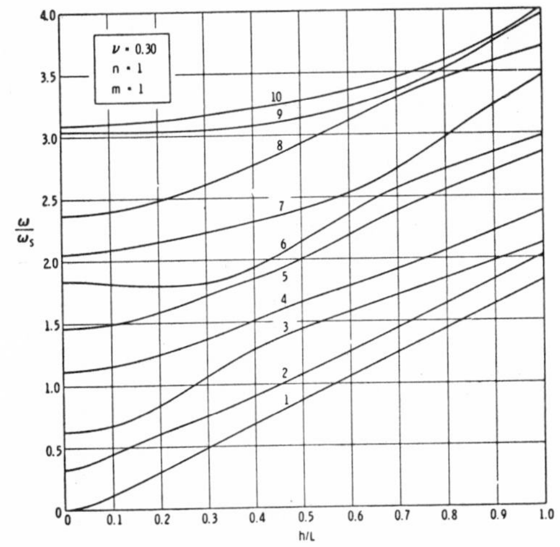

FIG. 4. Frequency *versus* wavelength for $n=1$ and $m=h/R=1$.

The various modes shall be identified from their displacement field corresponding to very long wavelengths. Thus the lowest mode for $n=1$ is the bar-flexural mode of the cylinder, corresponding essentially to a uniform translation sinusoidal in z of the entire cross section. For $n=2$ the lowest mode is the ring-flexural mode corresponding to radial displacements symmetric about two diametric planes. For both $n=1$ and $n=2$, the second and third modes are identified as the lowest longitudinal shear mode[4] and the ring-extensional ("breathing") mode, respectively. All modes higher than the third involve at least one nodal cylindrical surface across the thickness of the cylinder and shall be generally referred to as thickness modes. In particular they shall be designated as thickness-stretch or thickness-shear modes depending on whether the displacements are predominantly normal or parallel to the nodal surfaces. Thus, for example, for $\nu=0.3$, the fourth, fifth, and sixth modes are, respectively, a longitudinal thickness-

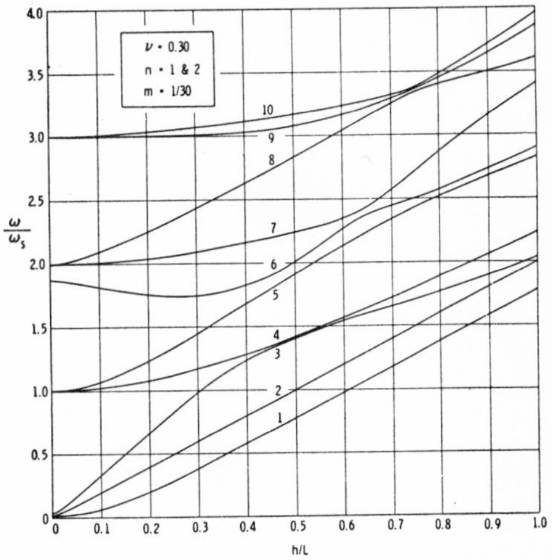

FIG. 6. Frequency *versus* wavelength for $n=1$ (and 2) and $m=h/R=1/30$.

[3] D. C. Gazis, J. Acoust. Soc. Am. 30, 786–794 (1958).

[4] See also Part I and reference 3 for definitions and descriptions of these modes.

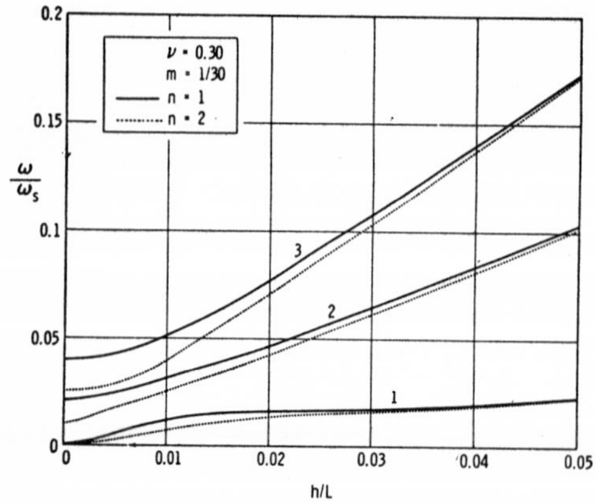

FIG. 7. Detail of the three lowest modes near the origin for $n=1$ and 2 and $m=h/R=1/30$.

shear mode, a circumferential thickness-shear mode, and a thickness-stretch mode.

Of particular interest is the lowest mode for $n=1$ which is identified as the bar-flexural mode. When the wavelength is very long in comparison with the outer diameter of the cylinder the frequency, phase velocity, and displacement field are quite adequately described by the classical beam theory. As the wavelength decreases the frequency rises until it approaches the frequency of the lowest longitudinal shear mode, at which point strong coupling takes place manifested by a large component of shearing deformation of the cylinder across its diameter, and in the axial direction. In Fig. 11, the coupling becomes apparent a little before the maximum of the phase velocity, as the variation of the phase velocity with h/L deviates from the linear law of the beam theory. If the wavelength is further

reduced, first the effect of coupling with the ring-extensional mode manifests itself and then, for wavelengths comparable to the wall thickness of the cylinder, the effect of coupling with the lowest longitudinal thickness-shear mode. Finally the phase velocity tends to the velocity of the Rayleigh surface waves, as in the case of the cylindrical rod.[5]

Similar observations may be made on the lowest (ring-flexural) mode for $n=2$ which, however, tends to a nonzero frequency and hence infinite phase velocity as $\delta \to 0$. At the other end of the spectrum, i.e., for $\delta \to \infty$ the phase velocity again tends to the velocity of the Rayleigh surface waves. Thus the dispersion curve of the lowest mode for $n=2$ coincides with the corresponding curve for $n=1$, in the region of small wavelengths. In this connection the following observation may be made on the character of all the curves of the spectrum: If the circumferential wavelength, which is

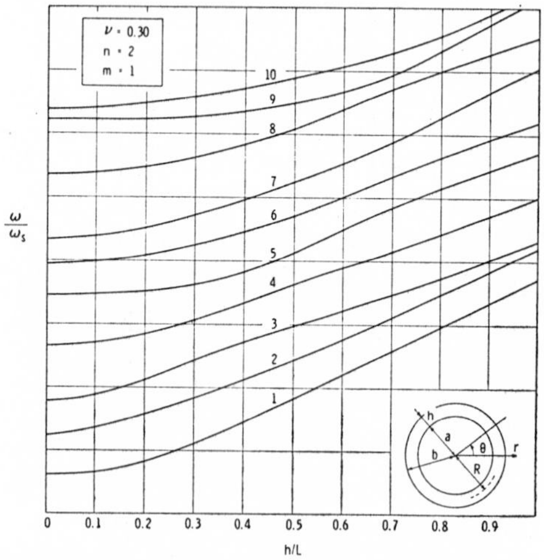

FIG. 9. Frequency *versus* wavelength for $n=2$ and $m=h/R=1$.

equal to $2\pi R/n$, is very large in comparison with the wall thickness of the cylinder (i.e., for small values of n and a relatively thin shell) the dispersion curves of all the modes which involve predominantly a deformation of the cylinder across its thickness tend to become almost independent of n as $\delta \to \infty$. This is particularly true for the fourth and all higher modes which have been identified as thickness modes involving one or more nodal cylindrical surfaces. The dispersion curves of these modes become essentially independent of n for quite small values of δ, as may be ascertained by an examination of Fig. 6 or a comparison of the Figs. 5 and 10. It should be remarked that the variation of these curves with a varying n appears greater in the neighborhood of the intersection points of the uncoupled spectra of the

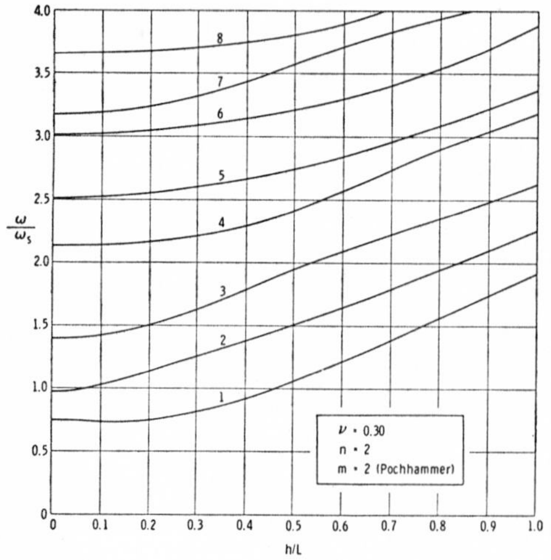

FIG. 8. Frequency *versus* wavelength for $n=2$ and $m=h/R=2$ (Pochhammer modes).

[5] See, for example, H. Kolsky, *Stress Waves in Solids* (Clarendon Press, Oxford, 1953), p. 72.

longitudinal and torsional modes, which are obtained for $n=0$. Thus the "bunching" of the dispersion curves, which is characteristic of a coupled spectrum ($n\neq 0$), is more pronounced, i.e., the coupling is weaker, for smaller values of n. The preceding observations may be utilized in order to expedite the evaluation of the frequency spectrum of relatively thin shells for a number of values of $n\geqslant 1$.

If now the value of n is kept constant and m is varied, it is observed that a reduction of the value of m has a similar effect on the spectrum as an increase of the value of n, since they both correspond to a reduction of the circumferential wavelength relative to the wall thickness of the cylinder. Thus the dispersion curves of the thickness modes of relatively thin shells appear almost independent of m, but their "bunching" is more pronounced for smaller values of m. In the limit, for $m\rightarrow 0$, one would obtain two uncoupled and hence intersecting families of dispersion curves, namely, those of the longitudinal straight-crested waves and the face-shear waves in an infinite plate, which are limiting forms of the pure longitudinal waves and the torsional waves of the hollow cylinder, respectively. The intersecting curves separate when $m\neq 0$ and their separation is accentuated by an increase in the value of m. At the same time, the cutoff ratios ω/ω_s keep rising as shown in Figs. 1 and 2, and the whole character of the spectrum is altered approaching asymptotically the Pochhammer spectrum ($m=2$) as is exemplified by the spectrum for the intermediate value $m=1$ (Figs. 2 and 9). It should be remarked, however, that this asymptotic behavior does not apply to the stress and strain field. As the internal radius is reduced the usual stress concentration effects appear in the neighborhood of the internal boundary. However, the stress concentration tends to become highly localized, in the limit $m\rightarrow 2$, $(a/h\rightarrow 0)$, and hence it does not affect the asymptotic values of the frequency.

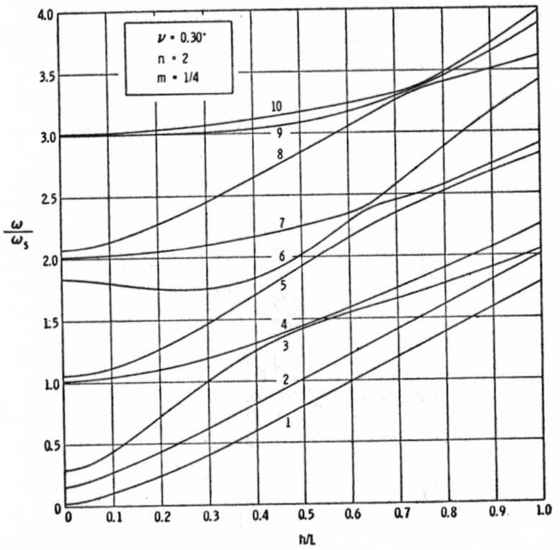

FIG. 10. Frequency *versus* wavelength for $n=2$ and $m=h/R=1/4$.

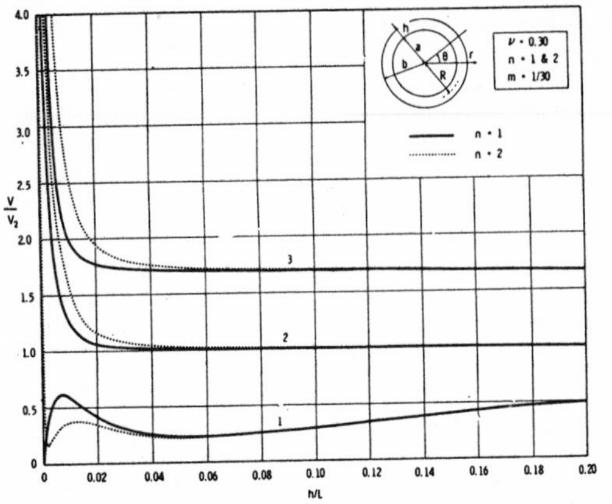

FIG. 11. Axial phase velocity *versus* wavelength for $m=h/R=1/30$ and $n=1$ and 2.

COMPARISON WITH THE RESULTS OF A SHELL THEORY

To this date a number of shell theories have been developed on the basis of various approximations. They range from Love's first approximation[6] to recent "Timoshenko-type" theories by Lin and Morgan,[7] Mirsky and Herrmann,[1] and Naghdi and Cooper,[8] which take into account the effect of transverse shear deformation and/or rotatory inertia on the natural frequencies of a moderately thick shell. The Mirsky-Herrmann theory, which is a five-mode theory, was used as representative of the Timoshenko-type shell theories for a comparison with the results of the three-

TABLE I. Percent deviation of the Mirsky-Herrmann phase velocities from the corresponding values of the three-dimensional solution.

		$m=1/30$			$m=1/4$		
	Mode	$\delta=0.1$	$\delta=0.5$	$\delta=1.0$	$\delta=0.1$	$\delta=0.5$	$\delta=1.0$
	1st	~0	− 2.4	− 3.2	−0.2	− 1.7	− 3.2
	2nd	~0	~0	~0	+0.4	~0	~0
$n=1$	3rd	+0.3	+ 0.6	+ 9.9	+0.5	+ 0.5	+10.0
	4th	~0	+19.2	+51.4	+0.1	+18.6	+50.9
	5th	+0.7	+ 6.4	+26.6	+0.7	+ 6.3	+26.8
	1st	−0.3	− 2.4	− 3.2	−0.4	− 2.2	− 3.2
	2nd	~0	~0	~0	~0	~0	~0
$n=2$	3rd	+0.5	+ 0.7	+ 9.9	+0.5	+ 1.1	+10.0
	4th	~0	+19.3	+51.4	+0.1	+18.8	+50.9
	5th	+0.7	+ 6.4	+26.6	+0.7	+ 6.4	+27.1

[6] A. E. H. Love, *A Treatise on the Mathematical Theory of Elasticity* (Dover Publications, New York, 1944), fourth edition, Chap. 24, p. 543.

[7] T. C. Lin and G. W. Morgan, Trans. Am. Soc. Mech. Engrs. **78**, 255–261 (1956).

[8] P. M. Naghdi and R. M. Cooper, J. Acoust. Soc. Am. **29**, 1365–1373 (1957).

dimensional theory.[9] The comparison is shown in Table I.

It is seen that the shell theory provides a very good description of the first three modes even for quite thick shells ($m=1/4$). In particular, for the lowest (flexural) mode, the agreement with the three-dimensional results is striking in the region of coupling with the lowest longitudinal shear mode. The agreement is still very good for wavelengths as small as the wall thickness of the cylinder. For example, the deviation of the phase velocity from the value given by the three-dimensional theory is only -3.2% for both $m=1/30$ and $m=1/4$, when $\delta=1.0$. It is possible to make the asymptotic value of the phase velocity of the first mode exactly equal to the velocity of the Rayleigh surface waves, by an appropriate selection of the correction constant κ_z of the shell theory.[10] However, this will damage the agreement of the shell theory with the three-dimensional theory in the region of long wavelengths, particularly for the fourth mode whose cutoff frequency was used by Mirsky and Herrmann for the determination of κ_z.

The second and third shell modes are in general agreement with the corresponding modes of the three-dimensional theory. However, the fourth and fifth modes are described with sufficient accuracy by the shell theory only for very small values of δ. This is because in the region of large δ these modes are coupled with higher thickness modes which are not included in the shell theory. For the same reason the asymptotic value of v/v_2 of the fourth and fifth shell mode for $\delta\rightarrow\infty$ is $[2/(1-\nu)]^{\frac{1}{2}}$, as against 1 for the three-dimensional theory.

In conclusion, a Timoshenko-type shell theory provides a very good description of the lowest (flexural) mode even for relatively thick shells. It also provides a fairly good description of the second and third modes, but its results for the fourth and fifth modes are good only for wavelengths long in comparison with the wall thickness of the cylinder.

At this point it may be mentioned that in the Mirsky-Herrmann theory two correction constants κ_z and κ_θ have been used in order to match the cutoff frequencies of the two lowest axially symmetric thickness-shear modes with the corresponding frequencies of the three-dimensional theory. These cutoff frequencies are exactly matched, even in the nonaxially symmetric case, for $h/a=0$, as seen in Figs. 1 and 2 (fourth and fifth modes). They deviate if $h/a\neq 0$ but the deviation is within 6.2% for values of h/a as large as 0.3. However, it may be interesting to attempt an even better "matching" of the three-dimensional theory by a shell theory. The constants κ_z and κ_θ may be determined so as to match exactly the cutoff frequencies of the fourth and fifth modes for a particular m. Another possibility may be the introduction of two additional correction constants multiplying the inertia terms of the equations of motion, as has been done by Mindlin in his second-order theory of plates.[11] The four correction constants may permit a better matching of a major portion of the spectrum of the shell theory and the three-dimensional solution.

ACKNOWLEDGMENTS

The author wishes to thank Professor R. D. Mindlin, Columbia University, for many valuable suggestions and Professor G. Herrmann, Columbia University, for making available a tabulation of all of the numerical results of the Mirsky-Herrmann shell theory. He also wishes to thank Dr. H. W. Milnes, General Motors Research Laboratory, for his advice in the course of the numerical computations. And, in particular, he wishes to express his appreciation to Dr. Robert Herman, General Motors Research Laboratory, for his very active interest and a number of very valuable discussions and suggestions during the course of this work.

[9] In the original paper of Mirsky and Herrmann their numerical results contained a computational error. For the present comparison the Mirsky-Herrmann frequency equation has been re-evaluated. See J. Acoust. Soc. Am. **31**, 250 (1959).

[10] Reference 1, p. 1123.

[11] R. D. Mindlin and M. A. Medick, ONR Contract Nonr-266(09), NR-064-388, Department of Civil Engineering and Engineering Mechanics, Columbia University, Tech. Rept. No. 28, J. Appl. Mechanics (to be published).

Errata: Three-Dimensional Investigation of the Propagation of Waves in Hollow Circular Cylinders. II

[J. Acoust. Soc. Am. 31, 573–578 (1959)]

D. C. GAZIS
Research Laboratories, General Motors Corporation, Detroit 2, Michigan

(Received January 29, 1960)

IN Fig. 7, p. 576, the second and third modes were erroneously drawn with dotted lines for $n=1$ and solid lines for $n=2$ instead of vice versa. As a result these two modes have the wrong n assigned to them. Actually both these modes for $n=2$ lie above the corresponding modes for $n=1$, as shown also in Fig. 11, p. 577.

27

Reprinted from *J. Acoust. Soc. Amer.*, **38**(6), 994–1002 (1965)

Transient Vibration of Thin Elastic Shells

H. KRAUS*

Pratt & Whitney Aircraft Division, United Aircraft Corporation, East Hartford, Connecticut

A. KALNINS†

Department of Mechanics, Lehigh University, Bethlehem, Pennsylvania

A complete solution for the response of an arbitrary shell subjected to time-dependent surface loads is derived by means of the classical method of spectral representation. The solution is expanded in terms of the modes of free vibration, and their orthogonality is proved for an arbitrary shell. As an example, the response of a spherical shell to a suddenly applied uniform normal load is calculated in detail.

INTRODUCTION

RECENT progress in the analysis of free vibration of thin elastic shells has made available a wealth of information regarding the normal modes and natural frequencies of a large class of shell configurations. Among the numerous papers dealing with free vibration of shells, we cite here only a few representative examples of detailed treatments of shells of revolution, which are based on the bending theory of shells. The natural frequencies and mode shapes of a cylindrical shell have been analyzed in detail by Arnold and Warburton[1] and those of a conical shell by Goldberg, Bogdanoff, and Marcus.[2] Shallow spherical shells have been treated by Reissner[3] and by Kalnins and Naghdi,[4] while a detailed examination of the modes of free vibration of nonshallow spherical shells has been carried out by Kalnins.[5] In addition to the many other papers listed in the above references as well as in a bibliography compiled by Gros and Forsberg,[6] we mention here the method for the calculation of the natural frequencies and mode shapes of arbitrary rotationally symmetric shells given by Kalnins.[7]

The purpose of the present paper is to investigate transient vibration of shells for which the free-vibration characteristics are known. Employing the classical method of spectral representation, which involves the expansion of the solution in terms of the normal modes of free vibration, we derive the complete solution for the transient response of an arbitrary shell subjected to time-dependent surface loads.[8] The analysis in this paper is carried out only for homogeneous boundary conditions, because the case of time-dependent nonhomogeneous boundary conditions can be treated in a similar manner by means of well-known methods.[9,10]

As an example of the application of the complete solution derived in this paper, we calculate the transient response of open spherical shells on a viscoelastic foundation subjected to a suddenly applied uniform pressure. The natural frequencies and normal modes are obtained by means of the methods of analysis utilized earlier in Refs. 5 and 7.

* Also: Rensselaer Polytechnic Inst. of Conn., E. Windsor Hill, Conn.

† Formerly: Dept. Eng. & Appl. Sci., Yale Univ., New Haven, Conn.

[1] R. N. Arnold and G. B. Warburton, Inst. Mech. Eng. Proc. **A 167**, 62–80 (1953).

[2] J. E. Goldberg, J. L. Bogdanoff, and L. Marcus, J. Acoust. Soc. Am. **32**, 738–742 (1960).

[3] E. Reissner, Quart. Appl. Math. **13**, 279–290 (1955).

[4] A. Kalnins and P. M. Naghdi, J. Acoust. Soc. Am. **32**, 342–347 (1960).

[5] A. Kalnins, J. Acoust. Soc. Am. **36**, 74–81 (1964).

[6] C. G. Gros and K. Forsberg, "Vibrations of Thin Shells: A Partially Annotated Bibliography," Missiles & Space Co., Lockheed Aircraft Corp., Sunnyvale, Calif., Rept. SB-63-43 (Apr. 1963).

[7] A. Kalnins, J. Acoust. Soc. Am. **36**, 1355–1365 (1964).

[8] Similar transient solutions have been previously employed for some simple shell configurations, such as cylindrical and shallow spherical shells. See, for example, J. Sheng, AIAA J. **3**, 701–709 (1965); L. R. Koval and P. G. Bhuta, Space Technol. Lab., Inc., Rept. No. EM 13-21 (Oct. 1963). The emphasis in the present paper is on shells of arbitrary shape.

[9] R. D. Mindlin and L. E. Goodman, J. Appl. Mech. **17**, 377–380 (1950).

[10] J. G. Berry and P. M. Naghdi, Quart. Appl. Math. **14**, 43–50 (1956).

I. SOLUTION FOR TRANSIENT VIBRATION OF A SHELL

In this section, the complete solution is derived for an elastic shell of arbitrary shape subjected to time-dependent surface loads. The derivation is based on the classical method of spectral representation, which is commonly used in transient analyses of elastic systems but has not been applied specifically to the general case of an elastic shell.

Let a point of the shell be defined by the coordinates ξ_1 and ξ_2, coinciding with the lines of curvature of the middle surface, and the distance ζ, measured along the normal of the middle surface. An element of the shell is subjected to the time-dependent load vector $\mathbf{P}(\xi_1,\xi_2,t)$ that produces the displacement vector $\mathbf{U}(\xi_1,\xi_2,t)$. For the classical theory of shells,[11] **P** and **U** are column vectors containing three elements in the form

$$\mathbf{P}=\begin{Bmatrix} p_1 \\ p_2 \\ p \end{Bmatrix}, \tag{1}$$

$$\mathbf{U}=\begin{Bmatrix} u_1 \\ u_2 \\ w \end{Bmatrix}, \tag{2}$$

which designate the load and displacement components of the middle surface in the ξ_1, ξ_2, and ζ directions, respectively.

The partial differential equations, which govern the motion of an element of the shell, can be found in many references. The derivation given here can be used with any one of the shell theories proposed in the literature. For our purposes, we need only to observe that the differential equations of motion can be written in the form

$$\mathbf{LU}=-\mathbf{P}+\rho h\frac{\partial^2\mathbf{U}}{\partial t^2}+\lambda\frac{\partial\mathbf{U}}{\partial t}+k\mathbf{U}, \tag{3}$$

where ρ and h denote the density and thickness of the shell, respectively, t is time, and **L** is a (3,3) differential operator matrix, dependent on ξ_1 and ξ_2 only, which is given by a specific theory of shells.[12] We have included in the present analysis the effect of a viscoelastic foundation, characterized by an elastic parameter k and a viscous-damping parameter λ, in the same way as it was done by Reissner[13] for a plate resting on a viscoelastic foundation. It is assumed here that both k and λ are the same in the normal and tangential directions of the coordinate curves. With this assumption, it is possible to employ the undamped spectrum of the free-vibration modes. If k and λ in the three directions are not equal, then the spectral representation in terms of the damped modes is necessary.

We now assume that a required number of normal modes of free-vibration of the elastic shell, for which the transient response is sought, are known; i.e., we know the natural frequency ω_i and the mode shapes of the displacement vector $\mathbf{U}_i(\xi_1,\xi_2)$ of the ith mode for any required value of i. We recall that the free-vibration displacements are governed by

$$\mathbf{LU}_i=-\rho h\omega_i^2\mathbf{U}_i, \tag{4}$$

and the same boundary conditions as specified for the transient problem.

Expanding the transient displacements in terms of the normal modes in the form

$$\mathbf{U}(\xi_1,\xi_2,t)=\sum_{i=1}^{\infty}\mathbf{U}_i(\xi_1,\xi_2)q_i(t), \tag{5}$$

where $q_i(t)$ denote the generalized coordinates, and substituting into Eq. 3, we get

$$\sum_{i=1}^{\infty}D_i\mathbf{U}_i=\mathbf{P}, \tag{6}$$

where

$$D_i=\rho h\frac{d^2q_i}{dt^2}+\lambda\frac{dq_i}{dt}+(k+\rho h\omega_i^2)q_i. \tag{7}$$

At this point, it is necessary to employ the Green's identity that is associated with the system of differential equations of a given theory of shells. We have derived this identity in detail for one particular theory in Appendix A, but we wish to emphasize that, whatever theory is chosen for the free-vibration analysis, it is necessary to satisfy the Green's identity in the form

$$\int\!\!\int_S\left[\left(\mathbf{P}^T-\rho h\frac{\partial^2\mathbf{U}^T}{\partial t^2}-\lambda\frac{\partial\mathbf{U}^T}{\partial t}-k\mathbf{U}^T\right)\mathbf{U}' -\left(\mathbf{P}'^T-\rho h\frac{\partial^2\mathbf{U}'^T}{\partial t^2}-\lambda\frac{\partial\mathbf{U}'^T}{\partial t}-k\mathbf{U}'^T\right)\mathbf{U}\right]\alpha_1\alpha_2 d\xi_1 d\xi_2 =\int_B W(\mathbf{y},\mathbf{y}')ds, \tag{8}$$

where the primed and unprimed quantities represent two different solution states, each of which satisfies the governing equations of the theory of shells. Thus, the load system **P** produces the displacement vector **U** and **y**, which designates any relevant dependent variable of the unprimed solution state. Similarly, **P′** produces **U′** and **y′**. In Eq. 8, we have denoted by ds the arc length along the boundary B of a shell with a middle surface S, and a superscript T refers to the transpose of a matrix.

[11] See, for example, the version of the classical theory given by E. Reissner, Am. J. Math. **63**, 177–184 (1941). It is of interest to remark that the transient solution derived later in the paper is also valid, in the same form, for the improved shell theory in which the transverse shear effects are accounted for. For this theory, the vectors **P** and **U** must each contain two more elements that are related to the load moments and rotations of the normal.

[12] The differences in the theories are reflected exclusively in the operator matrix **L**, which, as it becomes evident in the following, is employed only in the free-vibration analysis but not in the transient solution.

[13] E. Reissner, J. Appl. Mech. **25**, 144–145 (1958).

For the classical theory of shells, the integrand in the line integral in Eq. 8 is given by

$$W(\mathbf{y},\mathbf{y}')=N_{nn}'u_n-N_{nn}u_n'+N'u_\theta-Nu_\theta' +M_{nn}'\beta_n-M_{nn}\beta_n'+Q'w-Qw', \quad (9)$$

where

$$N=N_{n\theta}+(1/R_\theta)M_{n\theta}, \quad (10)$$

$$Q=Q_n+(1/\alpha_\theta)M_{n\theta,\theta}. \quad (11)$$

Here, N_{nn} and $N_{n\theta}$ denote the membrane stress resultants, M_{nn} and $M_{n\theta}$ are the stress couples, and Q_n is the transverse shear resultant as defined in the usual way in any theory of shells; R_θ denotes one of the principal radii of curvature and α with subscripts is the square root of the metric. The subscripts n and θ refer to the normal and tangential directions with respect to the edge of the shell as defined by the boundary B.

Let us identify in Eq. 8 the unprimed solution state with an undamped free-vibration state having a natural frequency ω_i and the primed solution state with another undamped free-vibration state having a frequency ω_j, and let us denote the dependent variables of each state by subscripts i and j, respectively. Assuming that both states satisfy the same homogeneous boundary conditions on the boundary B, then it follows directly from Eq. 8 that

$$(\omega_i^2-\omega_j^2)\int\int_S \mathbf{U}_i^T\mathbf{U}_j\alpha_1\alpha_2 d\xi_1 d\xi_2=0, \quad (12a)$$

or

$$\int\int_S \mathbf{U}_i^T\mathbf{U}_j\alpha_1\alpha_2 d\xi_1 d\xi_2=\delta_{ij}N_i, \quad (12b)$$

where δ_{ij} is the Kronecker delta and

$$N_i=\int\int_S \mathbf{U}_i^T\mathbf{U}_i\alpha_1\alpha_2 d\xi_1 d\xi_2. \quad (12c)$$

Equation 12b represents the orthogonality condition of the modes of undamped free vibration of an arbitrary shell for the case when ρ and h are constant over the middle surface of the shell.

Multiplying the transpose of Eq. 6 by the displacement vector $\mathbf{U}_j$, integrating over the middle surface of the shell, and making use of Eq. 12b, we get

$$\frac{d^2q_i}{dt^2}+\frac{\lambda}{\rho h}\frac{dq_i}{dt}+\left(\frac{k}{\rho h}+\omega_i^2\right)q_i=\frac{1}{\rho h}G_i(t), \quad (13)$$

where

$$G_i(t)=1/N_i\int\int_S \mathbf{P}^T\mathbf{U}_i\alpha_1\alpha_2 d\xi_1 d\xi_2. \quad (14)$$

The complete solution of Eq. 13 for the underdamped case $(\lambda/2\rho h)^2<k/\rho h+\omega_i^2$ is given by

$$q_i(t)=e^{-\lambda t/2\rho h}(A_i\cos\gamma_i t+B_i\sin\gamma_i t) +(1/\rho h\gamma_i)\int_0^t G_i(\tau)e^{-\lambda(t-\tau)/2\rho h}\sin\gamma_i(t-\tau)d\tau, \quad (15)$$

where

$$\gamma_i=[(k/\rho h)+\omega_i^2-(\lambda/2\rho h)^2]^{\frac{1}{2}}, \quad (16)$$

and the arbitrary constants A_i and B_i are determined from the initial conditions. If we assume that the initial displacements and velocities of the shell are given by

$$\mathbf{U}(\xi_1,\xi_2,0)=\mathbf{U}_0, \quad (17a)$$

$$(\partial\dot{\mathbf{U}}/\partial t)(\xi_1,\xi_2,0)=\dot{\mathbf{U}}_0, \quad (17b)$$

then it can be shown, with the use of the orthogonality condition, that

$$A_i=1/N_i\int\int_S \mathbf{U}_0^T\mathbf{U}_i\alpha_1\alpha_2 d\xi_1 d\xi_2, \quad (18a)$$

$$B_i=1/N_i\gamma_i\int\int_S \dot{\mathbf{U}}_0^T\mathbf{U}_i\alpha_1\alpha_2 d\xi_1 d\xi_2+(\lambda/2\rho h\gamma_i)A_i. \quad (18b)$$

Having determined the $q_i(t)$, we have obtained the complete solution of the problem of transient vibration of a general shell subjected to arbitrary time-dependent surface loads, because all the physical variables of the response are determined from the displacement field given by Eq. 5. We note that the solution as given by Eq. 5 and 15 requires the explicit values of the natural frequencies and the corresponding normal modes of free vibration for the shell configuration for which the transient response is sought. As seen from the bibliography of Ref. 6, the class of shells for which the free-vibration characteristics can be found is rapidly growing. For example, if the calculation of the foregoing transient solution is coupled with the method of free-vibration analysis given in Ref. 7, then such an arrangement enables one to find the transient response of an arbitrary rotationally symmetric shell when subjected to any time-dependent surface loads.

II. MODE PARTICIPATION

Let us explore the contribution that is made by the individual modes of free-vibration to the transient solution given by Eq. 5. For this purpose, it is convenient to write all variables and parameters in a nondimensional form. After the introduction of a general reference length L and a reference pressure p_0 for each shell, as well as a reference length l_i for the ith mode of free vibration, we define the following nondimensional quantities:

$$\begin{aligned}\{U_1,U_2,W\}&=(D/L^4p_0)\{u_1,u_2,w\}\\ T&=(D/L^4\rho h)^{\frac{1}{2}}t\\ \mathrm{K}&=(L^4/D)k,\\ \Lambda&=(L^4/\rho hD)^{\frac{1}{2}}\lambda/2,\\ \Omega&=(\rho/E)^{\frac{1}{2}}L\omega.\end{aligned} \quad (19)$$

Assuming that the surface loads are separable in the form

$$\begin{aligned} p_1(\xi_1,\xi_2,t)&=f_1(t)p_1^*(\xi_1,\xi_2),\\ p_2(\xi_1,\xi_2,t)&=f_2(t)p_2^*(\xi_1,\xi_2),\\ p(\xi_1,\xi_2,t)&=f(t)p^*(\xi_1,\xi_2), \end{aligned} \quad (20)$$

we can introduce the mode participation factors of the ith mode[14] defined by

$$\begin{Bmatrix} m_{1i} \\ m_{2i} \\ m_i \end{Bmatrix} = (l_i/p_0N_i)\int\int_S \begin{Bmatrix} p_1^*u_{1i} \\ p_2^*u_{2i} \\ p^*w_i \end{Bmatrix} \alpha_1\alpha_2 d\xi_1 d\xi_2, \quad (21)$$

and the time-dependent parameters given by

$$\begin{Bmatrix} F_{1i}(T) \\ F_{2i}(T) \\ F_i(T) \end{Bmatrix} = \int_0^T \begin{Bmatrix} f_1(\tau) \\ f_2(\tau) \\ f(\tau) \end{Bmatrix} e^{-(T-\tau)\Lambda}\sin\Gamma_i(T-\tau)d\tau. \quad (22)$$

With this notation, Eq. 15 can be written as

$$q_i(T)=e^{-T\Lambda}(A_i\cos\Gamma_iT+B_i\sin\Gamma_iT) + (L^4p_0/Dl_i\Gamma_i)(m_{1i}F_{1i}+m_{2i}F_{2i}+m_iF_i), \quad (23)$$

where

$$\Gamma_i=[\mathrm{K}+12(1-\nu^2)\Omega_i^2(L/h)^2-\Lambda^2]^{\frac{1}{2}}. \quad (24)$$

It follows from Eqs. 5 and 23 that, for large values of T, any dependent variable $\mathbf{y}$ of the transient response is given by

$$(D/L^4p_0)\mathbf{y}(\xi_1,\xi_2,t)=\sum_{i=1}^{\infty}\frac{\mathbf{y}_i(\xi_1,\xi_2)}{l_i\Gamma_i} \times(m_{1i}F_{1i}+m_{2i}F_{2i}+m_iF_i). \quad (25)$$

We can always consider the response of the shell produced by the three components of the surface load p_1, p_2, p acting separately (one at a time), because the total response can then be obtained by superposition. For example, consider the case when $p_1=p_2=0$. Then for large values of T, the ith term in the series given by Eq. 25 is proportional to m_i. Clearly, if a mode of free vibration has w_i much smaller than the other displacements, then the m_i calculated from Eq. 21 will be much smaller than the m_j of some of the other modes, and the participation of this mode in the solution produced by p will be small. By a similar argument, those modes that have a small u_{1i} (or u_{2i}) will not affect appreciably the response produced by p_1 (or p_2).

These conclusions are of importance in the theory of transient vibration of shallow shells, where the analysis can be simplified by neglecting the longitudinal-inertia terms and considering only the transverse vibration, as it was done by Reissner.[15] It was shown by Kalnins[16] that, if the longitudinal-inertia terms for a shallow spherical shell are included, then (1) the transverse modes are almost the same as those predicted by Reissner's equations and (2) an additional set of modes, called longitudinal modes, is obtained which is not predicted by the equations of Ref. 15. Moreover, the analysis of the modes of free vibration carried out in Ref. 16 shows that the absolute values of the displacement components obey the following rule: for the transverse modes $w>u_1, u_2$ and for the longitudinal modes $u_1, u_2>w$. From Eq. 25, it follows that, in the presence of a normal load p, the longitudinal modes will have a small participation in the response if the ratio of the m_i of the longitudinal modes over the m_i of the transverse modes is a small number. If this is true, then in the transient reponse produced by p the longitudinal modes can indeed be neglected and only the frequency spectrum predicted by Ref. 15 need by employed. Similarly, if in the response produced by p_1 (or p_2) the factors m_{1i} (or m_{2i}) of the transverse modes are much smaller than those of the longitudinal modes, then it would be possible to use only the longitudinal modes as predicted by the membrane theory and omit the transverse (or bending) modes. Judging from the results of Refs. 5 and 16, this, however, may not be true, and therefore in the calculation of the response of a shallow shell produced by p_1 or p_2 the complete frequency spectrum (longitudinal and transverse modes) may have to be employed.

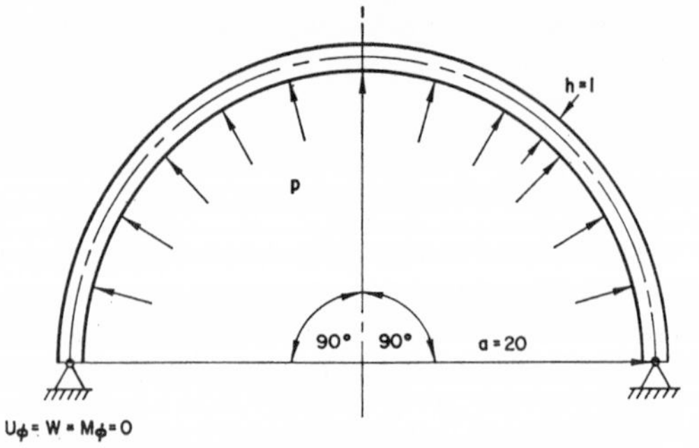

FIG. 1. Geometry of spherical shell under consideration.

III. EXAMPLE: TRANSIENT RESPONSE OF A SPHERICAL SHELL

As an application of the results given in Sec. II, we consider the transient response of a simply supported spherical shell on a viscoelastic foundation subjected to a sudden change of radial pressure applied uniformly over its entire surface (see Fig. 1). For such a shell, the relevant parameters are given by $\xi_1=\phi$, $\xi_2=\theta$, $\alpha_1=a$, $\alpha_2=a\sin\phi$, where we have taken a to be the radius of the middle surface of the shell. The components of the

[14] This factor is similar to the "mode participation of load" factor employed by W. T. Thomson and M. V. Barton, J. Appl. Mech. 24, 248–251 (1957).

[15] E. Reissner, Quart. Appl. Math. 13, 169–176 (1955).

[16] A. Kalnins, Proc. Natl. Congr. Appl. Mech. 4th, USA 1962, 225–233 (1963).

displacement vector are $u_1=u_\phi(\phi)$, $u_2=u_\theta=0$, $w=w(\phi)$, while the applied loads take on the values $p_1=p_\phi=0$, $p_2=p_\theta=0$, $p=p^*f(t)$.

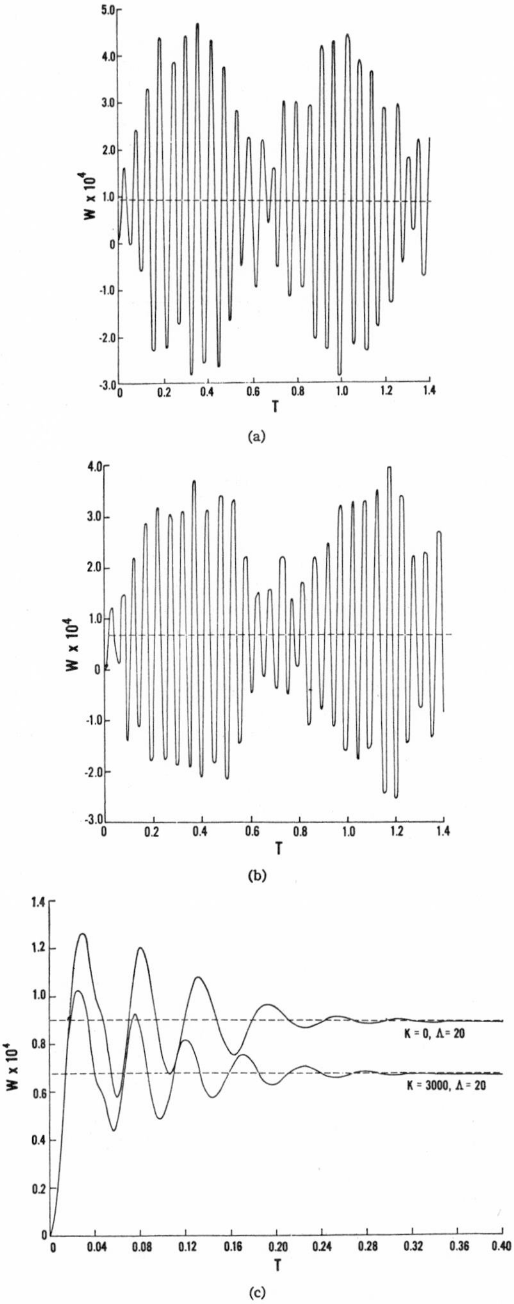

FIG. 2. Response of normal deflection at apex of spherical shell shown in Fig. 1 resting on a foundation. (a) $K=\Lambda=0$. (b) $K=3000$, $\Lambda=0$. (c) $K=0$, $\Lambda=20$, and $K=3000$, $\Lambda=20$.

We choose the reference parameters $L=a$, $l_i=w_i(0)$, $p_0=1$, and assume that the shell is initially at rest. From Eqs. 18 it follows that $A_i=B_i=0$. The prescribed pressure at the middle surface of the shell is

$$p^*=\text{const}; \tag{26}$$

$$f(t)=\begin{matrix} 0 & \text{for} & t<0 \\ 1 & \text{for} & t\geqslant 0 \end{matrix}. \tag{27}$$

From Eq. 22, after integration, we get $F_{1i}(T)=F_{2i}(T)=0$, and

$$F_i(T)=\{\Gamma_i/[K+12(1-\nu^2)\Omega_i^2(a/h)^2]\} \times[1-(e^{-T\Lambda}/\Gamma_i)(\Lambda \sin\Gamma_i T+\Gamma_i \cos\Gamma_i T)]. \tag{28}$$

The final solution can be written in the form

$$W(\phi,T)=\sum_{i=1}^{\infty}\frac{w_i(\phi)}{w_i(0)}\frac{m_iF_i(T)}{\Gamma_i}, \tag{29a}$$

$$U_\phi(\phi,T)=\sum_{i=1}^{\infty}\frac{u_{\phi i}(\phi)}{w_i(0)}\frac{m_iF_i(T)}{\Gamma_i}. \tag{29b}$$

For a large value of T, the transverse displacement of the response at the apex ($\phi=0$) of the shell is given by

$$W(0,T)=\sum_{i=1}^{\infty}\frac{m_i}{K+12(1-\nu^2)\Omega_i^2(a/h)^2}. \tag{30}$$

From Eq. 30, we can observe that the convergence of the expansion in terms of normal modes depends on the convergence of m_i. We also note that the series of Eq. 30 should converge to the solution of the corresponding static problem. By knowing the static solution, we can easily establish the number of modes that are necessary in the transient solution.

The foregoing analysis was applied to spherical shells with half-angles of opening of 15° and 90° resting on a variety of foundations. The natural frequencies and the individual modes of free vibration required for the mode participation factors in Eq. 21 and also in Eqs. 29 were obtained by means of the computer programs employed

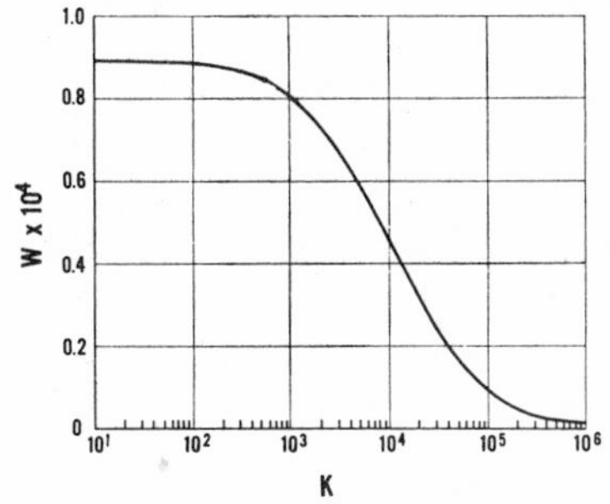

FIG. 3. Effect of elastic foundation parameter on the static deflection at the apex of the shell shown in Fig. 1.

TABLE I. Results for a typical shallow spherical shell. The notation $AE\pm N$ stands for $A(10^{\pm N})$. $\phi_0=15°$, $a/h=20$, $\nu=0.3$, $k=0$, simply supported edge.

	Mode participation factors			Static deflection ($\phi=0$, $T\rightarrow\infty$)	
i	m_i	$m_{\phi i}$	Ω_i	q_i	W
1	0.15924 E 01	−0.85479 E−01	1.863	0.10536 E−03	0.10536 E−03
2	−0.10444 E 01	−0.19336 E 00	6.618	−0.54763 E−05	0.99889 E−04
3	0.63996 E−01	0.93213 E 00	15.35	0.62351 E−07	0.99951 E−04
4	0.76549 E 00	−0.65718 E 00	16.43	0.65117 E−06	0.10060 E−03
5	−0.13295 E−01	−0.10480 E 00	28.15	−0.38542 E−08	0.10059 E−03
6	−0.69489 E 00	0.30452 E−01	30.56	−0.17090 E−06	0.10042 E−03
7	0.15599 E−02	0.13750 E 00	40.84	0.21479 E−09	0.10042 E−03
8	0.61725 E 00	−0.13740 E−01	49.01	0.59038 E−07	0.10048 E−03
9	0.10651 E−01	−0.94045 E−01	53.56	0.85282 E−09	0.10048 E−03
10	−0.10017 E−01	−0.69344 E−01	65.98	−0.52845 E−09	0.10048 E−03
11	−0.55460 E 00	0.17924 E 00	72.06	−0.24530 E−07	0.10046 E−03
12	−0.72507 E−02	−0.26795 E−01	78.73	−0.26872 E−09	0.10046 E−03
				Static deflection using Ref. 17:	0.10048 E−03

in Refs. 5 and 7. The integrals occurring in Eq. 21 were evaluated by means of Simpson's rule.

For the calculation of the free-vibration characteristics, the shells were assumed to be simply supported at their edges and to have the following properties: $a=20$, $h=1$, $\nu=0.3$, $E=3\times10^7$, $p^*=1$ (see Fig. 1). The results are summarized in Tables I and II and Figs. 2 and 3. We note that, although the Tables give data for both the 15° and 90° shells, the Figures give information only for the case of 90°. It was found that the results for the case of 15° were not sufficiently different to warrant their presentation.

In Figs. 2(a)–(c), we have plotted the response of the normal deflection at the apex of a hemispherical shell resting on foundations, described by the following parameter combinations: $\Lambda=K=0$; $\Lambda=0$, $K=3000$; $\Lambda=20$, $K=0$; and $\Lambda=20$, $K=3000$, respectively. In the Figures, we have also indicated, with horizontal lines, the appropriate static deflection for each case as obtained from Eq. 30 (see also Tables I and II). For the two cases where $K=0$, the static deflections agree with the results obtained by means of the computer program developed by Kalnins[17] for the static analysis of general shells of revolution. No direct check of the static deflection for those cases where $K=3000$ is available.

It is seen from Figs. 2 (a) and (b) that the response of a shell resting on a purely elastic foundation consists of an oscillation about the appropriate static deflection and that this oscillation continues for all values of time. For the case of a shell resting on a viscoelastic foundation, Fig. 2(c) shows that the period of the oscillation is greatly increased while its amplitude decays and the deflection approaches the appropriate static result as time goes on.

Figures 2(a) and (b) suggest that the phenomenon of beating is taking place between two of the more predominant modes in the transient response at the apex of the hemispherical shell under consideration (the

TABLE II. Results for a typical nonshallow spherical shell. The notation $AE\pm N$ stands for $A(10^{\pm N})$. $\phi_0=90°$, $a/h=20$, $\nu=0.3$, $k=0$, simply supported edge.

	Mode participation factors			Static deflection ($\phi=0$, $T\rightarrow\infty$)	
i	m_i	$m_{\phi i}$	Ω_i	q_i	W
1	0.48079 E−01	−0.10020 E 01	0.773	0.18454 E−04	0.18454 E−04
2	−0.10486 E 00	0.23184 E 00	1.010	−0.23610 E−04	−0.51561 E−05
3	0.35821 E 00	−0.43871 E 00	1.242	0.53323 E−04	0.48166 E−04
4	−0.15912 E 01	0.25153 E 00	1.578	−0.14652 E−03	−0.98362 E−04
5	0.27978 E 01	0.88919 E−01	1.720	0.21689 E−03	0.11852 E−03
6	−0.87944 E 00	−0.55614 E 00	2.192	−0.41994 E−04	0.76534 E−04
7	0.37398 E 00	0.26197 E 01	2.793	0.10999 E−04	0.87534 E−04
8	0.26195 E 00	−0.14683 E 01	2.941	0.69521 E−05	0.94486 E−04
9	−0.48004 E 00	0.13182 E 00	3.733	−0.79110 E−05	0.86575 E−04
10	0.35957 E 00	0.18832 E 00	4.682	0.37684 E−05	0.90344 E−04
11	0.47701 E−01	−0.30506 E 00	4.843	0.46709 E−06	0.90811 E−04
12	−0.37829 E 00	−0.52766 E−01	5.804	−0.25789 E−05	0.88232 E−04
13	0.55733 E−01	0.67072 E 00	6.876	0.27078 E−06	0.88503 E−04
14	0.30631 E 00	−0.45152 E 00	7.047	0.14172 E−05	0.89920 E−04
15	−0.34113 E 00	−0.21393 E−01	8.366	−0.11158 E−05	0.88804 E−04
				Static deflection using Ref. 17:	0.89751 E−04

17 A. Kalnins, J. Appl. Mech. **31**, 467–476 (1964).

15° shell considered here does not respond in this manner). To investigate this possibility, Tables I and II can be searched to find two consecutive modes that are predominant in the solution and that have roughly the same frequency and order of magnitude. Table I reveals no such cases for the 15° shell, as could be expected. Table II indicates several cases where consecutive modes have almost the same frequency, but only in the case of the 4th and 5th modes are the modes predominant in the solution and also of roughly the same amplitude q_i. Recalling that when beating occurs the frequency of the beats is equal to the difference in the frequencies of the modes involved, we find in the present situation that the frequency of beating should be $\Omega_5-\Omega_4=0.142$ and the period of one beat should be $2\pi/(\Gamma_5-\Gamma_4)=0.668$. It is seen in Figs. 2(a) and (b) that the period of the envelope surrounding the response is just this value, within the accuracy of the plot. Furthermore, the period of the more rapid vibrations should be, according to the present reasoning, $2\pi/\Gamma_5=0.056$ and this is confirmed by Figs. 2(a) and (b). Undoubtedly, beating is also taking place between other modes in the response. However, in the other possible cases such as modes 10 and 11 in Table II, the amplitudes of the two modes differ by an order of magnitude and are small in comparison to the over-all response. As a result, the effect of these beats is not as pronounced in Figs. 2(a) and (b) as is the beating between modes 4 and 5.

As could be expected from Eqs. 29 and 30, Figs. 2(a)–(c) indicate that one of the main effects of the elasticity of the foundation, on which the shell rests, is an alteration of its static deflection. The period and amplitude of its response are also affected. The extent of the influence of the elastic foundation parameter K on the static deflection as obtained from Eq. 30 is shown in Fig. 3 for a 90° shell. The static solution in Fig. 3 is independent of the viscosity of the foundation.

To obtain information on the convergence of the method of spectral representation utilized in this paper, we have included Tables I and II. In particular, in the damped case we may study the convergence toward the static solution obtained with the use of the method of Ref. 17. It is seen in the last column of each of the Tables that the convergence of the static solution is fairly rapid, especially in the case of the 15° shell, where it appears that the first mode alone gives virtually the correct result. The solution of the 90° shell converges less rapidly; this is felt to be attributable to the close spacing of the natural frequencies of that shell. Note that the first five natural frequencies of the 90° shell occur before the first natural frequency of the 15° shell. It is reasonable to assume that the convergence of the static solution toward the known result as demonstrated above is representative of the convergence of the solution at any other value of time. It is seen, therefore, that the transient-vibration analysis of a shell presents no computational difficulties.

IV. RÔLE OF LONGITUDINAL INERTIA

Finally, it is appropriate to make an assessment of the rôle of the longitudinal inertia in the transient response of a shell to a normal load. To facilitate this discussion, we have presented in Tables I and II the mode participation factors m_i and $m_{\phi i}$ for the two spherical shells that have been studied. As can be seen from Eqs. 29, only the m_i are used in the determination of the response to a normal pressure, and, therefore, the magnitude of m_i for a given mode of free vibration will give an indication of the participation of that mode in the response to p. For a shallow shell, as exemplified by the 15° case, it can be seen from Table I that those modes that have $m_{\phi i}$ greater than m_i have very little participation in the response to a normal load. According to a node count, these are the longitudinal modes discussed in Ref. 16. The contribution of these modes can be safely ignored in the calculation of the response of a normally loaded shallow shell.

For a nonshallow configuration, as exemplified by a hemispherical shell, Table II indicates that for many modes, where $m_{\phi i}$ is greater than m_i, the two values are nevertheless of the same order of magnitude, and consequently the contributions of none of these modes can be neglected. A striking example is that of the first mode in Table II, where $m_{\phi 1}$ is much greater than m_1, and yet the contribution of that mode to the final result is significant. It is thus seen that for the hemispherical shell the participation of all the modes is important and none can be omitted in the determination of the response to a normal load.

ACKNOWLEDGMENTS

The work of H. Kraus was supported by the U. S. Army Research Office (Durham) and the work of A. Kalnins was supported by the National Science Foundation. The authors are indebted to D. Lombari, Pratt & Whitney Aircraft, for writing a computer program that was used in the sample calculation.

Appendix A

For the derivation of the orthogonality condition of a general shell, we make use of an identity that can be regarded as the Green's identity of the system of partial differential Eqs. 3. The derivation of this identity parallels that of the reciprocal theorem that, for the static theory of shells, has been given in many papers.[A1] Since, in the present paper, we need the Green's

[A1] See, e.g., P. M. Naghdi, in *Progress in Solid Mechanics*, I. N. Sneddon and R. Hill, Eds. (North-Holland Publ. Co., Amsterdam, 1963), Vol. 4, p. 60.

identity in the presence of the inertia terms and a viscoelastic foundation, we include here a brief outline of its derivation for one particular classical theory of shells given by Reissner.[11]

Consider a solution state **y** whose variables satisfy the membrane equation of equilibrium (Ref. 11, Eq. 18) in the form

$$-\alpha_1\alpha_2\left(p_1-ku_1-\lambda\frac{\partial u_1}{\partial t}-\rho h\frac{\partial^2 u_1}{\partial t^2}\right)=(\alpha_2N_{11})_{,1}+(\alpha_1N_{12})_{,2}+N_{12}\alpha_{1,2}-N_{22}\alpha_{2,1}+\left(\frac{\alpha_2M_{11}}{R_1}\right)_{,1}-\alpha_2\left(\frac{1}{R_1}\right)_{,1}M_{11}$$

$$+\left(\frac{\alpha_1M_{12}}{R_1}\right)_{,2}-\alpha_1\left(\frac{1}{R_1}\right)_{,2}M_{12}+\frac{\alpha_{1,2}}{R_1}M_{12}-\frac{\alpha_{2,1}}{R_1}M_{22},\qquad\text{(A1)}$$

where Q_1 has been eliminated by the moment equilibrium equation. Multiplication of Eq. A1 by u_1' of another solution state $\mathbf{y}'$, integration with respect to ξ_1 and ξ_2 over the middle surface of the shell, followed by integration by parts, leads to

$$-\iint_S\left(p_1-ku_1-\lambda\frac{\partial u_1}{\partial t}-\rho h\frac{\partial^2 u_1}{\partial t^2}\right)u_1'd\sigma=\iint_S\left\{-\frac{u_{1,1}'}{\alpha_1}N_{11}-\left(\frac{u_{1,2}'}{\alpha_2}-\frac{\alpha_{1,2}}{\alpha_1\alpha_2}u_1'\right)N_{12}-\frac{\alpha_{2,1}}{\alpha_1\alpha_2}u_1'N_{22}\right.$$

$$\left.-\left[\frac{u_{1,1}'}{\alpha_1R_1}+\frac{1}{\alpha_1}\left(\frac{1}{R_1}\right)_{,1}u_1'\right]M_{11}-\left[\frac{u_{1,2}'}{\alpha_2R_1}+\frac{1}{\alpha_2}\left(\frac{1}{R_1}\right)_{,2}u_1'-\frac{\alpha_{1,2}}{\alpha_1\alpha_2R_1}u_1'\right]M_{12}-\frac{\alpha_{2,1}}{\alpha_1\alpha_2R_1}u_1'M_{22}\right\}d\sigma$$

$$+\int_B\left[\left(N_{11}u_1'+\frac{u_1'}{R_1}M_{11}\right)\alpha_2d\xi_2+\left(N_{12}u_1'+\frac{u_1'}{R_1}M_{12}\right)\alpha_1d\xi_1\right],\qquad\text{(A2)}$$

where we have denoted the surface element $\alpha_1\alpha_2d\xi_1d\xi_2$ of the middle surface by $d\sigma$. Performing the same operations on the remaining membrane equilibrium equation, after multiplication by u_2', we obtain an expression that is identical to Eq. A2, except that the indices 1 and 2 are interchanged.

Similarly, consider the transverse equilibrium equation in the form

$$-\alpha_1\alpha_2\left(p-kw-\lambda\frac{\partial w}{\partial t}-\rho h\frac{\partial^2 w}{\partial t^2}\right)=(\alpha_2Q_1)_{,1}+(\alpha_1Q_2)_{,2}-\alpha_1\alpha_2\left(\frac{N_{11}}{R_1}+\frac{N_{22}}{R_2}\right).\qquad\text{(A3)}$$

Again, multiplication by w', integration with respect to ξ_1 and ξ_2, and integration by parts, leads to

$$-\iint_S\left(p-kw-\lambda\frac{\partial w}{\partial t}-\rho h\frac{\partial^2 w}{\partial t^2}\right)w'd\sigma=\iint_S\left\{-\frac{w_{,1}'}{\alpha_1}Q_1-\frac{w_{,2}'}{\alpha_2}Q_2-w'\left(\frac{N_{11}}{R_1}+\frac{N_{22}}{R_2}\right)\right\}d\sigma$$

$$+\int_B\{Q_1w'\alpha_2d\xi_2+Q_2w'\alpha_1d\xi_1\}.\qquad\text{(A4)}$$

We now eliminate Q_1 and Q_2 by means of the moment-equilibrium equations, integrate once more by parts, and get

$$-\iint_S\left(p-kw-\lambda\frac{\partial w}{\partial t}-\rho h\frac{\partial^2 w}{\partial t^2}\right)w'd\sigma=\iint_S\left\{\frac{1}{\alpha_1}\left(\frac{w_{,1}'}{\alpha_1}\right)_{,1}M_{11}+\frac{1}{\alpha_2}\left(\frac{w_{,1}'}{\alpha_1}\right)_{,2}M_{12}-\frac{w_{,1}'}{\alpha_1^2\alpha_2}\alpha_{1,2}M_{12}+\frac{w_{,1}'}{\alpha_1^2\alpha_2}\alpha_{2,1}M_{22}\right.$$

$$\left.+\frac{1}{\alpha_1}\left(\frac{w_{,2}'}{\alpha_2}\right)_{,1}M_{12}+\frac{1}{\alpha_2}\left(\frac{w_{,2}'}{\alpha_2}\right)_{,2}M_{22}-\frac{\alpha_{2,1}}{\alpha_1\alpha_2^2}w_{,2}'M_{12}+\frac{\alpha_{1,2}}{\alpha_1\alpha_2^2}w_{,2}'M_{11}-\frac{w'}{R_1}N_{11}-\frac{w'}{R_2}N_{22}\right\}d\sigma$$

$$+\int_B\left\{\left(Q_1w'-\frac{w_{,1}'}{\alpha_1}M_{11}-\frac{w_{,2}'}{\alpha_2}M_{12}\right)\alpha_2d\xi_2+\left(Q_2w'-\frac{w_{,2}'}{\alpha_2}M_{22}-\frac{w_{,1}'}{\alpha_1}M_{12}\right)\alpha_1d\xi_1\right\}.\qquad\text{(A5)}$$

Adding Eq. A2 to the corresponding equation with indices 1 and 2 exchanged and then to Eq. A5, we find that

$$\int\int_S\left[\left(p_1-ku_1-\lambda\frac{\partial u_1}{\partial t}-\rho h\frac{\partial^2 u_1}{\partial t^2}\right)u_1'+\left(p_2-ku_2-\lambda\frac{\partial u_2}{\partial t}-\rho h\frac{\partial^2 u_2}{\partial t^2}\right)u_2'+\left(p-kw-\lambda\frac{\partial w}{\partial t}-\rho h\frac{\partial^2 w}{\partial t^2}\right)w'\right]d\sigma$$

$$+\int_B(Q_n w'+N_{nn}u_n'+N_{n\theta}u_\theta'+M_{nn}\beta_n'+M_{n\theta}\beta_\theta')ds$$

$$=\int\int_S(N_{11}\epsilon_1'+N_{12}2\epsilon_{12}'+N_{22}\epsilon_2'+M_{11}k_1'+M_{12}2k_{12}'+M_{22}k_2')d\sigma, \quad \text{(A6)}$$

where we have employed the strain–displacement relations given in Ref. 11.

Since the stress resultants in the surface integral can be replaced in terms of strains, we note that the expression on the right-hand side of Eq. A6 is symmetric with respect to the primed and unprimed quantities. Therefore, the left-hand side of Eq. A6 can be equated to the same expression with the primed and unprimed variables interchanged, and the identity given by Eq. 8 follows if, as it is required in the classical theory of shells, we make use of the effective transverse and tangential shear resultants as defined by Eqs. 10 and 11.

It should be remarked, that Eq. A6 is merely an identity that is satisfied by any two solution states associated with the governing equations of Ref. 11. As shown in the main body of the paper, the orthogonality of the modes of free vibration follows directly from Eq. A6.

28

Reprinted from *J. Acoust. Soc. Amer.*, **39**(5) Part I, 895–898 (1966)

Transient Response of Thin Elastic Shells

J. P. Wilkinson

North American Aviation, Inc., Downey, California

A solution for the transient response of a thin elastic shell that rests on a viscoelastic foundation is found by means of the mode-acceleration method. For a uniform load, this solution is compared with a similar solution obtained by means of the mode-displacement method. A numerical example for a hemispherical shell illustrates that the mode-acceleration method converges slightly more rapidly than the mode-displacement method during the time of application of the transient pulse.

LIST OF SYMBOLS

ξ_1,ξ_2	orthogonal coordinates of shell middle surface
α_1,α_2	Lamé parameters
u_1,u_2,u_3	displacement of middle surface in ξ_1, ξ_2, and normal directions
p_1,p_2,p_3	applied surface loads
ρ	mass density of shell
h	shell thickness
a	characteristic length of shell
E	Young's modulus of shell
k	elastic parameter of foundation
λ	viscous damping parameter of foundation
t	time
Ω_i	ith natural frequency of undamped shell
ω_i	$a(\rho/E)^{\frac{1}{2}}\Omega_i$
q_i	generalized coordinate
m_i	generalized mass
P_i	generalized force
u_{1i},u_{2i},u_{3i}	normal modes of free vibration
u_{1s},u_{2s},u_{3s}	pseudostatic displacements
$(\ )_{,x}$	$d(\ )/dx$
$(\dot{\ })$	$d(\ )/dt$

Other symbols are defined as they appear in the text.

INTRODUCTION

The transient response of thin elastic shells has been discussed recently by Kraus and Kalnins,[1] who used the well-known mode-displacement method of solution in which the response is written as an infinite sum of the normal modes of free vibration. It is the purpose of this paper to develop an alternate solution by means of the mode-acceleration method in which the response is expressed in two portions: one is the response to a pseudostatic load and the other is a sum of normal modes, which represents the response of the shell to the inertia loads. The advantage of the mode-acceleration method is that its convergence during the time of action of the transient loads is slightly more rapid than that obtained by means of the mode-displacement method.

In the analysis, it is assumed that the shell rests on a viscoelastic foundation whose parameters k and λ are the same in the tangential and normal directions of the coordinate curves. With this assumption, it is possible to express the response in terms of the normal modes of free undamped vibrations. Furthermore, attention is limited to the case where the lines of curvature of the shell middle surface coincide with an orthogonal coordinate system (ξ_1,ξ_2). In general, the pseudostatic response and the mode shapes of free vibration are not known in closed form; however, for a shell of revolution of arbitrary meridional shape, they may be evaluated numerically by using, for instance, a technique described by Kalnins.[2] For the purposes of this paper, it is, therefore, assumed that the free vibration modes and pseudostatic response are known.

[1] H. Kraus and A. Kalnins, "Transient Vibration of Thin Elastic Shells," J. Acoust. Soc. Am. 38, 994–1002 (1965).

[2] A. Kalnins, "Analysis of Shells of Revolution Subjected to Symmetrical and Nonsymmetrical Loads," J. Appl. Mech. 31, 467–476 (1964); "Free Vibration of Rotationally Symmetric Shells," J. Acoust. Soc. Am. 36, 1355–1365 (1964).

I. TRANSIENT RESPONSE

According to the classical linear theories of thin elastic shells, such as Love's first approximation[3] or Sanders' theory,[4] the equations of motion of a shell may be written in terms of the middle-surface displacements as a set of three coupled linear differential equations. These equations are of the form

$$[H]\{u_j\}-k\{u_j\}-\lambda\{u_j\}_{,t}-\rho h\{u_j\}_{,tt}=-\{p_j\}, \quad (1)$$

where $[H]$ is a matrix differential operator, $\{u_j\}$ is the column vector of displacements, and $\{p_j\}$ is the column vector of applied surface loads. The analysis that follows can be used in conjunction with any classical linear shell theory providing that the orthogonality conditions of the normal modes of free vibration are satisfied.

A solution to Eq. 1 subject to the initial conditions

$$\begin{aligned}\{u_j(\xi_1,\xi_2,0)\}&=\{u_{j0}(\xi_1,\xi_2)\},\\ \{u_j(\xi_1,\xi_2,0)\}_{,t}&=\{\dot u_{j0}(\xi_1,\xi_2)\}\end{aligned} \quad (2)$$

may be obtained by means of the mode-acceleration method first introduced by Williams[5] in connection with the transient vibrations of linear elastic systems. The method has been extended by Sheng[6] to the transient vibrations of elastic cylindrical shells. Let

$$\{u_j(\xi_1,\xi_2,t)\}=\{u_{js}(\xi_1,\xi_2,t)\}+\sum_{i=1}^{\infty} q_i(t)\{u_{ji}(\xi_1,\xi_2)\}, \quad (3)$$

where we define the subscripted quantities u_{ji} and u_{js} $(j=1,2,3)$ such that

$$[H]\{u_{ji}\}+\rho h\Omega_i^2\{u_{ji}\}=0, \quad (4)$$

$$[H]\{u_{js}\}-k\{u_{js}\}=-\{p_j(\xi_1,\xi_2,t)\}. \quad (5)$$

Here, u_{ji} are the ith normal modes of free undamped vibration corresponding to the natural frequency Ω_i, and u_{js} are the pseudostatic displacements resulting from the surface loads as if they were applied statically. Physically, the solution, Eq. 3, is to be interpreted as being made up of a system of displacements due to the pseudostatic loads at any given time, to which is added a displacement due to the dynamic loading. This concept is commonly used in aircraft design.[7]

Introduction of Eq. 3 into Eq. 1 results in

$$\sum_{i=1}^{\infty}[(\rho h\Omega_i^2+k)q_i+\lambda q_{i,t}+\rho h q_{i,tt}]\{u_{ji}\} =-\lambda\{u_{js}\}_{,t}-\rho h\{u_{js}\}_{,tt}. \quad (6)$$

We multiply Eq. 6 by the row vector $\alpha_1\alpha_2\lfloor u_{1l}u_{2l}u_{3l}\rfloor$, integrate over the shell middle surface, and use the following orthogonality relation for the undamped system derived by Kraus and Kalnins[1]:

$$\iint_S \sum_{j=1}^{3} u_{ji}u_{jl}d\sigma=\delta_{il}m_i, \quad (7)$$

where δ_{il} is the Kronecker delta, and

$$\begin{aligned}m_i&=\iint_S \sum_{j=1}^{3} u_{ji}^2 d\sigma,\\ d\sigma&=\alpha_1\alpha_2 d\xi_1 d\xi_2.\end{aligned} \quad (8)$$

This operation results in

$$q_{i,tt}+(\lambda/\rho h)q_{i,t}+(k/\rho h+\Omega_i^2)q_i =-(1/m_i)[(\lambda/\rho h)P_{i,t}+P_{i,tt}], \quad (9)$$

where

$$P_i(t)=\iint_S \sum_{j=1}^{3} u_{ji}u_{js}d\sigma. \quad (10)$$

A similar operation on the initial conditions given by Eq. 2 results in a new set of initial conditions on the generalized coordinate q_i:

$$\begin{aligned}m_iq_i(0)&=-P_i(0)+y_{0i},\\ m_iq_{i,t}(0)&=-\dot P_i(0)+\dot y_{0i},\end{aligned} \quad (11)$$

where

$$\begin{aligned}y_{0i}&=\iint_S \sum_{j=1}^{3} u_{j0}u_{ji}d\sigma,\\ \dot y_{0i}&=\iint_S \sum_{j=1}^{3} \dot u_{j0}u_{ji}d\sigma.\end{aligned} \quad (12)$$

The solution to Eqs. 9 and 11 for the underdamped case when $(\lambda/2\rho h)^2<k/\rho h+\Omega_i^2$ is

$$\begin{aligned}m_iq_i(t)&=e^{-\lambda t/2\rho h}\{y_{0i}\cos\gamma_i t+(1/\gamma_i)\\ &\quad\times[\dot y_{0i}+(\lambda/2\rho h)y_{0i}]\sin\gamma_i t\}\\ &\quad-P_i(t)+(1/\gamma_i)[(\lambda/2\rho h)^2+\gamma_i^2]\\ &\quad\times\int_0^t P_i(\tau)e^{-\lambda(t-\tau)/2\rho h}\sin\gamma_i(t-\tau)d\tau, \quad (13)\end{aligned}$$

[3] E. Reissner, "A New Derivation of the Equations for the Deformation of Elastic Shells," Am. J. Math. **63**, 177–184 (1941).
[4] J. L. Sanders, Jr., "An Improved First-Approximation Theory for Thin Shells," NASA Rept. **24** (June 1959).
[5] D. Williams, "Displacements of a Linear Elastic System under a Given Transient Load," Aeron. Quart. **1**, 123–136 (1949).
[6] J. Sheng, "The Response of a Thin Cylindrical Shell to a Transient Surface Loading," AIAA J. **3**, 701–709 (1965).
[7] R. L. Bisplinghoff, H. Ashley, and R. L. Halfman, *Aeroelasticity* (Addison-Wesley Publ. Co., Inc., Reading, Mass., 1957), pp. 642–643; J. W. Mar, T. H. H. Pian, and J. M. Calligeros, "A Note on Methods for the Determination of Transient Stresses," J. Aeron. Sci. **23**, 94–95 (1956).

where

$$\gamma_i=[k/\rho h+\Omega_i^2-(\lambda/2\rho h)^2]^{\frac{1}{2}}. \quad (14)$$

The dynamic reciprocity theorem given by Kraus and Kalnins[1] requires that the displacement fields of a primed and an unprimed solution state satisfy the relation

$$\int_S\int\sum_{j=1}^{3}(p_j-ku_j-\lambda u_{j,t}-\rho hu_{j,tt})u_j'd\sigma$$
$$=\int_S\int\sum_{j=1}^{3}(p_j'-ku_j'-\lambda u_{j,t}'-\rho hu_{j,tt}')u_jd\sigma. \quad (15)$$

Thus, if the pseudostatic deformation state is regarded as the primed state, and the modal portion of the solution is regarded as the unprimed state, we find that

$$\int_S\int\sum_{j=1}^{3}u_{ji}u_{js}d\sigma=(\rho h\Omega_i^2)^{-1}\int_S\int\sum_{j=1}^{3}(p_j-ku_{js})u_{ji}d\sigma. \quad (16)$$

Hence, from Eq. 10, the function $P_i(t)$ may be written in a more convenient form as

$$P_i(t)=(\rho h\Omega_i^2)^{-1}\int_S\int\sum_{j=1}^{3}(p_j-ku_{js})u_{ji}d\sigma. \quad (17)$$

II. COMPARISON OF MODE-ACCELERATION AND MODE-DISPLACEMENT METHODS

In order to obtain an insight into the mode-acceleration method, we compare the response of a shell obtained by this method (denoted by R_A) with that obtained by means of the mode-displacement method (denoted by R_D). We recall that R_A and R_D are given by

$$R_A=R_s(\xi_1,\xi_2,t)+\sum_{i=1}^{\infty}q_{iA}(t)R_i(\xi_1,\xi_2), \quad (18)$$

$$R_D=\sum_{i=1}^{\infty}q_{iD}(t)R_i(\xi_1,\xi_2), \quad (19)$$

where q_{iA} and q_{iD} are the ith generalized coordinates according to the mode-acceleration and mode-displacement methods, respectively, while R_s is the pseudostatic response and R_i the normal mode of free vibration.

For an elastic shell under a uniform load $p_3(t)$ normal to its middle surface, it may be shown[1] that, when the foundation is absent (i.e., when $k=\lambda=0$),

$$q_{iD}=(y_{0i}/m_i)\cos\Omega_it+(\dot{y}_{0i}/m_i\Omega_i)\sin\Omega_it+A_iI_i/\Omega_i, \quad (20)$$

where

$$A_i=(\rho hm_i)^{-1}\int_S\int u_{3i}d\sigma,$$
$$I_i=\int_0^t p_3(\tau)\sin\Omega_i(t-\tau)d\tau. \quad (21)$$

From Eq. 13, we find that

$$q_{iA}=(y_{0i}/m_i)\cos\Omega_it+(\dot{y}_{0i}/m_i\Omega_i)\sin\Omega_it$$
$$+A_iI_i/\Omega_i-A_ip_3(t)/\Omega_i^2. \quad (22)$$

Incidentally, because $R_A=R_D$, we note that the pseudostatic response is given in terms of the normal modes of free vibration by

$$R_s=p_3(t)\sum_{i=1}^{\infty}R_iA_i/\Omega_i^2. \quad (23)$$

Furthermore, when a steady load acts on the shell, the responses R_A and R_D are of exactly the same form.

Kraus and Kalnins[1] have presented a Table of nondimensional natural frequencies ω_i of a simply supported hemispherical shell of radius a, and the nondimensional products corresponding to A_iR_i (where R_i now denotes the normal displacement u_{3i} at the shell apex). A portion of their Table is reproduced here as Table I. Using these results, we have computed by means of Eqs. 20 and 22 the response of the hemisphere initially at rest during the time of application of the following pulses using both the mode-displacement and mode-acceleration methods:

Half-sine pulse

$$p_3(T)=P_0\sin pT/2,\quad 0<T<2\pi/p,$$
$$=0,\quad 2\pi/p<T. \quad (24)$$

Exponential pulse

$$p_3(T)=P_0(p/\pi)eTe^{-pT/\pi}, \quad (25)$$

where T is a nondimensional time, $a^{-1}(\rho/E)^{-\frac{1}{2}}t$. The responses are illustrated for a particular value of p in Figs. 1 and 2, where the rate of convergence is indicated by numbering consecutively the effect of each mode as it is added to the converging sum.

We observe that in both cases the response converges

TABLE I. Natural frequencies of a simply supported hemisphere.

i	A_iR_i	ω_i
1	0.048 079	0.773
2	−0.104 86	1.010
3	0.358 21	1.242
4	−1.591 2	1.578
5	2.797 8	1.720
6	−0.879 44	2.192
7	0.373 98	2.793
8	0.261 95	2.941
9	−0.480 04	3.733
10	0.359 57	4.682

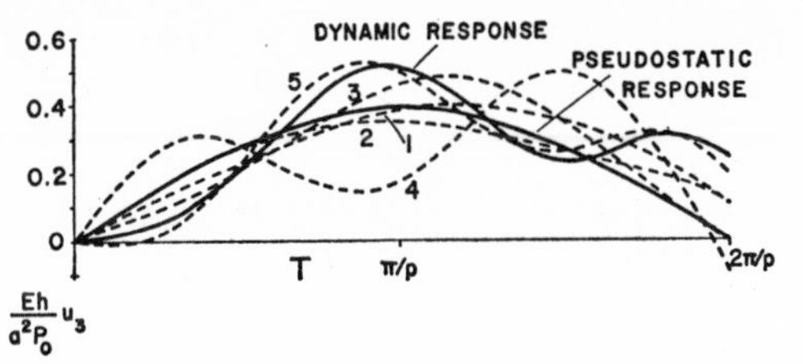

(a)

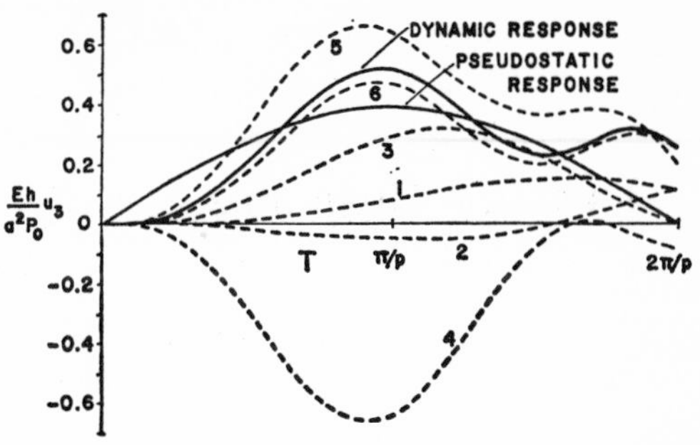

(b)

FIG. 1. Displacement of hemisphere at apex owing to half-sine pulse, $p=1.01$. (a) Response by mode-acceleration method. (b) Response by mode-displacement method.

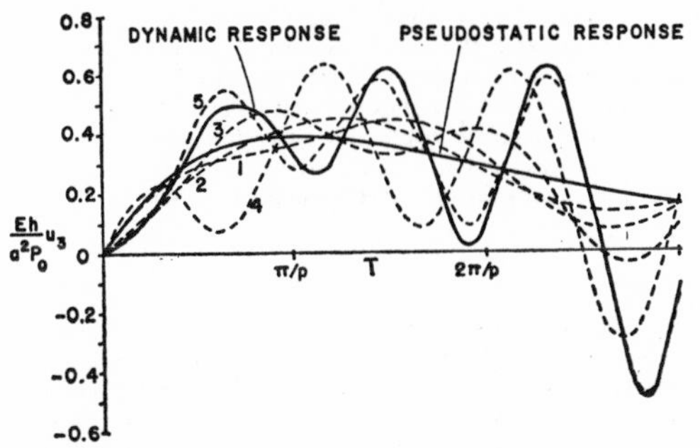

(a)

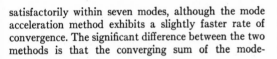

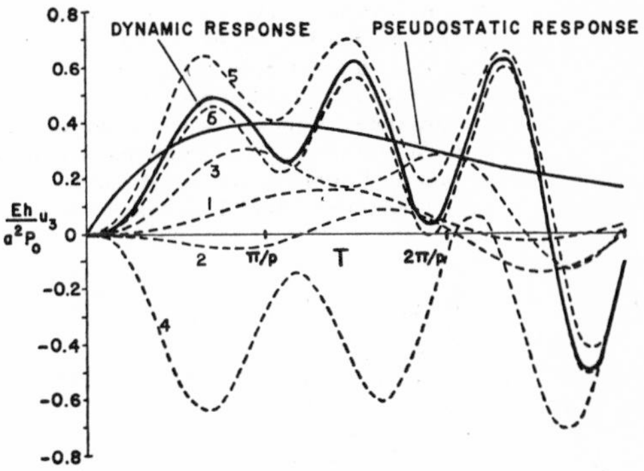

(b)

FIG. 2. Displacement of hemisphere at apex owing to exponential pulse, $p=0.789$. (a) Response by mode-acceleration method. (b) Response by mode-displacement method.

satisfactorily within seven modes, although the mode acceleration method exhibits a slightly faster rate of convergence. The significant difference between the two methods is that the converging sum of the mode-displacement method varies widely from the final response, whereas the converging sum of the mode-acceleration method tends to follow the final response more closely.

29

Reprinted from *AIAA J.*, **4**(3), 486–494 (1966)

Dynamic Response of a Cylindrical Shell: Two Numerical Methods

DONALD E. JOHNSON* AND ROBERT GREIF*
Avco Corporation, Wilmington, Mass.

The dynamic response of a cylindrical shell subjected to arbitrarily time-varying load distributions is treated within the framework of first-order linear elastic thin shell theory by use of two different numerical methods of timewise integration: an explicit method and an implicit (Houbolt) method. In both of these methods the dependent variables are expanded in Fourier series in the circumferential direction, and the resulting equations are expressed in finite difference form. Although the problem considered is limited to cylindrical shells with constant geometric and material properties, this study is intended to provide the groundwork for the development of a general computer program applicable to a wide class of shell problems. The relative efficiencies of the two numerical methods, as encountered in practical applications, are discussed. Results for a typical example of a blast loading are presented for the first three Fourier harmonics.

Nomenclature

ϵ	= dimensionless time increment $(E_0/\rho)^{1/2}\Delta t/R$
ζ	= normal outward distance from shell middle surface
θ	= circumferential angle
ξ	= dimensionless longitudinal coordinate, s/R
Δ	= dimensionless increment of longitudinal coordinate, $\bar{s}/RN$
Φ_ξ, Φ_θ	= rotations of the middle surface
σ_ξ, σ_θ, $\sigma_{\xi\theta}$	= longitudinal, circumferential, and shear stresses
g	= acceleration due to gravity
q, q_ξ, q_θ	= external loads per unit area
s	= dimensional longitudinal coordinate
$\bar{s}$	= length of cylindrical shell
Δt	= dimensional time increment
M_ξ, M_θ, $M_{\xi\theta}$	= bending moments per unit length
N_ξ, N_θ, $N_{\xi\theta}$	= membrane force per unit length
N	= number of longitudinal space increments
N_t	= number of time increments
Q_ξ, Q_θ	= transverse forces per unit length
U_ξ, U_θ, W	= longitudinal, circumferential, and normal displacement

1. Introduction

THE determination of the dynamic response of shell structures to arbitrary external loads is an important problem in the design of missiles and space vehicles. Although analytical techniques have been applied successfully in certain cases,[6] the general problem of predicting the dynamic response of shell structures is treated most effectively by the direct use of numerical methods. One frequently used numerical method involves the determination of the modes of vibration and the subsequent combining of these modes to yield the total response. Because of certain limitations inherent in this modal approach, efforts have been made recently to apply numerical integration schemes directly to the equations of motion obtained from shell theory.[3,9] This paper is concerned with this latter approach.

The problem considered here is treated within the framework of linear, elastic, first-order shell theory. The purpose of this work is to show that the problem of predicting the dynamic response of shells to arbitrary external loads can be effectively treated by applying either of two different numerical integration schemes directly to the equations of motion. The relative merits of each of these methods of numerical integration also are investigated and discussed. In this work, only a cylindrical shell (with constant geometric and material properties and subjected to isothermal conditions) is considered; however, the conclusions drawn should provide the groundwork for the development of a general computer program for the dynamic response of complicated shells and shell structures.

The two numerical methods used in this work are both based on Sanders' shell equations.[1] In both methods, the expansion of the dependent variables in Fourier series in the circumferential variable and the subsequent representation of the spatial derivatives in difference form are essentially the same as used by Budiansky and Radkowski[2] in their static analysis. These two methods differ only† in the manner in which the acceleration terms are expressed in finite

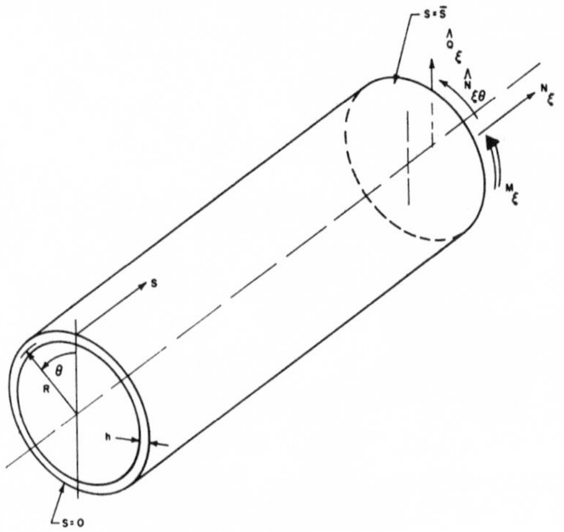

Fig. 1 Right circular cylindrical shell.

Received June 2, 1965; revision received October 29, 1965. The authors wish to thank B. Budiansky of Harvard University for his helpful advice and contributions during the course of this investigation.

* Staff Scientist, Research and Development Division. Member AIAA.

† An additional formulation of the explicit method, which differs in other ways from the formulation used for the Houbolt method, also was programed.

difference form. In the first method, the Houbolt[4] method (an implicit method), a backwards difference formula is used to represent the acceleration terms. In the second method, an explicit method, a central difference formula is used to represent the acceleration terms. The time increment used in the explicit method is restricted by certain numerical stability considerations, whereas in the Houbolt method, the time increment need only be chosen small enough to provide the desired accuracy for a given problem.

2. Continuum Theory

The material presented in this section is treated briefly since it closely parallels the work of Sanders[1] and Budiansky and Radkowski.[2] The notation used is identical to that of Ref. 2, except where otherwise noted.

The circular cylindrical shell is shown in Fig. 1. The forces, moments, and loads acting on a section of the shell parallel to the coordinate lines are shown in Fig. 2. With the use of Sanders' theory, the governing equations of motion for the circular cylindrical shell are

$$\left.\begin{aligned}
&\frac{\partial N_\xi}{\partial \xi} + \frac{\partial \bar{N}_{\xi\theta}}{\partial \theta} - \frac{1}{2R}\frac{\partial \bar{M}_{\xi\theta}}{\partial \theta} + R\left[q_\xi - \frac{E_0 h}{R^2}\frac{\partial^2 U_\xi}{\partial \tau^2}\right] = 0\\
&\frac{\partial N_\theta}{\partial \theta} + \frac{\partial \bar{N}_{\xi\theta}}{\partial \xi} + \frac{3}{2R}\frac{\partial \bar{M}_{\xi\theta}}{\partial \xi} + \frac{1}{R}\frac{\partial M_\theta}{\partial \theta} + \\
&\qquad R\left[q_\theta - \frac{E_0 h}{R^2}\frac{\partial^2 U_\theta}{\partial \tau^2}\right] = 0\\
&-N_\theta + \frac{1}{R}\left[\frac{\partial^2 M_\xi}{\partial \xi^2} + 2\frac{\partial^2 \bar{M}_{\xi\theta}}{\partial \xi \partial \theta} + \frac{\partial^2 M_\theta}{\partial \theta^2}\right] + \\
&\qquad R\left[q - \frac{E_0 h}{R^2}\frac{\partial^2 W}{\partial \tau^2}\right] = 0
\end{aligned}\right\} \quad (2.1)$$

where the nondimensional time and space variables τ and ξ are related to the respective dimensional time and space quantities t and s by the expressions

$$\tau = (E_0/\rho)^{1/2} t/R \qquad \xi = s/R \qquad (2.2)$$

In these formulas, h and R are the respective thickness and radius of the shell, ρ is the mass density, and E_0 is a reference Young's modulus. The modified shearing force $\bar{N}_{\xi\theta}$ and the modified twisting moment $\bar{M}_{\xi\theta}$ are defined as

$$\begin{aligned}
\bar{N}_{\xi\theta} &= \tfrac{1}{2}(N_{\xi\theta} + N_{\theta\xi}) + [1/(4R)](M_{\xi\theta} - M_{\theta\xi})\\
\bar{M}_{\xi\theta} &= \tfrac{1}{2}(M_{\xi\theta} + M_{\theta\xi})
\end{aligned} \qquad (2.3)$$

Quantities shown in Eqs (2.1) now will be expanded in Fourier series in the circumferential variable θ as follows:

$$\left.\begin{aligned}
N_\xi &= \sigma_0 h \sum_{n=0}^{\infty} t_\xi^{(n)} \cos n\theta\\
N_\theta &= \sigma_0 h \sum_{n=0}^{\infty} t_\theta^{(n)} \cos n\theta\\
\bar{N}_{\xi\theta} &= \sigma_0 h \sum_{n=1}^{\infty} t_{\xi\theta}^{(n)} \sin n\theta
\end{aligned}\right\} \qquad (2.4)$$

$$\left.\begin{aligned}
M_\xi &= \frac{\sigma_0 h^3}{R} \sum_{n=0}^{\infty} m_\xi^{(n)} \cos n\theta\\
M_\theta &= \frac{\sigma_0 h^3}{R} \sum_{n=0}^{\infty} m_\theta^{(n)} \cos n\theta\\
\bar{M}_{\xi\theta} &= \frac{\sigma_0 h^3}{R} \sum_{n=1}^{\infty} m_{\xi\theta}^{(n)} \sin n\theta
\end{aligned}\right\} \qquad (2.5)$$

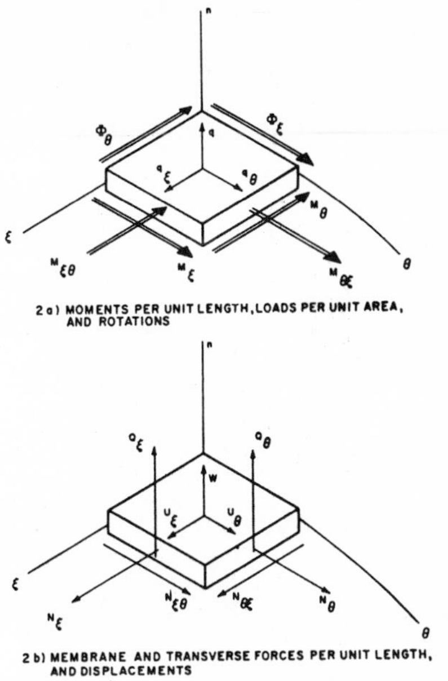

Fig. 2 Sign convention and coordinates, moments, forces, loads, displacements, and rotations.

$$\left.\begin{aligned}
U_\xi &= \frac{R\sigma_0}{E_0} \sum_{n=0}^{\infty} u_\xi^{(n)} \cos n\theta\\
U_\theta &= \frac{R\sigma_0}{E_0} \sum_{n=1}^{\infty} u_\theta^{(n)} \sin n\theta\\
W &= \frac{R\sigma_0}{E_0} \sum_{n=0}^{\infty} w^{(n)} \cos n\theta
\end{aligned}\right\} \qquad (2.6)$$

$$\left.\begin{aligned}
q &= \frac{\sigma_0 h}{R} \sum_{n=0}^{\infty} p^{(n)} \cos n\theta\\
q_\xi &= \frac{\sigma_0 h}{R} \sum_{n=0}^{\infty} p_\xi^{(n)} \cos n\theta\\
q_\theta &= \frac{\sigma_0 h}{R} \sum_{n=1}^{\infty} p_\theta^{(n)} \sin n\theta
\end{aligned}\right\} \qquad (2.7)$$

The term σ_0 refers to a reference stress level. The Fourier coefficients used in this report are all nondimensional functions of ξ and τ. For future convenience, the superscripts n on all Fourier coefficients usually will be suppressed.

If the forces and moments are expressed in terms of the four variables u_ξ, u_θ, w, and m_ξ, as was done by Budiansky and Radkowski[2], Eqs. (2.1) become

$$\left.\begin{aligned}
&\bar{a}_1 u_\xi'' + \bar{a}_3 u_\xi + \bar{a}_4 u_\theta' + \bar{a}_6 w' = -p_\xi + \ddot{u}_\xi\\
&\bar{a}_{10} u_\xi' + \bar{a}_{12} u_\theta'' + \bar{a}_{14} u_\theta + \bar{a}_{15} w'' + \bar{a}_{17} w +\\
&\qquad \bar{a}_{18} m_\xi = -p_\theta + \ddot{u}_\theta\\
&\bar{a}_{19} u_\xi' + \bar{a}_{21} u_\theta'' + \bar{a}_{23} u_\theta + \bar{a}_{24} w'' + \bar{a}_{26} w +\\
&\qquad \bar{a}_{27} m_\xi'' + \bar{a}_{29} m_\xi = -p + \ddot{w}
\end{aligned}\right\} \qquad (2.8)$$

The definition of m_ξ may be written in a similar form in terms of u_θ, w'', and w:

$$\bar{a}_{32} u_\theta + \bar{a}_{33} w'' + \bar{a}_{35} w + \bar{a}_{36} m_\xi = 0 \qquad (2.9)$$

The coefficients $\bar{a}_i$ are given in Appendix A in terms of the Fourier index n, Poisson's ratio ν, Young's modulus E, and the ratio of shell thickness to radius $\lambda = h/R$. The derivatives with respect to ξ and τ are denoted by primes and dots, respectively.

Eqs. (2.8 and 2.9) may be combined and put in the following matrix form:

$$Kz'' + Fz' + Gz = y + D\ddot{z} \tag{2.10}$$

where

$$z = \begin{bmatrix} u_\xi \\ u_\theta \\ w \\ m_\xi \end{bmatrix} \quad y = \begin{bmatrix} -p_\xi \\ -p_\theta \\ -p \\ 0 \end{bmatrix} \quad D = \begin{bmatrix} 1 & 0 & 0 & 0 \\ 0 & 1 & 0 & 0 \\ 0 & 0 & 1 & 0 \\ 0 & 0 & 0 & 0 \end{bmatrix} \tag{2.11}$$

$$K = \begin{bmatrix} \bar{a}_1 & 0 & 0 & 0 \\ 0 & \bar{a}_{12} & \bar{a}_{15} & 0 \\ 0 & \bar{a}_{21} & \bar{a}_{24} & \bar{a}_{27} \\ 0 & 0 & \bar{a}_{33} & 0 \end{bmatrix} \quad F = \begin{bmatrix} 0 & \bar{a}_4 & \bar{a}_6 & 0 \\ \bar{a}_{10} & 0 & 0 & 0 \\ \bar{a}_{19} & 0 & 0 & 0 \\ 0 & 0 & 0 & 0 \end{bmatrix}$$

$$G = \begin{bmatrix} \bar{a}_3 & 0 & 0 & 0 \\ 0 & \bar{a}_{14} & \bar{a}_{17} & \bar{a}_{18} \\ 0 & \bar{a}_{23} & \bar{a}_{26} & \bar{a}_{29} \\ 0 & \bar{a}_{32} & \bar{a}_{35} & \bar{a}_{36} \end{bmatrix}$$

The boundary conditions applicable to Sanders' theory reduce to prescribing the following quantities at the boundaries $s = 0, \bar{s}$ of the cylindrical shell

$$\left.\begin{matrix} N_\xi & \text{or} & U_\xi \\ \hat{N}_{\xi\theta} & \text{or} & U_\theta \\ \hat{Q}_\xi & \text{or} & W \\ M_\xi & \text{or} & \Phi_\xi \end{matrix}\right\} s = 0, \bar{s} \tag{2.12}$$

where

$$\left.\begin{matrix} \hat{N}_{\xi\theta} = \bar{N}_{\xi\theta} + [3/(2R)]\bar{M}_{\xi\theta} \\ \hat{Q}_\xi = \dfrac{1}{R}\left[\dfrac{\partial M_\xi}{\partial \xi} + 2\dfrac{\partial \bar{M}_{\xi\theta}}{\partial \theta}\right] \qquad \Phi_\xi = -\dfrac{1}{R}\dfrac{\partial W}{\partial \xi} \end{matrix}\right\} \tag{2.13}$$

The "effective" forces per unit length $\hat{N}_{\xi\theta}$ and $\hat{Q}_\xi$ are shown in Fig. 1, and the rotation of the middle surface Φ_ξ is shown in Fig. 2. More general boundary conditions than those of (2.12) could, of course, be prescribed. For example, N_ξ and U_ξ (as well as $\hat{N}_{\xi\theta}$ and U_θ, etc.) could be related through an elastic constraint condition. However, in this paper, only physical conditions leading to the boundary conditions of (2.12) were examined.

For any Fourier index n, the boundary conditions can be written entirely in terms of the elements (and the derivatives of the elements) of the column matrix z defined in (2.11). Letting

$$\left.\begin{matrix} \hat{N}_{\xi\theta} = \sigma_0 h \sum_{n=1}^{\infty} \hat{t}_{\xi\theta}^{(n)} \sin n\theta \\ \hat{Q}_\xi = \sigma_0 h \sum_{n=0}^{\infty} \hat{f}_\xi^{(n)} \cos n\theta \\ \Phi_\xi = \dfrac{\sigma_0}{E_0} \sum_{n=0}^{\infty} \phi_\xi^{(n)} \cos n\theta \end{matrix}\right\} \tag{2.14}$$

and substituting into (2.13) yields the relations between z and the Fourier coefficients defined in (2.14). The boundary conditions now can be rewritten for the nth Fourier component, as prescribing the following dimensionless quantities at the boundaries $s = 0, \bar{s}$:

$$\left.\begin{matrix} t_\xi & \text{or} & u_\xi \\ \hat{t}_{\xi\theta} & \text{or} & u_\theta \\ \hat{f}_\xi & \text{or} & w \\ m_\xi & \text{or} & \phi_\xi \end{matrix}\right\} s = 0, \bar{s} \tag{2.15}$$

Following Ref. 2, the boundary conditions finally may be recast in nondimensional matrix form as

$$\Omega H z' + (I - \Omega + \Omega J)z = l \tag{2.16}$$

where

$$H = \begin{bmatrix} \bar{b}_1 & 0 & 0 & 0 \\ 0 & \bar{b}_6 & \bar{b}_8 & 0 \\ 0 & \bar{b}_{11} & \bar{b}_{13} & \bar{b}_{15} \\ 0 & 0 & -1 & 0 \end{bmatrix} \quad J = \begin{bmatrix} 0 & \bar{b}_3 & \bar{b}_4 & 0 \\ \bar{b}_5 & 0 & 0 & 0 \\ \bar{b}_{10} & 0 & 0 & 0 \\ 0 & 0 & 0 & 0 \end{bmatrix} \tag{2.17}$$

The elements $\bar{b}_i$ are given in Appendix A, Ω is a diagonal matrix, l is a given column matrix, and I is a unit matrix. As explained in Ref. 2, the matrices Ω and l indicate what type of boundary conditions are being considered. For example, if u_ξ is known on the boundary, the first diagonal element of Ω is zero, and the first element of l is the prescribed value of u_ξ.

Formulas used to calculate the stresses at any point in the shell are given in Ref. 2.

3. Discrete Theory

Discrete values of the nondimensional space variable ξ are denoted by

$$\xi_i = i\Delta \qquad i = -1, 0, 1, 2, \ldots, N, N+1 \tag{3.1}$$

where the dimensionless space increment is defined to be $\Delta = \bar{s}/RN$. The "fictitious" space points adjacent to the ends of the cylinder are denoted by ξ_{-1} and ξ_{N+1}. Similarly, discrete values of the nondimensional time variable τ are denoted by

$$\tau_j = j\epsilon \qquad j = -2, -1, 0, 1, 2, \ldots \tag{3.2}$$

The values of ϵ, employed in the Houbolt method and in the explicit method, are discussed in subsequent sections. The fictitious times τ_{-2} and τ_{-1} are used in the Houbolt method to describe the initial conditions and will be explained in more detail later.

The difference formulas used to represent the derivatives with respect to ξ (at any time τ_j at position ξ_i) are

$$\left.\begin{matrix} z_{i,j}'' = (z_{i+1,j} - 2z_{i,j} + z_{i-1,j})/\Delta^2 \\ z_{i,j}' = (z_{i+1,j} - z_{i-1,j})/2\Delta \end{matrix}\right\} i = 1, 2, \ldots, N-1 \tag{3.3}$$

$$\begin{matrix} z_{0,j}' = (-\tfrac{3}{2}z_{0,j} + 2z_{1,j} - \tfrac{1}{2}z_{2,j})/\Delta \\ z_{N,j}' = (\tfrac{3}{2}z_{N,j} - 2z_{N-1,j} + \tfrac{1}{2}z_{N-2,j})/\Delta \end{matrix} \tag{3.4}$$

The third elements in the column matrices $z_{0,j}$ and $z_{N,j}$ are $w_{0,j}$ and $w_{N,j}$, respectively [Eq. (2.11)]. In order to have the same spatial finite difference formulas for both the Houbolt method and the explicit method, it was necessary to represent the derivatives of $w_{0,j}$ and $w_{N,j}$ by central differences instead of the backward differences shown in (3.4). This central difference representation of the derivatives of $w_{0,j}$ and $w_{N,j}$ normally involves the values at the fictitious points $w_{-1,j}$ and $w_{N+1,j}$; however, it is most convenient to first algebraically eliminate $w_{-1,j}$ and $w_{N+1,j}$ by using the definitions of $m_{\xi_{0,j}}$ and $m_{\xi_{N,j}}$ (i.e. the difference equations equivalent to 2.9). The resulting equations for $w_{0,j}'$ and $w_{N,j}'$ are

$$\begin{aligned} w_{0,j}' &= \frac{w_{1,j}}{\Delta} + \left(\frac{\Delta}{2}\frac{\bar{a}_{35}}{\bar{a}_{33}} - \frac{1}{\Delta}\right) w_{0,j} - \frac{\Delta}{2\bar{a}_{33}} m_{\xi_{0,j}} + \frac{\Delta \bar{a}_{32}}{2\bar{a}_{33}} u_{\theta_{0,j}} \\ w_{N,j}' &= -\frac{w_{N-1,j}}{\Delta} - \left(\frac{\Delta}{2}\frac{\bar{a}_{35}}{\bar{a}_{33}} - \frac{1}{\Delta}\right) w_{N,j} + \frac{\Delta}{2\bar{a}_{33}} m_{\xi_{N,j}} - \frac{\Delta \bar{a}_{32}}{2\bar{a}_{33}} u_{\theta_{N,j}} \end{aligned} \tag{3.5}$$

Equations (3.5) retain the accuracy and advantages associated with the use of fictitious points without introducing the values of any of the variables evaluated at the fictitious points. The spatial finite difference formulas given previously were used for both the Houbolt method and the explicit method, thus allowing attention to be focused on the only difference between these two methods, namely the manner in which the acceleration terms are represented in finite difference form.

A general formula for the acceleration terms, which encompasses the difference formulations involved in both methods, is

$$\ddot{z}_{i,j} = \bar{\alpha}_j z_{i,j+1} + \alpha_j z_{i,j} + \beta_j z_{i,j-1} + \gamma_j z_{i,j-2} + \delta_j z_{i,j-3} \quad (j = 0, 1, 2, \ldots) \tag{3.6}$$

As indicated by the subscript j, the coefficients $\bar{\alpha}_j$, α_j, β_j, γ_j, and δ_j may, in general, depend on time; the actual values of these coefficients are given in the next two sections.

By making use of Eqs. (3.3–3.6) to represent Eqs. (2.10) and (2.16) in difference form, one can reduce the equations of motion and the boundary conditions to the following sets of matrix equations:

$$V_0 z_{2,j} + A_0 z_{1,j} + B_0 z_{0,j} = g_{0,j} \tag{3.7}$$

$$A z_{i+1,j} + B_j z_{i,j} + C z_{i-1,j} = g_{i,j} + \bar{D}_j z_{i,j+1} \quad (i = 1, 2, \ldots, N-1) \tag{3.8}$$

$$B_N z_{N,j} + C_N z_{N-1,j} + V_N z_{N-2,j} = g_{N,j} \tag{3.9}$$

Here, by introducing the two matrices

$$Q = \begin{bmatrix} 0 & 0 & 0 & 0 \\ 0 & 0 & 0 & 0 \\ 0 & 0 & 1 & 0 \\ 0 & 0 & 0 & 0 \end{bmatrix}$$

$$T = \begin{bmatrix} 0 & 0 & 0 & 0 \\ 0 & 0 & 0 & 0 \\ 0 & \dfrac{\Delta}{2}\dfrac{\bar{a}_{32}}{\bar{a}_{33}} & \left(\dfrac{1}{2\Delta} + \dfrac{\Delta \bar{a}_{35}}{2\bar{a}_{33}}\right) & -\dfrac{\Delta}{2\bar{a}_{33}} \\ 0 & 0 & 0 & 0 \end{bmatrix} \tag{3.10}$$

one obtains

$$\left.\begin{aligned} g_{0,j} &= l_{0,j} \qquad V_0 = [1/(2\Delta)]\Omega_0 H[-I + Q] \\ A_0 &= (1/\Delta)\Omega_0 H[2I - Q] \\ B_0 &= I - \Omega_0 + \Omega_0\{J + H[(-\tfrac{3}{2}I/\Delta) + T]\} \end{aligned}\right\} \tag{3.11}$$

where the zero subscript refers to the end $s = 0$:

$$\left.\begin{aligned} A &= (2K/\Delta) + F \\ B_j &= (-4K/\Delta) + 2\Delta[G - \alpha_j D] \\ C &= (2K/\Delta) - F \\ g_{i,j} &= 2\Delta y_{i,j} + 2\Delta D[\beta_j z_{i,j-1} + \gamma_j z_{i,j-2} + \delta_j z_{2,j-3}] \\ \bar{D}_j &= 2\Delta \bar{\alpha}_j D \end{aligned}\right\} (i = 1, 2, \ldots, N-1) \tag{3.12}$$

For the end $s = \bar{s}$:

$$\left.\begin{aligned} g_{N,j} &= l_{N,j} \qquad V_N = [1/(2\Delta)]\Omega_N H[I - Q] \\ B_N &= I - \Omega_N + \Omega_N\{J + H[(\tfrac{3}{2}I/\Delta) - T]\} \\ C_N &= (1/\Delta)\Omega_N H[-2I + Q] \end{aligned}\right\} \tag{3.13}$$

4. Houbolt (Implicit) Method of Numerical Integration

The Houbolt method[4] of numerical integration is an implicit method that has been used effectively for the transient solution to the equations of motion of elastic systems.[5,7] It is a stable method, although the time increment ϵ must be chosen small enough to provide the desired accuracy for a given problem. For future reference, the general expression for the acceleration given by (3.6) is written below:

$$\ddot{z}_{i,j} = \bar{\alpha}_j z_{i,j+1} + \alpha_j z_{i,} + \beta_j z_{i,j-1} + \gamma_j z_{i,j-2} + \delta_j z_{i,j-3} \quad (j = 0, 1, 2, \ldots) \tag{4.1}$$

The Houbolt method, as applied here, uses backward time differences based on a third degree polynomial (in τ), which is fitted through the four discrete points τ_j, τ_{j-1}, τ_{j-2}, τ_{j-3}. This gives

$$\left.\begin{aligned} &\bar{\alpha}_j = 0 \text{ (all } j) && \alpha_0 = \beta_0 = \gamma_0 = \delta_0 = 0 \\ &\alpha_1 = 6/\epsilon^2 && \beta_1 = \gamma_1 = \delta_1 = 0 \\ &\alpha_2 = 2/\epsilon^2 && \beta_2 = -4/\epsilon^2 \qquad \gamma_2 = \delta_2 = 0 \\ &\alpha_j = 2/\epsilon^2 && \beta_j = -5/\epsilon^2 \qquad \gamma_j = 4/\epsilon^2 \\ &\delta_j = -1/\epsilon^2 && (j = 3, 4, \ldots) \end{aligned}\right\} \tag{4.2}$$

The coefficients α_j, β_j, γ_j, δ_j take the foregoing special values during the initial increments of calculation ($j = 0, 1, 2$) since functions evaluated at "negative values of time" $\tau < 0$ otherwise would be involved. These special values for $j = 0, 1, 2$ correspond to the initial conditions $z_{i,0} = \dot{z}_{i,0} = \ddot{z}_{i,0} = 0$. This is explained further in Appendix B.

From a computational point of view, the significant feature of the Houbolt method is the fact that $\bar{\alpha}_j = 0$ as is shown previously in (4.2). This fact implies that $\bar{D}_j = 0$ [Eqs. (3.12)], and, consequently, it also implies that the $z_{i,j+1}$ term does not appear in Eqs. (3.8). It is therefore necessary to solve Eqs. (3.7–3.9) for $z_{i,j}$ [the values of $z_{i,j-1}$, $z_{i,j-2}$, $z_{i,j-3}$ required for (3.8) by Eq. (3.12) are assumed to be known], and this was done in a manner similar to that used in Ref. 2 for the static problem. The method actually used differs somewhat from that shown in Ref. 2, because Eqs. (3.7) and (3.9) involve second-order difference equations, whereas the analogous equations in Ref. 2 involve only first-order difference equations. The method used is a modified Gaussian elimination technique[2,8] (Potters Method) and is described in detail in Appendix B. By repeating this elimination procedure for successive values of j, one can construct a numerical solution to the problem.

It is of importance to note that as ϵ becomes unbounded, the equations of motion (3.8) reduce to the static equilibrium equations. In addition, the modified Gaussian elimination procedure described in Appendix B remains valid and, indeed, becomes indentical to the procedure used in the static solution of the shell.[2] Consequently, it may be seen that the Houbolt method of solution for dynamic response reduces to the static solution as ϵ becomes unbounded.

5. Explicit Method

Basic Iteration

The explicit method of iteration used here is similar to that used in Ref. 3 for the axisymmetric nonlinear dynamic response of shells. The explicit method is characterized by the following central difference representation of the acceleration term:

$$\ddot{z}_{i,j} = (z_{i,j+1} - 2z_{i,j} + z_{i,j-1})/\epsilon^2 \tag{5.1}$$

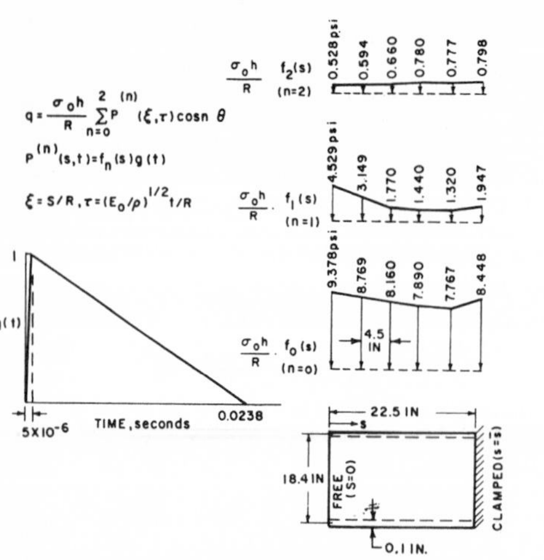

Fig. 3 Blast loading on a clamped-free circular cylindrical shell.

In terms of the general formula (3.6), this implies that

$$\bar{\alpha}_j = 1/\epsilon^2 \quad \alpha_j = -2/\epsilon^2 \quad \beta_j = 1/\epsilon^2 \quad \gamma_j = 0 \quad \delta_j = 0 \quad \text{(all } j) \tag{5.2}$$

The fact that $\bar{\alpha}_j \neq 0$ ($\bar{\alpha}_j = 0$ in the Houbolt method) allows $z_{i,j+1}$ to be computed from $z_{i,j}$ and $z_{i,j-1}$ in the explicit manner as follows:

To begin with, the first three rows of matrix Eqs. (3.8) are used to evaluate directly the first three elements of $z_{i,j+1}$ (i.e., $u_{\xi_{i,j+1}}$, $u_{\theta_{i,j+1}}$, and $w_{i,j+1}$) for $i = 1, 2 \ldots N-1$. Then by replacing j by $j+1$ in the last row of Eq. (3.8), the last element of $z_{i,j+1}$ (i.e. $m_{\xi_{i,j+1}}$) is directly evaluated for $i = 2, 3, \ldots, N-2$. This step is actually the evaluation of m_ξ from u_ξ, u_θ, w, and their derivatives.

The remaining unknown elements of $z_{i,j+1}$ are found, as follows. The four elements of $z_{0,j+1}$ and the last element of $z_{1,j+1}$ are determined by using matrix Eq. (3.7) and the last row of matrix Eq. (3.8) in which i is set equal to 1. Similarly, the four elements of $z_{N,j+1}$ and the last element of $z_{N-1,j+1}$ are determined by using matrix Eq. (3.9) and the last row of matrix Eq. (3.8) in which i is set equal to $N-1$.

The process described previously may be repeated for each successive time step. It is necessary to start the process by determining $z_{i,0}$ and $z_{i,1}$ from the initial conditions. The initial velocity may be represented in forward difference form and used to compute $z_{i,1}$ from $z_{i,0}$. For the problems discussed in this paper, the initial deformation and velocity are zero.

From the preceding description one can see that the explicit method can be carried out by using the same dependent variables ($z_{i,j}$, i.e., $u_{\xi_{i,j}}$, $u_{\theta_{i,j}}$, $w_{i,j}$ and $m_{\xi_{i,j}}$) and the same spatial finite difference representations as used in the Houbolt method. One advantage of this is that comparisons of results obtained by the Houbolt and explicit methods will show only the effect of the different representations of the acceleration term. The more usual approach (for the explicit method) lacks this advantage and involves the use of the dependent variables $u_{\xi_{i,j}}$, $u_{\theta_{i,j}}$, and $w_{i,j}$ (but not $m_{\xi_{i,j}}$); this approach was also used to develop an operational computer program.

Numerical Stability

The dimensionless time interval ϵ, used in the explicit method, must be chosen small enough to insure numerical stability. The following stability criteria for the cylindrical shell were developed from physical arguments and were substantiated by numerical experiments:

$$\left.\begin{aligned}
&(\epsilon^2/\Delta^2)\{1 + [(1-\nu)/8]n^2\Delta^2\} \le (1-\nu^2)\\
&(\epsilon^2/\Delta^2)\{[(1-\nu)/2] + \tfrac{1}{4}n^2\Delta^2\} \le (1-\nu^2)\\
&(\epsilon/\Delta^2)\{[h/(3^{1/2}R)]\}[1 + \tfrac{1}{4}n^2\Delta^2] \le (1-\nu^2)^{1/2}\\
&(\Delta = \bar{s}/RN)
\end{aligned}\right\} \tag{5.3}$$

The physical arguments used to develop (5.3) are briefly outlined below. 1) For a given n and a given nonzero value of Δ, the discrete system described by the difference equation has a finite number of modes of vibration. (The general response of the shell to an arbitrary loading can be obtained by superimposing modes of vibration.) 2) It is assumed that the modes of vibration of the discrete system, which have the highest frequencies, also have the most rapid variation with respect to ξ. 3) Since relatively small values of Δ usually are required to provide accurate solutions, the highest natural frequencies of the discrete system are estimated by neglecting the curvature effects and by considering the vibrations of an equivalent flat plate, i.e., vibrations associated with plate bending and those associated with the plane stress problem. 4) In order to analytically determine the criteria for the equivalent plate, the $s = 0$ and $s = \bar{s}$ boundaries are assumed to be simply supported. (For relatively small values of Δ, the boundary conditions appear to have little influence on the stability criteria.)

The previously listed arguments 2–4 appear to be most valid for relatively small values of Δ. This is a fortunate circumstance, because the limitations imposed on the value of ϵ by the criteria are of practical significance only when Δ is relatively small. The limits of stability given by (5.3) were found to agree closely with the limits of stability observed by running the computer program for a large number of practical problems. (Instabilities in the computer calculations were detected readily by considering the total energy balance of the system.)

The first two inequalities (5.3) are associated with the plane stress problem; the last inequality (5.3) is associated with the plate bending problem. For $n = 0$ these criteria are very similar to those given by Balmer and Witmer[9] for beams and rings.

In many practical applications of the explicit method, the restrictions imposed on ϵ by the criteria (5.3) play a much simpler role than is immediately apparent. First, the values of Δ required by many practical problems are such that

$$n^2\Delta^2 \ll 1 \tag{5.4}$$

In such cases, criteria (5.3) reduce approximately to

$$\epsilon/\Delta \le (1-\nu^2)^{1/2} \tag{5.5}$$

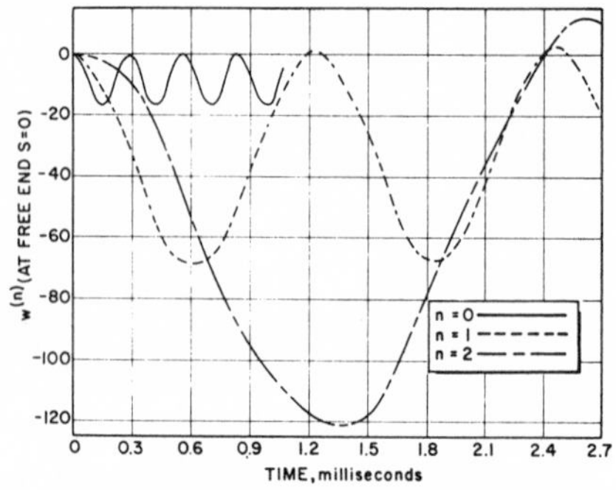

Fig. 4 Nondimensional Fourier coefficient of normal displacement at free end vs time.

$$(\epsilon/\Delta^2)[h/(3^{1/2}R)] \leq (1 - \nu^2)^{1/2} \quad (5.6)$$

Furthermore, in many applications for which (5.4) is valid, one also has

$$\Delta > h/(3^{1/2}R) \quad (5.7)$$

In such applications only the restriction on ϵ given by (5.5) need be considered. For example, in the numerical solution of the typical problem discussed in the next section, only criteria (5.5) had to be used.

6. Numerical Results

Computer programs based on both the Houbolt method and the explicit method were applied to a variety of cylindrical shell problems. These programs were checked by applying both methods to the same problem and also by comparison of results to certain analytical solutions.

In this section a typical problem, defined in Fig. 3, is presented in detail. The geometric and material properties of the shell considered are $R = 9.2$ in., $h = 0.1$ in., $\bar{s} = 22.5$ in., $\rho = 2.4 \times 10^{-4}$ lb-sec²/in.⁴ ($\rho g = 0.0926$ lbs/in.³), $E = 10.5 \times 10^6$ psi, and $\nu = 0.3$. The reference quantities used are $\sigma_0 = 100.0$ psi and $E_0 = E = 10.5 \times 10^6$ psi.

The shell geometry and material properties are typical of those used in the missile industry. The clamped-free boundary conditions give rise to two kinds of boundary effects. [The term clamped, used in this paper, implies that the boundary is constrained completely, so that $u_\xi(\bar{s}, t) = u_\theta(\bar{s}, t) = w(\bar{s}, t) = w'(\bar{s}, t) = 0$.] The loading shown in Fig. 3 is typical of blast loadings, except possibly for the fact that only the harmonics $n = 0, 1, 2$ are considered. The computer programs can be used to treat large integral values of n; however, for the purposes of illustration, only the harmonics $n = 0, 1, 2$ are considered. (The harmonics $n = 0, 1, 2$ are sufficient to describe the pressures obtained from Newtonian flow theory).

For the purposes of separating the various types of response that occur for this problem, each of the harmonics $n = 0, 1,$ and 2 are considered separately. Results for these harmonics

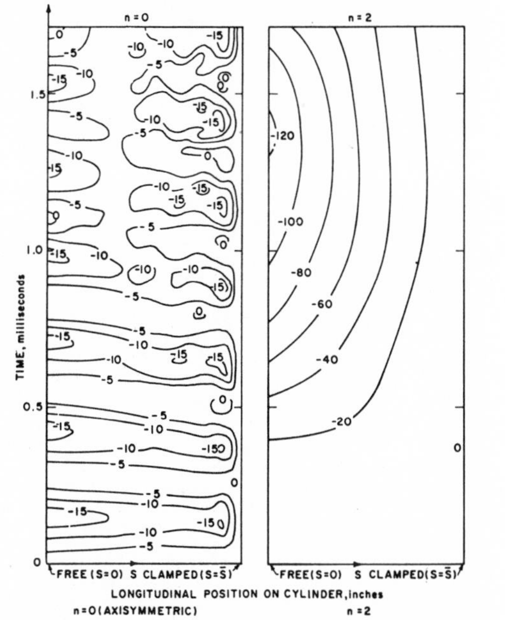

Fig. 5 Contour plot of $w^{(n)}$ for $n = 0$ and $n = 2$ (nondimensional Fourier coefficient of normal displacement).

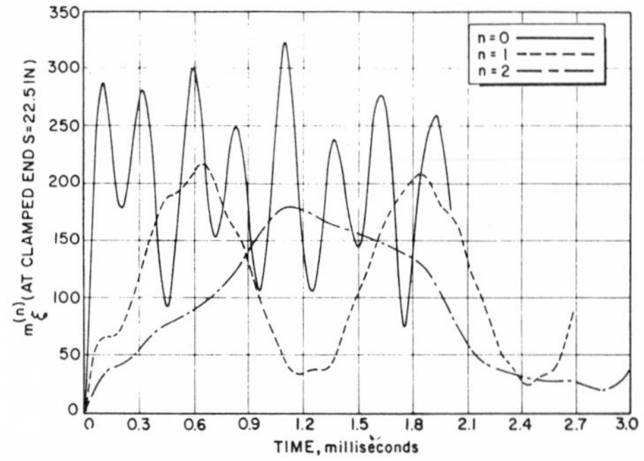

Fig. 6 Nondimensional Fourier coefficient of axial bending moment at clamped end vs time.

are shown in Figs. 4–6. These same results were obtained by both numerical methods. As will be discussed later, Figs. 7 and 8 show how some of these results are approached as the number of discrete points used in the numerical integration is increased.

Figure 4 shows the time history of the Fourier coefficients of normal displacement $w^{(n)}$ at the free end. This figure reveals an almost periodic response in the neighborhood of the free end. A more complete picture of the time history of $w^{(n)}$ for all points on the cylinder and for $n = 0$ and $n = 2$ is shown in Fig. 5. From Fig. 5, it is clearly evident that for $n = 0$ the response is almost periodic only near the free end. Figure 5 also reveals several other significant phenomena. First, the displacement pattern varies much more rapidly near the clamped end for the $n = 0$ case than for the $n = 2$ case. This is largely because in the $n = 0$ case, the load is resisted primarily by the action of hoop stresses except near the clamped end, whereas in the $n = 2$ case the load is resisted primarily by the action of bending. In general, the rapid variation of $w^{(0)}$ with respect to ξ near the clamped boundary requires that small values of the spatial increment Δ be used in the $n = 0$ case. Secondly, the over-all displacement pattern shown in Fig. 5 is much smoother for $n = 2$ than for $n = 0$. The fact that the character of the displacement pattern for the $n = 0$ case becomes increasingly more rough as time increases perhaps may be explained, as follows: The free end appears to oscillate with the frequency of a ring without lateral constraint, whereas the material near the clamped end appears to oscillate with the frequency of a ring with lateral constraint, i.e., at a slightly higher frequency. As the response proceeds, the behavior at the two ends be-

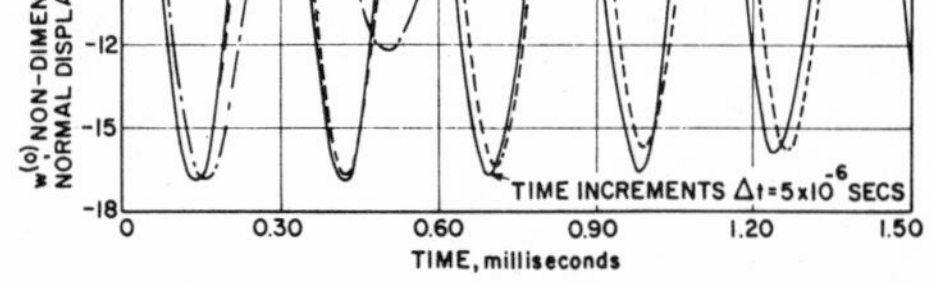

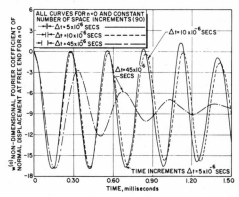

Fig. 7 Dependence of decay factor in Houbolt method on coarseness of time grid.

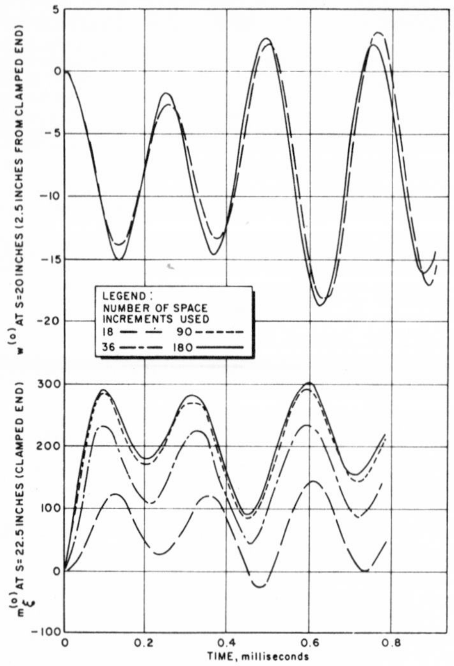

Fig. 8 Convergence of explicit method.

comes increasingly out of phase, and a very rough over-all displacement pattern is developed.

Figure 6 shows the time histories of $m_\xi^{(n)}$ at the clamped end of the cylinder. These time histories of $m_\xi^{(n)}$ differ from the time histories of $w^{(n)}$ at the free end (Fig. 4) in several ways. First, it may be noticed that for the values of n considered, the absolute maximum values of $w^{(n)}$ at the free end increase with n, whereas the absolute maximum values of $m_\xi^{(n)}$ at the clamped end decrease with n. This is not surprising in view of the nature of $w^{(n)}$ (ξ, τ) near the clamped edge shown in Fig. 5 and the fact that at the clamped end

$$m_\xi^{(n)} \sim -(\partial^2 w^{(n)})/(\partial \xi^2)$$

Secondly, the time histories shown in Fig. 6 are more irregular in shape than the time histories shown in Fig. 4. This is probably because of the previously mentioned relationship between $m_\xi^{(n)}$ and $w^{(n)}$, but it also appears to be due to some sort of wave action. For the $n = 1$ curves shown in Fig. 6, there appear to be local irregularities occurring at distinct times. The time between these local irregularities roughly corresponds to the time required for a shear wave to travel twice the length of the shell. It is quite possible that these local irregularities are caused by the reflection of waves from the clamped end.

The results shown in Figs. 4–6 were obtained by successively decreasing the values of ϵ and Δ until the results appeared to converge to a limiting solution. It is important to observe just how this limiting solution was approached as ϵ and Δ were decreased. Figure 7 shows the numerical solutions obtained by the Houbolt method for $w^{(0)}$ at the free end for various values of the time increment Δt, $[\epsilon = (E_0/\rho)^{1/2}\Delta t/R]$. As previously noted by Levy,[5] if the time increment is too large, the Houbolt method has a damping effect on the solution. It was found that the time increment Δt has to be smaller than about 1/50th of the period of a particular mode of vibration in order that important output quantities in that particular mode not be significantly damped. Inasmuch as the total response may be considered as a superposition of many modes of vibration, it is evident that the Houbolt method will tend to damp out the higher modes unless the foregoing criterion is obeyed.

Figure 8 shows the numerical solutions obtained by the explicit method for $m_\xi^{(0)}$ at the clamped end and for $w^{(0)}$ at 2.5 in. from the clamped end for various values of ϵ where the ratio ϵ/Δ is held fixed at a value of 0.837. This insures stability in this case ($n = 0$), because the values of Δ required for accuracy are such that

$$\Delta > h/(3^{1/2}R) \tag{6.1}$$

and hence stability criteria (5.3) reduce to (5.5)

$$\epsilon/\Delta \leq (1 - \nu^2)^{1/2} \tag{6.2}$$

which for this case ($\nu = 0.3$) becomes

$$\epsilon/\Delta \leq 0.954$$

Although any ratio ϵ/Δ less than 0.954 would have provided stability, a value of 0.837 was used to provide a convenient time spacing for the output. In contrast to the Houbolt method, numerically stable solutions obtained from the explicit method do not exhibit a damping effect if the time increment is large. Figure 8 also shows that much smaller values of Δ are required to give accurate plots of $m_\xi^{(0)}$ than are required to give accurate plots of $w^{(0)}$. This phenomenon, which is also noticed in the Houbolt method, is not surprising in view of the fact that $m_\xi^{(n)}$ depends in part on derivatives of $w^{(n)}$ with respect to ξ.

The remaining part of this section deals with the relative efficiencies of the two numerical methods. For the purpose of this paper, efficiency will be measured solely in terms of the computer time required to yield an accurate solution for a given problem. To provide a reasonable basis for comparison, only the computer time required for the basic iterations is considered here, i.e., the computer time required for such operations as processing input and output is excluded.

Let us consider the numerical solution of a particular problem for which it is desired to calculate the dynamic response up to a time t_{max}. Let N_t represent the number of time increments,

$$N_t = t_{max}/\Delta t$$

and, as previously defined, let N represent the number of space increments. Let T represent the computer time (in minutes) required to calculate (excluding input, output, etc.) the numerical solution, then

$$T = C\, N N_t \tag{6.3}$$

For the programs written,‡ one has

$$C = 8.0 \times 10^{-6} \text{ min/(increment)}^2 \text{ (for explicit method)}$$
$$(6.4)$$
$$C = 5.0 \times 10^{-5} \text{ min/(increment)}^2 \text{ (for Houbolt method)}$$

The values of N_t, required by each of the two numerical methods to give accurate results for a given problem, are not the same, primarily because of the stability requirements. Consequently, the relative efficiencies of the two methods cannot be determined from the values of C alone. For the problem considered in this section, the values of T for com-

‡ 1) These programs were written in Fortran, were run on an IBM 7094 computer, and did not utilize tape for storage of variables. 2) The value of C for the Houbolt method is taken from a program that stores several of the inverses shown in Eq. B7. For cases in which the storage capacity is insufficient to store these inverses, the value of C shown in (6.4) must be increased by a factor of 4.5. 3) The value of C for the explicit method is taken from a program that utilizes the variables u_ξ, u_θ, and w (Sec. 5). The explicit method (described in Sec. 5), which also uses the variable m_ξ, gives a higher value of C than shown in (6.4).

puter runs, which gave accurate results, are presented below in Table 1.

In Table 1, the values of $w^{(n)}$ at the free end are considered to be accurate if they change less than 2% when the values of N and N_t are doubled. In Table 2, accuracy with respect to $m_\xi^{(n)}$ at the clamped end implies a change of less than 4% when N and N_t are doubled. The values of T listed in Tables 1 and 2 correspond to a t_{max} of 1 msec.

Various aspects of the relative efficiencies of the two methods may be discerned by observing the values of N_t and T shown in the tables. (The value of N is the same for all of the cases shown.) The values of N_t used in the explicit method for these cases are limited by stability criteria (5.5) and are the same for each case. On the other hand, the fact that smaller values of N_t are sufficient to yield accurate output is revealed by the smaller values of N_t used in the Houbolt method. The two tables show that the values of N_t required to yield accurate output depend on the nature of the output considered. For $n = 0$ and $n = 1$, the values of $m_\xi^{(n)}$ at the clamped end vary much more rapidly with time than do the values of $w^{(n)}$ at the free end (Figs. 4 and 6), consequently, larger values of N_t are required to give accurate values of $m_\xi^{(n)}$. For the $n = 2$ case, the response is slower than for the $n = 0$ and $n = 1$ case, and smaller values of N_t give an accurate solution. On the basis of these observations, we arrive at the generalization: the explicit method tends to be more efficient for problems in which the output quantities vary relatively rapidly with time, whereas the Houbolt method tends to be more efficient for problems in which the important output quantities vary relatively slowly with time.

In generalizing the results shown in the tables to other problems, two additional factors should be considered. First, in problems that require very small values of Δ [i.e., Δ of order $h/(R\cdot 3^{1/2})$ or less] to give the desired accuracy, the explicit method generally becomes limited by the last of inequalities (5.3) instead of the first. In such problems, such small values of ϵ (large N_t) are required by the explicit method that the method becomes very inefficient. Secondly, although the spatial stations $i = -1, 0, 1, 2, \ldots$ (Eq. 3.1) were equally spaced in this work, it would be, in general, advantageous to make use of a relatively larger number of stations in certain important regions, such as near the boundaries. A comparison of the efficiencies of the explicit and Houbolt methods as applied to unequal spacings is not considered here, but the Houbolt method appears to be more adaptable to these unequal spacings.

7. Conclusions

In conclusion, computer programs were written that can be used to predict the linear dynamic response of a cylindrical shell. These programs were based on two numerical methods of integration: an explicit method and an implicit (Houbolt) method. It appears that the explicit method tends to be more efficient when the response varies rapidly, whereas the Houbolt method tends to be more efficient for the prediction of the slower responses. The Houbolt method is generally the more flexible method, because, unlike the explicit method, it is always numerically stable. The general methods developed here may be extended to include other shell shapes and more complicated physical models, such as orthotropic shells, multilayered shells, etc. With the ever increasing speeds and storage capacities of modern digital computers, such extensions appear to be both feasible and of practical importance.

Table 1 Computer time for accurate prediction of $w^{(n)}$ at free end

	Explicit method			Houbolt method		
n	N	N_t	T (min)	N	N_t	T (min)
0	90	1000	0.72	90	100	0.45
1	90	1000	0.72	90	50	0.23
2	90	1000	0.72	90	34	0.15

Table 2 Computer time for accurate prediction of $m_\xi^{(n)}$ at clamped end

	Explicit			Houbolt		
n	N	N_t	T (min)	N	N_t	T (min)
0	90	1000	0.72	90	200	0.90
1	90	1000	0.72	90	133	0.60
2	90	1000	0.72	90	67	0.30

Appendix A

Formulas for Coefficients

From Eqs. (2.8) and (2.9):

$$\bar{a}_1 = (E/E_0)[1/(1-\nu^2)] \qquad \bar{a}_{19} = -\bar{a}_6$$

$$\bar{a}_3 = \frac{-n^2}{2(1+\nu)}\left(\frac{E}{E_0}\right)\left[1 + \frac{\lambda^2}{48}\right] \qquad \bar{a}_{21} = \bar{a}_{15}$$

$$\bar{a}_4 = \frac{n}{2}\left(\frac{E}{E_0}\right)\left[\frac{1}{1-\nu} - \frac{\lambda^2}{16(1+\nu)}\right] \qquad \bar{a}_{23} = \bar{a}_{17}$$

$$\bar{a}_6 = \frac{1}{1+\nu}\left(\frac{E}{E_0}\right)\left[\frac{\nu}{1-\nu} - \frac{n^2\lambda^2}{24}\right] \qquad \bar{a}_{24} = \frac{n^2\lambda^2/6}{1+\nu}\left(\frac{E}{E_0}\right)$$

$$\bar{a}_{10} = -\bar{a}_4 \qquad \bar{a}_{26} = -\left(\frac{E}{E_0}\right)\left[\frac{1}{1-\nu^2} + \frac{\lambda^2 n^4}{12}\right]$$

$$\bar{a}_{12} = \frac{1}{2(1+\nu)}\left(\frac{E}{E_0}\right)\left[1 + \frac{3\lambda^2}{16}\right] \qquad \bar{a}_{27} = \lambda^2$$

$$\bar{a}_{14} = -n^2\left(\frac{E}{E_0}\right)\left[\frac{1}{1-\nu^2} + \frac{\lambda^2}{12}\right] \qquad \bar{a}_{29} = n\bar{a}_{18}$$

$$\bar{a}_{15} = \left(\frac{E}{E_0}\right)\frac{n\lambda^2/8}{1+\nu} \qquad \bar{a}_{32} = \left(\frac{E}{E_0}\right)\frac{\nu n/12}{1-\nu^2}$$

$$\bar{a}_{17} = -n\left(\frac{E}{E_0}\right)\left[\frac{1}{1-\nu^2} + \frac{\lambda^2 n^2}{12}\right] \qquad \bar{a}_{33} = -\frac{1}{12}\bar{a}_1$$

$$\bar{a}_{18} = -\nu\lambda^2 n \qquad \bar{a}_{35} = n\bar{a}_{32} \qquad \bar{a}_{36} = -1$$

From Eq. (2.17):

$$\bar{b}_1 = \left(\frac{E}{E_0}\right)\frac{1}{1-\nu^2} \qquad \bar{b}_3 = \nu n\bar{b}_1 \qquad \bar{b}_4 = \nu\bar{b}_1$$

$$\bar{b}_5 = \frac{n/2}{1+\nu}\left(\frac{E}{E_0}\right)\left[-1 + \frac{\lambda^2}{16}\right]$$

$$\bar{b}_6 = \frac{1/2}{1+\nu}\left(\frac{E}{E_0}\right)\left[1 + \frac{3\lambda^2}{16}\right] \qquad \bar{b}_8 = \frac{\lambda^2 n/8}{1+\nu}\left(\frac{E}{E_0}\right)$$

$$\bar{b}_{10} = \frac{n}{3}\bar{b}_8 \qquad \bar{b}_{11} = \bar{b}_8 \qquad \bar{b}_{13} = \frac{4}{3}n\bar{b}_8 \qquad \bar{b}_{15} = \lambda^2$$

Appendix B

Starting Methods in the Houbolt Procedure

Houbolt[4] suggests using the following difference formulas to start the recurrence process:

$$\begin{aligned}\dot{z}_{i,0} &= (2z_{i,1} + 3z_{i,0} - 6z_{i,-1} + z_{i,-2})/6\epsilon \\ \ddot{z}_{i,0} &= (z_{i,1} - 2z_{i,0} + z_{i,-1})/\epsilon^2\end{aligned} \tag{B1}$$

The initial conditions for the problems analyzed in this paper are $z_{i,0} = \dot{z}_{i,0} = \ddot{z}_{i,0} = 0$. (The condition $\ddot{z}_{i,0} = 0$ used here implies that the forces initially applied to the shell are zero. The method shown here could be extended to the general case $\ddot{z}_{i,0} \neq 0$ by considering the magnitude and distribution of the initially applied forces.) Applying the initial conditions to the terms in (B1) yields the proper relations between the z's at negative times z_{i-1} and $z_{i,-2}$ and the z's at positive times $z_{i,1}$, namely

$$z_{i,-1} = -z_{i,1} \qquad z_{i,-2} = -8z_{i,1} \tag{B2}$$

Substituting these relations and the initial conditions into the standard Houbolt expression

$$\ddot{z}_{i,j} = (2/\epsilon^2)z_{i,j} - (5/\epsilon^2)z_{i,j-1} + (4/\epsilon^2)z_{i,j-2} - (1/\epsilon^2)z_{i,j-3} \tag{B3}$$

gives certain formulas for the first two time increments of calculation ($j = 1, 2$) in terms of $\ddot{z}_{i,1}$, $z_{i,1}$, $\ddot{z}_{i,2}$, and $z_{i,2}$. Identifying the coefficients in these formulas with the general coeffients of the Houbolt formula

$$\ddot{z}_{i,j} = \alpha_j z_{i,j} + \beta_j z_{i,j-1} + \gamma_j z_{i,j-2} + \delta_j z_{i,j-3} \tag{B4}$$

leads to the coefficients listed in (4.2) for $j = 1, 2$.

Details of Elimination Procedure Used for Houbolt Method

The set of equations specified by (3.7–3.9) using the values in (4.2) were solved for $z_{i,j}$ by a modified Gaussian elimination technique (Potters Method) similar to that used in Ref. 2. In this procedure, a general expression is found relating $z_{i,j}$ to $z_{i+1,j}$. Then using (3.9), together with the relationship between $z_{N-1,j}$ and $z_{N,j}$, the correct value of $z_{N,j}$ can be calculated. All of the remaining z's are then calculated in reverse order, starting from this value of $z_{N,j}$. In detail, $z_{1,j}$ can be written in terms of $z_{2,j}$ by eliminating $z_{0,j}$ from (3.7) and (3.8):

$$z_{1,j} = -[B_0C^{-1}B_j - A_0]^{-1}[(B_0C^{-1}A - V_0)z_{2,j} - B_0C^{-1}g_{1,j} + g_{0,j}] \tag{B5}$$

If the relationship between $z_{i,j}$ and $z_{i+1,j}$ is denoted by

$$z_{i,j} = -P_{i,j}z_{i+1,j} + x_{i,j} \qquad (i = 1, 2, \ldots, N-1) \tag{B6}$$

then (B5) and (3.8) yield

$$\left.\begin{aligned} P_{1,j} &= [B_0C^{-1}B_j - A_0]^{-1}[B_0C^{-1}A - V_0] \\ x_{1,j} &= [B_0C^{-1}B_j - A_0]^{-1}[B_0C^{-1}g_{1,j} - g_{0,j}] \\ P_{i,j} &= [B_j - CP_{i-1,j}]^{-1}A \\ x_{i,j} &= [B_j - CP_{i-1,j}]^{-1} \times [g_{i,j} - Cx_{i-1,j}] \end{aligned}\right\} \quad (i = 2, 3, \ldots, N-1) \tag{B7}$$

These recurrence relations enable all the P's and x's up to $P_{N-1,j}$ and $x_{N-1,j}$ to be calculated. Then from (B6), letting $z_{N-1,j} = -P_{N-1,j}z_{N,j} + x_{N-1,j}$ and $z_{N-2,j} = -P_{N-2,j}z_{N-1,j} + x_{N-2,j}$, Eq. (3.9) can be manipulated to give the final result for z_N:

$$z_{N,j} = -[B_N - C_NP_{N-1,j} + V_NP_{N-2,j}P_{N-1,j}]^{-1} \times [C_Nx_{N-1,j} - V_NP_{N-2,j}x_{N-1,j} + V_Nx_{N-2,j} - g_{N,j}] \tag{B8}$$

Using this value of $z_{N,j}$, in conjunction with the relations from (B6), gives the final values of $z_{N-1,j}, z_{N-2,j}, \ldots, z_{1,j}$. Finally, $z_{0,j}$ is calculated from the equation of motion (3.8) with $i = 1$, as

$$z_{0,j} = C^{-1}[g_{1,j} - Az_{2,j} - B_jz_{1,j}] \tag{B9}$$

This entire process is repeated for each successive time step. The solution for the z's involves only inversions of 4×4 matrices.

References

[1] Sanders, J. L., Jr., "An improved first-approximation theory for thin shells," NASA Rept. 24 (June 1959).

[2] Budiansky, B. and Radkowski, P. P., "Numerical analysis of unsymmetrical bending of shells of revolution," AIAA J. 1, 1833–1842 (1963).

[3] Witmer, E. A., Balmer, H. A., Leech, J. W., and Pian, T. H. H., "Large dynamic deformations of beams, rings, plates, and shells," AIAA J. 1, 1848–1857 (1963).

[4] Houbolt, J. C., "A recurrence matrix solution for the dynamic response of elastic aircraft," J. Aeronaut. Sci. 17, 540–550 (1950).

[5] Levy, S. and Kroll, W. D., "Errors introduced by finite space and time increments in dynamic response computation," *Proceedings of the First U. S. National Congress on Applied Mechanics* (American Society of Mechanical Engineers, New York, 1951), Vol. 1.

[6] Yao, J. C., "An analytical and experimental study of cylindrical shells under localized impact loads," AIAA Preprint 65-111 (January 1965).

[7] Bisplinghoff, R. L., Ashley, H., and Halfman, R. L., *Aeroelasticity* (Addison Wesley Publishing Co., Inc., Reading Mass., 1957).

[8] Radkowski, P. P., Davis, R. M., and Bolduc, M. R., "Numerical analysis of equations of thin shells of revolution," ARS J. 32, 36–41 (1962).

[9] Balmer, H. A. and Witmer, E. A., "Theoretical-experimental correlation of large dynamic and permanent deformations of impulsively loaded simple structures," Tech. Doc. Rept. FDL-TDR-64-108 (July 1964).

30

Reprinted from *J. Acoust. Soc. Amer.*, **45**(1), 144–149 (1969)

Waves on a Spherical Shell

MICHAEL P. MORTELL

Center for the Applications of Mathematics, Lehigh University, Bethlehem, Pennsylvania 18015

Within the framework of classical linear elasticity theory, the dynamic equations governing the motion of a spherical shell are given. These include the effects of transverse shear and rotatory inertia, and can be derived on the basis of three assumptions additional to those of linear elasticity. The response of an incomplete spherical shell subject to a suddenly applied, constant moment M_θ is studied. The wavefront behavior of the solution is found by taking the Laplace transform of the equations, and then making an asymptotic expansion for large values of p, the Laplace variable. The solution is given in the form of a traveling wave, which is followed into the focus point and on its reflection from there. The applied discontinuity at the wavefront grows as $\theta \to \pi$; it gives a square-root singularity at $\theta = \pi$, and a logarithmic singularity on the reflected wavefront. These results are synthesized by an elliptic function.

INTRODUCTION

THE emphasis in the construction of linear shell theories has been to model more accurately the behavior at higher frequencies and shorter wavelengths, with the exact linear theory of elasticity as the standard. Classical membrane theory makes no pretense at including bending effects, and the thickness of the shell is neglected. Euler–Bernoulli theory includes bending but predicts infinite propagation speeds. Using a variational principle due to Reissner,[1] Naghdi[2] gave a set of equations that accounted for the effects of bending, transverse shear, and rotatory inertia, and also predicted finite propagation speeds.

The analysis of the transient response of a shell using these equations is no simple task. One classical approach is by means of the normal modes of vibration. This method has been exploited by Kraus and Kalnins,[3] for example, to find the response of a spherical shell subject to a suddenly applied uniform load. The normal-mode approach is not very illuminating if one wishes to examine progressing waves. Another familiar technique in this area is that of the integral transform. In many cases of interest, inversion of the transform presents serious difficulties, and one is satisfied with partial answers found by applying various asymptotic procedures to the inversion integral. The two methods have it in common that first an *exact* solution is constructed and then the approximations are made.

The problem of determining the stresses in a membrane spherical shell due to a blast-wave pressure loading was considered in Ref. 4. Owing to the complexity of the problem, the approach used was both analytical and numerical. In the analysis presented here, the inclusion of bending effects adds to the complexity, but we treat the more idealized problem of a suddenly applied moment at the lip ($\theta = \theta_0$) of a spherical shell. Also, attention is confined to the wavefront as it moves into the pole ($\theta = \pi$) and is then reflected back. The solution would be expected to be valid during the first few instants of time after the passage of the wavefront. A modal approach gives good results after several reflections of the waves. This type of analysis has been given in Ref. 5, and the work here complements these results.

The simplest concept of wave propagation is that of a wave progressing into a region of quiet. An example is the D'Alembert solution for a stretched string. One is struck by the dearth of such transparent representations in transient shell analysis. Since our attention is confined to the wavefronts, there should be no need for the exact solution under any guise. Accordingly, we perform the necessary asymptotic analysis directly on the governing equations, and in so doing we bypass many extraneous details and keep the analysis simple.

[1] E. Reissner, J. Math. Phys. 29, 90–95 (1950).
[2] P. M. Naghdi, Quart. Appl. Math. 14, 369–380 (1957).
[3] H. Kraus and A. Kalnins, J. Acoust. Soc. Amer. 38, 994–1002 (1965).
[4] J. H. Huth and J. D. Cole, J. Appl. Mech. 4, 473–478 (1955).
[5] J. P. Wilkinson and A. Kalnins, J. Appl. Mech. 3, 525–532 (1965).

The traveling waveform of the solution is then explicitly exhibited.

In Sec. I, the equations for the spherical shell are given on the basis of three assumptions, and an interesting reduction is noted that shows how bending effects have been incorporated. The problem is stated in Sec. II and some preliminary transformations are made. In Sec. III, a uniformly valid first approximation to the Laplace transform of the solution is given and it is also indicated how further approximations are constructed. Section IV gives the Laplace inversion and the final results.

The standard notation of shell theory is used throughout, and so it is not explained in any great detail.

I. GOVERNING EQUATIONS

The spherical polar coordinates (r,θ,ϕ) are related to the (x,y,z) Cartesian coordinate system by

$$x=r\sin\theta\cos\phi,$$
$$y=r\sin\theta\sin\phi,$$
$$z=r\cos\theta,$$
$$0\leq\theta\leq\pi,\quad 0\leq\phi<2\pi,\quad r\geq 0.$$

The spherical shell under consideration has constant thickness, h, and its midsurface is defined by $r=R$, constant. The theory is based on the three assumptions:

(a) $h/R\ll 1,$

(b) $t_{rr}\ll t_{\theta\theta},\ t_{\phi\phi},$

(c) $u_r=w'(\theta,\phi,t),$
$u_\theta=u'(\theta,\phi,t)+(r-R)\beta_\theta(\theta,\phi,t),$
$u_\phi=v'(\theta,\phi,t)+(r-R)\beta_\phi(\theta,\phi,t),$

where u_r, u_θ, u_ϕ are the three components of the displacement vector and β_θ, β_ϕ are the rotations of the normal to the midsurface. Furthermore, we confine attention to problems that are symmetrical in ϕ, and in which body and surface forces are absent. Following the usual method of integrating the exact equations of linear elasticity through the thickness of the shell, the momentum equations in dimensionless form are

$$\nabla^2u-[(1+\nu)/2+\cot^2\theta]u+c^{-2}\beta+[(3+\nu)/2]\times w_{,\theta}=u_{,tt}+2\epsilon^2\beta_{,tt}, \quad (1)$$

$$\nabla^2w-[2(1+\nu)/c^2]w+\beta_{,\theta}+\beta\cot\theta-[(3+\nu)/2c^2]\times(u_{,\theta}+u\cot\theta)=c^{-2}w_{,tt}, \quad (2)$$

and

$$\nabla^2\beta-(\nu+\cot^2\theta)\beta-c^2\lambda_0^2(\beta+w_{,\theta}-u)=\beta_{,tt}+2u_{,tt}, \quad (3)$$

where

$$c^2=(1-\nu)/2,\ \lambda_0^2=\epsilon^{-2}=12(R/h)^2$$

and

$$\nabla^2\equiv\partial^2/\partial\theta^2+\cot\theta(\partial/\partial\theta).$$

Here $c_p=[E/\rho(1-\nu^2)]^{\frac{1}{2}}$ is the speed of propagation of longitudinal waves in a plate, and with this normalization, c is recognized as the speed of distortion waves in an unbounded medium. In Eqs. 1–3, the dimensionless variables are

$$t=(c_p/R)t',\ u=u'/R,\ w=w'/R,\ \beta_\theta=\beta.$$

The dimensionless stress-displacement relations are

$$N_\theta=u_{,\theta}+\nu u\cot\theta+(1+\nu)w, \quad (4)$$
$$N_\phi=\nu u_{,\theta}+u\cot\theta+(1+\nu)w, \quad (5)$$
$$M_\theta=\beta_{,\theta}+\nu\beta\cot\theta, \quad (6)$$
$$M_\phi=\nu\beta_{,\theta}+\beta\cot\theta, \quad (7)$$

and

$$Q_\theta=c^2(\beta+w_{,\theta}-u), \quad (8)$$

where

$$(N_\theta,N_\phi,Q_\theta)=[(1-\nu^2)/Eh](N_\theta',N_\phi',Q_\theta')$$

and

$$(M_\theta,M_\phi)=[12(1-\nu^2)R/Eh^3](M_\theta',M_\phi').$$

We have chosen not to insert an averaging shear factor in Eq. 8 for Q_θ, since it does not arise naturally in our derivation and makes no essential difference to our results. Except for this, Eqs. 1–8 were given by Naghdi,[2] and Kalnins and Wilkinson.[5]

An order of magnitude analysis shows that it is consistent to neglect the term $2u_{,tt}$ on the right-hand side of Eq. 3. Then the system of Eqs. 1–3 is hyperbolic with double characteristics ±1 and single characteristics $\pm c$. For computational convenience, we retain the term $2u_{,tt}$ as it has the effect of splitting the double characteristics. The final results are obtained by allowing the characteristics to coalesce once again.

If we finally set $\epsilon=0$ in the momentum equations, Eq. 3 implies that

$$\beta=u-w_{,\theta}. \quad (9)$$

When account is taken of the scaling of the variables, Assumption (c) on u_θ becomes

$$u_\theta=u(\theta,t)+[r/R-1](u-w_{,\theta}),$$

and then

$$e_{r\theta}=\tfrac{1}{2}[(1/r)u_{r,\theta}-(1/r)u_\theta+u_{\theta,r}]=0. \quad (10)$$

Equation 10 is just the Love–Kirchoff hypothesis that points lying on the normal to the undeformed midsurface remain on the normal to the deformed midsurface. We also note that Eq. 9 implies $Q_\theta=0$.

If Eq. 9 is used to eliminate β from Eqs. 1 and 2, they reduce to

$$\nabla^2u-(\nu+\cot^2\theta)u+(1+\nu)w_{,\theta}=u_{,tt} \quad (11)$$

and

$$u_{,\theta}+u\cot\theta+2w=[-1/(1+\nu)]w_{,tt}. \quad (12)$$

These are the dynamical "membrane" equations for a spherical shell. We note that the shear speed c has been lost in the reduction to Eqs. 11 and 12.

II. STATEMENT OF THE PROBLEM

We consider the incomplete spherical shell $0<\theta_0\leq\theta\leq\pi$ when a constant moment $M_\theta=M_0$ is applied at the lip $\theta=\theta_0$. The ensuing wavefront behavior is examined as the disturbance travels to $\theta=\pi$, and is then reflected. The objective is achieved by the appropriate asymptotic procedure without recourse to any representation of the exact solution.

The sphere is initially at rest in its undeformed state corresponding to the initial conditions

$$t=0:\quad u=u_{,t}=\beta=\beta_{,t}=w=w_{,t}=0,\quad \theta_0\leq\theta\leq\pi. \tag{13}$$

It is loaded by a suddenly applied bending moment at the edge $\theta=\theta_0$:

$$\theta=\theta_0:\quad M_\theta=\beta_{,\theta}+\nu\beta\cot\theta=M_0H(t), \tag{14}$$

$$N_\theta=u_{,\theta}+\nu u\cot\theta+(1+\nu)w=0, \tag{15}$$

$$Q_\theta=c^2(\beta+w_{,\theta}-u)=0, \tag{16}$$

where $H(t)$ is the Heaviside unit step function. Furthermore, the radiation condition that the leading wave be moving into an undisturbed region on its first traversal from $\theta=\theta_0$ to $\theta=\pi$ is enforced.

We wish to solve Eqs. 1–3 subject to the initial conditions of Eq. 13, the boundary conditions of Eqs. 14–16, and the radiation condition.

We define the Laplace transform of $u(\theta,t)$ by

$$\bar{u}(\theta,p)=\int_0^\infty u(\theta,t)e^{-pt}dt, \tag{17}$$

with corresponding definitions for $\bar{w}(\theta,p)$ and $\bar{\beta}(\theta,p)$. The variable $u^*(\theta,p)$ is introduced by the relation

$$u^*(\theta,p)=\sin^{\frac{1}{2}}\theta\,\bar{u}(\theta,p); \tag{18}$$

$w^*(\theta,p)$ and $\beta^*(\theta,p)$ are similarly defined. It is now noted that

$$\nabla^2\bar{u}=\bar{u}_{,\theta\theta}+\cot\theta\,\bar{u}_{,\theta}=\sin^{-\frac{1}{2}}\theta \times[u^*_{,\theta\theta}-\tfrac{1}{4}u^*\cot^2\theta-\tfrac{1}{2}u^*(d/d\theta)(\cot\theta)], \tag{19}$$

and thus the transformation of Eq. 18 eliminates the first derivative term in the operator ∇^2.

When account is taken of Eq. 17, the initial conditions of Eq. 13, and the transformation of Eq. 18, the Eqs. 1–3 become

$$u^*_{,\theta\theta}-[\tfrac{3}{4}\cot^2\theta+\tfrac{1}{2}\nu+p^2]u^*+(c^2-2\epsilon^2p^2)\beta^* +\tfrac{1}{2}(3+\nu)(w^*_{,\theta}-\tfrac{1}{2}w^*\cot\theta)=0, \tag{20}$$

$$w^*_{,\theta\theta}+[\tfrac{1}{4}\cot^2\theta+\tfrac{1}{2}-(2a^2/c^4)-c^{-2}p^2]w^*+\beta^*_{,\theta} +\tfrac{1}{2}\beta^*\cot\theta-[(3+\nu)/2c^2](u^*_{,\theta}+\tfrac{1}{2}u^*\cot\theta)=0, \tag{21}$$

and

$$\beta^*_{,\theta\theta}-[\tfrac{3}{4}\cot^2\theta-\tfrac{1}{2}+\nu+\lambda_0^2c^2+p^2]\beta^*-\lambda_0^2c^2 \times[w^*_{,\theta}-\tfrac{1}{2}w^*\cot\theta-u^*]-2p^2u^*=0, \tag{22}$$

where $a=(1-\nu^2)^{\frac{1}{2}}$—under the scalings here—represents the speed of very long waves in a bar.

The boundary conditions of Eqs. 14–16 become

$$(1/p)M_0\sin^{\frac{1}{2}}\theta_0=\beta^*_{,\theta}-(\tfrac{1}{2}-\nu)\beta^*\cot\theta_0, \tag{23}$$

$$0=u^*_{,\theta}-(\tfrac{1}{2}-\nu)u^*\cot\theta_0+(1+\nu)w^*, \tag{24}$$

and

$$0=\beta^*-u^*+w^*_{,\theta}-\tfrac{1}{2}w^*\cot\theta_0. \tag{25}$$

III. UNIFORMLY VALID FIRST APPROXIMATION FOR θ IN $\theta_0\leq\theta\leq\pi$

In Eqs. 20–25, the parameter p is taken to be large, while ϵ is held fixed and not equal to zero. Large p ensures that we get the wavefront behavior, while $\epsilon\neq0$ allows the bending effects to be taken into account.

We define the variable $\bar{\theta}=p(\theta-\theta_0)$ and then assume that $u^*(\theta,p)$, $w^*(\theta,p)$, and $\beta^*(\theta,p)$ have the following expansions:

$$u^*(\theta,p)=u_0(\bar{\theta})+(1/p)u_1(\bar{\theta})+\cdots,$$

$$w^*(\theta,p)=w_0(\bar{\theta})+(1/p)w_1(\bar{\theta})+\cdots,$$

$$\beta^*(\theta,p)=\beta_0(\bar{\theta})+(1/p)w_1(\bar{\theta})+\cdots.$$

The equations for u_0, w_0, β_0 then are

$$\frac{\partial^2u_0}{\partial\bar{\theta}^2}-u_0-2\epsilon^2\beta_0=0, \tag{26}$$

$$\frac{\partial^2w_0}{\partial\bar{\theta}^2}-c^{-2}w_0=0, \tag{27}$$

and

$$\frac{\partial^2\beta_0}{\partial\bar{\theta}^2}-\beta_0-2u_0=0, \tag{28}$$

with the following boundary conditions at $\bar{\theta}=0$:

$$\frac{\partial\beta_0}{\partial\bar{\theta}}=\left(\frac{1}{p^2}\right)M_0\sin^{\frac{1}{2}}\theta_0, \tag{29}$$

$$\frac{\partial u_0}{\partial\bar{\theta}}=0, \tag{30}$$

and

$$\frac{\partial w_0}{\partial\bar{\theta}}=0. \tag{31}$$

The solutions to Eqs. 26–28 which satisfy Eqs. 29–31 and also the radiation condition are

$$w_0=0 \tag{32}$$

$$\beta_0=\frac{M_0}{p^2}\frac{q_2^2-1}{q_1(q_1^2-q_2^2)}\sin^{\frac{1}{2}}\theta_0e^{-q_1\bar{\theta}} -\frac{M_0}{p^2}\frac{q_1^2-1}{q_2(q_1^2-q_2^2)}\sin^{\frac{1}{2}}\theta_0e^{-q_2\bar{\theta}} \tag{33}$$

and

$$u_0=\frac{M_0}{p^2}\frac{(q_1^2-1)(q_2^2-1)}{2q_1(q_1^2-q_2^2)}\sin^{\frac{1}{2}}\theta_0 e^{-q_1\bar{\theta}}$$

$$-\frac{M_0}{p^2}\frac{(q_1^2-1)(q_2^2-1)}{2q_2(q_1^2-q_2^2)}\sin^{\frac{1}{2}}\theta_0 e^{-q_2\bar{\theta}}, \quad (34)$$

where

$$q_1^2=1+2\epsilon \quad \text{and} \quad q_2^2=1-2\epsilon.$$

The first approximation to $\bar{u}(\theta,p)$ is $\sin^{-\frac{1}{2}}\theta u_0(\bar{\theta})$ and this has a singularity at $\theta=\pi$. This nonuniformity could have been foreseen by noticing that in Eqs. 20–22 the terms involving $\cot^2\theta$ were neglected in comparison with p^2. This is obviously not valid as $\theta\to\pi$. We now construct an expansion which is valid in the neighborhood of $\theta=\pi$, and match it to the expansion valid near $\theta=\theta_0$.

We define the variable $\tilde{\theta}$ by

$$\tilde{\theta}=p(\pi-\theta)$$

and assume expansions of the form

$$w^*(\theta,p)=\tilde{w}_0(\tilde{\theta})+(1/p)\tilde{w}_1(\tilde{\theta})+\cdots,$$

$$u^*(\theta,p)=\tilde{u}_0(\tilde{\theta})+(1/p)\tilde{u}_1(\tilde{\theta})+\cdots,$$

and

$$\beta^*(\theta,p)=\tilde{\beta}_0(\tilde{\theta})+(1/p)\tilde{\beta}_1(\tilde{\theta})+\cdots.$$

The ordinary differential equations satisfied by $\tilde{w}_0$, $\tilde{u}_0$, $\tilde{\beta}_0$ are

$$\frac{\partial^2 u_0}{\partial\tilde{\theta}^2}-\left(\frac{3}{4\tilde{\theta}^2}+1\right)\tilde{u}_0-2\epsilon^2\tilde{\beta}_0=0, \quad (35)$$

$$\frac{\partial^2 w_0}{\partial\tilde{\theta}^2}+\left(\frac{1}{4\tilde{\theta}^2}-c^{-2}\right)\tilde{w}_0=0, \quad (36)$$

and

$$\frac{\partial^2\beta_0}{\partial\tilde{\theta}^2}-\left(\frac{3}{4\tilde{\theta}^2}+1\right)-2\tilde{u}_0=0. \quad (37)$$

The solutions of Eqs. 35–37 that are bounded at $\tilde{\theta}=0$ are

$$\tilde{w}_0=\tilde{A}_0\tilde{\theta}^{\frac{1}{2}}I_0[(1/c)\tilde{\theta}], \quad (38)$$

$$\tilde{\beta}_0=\tilde{B}_1\tilde{\theta}^{\frac{1}{2}}I_1(q_1\tilde{\theta})+\tilde{B}_2\tilde{\theta}^{\frac{1}{2}}I_1(q_2\tilde{\theta}), \quad (39)$$

and

$$\tilde{u}_0=\tfrac{1}{2}(q_1^2-1)\tilde{B}_1\theta^{\frac{1}{2}}I_1(q_1\tilde{\theta})+\tfrac{1}{2}(q_2^2-1)\tilde{B}_2\tilde{\theta}^{\frac{1}{2}}I_1(q_2\tilde{\theta}), \quad (40)$$

where I_0, I_1 are modified Bessel functions. We fix $\theta\neq\pi$ and let $p\to\infty$, then on using the asymptotic form of $I_1(q\tilde{\theta})$, we find from Eq. 39

$$\tilde{\beta}_0\sim\tilde{B}_1[1/(2\pi q_1)^{\frac{1}{2}}]e^{q_1p(\pi-\theta}+\tilde{B}_2[1/(2\pi q_2)^{\frac{1}{2}}]e^{q_2p(\pi-\theta)}. \quad (41)$$

On comparing Eq. 41 with Eq. 33, it is evident that

$$\tilde{B}_1=\frac{M_0}{p^2}\left(\frac{2\pi}{q_1}\right)^{\frac{1}{2}}\frac{q_2^2-1}{q_1^2-q_2^2}\sin^{\frac{1}{2}}\theta_0 e^{-q_1p(\pi-\theta_0)} \quad (42)$$

and

$$\tilde{B}_2=-\frac{M_0}{p^2}\left(\frac{2\pi}{q_2}\right)^{\frac{1}{2}}\frac{q_1^2-1}{q_1^2-q_2^2}\sin^{\frac{1}{2}}\theta_0 e^{-q_2p(\pi-\theta_0)}. \quad (43)$$

By a similar procedure

$$\tilde{A}_0=0.$$

Thus, the first-order approximation, which is uniformly valid in θ for $\theta_0\leq\theta\leq\pi$ as $p\to\infty$, is given by

$$\bar{w}_0(\theta,p)=0,$$

$$\bar{\beta}_0(\theta,p)=\frac{M_0}{p^{\frac{3}{2}}}\frac{q_2^2-1}{q_1^2-q_2^2}(2\pi\sin\theta_0)^{\frac{1}{2}}\left(\frac{\pi-\theta}{\sin\theta}\right)^{\frac{1}{2}}$$

$$\times e^{-q_1p(\pi-\theta_0)}I_1[q_1p(\pi-\theta)]$$

$$-\frac{M_0}{p^{\frac{3}{2}}}\frac{q_2^2-1}{q_1^2-q_2^2}(2\pi\sin\theta_0)^{\frac{1}{2}}\left(\frac{\pi-\theta}{\sin\theta}\right)^{\frac{1}{2}}$$

$$\times e^{-q_2p(\pi-\theta_0)}I_1[q_2p(\pi-\theta)],$$

and

$$\bar{u}_0(\theta,p)=\frac{M_0}{p^{\frac{3}{2}}}\frac{(q_1^2-1)(q_2^2-1)}{q_1^2-q_2^2}\left(\frac{\pi}{2}\sin\theta_0\right)^{\frac{1}{2}}\left(\frac{\pi-\theta}{\sin\theta}\right)^{\frac{1}{2}}$$

$$\times e^{-q_1p(\pi-\theta_0)}I_1[q_1p(\pi-\theta)]$$

$$-\frac{M_0}{p^{\frac{3}{2}}}\frac{(q_1^2-1)(q_2^2-1)}{q_1^2-q_2^2}\left(\frac{\pi}{2}\sin\theta_0\right)^{\frac{1}{2}}\left(\frac{\pi-\theta}{\sin\theta}\right)^{\frac{1}{2}}$$

$$\times e^{-q_2p(\pi-\theta_0)}I_1[q_2p(\pi-\theta)].$$

If we now calculate the second terms $u_1(\bar{\theta})$, $\beta_1(\bar{\theta})$ of the expansion valid near $\theta=\theta_0$, we find, for example, that $\beta_1(\bar{\theta})$ has a term of the form $\bar{\theta}e^{-q_1\bar{\theta}}$. By the definition of $\bar{\theta}$,

$$(1/p)\bar{\theta}e^{-q_1\bar{\theta}}=(\theta-\theta_0)e^{-q_1\bar{\theta}},$$

and so the second term in a naive expansion has the same order of magnitude as the first. The source of the difficulty is seen most simply by considering the equation

$$y_{,tt}+(1+\epsilon)y=\sin t, \quad \epsilon\ll 1.$$

A first approximation is given by

$$y_{,tt}+y=\sin t,$$

and solutions of this equation are unbounded as $t\to\infty$, which is not so for the exact equation. This difficulty is overcome by introducing a "slow" variable to correct the argument to the solution of the homogeneous equation. The two-variable expansion technique of Cole and Kevorkian[6] provides just such a technique. The results of this calculation (which yields the effects of the higher-order coupling terms) for the Eqs. 20–22 are

[6] J. D. Cole and J. Kevorkian, "Non-linear Differential Equations and Non-linear Mechanics," *Proceedings of the International Symposium* (Academic Press, Inc., New York, 1963), pp. 113–120.

given in the author's thesis,[7] but are not needed here. This method of correcting the wavefront approximation could be contrasted with that of Flügge and Zajac.[8]

Finally, the characteristics are allowed to coalesce, i.e., $q_1 \rightarrow 1$ and $q_2 \rightarrow 1$. Then we find

$$\bar{w}_0(\theta,p)=0, \qquad (44)$$

$$\bar{u}_0(\theta,p)=0, \qquad (45)$$

and

$$\bar{\beta}_0(\theta,p)=-(M_0/p^{3/2})(2\pi\sin\theta_0)^{\frac{1}{2}} \times[(\pi-\theta)/\sin\theta]^{\frac{1}{2}}e^{-p(\pi-\theta_0)}I_1[p(\pi-\theta)]. \qquad (46)$$

IV. WAVEFRONT BEHAVIOR

We now fix our attention on the value of the moment M_θ at the leading wavefront. If $\bar{M}_\theta(\theta,p)$ denotes the Laplace transform of $M_\theta(\theta,t)$, and if we assume that $\bar{M}_\theta$ may be written as

$$\bar{M}_\theta(\theta,p)=\bar{M}_{0\theta}+(1/p)\bar{M}_{1\theta}+\cdots,$$

then

$$\bar{M}_{0\theta}(\theta,p)=\bar{\beta}_{0,\theta}+\nu\bar{\beta}_0\cot\theta. \qquad (47)$$

On using Eq. 46, it is found that

$$M_{0\theta}(\theta,p)=M_0(2\pi\sin\theta_0)^{\frac{1}{2}}\left(\frac{\pi-\theta}{\sin\theta}\right)^{\frac{1}{2}}e^{-p(\pi-\theta_0)}p^{-\frac{1}{2}}I_0[p(\pi-\theta)]+M_0\left(\frac{\pi}{2}\sin\theta_0\right)^{\frac{1}{2}}\left(\frac{1}{\pi-\theta}+\cot\theta\right)\left(\frac{\pi-\theta}{\sin\theta}\right)^{\frac{1}{2}}$$
$$\times e^{-p(\pi-\theta_0)}p^{-\frac{3}{2}}I_1[p(\pi-\theta)]-\nu M_0(2\pi\sin\theta_0)^{\frac{1}{2}}\left(\frac{\pi-\theta}{\sin\theta}\right)^{\frac{1}{2}}\cot\theta e^{-p(\pi-\theta_0)}p^{-\frac{3}{2}}I_1[p(\pi-\theta)]$$
$$-M_0(2\pi\sin\theta_0)^{\frac{1}{2}}\left(\frac{\pi-\theta}{\sin\theta}\right)^{\frac{1}{2}}e^{-p(\pi-\theta_0)}\frac{1}{p^{\frac{3}{2}}(\pi-\theta)}I_1[p(\pi-\theta)]. \qquad (48)$$

We need the following Laplace inversion formulas (see Ref. 9):

$$e^{-a_1p}I_0(b_1p)\sim\begin{cases}[\pi(t-\alpha)^{\frac{1}{2}}(\gamma-t)^{\frac{1}{2}}]^{-1}, & \alpha<t<\gamma\\ 0, & \text{otherwise}\end{cases} \qquad (49)$$

and

$$e^{-a_1p}I_1(b_1p)\sim\begin{cases}(t-a_1)[b_1\pi(t-\alpha)^{\frac{1}{2}}(\gamma-t)^{\frac{1}{2}}]^{-1}, & \alpha<t<\gamma\\ 0, & \text{otherwise,}\end{cases} \qquad (50)$$

where

$$a_1=\pi-\theta_0, \quad b_1=\pi-\theta$$

$$\alpha=\theta-\theta_0, \quad \gamma=2\pi-\theta-\theta_0.$$

On using Eq. 49, we see that

$$p^{-\frac{1}{2}}e^{-a_1p}I_0(b_1p)\sim\pi^{-\frac{3}{2}}\int_\alpha^t\frac{du}{[(t-u)(u-\alpha)(\gamma-u)]^{\frac{1}{2}}}. \qquad (51)$$

The substitution

$$(u-\alpha)/(t-\alpha)=\sin^2\phi$$

reduces the right-hand side of Eq. 51 to

$$2\pi^{-\frac{3}{2}}(\gamma-\alpha)^{-\frac{1}{2}}\int_0^{\pi/2}\frac{d\phi}{(1-m\sin^2\phi)^{\frac{1}{2}}}, \quad m=\frac{(t-\alpha)}{(\gamma-\alpha)}.$$

Thus

$$p^{-\frac{1}{2}}e^{-a_1p}I_0(b_1p)\sim2\pi^{-\frac{3}{2}}(\gamma-\alpha)^{-\frac{1}{2}}K(m), \qquad (52)$$

where $K(m)$ is the complete elliptic integral of the first kind. The use of Eq. 50 yields

$$p^{-\frac{3}{2}}e^{-a_1p}I_1(b_1p)\sim2^{\frac{1}{2}}\pi^{-\frac{3}{2}}b_1^{-1}\int_\alpha^t\frac{(t-u)^{\frac{1}{2}}(u-a_1)}{[(u-\alpha)(\gamma-u)]^{\frac{1}{2}}}du$$
$$=\frac{2^{\frac{3}{2}}}{\pi^{\frac{3}{2}}b_1}\frac{(t-\alpha)(\alpha-a_1)}{(\gamma-\alpha)^{\frac{1}{2}}}\int_0^{\pi/2}\frac{\cos^2\phi}{(1-m\sin^2\phi)^{\frac{1}{2}}}d\phi, \qquad (53)$$

where $(t-\alpha)^2$ has been neglected, since $t=\alpha$ is the wavefront.

We denote by $M_\theta(wf)$ the dominant contribution to the wavefront behavior of $M_\theta(\theta,t)$. On examining Eq. 48 in the light of Eqs. 52 and 53, we find

$$M_\theta(wf)=M_0(2\pi\sin\theta_0)^{\frac{1}{2}}\left(\frac{\pi-\theta}{\sin\theta}\right)^{\frac{1}{2}}\left[\frac{2}{\pi^{\frac{3}{2}}(\gamma-\alpha)^{\frac{1}{2}}}K(m)\right.$$
$$-\left(\nu\cot\theta+\frac{1}{\pi-\theta}\right)\frac{2^{\frac{3}{2}}}{\pi^{\frac{3}{2}}b_1}\frac{(t-\alpha)(\alpha-a_1)}{(\gamma-\alpha)^{\frac{1}{2}}}$$
$$\left.\times\int_0^{\pi/2}\frac{\cos^2\phi}{(1-m\sin^2\phi)^{\frac{1}{2}}}d\phi\right]. \qquad (54)$$

[7] M. P. Mortell, "Some Approximate Solutions of Dynamic Problems in the Linear Theory of Thin Elastic Shells," thesis, California Inst. Technol. (October 1967).

[8] W. Flügge and E. E. Zajac, Ing.-Arch. **28**, 59–70 (1959).

[9] A. Erdélyi, W. Magnus, F. Oberhettinger, and F. Tricomi, *Tables of Integral Transforms* (Bateman Manuscript Project) (McGraw-Hill Book Co., New York, 1954), Vol. 1, p. 276.

For $\theta_0<\theta<\pi$, the wavefront is $t-(\theta-\theta_0)=0$ and then $m=0$. Then Eq. 54 reduces to

$$M_\theta(wf)=M_0(\sin\theta_0/\sin\theta)^{\frac{1}{2}}. \tag{55}$$

For $\theta=\pi$, the wavefront is $t-(\pi-\theta_0)=0$, and $m=0$. In this case, Eq. 54 becomes

$$M_\theta(wf)=M_0\frac{\sin^{\frac{1}{2}}\theta_0}{(\pi-\theta)^{\frac{1}{2}}}+\frac{1}{\sqrt{2}}M_0\sin^{\frac{1}{2}}\theta_0 \times\left(\nu\cot\theta+\frac{1}{\pi-\theta}\right)\frac{t-(\pi-\theta_0)}{(\pi-\theta)^{\frac{1}{2}}}.$$

Then

$$\lim_{\theta\to\pi} M_\theta(wf)=C(\pi-\theta)^{-\frac{1}{2}}, \tag{56}$$

where

$$C=M_0\sin^{\frac{1}{2}}\theta_0[(1+\sqrt{2})/\sqrt{2}-\nu/\sqrt{2}].$$

Thus, the effect of the focusing is to give a square-root singularity at the arrival time of the wavefront.

It now remains only to get the reflected wave. The speed of the leading wavefront is unity, and thus the time for a wave to go from θ_0 to π and back to a general point θ is $2\pi-\theta_0-\theta$. Now

$$m=(t-\alpha)/(\gamma-\alpha)=(t-\alpha)/(2\pi-\theta_0-\theta-\alpha)$$

and thus

$$m=1$$

is the determination of a once-reflected wavefront. Then, from Eq. 54

$$M_\theta(wf)=M_0(2\pi\sin\theta_0)^{\frac{1}{2}}[(\pi-\theta)/(\sin\theta)]^{\frac{1}{2}} \times(2/\pi^{\frac{1}{2}}2^{\frac{1}{2}}(\pi-\theta)^{\frac{1}{2}})K(1),$$

and so for $\pi<\theta<2\pi-\theta_0$,

$$M_\theta(wf)=-(M_0/\pi)(\sin\theta_0/\sin\theta)^{\frac{1}{2}}\log(1-m),\ m\uparrow 1. \tag{57}$$

Thus, there is a logarithmic singularity on the reflected wavefront.

Equations 55–57 constitute the final results. It is interesting to note that the function $K(m)$ synthesizes the results for the three different regions, when m is given the proper interpretation. That the interpretation given here is justified is confirmed in Tables 3 and 4 of Ref. 10, and the comparative simplicity of the analysis here should be noted.

Furthermore, even though the displacements u and w and the rotation β are small at the leading wavefront, the linear theory is not adequate to describe the situation there, when $\theta=\pi$, or on the reflected wavefront.

V. DISCUSSION

The complex system of equations governing the motion have been quickly reduced to tractable form by the appropriate asymptotic procedure. The resulting equations involve mainly higher-derivative terms, and so the results reflect the inclusion of bending effects. One could anticipate the square-root singularity at the focusing point from energy considerations. The procedure given here yields results that are valid for short times after the passage of the wavefront, and a systematic method of improving the approximation is indicated. One can easily envisage situations where a major part of the disturbance lies well away from the characteristic fronts, and then one would hope to find an asymptotic procedure that would yield both the region of the main disturbance and the structure of the solution there. A pertinent example is that of a blast wave enveloping a spherical shell. The author, guided by some results for a cylindrical shell, hopes to solve this problem.

ACKNOWLEDGMENTS

The author wishes to thank Dr. J. K. Knowles of the California Institute of Technology for his invaluable guidance during the course of this work, which was supported in the main by a contract of the U. S. Office of Naval Research at the California Institute of Technology, and in part by a contract of the U. S. Office of Naval Research at Lehigh University.

[10] G. N. Ward, Quart. J. Mech. Appl. Math. 1, 225–245 (1948).

31

Reprinted from *J. Appl. Mech.*, **32,** 346–350 (1965)

WILLIAM R. SPILLERS
Department of Civil Engineering and Engineering Mechanics, Columbia University, New York, N. Y.

Wave Propagation in a Thin Cylindrical Shell

A solution is presented for a semi-infinite cylindrical (Timoshenko type) shell subjected to dynamic loading at one end using the method characteristics. Explicit results are obtained for the propagation of discontinuities. These results are combined with a simple numerical procedure to obtain the solution in all regions. Numerical examples are given for an elastic and a viscoelastic material and comparisons are made to previous results concerning wave propagation in shells.

THE problem of the cylindrical shell loaded dynamically at its end which is fundamental to several fields of interest including space structures has been the subject of recent studies [1, 2].[1] These studies have been based on membrane theory, leaving open to a large extent the question of the response of a shell to very rapidly applied loads in which shear effects may play a significant role, as they do in wave propagation in a beam. It is the purpose of this paper to extend the foregoing studies to include the effects of bending and shear.

The problem is posed in terms of simultaneous partial differential equations with constant coefficients which are frequently treated using integral transforms. However, when the number of equations becomes large, as it does here, the characteristic equation of the system becomes intractable while it is still possible to determine the characteristic directions which depend only on the higher order terms in the characteristic equation. This fact itself suggests the method of characteristics. Another aspect to be considered is that if it is only possible to obtain explicit results for the wave fronts of a particular problem and numerical techniques are required to obtain the entire field, then the method of characteristics is ideal; the data at the wave fronts may be obtained by inspecting the equations and extremely simple procedures are available for numerical integration of the equations.

The method of characteristics has previously been applied by Leonard and Budiansky [3] to the Timoshenko beam.

Elastic Thin Shell

The equations for a Timoshenko-type cylindrical shell with axial symmetry have been derived by Herrmann and Mirsky [4]. These equations are as follows:

Equations of motion

$$N_{xx}' = \rho h(\ddot{u} + h^2/(12R)\ddot{\psi}) \tag{1}$$

$$Q_x' - N_{\theta\theta}R^{-1} = \rho h\ddot{w} \tag{2}$$

$$M_{xx}' - Q_x = \rho h^3(\ddot{\psi} + \ddot{u}/R)/12 \tag{3}$$

and the stress-displacement equations

$$N_{xx} = Eh(1 - \nu^2)^{-1}(u' + \nu wR^{-1} + h^2/(12R)\psi') \tag{4}$$

$$N_{\theta\theta} = Eh(1 - \nu^2)^{-1}[\nu u' + wR^{-1}(1 + h^2/(12R^2))] \tag{5}$$

$$Q_x = \kappa^2 Gh(\psi + w') \tag{6}$$

$$M_{xx} = Eh^3[12(1 - \nu^2)]^{-1}(\psi' + u'/R) \tag{7}$$

in which N_{xx}, Q_x, M_{xx}, and $N_{\theta\theta}$ are the stress resultants, Fig. 1; u, w, and ψ the shell displacements, Fig. 2; ρ the mass density; R the mean radius of the shell; h the shell thickness; E, ν, and G Young's modulus, Poisson's ratio, and the shear modulus, respectively; κ^2 is the Timoshenko shear coefficient and the prime and dot represent differentiation with respect to the space coordinate, x, and to time, t. Equations (1) to (7) constitute a system of linear partial differential equations which will be treated using the method given by Courant [5]. First, the nondimensional variables

$$U = uR^{-1} \tag{8}$$

$$W = wR^{-1} \tag{9}$$

$$n_x = N_{xx}[Eh/(1 - \nu^2)]^{-1} \tag{10}$$

$$n_\theta = N_{\theta\theta}[Eh/(1 - \nu^2)]^{-1} \tag{11}$$

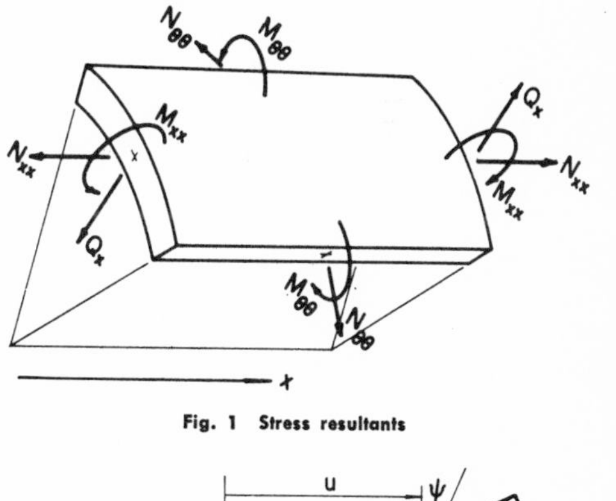

Fig. 1 Stress resultants

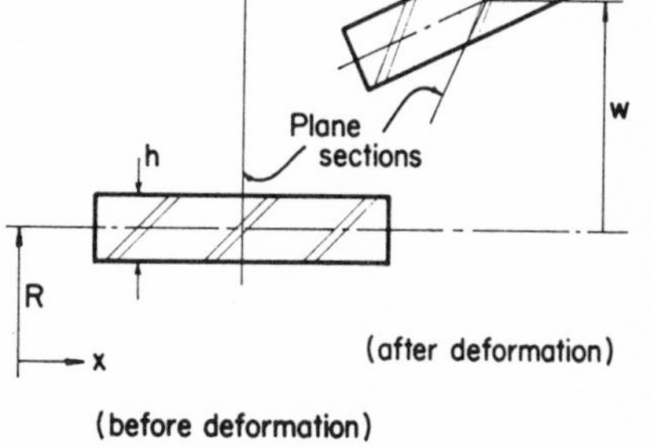

Fig. 2 Longitudinal shell element

[1] Numbers in brackets designate References at end of paper.

Contributed by the Applied Mechanics Division for presentation at the Winter Annual Meeting, New York, N. Y., November 29–December 4, 1964, of THE AMERICAN SOCIETY OF MECHANICAL ENGINEERS.

Discussion of this paper should be addressed to the Editorial Department, ASME, United Engineering Center, 345 East 47th Street, New York, N. Y. 10017, and will be accepted until one month after final publication of the paper itself in THE JOURNAL OF APPLIED MECHANICS. Manuscript received by ASME Applied Mechanics Division, January 7, 1964. Paper No. 64—WA/APM-23.

$$q_x = Q_x[Eh/(1-\nu^2)]^{-1} \qquad (12)$$

$$m_x = M_{xx}[REh/(1-\nu^2)]^{-1} \qquad (13)$$

$$\xi = xR^{-1} \qquad (14)$$

$$\tau = c_p t R^{-1} \qquad (15)$$

are introduced ($c_p = [E/(\rho(1-\nu^2))]^{1/2}$). Then, if equations (4) to (7) are differentiated with respect to time, the entire system may be written as a system of first-order equations (linear combinations of equations (1) and (3) have been used)

$$\frac{\partial}{\partial \xi}\begin{bmatrix} 0 & 0 & 0 & 0 & -1 & 0 & -\delta^2 \\ 0 & 0 & 0 & 0 & -\nu & 0 & 0 \\ 0 & 0 & 0 & 0 & 0 & -k^2 & 0 \\ 0 & 0 & 0 & 0 & -\delta^2 & 0 & -\delta^2 \\ -(1-\delta^2)^{-1} & 0 & 0 & (1-\delta^2)^{-1} & 0 & 0 & 0 \\ 0 & 0 & -1 & 0 & 0 & 0 & 0 \\ (1-\delta^2)^{-1} & 0 & 0 & -\delta^{-2}(1-\delta^2)^{-1} & 0 & 0 & 0 \end{bmatrix}\begin{bmatrix} n_x \\ n_\theta \\ q_x \\ m_x \\ \dot U \\ \dot W \\ \dot\psi \end{bmatrix} + \frac{\partial}{\partial \tau}\begin{bmatrix} n_x \\ n_\theta \\ q_x \\ m_x \\ \dot U \\ \dot W \\ \dot\psi \end{bmatrix} = \begin{bmatrix} \nu \dot W \\ (1+\delta^2)\dot W \\ k^2\dot\psi \\ 0 \\ (1-\delta^2)^{-1}q_x \\ -n_\theta \\ -\delta^{-2}(1-\delta^2)^{-1}q_x \end{bmatrix} \qquad (16)$$

where $k^2 = G\kappa^2/[E/(1-\nu^2)] \qquad \delta^2 = h^2/(12R^2)$

$$\dot U, \dot W, \dot\psi = \partial/\partial\tau(U, W, \psi)$$

The foregoing may be written formally

$$A\frac{\partial}{\partial \xi}v + \frac{\partial}{\partial \tau}v = c \qquad (17)$$

where A, v, and c are the appropriate matrices given by equation (16). This system is hyperbolic since there exist 7 linearly independent eigenvectors l^s for which

$$l^s A = \tau_s l^s \qquad (18)$$

and τ_s are the roots of the algebraic equation

$$|A - \tau I| = 0 \qquad (19)$$

The roots of equation (19) are $\tau_s = 1, -1, 1, -1, k, -k, 0$ for which the corresponding eigenvectors are

These equations are called the normal form of the system. In general the wave fronts depend on the boundary conditions but it can be seen immediately that there are two possible velocities for forward-moving waves: c_p, the velocity of compressional waves in a plate, and $c_s = (\kappa^2 G/\rho)^{1/2}$, the velocity of shear waves (corrected by κ^2). It can be seen that discontinuities propagate with undiminished jumps; discontinuities in n_x, $\dot U$, m_x, and $\dot\psi$ propagating with the velocity c_p and discontinuities in q_x and $\dot W$ propagating with the velocity c_s. Further equations (21)–(27) may be written

$$\frac{d}{ds_1}U_1 = \tfrac{1}{2}[\nu(U_6 - U_5)/k - \delta^{-2}(U_6 + U_5)] \qquad (28)$$

$$\frac{d}{ds_2}U_2 = \tfrac{1}{2}[\nu(U_6 - U_5)/k + \delta^{-2}(U_6 + U_5)] \qquad (29)$$

$$\frac{d}{ds_1}U_3 = \tfrac{1}{2}(U_6 + U_5) \qquad (30)$$

$$\frac{d}{ds_2}U_4 = -\tfrac{1}{2}(U_6 + U_5) \qquad (31)$$

$$\begin{aligned} l^1 &= [\;1 \quad 0 \quad 0 \quad -\delta^{-2} \quad 0 \quad 0 \quad (1-\delta^2)] \\ l^2 &= [\;1 \quad 0 \quad 0 \quad -\delta^{-2} \quad 0 \quad 0 \quad -(1-\delta^2)] \\ l^3 &= [\;0 \quad 0 \quad 0 \quad -\delta^{-2} \quad 1 \quad 0 \quad 1\;] \\ l^4 &= [\;0 \quad 0 \quad 0 \quad \delta^{-2} \quad 1 \quad 0 \quad 1\;] \\ l^5 &= [\;0 \quad 0 \quad 1 \quad 0 \quad 0 \quad -k \quad 0\;] \\ l^6 &= [\;0 \quad 0 \quad 1 \quad 0 \quad 0 \quad k \quad 0\;] \\ l^7 &= [-\nu(1-\delta^2)^{-1} \quad 1 \quad 0 \quad \nu(1-\delta^2)^{-1} \quad 0 \quad 0 \quad 0\;] \end{aligned} \qquad (20)$$

Multiplying equation (17) on the left by each l^s in turn gives a set of equations in which the differentiation occurs along the characteristic directions

$$\frac{\partial}{\partial \xi}[n_x - \delta^{-2}m_x + (1-\delta^2)\dot\psi] + \frac{\partial}{\partial \tau}[n_x - \delta^{-2}m_x + (1-\delta^2)\dot\psi] = \nu\dot W - \delta^{-2}q_x \qquad (21)$$

$$-\frac{\partial}{\partial \xi}[n_x - \delta^{-2}m_x - (1-\delta^2)\dot\psi] + \frac{\partial}{\partial \tau}[n_x - \delta^{-2}m_x - (1-\delta^2)\dot\psi] = \nu\dot W + \delta^{-2}q_x \qquad (22)$$

$$\frac{\partial}{\partial \xi}[m_x - \delta^2(\dot U + \dot\psi)] + \frac{\partial}{\partial \tau}[m_x - \delta^2(\dot U + \dot\psi)] = q_x \qquad (23)$$

$$-\frac{\partial}{\partial \xi}[m_x + \delta^2(\dot U + \dot\psi)] + \frac{\partial}{\partial \tau}[m_x + \delta^2(\dot U + \dot\psi)] = -q_x \qquad (24)$$

$$k\frac{\partial}{\partial \xi}[q_x - k\dot W] + \frac{\partial}{\partial \tau}[q_x - k\dot W] = k^2\dot\psi + kn_\theta \qquad (25)$$

$$-k\frac{\partial}{\partial \xi}[q_x + k\dot W] + \frac{\partial}{\partial \tau}[q_x + k\dot W] = k^2\dot\psi - kn_\theta \qquad (26)$$

$$\frac{\partial}{\partial \tau}[n_\theta + \nu(1-\delta^2)^{-1}(m_x - n_x)] = (1 - \nu^2 - \delta^4)(1-\delta^2)^{-1}\dot W \qquad (27)$$

$$\frac{d}{ds_3} U_5 = \tfrac{1}{2}k^2(U_1 - U_2)(1 - \delta^2)^{-1} + k[U_7 - \nu(1 - \delta^2)^{-1} \cdot \{(1 - \delta^{-2})(U_3 + U_4) - (U_1 + U_2)\}] \quad (32)$$

$$\frac{d}{ds_4} U_6 = \tfrac{1}{2}k^2(U_1 - U_2)(1 - \delta^2)^{-1} - k[U_7 - \nu(1 - \delta^2)^{-1} \cdot \{(1 - \delta^{-2})(U_3 + U_4) - (U_1 + U_2)\}] \quad (33)$$

$$\frac{\partial}{\partial \tau} U_7 = \tfrac{1}{2}(1 - \nu^2 - \delta^4)(1 - \delta^2)^{-1}(U_6 - U_5)/k \quad (34)$$

where

$$\begin{aligned} U_1 &= n_x - \delta^{-2}m_x + (1 - \delta^2)\dot{\psi} \\ U_2 &= n_x - \delta^{-2}m_x - (1 - \delta^2)\dot{\psi} \\ U_3 &= m_x - \delta^2(\dot{U} + \dot{\psi}) \\ U_4 &= m_x + \delta^2(\dot{U} + \dot{\psi}) \\ U_5 &= q_x - k\dot{W} \\ U_6 &= q_x + k\dot{W} \\ U_7 &= n_\theta + \nu(1 - \delta^2)^{-1}(m_x - n_x) \end{aligned} \quad (35)$$

which are convenient for numerical integration.

An Initial Value Problem for Elastic Thin Shell

Consider the problem of a smooth wall moving at constant velocity striking the end of a semi-infinite ($\xi \geq 0$) shell which is initially at rest. Formally

$$n_x = n_\theta = q_x = m_x = \dot{U} = \dot{W} = \dot{\psi} = 0 \qquad \text{for} \quad \tau = 0, \quad \xi > 0$$

$$q_x = \dot{\psi} = 0, \qquad \dot{U} = 1 \quad \text{for} \quad \tau > 0, \quad \xi = 0$$

From the "range of influence," Fig. 3, it follows that the first response travels with the velocity c_p. Since the solution is continuous along the boundary except at the point $\tau = \xi = 0$, the only possible lines of discontinuity are the two characteristics through the origin. Let $[U_s]$ represent the discontinuity in U_s. From equations (22), (24), (26) and (27) it follows that

$$[U_2] = [U_4] = [U_6] = [U_7] = 0$$

in the vicinity of the origin and that

$$\left.\begin{aligned} m_x &= -\delta^2 \\ \dot{W} &= 0 \\ n_x &= -1 \\ n_\theta &= -\nu \end{aligned}\right\} \quad \text{for } \xi = 0, \quad \tau = \epsilon, \quad \epsilon \ll 1 \quad (36)$$

Therefore, the only discontinuities for this problem occur along the line $\xi = \tau$; the solution is known explicitly at the wave front and can be found for the remainder of the domain by integrating equations (28)–(34) numerically.

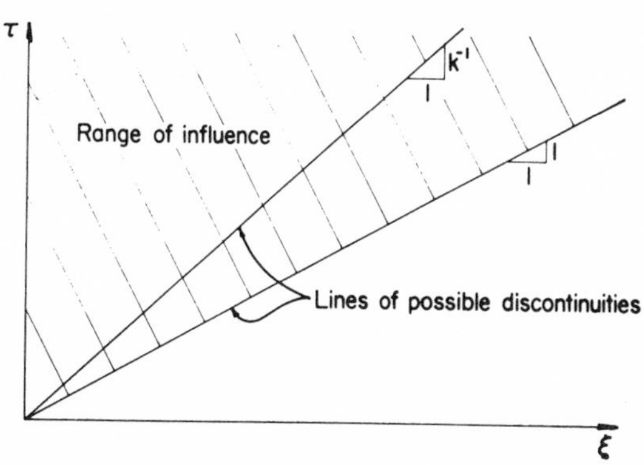

Fig. 3

Viscoelastic Thin Shell

Viscoelasticity can be introduced into the system by replacing the elastic constants in equations (4)–(7) by appropriate differential operators [6]. In general, these operators will be of arbitrary order and the simple procedure used to arrive at a set of first-order equations for the elastic shell will have to be modified. However, if the various viscoelastic models are broken down into springs and dashpots, a system of first-order equations describing these elements results. Of course the number of unknowns increases over the number required for the elastic shell. In order to avoid this problem but still be able to study a shell in which the viscoelasticity is introduced in a realistic manner, it will be assumed here that the material of the shell is incompressible and of a Maxwell type in shear. In this case

$$\mathbf{G} = G\frac{\partial/\partial\tau}{\partial/\partial\tau + \alpha}, \qquad (\alpha = \tau_0^{-1}R/c_p); \ \boldsymbol{\nu} = \tfrac{1}{2}; \ \mathbf{G} = \mathbf{E}/3$$

where τ_0 is the relaxation time associated with the Maxwell element (the bold-face letters indicate differential operators) and the equations of the system, equation (17), remain the same except for the matrix c which becomes

$$\begin{bmatrix} \nu\dot{W} - \alpha n_x \\ \dot{W} - \alpha n_\theta \\ k^2\dot{\psi} - \alpha q_x \\ -\alpha m_x \\ (1 - \delta^2)^{-1}q_x \\ -n_\theta \\ -\delta^{-2}(1 - \delta^2)^{-1}q_x \end{bmatrix} \quad (37)$$

The additional terms to be added to the right-hand sides of equations (28)–(34) are

$-\frac{1}{2}\alpha(U_1 + U_2)$	for equation (28)
$-\frac{1}{2}\alpha(U_1 + U_2)$	for equation (29)
$-\frac{1}{2}\alpha(U_3 + U_4)$	for equation (30)
$-\frac{1}{2}\alpha(U_3 + U_4)$	for equation (31)
$-\frac{1}{2}\alpha(U_5 + U_6)$	for equation (32)
$-\frac{1}{2}\alpha(U_5 + U_6)$	for equation (33)
$-\alpha U_7$	for equation (34)

As in the case of the elastic shell there are two possible velocities for forward-moving waves, c_x and c_s, but the discontinuities are no longer propagated with undiminished jumps.

With regard to more general viscoelastic models, it can be shown that for the model which has an initial elastic response, no new wave velocities are introduced when viscoelasticity is introduced. The viscoelastic operators which represent models with an initial elastic response can be written as the ratio of two polynomials of the same degree in the time operator. It follows that the higher order terms in the single higher order differential equation which corresponds to the first-order system used in the foregoing only differ by a factor of the time operator to some power for various viscoelastic materials. Therefore no new wave velocities are introduced as this kind of viscoelasticity is introduced or varied.

It may also be noted that, since the operator corresponding to a standard solid is the ratio of two linear functions of the time operator, the equations for an incompressible standard solid shell may be obtained from equation (17) simply by adding additional terms to the matrix c over those given in equation (37). Models more complex than a Maxwell model have not been pursued here in any detail since they offer no theoretical difficulties and the additional parameters would tend to obscure possible physical interpretations.

An Initial Value Problem for Viscoelastic Thin Shell

Consider the same problem as previously posed for the elastic shell; again, the first response travels with the velocity c_p and the same lines of possible discontinuity exist. Equation (36) is still valid so that discontinuities only occur along the line $\xi = \tau$. Again the only discontinuous quantities are n_x, n_θ, m_x and $\dot{U}$. From equations (28) and (30) (corrected) it follows that

$$d/ds_1[U_3] = -0.5\alpha[U_3] \tag{38}$$

or that the jumps decay exponentially along the characteristic.

Numerical Examples

The most simple example considered is the membrane shell version of the initial value problem discussed in the foregoing. Fig. 4 shows the form of the longitudinal stress resultant n_x for several values of time. The waveform predicted in [1] can be seen developing. These results were obtained by integrating equations (40)–(43) numerically using a time interval of 0.1 and $\nu = 1/3$. Figs. 5 to 11 show the results of similar calculations for the Timoshenko-type shell with comparisons with the membrane solution. These figures show that motions of a higher frequency than those which occur in the membrane are excited and numerical integration becomes correspondingly more difficult. The integration, which was performed on an IBM 7090, required about 2 min to reach the time $\tau = 2$ and it is estimated that it would require about 45 min to reach the time $\tau = 10$ as was done for the membrane. The time interval used was 0.01 with $\nu = 1/3$, $h/R = 0.1$, and $\kappa^2 = 0.87$. Finally, Fig. 12 shows the effect of viscoelasticity for a typical stress resultant. In this case the time interval was 0.01 with $\nu = 1/2$, $h/R = 0.1$, $\kappa^2 = 0.91$, and $\alpha = 0.2$.

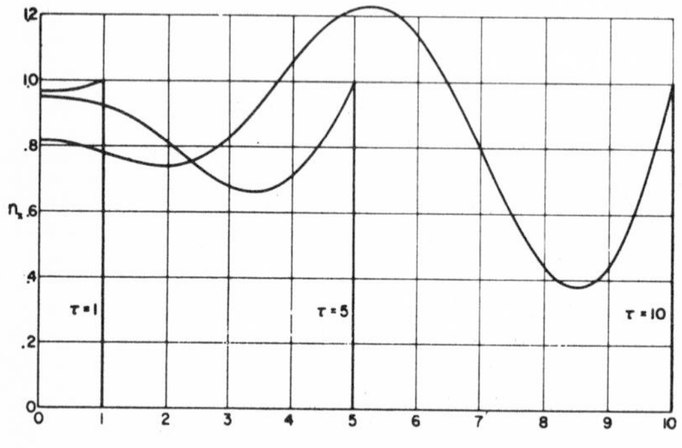

Fig. 4 Longitudinal stress in membrane shell

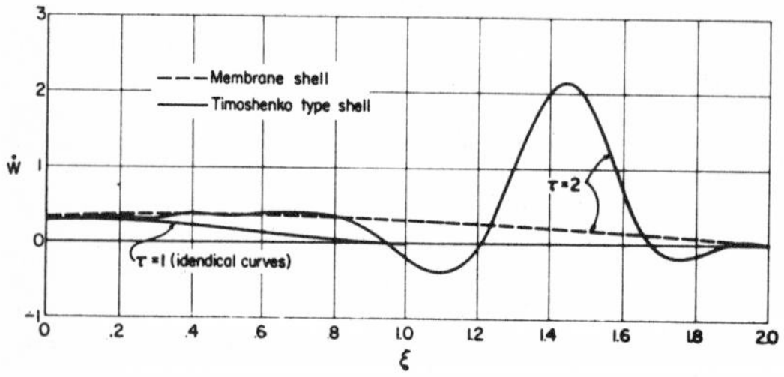

Fig. 5 Radial velocity

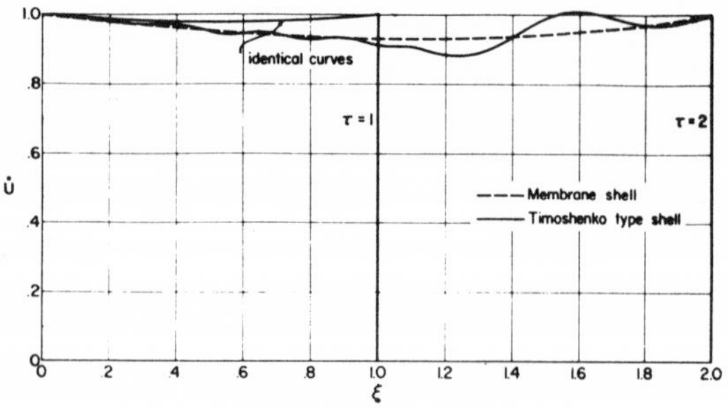

Fig. 6 Longitudinal velocity

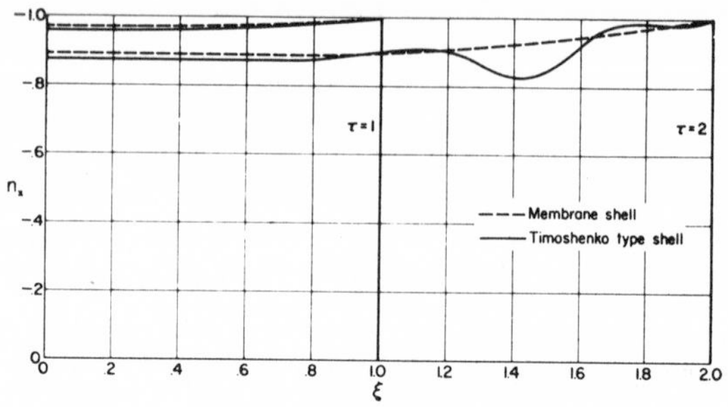

Fig. 7 Longitudinal stress

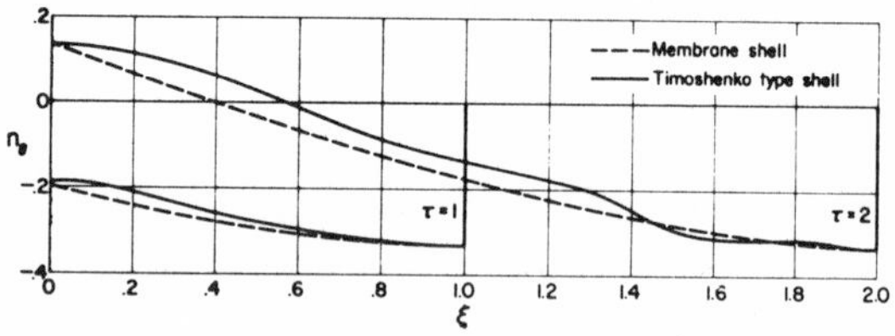

Fig. 8 Circumferential stress

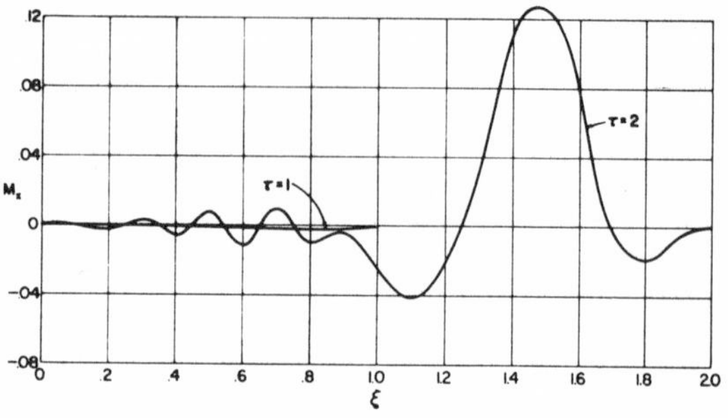

Fig. 9 Moment

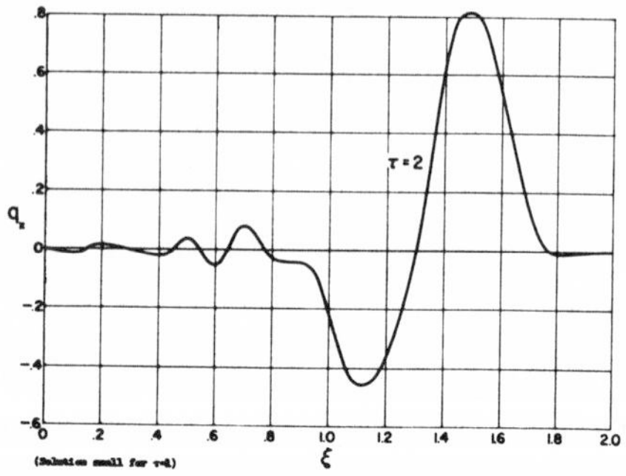

Fig. 10 Shear stress

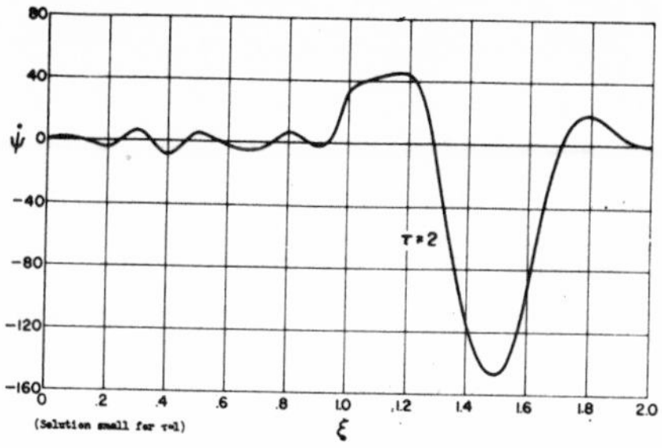

Fig. 11 Velocity of rotation

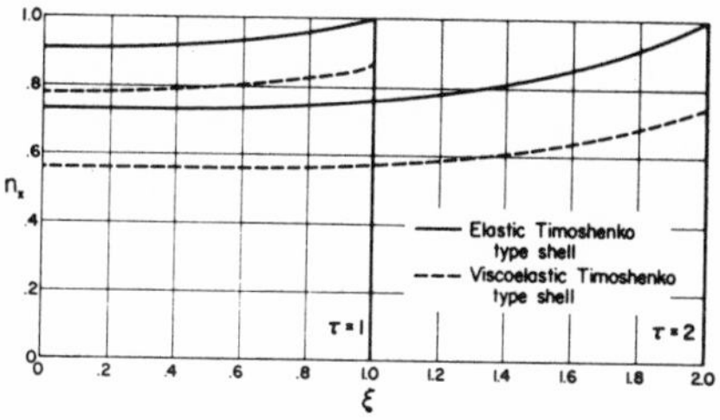

Fig. 12

Discussion of Results

While the equations for the shell allow discontinuous moments, they were not excited in the example considered and therefore the results in the vicinity of the wave front, i.e., the first response felt at any point, were quite similar to those predicted from membrane analysis. Since the initial response is frequently of primary interest, the use of membrane analysis in these problems is reinforced by the work here. However, in areas in which membrane analysis predicts high gradients, for example, areas directly behind the wave front for longer times, it is expected that the two theories will give appreciably different results.

Acknowledgment

The author is indebted to Prof. A. M. Freudenthal under whose direction this research was done. The research was sponsored by the Office of Naval Research, Contract Nonr 266(78).

References

1 Harvey M. Berkowitz, "Longitudinal Impact of a Semi-Infinite Elastic Cylindrical Shell," JOURNAL OF APPLIED MECHANICS, vol. 30, TRANS. ASME, vol. 85, Series E, September, 1963, pp. 347–355.

2 R. B. Testa and H. H. Bleich, "Longitudinal Impact of the Semi-Infinite Cylindrical Viscoelastic Shell," Office of Naval Research, Project NR 064-417, Technical Report 16, Columbia University, June, 1963.

3 R. W. Leonard and B. Budiansky, "On Traveling Waves in Beams," NACA Technical Note 2874, January, 1953.

4 G. Herrmann and I. Mirsky, "Three-Dimensional and Shell-Theory Analysis of Axially Symmetric Motions of Cylinders," JOURNAL OF APPLIED MECHANICS, vol. 23, TRANS. ASME, vol. 78, Series E, 1956, pp. 563–568.

5 R. Courant, *Methods of Mathematical Physics*, vol. 2, Partial Differential Equations, Interscience, 1962, p. 425.

6 A. M. Freudenthal and H. Geiringer, "The Mathematical Theories of the Inelastic Continuum," *Handbuch der Physik*, vol. 6, Springer-Verlag, Berlin, Germany, 1958, p. 695.

APPENDIX

Characteristic Solution for Membrane Shell Theory

In order to facilitate the comparison of the present work to that of [1], the membrane equations corresponding to equations (28)–(34) are given.

If shear and bending effects are neglected, equation (16) reduces to

$$\frac{\partial}{\partial \xi}\begin{bmatrix} 0 & 0 & -1 & 0 \\ 0 & 0 & -\nu & 0 \\ -1 & 0 & 0 & 0 \\ 0 & 0 & 0 & 0 \end{bmatrix}\begin{bmatrix} n_x \\ n_\theta \\ \dot{U} \\ \dot{W} \end{bmatrix} + \frac{\partial}{\partial \tau}\begin{bmatrix} n_x \\ n_\theta \\ \dot{U} \\ \dot{W} \end{bmatrix} = \begin{bmatrix} \nu \dot{W} \\ \dot{W} \\ 0 \\ -n_\theta \end{bmatrix} \quad (39)$$

The normal form of equation (39) is

$$\frac{\partial}{\partial \xi}(n_x - \dot{U}) + \frac{\partial}{\partial \tau}(n_x - \dot{U}) = \nu \dot{W} \quad (40)$$

$$-\frac{\partial}{\partial \xi}(n_x + \dot{U}) + \frac{\partial}{\partial \tau}(n_x + \dot{U}) = \nu \dot{W} \quad (41)$$

$$\frac{\partial}{\partial \tau}(-\nu n_x + n_\theta) = (1 - \nu^2)\dot{W} \quad (42)$$

$$\frac{\partial}{\partial \tau}\dot{W} = -n_\theta \quad (43)$$

From inspection of equations (40)–(43) it is seen that for the membrane there is only one velocity of forward waves, c_p, and that discontinuities are propagated with constant magnitude.

Part V

NONLINEAR VIBRATION

Editor's Comments on Papers 32 Through 37

All of the papers previously displayed and discussed in this volume are based on *linear* theories of behavior. This means, in kinematic terms, that the transverse motion of a beam, plate, or shell has been assumed to be small when compared to the appropriate length dimension of these structural elements. This in turn means that in a free-vibration analysis, the frequency and the amplitude of any motion cannot be linked to each other, since a linear eigenvalue analysis produces no information on the frequency–amplitude relation. It is true, of course, that relative modal amplitudes can be connected with particular frequencies, a fact that has been exploited particularly well in some shell analyses (see Paper 19); however, no absolute magnitude can be assigned to an amplitude in a linear problem, other than by applying some arbitrary normalization condition.

The burden of the papers presented in this section, then, can be viewed in one of two ways. That is, we can seek solutions to problems where we anticipate that the deflections will be of sufficient magnitude that some kinematic nonlinearity will enter the problem. This would appear to be an approach that is very physical in its motivation. Somewhat more mathematically, we could inquire as to the existence of amplitude–frequency relations, and their meaning. In any event, we are led to consider higher-order terms, their complications, and their implications for an understanding of the vibrations of structures. And as we shall see in the last paper to be presented here, there is a very strong analogy between linear and nonlinear vibrations and between the buckling and postbuckling of elastic structures.

Carrier's paper, generally accepted as the first contribution to the theory of nonlinear vibrations of elastic structures, is concerned with the motion of strings whose tensions vary significantly from their initial values. Although the motivation for the paper is described in physical terms, the analysis itself is a very neat piece of applied mathematics. The heart of the paper lies in the perturbation expansions introduced in equations (5), and in the following discussion. Here we see immediately that physical reasoning has dictated the (mathematical) form of the expansions, and once this is done "it is now a simple matter" to complete the problem of determining the perturbation coefficients. However, Carrier also carried the analysis a step further and showed that the results could be regrouped in a different form, so as to be presentable in terms of elliptic functions [equations (15)]. The numerical results portray very graphically the dependence of the period on the change in tension (Figure 2), which is of course a direct analogue of a frequency–amplitude plot.

Chu and Herrmann, in a well-known paper, make use of the same type of perturbation procedure to analyze the large deflections of a vibrating plate, as did Eringen for a beam and a circular membrane (see Refs. 4 and 5 of Paper 33). In addition to showing that for plates, too, there is a significant relation between the fundamental period of motion and the amplitude of vibration, they also indicated [equations (33) through (38)] how energy principles could be effectively applied to a nonlinear vibration problem. As a final point, Chu and Herrmann considered the special case of a beam, and showed that they could replicate the result obtained from their perturbation analysis by using a simpler method due to Duffing. Inasmuch as the perturbation analysis led to elliptic functions, as in Carrier's work, and Duffing's method did not, this was a happy conclusion.

Now, at about the same time during which the above work was being completed, Eric Reissner completed the first investigation into the nonlinear vibrations of shells, in particular, circular cylindrical shells (E. Reissner, "Non-linear effects in vibrations of cylindrical shells," *The Ramo-Woolridge Corporation* Report AM 5-6, September 1955). As was noted in a recent survey paper by Evensen [D. A. Evensen, "Nonlinear vibrations of circular cylindrical shells," in Y. C. Fung and E. E. Sechler (eds.), *Thin-Shell Structures,* Englewood Cliffs, N.J.: Prentice-Hall, pp. 133–155, 1974], Reissner set the pattern for much of the nonlinear analysis of shells that would be done over the next two decades. Although not all the details were precisely correct, Reissner was the first to use shallow shell theory and to neglect the in-plane inertia terms. Further, he initiated in this context the use of the Lindstedt perturbation technique (R. E. Bellman, *Perturbation Techniques in Mathematics, Physics, and Engineering,* New York: Holt, Rinehart and Winston, pp. 57–58, 1964) to eliminate secular terms so as to obtain a periodic solution. It is worth noting, as a side point, that the Lindstedt perturbation scheme can be viewed as an extended, formal version of the method of Duffing used by Chu and Herrmann. It is unfortunate that, due to length limitations, we are able to reproduce in our collection neither Reissner's report nor Evensen's comprehensive survey.

Nowinski's paper represents a contribution in that it extended some of Reissner's work to include the effects of orthotropy and, more importantly, he used a mode shape for the radial deflection that, unlike Reissner's assumption of the checkerboard pattern, allowed the circumferential displacement to be periodic around the shell circumference. One significant difference in the outcome that resulted from this analysis was that the nonlinear spring effect turned out to always be that of a hardening spring [see equations (4.3) and (4.4)]. A particular set of Nowinski's results had been anticipated, using somewhat different methodology, by Chu [H.-N. Chu, *J. Aerospace Scis.,* **28,** 602–609 (1961)].

Evensen's paper on ring vibrations is of particular interest because it includes the results of the first experiments on nonlinear vibrations of shells or rings. Further, the experiments revealed a definite nonlinear softening behavior, which stood in distinct contrast to the works of Chu and Nowinski. A companion analysis showed that an axisymmetric mode had to be included in the displacement assumption [equation (8) of Evensen's paper] to reproduce the softening behavior.

Another result that Evensen uncovered was the presence, in his experiments, of traveling wave behavior in the ring oscillations. In order to account for this phenomenon, a companion mode was also included in the analysis. Even for the case where a forcing function distributed in a simple checkerboard fashion was included, both the axisymmetric and companion modes were required to reproduce the experimental results through an analysis.

In the decade that followed Evensen's paper, a number of investigations were conducted along the lines laid out by Reissner and Evensen. These investigations include the work of Mayers and Wrenn, who used a more elaborate shell theory to investigate transient nonperiodic response, and that of Leissa and Kadi, who examined the nonlinear response of shallow shell segments of varying curvatures. (These last two works are, respectively, Refs. 2 and 1 of the paper by El-Zaouk and Dym.) There were also a number of related investigations of nonlinear flutter and aeroelasticity problems during that time; the reader should refer to Evensen's survey for further details.

The paper by El-Zaouk and Dym breaks new ground in the nonlinear vibration of shells in that it includes the effects of orthotropy, varying curvature, and internal pressurization for complete shells. The analysis does account for circumferential continuity, and it goes on to tentatively suggest a new criterion for the jumps that are typical of nonlinear vibrational behavior. Of particular interest is the effect of the varying curvature ratio (Figure 9) and the effects of internal pressurization and orthotropy (Figures 10 and 11).

The final paper presented in this section represents something of a departure from the corpus of work discussed above. Rehfield has successfully transplanted the asymptotic analysis of postbuckling behavior developed by Koiter to the nonlinear vibration problem. Just as the linear eigenvalue problems of free vibrations and of linearized buckling are known to be equivalent in some sense, we can now infer a strong relation between the nonlinear vibration problem and postbuckling analysis. Rehfield has applied his asymptotic analysis to two problems whose results are well known, and has achieved the appropriate asymptotic frequency–amplitude relation in both cases. Thus, he demonstrates that the initial nonlinear hardening or softening behavior is amenable to a systematic asymptotic analysis.

32

Reprinted from *Quar. Appl. Math.*, **3**(2), 157–165 (1945)

ON THE NON-LINEAR VIBRATION PROBLEM OF THE ELASTIC STRING*

BY

G. F. CARRIER

Harvard University

1. Introduction. It is well known that the classical linearized analysis of the vibrating string can lead to results which are reasonably accurate only when the minimum (rest position) tension and the displacements are of such magnitude that the relative change in tension during the motion is small. The following analysis of the free vibrations of the string with fixed ends leads to a solution of the problem which adequately describes those motions for which the changes in tension are not small. The perturbation method is adopted, using as a parameter a quantity which is essentially the amplitude of the motion. The periodic motions arising from initial sinusoidal deformations are closely approximated in closed form. The method is applied to motions not restricted to a single plane and finally the exact solution for the transmission of a localized deformation is indicated.

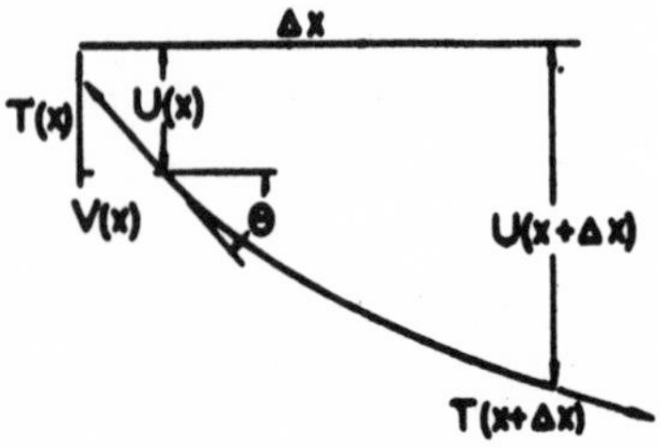

FIG. 1. Displaced element of string.

2. The equations of motion. The equations of dynamic equilibrium of an element of the string, deformed into a plane curve as shown in Fig. 1, are

$$\frac{\partial}{\partial x}[T\sin\theta] = \rho A\frac{\partial^2 u}{\partial t^2}, \qquad \frac{\partial}{\partial x}[T\cos\theta] = \rho A\frac{\partial^2 v}{\partial t^2}, \tag{1}$$

where ρ denotes the mass per unit volume, A the cross-sectional area of the string in the rest position, and $\theta = \arctan[u'/(1+v')]$, the primes indicating differentiation with respect to x. The condition of fixed ends implies that,

$$\int_0^l v'dx = 0 \quad \text{for all } t. \tag{2}$$

The stress-strain relation of the string is assumed in the form,

$$T - T_0 = EA\{[(1+v')^2 + (u')^2]^{1/2} - 1\}, \tag{3}$$

where T_0 is the tension in the rest position and E is a constant characteristic of the string material. The following dimensionless quantities are introduced to simplify the algebraic work

$$\alpha^2 = \frac{T_0}{EA}, \qquad \tau = \frac{T - T_0}{T_0}, \qquad \xi = \frac{\pi x}{l}, \qquad \eta = \frac{\pi}{l}\left(\frac{T_0}{\rho A}\right)^{1/2} t.$$

After differentiating Eqs. (1) with respect to x, setting

* Received Jan. 3, 1945.

$$\sin\theta = \frac{u'}{[(u')^2 + (1 + v')^2]^{1/2}} = \frac{u'}{1 + \alpha^2\tau} = \varphi,$$
$$\cos\theta = (1 - \varphi^2)^{1/2},$$

and eliminating v' between Eqs. (1) and (2), we obtain

$$\frac{\partial^2}{\partial\xi^2}\left[(1 + \tau)\varphi\right] = \frac{\partial^2}{\partial\eta^2}\left[(1 + \alpha^2\tau)\varphi\right], \tag{4a}$$

$$\frac{\partial^2}{\partial\xi^2}\left[(1 + \tau)(1 - \varphi^2)^{1/2}\right] = \frac{\partial^2}{\partial\eta^2}\left[(1 + \alpha^2\tau)(1 - \varphi^2)^{1/2}\right], \tag{4b}$$

$$\int_0^\pi (1 + \alpha^2\tau)(1 - \varphi^2)^{1/2}d\xi = \pi. \tag{4c}$$

These equations rigorously define the motion of the string which is acted on by on external forces.

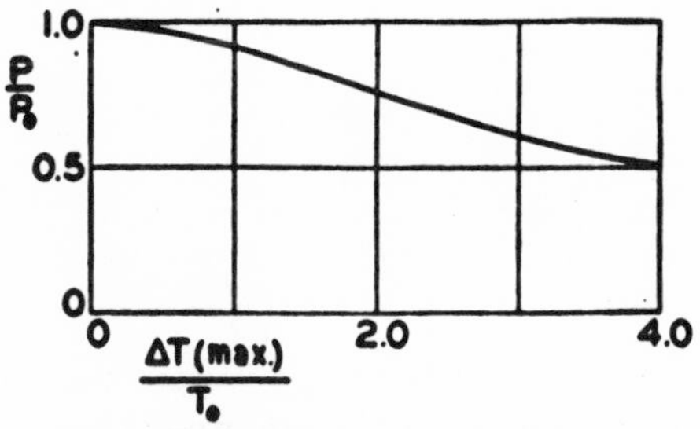

FIG. 2. Comparison of periods obtained by linear and non-linear theories.

$$\frac{P}{P_0} = \frac{\text{non-linear period}}{\text{linear period}};$$

$$\frac{\Delta T_{max}}{T_0} = \frac{\epsilon^2}{4} = \left(\frac{\text{"amplitude"}}{2\alpha}\right)^2.$$

Motion defined by Eqs. (15). $\alpha \neq 0$.

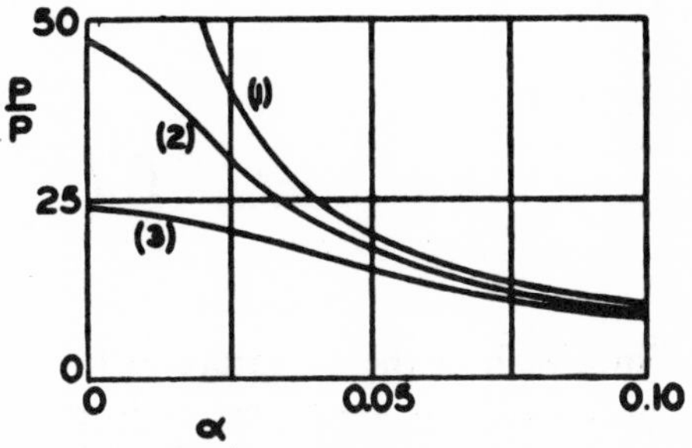

FIG. 3. Period v.s. initial tension.*

$$\frac{P}{P^*} = \frac{\text{non-linear period}}{(E\pi^2/\rho l^2)^{1/2}}; \quad \alpha^2 = T_0/EA$$

(1) vanishing amplitude (linear theory)
(2) $a = \alpha\epsilon =$ "amplitude" $= 0.05$
(3) $a = 0.10$

3. **The perturbation procedure.** It is convenient to choose, as the perturbation parameter of the problem, a number ϵ which is essentially the amplitude of the motion.** The two functions φ and τ are therefore expanded in powers of this parameter as follows:

$$\varphi = \alpha[\epsilon\varphi_1 + \epsilon^3\varphi_3 + \epsilon^5\varphi + \cdots], \qquad \tau = \epsilon^2\tau_2 + \epsilon^4\tau_4 + \cdots. \tag{5}$$

It is easily seen that a reversal of the sign of ϵ should merely reverse the sign of φ. Hence the omission of the even powers of ϵ is justified. In a similar manner the funtions τ_1, τ_3, $\cdots$ can be seen to vanish. That τ_0 vanishes is seen by inspection of Eq. (4c). The expressions for φ and τ are now substituted into Eqs. (4), the coeffi-

* In Fig 3. the ordinates should be labeled P/P^*.

** Equation (15) indicates more precisely the meaning of ϵ.

cients of each power of ϵ are equated to zero, and the following system of equations is obtained:

$$L_0(\varphi_1) = 0, \quad (6a), \qquad L_1(\tau_2) = -\alpha^2 L_0\left(\frac{\varphi_1^2}{2}\right), \quad (6b)$$

$$L_0(\varphi_3) = L_1(\tau_2\varphi_1), \quad (6c), \qquad L_1(\tau_4) = -\alpha^2 L_0\left(\varphi_1\varphi_3 - \frac{\alpha^2\varphi_1^4}{8}\right), \quad (6d)$$

$$L_0(\varphi_5) = L_1(\tau_4\varphi_1 + \tau_2\varphi_3), \quad (6e), \qquad L_1(\tau_6) = -\alpha^2 L_0\left(\varphi_1\varphi_5 + \frac{\varphi_3^2}{2} + \alpha^2\frac{\varphi_1\varphi_3^3}{2} + \alpha^4\frac{\varphi_1^6}{48}\right), \quad (6f)$$

. . . . ,

where

$$L_0 = \frac{\partial^2}{\partial\xi^2} - \frac{\partial^2}{\partial\eta^2}, \qquad L_1 = \alpha^2\frac{\partial^2}{\partial\eta^2} - \frac{\partial^2}{\partial\xi^2},$$

and

$$\int_0^\pi \left(\tau_2 - \frac{\varphi_1^2}{2}\right) d\xi = 0, \quad (7a)$$

$$\int_0^\pi \left(\tau_4 - \varphi_1\varphi_3 + \frac{\alpha^2\varphi_1^4}{8}\right) d\xi = 0, \quad (7b)$$

. . . .

Since each of the operators in the foregoing equations is linear, it is now a simple matter to evaluate successively the φ_i and the τ_i. For the moment, we confine our attention to the motion defined by choosing as a solution to Eq. (6a) the function,

$$\varphi_1 = \cos\xi\cos\eta. \quad (8a)$$

Note that for $\varphi_1 = \cos n\xi \cos n\eta$ the same solution will exist when l is replaced by l/n in the definitions of ξ and η. Solving successively Eqs. (6), starting with the foregoing definition of φ_1, and using Eqs. (7) to determine the arbitrary terms appearing in the τ_i, we obtain

$$\tau_2 = \frac{1}{4}\cos^2\eta + \frac{\alpha^2}{8}\cos 2\xi, \quad (8b)$$

$$\varphi_3 = \cos\xi\left[-\frac{3 - 2\alpha^2 - \alpha^4}{32}\eta\sin\eta + \frac{1 - 9\alpha^2}{128}\cos 3\eta - \frac{1}{128}\cos\eta\right] - \frac{\alpha^2(9 - \alpha^2)}{128}\cos 3\xi\cos\eta, \quad (8c)$$

$$\tau_4 = \left[-\frac{3 - 2\alpha^2 - \alpha^4}{128}\eta\sin 2\eta + \frac{1 - 9\alpha^2}{512}(\cos 2\eta + \cos 4\eta) - \frac{3\alpha^2}{512}\cos^4\eta - \frac{1}{256}\cos^2\eta\right] + \frac{\alpha^4(21 - \alpha^2)}{512}\cos 2\xi + \cdots - \frac{3\alpha^2}{2048}\frac{13 - \alpha^2}{4 - \alpha^2}\cos 4\xi\cos 2\eta, \quad (8d)$$

$$\varphi_5 = \cos\xi\left[-\frac{9(1+\alpha^2+\cdots)}{2048}\eta^2\cos\eta - \frac{3}{512}\eta\sin\eta - \frac{9}{4096}\eta\sin 3\eta + 2^{-14}\cos 5\eta\right] + \cdots, \tag{8e}$$

. . . .

The arbitrary solutions of Eq. (6a) which may be added to each of the φ_i as they are evaluated have been chosen in such a manner that lim $\alpha\epsilon^i\varphi_i$ exists when α tends to zero and $\alpha\epsilon = \text{constant} = \alpha$.* This limiting process, of course, defines the motion wherein the initial tension T_0 is zero and the amplitude a is non-vanishing. An investigation of this problem will simplify the question of the convergence of the functions φ and τ as defined by Eqs. (5) and (8). When α tends to zero as specified above, the symbols τ and η become meaningless. Hence, we replace them by

$$\sigma = \frac{T}{EA} = \alpha^2\tau, \quad \text{and} \quad \eta = \alpha s.$$

The limiting process then yields the following expressions for the φ_j and the σ_j

$$\left.\begin{aligned} \alpha\epsilon\varphi_1 &= a\cos\xi, \\ \alpha\epsilon^3\varphi_3 &= a^3\left[-\frac{1}{2!}\left(\frac{s}{2}\right)^2\cos\xi + \frac{9}{128}\cos\xi - \frac{9}{128}\cos 3\xi\right], \\ \alpha\epsilon^5\varphi_5 &= a^5\left[\frac{3}{4!}\left(\frac{s}{2}\right)^4\cos\xi + \frac{5}{2^8}\left(\frac{s}{2}\right)^2\cos\xi - \frac{45}{128}\left(\frac{s}{2}\right)^2\cos 3\xi + f(\xi)\right], \\ \alpha\epsilon^7\varphi_7 &= a^7\left[-\frac{27}{6!}\left(\frac{s}{2}\right)^6\cos\xi + \cdots\right], \\ &\cdots\cdots\cdots\cdots , \end{aligned}\right\} \tag{9}$$

$$\left.\begin{aligned} \epsilon^2\sigma_2 &= \frac{1}{4}a^2, \\ \epsilon^4\sigma_4 &= a^4\left[-\frac{1}{16}s^2 + \frac{\cos 2\xi}{32} + \frac{37}{256}\right], \\ \epsilon^6\sigma_6 &= a^6\left[\frac{s^4}{128} - \frac{13s^2}{512} + \frac{3s^2}{128}\cos 2\xi\right] \\ &\quad + \left[\sigma_4\frac{\varphi_1^2}{2} + \sigma_2\varphi_1\varphi_3 + \sigma_2\frac{\varphi_1^4}{8}\right]\alpha^2\epsilon^6 + g(\xi), \\ \epsilon^8\sigma_8 &= a^8\left[-\frac{s^6}{1280} + \cdots\right], \\ &\cdots\cdots\cdots\cdots . \end{aligned}\right\} \tag{10}$$

These solutions may also be obtained, of course, by assuming α equal to zero at

* Such complementary solutions are usually chosen to be consistent with a given set of initial conditions. However, it is convenient here to choose them so that the solution does not become meaningless when $\alpha\to 0$, $\alpha\epsilon = a$. Equations (15) indicate that this choice leads to a solution corresponding to a nearly sinusoidal initial deformation.

the outset, expanding φ and σ in powers of a parameter a, and proceeding in the foregoing manner.

Note that the leading terms of the φ_j define an absolutely converging series for all a. Note also that the remaining terms of each φ_j are dominated by this leading term. In fact, for sufficiently large s, the sum of the remaining terms in each φ_j is as small as we please compared to this leading term. Although this dominance has not been shown to occur uniformly, it is to be expected that the series defined by Eqs. (9) and (10) will converge over some range of a. The requirement, "sufficiently large s" introduces no difficulty since the initial value of s may be chosen arbitrarily large.

The functions φ and σ are now most conveniently written in the forms

$$\begin{aligned} \varphi(\xi, s, a) &= af_1(as, \xi) + a^3 f_3(as, \xi) + \cdots, \\ \sigma(\xi, s, a) &= a^2 g_2(as, \xi) + a^4 g_4(as, \xi) + \cdots, \end{aligned} \tag{11}$$

where the terms of the series defining the f_j and the g_j are easily chosen from Eqs. (9) and (10). f_1 and g_2 are composed of the previously mentioned leading terms, and it is easily established that they converge to the values

$$f_1 = \text{cn}\left(\frac{as}{2}, \frac{1}{\sqrt{2}}\right)\cos\xi, \qquad g_2 = \frac{1}{4}\,\text{cn}^2\left(\frac{as}{2}, \frac{1}{\sqrt{2}}\right), \tag{12}$$

where cn denotes the elliptic cosine. Energy considerations may be used to show that the remaining f_j and g_j are bounded, and it is to be expected that the motion is closely described by $\varphi = af_1$ and $\sigma = a^2 g_2$ when a is sufficiently small. For most materials, a value of a^2 greatly in excess of 10^{-3} will lead to plastic deformations; hence, the motion of such strings is well defined.

The motions arising when T_0 is arbitrary, as defined by Eqs. (5) and (8), can also be written in the form,

$$\begin{aligned} \varphi &= \alpha\epsilon F_1(\xi, \eta, \epsilon) + \alpha^3\epsilon^3 F_3(\xi, \eta, \epsilon) + \cdots + P(\xi, \eta, \alpha, \epsilon), \\ \tau &= \epsilon^2[G_2(\xi, \eta, \epsilon) + \alpha^2\epsilon^2 G_4(\xi, \eta, \epsilon + \cdots] + Q(\xi, \eta, \alpha, \epsilon), \end{aligned} \tag{13}$$

where P and Q are those parts of φ and τ which vanish when α tends to zero and $\alpha\epsilon = a$. For this case,

$$F_1 = \text{cn}\left(\sqrt{1 + \frac{\epsilon^2}{4}}\,\eta, k\right)\cos\xi, \qquad G_2 = \frac{1}{4}\,\text{cn}\left(\sqrt{1 + \frac{\epsilon^2}{4}}\,\eta, k\right), \tag{14}$$

where $k = \epsilon[2(4+\epsilon^2)]^{-1/2}$. It is evident, in view of the foregoing results, that

$$\lim_{\epsilon \to \infty} F_j(\xi, \eta, \epsilon) = f_j(\xi, as)$$

and it is to be concluded that since the series defining the F_i converge as ϵ tends to infinity, they will also converge for the smaller values of ϵ. Both α and $\alpha\epsilon$ must be small because of elastic considerations, which indicates that P and Q will also exist. We conclude therefore that the motion of the string, whose "amplitude" $\alpha\epsilon$ is of the order of magnitude required by elastic considerations, is adequately defined by the leading terms of Eq. (13). That is, in the first approximation,

$$\varphi = \alpha\epsilon \, \mathrm{cn}\left[\sqrt{1+\frac{\epsilon}{4}}\,\eta, k\right]\cos\xi$$
$$\tau = \frac{\epsilon^2}{4}\,\mathrm{cn}^2\left[\sqrt{1+\frac{\epsilon^2}{4}}\,\eta, k\right]. \tag{15}$$

Figs. 2 and 3 compare the results of this analysis with those of the linear theory.

4. The motion following an arbitrary initial deformation. The motions derived in the preceding section are obviously those corresponding to initial sinusoidal deformations. If the perturbation procedure is again carried out, and if for φ_1 the function $\varphi_1 = \sum_j b_j \cos j\xi \cos j\eta$ is selected, a solution will be obtained, the leading terms of which contain no powers of α greater than unity. The solution so obtained will correspond to an initial deformation, $\varphi_1(\xi, 0) = \sum_j b_j \cos j\xi$. This predominating part of the solution may, however, be obtained by a simpler, less rigorous, procedure which nevertheless leads to identical results. We merely expand $(1-\varphi^2)^{1/2}$ in the conventional power series and omit in Eqs. (4), φ^{n+2} as compared to φ^n, and α^2 as compared to 1. We thus obtain as replacement for Eqs. (4)

$$\left.\begin{aligned}
&\frac{\partial^2}{\partial\xi^2}\left[(1+\tau)\varphi\right] = \frac{\partial^2\varphi}{\partial\eta^2},\\
&\frac{\partial^2\tau}{\partial\xi^2} = 0, \quad \text{hence} \quad \tau = \tau(\eta),\\
&\int_0^{\pi}\left(\alpha^2\tau - \frac{\varphi^2}{2}\right)d\xi = 0.
\end{aligned}\right\} \tag{16}$$

Finally the first of these becomes

$$\left[1 + \frac{\alpha^{-2}}{2\pi}\int_0^{\pi}\varphi^2(\xi, \eta)d\xi\right]\frac{\partial^2\varphi}{\partial\xi^2} = \frac{\partial^2\varphi}{\partial\eta^2}\cdot \tag{17}$$

The solution corresponding to the initial conditions specified at the beginning of this section is found by considering that solution of the form $\varphi = \alpha\sum_j b_j \cos j\xi\psi_j(\eta)$, where $\psi_j(0) = 1$ for each j.

Upon substitution of this function, Eq. (17) yields the following set of ordinary differential equations

$$\psi_n'' + n^2\psi_n\left[1 + \frac{1}{4}\sum_j b_j^2\psi_j^2\right] = 0. \tag{18}$$

These may be written in the conventional operational form

$$(D^2 + n^2)\psi_n = -\frac{n^2}{4}\psi_n\sum_j b_j^2\psi_j^2, \tag{19}$$

and standard integration procedure leads immediately to the integral equation

$$\psi_n(\eta) = \cos n\eta - \frac{n}{4}\int_0^{\eta}\sin n(s-\eta)\psi_n(s)\sum_j b_j^2\psi_j^2(s)ds. \tag{20}$$

The method of successive approximations when applied to this equation will produce a converging sequence of solutions. This method is obviously preferable to the direct application of the perturbation method, once the equivalence of the results has been established, since no minor terms are carried along in the algebraic work, no complimentary solutions need be added as the integration proceeds,* and the τ_i do not appear when the function φ is evaluated.

It is of interest to note that when $b_j = 0$ for $j \neq 1$, Eq. (20) assumes the form

$$\psi_1 = \cos \eta - \frac{b_1^2}{4} \int_0^\eta \sin (z - \eta) \psi_1^3(z) dz, \tag{21}$$

and that this equation must generate the elliptic function previously encountered. When the method of successive approximations is applied to this equation, the series obtained is that one found in the first solution obtained in this paper. This function may be obtained more directly by solving Eq. (18) for this particular set of initial conditions.

Perhaps the quickest way to obtain an approximation to the motion for non-sinusoidal initial deformation is to be found in the application of a numerical procedure using finite differences. Equation (17) lends itself readily to such a treatment and the results are considerably easier to interpret than those found by the more rigorous integral equation treatment.

5. The three dimensional problem. If we now allow deflections w normal to the plane of u, the procedure of the foregoing sections of this paper leads to the equation

$$\left\{1 + \frac{\alpha^{-2}}{2\pi} \int_0^\pi [\varphi^2(\xi, \eta) + \chi^2(\xi, \eta)] d\xi \right\} \frac{\partial^2 \varphi}{\partial \xi^2} = \frac{\partial^2 \varphi}{\partial \eta^2} \tag{22}$$

and to the equation obtained by interchanging φ and χ in (22). τ is given by the integral on the left side of this equation and $\chi = w/(1+\alpha^2\tau)$. It follows immediately from the similarity of Eq. (17) and that given above that the integral equation method previously described will provide the solutions to problems of this nature. In particular, however, the motions wherein the string at any instant lies in a single plane and wherein each particle describes a quasi-elliptical path is easily determined in closed form by considering the deformation expressed in the complex form

$$\varphi = \epsilon\alpha\, \psi(\eta) e^{i\mu(\eta)} \cos \xi,$$

where ψ and μ are each real. Equation (22) assumes the form,

$$\left[1 + \frac{\epsilon^2}{2} \int_0^\pi | \varphi(\xi, \eta) |^2 d\xi \right] \frac{\partial^2 \varphi}{\partial \xi^2} = \frac{\partial^2 \varphi}{\partial \eta^2} \tag{23}$$

which, when separated into its real and imaginary parts, implies,

$$\mu'(\eta) = c/\psi^2(\eta)$$

* When dealing with the differential equations leading to Eq. (8), it was necessary to choose complementary solutions to conform to given initial (or other auxiliary) conditions of the problem. In the integral equation approach, such conditions are always included in the equations.

and

$$\psi'' + \psi + \frac{\epsilon^2}{4}\psi^3 - c^2\psi^{-3} = 0. \tag{24}$$

Here, c is a constant defined by the initial conditions as follows;

$$\psi(0) = 1, \quad \psi'(0) = 0, \quad \mu(0) = 0, \quad \mu'(0) = c.$$

When $c<1+\epsilon^2/4$, these initial conditions lead to a solution of Eq. (24) given by

$$\psi = \left[1 - (1-\gamma)\,\mathrm{sn}^2\left\{\sqrt{\frac{1+\beta}{8}}\,\epsilon\eta,\ \sqrt{\frac{1-\gamma}{1+\beta}}\right\}\right]^{1/2} \tag{25}$$

$$\mu = c\int_0^\eta \psi^{-2}(s)ds, \qquad \tau = \frac{\epsilon^2}{4}\psi^2,$$

where

$$\beta = \frac{1}{2\epsilon^2}\left[\sqrt{(8+\epsilon^2)^2 + 32\epsilon^2c^2} + (8+\epsilon^2)\right],$$

$$\gamma = \frac{1}{2\epsilon^2}\left[\sqrt{(8+\epsilon^2)^2 + 32\epsilon^2c^2} - (8+\epsilon^2)\right].$$

Note that as c tends to $1+\epsilon^2/4$, ψ becomes identically unity and the motion of each particle is circular. That is,

$$\varphi = \alpha\epsilon\cos\xi e^{i\tau\sqrt{1+\epsilon^2/4}}. \tag{26}$$

When $c>1+\epsilon^2/4$, integration of Eq. (24) yields,

$$\psi^2 - 1 = \frac{(\beta+1)(\gamma-1)Z^2}{\gamma+\beta-(\gamma-1)Z^2}, \tag{27}$$

where

$$Z = \mathrm{sn}\left[\sqrt{\frac{\gamma+\beta}{8}}\,\epsilon\eta,\ \sqrt{\frac{\gamma-1}{\gamma+\beta}}\right].$$

It is interesting to observe that the string never passes through its rest position for values of c different from zero. This follows from the fact that ψ never vanishes.

The function which rigorously defines the transmission of a localized disturbance along the string is easily found by considering those solutions of Eqs. (4) which allow the function τ to assume a constant value. Equations (4) become, under this assumption,

$$p^2\frac{\partial^2 u'}{\partial\xi^2} = \frac{\partial^2 u'}{\partial\eta^2}, \qquad p^2\frac{\partial^2 v'}{\partial\xi^2} = \frac{\partial^2 v'}{\partial\eta^2}, \qquad \int_0^\pi [(1+\alpha^2\tau)^2 - (u')^2]^{1/2}d\xi = \pi, \tag{28}$$

where

$$p^2 = \frac{1+\tau}{1+\alpha^2\tau}.$$

If we now choose

$$u' = f(\xi - p\eta), \qquad v' = \{(1 + \alpha^2\tau)^2 - (u')^2\}^{1/2} - 1, \tag{29}$$

where τ is determined by

$$\frac{1}{\pi}\int_0^\pi \{[1 + \alpha^2\tau]^2 - |u'(\xi, 0)|^2\}^{1/2} d\xi = 1,$$

and where $f(\xi)$ is non-vanishing in a small region in ξ, all equations are satisfied. This solution is valid until the deformation reaches a fixed point in the string. When this occurs, the reflection phenomenon requires a change in τ. This solution is in agreement with that found by the linear theory except that p would assume the value unity in that theory.

33

Reprinted from *J. Appl. Mech.*, **23**, 532–540 (1956)

Influence of Large Amplitudes on Free Flexural Vibrations of Rectangular Elastic Plates[1]

By HU-NAN CHU[2] and GEORGE HERRMANN,[3] NEW YORK, N. Y.

In a recent paper (1)[4] a set of plate equations was derived, which governs motions with small elongations and shears, but moderately large rotations, valid for an isotropic material obeying Hooke's law. The resulting theory, which may be considered the dynamic analog of the von Karman plate theory, is applied presently to the study of free vibrations of a rectangular, elastic plate with hinged, immovable edges. The nonlinear equations are solved approximately by employing a perturbation procedure and also the principle of conservation of energy directly. The influence of large amplitudes on the period of free vibration and on the maximum normal stress is established. The free vibrations of a beam are studied as a special case and the resulting period compared with a previous investigation.

Nomenclature

The following nomenclature is used in the paper:

x, y, z = original co-ordinates of a particle (before deformation)
u, v, w = displacement components in the x, y, z-direction, respectively
u_0, v_0, w_0 = plate-displacement components (displacement of particle on middle surface of plate)
$\bar{u}, \bar{v}, \bar{w}$ = dimensionless displacement components defined by Equation [7]
h = plate thickness, beam depth
a, b = plate width and length, respectively
ξ, η = dimensionless space variables defined by Equation [6]
τ, ζ = dimensionless time defined by Equations [6] and [9], respectively
δ = perturbation parameter defined by Equation [5]
r = aspect ratio, $r = a/b$
ρ = density of plate material
ν = Poisson's ratio
c_p = phase velocity of compressional waves in a plate, $c_p = \dfrac{E}{\rho(1-\nu^2)}$
c_b = phase velocity of compressional waves in a beam, $c_b = \dfrac{E}{\rho}$
N_1, N_2, N_{12} = membrane forces
M_1, M_2, M_{12} = plate moments
D = flexural rigidity of plate
$u^{(2)}, v^{(2)}, w^{(1)}$ = lowest order perturbation components of displacement
$\bar{F}$ = stress function defined by Equation [15]
$\bar{H}(\zeta), H(t)$ = time functions
$\bar{T}$ = kinetic energy
$\bar{W}$ = strain energy
i, j, m, n = indices of summation
σ_{1m} = maximum membrane stress
σ_{1b} = maximum bending stress
B = amplitude of vibration
$\beta = B/h$ = dimensionless amplitude of vibration
Z = time function, of dimensionless time ζ
S = dimensionless membrane stress defined by Equation [29]
R = $h/\sqrt{12}$
E = Young's modulus
K = complete elliptic integral of first kind
T^* = nonlinear period
ω^* = nonlinear frequency
T = linear period
ω = linear frequency

Introduction

The linear theories of motion of bodies, one or two of whose dimensions are small as compared to the third (beams, plates, and so on) are based, among others, on the assumption that the deflections are small in comparison with the beam depth or plate thickness. In most practical cases this implies in turn that the slopes (or rotations of the cross section) are small as compared to unity.

In a variety of circumstances of practical importance, however, motions may be excited, where this basic assumption is no longer valid; that is, the deflections have the order of magnitude of the thickness. It is of interest, therefore, to establish the influence of such large amplitudes on the salient parameters of the problem; e.g., the frequency of free vibrations.

Free vibrations with large amplitudes of elastic bodies, one or two of whose dimensions are small as compared to the third, have previously been studied, for example, by G. F. Carrier (2, 3) who investigated strings and by A. C. Eringen who dealt with beams (4) and circular membranes (5).

In the present paper, we propose to investigate free flexural vibrations of a rectangular elastic plate, supported by immovable hinges along all edges. The plate equations used for this purpose are those discussed in a previous paper (1). They may be considered the dynamic analog of the von Karman large-deflection plate theory of equilibrium, and are valid for moderately large amplitudes. These equations are also quite similar to those pre-

[1] This investigation was supported by the National Advisory Committee for Aeronautics, under Contract NAw-6366.

[2] Department of Civil Engineering and Engineering Mechanics, Institute of Flight Structures, Columbia University.

[3] Associate Professor of Civil Engineering, Institute of Flight Structures, Columbia University. Mem. ASME.

[4] Numbers in parentheses refer to the Bibliography at the end of the paper.

Presented at the National Applied Mechanics Division Conference, Urbana, Ill., June 14–16, 1956, of The American Society of Mechanical Engineers.

Discussion of this paper should be addressed to the Secretary, ASME, 29 West 39th Street, New York, N. Y., and will be accepted until January 10, 1957, for publication at a later date. Discussion received after the closing date will be returned.

Note: Statements and opinions advanced in papers are to be understood as individual expressions of their authors and not those of the Society. Manuscript received by ASME Applied Mechanics Division, January 9, 1956. Paper No. 56—APM-27.

viously obtained by Saint Venant.[5] If the body forces in Saint Venant's equations were taken to be the negative of the inertia forces and the transverse load were identified with the negative of the transverse inertia term, then they can be seen to be the same as Equations [1] of the present paper.

As in previous work by Carrier and Eringen, a perturbation procedure is used in the solution of the nonlinear problem. Since all the nonlinear terms are considered to be small throughout, only the lowest order equations of motion are solved. In addition, the problem was treated by direct energy considerations for the purpose of a verification. Graphs, indicating the influence of large amplitudes of vibration on the periodic time and on the maximum normal stress, have been plotted for various aspect ratios of the plate. Finally, the vibrating beam was considered as a special case, employing again a perturbation procedure, and furthermore, a procedure corresponding to Duffing's method (6) for a single lumped mass. The computed values are compared with those by A. C. Eringen (4). All the results obtained in this study reduce readily to those of the classical linear plate theory when the amplitudes of vibration tend to zero.

Statement of Problem

Considered are free vibrations of a flat rectangular plate of thickness h and edge lengths a and b. The edges are assumed to be hinged and immovable. The plate is referred to a Cartesian co-ordinate system $Oxyz$, the x,y-plane being in the middle plane of the plate and the origin O at a corner of the plate. We designate by subscripts 1 and 2 the directions of those fibers of the deformed plate which were in the x and y-directions, respectively, before deformation. The stress equations of free flexural motions of plates with moderately large amplitudes are

$$\left.\begin{aligned} &\frac{\partial N_1}{\partial x} + \frac{\partial N_{12}}{\partial y} = ph\ddot{u}_0 \\ &\frac{\partial N_2}{\partial y} + \frac{\partial N_{12}}{\partial x} = ph\ddot{v}_0 \\ &\frac{\partial^2 M_1}{\partial x^2} + \frac{\partial^2 M_2}{\partial y^2} + 2\frac{\partial^2 M_{12}}{\partial x \partial y} + \frac{\partial}{\partial x}\left(N_1 \frac{\partial w_0}{\partial x}\right) \\ &\quad + \frac{\partial}{\partial y}\left(N_2 \frac{\partial w_0}{\partial y}\right) + \frac{\partial}{\partial x}\left(N_{12} \frac{\partial w_0}{\partial y}\right) \\ &\quad + \frac{\partial}{\partial y}\left(N_{12} \frac{\partial w_0}{\partial x}\right) = ph\ddot{w}_0 \end{aligned}\right\} \quad \ldots [1]$$

Dots indicate differentiation with respect to time t. In these equations u_0, v_0 designate the displacement of a particle in the middle plane of the plate in x and y-directions, respectively, while w_0 is the displacement of such particle normal to the middle plane. N_1, N_2, N_{12} are the membrane forces, characteristic of the state of stress in the middle plane of the plate. M_1, M_2, M_{12} are the usual moments of the classical linear plate theory. Subscripts 1 and 2 are used here to indicate that these forces and moments are taken with respect to fibers after deformation. If the inertia in the plane of the plate is neglected, that is, if the right-hand side of the first two equations of Set [1] is put equal to zero, the third equation may be simplified considerably. Set [1] becomes then

$$\left.\begin{aligned} &\frac{\partial N_1}{\partial x} + \frac{\partial N_{12}}{\partial y} = 0 \\ &\frac{\partial N_2}{\partial y} + \frac{\partial N_{12}}{\partial x} = 0 \\ &\frac{\partial^2 M_1}{\partial x^2} + \frac{\partial^2 M_2}{\partial y^2} + 2\frac{\partial^2 M_{12}}{\partial x \partial y} + N_1 \frac{\partial^2 w_0}{\partial x^2} + N_2 \frac{\partial^2 w_0}{\partial y^2} \\ &\quad + 2N_{12}\frac{\partial^2 w_0}{\partial x \partial y} = ph\ddot{w}_0 \end{aligned}\right\} \quad [1a]$$

[5] See, for example, reference (7), p. 301.

This set is identical to the static von Karman equations, if we consider the negative of the transverse load to be the transverse inertia. However, we note that a mere addition of longitudinal inertia terms to the first two equations of Set [1a] does not result in the original Set [1], because terms of the type

$$\frac{\partial N_1}{\partial x}\frac{\partial w_0}{\partial x}$$

the so-called "buoyancy" terms, would be missing.

The stress-displacement relations for an elastic, isotropic plate are given in (1, 7) as

$$\left.\begin{aligned} N_1 &= \frac{Eh}{1-\nu^2}\left[\frac{\partial u_0}{\partial x} + \frac{1}{2}\left(\frac{\partial w_0}{\partial x}\right)^2 + \nu\frac{\partial v_0}{\partial y} + \frac{\nu}{2}\left(\frac{\partial w_0}{\partial y}\right)^2\right] \\ N_2 &= \frac{Eh}{1-\nu^2}\left[\frac{\partial v_0}{\partial y} + \frac{1}{2}\left(\frac{\partial w_0}{\partial y}\right)^2 + \nu\frac{\partial u_0}{\partial x} + \frac{\nu}{2}\left(\frac{\partial w_0}{\partial x}\right)^2\right] \\ N_{12} &= Gh\left(\frac{\partial u_0}{\partial y} + \frac{\partial v_0}{\partial x} + \frac{\partial w_0}{\partial x}\frac{\partial w_0}{\partial y}\right) \\ M_1 &= -D\left(\frac{\partial^2 w_0}{\partial x^2} + \nu\frac{\partial^2 w_0}{\partial y^2}\right), \\ M_2 &= -D\left(\frac{\partial^2 w_0}{\partial y^2} + \nu\frac{\partial^2 w_0}{\partial x^2}\right), \\ M_{12} &= -D(1-\nu)\frac{\partial^2 w_0}{\partial x \partial y} \end{aligned}\right\} \quad \ldots [2]$$

The occurrence of nonlinear terms in the first three relations is due to the fact that an element of the plate is not only sheared and elongated but also rotated about the x, y-axes, respectively, such that a fiber originally in the x-direction is after deformation in the 1-direction, the angle of rotation being given by $\partial w_0/\partial x$. A more complete discussion of these equations may be found in reference (1). The stress Equations of Motion [1] are converted to displacement equations of motion, using Relations [2]

$$\left.\begin{aligned} &\frac{\partial^2 u_0}{\partial x^2} + \frac{\partial w_0}{\partial x}\frac{\partial^2 w_0}{\partial x^2} + \nu\left(\frac{\partial^2 v_0}{\partial x \partial y} + \frac{\partial w_0}{\partial y}\frac{\partial^2 w_0}{\partial x \partial y}\right) \\ &\quad + \frac{1-\nu}{2}\left(\frac{\partial^2 u_0}{\partial y^2} + \frac{\partial^2 v_0}{\partial x \partial y} + \frac{\partial w_0}{\partial x}\frac{\partial^2 w_0}{\partial y^2} + \frac{\partial w_0}{\partial y}\frac{\partial^2 w_0}{\partial x \partial y}\right) = \frac{\ddot{u}_0}{c_p^2} \\ &\frac{1-\nu}{2}\left(\frac{\partial^2 u_0}{\partial x \partial y} + \frac{\partial^2 v_0}{\partial x^2} + \frac{\partial w_0}{\partial x}\frac{\partial^2 w_0}{\partial x \partial y} + \frac{\partial w_0}{\partial y}\frac{\partial^2 w_0}{\partial x^2}\right) \\ &\quad + \frac{\partial^2 v_0}{\partial y^2} + \frac{\partial w_0}{\partial y}\frac{\partial^2 w_0}{\partial y^2} + \nu\left(\frac{\partial^2 u_0}{\partial x \partial y} + \frac{\partial w_0}{\partial x}\frac{\partial^2 w_0}{\partial x \partial y}\right) = \frac{\ddot{v}_0}{c_p^2} \end{aligned}\right\} \quad \ldots [3]$$

$$
\begin{aligned}
\frac{h^2}{12}\left(\frac{\partial^4 w_0}{\partial x^4} + 2\frac{\partial^4 w_0}{\partial x^2\partial y^2} + \frac{\partial^4 w_0}{\partial y^4}\right) &= \frac{\partial u_0}{\partial x}\frac{\partial^2 w_0}{\partial x^2} \\
&+ \frac{1}{2}\left(\frac{\partial w_0}{\partial x}\right)^2\frac{\partial^2 w_0}{\partial x^2} + \nu\left[\frac{\partial v_0}{\partial y}\frac{\partial^2 w_0}{\partial x^2} + \frac{1}{2}\left(\frac{\partial w_0}{\partial y}\right)^2\frac{\partial^2 w_0}{\partial x^2}\right] \\
&+ \frac{\partial v_0}{\partial y}\frac{\partial^2 w_0}{\partial y^2} + \frac{1}{2}\left(\frac{\partial w_0}{\partial y}\right)^2\frac{\partial^2 w_0}{\partial y^2} + \nu\left[\frac{\partial u_0}{\partial x}\frac{\partial^2 w_0}{\partial y^2}\right. \\
&+ \left.\frac{1}{2}\left(\frac{\partial w_0}{\partial x}\right)^2\frac{\partial^2 w_0}{\partial y^2}\right] + (1-\nu)\left(\frac{\partial u_0}{\partial y}\frac{\partial^2 w_0}{\partial x\partial y} + \frac{\partial v_0}{\partial x}\frac{\partial^2 w_0}{\partial x\partial y}\right. \\
&+ \left.\frac{\partial w_0}{\partial x}\frac{\partial w_0}{\partial y}\frac{\partial^2 w_0}{\partial x\partial y}\right) + \frac{\ddot{u}_0}{c_p^2}\frac{\partial w_0}{\partial x} + \frac{\ddot{v}_0}{c_p^2}\frac{\partial w_0}{\partial y} + \frac{\ddot{w}_0}{c_p^2}
\end{aligned}
\quad \ldots[3] \text{ (cont.)}
$$

where

$$c_p^2 = \frac{E}{\rho(1-\nu^2)}$$

The boundary conditions of a rectangular plate, all four of whose edges are hinged and immovable, are

$$
\left.
\begin{aligned}
&\text{at } x = 0, a: \quad u_0 = w_0 = \frac{\partial^2 w_0}{\partial x^2} = 0 \\
&\text{at } y = 0, b: \quad v_0 = w_0 = \frac{\partial^2 w_0}{\partial y^2} = 0
\end{aligned}
\right\} \ldots\ldots[4]
$$

Since only free vibrations are to be investigated, there is no need to specify the initial conditions.

Perturbation Procedure

A perturbation parameter δ is chosen, which depends only on the geometry of the plate and not on the type of motion. Letting, without loss of generality, $a \leq b$, we set

$$\delta = \frac{h}{a} \ldots\ldots\ldots\ldots[5]$$

For convenience, the independent variables of the problem, x, y, and t are transformed to dimensionless quantities ξ, η, and τ, respectively, by

$$\xi = \frac{x}{a}, \qquad \eta = \frac{y}{a}, \qquad \tau = \frac{c_p t}{\sqrt{12}\, a} \ldots\ldots\ldots[6]$$

and the dependent variables u_0, v_0, and w_0 are transformed to $\bar{u}$, $\bar{v}$, and $\bar{w}$ by

$$\bar{u} = \frac{u_0}{a}, \qquad \bar{v} = \frac{v_0}{a}, \qquad \bar{w} = \frac{w_0}{a} \ldots\ldots\ldots[7]$$

Furthermore, the notation is introduced

$$r = \frac{a}{b} \ldots\ldots\ldots\ldots[8]$$

$$\zeta = \delta\tau \ldots\ldots\ldots\ldots[9]$$

This last substitution is performed in order to make all the terms of the classical linear plate theory of the same order in the parameter δ.

The Equations of Motion [3] may then be written as

$$
\begin{aligned}
&\frac{\partial^2 \bar{u}}{\partial \xi^2} + \frac{\partial \bar{w}}{\partial \xi}\frac{\partial^2 \bar{w}}{\partial \xi^2} + \nu\left(\frac{\partial^2 \bar{v}}{\partial \xi\partial \eta} + \frac{\partial \bar{w}}{\partial \eta}\frac{\partial^2 \bar{w}}{\partial \xi\partial \eta}\right) \\
&\quad + \frac{1-\nu}{2}\left(\frac{\partial^2 \bar{u}}{\partial \eta^2} + \frac{\partial^2 \bar{v}}{\partial \xi\partial \eta} + \frac{\partial \bar{w}}{\partial \xi}\frac{\partial^2 \bar{w}}{\partial \eta^2} + \frac{\partial \bar{w}}{\partial \eta}\frac{\partial^2 \bar{w}}{\partial \xi\partial \eta}\right) = \frac{\delta^2}{12}\frac{\partial^2 \bar{u}}{\partial \zeta^2} \\
&\frac{1-\nu}{2}\left(\frac{\partial^2 \bar{u}}{\partial \xi\partial \eta} + \frac{\partial^2 \bar{v}}{\partial \xi^2} + \frac{\partial \bar{w}}{\partial \xi}\frac{\partial^2 \bar{w}}{\partial \xi\partial \eta} + \frac{\partial^2 \bar{w}}{\partial \xi^2}\frac{\partial \bar{w}}{\partial \eta}\right) \\
&\quad + \frac{\partial^2 \bar{v}}{\partial \eta^2} + \frac{\partial \bar{w}}{\partial \eta}\frac{\partial^2 \bar{w}}{\partial \eta^2} + \nu\left(\frac{\partial^2 \bar{u}}{\partial \xi\partial \eta} + \frac{\partial \bar{w}}{\partial \xi}\frac{\partial^2 \bar{w}}{\partial \xi\partial \eta}\right) = \frac{\delta^2}{12}\frac{\partial^2 \bar{v}}{\partial \zeta^2} \\
&\frac{\delta^2}{12}\left(\frac{\partial^4 \bar{w}}{\partial \xi^4} + 2\frac{\partial^4 \bar{w}}{\partial \xi^2\partial \eta^2} + \frac{\partial^4 \bar{w}}{\partial \eta^4}\right) \\
&\quad = \frac{\partial \bar{u}}{\partial \xi}\frac{\partial^2 \bar{w}}{\partial \xi^2} + \frac{1}{2}\left(\frac{\partial \bar{w}}{\partial \xi}\right)^2\frac{\partial^2 \bar{w}}{\delta \xi^2} \\
&\quad + \nu\left[\frac{\partial \bar{v}}{\partial \eta}\frac{\partial^2 \bar{w}}{\partial \xi^2} + \frac{1}{2}\left(\frac{\partial \bar{w}}{\partial \eta}\right)^2\frac{\partial^2 \bar{w}}{\partial \xi^2}\right] + \frac{\partial \bar{w}}{\partial \eta}\frac{\partial^2 \bar{w}}{\partial \eta^2} \\
&\quad + \frac{1}{2}\left(\frac{\partial \bar{w}}{\partial \eta}\right)^2\frac{\partial^2 \bar{w}}{\partial \eta^2} + \nu\left[\frac{\partial \bar{u}}{\partial \xi}\frac{\partial^2 \bar{w}}{\partial \eta^2} + \frac{1}{2}\left(\frac{\partial \bar{w}}{\partial \xi}\right)^2\frac{\partial^2 \bar{w}}{\partial \eta^2}\right] \\
&\quad + (1-\nu)\left[\frac{\partial \bar{u}}{\partial \eta}\frac{\partial^2 \bar{w}}{\partial \xi\partial \eta} + \frac{\partial \bar{v}}{\partial \xi}\frac{\partial^2 \bar{w}}{\partial \xi\partial \eta} + \frac{\partial \bar{w}}{\partial \xi}\frac{\partial \bar{w}}{\partial \eta}\frac{\partial^2 \bar{w}}{\partial \xi\partial \eta}\right] \\
&\quad + \frac{\delta^2}{12}\frac{\partial \bar{w}}{\partial \xi}\frac{\partial^2 \bar{u}}{\partial \zeta^2} + \frac{\delta^2}{12}\frac{\partial \bar{w}}{\partial \eta}\frac{\partial^2 \bar{v}}{\partial \zeta^2} - \frac{\delta^2}{12}\frac{\partial^2 \bar{w}}{\partial \zeta^2}
\end{aligned}
\quad \ldots[10]
$$

and the boundary conditions as

$$
\left.
\begin{aligned}
&\text{at } \xi = 0, 1: \quad \bar{u} = \bar{w} = \frac{\partial^2 \bar{w}}{\partial \xi^2} = 0 \\
&\text{at } \eta = 0, \frac{1}{r}: \quad \bar{v} = \bar{w} = \frac{\partial^2 \bar{w}}{\partial \eta^2} = 0
\end{aligned}
\right\} \ldots\ldots[11]
$$

Recognizing that the displacements $\bar{u}$ and $\bar{v}$, since they are omitted in classical linear theory, are of one order higher than the displacement $\bar{w}$ and noticing that $\bar{u}$ and $\bar{v}$ are even functions of δ, and that $\bar{w}$ is an odd function, the following perturbation series are used for these displacements

$$
\left.
\begin{aligned}
\bar{u} &= \delta^2 u^{(2)} + \delta^4 u^{(4)} + \ldots \\
\bar{v} &= \delta^2 v^{(2)} + \delta^4 v^{(4)} + \ldots \\
\bar{w} &= \delta w^{(1)} + \delta^3 w^{(3)} + \ldots
\end{aligned}
\right\} \ldots\ldots\ldots[12]
$$

Substituting the Series [12] into the Equations of Motion [10] and retaining terms of lowest order in δ only, the following equations of motion result

$$
\begin{aligned}
&\frac{\partial^2 u^{(2)}}{\partial \xi^2} + \frac{\partial w^{(1)}}{\partial \xi}\frac{\partial^2 w^{(1)}}{\partial \xi^2} + \nu\left(\frac{\partial^2 v^{(2)}}{\partial \xi\partial \eta} + \frac{\partial w^{(1)}}{\partial \eta}\frac{\partial^2 w^{(1)}}{\partial \xi\partial \eta}\right) \\
&\quad + \frac{1-\nu}{2}\left(\frac{\partial^2 u^{(2)}}{\partial \eta^2} + \frac{\partial^2 v^{(2)}}{\partial \xi\partial \eta} + \frac{\partial w^{(1)}}{\partial \xi}\frac{\partial^2 w^{(1)}}{\partial \eta^2} + \frac{\partial w^{(1)}}{\partial \eta}\frac{\partial^2 w^{(1)}}{\partial \xi\partial \eta}\right) = 0 \\
&\frac{1-\nu}{2}\left(\frac{\partial^2 u^{(2)}}{\partial \xi\partial \eta} + \frac{\partial w^{(1)}}{\partial \xi}\frac{\partial^2 w^{(1)}}{\partial \xi\partial \eta} + \frac{\partial w^{(1)}}{\partial \eta}\frac{\partial^2 w^{(1)}}{\partial \xi^2} + \frac{\partial^2 v^{(2)}}{\partial \xi^2}\right) \\
&\quad + \frac{\partial^2 v^{(2)}}{\partial \eta^2} + \frac{\partial w^{(1)}}{\partial \eta}\frac{\partial^2 w^{(1)}}{\partial \eta^2} + \nu\left(\frac{\partial^2 u^{(2)}}{\partial \xi\,\partial \eta} + \frac{\partial w^{(1)}}{\partial \xi}\frac{\partial^2 w^{(1)}}{\partial \xi\,\partial \eta}\right) = 0 \\
&\frac{1}{12}\left(\frac{\partial^4 w^{(1)}}{\partial \xi^4} + 2\frac{\partial^4 w^{(1)}}{\partial \xi^2\partial \eta^2} + \frac{\partial^4 w^{(1)}}{\partial \eta^4}\right)
\end{aligned}
\quad \ldots[13]
$$

$$
\begin{aligned}
= \frac{\partial u^{(2)}}{\partial \xi}\frac{\partial^2 w^{(1)}}{\partial \xi^2} + \frac{1}{2}\left(\frac{\partial w^{(1)}}{\partial \xi}\right)^2 \frac{\partial^2 w^{(1)}}{\partial \xi^2} \\
+ \nu\left[\frac{\partial v^{(2)}}{\partial \eta}\frac{\partial^2 w^{(1)}}{\partial \xi^2} + \frac{1}{2}\left(\frac{\partial w^{(1)}}{\partial \eta}\right)^2 \frac{\partial^2 w^{(1)}}{\partial \xi^2}\right] \\
+ \frac{\partial v^{(2)}}{\partial \eta^2}\frac{\partial^2 w^{(1)}}{\partial \eta^2} + \frac{1}{2}\left(\frac{\partial w^{(1)}}{\partial \eta}\right)^2 \frac{\partial^2 w^{(1)}}{\partial \eta^2} + \nu\left[\frac{\partial u^{(2)}}{\partial \xi}\frac{\partial^2 w^{(1)}}{\partial \eta^2} \right.\\
\left. + \frac{1}{2}\left(\frac{\partial w^{(1)}}{\partial \xi}\right)^2 \frac{\partial^2 w^{(1)}}{\partial \eta^2}\right] + (1-\nu)\left(\frac{\partial u^{(2)}}{\partial \eta}\frac{\partial^2 w^{(1)}}{\partial \xi \partial \eta}\right. \\
\left. + \frac{\partial v^{(2)}}{\partial \xi}\frac{\partial^2 w^{(1)}}{\partial \xi \partial \eta} + \frac{\partial w^{(1)}}{\partial \xi}\frac{\partial w^{(1)}}{\partial \eta}\frac{\partial^2 w^{(1)}}{\partial \xi \partial \eta}\right) - \frac{1}{12}\frac{\partial^2 w^{(1)}}{\partial \zeta^2}
\end{aligned}
\quad \ldots [13]\ (cont.)
$$

and, from Equations [11], the following boundary conditions

$$
\left.
\begin{aligned}
\text{at } \xi = 0, 1: \quad & u^{(2)} = w^{(1)} = \frac{\partial^2 w^{(1)}}{\partial \xi^2} = 0 \\
\text{at } \eta = 0, \frac{1}{r}: \quad & v^{(2)} = w^{(1)} = \frac{\partial^2 w^{(1)}}{\partial \eta^2} = 0
\end{aligned}
\right\} \ldots [14]
$$

Equations [13] represent a first-order approximation to the problem stated originally by Equation [1]. The only difference is the absence of longitudinal inertia terms, that is, Equations [13] are equivalent to Equations [1a]. Thus, Equations [13] may have been obtained by replacing in von Karman's equations the negative of the transverse loading term by transverse inertia. There would be no assurance, however, that all the terms of the same order are retained.

Similarly, as in the equilibrium case, a stress function $\bar{F}$ may now be introduced to replace the variables $u^{(2)}$ and $v^{(2)}$. Defining $\bar{F}$ as

$$
\left.
\begin{aligned}
\frac{\partial^2 \bar{F}}{\partial \eta^2} &= \frac{\partial u^{(2)}}{\partial \xi} + \frac{1}{2}\left(\frac{\partial w^{(1)}}{\partial \xi}\right)^2 + \nu\left[\frac{\partial v^{(2)}}{\partial \eta} + \frac{1}{2}\left(\frac{\partial w^{(1)}}{\partial \eta}\right)^2\right] \\
\frac{\partial^2 \bar{F}}{\partial \xi^2} &= \frac{\partial v^{(2)}}{\partial \eta} + \frac{1}{2}\left(\frac{\partial w^{(1)}}{\partial \eta}\right)^2 + \nu\left[\frac{\partial u^{(2)}}{\partial \xi} + \frac{1}{2}\left(\frac{\partial w^{(1)}}{\partial \xi}\right)^2\right] \\
-\frac{\partial^2 \bar{F}}{\partial \xi \partial \eta} &= \frac{1-\nu}{2}\left(\frac{\partial u^{(2)}}{\partial \eta} + \frac{\partial v^{(2)}}{\partial \xi} + \frac{\partial w^{(1)}}{\partial \xi}\frac{\partial w^{(1)}}{\partial \eta}\right)
\end{aligned}
\right\} \ldots [15]
$$

the first two of Equations [13] will be satisfied identically and the third equation becomes

$$
\begin{aligned}
\frac{1}{12}\left(\frac{\partial^4 w^{(1)}}{\partial \xi^4} + 2\frac{\partial^4 w^{(1)}}{\partial \xi^2 \partial \eta^2} + \frac{\partial^4 w^{(1)}}{\partial \eta^4} + \frac{\partial^2 w^{(1)}}{\partial \zeta^2}\right) = \frac{\partial^2 \bar{F}}{\partial \eta^2}\frac{\partial^2 w^{(1)}}{\partial \xi^2} \\
+ \frac{\partial^2 \bar{F}}{\partial \xi^2}\frac{\partial^2 w^{(1)}}{\partial \eta^2} - 2\frac{\partial^2 \bar{F}}{\partial \xi \partial \eta}\frac{\partial^2 w^{(1)}}{\partial \xi \partial \eta} \quad \ldots [16a]
\end{aligned}
$$

Moreover, $\bar{F}$ has to satisfy the equation

$$
\frac{\partial^4 \bar{F}}{\partial \xi^4} + 2\frac{\partial^4 \bar{F}}{\partial \xi^2 \partial \eta^2} + \frac{\partial^4 \bar{F}}{\partial \eta^4} = (1-\nu^2)\left[\left(\frac{\partial^2 w^{(1)}}{\partial \xi \partial \eta}\right)^2 - \frac{\partial^2 w^{(1)}}{\partial \xi^2}\frac{\partial^2 w^{(1)}}{\partial \eta^2}\right] \quad \ldots [16b]
$$

as a consequence of its definition. $\bar{F}$ is related to the stress function F (for example, equation [e] of reference (7), p. 343)

$$
\bar{F} = \frac{1-\nu^2}{Eh^2} F
$$

For the present purpose, it was found more convenient to continue to deal with Set [13] in three variables $u^{(2)}$, $v^{(2)}$, $w^{(1)}$, rather than with Set [16] in two variables $\bar{F}$ and $w^{(1)}$, largely because the Boundary Conditions [14] are in terms of $u^{(2)}$ and $v^{(2)}$.

Moreover, it appears that in a dynamic problem the use of the stress function $\bar{F}$ does not offer the advantages as in the corresponding static problem (8), because the time function is not separable from the space function. In fact, assuming the transverse displacement $w^{(1)}$ in the form

$$
w^{(1)} = \bar{H}(\zeta) \sum_{i,j \text{ odd}} a_{ij} \sin i\pi\xi \sin j\pi r\eta
$$

Equation [16b] determines the stress function $\bar{F}$ as

$$
\bar{F} = \bar{H}^2 \sum f_{mn} \cos m\pi\xi \cos n\pi r\eta
$$

where

$$
\begin{aligned}
f_{mn} = \frac{(1-\nu^2)r^2}{4(m^2+n^2r^2)^2}\Bigg\{ & \sum_{\substack{m>i\\ n>j}} [ij(m-i)(n-j) \\
& - i^2(n-j)^2]\, a_{ij}a_{(m-i)(n-j)} + \sum_{\substack{i>m\\ n>j}} [ij(i-m)(n-j) \\
& + i^2(n-j)^2] a_{ij}a_{(i-m)(n-j)} + \sum_{\substack{m>i\\ j>n}} [ij(m-i)(j-n) \\
& + i^2(j-n)^2] a_{ij}a_{(m-i)(j-n)} + \sum_{\substack{i>m\\ j>n}} [ij(i-m)(j-n) \\
& - i^2(j-n)^2] a_{ij}a_{(i-m)(j-n)} \Bigg\}
\end{aligned}
$$

Substitution of the foregoing expressions for $w^{(1)}$ and $\bar{F}$ into Equation [16a] results in

$$
\begin{aligned}
& \frac{1}{12}\left[\bar{H}\pi^4 \sum a_{ij}(i^2+j^2r^2)^2 \sin i\pi\xi \sin j\pi r\eta + \frac{d^2\bar{H}}{d\zeta^2}\sum a_{ij} \sin i\pi\xi \sin j\pi r\eta\right] \\
& = \frac{\bar{H}^3\pi^4 r^2}{4}\Bigg\{\sum_{\substack{i>m\\ j>n}} (i-m)^2 n^2 a_{(i-m)(j-n)} + \sum_{\substack{j>n\\ m>i}} (i+m)^2 n^2 a_{(i+m)(j-n)} + \sum_{\substack{i>m\\ n>j}} (i-m)^2 n^2 a_{(i-m)(j+n)} \\
& + \sum_{\substack{m>i\\ n>j}} (i+m)^2 n^2 a_{(i+m)(j+n)} + \sum_{\substack{i>m\\ j>n}} (j-n)^2 m^2 a_{(i-m)(j-n)} + \sum_{\substack{j>m\\ m>i}} (j-n)^2 m^2 a_{(i-m)(j-n)} + \sum_{\substack{i>m\\ n>j}} (j+n)^2 m^2 a_{(i-m)(j-n)} \\
& + \sum_{\substack{m>i\\ n>j}} (j+n)^2 m^2 a_{(i+m)(j+n)} - 2\sum_{\substack{i>m\\ j>n}} (i-m)(j-n) mn a_{(i-m)(j-n)} + 2\sum_{\substack{j>n\\ m>i}} (i+m)(j-n) mn a_{(i+m)(j-n)} \\
& + 2\sum_{\substack{i>m\\ n>j}} (i-m)(j+n) mn a_{(i-m)(j+n)} - 2\sum_{\substack{m>i\\ n>j}} (i+m)(j+n) mn a_{(i+m)(j+n)}\Bigg\} f_{mn} \sin i\pi\xi \sin j\pi r\eta
\end{aligned}
$$

The presence of the factors $(i^2 + j^2r^2)^2$ on the left-hand side of the foregoing expression precludes the possibility of separating the time function $\bar{H}(\zeta)$ from the space function.

Seeking now a solution of the perturbed problem, the transverse displacement $w^{(1)}$ is assumed in the form

$$w^{(1)} = \beta Z(\zeta) \sin \pi\xi \sin \pi r\eta \qquad [17]$$

where β is the dimensionless amplitude and $z(\zeta) \leq 1$. As may be verified, $\beta = B/h$, where B is the actual amplitude. The first two equations of Set [13] are satisfied if the displacements $u^{(2)}$, $v^{(2)}$ are of the form

$$\left.\begin{aligned} u^{(2)} &= \frac{\beta^2 Z^2\pi}{16}(\cos 2\pi r\eta - 1 + \nu r^2)\sin 2\pi\xi \\ v^{2} &= \frac{\beta^2 Z^2\pi}{16}\left(r\cos 2\pi\xi - r + \frac{\nu}{r}\right)\sin 2\pi r\eta \end{aligned}\right\} \qquad [18]$$

The Expressions [17] and [18] satisfy the Boundary Conditions [14].

Substitution of Equations [17] and [18] into the third equation of Set [13] yields

$$\left\{\frac{1}{12}\left[\pi^4(1+r^2)^2 Z + \frac{d^2Z}{d\zeta^2}\right]\beta + \frac{1}{8}Z^3\pi^4\beta^3\left[\left(\frac{3}{2} - \frac{\nu^2}{2}\right)(1+r^4) + 2\nu r^2\right]\right\}\sin \pi\xi \sin \pi r\eta$$
$$+ \frac{\beta^3 Z^3\pi^4}{16}(1-\nu^2)\left[\sin \pi\xi \sin 3\pi r\eta + r^4 \sin 3\pi\xi \sin \pi r\eta\right] = 0 \qquad [19]$$

It is noted that the nonlinearity of the problem introduces a coupling effect between the fundamental mode contained in the first term and the two higher modes contained in the second term of Equation [19]. Neglecting the effect of these higher modes, which is justified by observing that for small nonlinearities this effect will be small, we obtain the equation governing the time function $Z(\zeta)$

$$\frac{d^2Z}{d\zeta^2} + \pi^4(1+r^2)^2 Z + 3\beta^2\pi^4 Z^3\left[\left(\frac{3}{4} - \frac{\nu^2}{4}\right)(1+r^4) + \nu r^2\right] = 0 \qquad [20]$$

Measuring the time ζ such that

$$(Z)_{\zeta=0} = 1, \quad \left(\frac{dZ}{d\zeta}\right)_{\zeta=0} = 0$$

solution of Equation [20] will be given in the form of an elliptic cosine c_n

$$Z = c_n(\omega^*\zeta, k) \qquad [21]$$

where

$$\omega^* = \pi^2\left\{(1+r^2)^2 + 3\beta^2\left[\left(\frac{3}{4} - \frac{\nu^2}{4}\right)(1+r^4) + \nu r^2\right]\right\}^{1/2} \qquad [22]$$

$$k^2 = \left\{2 + \frac{2(1+r^2)^2}{3\beta^2\left[\left(\frac{3}{4} - \frac{\nu^2}{4}\right)(1+r^4) + \nu r^2\right]}\right\}^{-1} \qquad [23]$$

The period T^* of $c_n(\omega^*\zeta, k)$ is

$$T^* = 4K\frac{\pi^2}{\omega^*} = \frac{4K}{\left\{(1+r^2)^2 + 3\beta^2\left[\left(\frac{3}{4} - \frac{\nu^2}{4}\right)(1+r^4) + \nu r^2\right]\right\}^{1/2}} \qquad [24]$$

where $K(k)$ is the complete elliptic integral of the first kind.

The corresponding linear period is

$$T = \frac{2\pi}{(1+r^2)} \qquad [25]$$

T also may be deduced from Expression [24] for T^* by observing that as $\beta \to 0$, $\omega^* \to \pi^2(1+r^2)$, $K \to \frac{\pi}{2}$. The ratio T^*/T is given by

$$\frac{T^*}{T} = \frac{2}{\pi}\left\{(1+r^2)^2 + 3\beta^2\left[\left(\frac{3}{4} - \frac{\nu^2}{4}\right)(1+r^4) + \nu r^2\right]\right\}^{-1/2}(1+r^2)K \qquad [26]$$

In Fig. 1 this ratio is plotted versus the dimensionless amplitude β, for various values of the aspect ratio r and for Poisson's ratio $\nu = 0.318$, which is typical for aluminum alloys. It is seen that the period, for any aspect ratio, decreases rapidly with increasing amplitude.

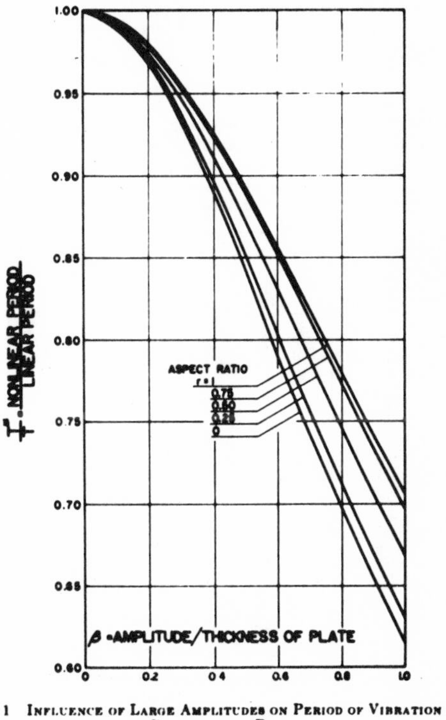

Fig. 1 Influence of Large Amplitudes on Period of Vibration of a Rectangular Plate

Turning now our attention to the state of stress in the plate, we note that the occurrence of moderately large rotations manifests itself in the presence of membrane stresses, one of whose components is, by the first equation of Set [2]

$$\sigma_{1m} = \frac{N_1}{h} = \frac{E}{1-\nu^2}\left\{\frac{\partial u_0}{\partial x} + \frac{1}{2}\left(\frac{\partial w_0}{\partial x}\right)^2 + \nu\left[\frac{\partial v_0}{\partial y} + \frac{1}{2}\left(\frac{\partial w_0}{\partial y}\right)^2\right]\right\} \qquad [27]$$

and which are absent in linear plate theory. The maximum value of σ_{1m} will be reached at the center of the plate, $x = a/2$, $y = b/2$. From Equations [17] and [18], the displacements u_0, v_0, w_0, are

$$\left.\begin{aligned} u_0 &= \frac{\beta^2h^2Z^2}{16a}\left(\cos\frac{2\pi y}{b} - 1 + \nu r^2\right)\sin\frac{2\pi x}{a} \\ v_0 &= \frac{\beta^2h^2Z^2}{16a}\left(r\cos\frac{2\pi x}{a} - r + \frac{\nu}{r}\right)\sin\frac{2\pi y}{b} \\ w_0 &= \beta hZ\sin\frac{\pi x}{a}\sin\frac{\pi y}{b} \end{aligned}\right\} \qquad [28]$$

and σ_{1m} becomes, with $Z_{max} = 1$

$$\sigma_{1m} = \frac{E\pi}{8(1-\nu^2)}\left(\frac{\beta h}{a}\right)^2(2 + \nu r^2 - \nu^2)$$

We define a dimensionless membrane stress S as

$$S = \frac{8\sigma_{1m}}{E\pi\left(\frac{h}{a}\right)^2} \qquad [29]$$

which may be expressed, with a value of $\nu = 0.318$, as

$$S = \beta^2(2.11 + 0.353r^2) \qquad [30]$$

For various values of aspect ratio r, this dimensionless membrane stress S is plotted versus the dimensionless amplitude β in Fig. 2. It is seen that the aspect ratio has a relatively small effect on the magnitude of the membrane stress.

It also may be of interest to establish the influence of the membrane stress on the total maximum stress, which will occur at the face of the plate at its center. This total maximum stress will be the sum of the membrane stress σ_{1m} computed above and the bending stress σ_{1b}

$$\left.\begin{aligned} \sigma_{1b} &= \frac{6M_1}{h} = -\frac{Eh}{2(1-\nu^2)}\left(\frac{\partial^2 w_0}{\partial x^2} + \nu\frac{\partial^2 w_0}{\partial y^2}\right) \\ &= \frac{E}{2}\left(\frac{h}{a}\right)^2\pi^2\beta(1.111 + 0.353r^2) \end{aligned}\right\} \qquad [31]$$

again with $Z_{max} = 1$ and $\nu = 0.318$.

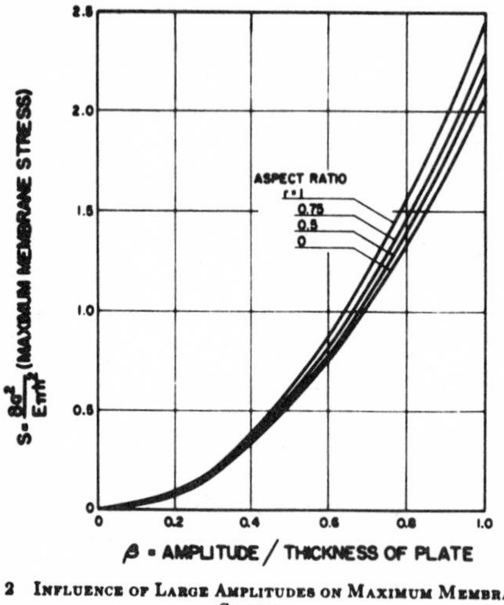

FIG. 2 INFLUENCE OF LARGE AMPLITUDES ON MAXIMUM MEMBRANE STRESS

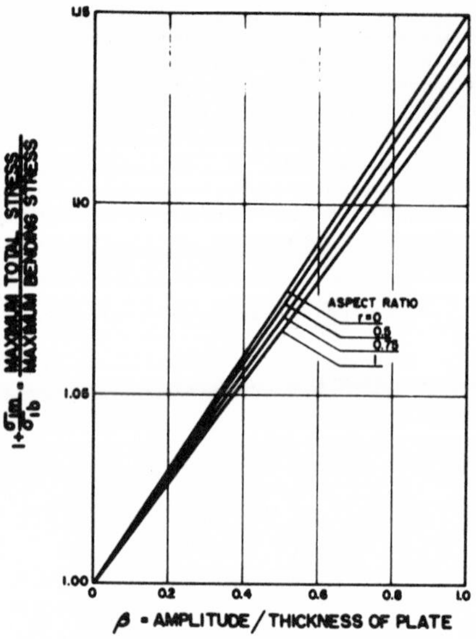

FIG. 3 INFLUENCE OF LARGE AMPLITUDES ON MAXIMUM TOTAL STRESS

For various aspect ratios r, the ratios of the total stress $\sigma_{1m} + \sigma_{1b}$ to the bending stress σ_{1b}, which is the only stress in classical plate theory, are plotted versus the dimensionless amplitude β in Fig. 3, in accordance with the relation

$$\frac{\sigma_{1b} + \sigma_{1m}}{\sigma_{1b}} = 1 + \frac{\beta(1.8989 + 0.318r^2)}{4\pi(1 + 0.318r^2)} \qquad [32]$$

It is evident from the graph that, as a consequence of moderately large amplitudes, the total stress may be considerably larger than the stress predicted by the classical linear plate theory. However, it appears that percentagewise the large amplitudes affect the stress markedly less than the period of vibration.

PRINCIPLE OF CONSERVATION OF ENERGY

As an alternative, the same problem is studied in this section, using the principle of conservation of energy directly. The strain energy $\overline{W}$ in the present theory (1) is given by the expression

$$2\overline{W} = \iint\left\{N_1\left[\frac{\partial u_0}{\partial x} + \frac{1}{2}\left(\frac{\partial w_0}{\partial x}\right)^2\right] + N_2\left[\frac{\partial v_0}{\partial y} + \frac{1}{2}\left(\frac{\partial w_0}{\partial y}\right)^2\right] + N_{12}\left(\frac{\partial u_0}{\partial y} + \frac{\partial v_0}{\partial x} + \frac{\partial w_0}{\partial x}\frac{\partial w_0}{\partial y}\right) - M_1\frac{\partial^2 w_0}{\partial x^2} - M_2\frac{\partial^2 w_0}{\partial y^2} - 2M_{12}\frac{\partial^2 w_0}{\partial x\partial y}\right\}dxdy \qquad [33]$$

where integration is to be extended over the area of the plate face. The kinetic energy T (1) is given by, neglecting rotatory inertia terms

$$2T = \rho \iint (h\dot{u}_0^2 + h\dot{v}_0^2 + h\dot{w}_0^2)dxdy \quad \ldots\ldots [34]$$

Assuming the displacements in the form

$$\left.\begin{aligned} u_0 &= \frac{B^2H^2(t)\pi}{16a}\left(\cos\frac{2\pi y}{b} - 1 + \nu\frac{a^2}{b^2}\right)\sin\frac{2\pi x}{a} \\ v_0 &= \frac{B^2H^2(t)\pi}{16a}\left(\frac{a}{b}\cos\frac{2\pi x}{a} - \frac{a}{b} + \nu\frac{b}{a}\right)\sin^2\frac{\pi y}{b} \\ w_0 &= BH(t)\sin\frac{\pi x}{a}\sin\frac{\pi y}{b} \end{aligned}\right\} \quad \ldots [35]$$

the potential and the kinetic energies become, respectively

$$\left.\begin{aligned} 2W = &\int_0^a\int_0^b \frac{Eh}{1-\nu^2}\left(\frac{B^2H^2\pi^2}{8}\right)^2\left\{\frac{1}{a^4}\left[1 - \cos\frac{2\pi y}{b} + \nu\frac{a^2}{b^2}\cos\frac{2\pi x}{a}\right]^2 + \frac{\nu}{a^2b^2}\left[1 - \cos\frac{2\pi x}{a} + \nu\frac{b^2}{a^2}\cos\frac{2\pi y}{b}\right]\right. \\ &\left.\left[1 - \cos\frac{2\pi y}{b} + \nu\frac{a^2}{b^2}\cos\frac{2\pi x}{a}\right]\right\}dxdy + \int_0^a\int_0^b \frac{Eh}{1-\nu^2}\left(\frac{B^2H^2\pi^2}{8}\right)^2\left\{\frac{1}{b^4}\left[1 - \cos\frac{2\pi x}{a} + \nu\frac{b^2}{a^2}\cos\frac{2\pi y}{b}\right]^2\right. \\ &\left. + \frac{\nu}{a^2b^2}\left[1 - \cos\frac{2\pi y}{b} + \nu\frac{a^2}{b^2}\cos\frac{2\pi x}{a}\right]\left[1 - \cos\frac{2\pi x}{a} + \nu\frac{b^2}{a^2}\cos\frac{2\pi y}{b}\right]\right\}dxdy \\ &+ \frac{Eh^3}{1-\nu^2}B^2H^2\pi^4\int_0^a\int_0^b \sin^2\frac{\pi x}{a}\sin^2\frac{\pi y}{b}\left(\frac{1}{a^4} + \frac{1}{b^4} + \frac{2}{a^2b^2}\right)dxdy = \frac{Eh}{1-\nu^2}\left(\frac{B^2H^2\pi^2}{8}\right)^2 \\ &\left\{\frac{b}{a^3}\left[\frac{3}{2} + \frac{\nu^2}{2}\frac{a^4}{b^4}\right] + \frac{2\nu}{ab}\left[1 - \frac{\nu}{2}\frac{b^2}{a^2} - \frac{\nu}{2}\frac{a^2}{b^2}\right] + \frac{a}{b^3}\left[\frac{3}{2} + \frac{\nu^2}{2}\frac{b^4}{a^4}\right]\right\} + \frac{Eh^3}{12(H\nu^2)}\frac{B^2\pi^4H^2}{4}\left(\frac{1}{a^2} + \frac{1}{b^2}\right)^2 ab \end{aligned}\right\} \ldots [36]$$

$$2T = \rho\left(\frac{B^2H\dot{H}\pi}{8}\right)^2\left[\frac{h}{a}b\left(\frac{3}{4} + \nu^2\frac{a^4}{b^4} - 2\nu\frac{a^2}{b^2}\right)\right.$$

$$\left. + \frac{h}{b}a\left(\frac{3}{4} + \nu^2\frac{b^4}{a^2} - 2\nu\frac{b^2}{a^4}\right)\right] + \rho\frac{B^2\dot{H}^2}{4}abh \quad \ldots [37]$$

The principle of conservation of energy is stated by requiring that

$$\frac{\partial}{\partial t}(T + W) = 0$$

Letting again $\delta = h/a$, $r = a/b$, and designating by p the corresponding linear frequency of the plate, given by (9)

$$p^2 = \frac{\pi^4}{a^4}\frac{E}{\rho}\frac{h^2}{12(1-\nu^2)}(1 + r^2)^2$$

the result is, retaining terms of lowest power in δ only

$$\ddot{H} + p^2H + \frac{3}{2}p^2\left(\frac{B}{h}\right)^2 H^3\left[\left(\frac{3}{2} - \frac{\nu^2}{2}\right)(1 + r^4) + 2\nu r^2\right]$$

$$(1 + r)^{-2} = 0 \quad \ldots [38]$$

This equation on $H(t)$ is identical to Equation [20] on $Z(\zeta)$ if the appropriate transformation of variables is performed.

A Special Case: Beam Vibrations

The stress equations of motion of a beam are obtained from the corresponding plate Equations [1] by letting $v_1 = 0$ and $\partial(\,)/\partial y = 0$

$$\left.\begin{aligned} \frac{\partial N_1}{\partial x} &= \rho h\ddot{u}_0 \\ \frac{\partial^2 M_1}{\partial x^2} + \frac{\partial}{\partial x}\left(N_1\frac{\partial w_0}{\partial x}\right) &= \rho h\ddot{w}_0 \end{aligned}\right\} \ldots\ldots [39]$$

For the sake of simplicity, but without loss of generality, a beam of rectangular cross section, of height h and unit width, is considered. The area of the cross section is thus given by h and the moment of inertia by $h^3/12$.

For the case of such a beam, the stress-displacement Relations [2] simplify to

$$N_1 = Eh\left[\frac{\partial u_0}{\partial x} + \frac{1}{2}\left(\frac{\partial w_0}{\partial x}\right)^2\right]$$

$$M_1 = -\frac{Eh^3}{12}\frac{\partial^2 w_0}{\partial x^2}$$

Using the same Transformations [6] and [7], where, however, the velocity c_p has to be replaced by the velocity of compressional waves in a beam

$$c_b = \sqrt{\frac{E}{\rho}}$$

the dimensionless equations result

$$\frac{\partial^2\bar{u}}{\partial\xi^2} + \frac{\partial\bar{w}}{\partial\xi}\frac{\partial^2\bar{w}}{\partial\xi^2} = \frac{\delta^2}{12}\frac{\partial^2\bar{u}}{\partial\zeta^2}$$

$$\frac{\delta^2}{12}\left(\frac{\partial^4\bar{w}}{\partial\xi^4} + \frac{\partial^2\bar{w}}{\partial\zeta^2}\right) = \frac{3}{2}\frac{\partial^2\bar{w}}{\partial\xi^2}\left(\frac{\partial\bar{w}}{\partial\xi}\right)^2 + \frac{\partial^2\bar{w}}{\partial\xi^2}\frac{\partial\bar{u}}{\partial\xi}$$

$$+ \frac{\partial\bar{w}}{\partial\xi}\frac{\partial^2\bar{u}}{\partial\xi^2} \quad \ldots [40]$$

The boundary conditions for immovable, hinged, beam ends are

$$\bar{u} = \bar{w} = \frac{\partial^2\bar{w}}{\partial\xi^2} = 0 \qquad \text{at } \xi = 0, 1 \quad \ldots\ldots [41]$$

Similarly, as in the plate case, the perturbation yields in first approximation the equations of motion

$$\frac{\partial^2 u^{(2)}}{\partial\xi^2} + \frac{\partial w^{(1)}}{\partial\xi}\frac{\partial^2 w^{(1)}}{\partial\xi^2} = 0$$

$$\frac{1}{12}\left(\frac{\partial^4 w^{(1)}}{\partial\xi^4} + \frac{\partial^2 w^{(1)}}{\partial\zeta^2}\right) = \frac{3}{2}\frac{\partial^2 w^{(1)}}{\partial\xi^2}\left(\frac{\partial w^{(1)}}{\partial\xi}\right)^2 + \frac{\partial^2 w^{(1)}}{\partial\xi^2}\frac{\partial u^{(2)}}{\partial\xi}$$

$$+ \frac{\partial w^{(1)}}{\partial\xi}\frac{\partial^2 u^{(2)}}{\partial\xi^2} \quad \ldots [42]$$

and the boundary conditions

$$u^{(2)} = w^{(1)} = \frac{\partial^2 w^{(1)}}{\partial \xi^2} = 0 \quad \text{at } \xi = 0, 1 \quad \ldots\ldots [43]$$

With the aid of the boundary condition on $u^{(2)}$, the variable $u^{(2)}$ may be eliminated from the Equations of Motion [42], such that only one equation governing $w^{(1)}$ results

$$\frac{1}{12}\left(\frac{\partial^4 w^{(1)}}{\partial \xi^4} + \frac{\partial^2 w^{(1)}}{\partial \zeta^2}\right) = \frac{\partial^2 w^{(1)}}{\partial \xi^2}\int_0^1 \frac{1}{2}\left(\frac{\partial w^{(1)}}{\partial \xi}\right)^2 d\xi \ldots [44]$$

Assuming $w^{(1)}$ in the form

$$w^{(1)} = \beta Z(\zeta) \sin \pi\xi \ldots\ldots [45]$$

the time function will be governed by

$$\frac{d^2 Z}{d\zeta^2} + \pi^4 Z + 3\,\beta^2\pi^4 Z^3 = 0 \ldots\ldots [46]$$

whose solution may be given again in terms of elliptic functions. Similarly as in the case of plates, we evaluate the ratio of the nonlinear period T^* to the linear period T

$$\frac{T^*}{T} = \frac{2}{\pi}(1 + 3\beta^2)^{-1/4}K \ldots\ldots [47]$$

where K is again the complete elliptic integral of the first kind. This result, plotted in Fig. 4, is identical with the one obtained by A. C. Eringen (4), who, however, based his analysis on more exact equations than those used presently. The theory discussed in (4) describes rotations of any magnitude and strains which may be large. The present theory, on the contrary, deals with merely moderately large rotations and with small strains. It is thus of interest to observe that both theories predict the same period from first-order perturbations.

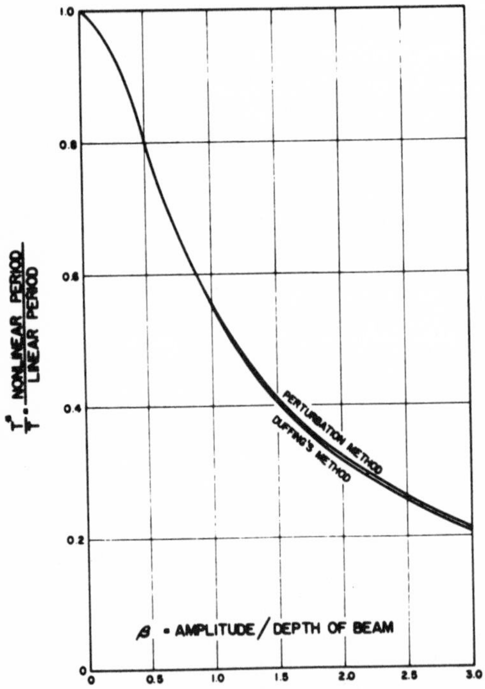

FIG. 4 INFLUENCE OF LARGE AMPLITUDES ON PERIOD OF VIBRATION OF A BEAM

Furthermore, it may be observed that the perturbed equations of motion can be solved exactly in the case of a beam, Equation [42], but not in the case of a plate, Equation [13]. It is also of interest to note that the ratio of the nonlinear period to the linear period of the plate, given by Equation [26], does not reduce to the corresponding relation of the beam, Equation [47], by letting the aspect ratio $r = 0$. This behavior may be made plausible by recognizing that the period is affected by terms containing $\partial v_0/\partial y$ which do not exist in the beam case but are present in the plate problem, even as r approaches zero.

As an alternative, a different solution of the same beam problem will be obtained, in the following, employing a method which is essentially the one due to Duffing (6).

The two equations of motion for the beam, from Equation [3], in dimensional form, are

$$\left.\begin{aligned} &\frac{\partial^2 u_0}{\partial x^2} - \frac{\ddot{u}_0}{c_b^2} = -\frac{\partial w_0}{\partial x}\frac{\partial^2 w_0}{\partial x^2} \\ &\frac{\partial^4 w_0}{\partial x^4} - \frac{1}{R^2}\left[\frac{3}{2}\left(\frac{\partial w_0}{\partial x}\right)^2\frac{\partial^2 w_0}{\partial x^2} + \frac{\partial w_0}{\partial x}\frac{\partial^2 u_0}{\partial x^2} + \frac{\partial^2 w_0}{\partial x^2}\frac{\partial u_0}{\partial x}\right] \\ &\qquad = -\frac{1}{c_b^2 R^2}\ddot{w}_0 \end{aligned}\right\} \ldots [48]$$

where

$$c_b^2 = \frac{E}{\rho}, \qquad R^2 = \frac{h^2}{12}$$

The transverse displacement w_0 is assumed in the form

$$w_0 = B(\omega^*) \sin\frac{\pi x}{a}\cos\omega^* t \ldots\ldots [49]$$

ω^* being the nonlinear frequency, and the longitudinal displacement u_0 is obtained from the first of Equations [48] as

$$u_0 = -\frac{1}{16}\frac{\pi}{a}B^2\left[1 + \frac{\cos 2\omega^* t}{1 - \left(\frac{\omega^* a}{c_b \pi}\right)^2}\right]\sin\frac{2\pi x}{a} \ldots [50]$$

From the second of Equations [48] we obtain, after equating the coefficients of $\sin(\pi x/a)\cos\omega^* t$

$$\frac{B^2}{32R^2}\left[7 - \frac{1}{1 - \left(\frac{\omega^* a}{c_b\pi}\right)^2}\right]\left(\frac{\pi}{a}\right)^4 = \left(\frac{1}{c_b R}\right)^2\omega^{*2} - \left(\frac{\pi}{a}\right)^4 \ldots [51]$$

Noting that the frequency of the linear problem ω is (9)

$$\omega = \left(\frac{\pi}{a}\right)^2 c_b R \ldots\ldots [52]$$

the ratio of the squares of the frequencies may be written as

$$\left(\frac{\omega^*}{\omega}\right)^2 = 1 + \frac{3}{8}\left(\frac{B}{h}\right)^2\left[7 - \frac{1}{1 - \frac{\pi^2}{12}\left(\frac{\omega^*}{\omega}\frac{h}{a}\right)^2}\right] \ldots [53]$$

Since for a beam $h/a \ll 1$, we may neglect the term $[(\omega^*/\omega)(h/a)]^2$ in comparison to unity. Relation [53] is then simplified to

$$\left(\frac{\omega^*}{\omega}\right)^2 = 1 + \frac{9}{4}\beta^2 \qquad [54]$$

β stands, as before, for B/h, and the ratio of periods is given by

$$\frac{T^*}{T} = \left[1 + \left(\frac{3}{2}\beta\right)^2\right]^{-1/2} \qquad [55]$$

This ratio is also plotted in Fig. 4 and it is seen that the agreement between the two solutions is excellent. Thus the conclusion is reached that in the present case the more elementary Duffing's approach yields practically the same result as a perturbation procedure followed by an integration involving elliptic functions.

In the case of a plate, however, Duffing's method would be rather tedious and no attempt was made to develop a solution by that method.

BIBLIOGRAPHY

1 "Influence of Large Amplitudes on Flexural Motions of Elastic Plates," by G. Herrmann, NACA TN 3578.

2 "On the Nonlinear Vibration Problem of the Elastic String," by G. F. Carrier, *Quarterly of Applied Mathematics*, vol. 3, 1945, p. 157.

3 "A Note on the Vibrating String," by G. F. Carrier, *Quarterly of Applied Mathematics*, vol. 7, 1949, p. 97.

4 "On the Nonlinear Vibration of Elastic Bars," by A. C. Eringen, *Quarterly of Applied Mathematics*, vol. 10, 1952, p. 361.

5 "On the Nonlinear Vibration of Circular Membrane," by A. C. Eringen, Proceedings of the First U. S. National Congress of Applied Mechanics, Chicago, Ill., 1951, p. 139.

6 "Nonlinear Vibrations," by J. J. Stoker, Interscience Publishers, Inc., New York, N. Y., 1950.

7 "Theory of Plates and Shells," by S. Timoshenko, McGraw-Hill Book Company, Inc., New York, N. Y., 1940.

8 "Large Deflection Theory for Rectangular Plates," by S. Levy, Proceedings of Symposia in Applied Mathematics, American Mathematical Society, New York, N. Y., vol. 1, 1949, p. 197.

9 "Vibration Problems in Engineering," by S. Timoshenko, D. Van Nostrand Company, Inc., New York, N. Y., 1937.

34

Reprinted from *AIAA J.*, **1**(3), 617–620 (1963)

Nonlinear Transverse Vibrations of Orthotropic Cylindrical Shells

J. L. Nowinski*
University of Delaware, Newark, Del.

Nonlinear transverse vibrations of elastic orthotropic shells are investigated using von Kármán-Tsien equations generalized to dynamic and orthotropic case. The deflection function is chosen in a simple separable form and the stress function is determined from the compatibility equation. The governing equation for the time function is derived by Galerkin's procedure, and its solution discussed for two types of orthotropy and for the isotropic case. A sharp decrease of the period of nonlinear vibrations with an increasing amplitude is corroborated, the mode pattern influencing the period more than the degree of anisotropy. Parenthetically, the influence of anisotropy on free linear vibrations and on the buckling under normal pressure is discussed.

Nomenclature

x,y	= axial and circumferential coordinates
u,v,w	= displacements of the median surface in the axial, circumferential, and transverse directions
t	= time
$\epsilon_x,\epsilon_y,\gamma_{xy}$	= components of strain in the median surface
$\sigma_x,\sigma_y,\tau_{xy}$	= components of stress in the median surface
E_1,E_2,G_{12}	= Young's moduli in the axial and circumferential directions, and shear modulus
ν_1,ν_2	= Poisson's ratios
k^2,m^2,p^2	= defined by Eq. (1.4)
h,a,l	= thickness, radius, and length of the shell
q_0	= lateral normal pressure
Φ	= stress function [see Eqs. (1.1)]
ρ	= mass density of the wall of the shell
i,j	= numbers of longitudinal half-waves and circumferential waves
α_i	= $i\pi/l$
β_j	= j/a
$f(t),f_0(t)$	= defined by Eq. (1.10)
$\delta_1,\delta_2,\delta_3,\delta^*$	= defined by Eq. (1.12)
$\epsilon_1,\epsilon_2,\delta$	= defined by Eq. (1.18)
ω	= frequency of linear free vibrations
Λ	= defined by Eq. (3.2)
$A,\tau(t)$	= defined by Eq. (4.2)
Λ_1,Λ_2	= defined by Eq. (4.4)
$K(k^*)$	= complete elliptic integral of the first kind
k^*,ω^*	= defined by Eq. (4.5.1)
$T,T^* = 4K/\omega^*$	= periods of linear and nonlinear free vibrations

Introduction

In June 1962, the author of the present note submitted for publication in the Journal of Aerospace Sciences an article on transverse nonlinear vibrations of cylindrical orthotropic shells prepared earlier as a technical report for the University of Delaware.[1] At this opportunity, one of the reviewers of the article drew the author's attention to a paper by Chu published in the October 1961 issue of the Journal of Aerospace Sciences.[2] In this paper, a problem similar to that mentioned was solved independently, confined, however, to the isotropic material. Again the field equations derived in Ref. 2 are similar to those derived in Ref. 1 if one specifies the latter to the isotropic case. Since the contribution of Chu represents a particular case of the solution given in Ref. 1, the present author was glad to note a close agreement between the final results obtained in both papers for the isotropic case. However, a publication of the full text of Ref. 1, after the previous publication of Ref. 2, did not seem necessary. In this connection, the writer has prepared the present note as an abstract of Ref. 1, quoting only those results that may possess a self-dependent interest. To make the note self-contained, the author starts with derivation of the general field equations, using, in contrast to Ref. 2 (in which these equations are derived as Euler-Lagrange equations from Hamilton's principle), the balance of momenta and the compatibility condition. As another difference, the derivation of the crucial differential time equation is carried on using Galerkin's method.

Received by IAS June 11, 1962; revision received December 19, 1962.

* Professor of Mechanical Engineering.

1. Fundamentals

Consider a thin circular shell with the dimensions shown in Fig. 1. Locate the origin of curvilinear coordinates x, y at an arbitrary point of one edge of the shell, and measure x in the axial and y in the circumferential direction in the median surface of the undeformed cylinder (of mean radius a). Assume that the wall of the shell is made of an orthotropic material, the principal elastic directions of which coincide with the axes x and y. The external (lateral) face of the shell is subjected to a normal pressure $q(x, y)$, which, for definiteness, is assumed to be constant and equal to q_0. Transverse vibrations of the wall of the shell are supposed to be large, that is, of the order of magnitude of the thickness of the plate. In problems such as this in which vibrations take place principally in the direction of the least stiffness (perpendicular to the median surface), it is reasonable and customary to neglect the inertia terms inherent with the motion in the median surface. One then may represent the stress components in terms of a stress function $\Phi(x,y,t)$ in the form

$$\sigma_x = \partial^2\phi/\partial y^2 \qquad \sigma_y = \partial^2\phi/\partial x^2 \qquad \tau_{xy} = -\partial^2\phi/\partial x\partial y \tag{1.1}$$

thus automatically satisfying the equations of equilibrium in the planes tangent to the median surface of the shell.

Following von Kármán and Tsien,[3] the strain components, including terms up to the second order, are expressed in the following form:

$$\begin{aligned}\epsilon_x &= (\partial u/\partial x) + \tfrac{1}{2}(\partial w/\partial x)^2 \\ \epsilon_y &= (\partial v/\partial y) + \tfrac{1}{2}(\partial w/\partial y)^2 - (w/a) \\ \gamma_{xy} &= (\partial u/\partial y) + (\partial v/\partial x) + (\partial w/\partial x)(\partial w/\partial y)\end{aligned} \tag{1.2}$$

where u, v, w are the components of displacement in the x, y, and transverse directions, respectively.

The stress and strain components in the median surface of the orthotropic wall of the shell then are related to each other by the familiar equations

$$\begin{aligned}\epsilon_x &= (\sigma_x/E_1) - \nu_2(\sigma_y/E_2) \\ \epsilon_y &= (\sigma_y/E_2) - \nu_1(\sigma_x/E_1) \\ \gamma_{xy} &= \tau_{xy}/G_{12}\end{aligned} \tag{1.3}$$

where E_1, E_2, and G_{12} denote Young's moduli in the x and y

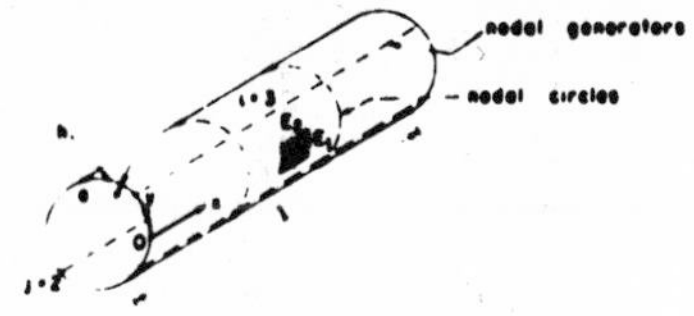

Fig. 1 Dimensions of the shell

directions and the shear modulus, respectively; ν_1 represents the relative contraction in the y direction influenced by the tension in the x direction. Apparently, the relation $E_1\nu_2 = E_2\nu_1$ holds.

In what follows, the following symbolism is used:

$$
\begin{aligned}
k^2 &= E_2/E_1 = D_2/D_1 \\
p^2 &= 2(G_{12}/E_1)(1 - k^2\nu_1^2) + \nu_2 = D_3/D_1 \\
m^2 &= (E_2/G_{12}) - 2\nu_2
\end{aligned}
\qquad (1.4)
$$

with

$$
D_1 = E_1I/(1 - \nu_1\nu_2) \qquad D_2 = E_2I/(1 - \nu_1\nu_2) \qquad D_3 = \tfrac{1}{2}(D_1\nu_2 + D_2\nu_1) + 2D_k \qquad (1.5)
$$

and $I = h^3/12$, $D_k = G_{12}I$, where D_1, D_2, and D_3 represent flexural and torsional rigidities of the wall of the shell, respectively.

The equation of motion in the direction normal to the median surface now yields

$$
\frac{\partial Q_x}{\partial x} + \frac{\partial Q_y}{\partial y} + \sigma_x h\frac{\partial^2 w}{\partial x^2} + \sigma_y h\left(\frac{1}{a} + \frac{\partial^2 w}{\partial y^2}\right) + 2\tau_{xy}h\frac{\partial^2 w}{\partial x\partial y} + q_0 - \rho h\frac{\partial^2 w}{\partial t^2} = 0 \qquad (1.6)
$$

where ρ denotes the mass density of the material of the wall, t the time, and

$$
\begin{aligned}
Q_x &= -(\partial/\partial x)[D_1(\partial^2 w/\partial x^2) + D_3(\partial^2 w/\partial y^2)] \\
Q_y &= -(\partial/\partial y)[D_3(\partial^2 w/\partial x^2) + D_2(\partial^2 w/\partial y^2)]
\end{aligned}
\qquad (1.7)
$$

are the shear forces in the transverse and axial cross sections. The variables u and v are now eliminated from Eqs. (1.1–1.3), thus establishing the following compatibility equation:

$$
\frac{\partial^4\phi}{\partial x^4} + m^2\frac{\partial^4\phi}{\partial x^2\partial y^2} + k^2\frac{\partial^4\phi}{\partial y^4} = E_2\left[\left(\frac{\partial^2 w}{\partial x\partial y}\right)^2 - \frac{\partial^2 w}{\partial x^2}\frac{\partial^2 w}{\partial y^2} - \frac{1}{a}\frac{\partial^2 w}{\partial x^2}\right] \qquad (1.8)
$$

first obtained by Donnell for the isotropic case. On the other hand, by virtue of (1.1) and (1.7), Eq. (1.6) yields

$$
\frac{\partial^4 w}{\partial x^4} + 2p^2\frac{\partial^4 w}{\partial x^2\partial y^2} + k^2\frac{\partial^4 w}{\partial y^4} = \frac{h}{D_1}\left[\frac{\partial^2 w}{\partial x^2}\frac{\partial^2\phi}{\partial y^2} + \frac{\partial^2 w}{\partial y^2}\frac{\partial^2\phi}{\partial x^2} - 2\frac{\partial^2\phi}{\partial x\partial y}\frac{\partial^2 w}{\partial x\partial y} + \frac{1}{a}\frac{\partial^2\phi}{\partial x^2} + \frac{q_0}{h} - \rho\frac{\partial^2 w}{\partial t^2}\right] \qquad (1.9)
$$

Equation (1.9) represents the second fundamental field equation, besides (1.8), governing the nonlinear dynamical problem for the circularly cylindrical orthotropic shell.

The deflection function $w(x, y, t)$ is chosen in the separable form

$$
w(x,y,t) = f(t)\sin\alpha_i x\cdot\sin\beta_j y + f_0(t) \qquad (1.10)
$$

where $\alpha_i = i\pi/l$ and $\beta_j = j/a$. Clearly, i represents the number of longitudinal half-waves and j the number of circumferential waves. If $i = 0$ or $j = 0$, a uniform radial oscillation is excited which, however, requires that the ends of the cylinder may "breath" without any hindrance. If $j = 1$, a translational motion of the cross sections, with no transverse distortion, occurs. To define a particular vibration pattern, it is necessary to prescribe the values both of i and of j (cf., Fig. 1, in which $i = 3$ and $j = 2$ have been chosen). Expression (1.10) is inserted into the compatibility equation (1.8), and a particular integral is obtained:

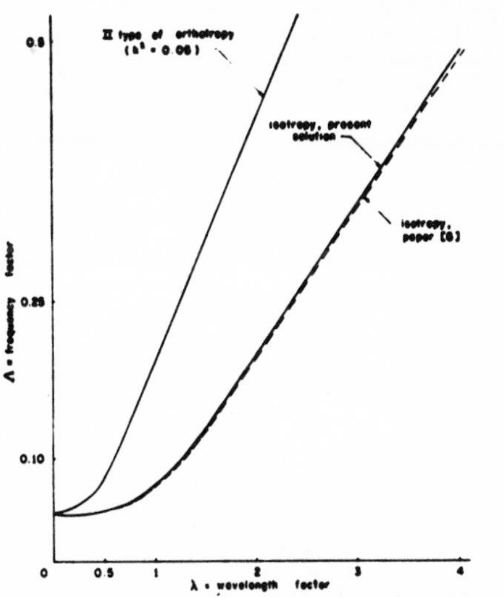

Fig. 2 Frequency factor vs wavelength factor for linear vibrations ($j = 4$, $h/a = 0.01$)

$$
\Phi(x,y,t) = E_2[(\delta_1\cos 2\alpha_i x + \delta_2\cos 2\beta_j y)f^2 + \delta_3\sin\alpha_i x\cdot\sin\beta_j y] - p_y(x^2/2) \qquad (1.11)
$$

where the following notation is used:

$$
\delta_1 = \frac{\beta_j^2}{32\alpha_i^2} \qquad \delta_2 = \frac{\alpha_i^2}{32\beta_j^2 k^2} \qquad \delta_3 = \frac{\alpha_i^2}{a\delta^*}
$$

$$
\delta^* = \alpha_i^4 + m^2\alpha_i^2\beta_j^2 + k^2\beta_j^4 \qquad (1.12)
$$

p_y is a constant equal to q_0a/h. The conditions that satisfy the particular solution (1.11) are now investigated.

Note first that in view of (1.11), Eqs. (1.1) for the stress components in the median surface admit the representation

$$
\begin{aligned}
\sigma_x &= -E_2\beta_j^2(4\delta_2 f^2\cos^2\beta_j y + \delta_3 f\sin\alpha_i x\cdot\sin\beta_j y) \\
\sigma_y &= -E_2\alpha_i^2(4\delta_1 f^2\cos 2\alpha_i x + \delta_3 f\sin\alpha_i x\cdot\sin\beta_j y) - p_y \\
\tau_{xy} &= -E_2\alpha_i\beta_j\delta_3 f\cos\alpha_i x\cdot\cos\beta_j y
\end{aligned}
\qquad (1.13)
$$

With these in mind, it can be shown that the axial resultant force of the normal stresses in any cross section of the shell vanishes. Since this result includes the ends of the shell, it follows that the self-equilibrated systems of axial normal stresses acting at the ends of the shell are associated with what one can interpret as an integral condition for the freely movable edges (in the axial direction). On the other hand, at the end cross sections of the shell, the following statical quantities vanish: 1) the transverse resultant force, 2) an overall twisting moment of the shear stresses, and 3) a resultant bending moment of the normal stresses with respect to an arbitrary diameter. It follows that the reactions at the ends of the shell conform to the equilibrium requirements associated with the prescribed uniform loading of the outer surface of the shell. Furthermore, an inspection reveals that in view of (1.10) the local bending moments at the edges of shell vanish identically. This result in combination with (1.10) conforms to the conditions of the free support of the edges of the shell in an integral sense. Note that the equilibrium of a portion of the shell singled out by an axial cross

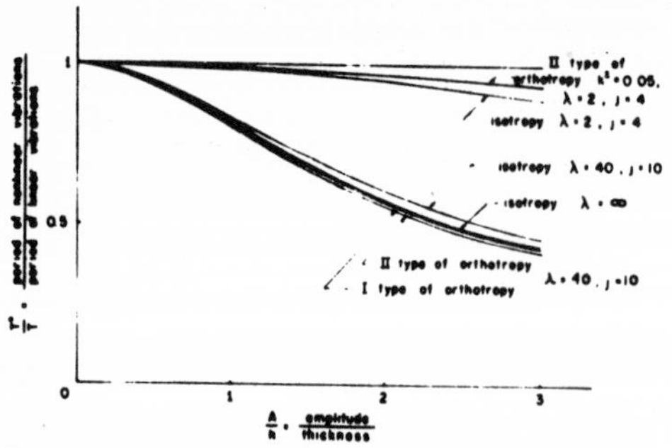

Fig. 3 Relative period vs relative amplitude for nonlinear vibrations

section $y = 0$ (or $y = \pi a$) yields the required equation for the membrane hoop stress:

$$p_y = q_0 a/h \tag{1.14}$$

induced by a uniform lateral pressure q_0.

The next task is to determine the time function $f_0(t)$ in terms of $f(t)$ by using the condition of closure (or periodicity of the circumferential displacement)

$$\int_0^{2\pi a} \frac{\partial v}{\partial y}\, dy = 0 \tag{1.15}$$

In view of (1.1–1.3), the desired relation appears as

$$f_0(t) = (\beta_j{}^2 a/8) f^2(t) + (q_0 a^2/hE_2) \tag{1.16}$$

The procedure of Galerkin is now applied to the only remaining equation of motion (1.9). To this end, one proceeds in the following way. Substitute expressions (1.10) and (1.13), representing in fact (with an appropriate sign) the second space derivatives of the stress function Φ, into Eq. (1.9). Then multiply both members of this equation by the spatial part $\sin \alpha_i x \cdot \sin \beta_j y$ of (1.10), and integrate the result over the domain of the median surface of the shell.

As a final result of a lengthy calculation, the following nonlinear differential equation of the second order is obtained for the time function:

$$(d^2 f/dt^2) + \epsilon_1 f + \epsilon_2 f^3 = 0 \tag{1.17}$$

where the following notation is used:

$$\epsilon_1 = \frac{D_1}{\rho h}\,\delta + \frac{E_2}{\rho}\left(\frac{\alpha_i{}^2}{a}\,\delta_2 + \frac{\beta_j{}^2 a}{E_2 h}\,q_0\right)$$
$$\epsilon_2 = 2\alpha_i{}^2\beta_j{}^2(\delta_1 + \delta_2)(E_2/\rho) \tag{1.18}$$
$$\delta = \alpha_i{}^4 + 2p^2\alpha_i{}^2\beta_j{}^2 + k^2\beta_j{}^4$$

This completes the general solution of the problem under consideration by reducing it to the solution of the governing time equation (1.17). The types of materials shown in Table 1 are now discussed.

2. Buckling under Uniform Pressure

Rejecting the inertia and nonlinear terms in (1.17), the equation for the uniform transverse pressure q_0 which provokes the buckling of the shell is thus obtained:

$$\frac{\delta h^2}{12k^2(1 - \nu_1\nu_2)} + \frac{\alpha_i{}^4}{\delta^* a^2} = \frac{a\beta_j{}^2}{E_1 h}\,q_0 \tag{2.1}$$

This equation, if specified to the isotropic case, confirms the result obtained in Ref. 4 and represents an orthotropic counterpart of the known equation of Mises, e.g., in Ref. 5.

3. Free Linear Vibrations

Rejecting the nonlinear term in Eq. (1.17), the equation is then reduced to the equation governing linear vibrations. If, moreover, the loading term is suppressed by posing $q_0 = 0$, the equation of free linear oscillations with the circular frequency is obtained:

$$\omega = (1/a)[E_2/\rho(1 - k^2\nu_1{}^2)]^{1/2}\Lambda \tag{3.1}$$

where

$$\Lambda = \left[\frac{\alpha^2}{12k^2}(\lambda^4 + 2p^2\lambda^2 j^2 + k^2 j^4) + \frac{(1 - k^2\nu_1{}^2)\lambda^4}{\lambda^4 + m^2\lambda^2 j^2 + k^2 j^4}\right]^{1/2}$$

is the frequency factor and $\lambda = i\pi a/l$, the wavelength factor introduced in Ref. 6. This result has been plotted in Fig. 2 for the second type of orthotropy and for the isotropic material (with $j = 4$ and $h/a = 0.01$) revealing, in the latter case, a close agreement with the corresponding results of Arnold and Warburton.[6] In the present case, however, for each chosen nodal pattern only one natural frequency (the lowest) exists instead of three possible, although fortunately the only one that appears to have practical significance (the others being well above the aural range). This fact is, apparently, the consequence of neglecting the inertia associated with in-plane motion. The ratio of dimensions chosen and $j = 4$ seem to result in a "primarily radial" mode of minimum frequency (see Ref. 7).

4. Free Nonlinear Vibrations

Turn now to the investigation of large vibrations of a shell to which is attributed one degree of freedom associated with the parametric function $f(t)$. To this purpose, a solution $f(t)$ of the complete time equation (1.17) has to be found. It is expedient to introduce the following representation:

$$f(t) = A\tau(t) \tag{4.1}$$

which permits the use of the normalized initial conditions:

$$\tau(0) = 1 \qquad \dot{\tau}(0) = 0 \tag{4.2}$$

Upon restriction to free vibrations, the loading term is suppressed in (1.18). With this in mind, Eq. (1.17) now is carried into

$$(\partial^2\tau/\partial t^2) + \Lambda_1\tau + \Lambda_3(A/h)^2\tau^3 = 0 \tag{4.3}$$

where the following notation is used:

$$\Lambda_1 = \frac{E_2}{\rho a^2}\left[\frac{\alpha^2(\lambda^4 + 2j^2p^2\lambda^2 + k^2j^4)}{12(1 - \nu_1\nu_2)k^2} + \frac{\lambda^4}{\lambda^4 + m^2j^2\lambda^2 + k^2j^4}\right] \tag{4.4}$$

$$\Lambda_3 = \frac{E_2}{\rho a^2}\,\alpha^2\,\frac{\lambda^4 + j^4k^2}{16k^2}$$

Table 1

Type	E_1	E_2	G_{12}	ν_1	ν_2	k^2	p^2	m^2
Orthotropy I	1×10^6	0.5×10^6	0.1×10^6	0.05	0.025	0.5	0.223	5
Orthotropy II	1×10^6	0.05×10^6	0.05×10^6	0.2	0.01	0.05	0.108	1
Isotropy	1×10^6	1×10^6	G	0.3	0.3	1	1	2

Obviously, $\Lambda_1 = \omega^2$, where ω is defined by (3.1).

A solution of the foregoing equation satisfying the initial conditions (4.2) is the cosine-type Jacobian elliptic function

$$\tau(t) = cn(\omega^* t, k^*) \tag{4.5}$$

with

$$\omega^* = [\Lambda_1 + \Lambda_3(A/h)^2]^{1/2}$$
$$k^{*2} = \frac{\Lambda_3(A/h)^2}{2[\Lambda_1 + \Lambda_3(A/h)^2]} \tag{4.5.1}$$

Since the period of the linear vibrations is $T = 2\pi/\Lambda_1^{1/2}$ the ratio of periods of nonlinear and linear vibrations is obtained as

$$\frac{T^*}{T} = \frac{2K}{\pi[1 + (\Lambda_3/\Lambda_1)(A/h)^2]^{1/2}} \tag{4.6}$$

In Fig. 3, the relation (4.6) is plotted for two types of orthotropy discussed previously and, for the sake of comparison, also for the isotropic case. For definiteness, it is assumed $\alpha = h/a = 0.01$, and posed $j = 10$ and $\lambda = 40$. An inspection reveals a known sharp decrease of the period of vibrations with an increasing amplitude. Also, graphs for isotropic case and for the II type of orthotropy are plotted assuming $j = 4$ and $\lambda = 2$. Apparently, the mode patterns influence the period of the nonlinear vibrations more than the degree of anisotropy (at least for the two types of orthotropy considered).

References

[1] Nowinski, J., "Response of a cylindrical shell to transverse nonlinear oscillations," TR 9, Dept. Mech. Eng., Univ. Delaware (May 1962).

[2] Chu, H.-N. "Influence of large amplitude on flexural vibrations of a thin circular cylindrical shell," J. Aerospace Sci. **28**, 602–609 (1961).

[3] von Kármán, T. and Tsien, H. S., "The buckling of thin cylindrical shells under axial compression," J. Aero/Space Sci. **8**, 303–312 (1941).

[4] Agamirov, V. L. and Volmir, A. S., "Behavior of cylindrical shells under hydrostatic dynamic loading and axial compression," Izv. Akad. Nauk SSSR, Otd. Tekhn. Nauk, Mekhan. i Mashinostr., no. 3, 78–83 (1959).

[5] Timoshenko, S. and Gere, J. M., *Theory of Elastic Stability* (McGraw-Hill Book Co. Inc., New York, 1961).

[6] Arnold, R. N. and Warburton, G. B., "The flexural vibrations of thin cylinders," Proc. Inst. Eng. **167**, 62–74 (1953).

[7] Baron, M. L. and Bleich, H., "Tables for frequencies and modes of free vibration of infinitely long thin cylindrical shells," J. Appl. Mech. **21**, 178–184 (1954).

Reprinted from *J. Appl. Mech.*, **33**, 553–560 (1966)

D. A. EVENSEN
Aerospace Engineer,
NASA Langley Research Center,
Langley Station, Hampton, Va.
Assoc. Mem. ASME

Nonlinear Flexural Vibrations of Thin Circular Rings[1]

The nonlinear flexural vibrations of thin circular rings are analyzed by assuming two vibration modes and then applying Galerkin's procedure on the equations of motion. The results show that vibrations involving either a single bending mode or two coupled bending modes can occur. Theory and experiment both indicate a nonlinearity of the softening type and the existence of these coupled-mode vibrations. Test results for the steady-state response are in good agreement with the calculated values, and the deflection modes used in the analysis agree with the experimental mode shapes. The analytical and experimental results exhibit several features that are characteristic of nonliner vibrations of axisymmetric systems in general and of circular cylindrial shells in particular.

The use of thin-walled cylinders in the design of missiles and launch vehicles has given rise to a number of theoretical studies on the nonlinear vibration of thin cylindrical shells [1, 2, 3].[2] Recent experimental results [4] suggest that such vibrations are still not fully understood. This prompted the author to investigate a simpler but related problem, namely, the nonlinear flexural vibrations of thin circular rings.

Linear vibrations of rings have been studied extensively, both theoretically [5, 6] and experimentally [7, 8]. Nonlinear flexural vibrations of rings have been examined briefly for free vibrations [9] and dynamic stability problems [10, 11]; however, no corresponding experiments are known to the author.

In the present work, only vibrations in the plane of the ring are considered, and the stress-strain law is taken to be linear. The nonlinearities which arise are geometric in nature, originating in the nonlinear terms of the strain-displacement relations. The ring is assumed to be of uniform rectangular cross section and to be relatively thin. With these restrictions, a detailed theoretical and experimental study of the vibrations was conducted.

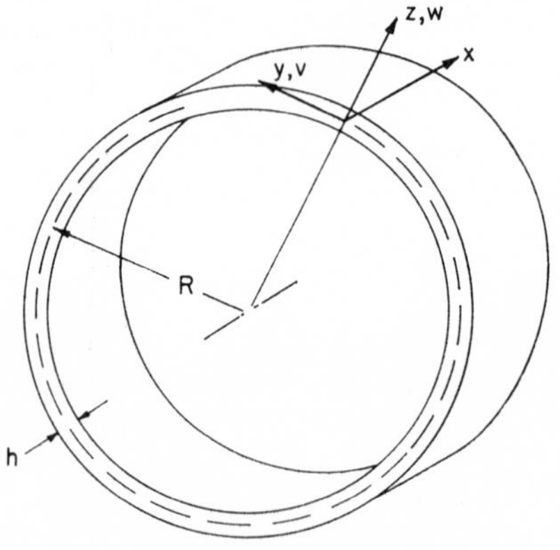

Fig. 1 Ring geometry and coordinate system

The Equations of Motion

The equations of motion for a thin ring can be obtained by a specialization of the analogous equations for cylindrical shells [1, 2]. This was accomplished using the following assumptions:

1 The radial displacement w, the tangential displacement v, and the applied load q are taken to be functions of only two variables: The circumferential coordinate y and time t. (See Fig. 1.)

2 The thickness of the ring h and its width are both taken to be constant. The ring is assumed to be thin, such that $(h/R)^2 \ll 1$.

3 The longitudinal in-plane force N_x and the shearing force N_{xy} are both taken to be zero throughout the ring.

Simplifying the cylinder equations as indicated in the foregoing yields the equations of motion for a thin ring:

$$\frac{\partial N_y}{\partial y} = \rho h \frac{\partial^2 v}{\partial t^2} \qquad (1a)$$

[1] Based on a doctoral dissertation of a similar title submitted by the author in May, 1964, to the faculty of the California Institute of Technology, Pasadena, Calif.

[2] Numbers in brackets designate References at end of paper.

Presented at the Fifth U. S. National Congress of Applied Mechanics, University of Minnesota, Minneapolis, Minn., June 14–17, 1966.

Discussion of this paper should be addressed to the Editorial Department, ASME, United Engineering Center, 345 East 47th Street, New York, N. Y. 10017, and will be accepted until October 15, 1966. Discussion received after the closing date will be returned. Manuscript received by ASME Applied Mechanics Division, March 29, 1965; final draft, September 7, 1965.

Nomenclature

$A_0(t), A_n(t), B_n(t)$ = generalized coordinates
$A(\tau), B(\tau), \phi(\tau), \psi(\tau)$ = slowly varying amplitudes and phases
D = bending stiffness, $\dfrac{Eh^3}{12(1-\nu^2)}$
E = Young's modulus
F_n = amplitude of the applied loading
G_n = dimensionless force
h = ring thickness
n = number of circumferential waves
N_x, N_{xy}, N_y = resultant forces/unit length
R = mean radius of the ring
t = time
v, w = displacements of a point on the middle surface in the circumferential and radial directions, respectively
x, y, z = coordinates in the axial, circumferential, and radial directions, respectively
β_c, β_s = percent critical damping in the cos (ny/R) and sin (ny/R) modes, respectively

(Continued on next page)

$$D\left(\frac{\partial^2}{\partial y^2}+\frac{1}{R^2}\right)\left(\frac{\partial^2 w}{\partial y^2}+\frac{w}{R^2}\right)+\frac{N_y}{R}-\frac{\partial}{\partial y}\left[N_y\frac{\partial w}{\partial y}\right]+\rho h\frac{\partial^2 w}{\partial t^2}=q(y,t) \quad (1b)$$

where the circumferential force is given by

$$N_y = Eh\left[\frac{\partial v}{\partial y}+\frac{w}{R}+\frac{1}{2}\left(\frac{\partial w}{\partial y}\right)^2\right] \quad (2)$$

and the bending term has been changed from $D(\partial^4 w/\partial y^4)$ to

$$D\left(\frac{\partial^2}{\partial y^2}+\frac{1}{R^2}\right)\left(\frac{\partial^2 w}{\partial y^2}+\frac{w}{R^2}\right)$$

to improve the accuracy of the equations [12, 13].

In addition to the equations of motion, the displacements must also satisfy the following continuity requirements:

$$w(y,t) = w(y+2\pi R, t)$$
$$\frac{\partial^k w}{\partial y^k}(y,t) = \frac{\partial^k w}{\partial y^k}(y+2\pi R,t) \quad (k=1,2,3) \quad (3a)$$

$$v(y,t) = v(y+2\pi R,t)$$
$$\frac{\partial v}{\partial y}(y,t) = \frac{\partial v}{\partial y}(y+2\pi R,t) \quad (3b)$$

These conditions insure that the displacements, slope, and strains will remain continuous in going around the circumference of the ring.

With respect to the accuracy of equations (1), it is well to note that they are subject to the limitations inherent in the cylindrical shell analyses [1, 2]. In the derivation of the shell equations, terms of order $1/n^2$ have been neglected in comparison to unity; as a result, equations (1) are slightly inaccurate for small values of n. These inaccuracies have been partially compensated for by modifying the bending term as indicated previously. Other corrections involving $1/n^2$ can be obtained by including additional nonlinear terms in the strain-displacement relations. [See equation (19).]

Simplification of the Equations for Flexural Vibrations

Previous work [12] has shown that, for flexural vibrations, the foregoing equations of motion can be simplified considerably without losing the essential features of the problem. This simplification is accomplished by

1 Assuming that the midsurface circumferential strain is zero.

2 Neglecting the effect of tangential inertia in equation (1*a*).

With the usual nonlinear strain-displacement relations [2], the first of these approximations requires

$$\epsilon_{yy} = \frac{\partial v}{\partial y}+\frac{w}{R}+\frac{1}{2}\left(\frac{\partial w}{\partial y}\right)^2 = 0 \quad (4)$$

and neglecting tangential inertia in (1*a*) gives

$$\frac{\partial N_y}{\partial y} = 0 \quad (5)$$

Combining these expressions with (2) gives $N_y = 0$; then equation (1*b*) reduces to

$$D\left(\frac{\partial^2}{\partial y^2}+\frac{1}{R^2}\right)\left(\frac{\partial^2 w}{\partial y^2}+\frac{w}{R^2}\right)+\rho h\frac{\partial^2 w}{\partial t^2}=q(y,t) \quad (6)$$

which is a linear equation for w. The problem itself is still nonlinear, however, since w must satisfy the nonlinear inextensionality condition of equation (4). Note that the continuity requirements (3) on w and v are not altered by the preceding discussion.

With the foregoing simplifications, the remaining problem is to solve equation (6) subject to the constraints (3) and (4). Despite the fact that (6) is a linear equation, the presence of the nonlinear inextensionality condition (4) obviates exact solutions and makes it necessary to resort to approximate techniques.

Reduction to Nonlinear Ordinary Differential Equations

Approximate solutions to equation (6) can be obtained by assuming the shape of the deflection in space. This approach is commonly used in nonlinear vibrations of structures, and it reduces the problem to one involving coupled nonlinear ordinary differential equations in t.

The most general radial deflection compatible with the continuity requirements is

$$w(y,t) = \sum_{n=0}^{\infty}\left[A_n(t)\cos\frac{ny}{R}+B_n(t)\sin\frac{ny}{R}\right]$$

where $A_n(t)$ and $B_n(t)$ are periodic in time. This case has been analyzed in detail elsewhere [12], but the main features of the results may be obtained by using the following two-mode approximation:

$$w(y,t) = A_n(t)\cos\frac{ny}{R}+B_n(t)\sin\frac{ny}{R}+A_0(t) \quad (n \geq 2) \quad (7)$$

Here $\cos(ny/R)$, $\sin(ny/R)$ are the linear vibration modes of the ring, and equation (7) is limited to $n \geq 2$ since only flexural motions are being considered. The $n = 0$ mode is (by itself) an axisymmetric mode involving extension of the midsurface of the ring, whereas the $n = 1$ modes correspond to motion of the ring as a rigid body. The $A_0(t)$-term in (7) is included to satisfy the continuity constraint on the circumferential displacement v.

Substituting the assumed deflection (7) into the inextensionality condition (4) and requiring continuity on the v-displacement determines $A_0(t)$ in terms of A_n and B_n. Then the expression for w becomes

$$w(y,t) = A_n(t)\cos\frac{ny}{R}+B_n(t)\sin\frac{ny}{R}-\frac{n^2}{4R}\left[A_n^2(t)+B_n^2(t)\right] \quad (n \geq 2) \quad (8)$$

Nomenclature

ϵ = dimensionless nonlinearity parameter, $(n^2h/R)^2$

ϵ_c = corrected nonlinearity parameter

ϵ_{yy} = circumferential strain on the midsurface of the ring

$\zeta_c(\tau)$, $\zeta_s(\tau)$ = nondimensional generalized coordinates associated with the cos (ny/R) and sin (ny/R) modes, respectively

ν = Poisson's ratio

ρ = mass density

τ = nondimensional time, $\omega_M t$

ω = vibration frequency, rad/sec

ω_M = approximate vibration frequency for linear vibrations, defined by $\omega_M^2 = \dfrac{E}{\rho R^2}\dfrac{(n^2-1)^2}{12(1-\nu^2)}\left(\dfrac{h}{R}\right)^2$

ω_L = vibration frequency for linear vibrations, defined by $\omega_L^2 = \dfrac{E}{\rho R^2}\dfrac{n^2}{(n^2+1)}\dfrac{(n^2-1)^2}{12(1-\nu^2)}\left(\dfrac{h}{R}\right)^2$

Ω = nondimensional frequency, ω/ω_M

A bar over a quantity indicates that it is an average value.

This deflection is compatible with the constraints (3), (4), and it can be used with Galerkin's procedure to approximately satisfy the remaining equation of motion (6). To do this, equation (8) is substituted for w in (6), and the resulting expression is multiplied by a weighting function associated with $A_n(t)$ and then integrated on y from 0 to $2\pi R$. This procedure yields an ordinary nonlinear differential equation involving primarily $A_n(t)$. An equation for $B_n(t)$ is obtained in a similar fashion; both equations are coupled in the nonlinear terms.

The weighting functions used are

$$\frac{\partial w}{\partial A_n} = \cos\frac{ny}{R} - \frac{n^2 A_n}{2R}$$

and

$$\frac{\partial w}{\partial B_n} = \sin\frac{ny}{R} - \frac{n^2 B_n}{2R}$$

respectively.

Carrying out the operations indicated in the foregoing and then nondimensionalizing the results yield the following coupled equations:

$$\frac{d^2\zeta_c}{d\tau^2} + 2\beta_c\frac{d\zeta_c}{d\tau} + \zeta_c + \frac{\epsilon\zeta_c}{2}\left[\zeta_c\frac{d^2\zeta_c}{d\tau^2} + \left(\frac{d\zeta_c}{d\tau}\right)^2 + \zeta_s\frac{d^2\zeta_s}{d\tau^2} + \left(\frac{d\zeta_s}{d\tau}\right)^2\right] = G_n\cos\Omega\tau \quad (9a)$$

$$\frac{d^2\zeta_s}{d\tau^2} + 2\beta_s\frac{d\zeta_s}{d\tau} + \zeta_s + \frac{\epsilon\zeta_s}{2}\left[\zeta_s\frac{d^2\zeta_s}{d\tau^2} + \left(\frac{d\zeta_s}{d\tau}\right)^2 + \zeta_c\frac{d^2\zeta_c}{d\tau^2} + \left(\frac{d\zeta_c}{d\tau}\right)^2\right] = 0 \quad (9b)$$

where the dimensionless variables are

$$\zeta_c = \frac{A_n}{h} \qquad \zeta_s = \frac{B_n}{h}$$

$$\tau = \omega_M t \qquad \Omega = \frac{\omega}{\omega_M}$$

$$\epsilon = \left(\frac{n^2 h}{R}\right)^2 \qquad G_n = \frac{F_n}{\pi R\rho h^2\omega_M^2}$$

and

(a) Two minor nonlinear terms of the form $\frac{\epsilon}{4n^4}\zeta_c(\zeta_c^2 + \zeta_s^2)$ have been discarded.

(b) For simplicity, the loading has been taken as

$$q(y, t) = \frac{F_n}{\pi R}\cos\frac{ny}{R}\cos\omega t \quad (10)$$

(c) Modal damping terms have been inserted to study the first-order effects of small viscous damping.

Although these results were derived using a Galerkin procedure, they can also be obtained by the Rayleigh-Ritz method. In general, these two approximate techniques are not equivalent; however, they can be made to coincide, as Singer [14] has indicated. This accounts for the choice of $\partial w/\partial A_n$ and $\partial w/\partial B_n$ as the weighting functions used to obtain equations (9).

Approximate Solutions to the Nonlinear Equations

Response of a Single Bending Mode

The applied loading [equation (10)] directly drives only one mode of the ring; thus a possible solution to equations (9) involves the response of only the driven mode. This response can be calculated approximately by applying the method of averaging [15]. To obtain such solutions, let

$$\zeta_c(\tau) = A(\tau)\cos[\Omega\tau + \phi(\tau)]$$
$$\zeta_s(\tau) = 0 \quad (11)$$

where $A(\tau)$ and $\phi(\tau)$ are presumed to be slowly varying functions of τ. When these expressions for ζ_c and ζ_s are substituted into equation (9a) and the appropriate averages are carried out, two equations for steady-state vibrations result (see Appendix 1):

$$[1 - \Omega^2]\bar{A} - \frac{\epsilon\Omega^2\bar{A}^3}{4} = G_n\cos\bar{\phi} \quad (12a)$$

$$-2\beta_c\Omega\bar{A} = G_n\sin\bar{\phi} \quad (12b)$$

Here $\bar{A}$ and $\bar{\phi}$ are average values (over one period) of $A(\tau)$ and

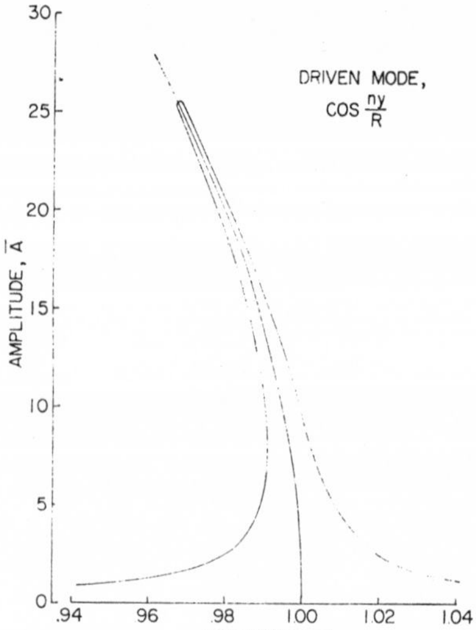

Fig. 2 Typical single-mode response: $\epsilon = 4.2 \times 10^{-4}$, $G_n = 0.10$, $\beta_c = 2 \times 10^{-3}$

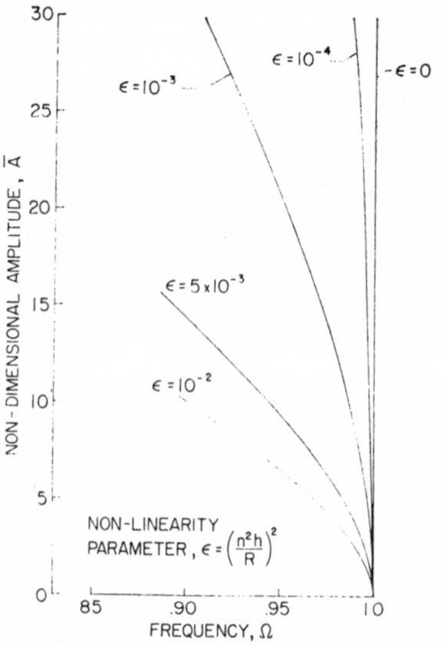

Fig. 3 Backbone curves for various values of ε

$\phi(\tau)$. For given values of G_n, β_c, and ϵ, the foregoing equations can be solved simultaneously for $\bar{A}$ and $\bar{\phi}$. The approximate solution to equations (9) then becomes

$$\zeta_c(\tau) = \bar{A} \cos [\Omega\tau + \bar{\phi}] \qquad \zeta_s(\tau) = 0 \tag{13}$$

for vibrations where only the driven mode [cos (ny/R)] responds. A typical resonance curve for this case is shown in Fig. 2, which demonstrates the nonlinearity of the softening type.

The case of free vibrations may be obtained by putting G_n and β_c equal to zero in the preceding results. Equations (12) then yield

$$\Omega = 1 - \frac{\epsilon \bar{A}^2}{8} + 0(\epsilon^2) \tag{14}$$

which is the so-called "backbone curve." This result is illustrated in Fig. 3 for various values of ϵ.

Stability of the One-Mode Response

The stability of the preceding solution was investigated by perturbing $\zeta_c(\tau)$ and $\zeta_s(\tau)$. A study of the resulting Mathieu-Hill equations indicated that

(*a*) Perturbations of ζ_c are unstable within the area bounded by

$$1 - \frac{3\epsilon\bar{A}^2}{8} + 0(\epsilon^2) < \Omega < 1 - \frac{\epsilon\bar{A}^2}{8} + 0(\epsilon^2)$$

(*b*) Perturbations of ζ_s are unstable within the region

$$1 - \frac{\epsilon\bar{A}^2}{8} + 0(\epsilon^2) < \Omega < 1 + \frac{\epsilon\bar{A}^2}{8} + 0(\epsilon^2)$$

(*c*) Both types of perturbations are unstable in narrow regions near $\Omega = 1/2, 1/3, 1/4, \ldots$.

The stability calculations are outlined in Appendix 2.

The boundaries of the first area (*a*) can be shown to coincide with the locus of vertical tangents to the response curves, where the jump phenomenon of nonlinear vibrations occurs. The narrow regions near $\Omega = 1/2, 1/3$, and so on, indicate the possibility of ultraharmonic responses in those areas. Within the second region (*b*), the companion mode [sin (ny/R), ζ_s] becomes parametrically excited and participates in the motion. This two-mode response occurs since nonlinear coupling exists between the driven mode [cos (ny/R)] and its companion mode [sin (ny/R)], which both have the same natural frequency.

Response of the Self-Coupled Bending Modes

When $\zeta_s(\tau)$ and $\zeta_c(\tau)$ both oscillate, let

$$\zeta_c(\tau) = A(\tau) \cos [\Omega\tau + \phi(\tau)] \qquad \zeta_s(\tau) = B(\tau) \sin [\Omega\tau + \psi(\tau)] \tag{15}$$

where A, B, ϕ, and ψ are presumed to be slowly varying functions of τ. Substituting the foregoing expressions for ζ_c and ζ_s into equations (9) and applying the method of averaging give

$$[1 - \Omega^2]\bar{A} - \frac{\epsilon\Omega^2\bar{A}}{4}[\bar{A}^2 - \bar{B}^2 \cos 2\bar{\Delta}] = G_n \cos \bar{\phi} \tag{16a}$$

$$-2\beta_c\Omega\bar{A} - \frac{\epsilon\Omega^2\bar{A}}{4}\bar{B}^2 \sin 2\bar{\Delta} = G_n \sin \bar{\phi} \tag{16b}$$

$$[1 - \Omega^2]\bar{B} - \frac{\epsilon\Omega^2\bar{B}}{4}[\bar{B}^2 - \bar{A}^2 \cos 2\bar{\Delta}] = 0 \tag{16c}$$

$$2\beta_s\Omega\bar{B} - \frac{\epsilon\Omega^2\bar{B}}{4}\bar{A}^2 \sin 2\bar{\Delta} = 0 \tag{16d}$$

for steady-state vibrations. Here $A(\tau)$ has been replaced by its average value over one cycle, denoted by $\bar{A}$, and so on, and the symbol $\bar{\Delta}$ is an average phase difference

$$\bar{\Delta} = \bar{\psi} - \bar{\phi} \tag{17}$$

Equations (16) and (17) can be solved simultaneously for $\bar{A}$, $\bar{B}$, $\bar{\phi}$, $\bar{\psi}$, and $\bar{\Delta}$; then the approximate two-mode solution becomes

$$\zeta_c(\tau) = \bar{A} \cos [\Omega\tau + \bar{\phi}] \qquad \zeta_s(\tau) = \bar{B} \sin [\Omega\tau + \bar{\psi}] \tag{18}$$

Details of the calculations are given in reference [12]. Some typical coupled-mode response curves are shown in Fig. 4.

An interesting feature of the two-mode response was the appearance of a "gap" or discontinuity in the solution. Slightly to the left of the resonance peak, the approximate solutions for ζ_s and ζ_c break down. The results of a stability analysis showed that the gap in the response coincides with a narrow region in which both the one-mode solution (13) and the two-mode solution (18) are unstable. Subsequent analog computer studies verified the calculated response curves, and nonsteady vibrations were found to occur in the gap region. Similar responses involving gaps have been observed in the case of fuel sloshing and in other nonlinear problems [16, 17].

Another interesting result demonstrated in Fig. 4 is that, in some cases, the companion mode [sin (ny/R), which is not driven by the forcing function] can vibrate to larger amplitudes than the driven mode [cos (ny/R)]. Responses of this type occur for $\Omega < 1$ and were detected experimentally. Finally, a comparison of Figs. 2 and 4 shows that, in the two-mode case, the maximum amplitude of the driven mode is considerably reduced from what it would be if [sin (ny/R)] did not respond.

Stability of the Coupled-Mode Response

As in the previous case, the stability of the vibrations can be found by perturbing ζ_c and ζ_s. The resulting analysis shows that real, nonzero values of $\bar{B}$ do not exist for $\Omega > 1 + \epsilon\bar{A}^2/8 + 0(\epsilon^2)$ and that the two-mode solution is unstable for

$$\Omega < 1 - \frac{2.38\epsilon\bar{A}^2}{8} + 0(\epsilon^2)$$

The calculations themselves are somewhat involved [12] and will not be repeated here. Comparing the foregoing results with

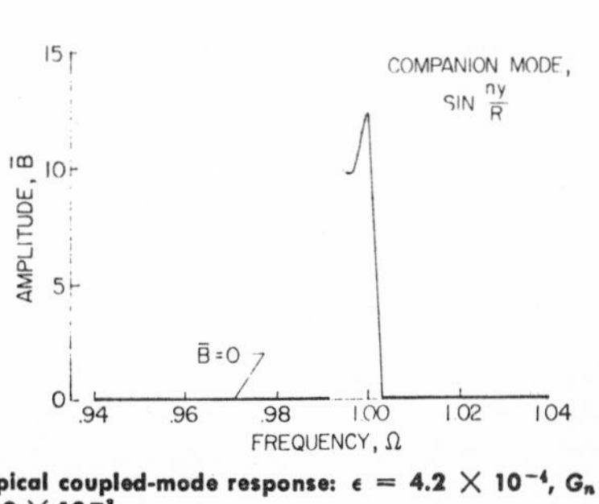

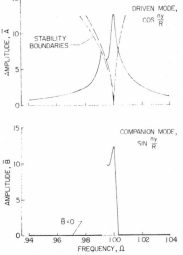

Fig. 4 Typical coupled-mode response: $\epsilon = 4.2 \times 10^{-4}$, $G_n = 0.10$: $\beta_c = \beta_s = 2 \times 10^{-3}$

the unstable regions given previously shows that the one-mode solution and the two-mode results are both unstable in a narrow region given by

$$1 - \frac{3\epsilon \bar{A}^2}{8} + 0(\epsilon^2) < \Omega < 1 - \frac{2.38\epsilon \bar{A}^2}{8} + 0(\epsilon^2)$$

This coincides with the gap in the two-mode response, as was previously noted.

Corrections for Additional Nonlinearities and Other Effects

The preceding analysis neglects shear deformation, rotary inertia, and extension of the midsurface of the ring. As shown by Buckens [6] and others [11], these effects are relatively minor as long as the wavelength of the vibration mode is large in comparison to the ring thickness.

A major improvement in the nonlinear analysis was obtained by employing the more exact strain-displacement relation [18]

$$\epsilon_{yy} = \frac{\partial v}{\partial y} + \frac{w}{R} + \frac{1}{2}\left[\left(\frac{\partial w}{\partial y} - \frac{v}{R}\right)^2 + \left(\frac{w}{R} + \frac{\partial v}{\partial y}\right)^2\right] \quad (19)$$

in the inextensionality condition. The improved deflection then becomes

$$w(y, t) = A_n(t) \cos \frac{ny}{R} + B_n(t) \sin \frac{ny}{R}$$

$$- \frac{n^2}{4R}\left(1 - \frac{1}{n^2}\right)^2 [A_n^2(t) + B_n^2(t)] \quad (20)$$

and the preceding analysis can be repeated using equation (20) in place of equation (8). It is unnecessary to repeat the actual computations, however, since both of these expressions for w have exactly the same form. That is, the results of the previous calculations can be modified to account for the additional nonlinearities by simply correcting the parameter ϵ.

It is also possible to retain tangential inertia in the analysis [12]. Both of these effects can be accounted for by replacing Ω by

$$\frac{\omega}{\omega_L} = \frac{\text{Frequency of vibration}}{\text{Linear vibration frequency}}$$

and ϵ by the corrected nonlinearity parameter

$$\epsilon_c = \left(\frac{n^2}{n^2 + 1}\right)\left(1 - \frac{1}{n^2}\right)^4\left(1 + \frac{1}{8n^2}\right)\epsilon \quad (21)$$

in the preceding analysis. These modifications result in significant improvement in the theory, bringing it into closer agreement with the experiments. The corrections are quite large for low values of n, with the effect of the additional nonlinearities predominating. Values of ϵ_c/ϵ for various values of n are given in Table 1.

Experiments

Apparatus

A thin copper ring was used in the experiments; it had a 4-in. radius and a thickness of 5.1×10^{-3} in. It was 0.988 in. long in the axial direction. The ring was supported by four very thin suspension threads, equally spaced around the circumference, as shown in Fig. 5.

Table 1

n, mode number	ϵ_c/ϵ
2	0.261
3	0.571
4	0.734
5	0.820
10	0.950

Radial motions of the ring were measured by two inductance-type pickups which operated in a push-pull fashion. These deflection sensors were mounted on a fixture with a large bearing that allowed them to move circumferentially around the ring. Vibrations of the ring were excited by means of an electrodynamic shaker. The shaker was connected to the ring by a fine tungsten drive wire (0.001 in. dia), which served as a very soft coupling spring between the shaker and the ring. (A block diagram of the apparatus is shown in Fig. 5.) This arrangement made it possible to estimate the amplitude of the force which was experimentally applied to vibrate the ring.

The response of the ring was analyzed, and the vibration modes were identified by means of Lissajous figures. Refined experimental techniques were used to minimize the nonlinearities introduced by the measuring system and by the suspension, shaker, and drive arrangement. (A detailed discussion of the experimental apparatus is given in reference [12].)

Measurement of the Mode Shapes

The mode shapes were measured by exciting one mode [cos (ny/R), $n = 4$] and recording the amplitude of the response at intervals along a half wavelength. Measurements were made separately on two different half waves; the amplitudes ranged from 1 to 27 times the thickness of the ring. The results are shown in Fig. 6, where the solid lines correspond to the deflection

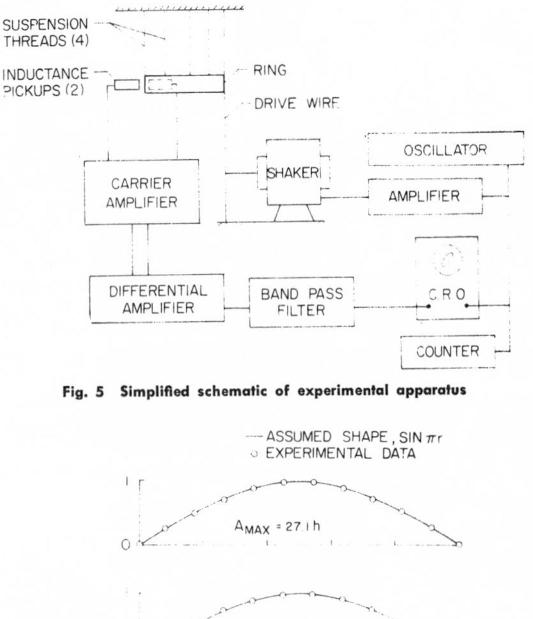

Fig. 5 Simplified schematic of experimental apparatus

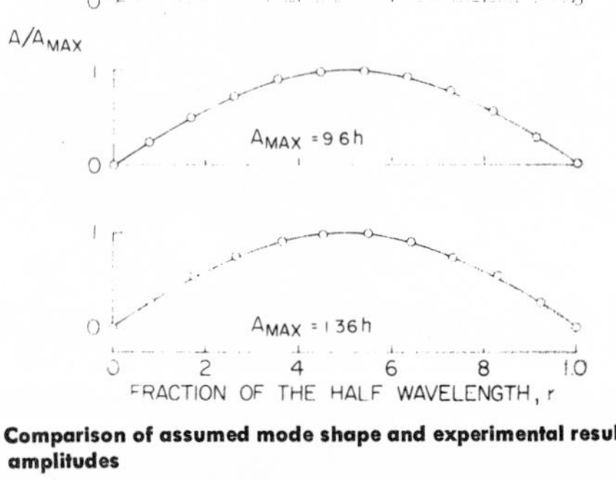

Fig. 6 Comparison of assumed mode shape and experimental results for various amplitudes

shape assumed in the analysis. As indicated by the figure, the main vibration shape was virtually independent of amplitude. Additional results [12] included a plot of the root-mean-square response around the circumference of the ring and measurement of the motions which occurred at the nodes of cos (ny/R).

Response Curves for the $n = 4$ Mode

Some typical experimental response curves are shown in non-dimensional form in Fig. 7, for two different values of the input force. The magnitude of the input force was held contant for the individual tests by maintaining a fixed displacement of the shaker (δ_0, peak to peak) during each run. An electronic counter was used to measure the period of the forcing function at each data point, which enabled the frequency ratio (ω/ω_L) to be very accurately determined. The amplitude of $A_n(t)$ was measured

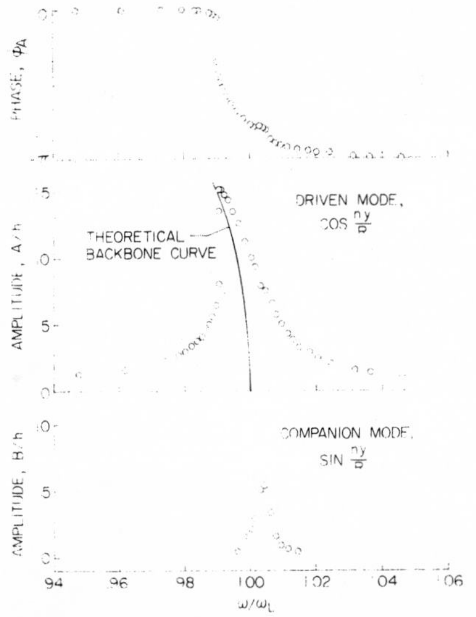

Fig. 7(a) Response curves for $n = 4$ modes: Shaker displacement $\delta_0 =$ 0.2 in. peak to peak

at one antinode of cos (ny/R), and the phase of the response (ϕ_A) was determined there with a standard phase meter; the magnitude of $B_n(t)$ was measured at an adjacent node of cos (ny/R). To facilitate comparison with the analysis, these experimental amplitudes were nondimensionalized on the ring thickness.

Corresponding theoretical results are shown by the solid lines in Fig. 7. The backbone curves were calculated using equation (14) with (ω/ω_L) in place of Ω and ϵ_c substituted for ϵ, both changes being made to account for the influence of tangential inertia and the additional nonlinearities discussed previously. The same corrections were used in computing the forced response, which was obtained from equations (12) for the undamped case. The input force used for these calculations was computed using the shaker displacement and the spring constant of the drive wire. A detailed comparison of the calculated and measured response for the driven mode is given in Fig. 8, which is for a peak-to-peak shaker displacement of $\delta_0 = 0.3$ in.

Discussion of the Experimental Results

As shown in Fig. 6, the experimentally determined mode shapes were nearly perfect sine waves along any half wavelength. This result was found to be independent of the maximum amplitude of the motion, as Fig. 6 indicates. For the single-mode response, the displacements at the nodes of cos (ny/R) were found to increase linearly with the square of the maximum amplitude [12]. This behavior is in agreement with the analytical results.

The response curves in Fig. 7 show good correlation with the theoretical backbone curves, and Fig. 8 shows a frequency difference of less than 1 percent between theory and experiment. This slight discrepancy was probably due to small errors in estimating the force applied to the ring.

The experimental response curves exhibited the jump phenomena and the appearance of a secondary resonance peak for the driven mode. The latter peak resulted from the fact that, in the experiment, the driven mode and the companion mode had slightly different natural frequencies. Because of this disparity, it was not possible to make a quantitative comparison of the coupled-mode response with the corresponding theory. Qualitative agreement with the analysis was obtained for the two-mode case, however, including the result that, in some cases, the companion mode vibrated to larger amplitudes than the driven mode.

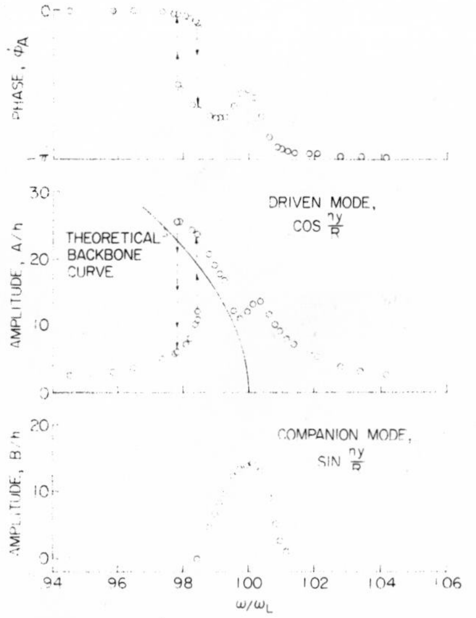

Fig. 7(b) Response curves for $n = 4$ modes: Shaker displacement $\delta_0 =$ 0.4 in. peak to peak

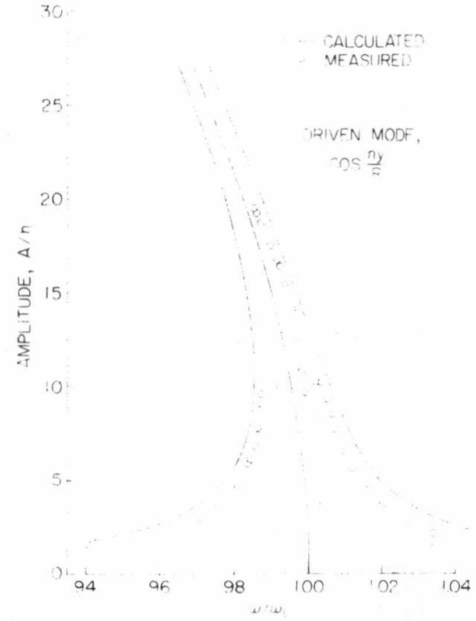

Fig. 8 Comparison of calculated response and experimental results, $n = 4$, $\epsilon_c = 3.08 \times 10^{-4}$, $\delta_0 = 0.3$ in. peak to peak

Concluding Remarks

The nonlinear flexural vibrations of thin rings were analyzed by choosing vibration modes and applying Galerkin's procedure. The vibrations were assumed to involve no stretching of the mid-surface of the ring; this was found to be adequate for the study of flexural vibrations. In the analysis, only one mode was directly driven by the forcing function; nevertheless, it was necessary to include two vibration modes in the calculations. This was due to the fact that under certain conditions nonlinear coupling caused the companion mode to respond and participate in the motion. In other cases, the single-mode response was sufficient.

Significant improvement in the theory was obtained by including the effect of additional nonlinearities in the strain-displacement relations. Retaining the effect of tangential inertia also improved the calculations, but to a lesser extent. Both these modifications lose their importance as the mode number increases; however, they combined to decrease the nonlinearity parameter (ϵ) by more than 25 percent for the $n = 4$ mode.

The experimental results were found to be in good agreement with the analysis, both qualitatively and quantitatively. Theory and experiment both exhibited the jump phenomena, nonlinearity of the softening type, and the appearance of the companion mode. The measured responses were in good agreement with the calculated values, and the experimental mode shapes demonstrated the appropriateness of the deflection form employed in the analysis.

The results of the present study are characteristic of the nonlinear forced vibrations of axisymmetric elastic bodies. In such structures, the nonlinear forced vibration of one mode often results in the response of both the driven mode and its companion [19]. Vibrations of this type occur because of the nonlinear coupling that exists between the modes involved. The nonlinear forced vibration of thin circular cylinders, thin circular cones, and other thin axisymmetric structures can be expected to exhibit similar behavior.

Acknowledgment

The author would like to express his appreciation to T. K. Caughey, E. E. Sechler, Y. C. Fung, and J. P. Raney for many helpful suggestions concerning this problem and its presentation.

References

1 E. Reissner, "Non-Linear Effects in the Vibration of Cylindrical Shells," Ramo-Wooldridge Corporation, Aeromechanics Report No. 5-6, 1955.

2 H. Chu, "Influence of Large Amplitudes on Flexural Vibrations of a Thin Circular Cylindrical Shell," *Journal of the Aerospace Sciences*, vol. 28, 1961, pp. 602–609.

3 B. E. Cummings, "Some Non-Linear Vibration and Response Problems of Cylindrical Panels and Shells," PhD thesis, California Institute of Technology, Pasadena, Calif., 1962.

4 D. A. Evensen, "Some Observations on the Non-Linear Vibration of Thin Cylindrical Shells," *AIAA Journal*, vol. 1, 1963, pp. 2857–2858.

5 J. W. Strutt, *The Theory of Sound*, second edition, Dover Publications, Inc., New York, N. Y., 1945.

6 F. Buckens, "Influence of the Relative Radial Thickness of a Ring on Its Natural Frequencies," *Journal of the Acoustical Society of America*, vol. 22, 1950, pp. 437–443.

7 E. R. Kaiser, "Acoustical Vibration of Rings," *Journal of the Acoustical Society of America*, vol. 25, 1953, pp. 617–623.

8 T. E. Lang, "Vibration of Thin Circular Rings," Technical Report No. 32–261, Jet Propulsion Laboratory, California Institute of Technology, Pasadena, Calif., 1962.

9 K. Federhofer, "Nicht-lineare Biegungsschwingungen des Kreisringes," *Ingenieur-Archiv*, vol. 28, 1959, pp. 53–58.

10 Y. S. Shkenev, "Non-Linear Vibrations of Circular Rings," *Inzhener Sbornik*, vol. 28, *Akad. Nauk SSSR*, 1960, pp. 82–86.

11 J. N. Goodier and I. K. McIvor, "Dynamic Stability and Non-Linear Oscillations of Cylindrical Shells (Plane Strain) Subjected to Impulsive Pressure," Stanford University Technical Report No. 132, Division of Engineering Mechanics, 1962.

12 D. A. Evensen, "Non-Linear Flexural Vibrations of Thin Circular Rings," PhD thesis, California Institute of Technology, Pasadena, Calif., 1964.

13 L. S. D. Morley, "An Improvement on Donnell's Approximation for Thin-Walled Circular Cylinders," *Quarterly Journal of Mechanics and Applied Mathematics*, vol. 12, Part 1, 1959, pp. 89–99.

14 J. Singer, "On the Equivalence of the Galerkin and Rayleigh-Ritz Methods," *Journal of the Royal Aeronautical Society*, vol. 66, 1962, p. 592.

15 N. W. McLachlan, *Ordinary Non-Linear Differential Equations*, Oxford University Press, London, England, 1955.

16 R. E. Hutton, "An Investigation of Resonant, Non-Linear, Non-Planar Free Surface Oscillations of a Fluid," NASA TN D-1870, 1963.

17 J. W. Miles, "Stability of Forced Oscillations of a Spherical Pendulum," *Quarterly of Applied Mathematics*, vol. 20, 1962, pp. 21–32.

18 G. Herrmann and A. E. Armenakas, "Dynamic Behavior of Cylindrical Shells Under Initial Stress," AFOSR TN 60-425, Columbia University, Institute of Flight Structures, 1960.

19 S. A. Tobias, "Free Undamped Non-Linear Vibrations of Imperfect Circular Disks," *Proceedings of The Institution of Mechanical Engineers*, vol. 171, 1957, pp. 691–715.

APPENDIX 1

The method of averaging (sometimes called the method of slowly varying amplitude and phase) can be demonstrated for equations (9) and (11) as follows:

Let $\chi = [\Omega\tau + \phi(\tau)]$. Then equation (11) gives

$$\zeta_c(\tau) = A \cos \chi \tag{22a}$$

$$\zeta_s(\tau) = 0 \tag{22b}$$

and

$$\frac{d\zeta_c}{d\tau} = -\Omega A \sin \chi + \frac{dA}{d\tau} \cos \chi - A \frac{d\phi}{d\tau} \sin \chi \tag{23}$$

In the method of averaging [15], equation (23) is replaced by two equations:

$$\frac{d\zeta_c}{d\tau} = -\Omega A \sin \chi \tag{24a}$$

and

$$\frac{dA}{d\tau} \cos \chi - A \frac{d\phi}{d\tau} \sin \chi = 0 \tag{24b}$$

Equation (24*a*) is then used to compute the second derivative:

$$\frac{d^2\zeta_c}{d\tau^2} = -\Omega^2 A \cos \chi - \frac{dA}{d\tau} \Omega \sin \chi - A \frac{d\phi}{d\tau} \Omega \cos \chi \tag{25}$$

These approximations for ζ_c, ζ_s, $d\zeta_c/d\tau$, and $d^2\zeta_c/d\tau^2$ are then substituted into equation (9*a*); the result is

$$\begin{aligned}[1 - \Omega^2]A \cos \chi - \frac{dA}{d\tau} \Omega \sin \chi - A\Omega \frac{d\phi}{d\tau} \cos \chi \\ - 2\beta_c \Omega A \sin \chi + \frac{\epsilon}{2} A \cos \chi \Big[-\Omega^2 A^2 \cos^2 \chi \\ -\Omega A \frac{dA}{d\tau} \sin \chi \cos \chi - \Omega A^2 \frac{d\phi}{d\tau} \cos^2 \chi + \Omega^2 A^2 \sin^2 \chi \Big] \\ - G_n \cos (\chi - \phi) = 0\end{aligned} \tag{26}$$

The following sequence of operations is then performed:

(*a*) Equation (26) is multiplied by $\cos \chi$.

(*b*) The result is added to equation (24*b*) after the latter has been multiplied by $\Omega \sin \chi$.

(*c*) These manipulations yield an equation involving squares and products of $\cos \chi$ and $\sin \chi$; the final equation is "averaged" by integrating both sides over one period on χ, say $\chi = 0$ to 2π. In the integration, the variables $A(\tau)$ and $\phi(\tau)$ are approximated by their average values, $\bar{A}$ and $\bar{\phi}$.

The result of these operations is an equation involving the average amplitude and phase:

$$[1 - \Omega^2]\bar{A} - 2\Omega\bar{A}\frac{d\bar{\phi}}{d\tau} - \frac{\epsilon\Omega^2\bar{A}^3}{4} - \frac{3\epsilon\Omega\bar{A}^3}{8}\frac{d\bar{\phi}}{d\tau} = G_n \cos\bar{\phi} \quad (27a)$$

A second equation for $\bar{A}$ and $\bar{\phi}$ can be obtained in a similar fashion by

(*a*) Multiplying equation (26) by $\sin\chi$.
(*b*) Multiplying equation (24*b*) by $-\Omega\cos\chi$.
(*c*) Adding the results and averaging on χ.

These operations give

$$-2\Omega\frac{d\bar{A}}{d\tau} - 2\beta_c\Omega\bar{A} - \frac{\epsilon\Omega\bar{A}^2}{8}\frac{d\bar{A}}{d\tau} = G_n \sin\bar{\phi} \quad (27b)$$

For steady-state vibrations, the average values $\bar{A}$ and $\bar{\phi}$ are constant; in this case, $d\bar{A}/d\tau = 0$, $d\bar{\phi}/d\tau = 0$, and equations (27) reduce to equations (12).

Equations (16) for the two-mode case can be derived in a similar manner.

APPENDIX 2

The stability of the approximate solution (13) for the one-mode response was examined by perturbing $\zeta_c(\tau)$ and $\zeta_s(\tau)$:

$$\zeta_c(\tau) = \bar{A}\cos[\Omega\tau + \bar{\phi}] + \xi_c(\tau) \quad (28a)$$

$$\zeta_s(\tau) = 0 + \xi_s(\tau) \quad (28b)$$

Substituting the foregoing expressions for ζ_c and ζ_s into equations (9) and retaining only first-order terms in the perturbations ξ_c and ξ_s give

$$\left[1 + \frac{\epsilon\bar{A}^2}{2}\cos^2(\Omega\tau + \bar{\phi})\right]\frac{d^2\xi_c}{d\tau^2} - \frac{\epsilon\bar{A}^2}{2}\sin 2(\Omega\tau + \bar{\phi})\frac{d\xi_c}{d\tau}$$
$$+\left[1 + \frac{\epsilon\bar{A}^2}{2}\{\sin^2(\Omega\tau + \bar{\phi}) - 2\cos^2(\Omega\tau + \bar{\phi})\}\right]\xi_c = 0 \quad (29a)$$

and

$$\frac{d^2\xi_s}{d\tau^2} + \left[1 - \frac{\epsilon\Omega^2\bar{A}^2}{2}\cos 2(\Omega\tau + \bar{\phi})\right]\xi_s = 0\,. \quad (29b)$$

where the damping constants β_c and β_s have been set equal to zero for simplicity. (The case of nonzero damping is discussed in reference [12].)

Equations (29) are Mathieu-Hill equations and, as such, they exhibit regions of instability. The first instability region of equation (29*b*) defines an area in which the perturbation ξ_s is unstable; this region is approximated by

$$1 - \frac{\epsilon\bar{A}^2}{8} + 0(\epsilon^2) < \Omega < 1 + \frac{\epsilon\bar{A}^2}{8} + 0(\epsilon^2)$$

and can be found from known results concerning the Mathieu equation [15].

Equation (29*a*) can be approximated by a Mathieu equation, and its first instability region defines an area in which the perturbation ξ_c is unstable:

$$1 - \frac{3\epsilon\bar{A}^2}{8} + 0(\epsilon^2) < \Omega < 1 - \frac{\epsilon\bar{A}^2}{8} + 0(\epsilon^2)$$

The second, third, and higher instability regions of equations (29*a*) and (29*b*) occur in narrow areas near $\Omega = 1/2, 1/3, 1/4, \ldots$. These regions denote possible ultraharmonic responses and are discussed in reference [12].

The stability of the coupled-mode solution (18) was examined by similar techniques.

36

Reprinted from *J. Sound Vibr.*, **31**(1), 89–103 (1973)

NON-LINEAR VIBRATIONS OF ORTHOTROPIC DOUBLY-CURVED SHALLOW SHELLS

B. R. EL-ZAOUK AND C. L. DYM

Department of Civil Engineering,
Carnegie Institute of Technology,
Carnegie-Mellon University,
Pittsburgh, Pennsylvania 15213, *U.S.A.*

(*Received* 9 *April* 1973)

The purpose of this study is the evaluation of the effects of curvature, material orthotropy and internal pressure upon the non-linear vibrations of shallow shells. The shells considered here are complete in the circumferential coordinate. A mode shape that leads to a continuous circumferential displacement is used as the basic solution in a Galerkin type procedure. Among the results of interest are the effects of the aforementioned physical parameters on the relative hardness of the shell response, as well as some interesting new perspectives on jump phenomena in non-linear shell vibrations.

1. INTRODUCTION

This study was prompted by the recent work of Leissa and Kadi [1] who investigated the effects of curvature on the free vibrations of shallow shell segments. As they have properly pointed out, almost all of the published studies of shell vibrations—and there are many—have dealt with cylinders or spheres. In the literature on non-linear shell vibrations one may go further and state that there appear to be no investigations that study the effects of orthotropy or of internal pressurization.

Extensive literature surveys on the topic of non-linear shell vibrations have been given by Leissa and Kadi [1] and by Mayers and Wrenn [2]. Of particular interest here are these two studies, and the paper of Evensen and Fulton [3]. The first of these studies was concerned with the linear and non-linear vibrations of isotropic, shallow shell segments, supported at the edges by shear diaphragms placed along the rectangular platform of the shell. Leissa and Kadi broke new ground by investigating the effects of curvature on the frequencies of free vibration. However, their work is restricted to shell segments because their choice of mode shape, even though it does satisfy segment boundary conditions exactly, does not satisfy the continuity requirement for a shell that is complete in the circumferential direction.

The analyses of Mayers and Wrenn and of Evensen and Fulton are concerned with the non-linear vibrations of isotropic circular cylinders. The first of these investigations is distinguished by a comparison of the results obtained with the widely used von Kármán–Donnell equations with results obtained by using a modified version of Sanders' non-linear shell equations. Mayers and Wrenn also presented some interesting plots of energy levels for various modal assumptions, and found that there were free, non-periodic oscillations for which the energy levels were lower than for corresponding periodic solutions. These non-periodic results were, however, almost periodic which, they suggested, would have interesting implications for experimental work in this area. However, according to Hayashi [4], when a periodic forcing function is applied, as is generally the case in experimental work, the non-periodic vibrations are always unstable and must give way to periodic oscillations.

The work of Evensen and Fulton includes some interesting results of a forced vibration analysis where a particular mode is directly excited by the choice of forcing function, and where a companion mode is excited and driven by the coupling inherent in this non-linear problem. They were able to correlate in a reasonable fashion with some experimental results presented by Olson [5], although they were not able to give an adequate explanation for certain jump phenomena observed both analytically [3] and experimentally [5, 6].

In the present analysis the effects of material orthotropy and of internal pressurization upon the non-linear vibrations are considered for the first time. This is in addition to a consideration of the interaction of the curvature changes, suggested by the work of Leissa and Kadi. The analysis is for shells complete in the circumferential directions, with the kinematic model based on the usual shallow shell approximations: i.e., the Marguerre–von Kármán–Donnell theory. Both free and forced vibrations will be considered, although only non-linear vibrations. It is assumed that the effects of in-plane inertia, pressurization, anisotropy and curvature on linear vibrations have been adequately explored: e.g., by Leissa and Kadi [1], Dym [7], and Oyler and Dym [8].

In addition to the investigation of the importance of the physical parameters mentioned above, this study will also be concerned with the stability of the non-linear oscillations, both free and forced. For the single mode vibration case, a stability criterion will be derived analytically, based on the Mathieu equation. For forced vibrations a criterion due to Klotter and Pinney [9] will be applied in such a way as to clarify hitherto unexplained jump phenomena.

2. NON-LINEAR SHELL ANALYSIS

Consider the shallow shell displayed in Figure 1. The meridional radius of curvature is denoted by R_1, while R is the radius of a parallel circle at the equator. The assumption of shallowness is here equivalent to requiring that $L/R_1 \ll 1$, where L is the length of the shell, measured along the axial coordinate. The strain-displacement relations are [10], for large rotations about lines in the shell middle surface,

$$\varepsilon_{xx} = \frac{\partial u}{\partial x} - \frac{w}{R_1} + \frac{1}{2}\left(\frac{\partial w}{\partial x}\right)^2 - z\frac{\partial^2 w}{\partial x^2},$$

$$\varepsilon_{yy} = \frac{\partial v}{\partial y} - \frac{w}{R} + \frac{1}{2}\left(\frac{\partial w}{\partial y}\right)^2 - z\frac{\partial^2 w}{\partial y^2},$$

$$\gamma_{xy} = \frac{\partial u}{\partial y} + \frac{\partial v}{\partial x} + \frac{\partial w}{\partial x}\frac{\partial w}{\partial y} - 2z\frac{\partial^2 w}{\partial x\,\partial y}. \tag{1}$$

Here x and y denote axial and circumferential coordinates, respectively, and u, v and w are the usual shell displacements, with the radial displacement taken as positive when directed towards the inside of the shell.

For a two-dimensional orthotropic continuum, the stress-strain relations can be written in the form [7]

$$\begin{Bmatrix} \tau_{xx} \\ \tau_{yy} \\ \tau_{xy} \end{Bmatrix} = c_{11}\begin{pmatrix} 1 & \alpha & 0 \\ \alpha & \beta & 0 \\ 0 & 0 & \gamma \end{pmatrix}\begin{Bmatrix} \varepsilon_{xx} \\ \varepsilon_{yy} \\ \gamma_{xy} \end{Bmatrix}, \tag{2}$$

where, then,

$$\alpha = c_{12}/c_{11},\ \beta = c_{22}/c_{11},\ \gamma = G_{12}/c_{11}. \tag{3}$$

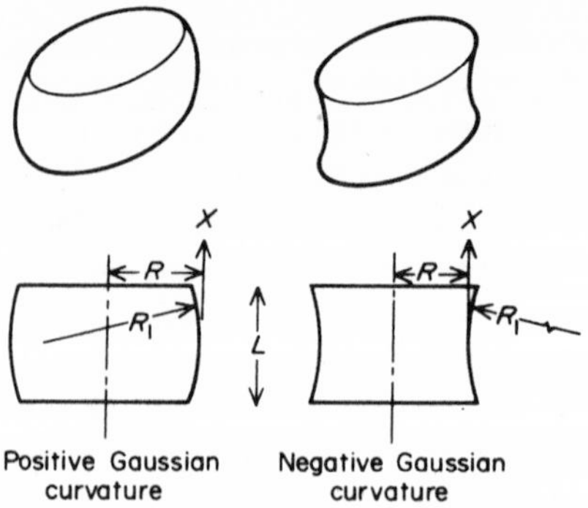

Figure 1. Geometry of various shells of revolution.

Now a Lagrangian can be formulated, with which Hamilton's Principle can be applied, so that the equations of motion for a shell described by equations (1) and (2) can be derived. The kinetic and strain energies are, respectively,†

$$T=\frac{\rho h}{2}\int_0^L\int_0^{2\pi R}\left[\left(\frac{\partial u}{\partial t}\right)^2+\left(\frac{\partial v}{\partial t}\right)^2+\left(\frac{\partial w}{\partial t}\right)^2\right]\mathrm{d}x\,\mathrm{d}y \tag{4}$$

and

$$U=\frac{1}{2}\int_0^L\int_0^{2\pi R}\int_{-h/2}^{h/2}(\tau_{xx}\,\varepsilon_{xx}+\tau_{yy}\,\varepsilon_{yy}+\tau_{xy}\,\gamma_{xy})\,\mathrm{d}z\,\mathrm{d}x\,\mathrm{d}y. \tag{5}$$

Finally, for an internal pressure p and an applied radial loading p_r, directed inward, the potential of applied loads is

$$V=-\int_0^L\int_0^{2\pi R}(p_r-p)\,w\,\mathrm{d}x\,\mathrm{d}y. \tag{6}$$

For the Lagrangian defined as $L=T-(U+V)$, a straightforward application of the variational calculus yields the equations of motion as follows:

$$\frac{\partial N_{xx}}{\partial x}+\frac{\partial N_{xy}}{\partial y}=0, \tag{7a}$$

$$\frac{\partial N_{xy}}{\partial x}+\frac{\partial N_{yy}}{\partial y}=0, \tag{7b}$$

$$\frac{\partial^2 M_{xx}}{\partial x^2}+2\frac{\partial^2 M_{xy}}{\partial x\,\partial y}+\frac{\partial^2 M_{yy}}{\partial y^2}+\frac{\partial}{\partial x}\left(N_{xx}\frac{\partial w}{\partial x}+N_{xy}\frac{\partial w}{\partial y}\right)+$$
$$+\frac{\partial}{\partial y}\left(N_{xy}\frac{\partial w}{\partial x}+N_{yy}\frac{\partial w}{\partial y}\right)+\frac{N_{xx}}{R_1}+\frac{N_{yy}}{R}=\rho h\frac{\partial^2 w}{\partial t^2}-(p_r-p). \tag{7c}$$

The N_{ij} and M_{ij} in equations (7) are the usual stress and moment resultants of their corresponding stress components τ_{ij}.

In the variational process described above a complete set of boundary force–displacement dualities can be obtained for any edge $x=\text{const.}$ or $y=\text{const.}$ For the present investigation,

† In the sequel the in-plane inertia terms in equation (4) will be deleted, in accordance with the earlier discussion.

the boundary conditions of greatest interest are the simple support conditions

$$v = w = N_{xx} = M_{xx} = 0 \text{ at } x = 0, L. \tag{8}$$

Now the first two of equations (7) can be satisfied identically by introducing an Airy stress function $F(x,y,t)$. In order to include the uniform pre-stress due to the internal pressure, the Airy function is taken here as

$$F(x,y,t) = \tau_{xo}\frac{x^2}{2} + \tau_{yo}\frac{y^2}{2} + \Phi(x,y,t). \tag{9}$$

Here τ_{xo} and τ_{yo} are the membrane stresses in a closed shell of revolution [8], viz.,

$$\tau_{xo} = \frac{pR}{2h}, \qquad \tau_{yo} = \frac{pR}{h}\left(1 - \frac{R}{2R_1}\right). \tag{10}$$

If equation (9) is appropriately substituted for the membrane stress resultants in equation (7c), together with the usual moment–curvature–displacement relations based on equations (1) and (2), then the radial equilibrium equation takes the form

$$-c_{11}\frac{h^2}{12}\left\{\frac{\partial^4 w}{\partial x^4} + 2(\alpha + 2\gamma)\frac{\partial^4 w}{\partial x^2\,\partial y^2} + \beta\frac{\partial^4 w}{\partial y^4}\right\} + \tau_{xo}\frac{\partial^2 w}{\partial x^2} + \tau_{yo}\frac{\partial^2 w}{\partial y^2} + \frac{1}{R_1}\frac{\partial^2 \Phi}{\partial y^2} + $$
$$+\frac{1}{R}\frac{\partial^2 \Phi}{\partial x^2} + S(\Phi, w) = \rho\frac{\partial^2 w}{\partial t^2} - \left(\frac{p_r}{h}\right), \tag{11}$$

where $S(a,b)$ is the non-linear partial differential operator

$$S(a,b) = \frac{\partial^2 a}{\partial x^2}\frac{\partial^2 b}{\partial y^2} - 2\frac{\partial^2 a}{\partial x\,\partial y}\frac{\partial^2 b}{\partial x\,\partial y} + \frac{\partial^2 a}{\partial y^2}\frac{\partial^2 b}{\partial x^2}. \tag{12}$$

The use of a stress function to satisfy membrane equilibrium requires also that compatibility of the membrane displacements be ensured through the satisfaction of the compatibility equation for the membrane strains. The latter equation is found by eliminating u and v between the middle surface components of equations (1). Then appropriate substitution yields the compatibility relation in the form

$$\frac{1}{c_{11}(\beta - \alpha^2)}\left\{\frac{\partial^4 \Phi}{\partial x^4} + \left(\frac{\beta - \alpha^2}{\gamma} - 2\alpha\right)\frac{\partial^4 \Phi}{\partial x^2\,\partial y^2} + \beta\frac{\partial^4 \Phi}{\partial y^4}\right\} = -\frac{1}{2}S(w,w) - \frac{1}{R}\frac{\partial^2 w}{\partial x^2} - \frac{1}{R_1}\frac{\partial^2 w}{\partial y^2}. \tag{13}$$

The pair of equations (11) and (13) thus forms a coupled system in the two dependent variables Φ and w. Their solution will be described in the next section.

3. NON-LINEAR VIBRATIONS

The solution to the coupled system above will be composed first of a radial displacement assumption that consists in part of linear vibrational mode shapes that satisfy the boundary conditions (8), to which are added terms such that the total radial displacement and the stress function Φ can be combined to satisfy continuity of the circumferential displacement around the shell circumference. Thus it is assumed that the radial displacement may be taken as

$$w(x,y,t) = \left[A_n(t)\cos\frac{ny}{R} + B_n(t)\sin\frac{ny}{R}\right]\sin\frac{m\pi x}{L} + \frac{n^2}{4R}[A_n^2(t) + B_n^2(t)]\sin^2\frac{m\pi x}{L}, \tag{14}$$

where $A_n(t)$ and $B_n(t)$ are unknown functions of time. With equation (14), whose linear terms satisfy the boundary conditions (8) exactly, and for which the radial displacement vanishes completely at the shell ends, a particular solution for the stress function Φ may be obtained from equation (13). Thus,

$$\Phi = \frac{1}{M_1}\left\{-\frac{1}{2}\left(\frac{m\pi}{L}\frac{n}{R}\right)^2\left[(A_n^2 - B_n^2)\cos\frac{2ny}{R} + 2A_n B_n \sin\frac{2ny}{R}\right]\right\} +$$
$$+ \frac{1}{M_2}\left\{\frac{1}{R}\left(\frac{m\pi}{L}\right)^2 + \frac{1}{R_1}\left(\frac{n}{R}\right)^2 - \frac{n^2}{4R}\left(\frac{m\pi}{L}\frac{n}{R}\right)^2 (A_n^2 + B_n^2)\right\}\left(A_n \cos\frac{ny}{R} + B_n \sin\frac{ny}{R}\right) \times$$
$$\times \sin\frac{m\pi x}{L} + \frac{1}{M_3}\left\{\frac{n^2}{4R}\left(\frac{m\pi}{L}\frac{n}{R}\right)^2 (A_n^2 + B_n^2)\left(A_n \cos\frac{ny}{R} + B_n \sin\frac{ny}{R}\right)\sin\frac{3m\pi x}{L}\right\}, \tag{15}$$

where

$$M_1 = \Psi\left(0, \frac{2n}{R}\right), \qquad M_2 = \Psi\left(\frac{m\pi}{L}, \frac{n}{R}\right), \qquad M_3 = \Psi\left(\frac{3m\pi}{L}, \frac{n}{R}\right), \tag{16}$$

and where

$$\Psi(a, b) = \frac{a^4 + (\beta - \alpha^2/\gamma - 2\alpha)\, a^2 b^2 + \beta b^4}{c_{11}(\beta - \alpha^2)}. \tag{17}$$

It is important to note here that equations (14) and (15) satisfy the periodicity (continuity) requirement $v(x,y,t) = v(x, y + 2\pi R, t)$. Or, in another form [11],

$$\int_0^{2\pi R} \frac{\partial v}{\partial y}\,\mathrm{d}y = 0 = \int_0^{2\pi R}\left[\varepsilon_{yy}\Big|_{z=0} + \frac{w}{R} - \frac{1}{2}\left(\frac{\partial w}{\partial y}\right)^2\right]\mathrm{d}y. \tag{18}$$

This requirement is satisfied by the solution of Evensen and Fulton [3], which the present solution contains, by some of the solutions of Mayers and Wrenn [2], but not by the solution of Leissa and Kadi [1].

Now the unknown coefficients $A_n(t)$ and $B_n(t)$ are determined by forcing the solution represented by equations (14) and (15) to satisfy the equilibrium equation in the approximate sense of Galerkin [12]. For the assumed surface loading $p_r(x,y,t)$, given as

$$p_r(x,y,t) = Q_{mn}\sin\frac{m\pi x}{L}\cos\frac{ny}{R}\cos\omega t, \tag{19}$$

the Galerkin solution of the equilibrium equation yields the following pair of non-linear ordinary differential equations:

$$\frac{\mathrm{d}^2\zeta_1}{\mathrm{d}\tau^2} + \zeta_1 + \frac{3\varepsilon}{16}\zeta_1\frac{\mathrm{d}^2}{\mathrm{d}\tau^2}(\zeta_1^2 + \zeta_2^2) - \varepsilon\eta\zeta_1(\zeta_1^2 + \zeta_2^2) + \varepsilon^2\delta\zeta_1(\zeta_1^2 + \zeta_2^2)^2 = G\cos\Omega\tau, \tag{20a}$$

$$\frac{\mathrm{d}^2\zeta_2}{\mathrm{d}\tau^2} + \zeta_2 + \frac{3\varepsilon}{16}\zeta_2\frac{\mathrm{d}^2}{\mathrm{d}\tau^2}(\zeta_1^2 + \zeta_2^2) - \varepsilon\eta\zeta_2(\zeta_1^2 + \zeta_2^2) + \varepsilon^2\delta\zeta_2(\zeta_1^2 + \zeta_2^2)^2 = 0. \tag{20b}$$

The following dimensionless parameters have been introduced:

$$\zeta_1 = \frac{A_n}{h}, \qquad \zeta_2 = \frac{B_n}{h}, \qquad \Omega = \frac{\omega}{\omega_s},$$
$$\tau = \omega_s t, \qquad G = \frac{Q_{mn}}{\rho h^2 \omega_s^2}, \tag{21}$$

as well as the non-linearity parameter

$$\varepsilon = \left(\frac{n^2 h}{R}\right)^2, \tag{22}$$

the aspect ratio of a vibration mode shape

$$\xi = \frac{m\pi}{L}\bigg/\frac{n}{R}, \tag{23}$$

and a dimensionless frequency factor

$$K^2 = \frac{\rho R^2 \omega_s^2}{c_{11}(\beta - \alpha^2)}. \tag{24}$$

The frequency of the linear vibration problem, ω_s, is given by

$$\omega_s^2 = \frac{c_{11}(\beta-\alpha^2)}{\rho R^2}\left\{\frac{(\xi^2 + R/R_1)^2}{\xi^4 + [(\beta-\alpha^2)/\gamma - 2\alpha]\,\xi^2 + B} + \frac{\varepsilon}{12}\,\frac{\xi^4 + 2(\alpha+2\gamma)\,\xi^2 + \beta}{(\beta-\alpha^2)} + \right.$$
$$\left. + \frac{n^2(\tau_{x0}\,\xi^2 + \tau_{y0})}{c_{11}(\beta-\alpha^2)}\right\}, \tag{25}$$

while the parameters η and δ are given by

$$\eta K^2 = \xi^4\left\{\frac{1 + R/R_1\,\xi^2}{\xi^4 + [(\beta-\alpha^2)/\gamma - 2\alpha]\,\xi^2 + \beta} - \frac{1}{16\beta} - \frac{\varepsilon}{12(\beta-\alpha^2)}\right\} - \frac{\tau_{x0}\,n^2\,\xi^2}{4c_{11}(\beta-\alpha^2)} \tag{26}$$

and

$$\delta K^2 = \frac{3\xi^4}{16}\left\{\frac{1}{\xi^4 + [(\beta-\alpha^2)/\gamma - 2\alpha]\,\xi^2 + \beta} + \frac{1}{(3\xi)^4 + [(\beta-\alpha^2)/\gamma - 2\alpha]\,(3\xi)^2 + \beta}\right\} \tag{27}$$

It is worth noting at this juncture that the sequence of equations (20) to (27) is identical in form to the results given by Evensen and Fulton [3], whose work is also based on the displacement assumption (14). Thus it is not only true that the present results reduce to those of Evensen and Fulton by allowing $\tau_{x0} = \tau_{y0} = 0$, $R_1 \to \infty$, $\beta = 1$, $\alpha = \nu$ and $\gamma = 1 - \nu$, but it is also true that the effects of the two curvatures, pressurization and orthotropy are all subsumed into the frequency ω_s and the extended parameters η and δ. The details of the effects of varying the physical parameters will be given later.

4. PERIODIC VIBRATIONS AND STABILITY

Two particular problems will be solved here, by using in both a solution technique which for these problems is known variously as "the method of averaging", "the method of Krylov and Bogoliubov", "the Ritz averaging method", "the method of slowly varying amplitude", among other names. For the first problem, the periodic response of a single bending mode, the solution proceeds as follows. It is assumed that a solution to equations (20) exists in the form

$$\zeta_1 = A\cos\Omega\tau, \qquad \zeta_2 = 0. \tag{28}$$

After substitution of the above into equation (20a)—the other is trivially satisfied—the equation is multiplied by $\cos\Omega\tau$ and integrated over one cycle of vibration: i.e., over the interval

$0 \leqslant \tau \leqslant 2\pi/\Omega$. This results in an algebraic equation for the amplitude A, to wit,

$$(1-\Omega^2)A - \frac{3\varepsilon}{4}\left(\frac{\Omega^2}{4}+\eta\right)A^3 + \frac{5\varepsilon^3}{8}\delta A^5 = G. \tag{29}$$

For free vibrations, of course, $G = 0$.

To investigate the stability of the single mode solution (29), the form of the solution (28) will be perturbed slightly: i.e., let

$$\zeta_1 = A\cos\Omega\tau + x_1, \qquad \zeta_2 = x_2, \tag{30}$$

where the $x_i(\tau)$ are very small perturbations. Substitution of equations (30) into the differential equations (20) and linearization with respect to the x_i yields a pair of equations close to the Mathieu type [13], here given up to order ε:

$$\ddot{x}_1 - \left(\frac{3\varepsilon}{8}\Omega A^2\sin 2\Omega\tau\right)\dot{x}_1 +$$

$$+ x_1\left[1 - \frac{3\varepsilon A^2}{16}(1+8\eta+\Omega^2) - \frac{3\varepsilon A^2}{16}(1+8\eta+3\Omega^2)\cos 2\Omega\tau\right]x_1 = 0,$$

$$\ddot{x}_2 + \left[1 - \frac{\varepsilon\eta}{2}A^2 - \frac{\varepsilon A^2}{16}(3\Omega^2+4\eta)\,2\cos 2\Omega\tau\right]x_2 = 0. \tag{31}$$

Through a standard analysis of the Mathieu stability criteria [14, 11], one can then show that the perturbations of $\zeta_1(\tau)$ are unstable within the region

$$1 - \frac{9\varepsilon A^2}{8}\left(\frac{1}{4}+\eta\right) < \Omega < 1 - \frac{3\varepsilon A^2}{8}\left(\frac{1}{4}+\eta\right), \tag{32}$$

while the perturbations of $\zeta_2(\tau)$ are not stable within the region

$$1 - \frac{3\varepsilon A^2}{8}\left(\frac{1}{4}+\eta\right) < \Omega < 1 + \frac{\varepsilon A^2}{8}\left(\frac{3}{4}-\eta\right). \tag{33}$$

It is simple enough to show that, within terms up to order ε, the instability region is bounded by the free vibration curve obtained from equation (29),

$$\Omega = 1 - \frac{3\varepsilon A^2}{8}\left(\frac{1}{4}+\eta\right), \tag{34}$$

and by the locus of the vertical tangents of the response diagrams which are found by requiring that $d\Omega/dA = 0$,

$$\Omega = 1 - \frac{9\varepsilon A^2}{8}\left(\frac{1}{4}+\eta\right). \tag{35}$$

The combination of equations (34) and (35) yields the quoted Mathieu result (32).

For the forced vibration problem the criterion of Klotter and Pinney [9] for stability is that dA/dG be greater than zero. From equation (29), to order ε, this condition implies that

$$\Omega > 1 - \frac{9\varepsilon A^2}{8}\left(\frac{1}{4}+\eta\right), \tag{36}$$

which agrees with the earlier results (32) and (35). It is also useful to note that this simple stability criterion of Klotter and Pinney will be a very powerful tool in explaining stability and the jump phenomena in both the single and the coupled mode problems.

A final point on the single mode solution (28) is that the points corresponding to $\Omega = 1/2$, $1/3$, $1/4 \ldots$, represent ultraharmonic frequencies, near which the perturbations x_1, x_2 above are also unstable [14].

For the coupled oscillations of a driven mode and a companion mode, it is assumed that

$$\zeta_1 = A\cos\Omega\tau, \qquad \zeta_2 = B\sin\Omega\tau. \tag{37}$$

It then follows that the amplitudes A and B are determined to be the solutions of the coupled algebraic system

$$(1-\Omega^2)A + \frac{3\varepsilon\Omega^2 A}{16}(B^2 - A^2) - \frac{\varepsilon\eta A}{4}(3A^2 + B^2) + \frac{\varepsilon^2\delta A}{8}(5A^4 + 2A^2B^2 + B^4) = G, \tag{38a}$$

$$(1-\Omega^2)B + \frac{3\varepsilon\Omega^2 B}{16}(A^2 - B^2) - \frac{\varepsilon\eta B}{4}(A^2 + 3B^2) + \frac{\varepsilon^2\delta B}{8}(A^4 + 2A^2B^2 + 5B^4) = 0. \tag{38b}$$

One root of equation (38b) is clearly $B = 0$. The substitution of this root into equation (38a) yields, not surprisingly, the single mode result, equation (29).

Aside from this root, equation (38b) can be used to eliminate B from equation (38a): i.e., to derive a single, tenth-order equation in A, Ω^2 and G, which is solved numerically. As with the single mode problems, the solutions are presented as plots of amplitude against frequency, which is the usual device in non-linear vibration problems. The results of the analysis will be detailed in the next section.

5. RESULTS AND DISCUSSION

The results of the above analysis are displayed in Figures 2–12. The first plot in this set (Figure 2) is for the single mode response problem ($B = 0$), and it shows the free vibration curve ($G = 0$) as well as two pairs of forced vibration curves. In Figure 3 a similar set of curves is given, only here the frequency scale has been expanded so that the jump phenomena may be clearly explained. Note first that there will be five (amplitude) roots for a given value of the frequency Ω, which is proper since equation (29) is a quintic equation. Generally these curves have been truncated when previously displayed (e.g., in the paper of Evensen and Fulton [3]) so that only three roots are visible: i.e., truncation at $A = 15$ in Figure 2. Up to this point the present results are in exact agreement with those in reference [3].

However, it is interesting to note what happens if the response curves are continued, as shown in Figures 2 and 3. First, it can be seen that the bandwidth due to the forcing coefficient G essentially vanishes. That is, for the larger amplitudes, the free response curves and the forced response curves coincide. Thus the forcing function, the applied load, has virtually no effect. The second point of interest here is that to go to the higher part of the response diagram, each curve must pass through a point where $\mathrm{d}G/\mathrm{d}A$ vanishes: that is, a point where the curves have vertical tangents [15]. According to the stability criterion of Klotter and Pinney [9], such points represent a set of stability boundaries for the response. Also, it is relevant to recall that if $\mathrm{d}G/\mathrm{d}A$ is positive along a portion of the response curve, then that portion represents a stable forced motion response.

Figure 3 shows a couple of possible response paths, including points where jumps may occur. Moving along the path AB, which is stable, one comes to the boundary point B where there exists a vertical tangent. Thus, to avoid continuing on the unstable portion of that curve, the solution must jump to the point C on the other curve for the same value of G, and then the path CD will be followed as Ω is increased. Now if decreasing frequencies are being

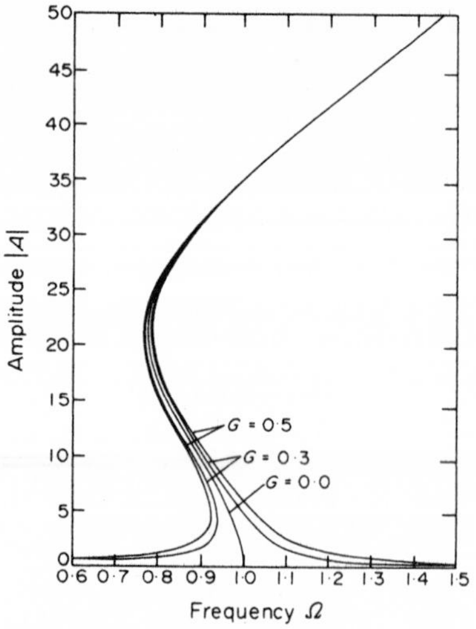

Figure 2. Resonance curves for varying forcing amplitude. Isotropic $\xi = 0{\cdot}1$; $\varepsilon = 0{\cdot}01$; $R/R_1 = 0{\cdot}0$.

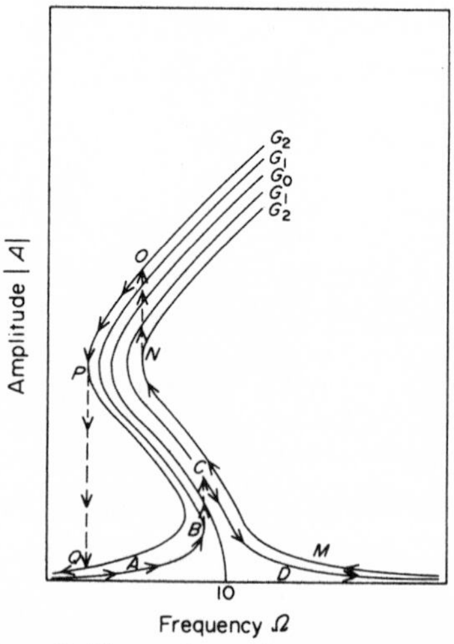

Figure 3. Illustration of jump phenomena.

followed, the stable portion of the curve MN is traversed until the boundary point N is reached. Then a jump occurs again to the point O, and as the frequency is further decreased another stability boundary is reached at the point P. There another jump must take place, to the point Q. Now if it is recalled that the bandwidth $G_2 - G_0 - G_2$ is greatly exaggerated in Figure 3 then it is clearly seen that this second path, with jumps NO and PQ, is akin to those seen in the experiments of Olson and of Kana, Lindholm and Abramson: i.e., see Figure 3 of reference [5] and Figure 5 of reference [6]. This possible explanation has been missed in earlier analyses because, as indicated previously, the related analytical investigations were truncated before the points N, O and P could be reached.†

† It must be conceded, as the referees have pointed out to the authors, that Olson's results appear to show the jump PQ before the reversal points, P and N, are reached, as do the analytical results of Evensen and Fulton for the same shell parameters. This is also true for unpublished results of Chen, which were kindly made available by one of the referees (J. C. Chen 1972 *Ph.D. Thesis, California Institute of Technology.* Non-linear vibration of cylindrical shells.). However, it is worth noting that in these cases the non-linearity parameter is quite small, which means that the curves G_2–G_0–G_2 will be very close together, even below the reversal point, and they will also be much steeper, which implies that the reversal point will be far less obvious than in the schematic of Figure 3. The precise nature of this behavior, then, would seem to require further study.

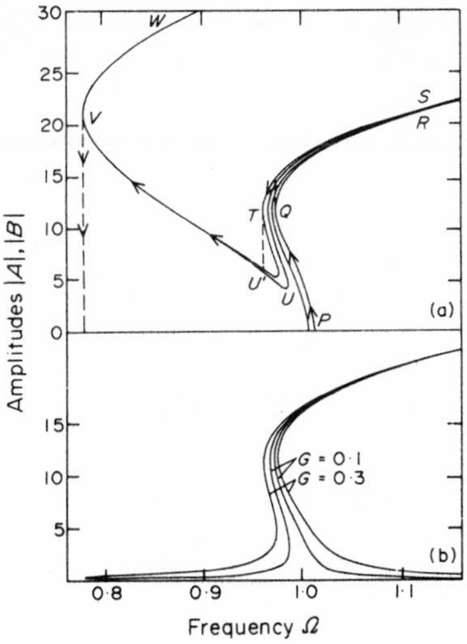

Figure 4. Coupled mode response curves. Isotropic $\varepsilon = 0{\cdot}01$; $\xi = 0{\cdot}1$; $R/R_1 = 0$. (a) Companion mode, (b) driven mode.

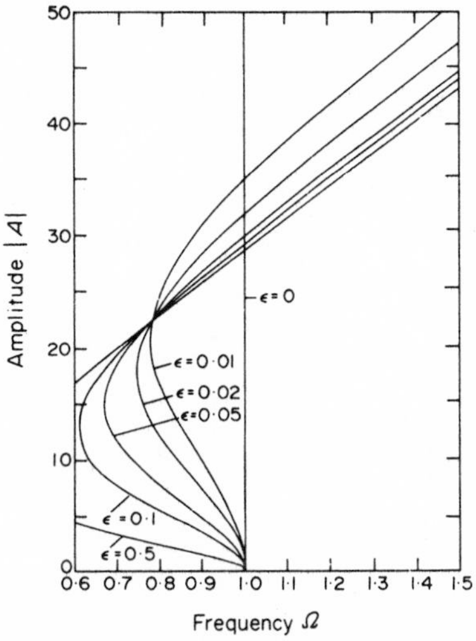

Figure 5. Resonance curves for varying non-linearity parameter. Isotropic $\xi = 0{\cdot}10$; $R/R_1 = 0{\cdot}0$; $G = 0{\cdot}0$.

In Figure 4, a coupled mode response is displayed. The linear terms in A are of the same spatial form as the applied load p_r; hence the terminology that calls A the driven mode and B the companion mode. Note that both modes are representative of soft springs for smaller amplitudes. This behavior then shifts to hard spring behavior as the amplitude gets sufficiently large. Such behavior has been previously observed for the cylinder [1, 2, 3], although the response of the companion mode has perhaps not been properly interpreted. In particular, Evensen and Fulton [3] claim that the entire portion of the curve UVW is unstable, as well the portions QR and TU. In fact this is not the case, as an application of the above stability and jump criteria clearly shows.† At Q there will be a jump to the stable portion of TS, followed by a jump down from T to U'. The curve $U'V$ represents stable response that is

† It has been assumed here, although a formal proof does not appear to be available, that the stability criterion derived by Klotter and Pinney, for a single-degree-of-freedom system, is applicable to the system analysed here.

terminated at the vertical tangent (stability boundary) V, where a jump down to the trivial solution ($B = 0$) for the companion mode takes place. Incidentally, the response of the companion mode is very much like that observed in vibration absorbers [16].

The remaining curves are resonance diagrams representing single mode response. In Figure 5 the effects of the non-linearity parameter ε are vividly displayed. Note that $\varepsilon = 0$ corresponds, by equation (22), to either of the cases $n = 0$ or $h/R = 0$. In the former case, that of axisymmetric deformation, the problem naturally becomes a linear one, which explains the given straight line of the linear eigenvalue problem. It is also clear from equations (20) that setting ε to zero produces a linear system. Further, as ε is increased from zero, the softening is increased as the linear term in ε dominates equation (29). After some point, however, the amplitude A becomes sufficiently large so that the term $\varepsilon^2 A^4$ dominates. Also of interest is the essentially linear amplitude–frequency relationship after the soft spring behavior has been dominated.

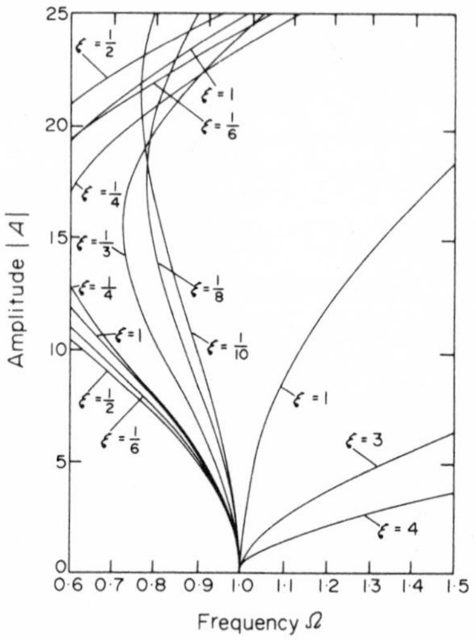

Figure 6. Resonance curves for varying aspect ratio. Isotropic $G = 0$; $\varepsilon = 0{\cdot}01$; $R/R_1 = 0$.

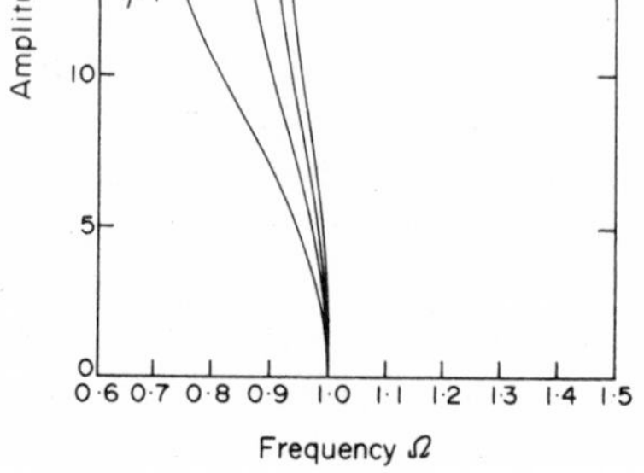

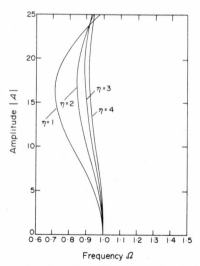

Figure 7. Resonance curves for varying circumferential wave number. Isotropic $\xi = 1{\cdot}0$; $G = 0{\cdot}0$; $\varepsilon = 0{\cdot}01$; $\tau_{x0} = 0{\cdot}1$.

From Figure 6 an appreciation of the effect of the aspect ratio may be gleaned. The most interesting point is the confirmation of the result of Evensen and Fulton [3] that $\xi = 1$ is a demarcation line between initially soft behavior ($\xi < 1$) and initially hard behavior ($\xi > 1$).

The curves in Figure 7 show that, while the response is generally soft, the shell stiffness does increase with the number of circumferential waves. Indeed, by the time the circumferential wave number reaches four, the response is virtually flat.

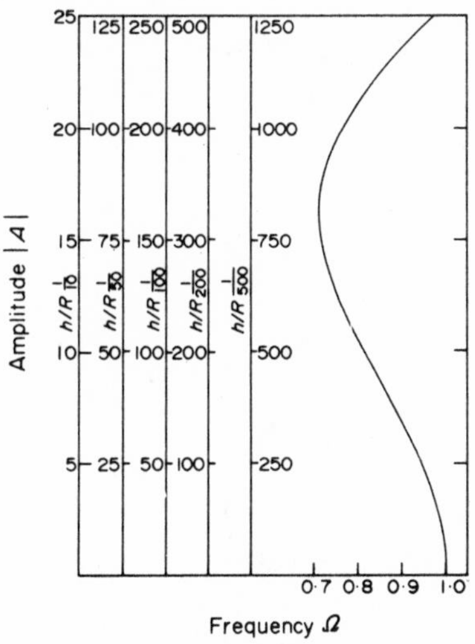

Figure 8. Resonance curves for varying thickness ratio. Isotropic $R/R_1 = 0{\cdot}0$; $\xi = 1{\cdot}0$; $G = 0{\cdot}0$; $\tau_{x0} = 0{\cdot}1$.

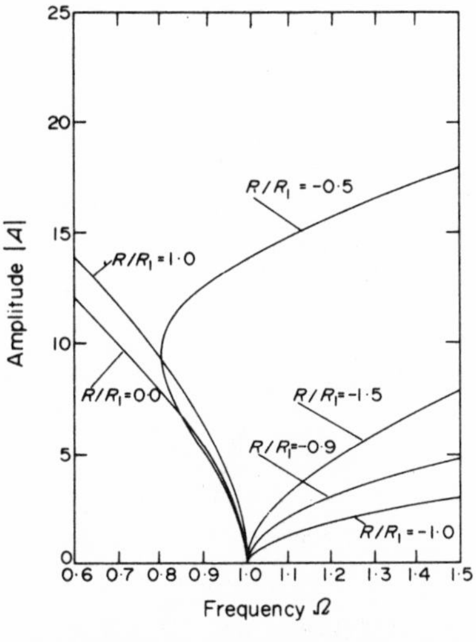

Figure 9. Resonance curves for varying curvature ratio. Isotropic $G = 0{\cdot}0$; $\varepsilon = 0{\cdot}01$; $\xi = 1{\cdot}0$.

The point of Figure 8 is rather simple: i.e., it would probably be more appropriate to render the amplitude of free vibration dimensionless with respect to the circumferential radius. There is no discernible effect of varying the thickness-to-radius ratio, other than the scaling of the amplitude A.

In Figure 9 are displayed the resonance curves for varied values of the curvature ratio, R/R_1. The results are in marked contrast to those given by Leissa and Kadi [1] for the shallow shell segment. In all but one of their results (see Figure 4 of reference [1]) a region of soft spring behavior is dominant before the hard spring region is reached. In the present analysis the initial response is that of a soft spring for $R/R_1 > -0{\cdot}5$, while for $R/R_1 < -0{\cdot}9$ the response is hard from the outset. The difference between the present results and those of Leissa and Kadi [1] is that the modal shape used by Leissa and Kadi would not satisfy periodicity requirements if it were applied to a complete shell. Their final solution contains terms that would allow asymmetric oscillations, terms that are multiplied by the interesting factor $(-1)^{m+n}$.

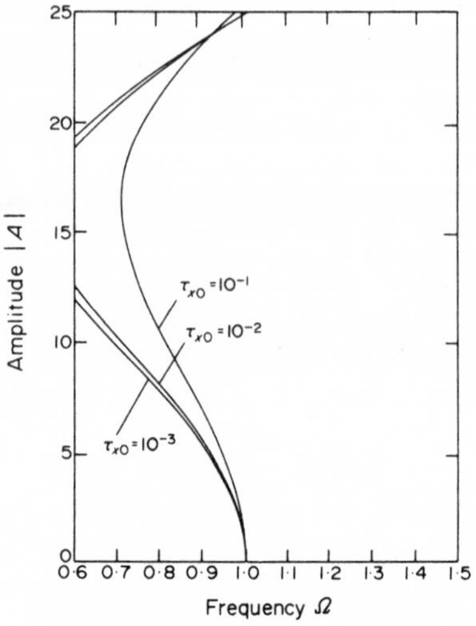

Figure 10. Resonance curves for varying pressurization. Isotropic $R/R_1 = 0{\cdot}0$; $\varepsilon = 0{\cdot}01$; $\xi = 1{\cdot}0$; $G = 0{\cdot}0$.

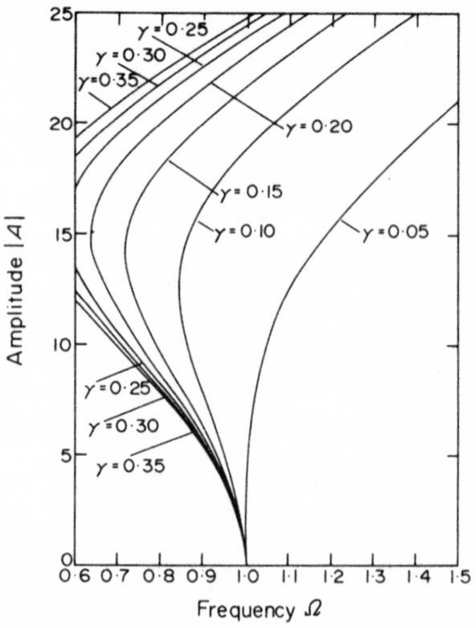

Figure 11. Resonance curves for varying shear ratio. Orthotropic $a = 0{\cdot}30$; $\beta = 1{\cdot}0$; $\xi = 1{\cdot}0$; $\varepsilon = 0{\cdot}01$; $G = 0{\cdot}0$; $R/R_1 = 0{\cdot}0$.

This factor indicates a sign change in the even order term in their differential equation—there are no even order terms in equations (20) here—every time a wave number changes by an integral value! This could imply a serious amplitude jump caused by a simple wave number change.

It is also interesting in discussing curvature effects to note that Vlasov [17] postulated that the lowest linear frequencies of free vibration would obtain in shells of negative Gaussian curvature. This suggestion is easily verified here in view of equation (25) that gives the formula for the linear frequency. Note in that equation the obvious dependence of the frequency on the term $(\xi^2 + R/R_1)^2$.

The effect of increasing the internal pressure in the shell is displayed in Figure 10. It is clear, and not very surprising, that increasing the internal pressure stiffens the shell markedly.

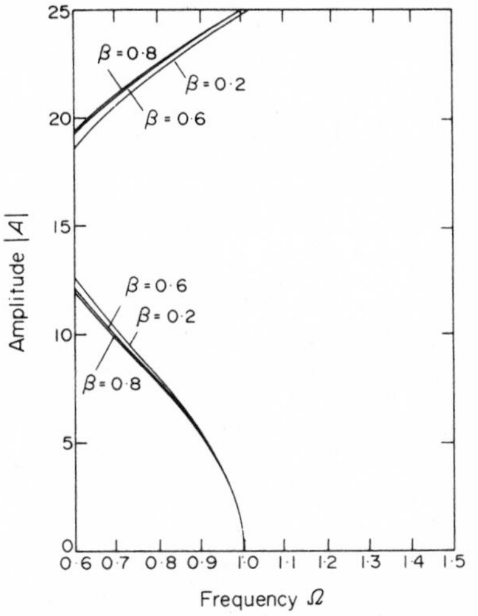

Figure 12. Resonance curves for varying circumferential modulus. Orthotropic $a = 0{\cdot}30$; $\gamma = 0{\cdot}35$; $\xi = 1{\cdot}0$; $\varepsilon = 0{\cdot}01$; $G = 0{\cdot}0$; $R/R_1 = 0{\cdot}0$.

The last two curves show the effect of varying the orthotropic material constants, in particular the in-plane shear modulus (Figure 11) and the circumferential normal modulus (Figure 12). The effect of increasing the shear modulus is to decrease the stiffness of the non-linear spring response. Thus as γ increases the spring gets quite soft. The variation of the circumferential stiffness, β, seems to have little or no effect on the shell response. It is interesting to note that varying β does create a large shift in the linear frequency ω_s (see, e.g., reference [7]), but perhaps because of the definition of the non-linear frequency this particular effect is entirely submerged in the non-linear analysis.

6. CONCLUSIONS

This paper has presented for the first time a study of the effects of internal pressurization, material orthotropy, and variable curvature, upon the non-linear vibrations of complete shells. It was shown that the curvature effects previously obtained for shallow shell segments are not replicated for complete shells, with initially hard behavior occurring uniformly for complete shells with sufficiently negative Gaussian curvature. Also the initially soft behavior of shells of positive (or not very negative) Gaussian curvature seemed to be relatively insensitive to the precise numerical value of the curvature ratio. The effects of changing the normal circumferential modulus or the in-plane shear modulus do not appear as drastic on the non-linear vibration frequencies as might have been expected from earlier linear studies.

This study also presented an explanation for certain jump phenomena that appear to have been seen experimentally, although they have not been properly analysed analytically. These jump phenomena occur because of higher order effects in the differential equations, due to terms that are generally accounted as negligible, but which actually dominate the large amplitude regime where the jumps actually take place. The application of the stability criterion of Klotter and Pinney to the shell vibration problem for the first time allowed a theoretically self-consistent explanation of the jump phenomena and a correction of certain errors of interpretation of previously obtained response diagrams.

REFERENCES

1. A. W. Leissa and A. S. Kadi 1971 *Journal of Sound and Vibration* **16**, 173–187. Curvature effects on shallow shell vibrations.
2. J. Mayers and B. G. Wrenn 1967 *Developments in Mechanics, Proceedings of the Tenth Midwestern Mechanics Conference* **4**, 819. On the nonlinear free vibrations of thin cylindrical shells. New York: Johnson Publishing Co.
3. D. A. Evensen and R. E. Fulton 1965 *International Conference on Dynamic Stability of Structures, Evanston, Illinois* (edited by G. Herrmann). Some studies on the nonlinear dynamic response of shell-type structures.
4. C. Hayashi 1953 *Forced Oscillations in Non-Linear Systems.* Osaka, Japan: Nippon Printing and Publishing Co.
5. M. D. Olson 1965 *American Institute of Aeronautics and Astronautics Journal* **3**, 1775–1777. Some experimental observations on the nonlinear vibrations of cylindrical shells.
6. D. D. Kana, U. S. Lindstrom and H. N. Abramson 1966 *Journal of Spacecraft and Rockets* **3**, 1183–1188. An experimental study of liquid instability in a vibrating elastic tank.
7. C. L. Dym 1970 *American Institute of Aeronautics and Astronautics Journal* **8**, 693–699. Vibrations of pressurized orthotropic cylindrical membranes.
8. J. F. Oyler and C. L. Dym 1973 *Developments in Mechanics, Proceedings of the Thirteenth Midwestern Mechanics Conference, Pittsburgh* **7**. The dynamics and stability of composite shells.
9. K. Klotter and E. Pinney 1953 *Journal of Applied Mechanics* **20**, 9–12. A comprehensive stability criterion for forced vibrations in nonlinear systems.
10. J. W. Hutchinson 1967 *International Journal of Solids and Structures* **3**, 97–115. Initial post-buckling behavior of toroidal shell segments.
11. B. R. El-Zaouk 1973 *Ph.D. Dissertation, Department of Civil Engineering of Carnegie-Mellon University.* Nonlinear vibrations of orthotropic doubly-curved shallow shells.
12. C. L. Dym and I. H. Shames 1973 *Solid Mechanics: A Variational Approach.* New York: McGraw-Hill Book Co.
13. N. McLachlan 1950 *Ordinary Nonlinear Differential Equations in Engineering and the Physical Sciences.* Oxford: Clarendon Press.
14. J. J. Stoker 1950 *Nonlinear Vibrations in Mechanical and Electrical Systems.* New York: Interscience Publishers.
15. C. Hayashi 1964 *Nonlinear Oscillations in Physical Systems.* New York: McGraw-Hill Book Co.
16. F. R. Arnold 1955 *Journal of Applied Mechanics* **22**, 487–492. Steady state behavior of systems provided with nonlinear dynamic vibration absorbers.
17. V. Z. Vlasov 1964 *NASA TT F*-99. General Theory of Shells and Its Application in Engineering. (Original book in Russian published in 1949.)

Reprinted from *Int. J. Solids Structures,* **9,** 581–590 (1973)

NONLINEAR FREE VIBRATIONS OF ELASTIC STRUCTURES

L. W. REHFIELD

Georgia Institute of Technology, Atlanta, Georgia

Abstract—An approach to nonlinear free vibrations of elastic structures is developed with the aid of Hamilton's principle and a perturbation procedure. The theory is analogous to the theory of initial postbuckling behavior due to Koiter. It provides information regarding the first order effects of finite displacements upon the frequency, period and dynamic stresses arising in the free, undamped vibration of structures. Attention is restricted to structures which are linearly elastic. The theory is illustrated by application to the free vibration of beams and rectangular plates.

INTRODUCTION

A THEORY of initial postbuckling behavior has been developed by Koiter [1, 2] which permits the first order effects of finite displacements and initial imperfections on the buckling process to be assessed. Although the original work was based upon potential energy considerations, Budiansky and Hutchinson [3] and Budiansky [4] rederived the essentials of the theory by another method which is based upon virtual work. In the present paper an approach is developed for the analysis of nonlinear free vibrations that is in much the same spirit. The approach is quite analogous to the treatment of Koiter's theory in Ref. [4].

A perturbation approach that is applicable to nonlinear partial differential equations which possess periodic solutions has been outlined by Keller [5] and applied by Keller and Ting [6]. The essential features of the perturbation approach are used in the present development. Solutions to the governing equations are sought as a power series in the amplitude of the linear vibration mode and higher order effects are systematically generated by successive perturbation equations.

OUTLINE OF THE THEORY

To facilitate a concise presentation of the theory, the functional notation used by Budiansky [4] will be employed. The motion of the structure produces generalized displacement $\mathbf{u}$, strain γ and stress $\boldsymbol{\sigma}$. The dynamics of the system is established by Hamilton's principle, which is symbolically written

$$\int_0^{2\pi/\omega} \left[\delta\left(\frac{1}{2}M\left(\frac{\partial \mathbf{u}}{\partial t}\right) \cdot \frac{\partial \mathbf{u}}{\partial t}\right) - \boldsymbol{\sigma} \cdot \delta\gamma \right] \mathrm{d}t = 0. \tag{1}$$

The "dot" operation signifies the appropriate inner multiplication of variables and integration of the result over the entire structure. The generalized mass operator M is assumed to be homogeneous and linear with the property that

$$M(\mathbf{u}) . \mathbf{v} = M(\mathbf{v}) . \mathbf{u} \tag{2}$$

for all **u** and **v**. Since only periodic motion is to be considered, the limits on the integral over time correspond to a single period of the motion. ω is the circular frequency of the vibration such that

$$\mathbf{u}(\mathbf{r}, t) = \mathbf{u}\left(\mathbf{r}, t+\frac{2\pi}{\omega}\right) \tag{3}$$

r is the position vector to an arbitrary point in the structure.

If a new time variable $\tau = \omega t$ is introduced in equation (1), this equation may be replaced by

$$\int_0^{2\pi} [\omega^2 \delta(\tfrac{1}{2}M(\dot{\mathbf{u}}) . \dot{\mathbf{u}}) - \boldsymbol{\sigma} . \delta\boldsymbol{\gamma}] \, d\tau = 0. \tag{4}$$

In the above, the notation $(\dot{\ }) = \partial(\ )/\partial\tau$ has been used. An integration by parts results in

$$\omega^2 M(\dot{\mathbf{u}}) . \delta\mathbf{u}|_0^{2\pi} - \int_0^{2\pi} [\omega^2 M(\ddot{\mathbf{u}}) . \delta\mathbf{u} + \boldsymbol{\sigma} . \delta\boldsymbol{\gamma}] \, d\tau = 0. \tag{5}$$

The boundary terms vanish by reason of periodicity and we are left with

$$\int_0^{2\pi} [\omega^2 M(\ddot{\mathbf{u}}) . \delta\mathbf{u} + \boldsymbol{\sigma} . \delta\boldsymbol{\gamma}] \, d\tau = 0 \tag{6}$$

$\delta\mathbf{u}$ is any virtual displacement that is consistent with all the kinematic boundary conditions imposed on the structure.

Equation (6) is supplemented by the strain–displacement relation

$$\boldsymbol{\gamma} = L_1(\mathbf{u}) + \tfrac{1}{2}L_2(\mathbf{u}) \tag{7}$$

where L_1 and L_2 are homogeneous linear and quadratic functionals, respectively. In addition, the homogeneous bilinear functional L_{11} is defined by the following equation:

$$L_2(\mathbf{u}+\mathbf{v}) = L_2(\mathbf{u}) + 2L_{11}(\mathbf{u}, \mathbf{v}) + L_2(\mathbf{v}). \tag{8}$$

It follows that $L_{11}(\mathbf{u}, \mathbf{v}) = L_{11}(\mathbf{v}, \mathbf{u})$ and $L_{11}(\mathbf{u}, \mathbf{u}) = L_2(\mathbf{u})$. If use is made of the above definitions, the variation of the generalized strain can be written as

$$\delta\boldsymbol{\gamma} = \delta\mathbf{e} + L_{11}(\mathbf{u}, \delta\mathbf{u}) \tag{9}$$

where

$$\mathbf{e} = L_1(\mathbf{u}) \tag{10}$$

is the linearized strain measure.

For Hookean (linear) elastic structures, the stress–strain relation can be written in the form

$$\boldsymbol{\sigma} = H(\boldsymbol{\gamma}) \tag{11}$$

where H is a homogeneous linear function. The following reciprocity relation

$$\boldsymbol{\sigma}^{(1)} . \boldsymbol{\gamma}^{(2)} = \boldsymbol{\sigma}^{(2)} . \boldsymbol{\gamma}^{(1)} \tag{12}$$

will be assumed also; "1" and "2" are any arbitrary states of stress and strain.

The vibration modes and frequencies of the linearized theory can be found by setting

$$\mathbf{u} = \xi\mathbf{u}_1 \qquad \boldsymbol{\gamma} = \xi\mathbf{e}_1 \qquad \boldsymbol{\sigma} = \xi\boldsymbol{\sigma}_1 \tag{13}$$

ξ is an amplitude parameter associated with the mode $\mathbf{u}_1$ which has natural frequency ω_0. If equation (13) is substituted into equation (6) and only linear terms are retained, we obtain

$$\int_0^{2\pi} [\omega_0^2 M(\ddot{\mathbf{u}}_1) \,.\, \boldsymbol{\delta}\mathbf{u} + \boldsymbol{\sigma}_1 \,.\, \boldsymbol{\delta}\mathbf{e}] \, d\tau = 0. \tag{14}$$

Equating the integrand to zero yields the linearized equation of motion.

If we now set $\boldsymbol{\delta}\mathbf{u} = \mathbf{u}_1$ in the above equation, we obtain an expression for ω_0^2.

$$\begin{aligned} \omega_0^2 &= \frac{-\int_0^{2\pi} \boldsymbol{\sigma}_1 \,.\, \mathbf{e}_1 \, d\tau}{\int_0^{2\pi} M(\ddot{\mathbf{u}}_1) \,.\, \mathbf{u}_1 \, d\tau} \\ &= \frac{\int_0^{2\pi} \boldsymbol{\sigma}_1 \,.\, \mathbf{e}_1 \, d\tau}{\int_0^{2\pi} M(\dot{\mathbf{u}}_1) \,.\, \dot{\mathbf{u}}_1 \, d\tau}. \end{aligned} \tag{15}$$

This is analogous to a Rayleigh quotient for the natural frequency ω_0.

We assume at this point that a single mode $\mathbf{u}_1$ is associated with the natural frequency ω_0. The case of multiple modes corresponding to the same natural frequency will be discussed later.

To discover how the structure behaves for finite amplitudes, we assume

$$\begin{aligned} \mathbf{u} &= \xi \mathbf{u}_1 + \xi^2 \mathbf{u}_2 + \ldots \\ \boldsymbol{\gamma} &= \xi \mathbf{e}_1 + \xi^2 (\mathbf{e}_2 + \tfrac{1}{2} L_2(\mathbf{u}_1)) + \ldots \\ \boldsymbol{\sigma} &= \xi \boldsymbol{\sigma}_1 + \xi^2 \boldsymbol{\sigma}_2 + \ldots \end{aligned} \tag{16}$$

where, in order to make the expansions unique, the displacement increments $\mathbf{u}_2$, $\mathbf{u}_3$, ... are orthogonalized with respect to $\mathbf{u}_1$ in the sense that

$$M(\dot{\mathbf{u}}_1) \,.\, \dot{\mathbf{u}}_k = M(\ddot{\mathbf{u}}_1) \,.\, \mathbf{u}_k = 0 \qquad (k \neq 1). \tag{17}$$

This relation, together with equation (12), also implies that

$$\boldsymbol{\sigma}_1 \,.\, \mathbf{e}_k = 0 \qquad (k \neq 1) \tag{18}$$

which by virtue of the reciprocity relation (12) implies further that

$$H(\mathbf{e}_k) \,.\, \mathbf{e}_1 = 0 \qquad (k \neq 1). \tag{19}$$

The substitution of (16) into (6) yields

$$\begin{aligned} \int_0^{2\pi} \{ &\xi(\omega^2 M(\ddot{\mathbf{u}}_1) \,.\, \boldsymbol{\delta}\mathbf{u} + \boldsymbol{\sigma}_1 \,.\, \boldsymbol{\delta}\mathbf{e}) + \xi^2(\omega^2 M(\ddot{\mathbf{u}}_2) \,.\, \boldsymbol{\delta}\mathbf{u} + \boldsymbol{\sigma}_2 \,.\, \boldsymbol{\delta}\mathbf{e} + \boldsymbol{\sigma}_1 \,.\, L_{11}(\mathbf{u}_1, \boldsymbol{\delta}\mathbf{u})) \\ &+ \xi^3(\omega^2 M(\ddot{\mathbf{u}}_3) \,.\, \boldsymbol{\delta}\mathbf{u} + \boldsymbol{\sigma}_3 \,.\, \boldsymbol{\delta}\mathbf{e} + \boldsymbol{\sigma}_1 \,.\, L_{11}(\mathbf{u}_2, \boldsymbol{\delta}\mathbf{u}) + \boldsymbol{\sigma}_2 \,.\, L_{11}(\mathbf{u}_1, \boldsymbol{\delta}\mathbf{u})) + \ldots \} \, d\tau = 0. \end{aligned} \tag{20}$$

If we now set $\boldsymbol{\delta}\mathbf{u} = \mathbf{u}_1$, $\boldsymbol{\delta}\mathbf{e} = \mathbf{e}_1$ in this expression and introduce (15), the following result is obtained:

$$\begin{aligned} \int_0^{2\pi} \Bigg\{ &\xi\left(1 - \frac{\omega^2}{\omega_0^2}\right) \boldsymbol{\sigma}_1 \,.\, \mathbf{e}_1 + \xi^2(\boldsymbol{\sigma}_2 \,.\, \mathbf{e}_1 + \boldsymbol{\sigma}_1 \,.\, L_{11}(\mathbf{u}_1, \mathbf{u}_1)) \\ &+ \xi^3(\boldsymbol{\sigma}_3 \,.\, \mathbf{e}_1 + \boldsymbol{\sigma}_1 \,.\, L_{11}(\mathbf{u}_1, \mathbf{u}_2) + \boldsymbol{\sigma}_2 \,.\, L_{11}(\mathbf{u}_1, \mathbf{u}_1)) + \ldots \Bigg\} \, d\tau = 0. \end{aligned}$$

However, the reciprocity relation (12) permits further simplification as

$$\boldsymbol{\sigma}_2 \cdot \mathbf{e}_1 = \boldsymbol{\sigma}_1 \cdot \boldsymbol{\gamma}_2 = \boldsymbol{\sigma}_1 \cdot (\mathbf{e}_2 + \tfrac{1}{2}L_{11}(\mathbf{u}_1, \mathbf{u}_1)) \\ = \tfrac{1}{2}\boldsymbol{\sigma}_1 \cdot L_{11}(\mathbf{u}_1, \mathbf{u}_1) \tag{21}$$

and

$$\boldsymbol{\sigma}_3 \cdot \mathbf{e}_1 = \boldsymbol{\sigma}_1 \cdot \boldsymbol{\gamma}_3 = \boldsymbol{\sigma}_1 \cdot (\mathbf{e}_3 + L_{11}(\mathbf{u}_1, \mathbf{u}_2)) \\ = \boldsymbol{\sigma}_1 \cdot L_{11}(\mathbf{u}_1, \mathbf{u}_2). \tag{22}$$

Consequently,

$$\int_0^{2\pi} \left\{ \xi\left(1 - \frac{\omega^2}{\omega_0^2}\right)\boldsymbol{\sigma}_1 \cdot \mathbf{e}_1 + \xi^2(\tfrac{3}{2}\boldsymbol{\sigma}_1 \cdot L_{11}(\mathbf{u}_1, \mathbf{u}_1)) \right. \\ \left. + \xi^3(2\boldsymbol{\sigma}_1 \cdot L_{11}(\mathbf{u}_1, \mathbf{u}_2) + \boldsymbol{\sigma}_2 \cdot L_{11}(\mathbf{u}_1, \mathbf{u}_1)) + \ldots \right\} d\tau = 0 \tag{23}$$

and we have the asymptotic relation

$$\frac{\omega^2}{\omega_0^2} = 1 + A\xi + B\xi^2 + \ldots \tag{24}$$

where

$$A = \frac{\int_0^{2\pi} \frac{3}{2}\boldsymbol{\sigma}_1 \cdot L_{11}(\mathbf{u}_1, \mathbf{u}_1) d\tau}{\int_0^{2\pi} \boldsymbol{\sigma}_1 \cdot \mathbf{e}_1 \, d\tau} \\ = \frac{\int_0^{2\pi} \frac{3}{2}\boldsymbol{\sigma}_1 \cdot L_{11}(\mathbf{u}_1, \mathbf{u}_1) \, d\tau}{\omega_0^2 \int_0^{2\pi} M(\dot{\mathbf{u}}_1) \cdot \dot{\mathbf{u}}_1 \, d\tau} \tag{25}$$

and

$$B = \frac{\int_0^{2\pi} (2\boldsymbol{\sigma}_1 \cdot L_{11}(\mathbf{u}_1, \mathbf{u}_2) + \boldsymbol{\sigma}_2 \cdot L(\mathbf{u}_1, \mathbf{u}_1)) \, d\tau}{\omega_0^2 \int_0^{2\pi} M(\dot{\mathbf{u}}_1) \cdot \dot{\mathbf{u}}_1 \, d\tau}. \tag{26}$$

If A is nonzero, the structure can exhibit a softening characteristic with the frequency decreasing for finite amplitudes for $A\xi$ negative. If A is zero, a negative value of B corresponds to softening (decreasing frequency) and a positive, nonzero value corresponds to hardening (increasing frequency). If both A and B are zero, higher order terms must be investigated to discover the nature of finite amplitude effects.

In the evaluation of B a solution for $\mathbf{u}_2$ and $\boldsymbol{\sigma}_2$ is required. It must be found from the second order term in equation (20). The variational equation of motion is, therefore,

$$\omega^2 M(\ddot{\mathbf{u}}_2) \cdot \delta\mathbf{u} + \boldsymbol{\sigma}_2 \cdot \delta\mathbf{e} + \boldsymbol{\sigma}_1 \cdot L_{11}(\mathbf{u}_1, \delta\mathbf{u}) = 0 \tag{27}$$

and

$$\boldsymbol{\gamma}_2 = \mathbf{e}_2 + \tfrac{1}{2}L(\mathbf{u}_1, \mathbf{u}_1) \qquad \boldsymbol{\sigma}_2 = H(\boldsymbol{\gamma}_2)$$

$\delta\mathbf{u}$ is orthogonal to $\mathbf{u}_1$ in the sense of equation (17).

MULTIPLE MODE SITUATIONS

If more than one mode corresponds to the same natural frequency found from the linearized theory, the above described solution process requires modification. If we assume that k modes correspond to the same frequency, with the linearly independent modes being identified as $\mathbf{u}_{11}, \mathbf{u}_{12}, \ldots, \mathbf{u}_{1k}$, then the expansion for displacement in equation (16) must be replaced by

$$\mathbf{u} = \sum_{i=1}^{k} \xi_i \mathbf{u}_{1i} + \mathbf{w} \tag{28}$$

where the modes are orthogonalized with respect to each other and to the additional displacement $\mathbf{w}$. Also we write

$$\boldsymbol{\sigma} = \sum_{i=1}^{k} \xi_i \boldsymbol{\sigma}_{1i} + \mathbf{s}. \tag{29}$$

It is possible to develop k equations of the same type as (23) which are obtained by setting $\delta\mathbf{u}$ equal to $\mathbf{u}_{11}, \mathbf{u}_{12}, \ldots, \mathbf{u}_{1k}$. These equations, retaining only quadratic terms in the ξ_i's as was done by Budiansky and Hutchinson [3] and thus neglecting the effects of $\mathbf{w}$, must be solved simultaneously. Since the equations are homogeneous, only amplitude ratios can be found. A higher order analysis can be made, but the solution process is far more difficult than the case of a single vibration mode.

APPLICATION TO FLEXURAL VIBRATIONS OF BEAMS

Consider the uniform, simply supported beam shown in Fig. 1. An axial force is induced in the beam due to finite amplitude vibrations because the supports are assumed to be immovable. Let U and W be the longitudinal and transverse components of displacement, N be the axial force and X and Z be the coordinates shown in the figure. The average axial strain ε is given by

$$\varepsilon = U_{,X} + \tfrac{1}{2}(W_{,X})^2 = \frac{N}{EA}, \tag{30}$$

where E is Young's modulus and A is the cross-sectional area. Since the supports cannot move apart

$$U(L) - U(0) = \int_0^L U_{,X}\,\mathrm{d}X = 0 \tag{31}$$

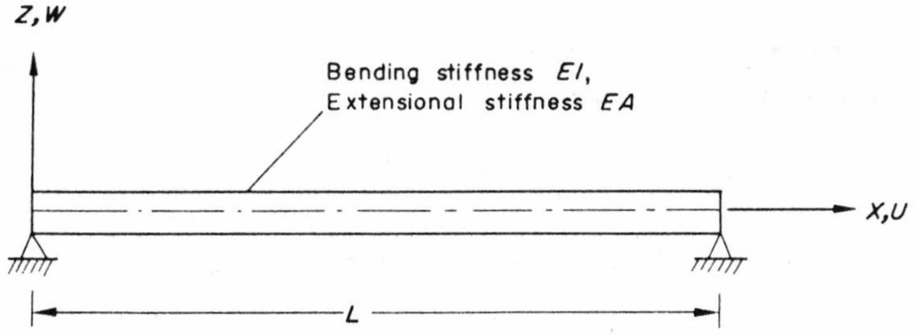

FIG. 1.

which implies

$$N = \frac{EA}{2L}\int_0^L (W_{,X})^2\,\mathrm{d}X. \tag{32}$$

The equation governing transverse vibrations of the beam is

$$mW_{,tt} + EIW_{,XXXX} - NW_{,XX} = 0, \tag{33}$$

where m is the mass per unit length of the beam and I is the second moment of area. If we introduce the following variables and parameters,

$$n = \frac{NL^2}{\pi^2 EI} \qquad x = \frac{\pi X}{L} \qquad w = \frac{W}{\sqrt{(\pi\rho)}} \qquad \rho = \sqrt{\left(\frac{I}{A}\right)} \qquad \omega_1^2 = \frac{\pi^4 EI}{mL^4} \tag{34}$$

then equations (32) and (33) can be written in a convenient dimensionless form. They become

$$n = \frac{1}{2}\int_0^\pi (w_{,x})^2\,\mathrm{d}x \tag{35}$$

$$\frac{1}{(\omega_1)^2}w_{,tt} + w_{,xxxx} - nw_{,xx} = 0. \tag{36}$$

Equation (36) is subject to the boundary conditions

$$\begin{aligned} w(0,t) &= w_{,xx}(0,t) = 0 \\ w(\pi,t) &= w_{,xx}(\pi,t) = 0. \end{aligned} \tag{37}$$

We now set $\tau = \omega t$, $\Omega = \omega^2/\omega_1^2$, and, in view of the structure of the equations,

$$\begin{aligned} w &= \xi w_1 + \xi^3 w_3 + \ldots \\ n &= \xi^2 n_2 + \xi^4 n_4 + \ldots. \end{aligned} \tag{38}$$

The first approximation involves only w_1. The linearized equation is

$$\Omega \ddot{w}_1 + w_{1,xxxx} = 0. \tag{39}$$

The solution is of the form

$$w_1 = \cos\tau \sin kx, \tag{40}$$

where k is an integer. This solution yields

$$\Omega_0 = k^4 \tag{41}$$

for the dimensionless frequency parameter.

The second approximation is simply

$$n_2 = \frac{1}{2}\int_0^\pi (w_{1,x})^2\,\mathrm{d}x = \frac{\pi k^2}{8}(1+\cos 2\tau). \tag{42}$$

The coefficient A in the expansion (24) is zero. B is determined to be

$$B = \frac{\int_0^{2\pi}\int_0^\pi n_2(w_{1,x})^2\,\mathrm{d}x\,\mathrm{d}\tau}{\Omega_0\int_0^{2\pi}\int_0^\pi (\dot{w}_1)^2\,\mathrm{d}x\,\mathrm{d}\tau} = \frac{3\pi}{16}. \tag{43}$$

Consequently, we have the asymptotic relation

$$\frac{\Omega}{\Omega_0} = 1+\frac{3\pi}{16}\xi^2+\ldots. \tag{44}$$

This asymptotic approximation can be shown to be in complete agreement with an asymptotic representation of the solution obtained by Woinowsky-Krieger [7] in terms of elliptic functions and with the solutions obtained by Chu and Herrmann [8] using two different methods.

APPLICATION TO FLEXURAL VIBRATIONS OF RECTANGULAR PLATES

A rectangular plate of length a, width b and thickness h is shown in Fig. 2, along with notation and a sign convention. The origin of surface coordinates (X, Y) is taken to be the corner of the plate, and U, V and W are midsurface displacement components. The plate is assumed to be simply supported on all four edges and relative motions of the edges are assumed to be prevented. Under these conditions, membrane stresses will be induced in the plate due to transverse flexural vibrations of finite amplitude.

The analysis is based upon the following equations, which are equivalent to the dynamic version of von Karman's equations used by Chu and Herrmann [8]:

$$\nabla^4 f = (w_{,xy})^2 - w_{,xx}w_{,yy} \tag{45}$$

$$\frac{4}{(\omega_{sp})^2}w_{,tt}+\nabla^4 w - f_{,yy}w_{,xx}+2f_{,xy}w_{,xy}-f_{,xx}w_{,yy} = 0 \tag{46}$$

$$u_{,x} = f_{,yy}-\nu f_{,xx}-\tfrac{1}{2}(w_{,x})^2 \tag{47}$$

$$v_{,y} = f_{,xx}-\nu f_{,yy}-\tfrac{1}{2}(w_{,y})^2. \tag{48}$$

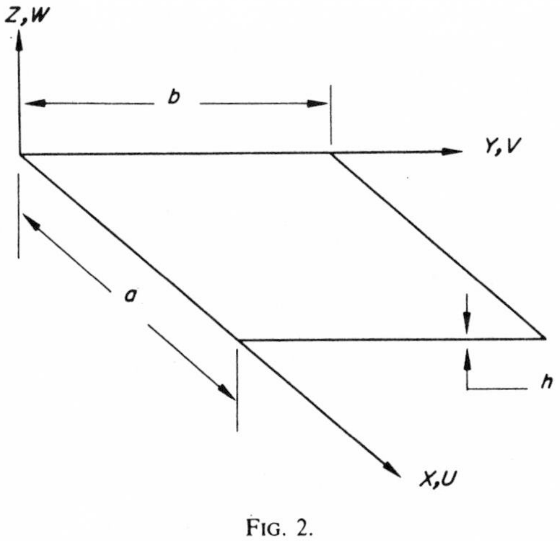

FIG. 2.

The dimensionless variables and parameters are listed below, along with any other important quantities.

$$x = \frac{\pi X}{b} \qquad y = \frac{\pi Y}{b} \qquad w = \frac{[12(1-\nu^2)]^{\frac{1}{2}} W}{h} \qquad u = \frac{4\pi E}{\sigma_p b} U \qquad v = \frac{4\pi E}{\sigma_p b} V$$

$$(\omega_{sp})^2 = \frac{Eh^3}{3(1-\nu^2)m}\left(\frac{\pi}{b}\right)^4 \qquad \mu = \frac{b}{a} \qquad \sigma_p = \frac{E}{3(1-\nu^2)}\left(\frac{\pi h}{b}\right)^2$$

$$\nabla^2(\ ) = \frac{\partial^2(\ )}{\partial x^2} + \frac{\partial^2(\ )}{\partial y^2} \qquad \nabla^4(\ ) = \nabla^2[\nabla^2(\ )]. \tag{49}$$

The stress parameter σ_p is the compressive buckling stress of long or square simply supported plates; ω_{sp} is the fundamental circular frequency of a square plate. ν is Poisson's ratio and f is a dimensionless Airy stress function defined such that the membrane stresses, σ_x^0, σ_y^0, and τ_{xy}^0, are given by the relation

$$\{\sigma_x^0, \sigma_y^0, \tau_{xy}^0\} = \frac{\sigma_p}{4}\{f_{,yy}, f_{,xx}, -f_{,xy}\}. \tag{50}$$

The above partial differential equations are subject to the following boundary conditions:

$$w(0, y) = w\left(\frac{\pi}{\mu}, y\right) = w(x, 0) = w(x, \pi) = 0 \tag{51}$$

$$w_{,xx}(0, y) = w_{,xx}\left(\frac{\pi}{\mu}, y\right) = w_{,yy}(x, 0) = w_{,yy}(x, \pi) = 0 \tag{52}$$

$$v(x, 0) = v(x, \pi) = 0 \tag{53}$$

$$u(0, y) = u\left(\frac{\pi}{\mu}, y\right) = 0. \tag{54}$$

$\mu = b/a$ is the aspect ratio of the plate.

We set $\tau = \omega t$, $\Omega = \dfrac{\omega^2}{(\omega_{sp})^2}$ and

$$\begin{aligned} w &= \xi w_1 + \xi^3 w_3 + \ldots \\ f &= \xi^2 f_2 + \xi^4 f_4 + \ldots. \end{aligned} \tag{55}$$

The first order equation involves only w_1 and is

$$4\Omega \ddot{w}_1 + \nabla^4 w_1 = 0. \tag{56}$$

We take the solution corresponding to the fundamental mode of the plate in the form

$$w_1 = \sin \mu x \sin y \sin \tau, \tag{57}$$

which leads to the linearized frequency ratio

$$\Omega_0 = \frac{(\mu^2+1)^2}{4}. \tag{58}$$

The second order equation is

$$\nabla^4 f_2 = (w_{1,xy})^2 - w_{1,xx}w_{1,yy} = \frac{\mu^2(1-\cos 2\tau)}{4}(\cos 2\mu x + \cos 2y). \tag{59}$$

We take the following solution for f_2:

$$f_2 = \tfrac{1}{64}(1-\cos 2\tau)\left[\frac{1}{\mu^2}\cos 2\mu x + \mu^2 \cos 2y + 2\alpha x^2 + 2\beta y^2\right] \tag{60}$$

α and β are constants which must be evaluated so as to satisfy the boundary conditions (53) and (54). The tangential displacement parameters can be determined with f_2 and w_1 known; they are

$$u_2 = \frac{1}{32\mu}(1-\cos 2\tau)\sin 2\mu x(\nu - \mu^2 + \mu^2 \cos 2y) \tag{61}$$

$$v_2 = \tfrac{1}{32}(1-\cos 2\tau)\sin 2y(\nu\mu^2 - 1 + \cos 2\mu x). \tag{62}$$

It can be verified that these expressions satisfy (53) and (54) and that we must require that

$$\alpha = \frac{1+\nu\mu^2}{(1-\nu^2)} \tag{63}$$

and

$$\beta = \frac{(\mu^2+\nu)}{(1-\nu^2)}. \tag{64}$$

Again, as for the beam, the coefficient A in the expansion (24) is zero. For the plate, B is found to be

$$B = \frac{\int_0^{2\pi}\int_0^{\pi/\mu}\int_0^{\pi}[f_{2,yy}(w_{1,x})^2 + f_{2,xx}(w_{1,y})^2]\,dx\,dy\,d\tau}{4\Omega_0\int_0^{2\pi}\int_0^{\pi/\mu}\int_0^{\pi}(\dot{w})^2\,dx\,dy\,d\tau} \tag{65}$$

$$= \frac{3}{32(\mu^2+1)}\left[\frac{\mu^2+2\nu\mu^2+1}{(1-\nu^2)} + \frac{1}{2}(\mu^4+1)\right].$$

Consequently, we have the asymptotic relation

$$\frac{\Omega}{\Omega_0} \cong 1 + \frac{3}{32(\mu^2+1)}\left[\frac{\mu^2+2\nu\mu^2+1}{(1-\nu^2)} + \frac{1}{2}(\mu^4+1)\right]\xi^2. \tag{66}$$

A solution to this problem has been obtained by Chu and Herrmann [8] in terms of elliptic functions. A careful study of their solution and an asymptotic representation of it for small amplitudes shows that there is complete agreement between it and the present solution, albeit a lengthy process.

CONCLUDING REMARKS

A general approach to nonlinear vibrations of elastic structures has been developed which provides information regarding the first order effects of finite displacements. It relies upon the use of Hamilton's principle and a perturbation procedure to obtain analytical results. The theory effectively reduces a nonlinear free vibration problem to a sequence of linear problems, only the first two of which usually need be solved to obtain an initial estimate of finite amplitude effects. The theory has been applied to beams and rectangular plates, structures for which solutions already exist, in order to illustrate the theory and demonstrate its usefulness.

REFERENCES

[1] W. T. KOITER, On the stability of elastic equilibrium. NASA TT F. **10**, 833 (1967).

[2] W. T. KOITER, Elastic stability and post-buckling behavior, in *Non-linear Problems*, edited by R. E. LANGER. University of Wisconsin Press (1963).

[3] B. BUDIANSKY and J. W. HUTCHINSON, Dynamic buckling of imperfection-sensitive structures, in *Applied Mechanics*, edited by H. GÖRTLER. Springer (1964).

[4] B. BUDIANSKY, Dynamic buckling of elastic structures: criteria and estimates, in *Dynamic Stability of Structures*, edited by G. HERRMANN. Pergamon Press (1966).

[5] J. B. KELLER, Nonlinear vibrations governed by partial differential equations, *Proc. Fifth U.S. Nat. Congr. appl. Mech.* American Society of Mechanical Engineers (1966).

[6] J. B. KELLER and L. TING, Periodic vibrations of systems governed by nonlinear partial differential equations. *Communications on Pure and Applied Mathematics*. **19**, 371 (1966).

[7] S. WOINOWSKY-KRIEGER, The effect of an axial force on the vibration of hinged bars. *J. appl. Mech.* **17**, 35 (1950).

[8] H. N. CHU and G. HERRMANN, Influence of large amplitudes on free flexural vibrations of rectangular elastic plates. *J. appl. Mech.* **23**, 532 (1956).

Part VI

DAMPING

Editor's Comments on Papers 38 Through 41

In this final section we shall be concerned with the effects of damping. It is a subject that is deeper than is widely believed, and not quite so well understood as is so often thought. Yet it is a subject in which a great many models have been developed, and, more importantly, a great deal of empirical evidence has been accumulated and evaluated. The four papers we have chosen for this section present some of the models and some of the experience.

Snowdon's paper is concerned with the representation of the damping of real materials in analytical terms. The foundation is laid in terms of linear viscoelasticity, from whence is derived the complex modulus with its real dynamic stiffness and the imaginary term which includes the damping factor [cf. equations (2.8) and (2.9)]. Then follows a characterization of a number of real materials as being low-damping or high-damping, depending upon the peak value of the damping factor. This, in turn, is also strongly related to the transition frequency, which is relatively low for high-damping materials (compare Figures 3 and 5). Snowdon points out that a high-damping material may lose its effectiveness as a vibration isolator because the dynamic stiffness increases rapidly with frequency, beyond the transition frequency. It is also noted that a high-damping rubber can be effectively mounted in parallel with a low-damping material to reduce the increase in dynamic stiffness, while retaining the high values of the damping factor. Snowdon's paper lays the groundwork for a series of subsequent papers which exploit the concept of the complex modulus. The results of much of that work can be found in Snowdon's book, *Vibration and Shock in Damped Mechanical Systems,* New York: John Wiley, 1968.

Crandall's survey was originally delivered as the first Fairey Lecture in the Institute of Sound and Vibration Research at the University of Southhampton. Although the bulk of the paper is concerned with various simple analytical models, the discussion is oriented throughout toward physical principles. For example, damping is defined as "the removal of energy from a vibratory system." Noting that energy can be lost from a system by radiation, or dissipated within a system, Crandall instructively presents some simple results to show that while much of the damping loss can be accounted for, and even properly apportioned among various mechanisms, it is not possible to account for all of the lost energy.

Among many interesting topics in Crandall's paper, there appears the relatively philosophical concept of *causality* (pp. 373–374), Figure 7). The point is that if we wish effect to follow cause, within a mathematical model of a forced, damped system, the damping model must itself obey certain restrictions. For example, if the loss factor is assumed to be independent of frequency, in a single-degree-of-freedom system, the response than precedes the excitation (Figure 7). In the same philosophic vein, Crandall also warns against using "nonequations" wherein time-domain variables and frequency-domain dependences are freely intermingled [see "equations" (21)]. Finally, it is of interest to note another warning, i.e., a system that is unstable in the undamped state can be destabilized by adding damping. This is illustrated with a simple model of a piece of rotating machinery (pp. 374–376) as a paradigm of nonconservative stability problems.

The first of Ungar's two papers is a definitive survey of damping in built-up structures. As the author notes, the damping of monolithic structures can be evaluated in terms of the materials of which they are made. However, structures that are manufactured by joining the various pieces together have, within these joints, additional capacity to dissipate energy. Thus the emphasis is placed here on the damping due to joints and on complex structures with many pieces joined together.

Among other points we note that for simple dry joints, the damping is dominated by the elastic interaction of rough surfaces. In lubricated joints, the lubricating fluid provides both the stiffness, through fluid entrapment, and the damping, through viscous shearing losses. It is worth noting that these shearing losses are generated by normal relative motion of the surfaces that entrap the fluid; this is the origin of the "pumping" model of damping in lubricated joints.

One of the interesting features of this paper is the continuing emphasis on the measurement of damping effects. Wherever possible, Ungar points to experimental verification and practical application, noting what can be measured and what cannot. This shows up particularly well in the discussion of measurement of damping in skin-stringer systems. For example, Ungar notes that decay rates as measured can be either too low or too high. It is of critical importance, here, that the paths of energy transport be well understood, so that one can evaluate the experimental data. This represents in part a caution often relayed to the analyst, but generally assumed to be understood by the experimentalist, that is, before one measures or calculates one should understand the physics of the problem!

In the final paper, Ungar presents a method for calculating stresses in a damped system, vibrating at resonance. The analysis is straightforward, and results in a general upper bound on the maximum principal stress [e.g., equation (25)] for rectangular, simply supported plates. The bound is intuitively pleasing as far as the damping is concerned, for if the damping factor is allowed to become vanishingly small, the (bending) stress at resonance can become infinitely large. Further, if the damping factor increases, we would expect that the stresses would decrease, as the bending of the plate is inhibited. We also note [see equation (24)] that the stresses at resonance decrease rapidly with increased thickness, and with increased wave number, both of which results are very reasonable. Similar results are shown to pertain to other configurations and geometries.

38

Reprinted from *J. Acoust. Soc. Amer.*, **35**(6), 821–829 (1963)

Representation of the Mechanical Damping Possessed by Rubberlike Materials and Structures*

J. C. Snowdon

Ordnance Research Laboratory, The Pennsylvania State University, University Park, Pennsylvania

(Received 17 October 1962)

Discussed in detail is the manner in which the damping possessed by rubberlike materials and structures experiencing sinusoidal vibration may be represented by the ratio of the imaginary to the real part of a complex elastic modulus. Examples are given of the way in which the damping possessed by low- and high-damping rubbers depends on frequency. Equations that predict the response to vibration of such distributed systems as damped rods and beams, and a simple structure comprised of two damped beams and a lumped element of mass are derived; and representative computations of input impedance and transmissibility are presented.

1. INTRODUCTION

It is becoming increasingly important and desirable to represent the damping possessed by rubberlike materials and structures in a realistic manner. High-damping materials are playing a widening role in such applications as backing layers for extended metal surfaces, and as materials for antivibration mountings. There is also much current interest in the use, as structural elements, of beams and plates composed of alternate bonded metal and dissipative laminations. Consequently, it is necessary that damping be included in theoretical analyses in a more adequate manner than has been customary in the past. It is the purpose of this paper to demonstrate that the realistic representation of damping in theoretical analyses of vibration problems is not a difficult task.

2. REPRESENTATION OF THE DAMPING POSSESSED BY RUBBERLIKE MATERIALS

The strain induced in a purely elastic linear material is proportional to the stress producing the deformation. Two basic types of deformation that the material may experience are described by two independent elastic constants. Thus, the shear modulus describes a shear deformation for which the material does not change in volume, and the bulk modulus describes a volume deformation for which the material does not change in shape. In most practical applications, a rubberlike material is utilized such that its behavior is governed by the shear modulus G, since this is many times smaller than the bulk modulus K. If a thin, flat sample of very large lateral dimensions is stressed in a direction perpendicular to the plane of its surfaces, there are changes in both shape and volume, the ratio of stress to strain being described by a modulus M given by

$$M=K+4G/3\approx K. \tag{2.1}$$

It is this modulus that governs the propagation of a longitudinal elastic wave in the material, provided the frequency is sufficiently high that the sample dimensions become large in comparison with the wavelength.

The strain induced in a *linear* viscoelastic material is also some function of the applied stress experienced by the material. Now, however, the two basic types of deformation are not related to the applied stress by a simple constant of proportionality—the elastic modulus G or K. Rather, the relation between stress and strain is generally represented by a *linear* partial differential equation of arbitrary order:

$$\left[A_0+A_1\frac{\partial}{dt}+A_2\frac{\partial^2}{\partial t^2}+\cdots A_n\frac{\partial^n}{\partial t^n}\cdots\right]\sigma$$

$$=\left[B_0+B_1\frac{\partial}{\partial t}+B_2\frac{\partial^2}{\partial t^2}+\cdots B_n\frac{\partial^n}{\partial t^n}\cdots\right]\epsilon, \tag{2.2}$$

* This paper was presented, essentially in the same form, at the National Aeronautic and Space Engineering Meeting of the Society of Automotive Engineers, Los Angeles, California, on 11 October 1962. It has not been published in the literature heretofore.

where σ is stress, ϵ is strain, and t is time; A_n and B_n are constants. The assumption of linearity is crucial in all subsequent analyses.

The foregoing equation may be derived by purely physical reasoning,[1] the derivation being completely rigorous when stress and strain vary sinusoidally with time—the case of interest in the present paper. It is instructive, however, to visualize the relation between stress and strain as being generated by an equivalent network of many springs and dashpots, the combination duplicating the mechanical behavior of the material under strain. Thus, consider the following simple examples:

1. Spring of stiffness λ_0 [Fig. 1(a)]:

$$\sigma=\kappa\lambda_0\epsilon, \tag{2.3}$$

where κ is a constant, A_0 and B_0 are finite, and all other values of A and B are zero.

2. Dashpot of viscosity η_0 [Fig. 1(b)]:

$$\sigma=\kappa\eta_0[\partial/\partial t]\epsilon. \tag{2.4}$$

(A_0 and B_1 are finite, all other values of A and B are zero.)

3. One spring and one dashpot [Fig. 1(c)]:

$$\sigma=\kappa[\lambda_0+\eta_0(\partial/\partial t)]\epsilon. \tag{2.5}$$

(A_0, B_0, and B_1 are finite, all other values of A and B are zero.)

4. Two springs and one dashpot [Fig. 1(d)]:

$$[\lambda_1+\eta_1(\partial/\partial t)]\sigma=\kappa[\lambda_0\lambda_1+(\lambda_0+\lambda_1)\eta_1(\partial/\partial t)]\epsilon. \tag{2.6}$$

(A_0, A_1, B_0, and B_1 are finite, all other values of A and B are zero.)

5. Three springs and three dashpots [Fig. 1(e)]:

$$\left\{\lambda_1\lambda_2+(\lambda_1\eta_2+\lambda_2\eta_1)\frac{\partial}{\partial t}+\eta_1\eta_2\frac{\partial^2}{\partial t^2}\right\}\sigma$$
$$=\kappa\Big\{\lambda_0\lambda_1\lambda_2+[\lambda_1\lambda_2(\eta_0+\eta_1+\eta_2)$$
$$+\lambda_0(\lambda_1\eta_2+\lambda_2\eta_1)]\frac{\partial}{\partial t}+[\eta_1\eta_2(\lambda_0+\lambda_1+\lambda_2)$$
$$+\eta_0(\lambda_1\eta_2+\lambda_2\eta_1)]\frac{\partial^2}{\partial t^2}+\eta_0\eta_1\eta_2\frac{\partial^3}{\partial t^3}\Big\}\epsilon. \tag{2.7}$$

(A_0, A_1, A_2, B_0, B_1, B_2, and B_3 are finite, all other values of A and B are zero.)

It is well to recall that a material which can be represented by a combination of springs and dashpots has no memory in the usual sense of the word.

When stress and strain vary sinusoidally with time, the nth partial derivative of stress or strain may be equated to $(j\omega)^n\sigma$ or $(j\omega)^n\epsilon$, respectively, where $j=(-1)^{\frac{1}{2}}$ and ω is angular frequency. The original partial differential equation then reduces to the following algebraic equation:

$$\{A_0+(j\omega)A_1+(j\omega)^2A_2+\cdots(j\omega)^nA_n\cdots\}\sigma$$
$$=\{B_0+(j\omega)B_1+(j\omega)^2B_2+\cdots(j\omega)^nB_n\cdots\}\epsilon,$$

which may be written more simply as

$$[a_1(\omega)+ja_2(\omega)]\sigma=[b_1(\omega)+jb_2(\omega)]\epsilon, \tag{2.8}$$

where $a_1(\omega)$, $a_2(\omega)$, $b_1(\omega)$, and $b_2(\omega)$ are functions of frequency.

It must be emphasized that the foregoing discussion relates only to the *linear* response of a viscoelastic material to an applied stress: the principle of superposition holds, Eq. (2.2) is linear, and the substitution of $(j\omega)^n$ for the operator $\partial^n/\partial t^n$ is permissible. It follows that alien frequencies will not be generated when the material is subjected to an applied force possessing one or more frequency components, as would be expected if the material were nonlinear.

Since the ratio of two complex numbers is itself a complex number, the ratio of stress to strain defined by Eq. (2.8) may be written as a single complex quantity. That is to say, the ratio of stress to strain in the material is represented not by a real but by a complex elastic modulus possessing real and imaginary parts that are, in general, functions of frequency. When the material is stressed in shear, for example, the complex ratio of stress to strain at any given temperature may be written

$$G^*=G_\omega(1+j\delta_\omega), \tag{2.9}$$

where G_ω and δ_ω are the real part and the ratio of the

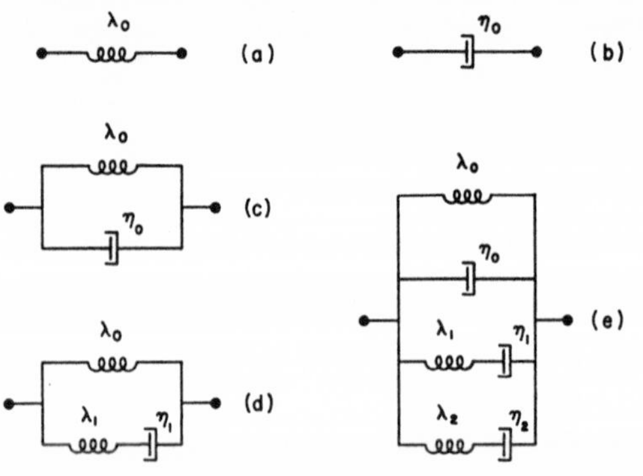

FIG. 1. Simple spring and dashpot models.

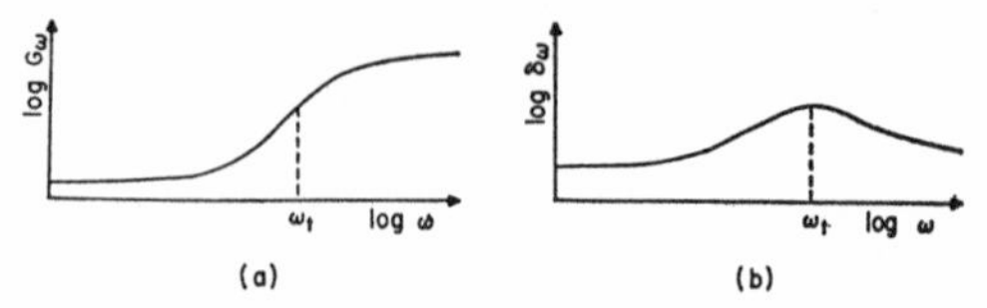

FIG. 2. Sketch of the frequency dependence of (a) the dynamic elastic modulus and (b) the damping factor characteristically possessed by rubberlike materials.

[1] E. J. Skudrzyk, Österr. Ing.-Arch. 3, 356 (1949).

imaginary to the real part of the complex shear modulus, respectively. The damping factor δ_ω describes the mechanical loss in the material, being analogous to the reciprocal of the quality factor Q employed in electrical circuit theory. The so-called dynamic modulus G_ω and damping factor δ_ω are functions of temperature as well as frequency, the mechanical properties of high-damping rubbers being far more temperature-sensitive than those of low-damping rubbers such as natural rubber. The results and discussion presented in this paper refer only to room temperature, the dependence of G_ω and δ_ω upon temperature being described in more detail elsewhere.[2,3]

3. FREQUENCY DEPENDENCE OF THE MECHANICAL PROPERTIES OF RUBBERLIKE MATERIALS

The dynamic modulus and damping factor of rubberlike materials are found experimentally to vary with frequency in the general manner shown by Fig. 2. The dynamic modulus increases over many decades in frequency at room temperature, at first slowly and then more rapidly. The damping factor exhibits a broad maximum value, which falls at the so-called rubber-to-glass transition frequency ω_t. In the neighborhood of this frequency, the dynamic modulus increases at a maximum rate and may be considered proportional to ω^α. The constant α appears to take some value between 0.5 and 1.0, which is characteristic of the material.

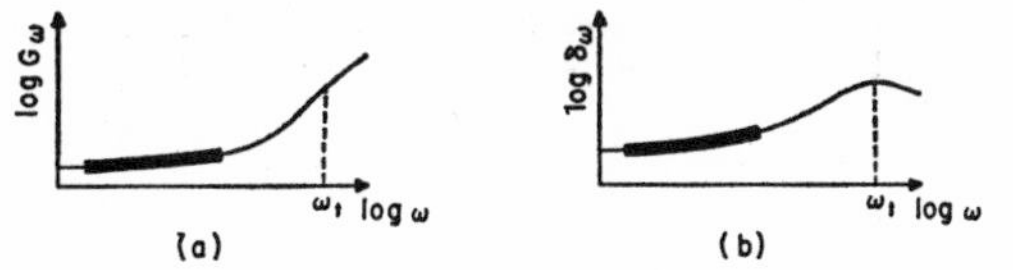

FIG. 3. Sketch of the frequency dependence of (a) the dynamic elastic modulus and (b) the damping factor of a low-damping rubber.

The transition frequency refers to the transition of rubberlike materials at sufficiently high frequencies to an "inextensible" or glasslike state, the dynamic modulus becoming so large that the characteristic resilience of the material is no longer apparent. The transition frequency of natural and other low-damping rubbers occurs at very high frequencies at room temperature, so that through the range of frequencies of concern in vibration problems (the shaded portions of the curves shown in Fig. 3) the damping factor is small and G_ω and δ_ω vary only slowly with frequency. For example, the dynamic shear modulus and damping factor of natural rubber filled with fifty parts by weight of carbon black are shown as a function of frequency in Fig. 4. These curves were deduced by the method of reduced variables from experimental data published by Fletcher and Gent.[4] (The method of reduced variables[5] enables the magnitude of G_ω or δ_ω to be predicted through a very broad frequency range, provided the experiment in which they are determined, through a restricted frequency range, is conducted at a number of temperatures.)

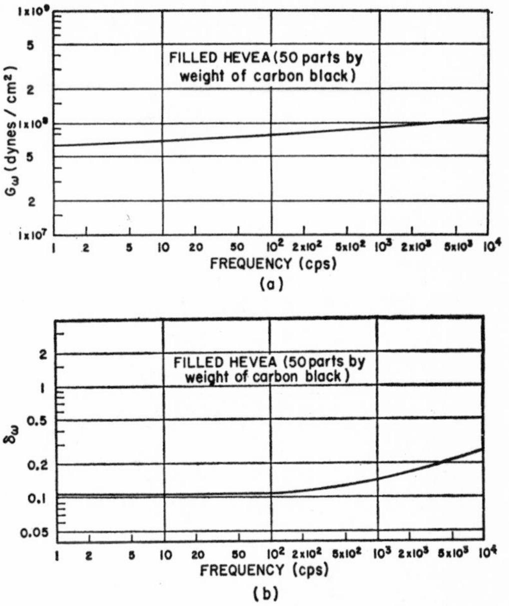

FIG. 4. Frequency dependence of (a) the dynamic shear modulus and (b) the damping factor possessed by natural rubber.

The transition frequencies of high-damping rubbers fall much below the transition frequency of natural rubber, occurring at frequencies that are normally of interest in vibration problems at room temperature. The dynamic modulus of the materials, therefore, increases very rapidly with frequency (refer to the shaded portions of the curves of Fig. 5); the damping factor is large and again varies slowly with frequency. The dynamic modulus and damping factor possessed by two high-damping materials are shown as functions of frequency in Figs. 6 and 7. The curves for plasticized

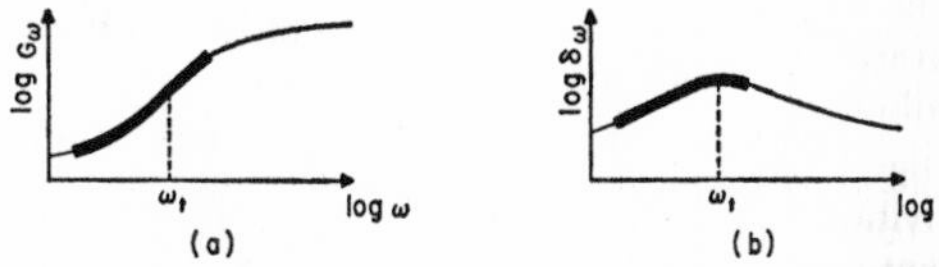

FIG. 5. Sketch of the frequency dependence of (a) the dynamic elastic modulus and (b) the damping factor of a high-damping rubber.

---

[2] E. E. Ungar and D. Kent Hatch, Prod. Eng. 32, No. 16, 44 (1961).

[3] J. C. Snowdon, Noise Control 6, No. 2, 18 (1960).

[4] W. P. Fletcher and A. N. Gent, Brit. J. Appl. Phys. 8, 194 (1957).

[5] J. D. Ferry, *Viscoelastic Properties of Polymers* (John Wiley & Sons, Inc., New York, 1961).

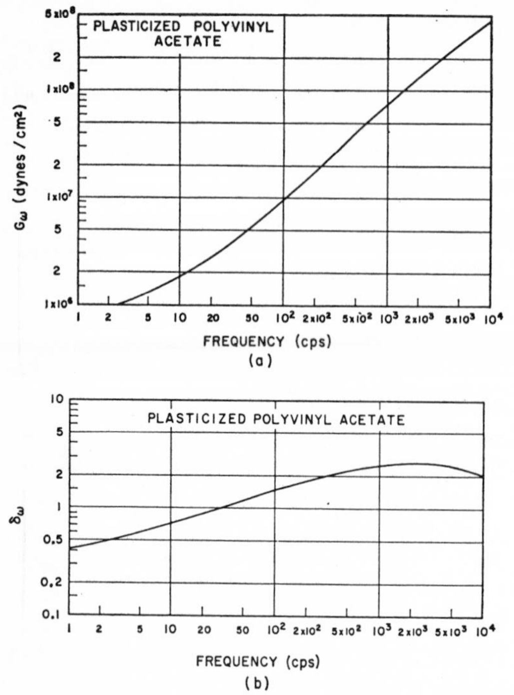

FIG. 6. Frequency dependence of (a) the dynamic shear modulus and (b) the damping factor possessed by plasticized polyvinyl acetate.

polyvinyl acetate (Fig. 6) have been deduced by the method of reduced variables from the experimental results of Williams and Ferry.[6] The value of the constant α describing the maximum slope of the dynamic shear modulus–frequency plot is 0.93, so that the modulus is essentially proportional to frequency through a certain frequency range neighboring the transition frequency. The data for plasticized polyvinyl butyral resin (Fig. 7), also obtained by the method of reduced variables, have been made available in their present form by Sinnott and Eby.[7] This material is of interest because the transition frequency—and, therefore, the maximum value of the damping factor—falls below 1 cps at room temperature, rather than at approximately 1000 cps as observed for plasticized polyvinyl acetate. In consequence, although the damping factor of the material is very large, it decreases as frequency increases through the usual frequency range of interest in vibration problems.

High-damping rubbers can effectively be utilized in antivibration mountings to reduce the vibration of the mounting system at its resonant frequencies, and to suppress the lower modes of vibration of any foundation that supports the mounting system.[8] Despite the common assumption that the inherent damping of such rubbers is responsible for the poor degree of isolation they afford at high frequencies, the adverse effects of high damping are really of second-order magnitude. Rather, the poor isolation is due to the manner in which the dynamic modulus increases rapidly with frequency.[3,8] Recently, it has been suggested[8,9] that the benefits resulting from the use of a high-damping rubber could be retained, yet the limitation of the large high-frequency stiffness avoided, if the rubber were mechanically placed in parallel with natural rubber of suitably greater cross section—each material being subject to the same strain. In this way, the dynamic modulus of the combination would increase only slowly with frequency, yet the associated damping factor would take values of significant magnitude intermediate to those possessed by the constituent rubbers. For example, the variation with frequency of the dynamic modulus and damping factor of a "parallel" mounting—composed of uniform elements of the high-damping rubber Thiokol RD, and natural rubber of five times greater cross-sectional area—is shown in Fig. 8. The performance of the mount-

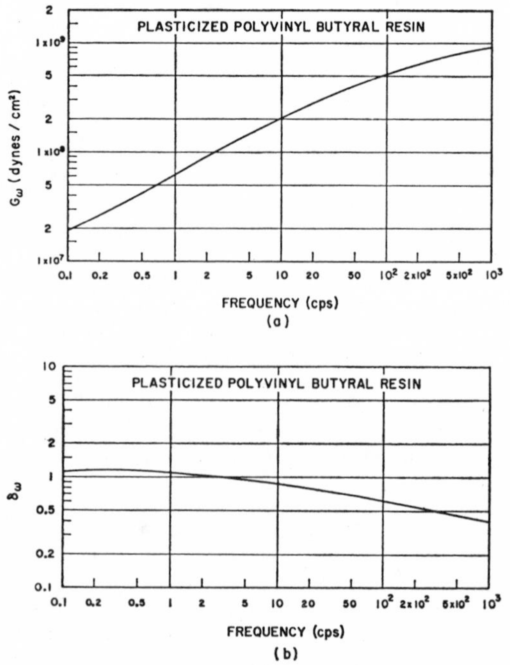

FIG. 7. Frequency dependence of (a) the dynamic shear modulus and (b) the damping factor possessed by plasticized polyvinyl butyral resin.

---

[6] M. L. Williams and J. D. Ferry, J. Colloid Sci. **10**, 1 (1955).

[7] K. M. Sinnott and R. K. Eby, private communication (to be published).

[8] J. C. Snowdon, J. Acoust. Soc. Am. **34**, 54 (1962).

[9] J. C. Snowdon, "Utilization of High-Damping Rubbers in Vibration Control," Paper P22, in *Proceedings of the Fourth International Congress on Acoustics, Copenhagen, August 1962* (to be published).

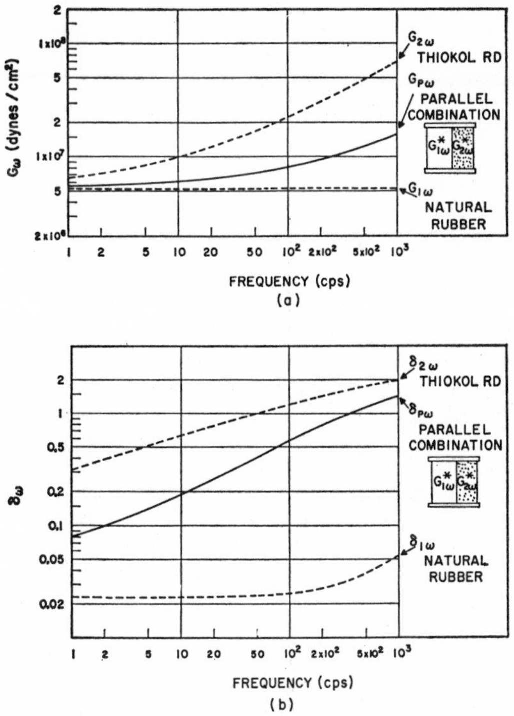

FIG. 8. Frequency dependence of (a) the dynamic shear modulus and (b) the damping factor possessed by a parallel combination of natural and Thiokol RD rubbers.

ing has been examined in detail elsewhere,[8,10] and compared with the performance of mountings composed entirely of Thiokol RD and entirely of natural rubber. Values of G_ω and δ_ω at a large number of frequencies were taken from the curves of Fig. 8 and inserted numerically into general equations describing the behavior of one- and two-stage mounting systems supported by a foundation possessing finite mechanical impedance. The over-all performance of the parallel mounting is found[8] to be superior to that of its constituent rubbers, as can be anticipated from the results presented in Fig. 8. Here, the damping factor of the parallel mounting is seen to possess significantly large values through three decades in frequency; yet the dynamic modulus increases by a factor of three only through the same frequency range—in contrast to the dynamic modulus of Thiokol RD, which increases by a factor greater than ten.

4. REPRESENTATION OF THE DAMPING POSSESSED BY STRUCTURES

In place of a real modulus of elasticity, it is now instructive to consider the substitution of a complex modulus in equations describing the forced vibration of a simple lumped system and the vibration of such simple distributed systems as a rod excited in its longitudinal modes and bars excited in their transverse or bending modes.

It is unnecessary, in general, to derive equations describing the response to vibration of damped systems from first principles. Rather, it is preferable to obtain solutions, or adopt solutions already available, that describe the response of systems with negligible damping. Thus, consider the well-known transmissibility equation for a one-stage, lumped, "mass–spring" system:

$$T=\left|\frac{1}{1-(\omega^2M/kG)}\right|, \tag{4.1}$$

where M is the mass of the mounted item, G is an elastic modulus describing the behavior of the undamped resilient mounting, and k is a constant (having the dimension of length) determined by the mount geometry. To obtain the transmissibility of a damped mounting, it is now only necessary that G be replaced by the complex modulus $G^*=G_\omega(1+j\delta_\omega)$; thus,

$$T=\left|\frac{1+j\delta_\omega}{1-(\omega^2M/kG_\omega)+j\delta_\omega}\right|. \tag{4.2}$$

Since the natural frequency ω_0 of the system is defined by the relation

$$\omega_0^2=kG_0/M, \tag{4.3}$$

where G_0 represents the value of G_ω at the frequency ω_0, the expression for transmissibility may finally be written

$$T^2=\frac{(1+\delta_\omega^2)}{\{[1-(\omega/\omega_0)^2(G_0/G_\omega)]^2+\delta_\omega^2\}}. \tag{4.4}$$

From this equation, the transmissibility of any material may be obtained,[8] provided the dependence of G_ω and δ_ω upon frequency is known.

Although the frequency dependence of G_ω and δ_ω can, in principle, be incorporated into any expression describing the response to vibration of distributed mechanical systems, it is simpler to assume that the quantities G_ω and δ_ω are independent of frequency. This is a valid assumption in the case of metals—for which δ is very small—and a good approximation to the behavior of thermosetting plastics such as Lucite, crystalline polymers such as Teflon and polyethylene, and relatively low-damping rubbers such as natural rubber.

It is possible to seek either a progressive-wave solution or a standing-wave solution to the wave equations that describe the vibration of distributed systems, depending upon the boundary conditions and nature of the particular system under consideration. For example, in the frequency range of interest in vibration problems and for the dimensions and damping normally possessed

[10] G. G. Parfitt, contribution to *Underwater Acoustics*, edited by V. M. Albers (Plenum Press, Inc., New York, 1963), Chap. 4, pp. 67–84.

by structures of engineering interest, pure wavepropagation is not often observed, so it is normally desirable to seek only the standing-wave solution. That is to say, systems of interest are seldom so large compared with the acoustic wavelength, or are seldom so heavily damped that reflections from the boundaries of the system—which give rise to opposing progressive waves and, therefore, standing waves—can be neglected. In consequence, emphasis will be placed upon the standing-wave solution to the wave equations of the simple distributed structures considered here. Equations describing the vibration of damped structures may again be obtained by substituting a complex elastic modulus in the appropriate wave equations, or in their solutions derived with the assumption that damping is negligible.

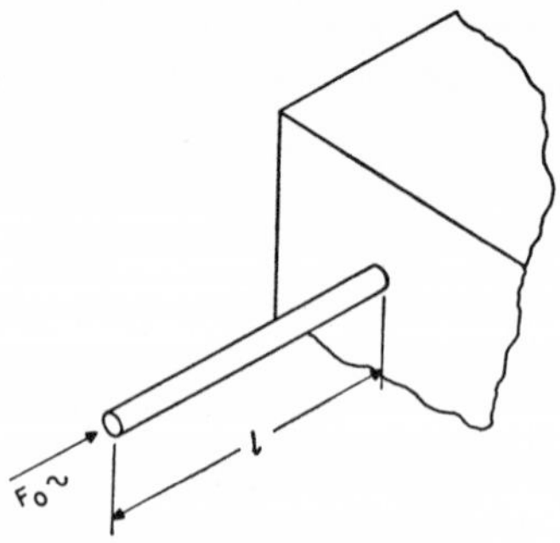

FIG. 9. A rod excited at its free end by a sinusoidally varying force.

5. LONGITUDINAL VIBRATION OF A DAMPED ROD

The wave equation for the longitudinal vibration of a rod of uniform (but not necessarily circular) cross section may be written directly in the following manner:

$$\partial^2\zeta/\partial x^2=(1/c^{*2})(\partial^2\zeta/\partial t^2), \tag{5.1}$$

provided that the wavelength of the vibration remains large compared with the cross-sectional dimensions of the rod. In this equation, ζ represents the displacement of an elementary section of the rod in the axial or "x" direction. The velocity of wavepropagation is given by $c^{*2}=E^*/\rho$, the quantities E^* and ρ representing the complex Young's modulus and density of the rod, respectively.

The standing-wave solution to the foregoing equation is of the form

$$\zeta=(P\sin m_dx+Q\cos m_dx)e^{j\omega t}, \tag{5.2}$$

where the parameter $m_d=\omega/c^*$ and P and Q are constants that may be determined from knowledge of the boundary conditions of the rod.

The standing-wave solution may equivalently be visualized as the sum of two periodic progressive-wave solutions. Thus,

$$\begin{aligned}\zeta&=\zeta_0[e^{j(\omega t-m_dx)}+(Re^{j\phi})e^{j(\omega t+m_dx)}]\\&=\zeta_0[e^{-jm_dx}+(Re^{j\phi})e^{jm_dx}]e^{j\omega t}.\end{aligned} \tag{5.3}$$

The first term of this equation refers to a wave traveling in the positive x direction, and the second term, to a wave that has been reflected from the end of the rod and is consequently traveling in the negative x direction. The quantity $Re^{j\phi}$ describes the relative magnitude of, and the phase difference between, the incident and reflected waves. The values of ζ_0, R, and ϕ may be determined from knowledge of the boundary conditions of the rod.

The solutions defined by Eqs. (5.2) and (5.3) are directly equivalent, of course; substitution of the boundary conditions for a rod that is clamped at one end and excited at the other by a force $F_0e^{j\omega t}$ (Fig. 9), for example, leads in both cases to the following results for the displacement at any point "x" and the impedance Z_0 at the driving point:

$$[\zeta]_x=\left(\frac{F_0}{\omega^2M_b}\right)(m_dl)\left[\left(\frac{\sin m_dl}{\cos m_dl}\right)\cos m_dx-\sin m_dx\right], \tag{5.4}$$

and

$$[Z_0/j\omega M_b]=-\cot m_dl/(m_dl), \tag{5.5}$$

where M_b is the mass of the rod. Since any coordinate "x" can be expressed as some fraction of the rod length, both ζ and Z_0 are functions of the parameter (m_dl), where

$$(m_dl)=\omega l/c^*=ml/(1+j\delta)^{\frac{1}{2}}. \tag{5.6}$$

It is now desirable to replace the parameter (m_dl) by the complex quantity $(p+jq)$, where

$$p=(ml/D)[\tfrac{1}{2}(D+1)]^{\frac{1}{2}}, \tag{5.7}$$

$$q=-(ml/D)[\tfrac{1}{2}(D-1)]^{\frac{1}{2}}, \tag{5.8}$$

and

$$D=(1+\delta^2)^{\frac{1}{2}}. \tag{5.9}$$

Then, for example, the expression for the driving-point impedance—normalized by division by the impedance of a lumped mass equal to that of the clamped rod—may be written

$$\frac{Z_0}{j\omega M_b}=-\frac{\cot(p+jq)}{(p+jq)}=\frac{(-\mathrm{c}p.\mathrm{ch}q.+j\mathrm{s}p.\mathrm{sh}q.)}{(p+jq)(\mathrm{s}p.\mathrm{ch}q.+j\mathrm{c}p.\mathrm{sh}q.)}, \tag{5.10}$$

so that the absolute value of Z_0 is simply

$$\left|\frac{Z_0}{j\omega M_b}\right|^2=\frac{[\mathrm{c}^2p.\mathrm{ch}^2q.+\mathrm{s}^2p.\mathrm{sh}^2q.]}{[p.\mathrm{s}p.\mathrm{ch}q.-q.\mathrm{c}p.\mathrm{sh}q.]^2+[q.\mathrm{s}p.\mathrm{ch}q.+p.\mathrm{c}p.\mathrm{sh}q.]^2}=\frac{(\mathrm{c}^2p.\mathrm{ch}^2q.+\mathrm{s}^2p.\mathrm{sh}^2q.)}{(p.^2+q.^2)(\mathrm{s}^2p.+\mathrm{sh}^2q.)}, \tag{5.11}$$

where such abbreviations as cp. and ch$q$. represent the quantities cos$p$ and cosh$q$, respectively.

The foregoing expression for the normalized driving-point impedance of the rod has been computed for values of the damping factor $\delta=0.01$, 0.1, and 1.0. The results are plotted in Fig. 10 as a function of the quantity ml, which is proportional to frequency. The highest damping factor is considered only as a hypothetical case since the real part of the complex Young's modulus associated with such high damping would not be a constant, as assumed here, but would be strongly dependent on frequency. At low frequencies, the impedance is springlike in nature, Z_0 being inversely proportional to frequency. The impedance subsequently exhibits minimum and maximum values where it changes from being predominantly springlike to predominantly masslike in character, and vice versa. It is apparent that, if the damping factor δ is changed by a certain factor, the maximum and minimum values of Z_0 change by a closely similar factor provided δ remains small.

In concluding this section, it is interesting to make reference to the equations governing wavepropagation in a damped rod. Thus, if reflections from the end of a rod such as that depicted in Fig. 9 may be neglected ($R=0$)—because the length of the rod is many times greater than the acoustic wavelength, or because the rod is heavily damped—then the expression for ζ presented previously [Eq. (5.3)] degenerates into an equation describing the propagation of a single progressive wave. This equation is usually stated in the following manner:

$$\zeta=\zeta_0 e^{-(\alpha+j\beta)x}e^{j\omega t}, \tag{5.12}$$

where α and β have replaced the quantities $-q/l$ and p/l, respectively. It follows that

$$\alpha=(m/D)[\tfrac{1}{2}(D-1)]^{\frac{1}{2}}, \tag{5.13}$$

and

$$\beta=(m/D)[\tfrac{1}{2}(D+1)]^{\frac{1}{2}}. \tag{5.14}$$

If the damping factor is very small compared with unity, the expressions for α and β simplify and may be written accurately as follows:

$$\alpha=\tfrac{1}{2}m\delta=\tfrac{1}{2}\omega\delta(\rho/E)^{\frac{1}{2}}, \tag{5.15}$$

and

$$\beta=m=\omega(\rho/E)^{\frac{1}{2}}. \tag{5.16}$$

When damping is small, therefore, the following solution to the progressive-wave equation is obtained:

$$\zeta=\zeta_0 e^{-\frac{1}{2}\delta(mx)}e^{j(\omega t-mx)}. \tag{5.17}$$

6. TRANSVERSE VIBRATION OF DAMPED BEAMS

The wave equation describing the transverse or bending vibration of a uniform damped beam is as follows:

$$\partial^4\zeta/\partial x^4=-(\rho/k^2E^*)(\partial^2\zeta/\partial t^2), \tag{6.1}$$

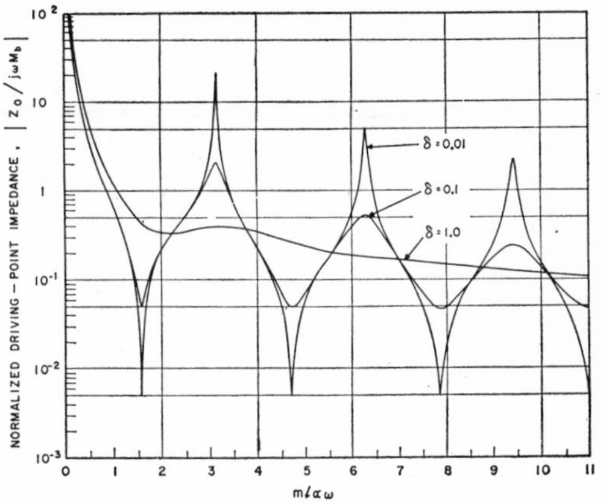

FIG. 10. Normalized driving-point impedance of the rod pictured in Fig. 9.

where ζ now represents the displacement of an element of the beam from its equilibrium position in a direction normal to the "x" axis, along which the beam lies. This equation embodies simplifying assumptions that are listed fully elsewhere,[11] examples being that the beam length is large compared with its cross-sectional dimensions, and that rotary inertia and shear displacement are negligible.

The standing-wave solution to Eq. (6.1) is

$$\zeta=(P\cosh m_d x+Q\cos m_d x+R\sinh m_d x+S\sin m_d x)e^{j\omega t}, \tag{6.2}$$

where the parameter m_d is now defined by the relation

$$m_d^4=\omega^2\rho/k^2E^*=m^4/(1+j\delta).$$

From knowledge of the boundary conditions of the beam, the arbitrary constants P, Q, R, and S may be determined. The constants ρ, k, and E^* represent the density, the radius of gyration of the cross section, and the complex Young's modulus of the beam, respectively.

To illustrate the complete solution to the damped-wave equation, two sets of constants, P, Q, R, and S, have been evaluated to satisfy the boundary conditions for a simply supported beam [Fig. 11(a)] and a simply clamped beam [Fig. 11(b)] driven at their midpoints by a force $F_0e^{j\omega t}$. The beams are assumed to possess identical half-lengths "a" and masses M_b, and to be identical in every other respect, apart from the manner in which they are terminated. It may readily be shown that the normalized driving-point impedance at the center of these beams is given by

$$\left[\frac{Z_0}{j\omega M_b}\right]_{\text{clamped}}=\left[\frac{\text{ch.s.}+\text{sh.c.}}{(m_da)(\text{ch.c.}-1)}\right]_{(m_da)}, \tag{6.3}$$

[11] N. W. McLachlan, *Theory of Vibrations* (Dover Publications, Inc., New York, 1951).

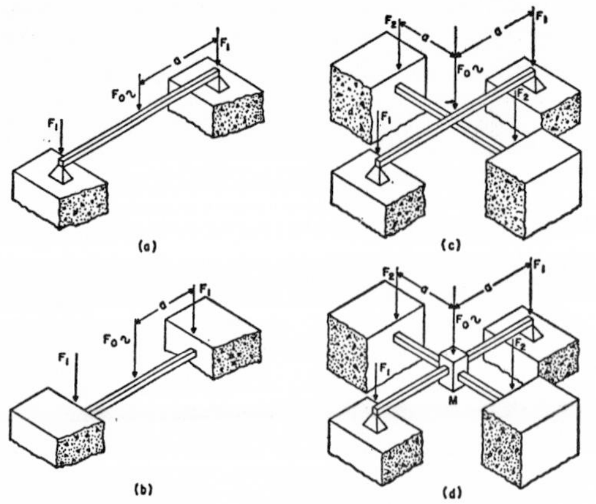

FIG. 11. Simply supported and simply clamped beams excited at their midpoints by a sinusoidally varying force.

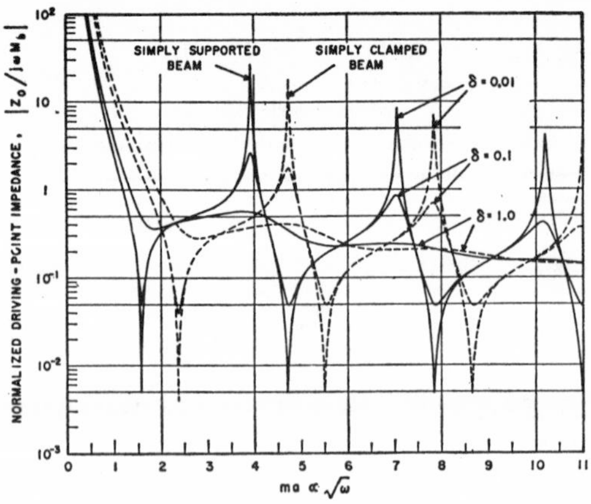

FIG. 12. Normalized driving-point impedance of the two beams shown in Figs. 11(a) and 11(b).

and

$$\left[\frac{Z_0}{j\omega M_b}\right]_{\text{supported}}=\left[\frac{2.\text{ch.c}}{(m_d a)(\text{sh.c.}-\text{ch.s.})}\right]_{(m_d a)} \tag{6.4}$$

The equations for Z_0, and all other equations describing the transverse vibration of the damped beams, are functions of the complex quantity $(m_d a)$; therefore, it is again desirable to express this parameter in the form $(p+jq)$, where p and q are now defined by the relations

$$p=(ma)\left[\frac{1}{2(D)^{\frac{1}{2}}}+\frac{(1+D)^{\frac{1}{2}}}{2\sqrt{2}D}\right]^{\frac{1}{2}}, \tag{6.5}$$

and

$$q=-(ma)\left[\frac{1}{2(D)^{\frac{1}{2}}}-\frac{(1+D)^{\frac{1}{2}}}{2\sqrt{2}D}\right]^{\frac{1}{2}}. \tag{6.6}$$

Circular and hyperbolic functions of the complex quantity $(p+jq)$ such as those found in Eqs. (6.3) and (6.4) may now be expanded by the "multiple angle" formulas and the sums and differences of their products grouped into real and imaginary parts. For example, Eq. (6.3) may be written

$$\left[\frac{Z_0}{j\omega M_b}\right]_{\text{clamped}}=\left[\frac{R_N+jI_N}{R_D+jI_D}\right], \tag{6.7}$$

where

$$R_N=[\text{ch}p.\text{s}p.(\text{ch}q.\text{c}q.+\text{sh}q.\text{s}q.)+\text{sh}p.\text{c}p.(\text{ch}q.\text{c}q.-\text{sh}q.\text{s}q.)], \tag{6.8}$$

$$I_N=[\text{ch}p.\text{c}p.(\text{ch}q.\text{s}q.+\text{sh}q.\text{c}q.)+\text{sh}p.\text{s}p.(\text{ch}q.\text{s}q.-\text{sh}q.\text{c}q.)], \tag{6.9}$$

$$R_D=[p(\text{ch}p.\text{c}p.\text{ch}q.\text{c}q.+\text{sh}p.\text{s}p.\text{sh}q.\text{s}q.-1)-q(\text{sh}p.\text{c}p.\text{ch}q.\text{s}q.-\text{ch}p.\text{s}p.\text{sh}q.\text{c}q.)], \tag{6.10}$$

and

$$I_D=[p(\text{sh}p.\text{c}p.\text{ch}q.\text{s}q.-\text{ch}p.\text{s}p.\text{sh}q.\text{c}q.)+q(\text{ch}p.\text{c}p.\text{ch}q.\text{c}q.+\text{sh}p.\text{s}p.\text{sh}q.\text{s}q.-1)]. \tag{6.11}$$

The absolute values of the impedances defined by Eqs. (6.3) and (6.4) have been computed, and are shown in Fig. 12 as the full and broken lines, respectively. As before, three values of the damping factor $\delta=0.01$, 0.1, and 1.0 are considered. Impedance is plotted as a function of ma, which is proportional to the square root of frequency. Since the natural frequencies of the beams are proportional, or closely so, to the squares of integers that form an arithmetic progression, the resonant peaks appear at regularly spaced intervals. The curves of Fig. 12 are very similar in appearance to those of Fig. 10. The impedance of both beams is again springlike at low frequencies, but the clamped beam appears "stiffer"; its fundamental resonance occurs at a higher frequency than that of the simply supported beam.

It is instructive now to examine the force transmissibility T_0 and T_m across the two-beam structures shown in Figs. 11(c) and 11(d), where the beams considered previously have been joined together at their midpoints. The beams shown in Fig. 11(d) are loaded by a lumped mass M at their point of juncture, this mass being larger by a factor γ than the mass of either beam. Both structures are excited by a force $F_0e^{j\omega t}$ as shown. Transmissibility may be determined from the following equations:

$$T_0=[T_m]_{\gamma=0}, \tag{6.12}$$

and

$$T_m=\left|\frac{2(F_1+F_2)}{F_0}\right|=\left|\left\{\frac{(1-\text{ch.c.})(\text{ch.}+\text{c.})+(\text{ch.s.}-\text{sh.c.})(\text{sh.}+\text{s.})}{[2\text{ch.c.}(1-\text{ch.c.})+(\text{ch}^2.\text{s}^2.-\text{sh}^2.\text{c}^2.)]-\gamma(m_d a)[(1-\text{ch.c.})(\text{ch.s.}-\text{sh.c.})]}\right\}\right|_{(m_d a)}. \tag{6.13}$$

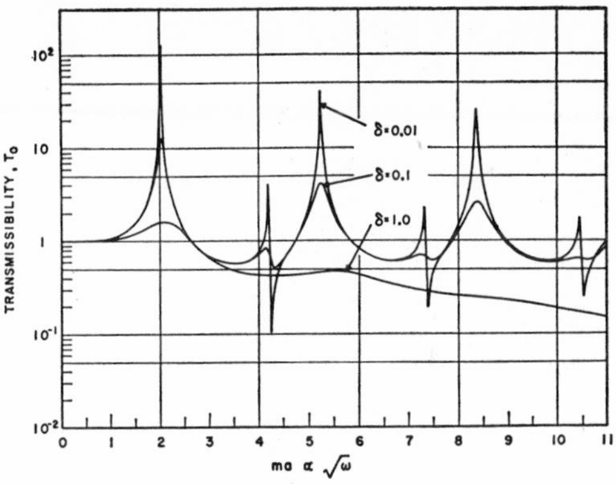

FIG. 13. Transmissibility across the two-beam structure shown in Fig. 11(c).

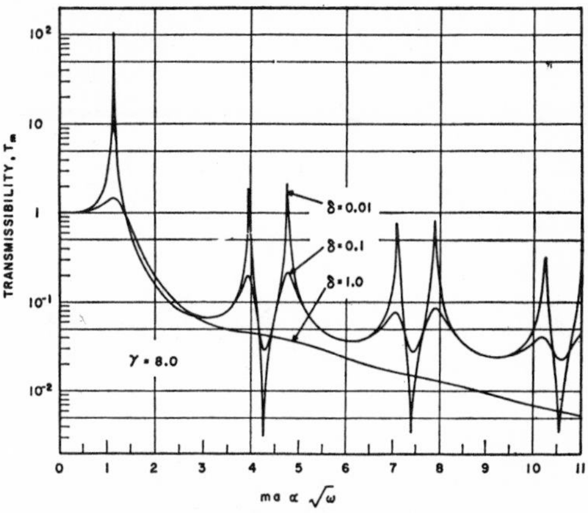

FIG. 14. Transmissibility across the mass-loaded two-beam structure shown in Fig. 11(d).

It is necessary to obtain the real and imaginary parts of the numerator and denominator of these equations, as before, in terms of the hyperbolic and circular functions of p and q. Transmissibility may then be determined when the damping factor δ and the mass ratio γ are specified. The results of one set of computations for T_0 and T_m are shown in Figs. 13 and 14, respectively. A value of $\gamma = 8.0$ was assumed in the computation of T_m.

The transmissibility T_0 exhibits a series of resonant peaks, every second peak being of relatively small magnitude. The frequencies at which the resonances of the system occur are specified by the values of ma for which the two sets of curves of Fig. 12 intersect. These curves refer to the individual input impedances of the simply supported and simply clamped beams. At the points of intersection, the impedance of one beam is equal in magnitude but opposite in sign to the impedance of the other beam, the required conditions for the generation of a resonance. The points of intersection fall at two distinct levels, the lower level (of the normalized impedance) being associated with the pronounced series of transmissibility peaks in Fig. 13.

The transmissibility T_m of the mass-loaded two-beam system [Fig. 11(d)] likewise exhibits resonant peaks at the frequencies for which the input impedances of the individual mass-loaded beams intersect. Now, however, neighboring resonant peaks are similar in appearance. The mass loading is sufficiently large to shift the resonant frequencies of the beams to their lower limit which, for any one resonant mode, corresponds to the next lower antiresonant frequency of the unloaded beam. The fundamental mode is not bounded in this manner, since there is not a further antiresonance at a lower frequency. Consequently, with the exception of the fundamental mode, the resonant frequencies of the mass-loaded system closely equal the antiresonant frequencies of the individual unloaded beams (Fig. 12).

ACKNOWLEDGMENT

This investigation was supported by the U. S. Department of the Navy, Bureau of Naval Weapons, under Contract NOrd 16597.

Reprinted from *J. Sound Vibr.*, **11**(1), 3–18 (1970)

THE ROLE OF DAMPING IN VIBRATION THEORY

S. H. CRANDALL

Department of Mechanical Engineering, Massachusetts Institute of Technology, Cambridge, Massachusetts, U.S.A.

(*Received* 28 *July* 1969)

In many applications of vibration and wave theory the magnitudes of the damping forces are small in comparison with the elastic and inertia forces. These small forces may, however, have very great influence under certain special circumstances. Damping arises from the removal of energy by radiation or dissipation. It is generally measured under conditions of cyclic or near-cyclic motion. The nature of some important damping mechanisms is discussed and an indication is given of how the damping depends on the amplitude and frequency of the cyclic motion. The idealized models of damping which are commonly employed in theoretical analyses are described and some limitations are noted. Damping is of primary importance in controlling vibration response amplitudes under conditions of steady-state resonance and stationary random excitation. Damping also plays a crucial role in fixing the borderline between stability and instability in many dynamical systems. Some examples which illustrate this are discussed, including shaft whirl, and pipeline flutter.

1. INTRODUCTION

The central phenomenon of vibration theory is cyclic oscillation. A major feature of oscillaation dynamics is the cyclic transformation of potential energy into kinetic energy and back again. This feature is clearly displayed by idealized models involving only elastic and inertial elements. For example the natural frequencies and natural modes of vibrating systems and the group and phase velocities of wave propagating systems are obtained from such idealized models. Secondary aspects of oscillation dynamics can be explained by accounting for a *damping* mechanism, i.e. a mechanism which removes energy from the oscillating system under consideration. Damping is responsible for the eventual decay of free vibrations and provides an explanation for the fact that the response of a vibratory system excited at resonance does not grow without limit.

One purpose of the present paper is to review some of the properties of actual damping mechanisms and to describe some of the mathematical models that are employed to represent these mechanisms. A second purpose is to draw attention to those special circumstances where small amounts of damping have an exaggerated importance in determining the dynamic behavior of a system.

2. THE NATURE OF DAMPING

Damping is the removal of energy from a vibratory system. The energy lost is either transmitted away from the system by some mechanism of radiation or dissipated within the system. Most measurements of damping are performed under conditions of cyclic or near cyclic oscillation. Commonly the decay of free oscillation is observed or measurements are made of steady-state forced vibration at (or in the vicinity of) resonance. In both cases the total energy W removed in a cycle can be inferred but the measurements are seldom precise enough to provide a detailed picture of how the instantaneous rate of energy removal fluctuates within a single cycle. If the oscillation of the system is in a single well-defined mode whose amplitude is characterized by a and whose frequency is ω, the energy lost per cycle generally varies with both a and ω, i.e. $W = W(a, \omega)$.

A convenient measure of damping is obtained by comparing the energy lost in a cycle with the peak potential energy V stored in the system during that cycle. The *loss factor* η is defined as

$$\eta = \frac{W}{2\pi V}. \tag{1}$$

If the energy could be removed at a uniform rate throughout a cycle of simple harmonic motion (such a mechanism is not actually very likely) then $W/2\pi$ could be interpreted as the energy loss per radian and η would be simply the energy loss per radian divided by the peak energy available.

In most dynamic systems which are of interest from the point of view of vibrations the damping is small. The values for loss factor that are encountered in practice range from about $\eta = 10^{-5}$ to $\eta = 2 \times 10^{-1}$; although larger values of η are found in instrument mechanisms, transducers and vehicle suspensions. In general the loss factor η depends on both the amplitude and frequency of the oscillation. If, however, the system is completely *linear*, then both W and V are proportional to a^2 and the loss factor η is *independent* of amplitude. For linear damping mechanisms the loss factor generally has an important frequency dependence.

These ideas can be illustrated by considering a bar of aluminum alloy suspended by strings as shown in Plate 1. Once the bar is set into bending vibration by a blow, it generates a sound with a fundamental frequency of 130 Hz which remains audible for several seconds as the vibration decays. The (measured) loss factor at this frequency is $\eta = 1{\cdot}5 \times 10^{-3}$. There are several damping mechanisms contributing to the total energy removed each cycle. The very fact that the vibration is audible is an indication that there is acoustic radiation away from the bar. There is undoubtedly some radiation of energy up the supporting strings even though they are attached at the nodal points of the fundamental bending node. There is also dissipation of energy within the aluminum. Even in a system as simple as this it is difficult to account quantitatively for all the damping.

Rough estimates can sometimes be made of the amounts of damping contributed by internal dissipation and by acoustic radiation. A great many linear and non-linear mechanisms of internal damping in metals have been identified [1, 2]. For aluminum in bending at room temperature the major contribution is made by transverse heat flow from the warmed compression fibers to the cooled tension fibers. This is a linear relaxation mechanism which depends on the temperature T, the thermal expansion coefficient α, the conductivity κ, and the specific heat c_v of the beam as well as on its thickness h, modulus E and frequency f of oscillation. In Figure 1 the loss factor predicted by this mechanism alone is compared with measured total internal loss factors of cantilever specimens. For the beam in Plate 1 the relaxation frequency is $f_0 = 7{\cdot}5$ Hz and the total internal loss factor for vibration in the fundamental mode can be estimated to be about $3{\cdot}5 \times 10^{-4}$.

The general nature of the acoustic radiation from a vibrating beam is known [5] but a quantitative analysis for the radiation from the lower modes appears to be beyond the present state of the art. A number of investigators [6, 7, 4, 8] have measured the air damping of thin cantilever beams. For large bending amplitudes air damping is non-linear, the loss factor increasing roughly in proportion to the amplitude. For small amplitudes the air damping loss factor appears to approach a limit which is independent of amplitude. This small amplitude loss factor appears to depend only on the ratio of the cantilever length L to the beam thickness h as shown in Figure 2. If the beam of Plate 1 had been supported as a cantilever

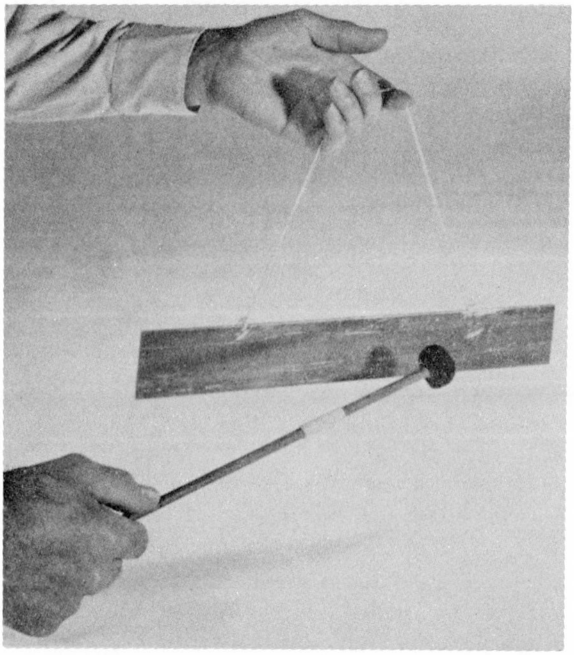

Plate 1. Vibrating beam suspended in air is damped by radiation up strings, acoustic radiation and internal damping within bar.

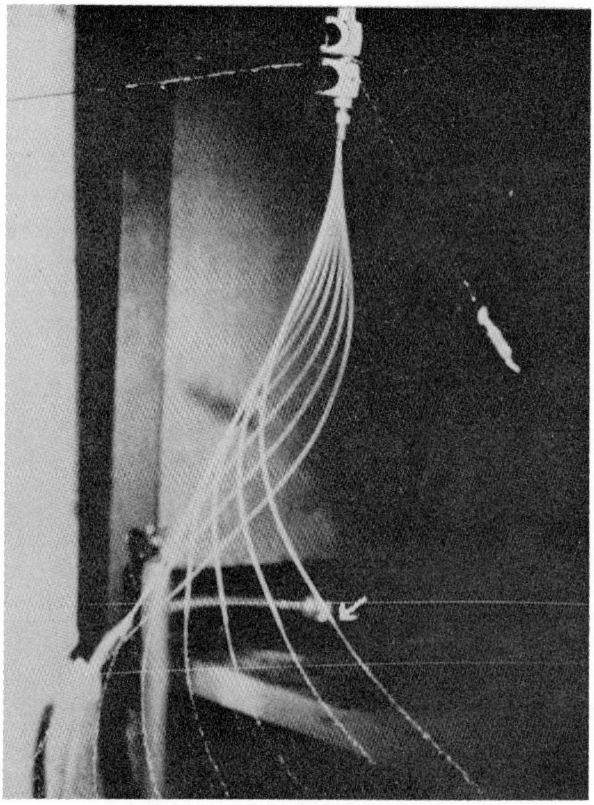

Plate 2. Strobe sequence of fluttering tube, eight flashes during half a cycle. Arrow marks end of tube. From [34].

beam, its fundamental frequency would have been 21 Hz with an acoustic radiation loss factor of $1{\cdot}6 \times 10^{-4}$ according to Figure 2. Unfortunately similar loss-factor information is not available for the fundamental free-free mode.

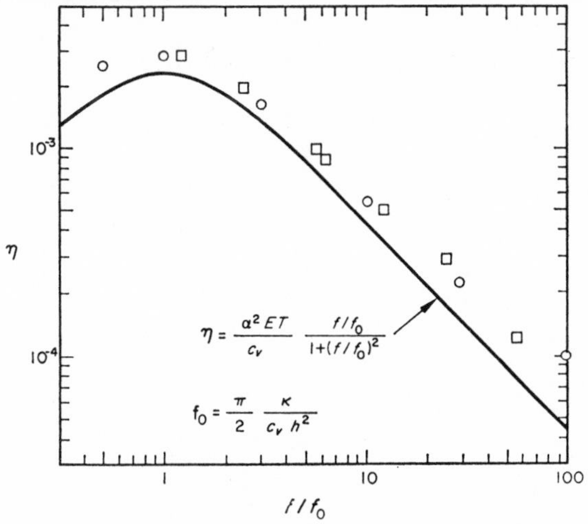

Figure 1. Loss factor as function of frequency. Comparison of measured values of internal damping for 2024-T4 cantilever beams with theoretical loss factor due to transverse heat flow. Adapted from [3].

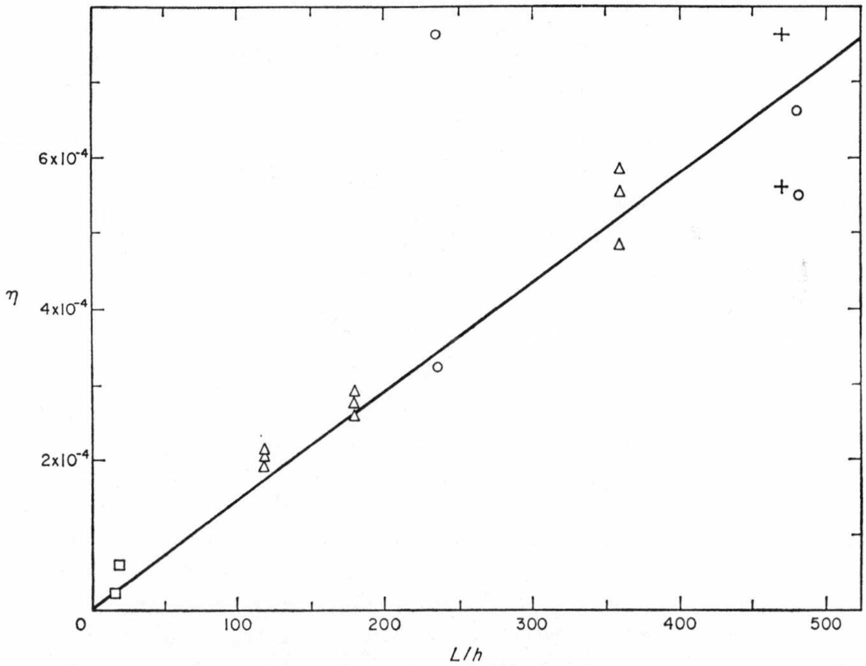

Figure 2. Low-amplitude air-damping loss factor for fundamental mode of thin cantilever beams as a function of length–thickness ratio. Adapted from [8].

The radiation damping of a system can be altered by changing the coupling with the exterior. The internal damping in a system can be altered by introducing energy absorbing devices. The beam in Plate 1 can be used to illustrate these statements. The radiation damping can be greatly increased if instead of supporting the beam by strings we clamp it to a desk.

Vibrational energy, which at first is localized within the beam, is rapidly transmitted throughout the larger structure. This energy transfer within the beam-desk system is counted as an energy loss for the beam considered as an isolated system. The internal damping of the beam can be increased by changing the beam material for aluminum to stainless steel (for example) or by adding damping tape or other acoustical coatings which utilize a viscoelastic material to dissipate energy. For example, Figure 3 shows how the loss factors of the first five modes

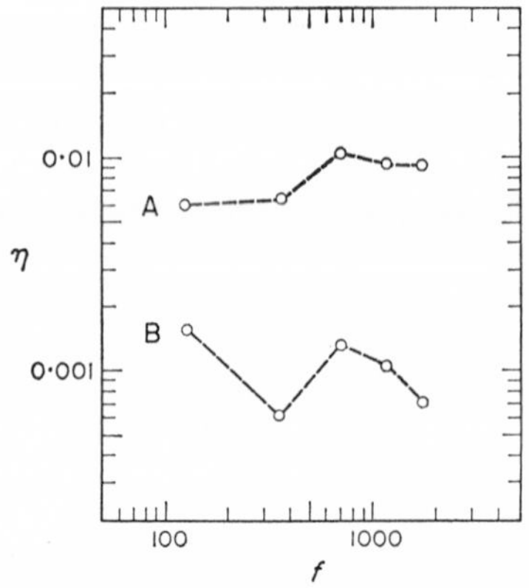

Figure 3. Loss factors for first five modes of free-free beam. A, After and B, before, applying damping tape.

of the beam in Plate 1 are increased by applying a $1\frac{3}{4}$ in. × 13 in. strip of damping tape (3-M No. 425 tape which consists of aluminum foil 0·003 in. thick and a layer of adhesive 0·002 in. thick) to one side of the $\frac{1}{8}$ in. thick beam whose face dimensions are 2 in. × $14\frac{5}{16}$ in.

Both radiation damping and internal dissipation can be expected to be frequency dependent. If the radiation is to a linear external system the loss factor will be affected by the frequency response of the external system. If the internal dissipation is due to a linear relaxation mechanism there will be a pronounced increase in loss factor when the oscillation frequency approaches the relaxation frequency.

3. MATHEMATICAL MODEL OF DAMPING

The prototype for a lossless vibration system is the simple spring-mass model shown in Figure 4(a). The natural free vibration is simple harmonic motion with frequency $\omega_n = \sqrt{k/m}$. When the exciting force is a steady-state sinusoid with frequency ω there is a steady-state sinusoidal solution for the motion which has the same frequency ω and has a finite amplitude if $\omega \neq \omega_n$. This model possesses many characteristics of actual physical systems but suffers from the following drawbacks. The free vibration, once excited, never decays. When excited at resonance ($\omega = \omega_n$) by a steady-state sinusoid applied at $t = 0$ the predicted response is an oscillation whose amplitude grows linearly with t, i.e. a steady state is never achieved. When excited by a stationary random force whose spectrum includes ω_n the response is a non-stationary random process [9] whose r.m.s. level grows essentially in proportion to $\sqrt{t}$; i.e. a stationary state is never achieved.

The classical remedy for these drawbacks is to introduce an ideal linear dashpot into the model as shown in Figure 4(b). It is assumed that the dashpot exerts a force $f_d = cv$ in opposition to a relative velocity v across its terminals. The constant c is called the dashpot parameter. The free vibration in this case is a damped oscillation and there is a finite steady-state response to any steady-state sinusoidal excitation. There is also a stationary random response process for any stationary excitation process.

Let us examine the ideal linear damper more closely. Suppose that a steady-state simple harmonic motion $x = a\cos\omega t$ is established in Figure 4(b). The energy W dissipated in a cycle is

$$W = \int_0^{2\pi/\omega} \left(c\frac{dx}{dt}\right)\frac{dx}{dt}\,dt = \pi c a^2|\omega| \tag{2}$$

while the peak potential energy stored in the spring during the cycle is

$$V = \tfrac{1}{2}ka^2. \tag{3}$$

The loss factor for the ideal damper in the system of Figure 4(a) is thus

$$\eta = \frac{c|\omega|}{k} \tag{4}$$

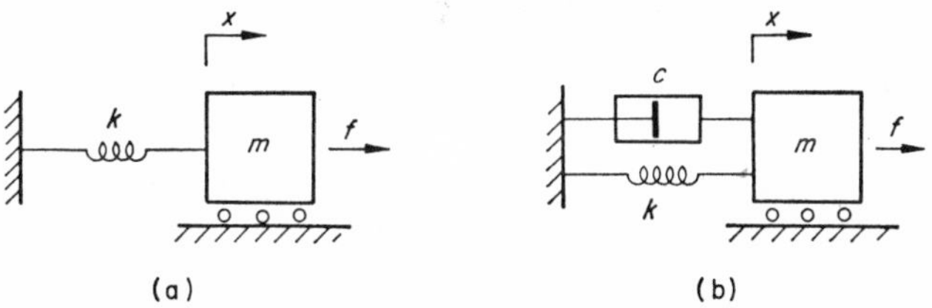

Figure 4. Ideal single-degree-of-freedom vibration models. (a) Lossless model, (b) model with ideal viscous damper.

according to (1). We may note that the absolute value signs on ω are required in (2) since the motion $x(t)$ itself, as well as the energy loss per cycle, is not affected by changing the sign of ω. Thus, considered as functions of ω, the energy loss per cycle W and the loss factor η are real, non-negative, and even.

In many applications where the damping is light the affect of damping is appreciable only for resonant or near-resonant motions. These effects can be described in terms of the loss factor at resonance ($\omega = \omega_n$)

$$\eta_n = \frac{c\omega_n}{k} = \frac{c}{\sqrt{km}}. \tag{5}$$

Thus the free vibration of the system of Figure 4(b) has the form

$$x(t) = a_0\,e^{-1/2\eta_n\omega_n t}\cos(\omega_d t + \phi_0) \tag{6}$$

where $\omega_d^2 = \omega_n^2(1 - \eta_n^2/4)$ and the amplitude a_0 and phase ϕ_0 depend on the initial conditions. Note that the rate of envelope decay depends on η_n. The logarithmic decrement is

$$\delta = \log\frac{x(t)}{x(t + 2\pi/\omega_d)} = \pi\eta_n\frac{\omega_n}{\omega_d} = \frac{\pi\eta_n}{\sqrt{1 - \eta_n^2/4}}. \tag{7}$$

When the exciting force in Figure 4(b) is a steady-state sinusoid $f = F\sin\omega t$ the steady-state response is

$$x = X\sin(\omega t + \phi) \tag{8}$$

where the response amplitude X depends on the frequency ω and on the resonant loss factor η_n as indicated in Figure 5. Note that for light damping the response amplitude is substantially independent of the damping except in the vicinity of $\omega = \omega_n$ where it depends critically on

η_n. We can also infer from Figure 5 that the spectral density of the response to a stationary random exciting force depends critically on the resonant loss factor.

In section 2 it was suggested that most linear damping mechanisms have considerable frequency dependence. In many cases, however, the frequency dependence of actual loss factors bear little relation to equation (4) for the ideal dashpot. Nevertheless for many purposes the actual damping can be modelled satisfactorily by an equivalent dashpot. This is indicated in Figure 6 where the frequency dependence of an actual loss factor in a single-degree-of-freedom oscillator is compared with that of an ideal dashpot. If the dashpot is

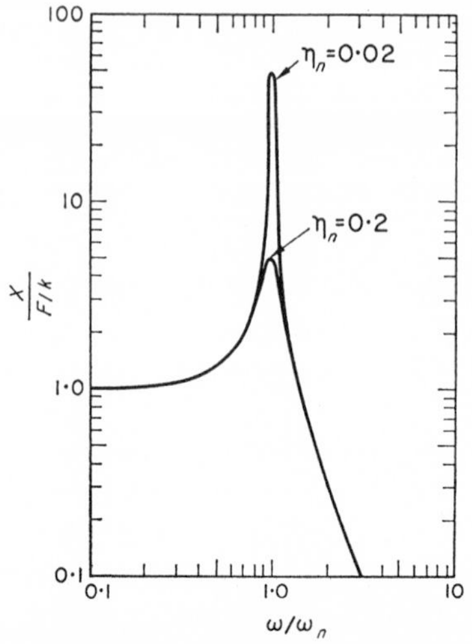

Figure 5. Steady-state frequency response of single-degree-of freedom oscillator with ideal linear dashpot.

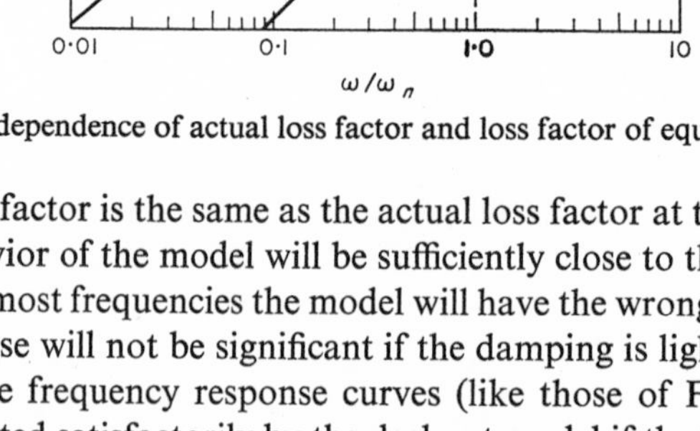

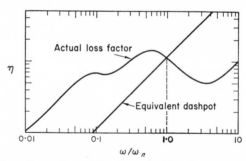

Figure 6. Frequency dependence of actual loss factor and loss factor of equivalent ideal dashpot.

selected so that its loss factor is the same as the actual loss factor at the natural frequency of the oscillator the behavior of the model will be sufficiently close to that of the actual system for most purposes. At most frequencies the model will have the wrong damping but the effect on the dynamic response will not be significant if the damping is light. The rate of decay of free oscillation and the frequency response curves (like those of Figure 5) for the actual damping can be estimated satisfactorily by the dashpot model if the loss factor at resonance is matched.

It is widely accepted that this procedure can be extended to model the damping in systems with many degrees of freedom. An equivalent dashpot is assigned to each natural mode in such a way that the dashpot loss factor matches the actual loss factor of the system when the system vibrates *in* that mode at the resonant frequency *of* that mode. Such a procedure neglects the possible coupling of modes due to damping. Some analytical studies of damping coupling have been made [10, 11, 12]. Further studies may show that there are practical applications where damping coupling plays an important role in establishing a stability borderline or in influencing the partition of energy during random vibration. For the purpose of predicting stationary response levels (random or deterministic) there does not, however, seem to be any compelling practical reason within the present state-of-the art for including the additional complication of damping coupling in the mathematical model.

A more serious discrepancy between model and reality occurs when the actual damping mechanism is non-linear. Analytical treatments of non-linear damping exist [13, 14, 15], but are of quite limited scope. In practice it is common to measure or estimate the actual loss factors for amplitudes which are representative of the expected application and then to use a model with linear dashpots having these loss factors.

4. THE FREQUENCY DEPENDENT DASHPOT

When the frequency at which damping is important is known in advance, it is usually adequate to model light damping by an equivalent ideal dashpot as described in the preceding section. A more faithful model of the damping is, however, required when it is not known initially at which frequency the damping will be critical. This latter case occurs in stability analyses where the critical frequency for onset of instability is itself a sensitive function of the damping. Some examples of cases of this sort are examined in the following sections.

In order to describe a model for an actual damping mechanism like that shown in Figure 6 we use the Fourier integral to transform the time domain into a frequency domain. The Fourier transform $X(\omega)$ of a time history $x(t)$ is defined by the integral

$$X(\omega) = \int_{-\infty}^{\infty} x(t) e^{-\iota\omega t} \, dt \tag{9}$$

provided that $x(t)$ satisfies modest convergence and smoothness requirements [16]. The time history can be recovered from the inverse transform

$$x(t) = \frac{1}{2\pi} \int_{-\infty}^{\infty} X(\omega) e^{\iota\omega t} \, d\omega. \tag{10}$$

The time-domain description of the ideal-dashpot relation

$$f_d = c \frac{dx}{dt} \tag{11}$$

can be transformed into a frequency-domain relation by multiplying both sides of (11) by $e^{-\iota\omega t}$ and integrating to obtain

$$F_d(\omega) = i\omega c X(\omega) \tag{12}$$

where $F_d(\omega)$ is the Fourier transform ot the dashpot force f_d(t). Equation (12) can be interpreted as the relation between the complex amplitudes of force and displacement during cyclic oscillation at frequency ω.

For an ideal dashpot the parameter c is constant and the loss factor is given by (4). For a damping mechanism which has a different frequency dependence we can use (4) to define a frequency dependent dashpot parameter

$$c(\omega) = \frac{k\eta(\omega)}{|\omega|}. \tag{13}$$

Since the loss factors of actual damping mechanisms are almost always measured under cyclic or nearly cyclic oscillation it is appropriate to generalize (12) by inserting the frequency dependent dashpot parameter (13) to get

$$F_d(\omega) = i\omega c(\omega)\, X(\omega) = \frac{i\omega k}{|\omega|}\eta(\omega)\, X(\omega) = ik\eta(\omega)\,\mathrm{sgn}\omega\; X(\omega) \tag{14}$$

where the signum function $\mathrm{sgn}\,\omega$ takes the value $+1$ for positive ω and (-1) for negative ω (and is zero for $\omega = 0$). Equation (14) defines the behavior of a useful analytical model: a linear frequency dependent dashpot. Note that the definition is made in the frequency domain. The corresponding time-domain description is obtained by multiplying (14) by $e^{i\omega t}/2\pi$ and integrating to obtain

$$\begin{aligned} f_d(t) &= \frac{1}{2\pi}\int_{-\infty}^{\infty} i\omega c(\omega)\, X e^{i\omega t}\, d\omega \\ &= \frac{1}{2\pi}\int_{-\infty}^{\infty} i\omega c(\omega)\, e^{i\omega t}\, d\omega \int_{-\infty}^{\infty} x(\tau)^{-i\omega\tau}\, d\tau \end{aligned} \tag{15}$$

or alternatively

$$x(t) = \frac{1}{2\pi}\int_{-\infty}^{\infty} \frac{e^{i\omega t}}{i\omega c(\omega)}\, d\omega \int_{-\infty}^{\infty} f_d(\tau)\, e^{-i\omega\tau}\, d\tau. \tag{16}$$

Unfortunately the integrals in (15) and (16) are not elementary except for rather special choices of $c(\omega)$. If $c(\omega)$ is a rational fraction of polynomials in ω^2 then the relation between $f_d(t)$ and $x(t)$ can be written as a (higher order) differential equation. This means that in place of the simple differential relation (11) for an ideal dashpot we now have a very complicated (although still linear) relationship between $f_d(t)$ and $x(t)$ for the frequency dependent dashpot.

When an ideal dashpot is placed in the oscillator in Figure 4(b) the time-domain description of the relation between the exciting force $f(t)$ and the displacement $x(t)$ is the differential equation

$$m\frac{d^2x}{dt^2} + c\frac{dx}{dt} + kx = f. \tag{17}$$

The corresponding frequency-domain description is

$$(-m\omega^2 + i\omega c + k)\, X(\omega) = F(\omega). \tag{18}$$

If now we replace the ideal dashpot (with constant c) by a frequency dependent dashpot with parameter $c(\omega)$ given by (13) the frequency-domain description becomes

$$(-m\omega^2 + i\omega c(\omega) + k)\, X(\omega) = F(\omega) \tag{19}$$

or

$$(-m\omega^2 + k[1 + i\eta(\omega)\,\mathrm{sgn}\,\omega])\, X(\omega) = F(\omega). \tag{20}$$

These frequency-domain relations are immediately useful for obtaining steady-state sinusoidal responses or stationary random response spectra. They can also be used to obtain more general response information by Fourier inversion. On occasion the following non-equations are employed to represent the inverses of (19) and (20):

$$\begin{aligned} m\frac{d^2x}{dt^2} + c(\omega)\frac{dx}{dt} + kx &= f \\ m\frac{d^2x}{dt^2} + kx[1 + i\eta(\omega)\,\mathrm{sgn}\,\omega] &= f. \end{aligned} \tag{21}$$

These are obtained from (19) and (20) by properly inverting the inertia, stiffness and excitation terms but leaving only a mnemonic indication of the damping inverse (15) by the dubious device of mixing time-domain and frequency-domain operations. The advantage gained from the mnemonic shorthand in such non-equations is not enough to compensate for the potential confusion which they may create. The difficulty is that they *look* like equations and there is a very real danger that they *will* be interpreted literally. In the past 20 years a number [17 to 26] of ingenous, *ad hoc* rationalizations have been used to interpret non-equations of the form (21).

There are certain limitations on the frequency dependence of dashpot models if they are to represent physically realizable damping mechanisms. As we have already noted for an ideal dashpot the loss factor $\eta(\omega)$ must be a real, non-negative and even function of ω if it is to represent a loss mechanism. The frequency-dependent dashpot parameter $c(\omega)$ must likewise be real, non-negative and even. Beyond this there is the considerably more subtle causality requirement: $c(\omega)$ must be such a function of ω that the corresponding time relation represented by (16) is *causal* in the sense that the response $x(t)$ at any instant t may depend on the *previous* history of the excitation $f_d(\tau)$ for $\tau < t$ but should be independent of the *future* behavior of $f_d(\tau)$ for $\tau > t$.

Although some general theorems exist [16] it is not always easy to decide whether a given frequency function $c(\omega)$ is causal or not. In some cases it is possible to demonstrate non-causality by exhibiting a particular excitation-response pair which satisfy (16) and for which the response anticipates the excitation. This is in fact the case for the widely employed model for linear hysteretic damping in which it is assumed that the loss factor $\eta(\omega)$ is a *constant* η_0 independent of frequency. Under this hypothesis

$$c(\omega) = \frac{k\eta_0}{|\omega|} = \frac{k\eta_0}{\omega} \operatorname{sgn} \omega \tag{22}$$

and (16) reduces to

$$x(t) = \frac{1}{2\pi k\eta_0} \int_{-\infty}^{\infty} \frac{e^{i\omega t}}{i \operatorname{sgn} \omega} \, d\omega \int_{-\infty}^{\infty} f_d(\tau)\, e^{-i\omega\tau} \, d\tau. \tag{23}$$

If we take $f_d(t)$ to be the unit impulse $\delta(t)$ and designate the corresponding response for $x(t)$ as the impulse response function $h(t)$ it is a simple exercise in generalized limits [16, 27] to show

$$h(t) = \frac{1}{\pi k\eta_0} \frac{1}{t} \quad (-\infty < t < \infty). \tag{24}$$

This excitation-response pair, which clearly violates the causality requirement, is shown in Figure 7.

The non-causal nature of the assumption of a frequency-independent loss factor was noted independently by Fraeijs de Veubeke [28], Caughey [29] and Crandall [30]. In [30] the impulse response of a single-degree-of-freedom oscillator with small frequency-independent loss factor was calculated. The post-impulse response has the general character of a damped sinusoid but there is a small negative pre-impulse response which slowly grows from zero of $t = -\infty$ to a maximum negative value which occurs shortly after the application of the impulse (the magnitude of this maximum precursor is somewhat larger than one per-cent of the peak post-impulse response for the case where the loss-factor is $\eta_0 = 0{\cdot}05$).

Although there is no realizable linear damping mechanism whose loss factor is strictly independent of frequency there are many mechanisms that have loss factors which remain substantially constant within certain ranges of frequency. An example of a combined energy storage and loss mechanism with very weak frequency dependence was given by Biot [31]. See also [29].

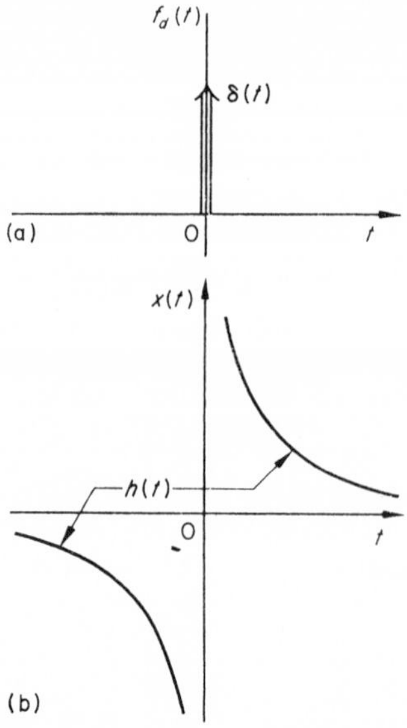

Figure 7. Example of non-causal behavior. Exciting force (a) and displacement response (b) for frequency dependent dashpot assumed to have constant loss factor at *all* frequencies.

5. DYNAMIC STABILITY OF NON-CONSERVATIVE SYSTEMS

The generally small effects of light damping become of major importance in the vicinity of resonance as we have seen. They are also of critical importance in determining stability borderlines for non-conservative systems such as flexible rotors under rotation and aero-elastic structures interacting with flowing fluid. In these systems there are generally regions (of rotational speed or fluid velocity) in which the system is stable in the sense that small disturbances decay. There may also be regions in which the system is unstable, i.e. small disturbances grow until either a non-linear limiting mechanism sets in or there is structural failure. At borderlines between stability and instability the system usually responds to a small disturbance by oscillating steadily with a small fixed amplitude. The frequency of this neutral oscillation and the location of the borderline (in terms of rotational speed or fluid velocity) can both be very sensitive to small damping in the system. In multi-moded systems the spatial distribution of damping (or equivalently the modal distribution) can have a strong influence on both of these. The frequency dependence of the damping also plays an important role in establishing the neutral stability conditions. These facts are illustrated in the examples which follow. For an extended treatment of stability of non-conservative systems see [32].

An excellent illustration is furnished by an idealized model of the classical problem of shaft whirl [33, 32]. In Figure 8 a mass point is suspended symmetrically by massless springs so that whenever the mass is displaced from the origin by a distance a there is a centrally directed restoring force of magnitude ka (independently of the rotation speed Ω of the frame). When the frame is stationary ($\Omega = 0$) the mass has two degrees of freedom in the plane of the sketch. All natural motions are linear combinations of two independent modes with the same natural frequency $\omega_n = \sqrt{k/m}$. The pair of basic modes may be taken as simple translation in the x and y directions, respectively, or (as is usually more convenient in whirling problems) as a pair of circularly polarized vibrations, one clockwise and one counterclockwise.

Furthermore, in the ideal undamped system these natural motions are completely independent of the frame rotation speed. These motions are described by the equations

$$m\begin{Bmatrix}\ddot{x}\\ \ddot{y}\end{Bmatrix}+k\begin{Bmatrix}x\\ y\end{Bmatrix}=0 \tag{25}$$

in stationary co-ordinates or by

$$m\begin{Bmatrix}\ddot{\xi}\\ \ddot{\eta}\end{Bmatrix}-m\Omega^2\begin{Bmatrix}\xi\\ \eta\end{Bmatrix}+2m\begin{bmatrix}0 & -\Omega\\ \Omega & 0\end{bmatrix}\begin{Bmatrix}\dot{\xi}\\ \dot{\eta}\end{Bmatrix}+k\begin{Bmatrix}\xi\\ \eta\end{Bmatrix}=0 \tag{26}$$

in rotating co-ordinates. Since small disturbances remain small the ideal system of Figure 8(a) is not unstable at any speed. There is, however, a latent source of instability in the centrifugal field set up by the steady rotation Ω.

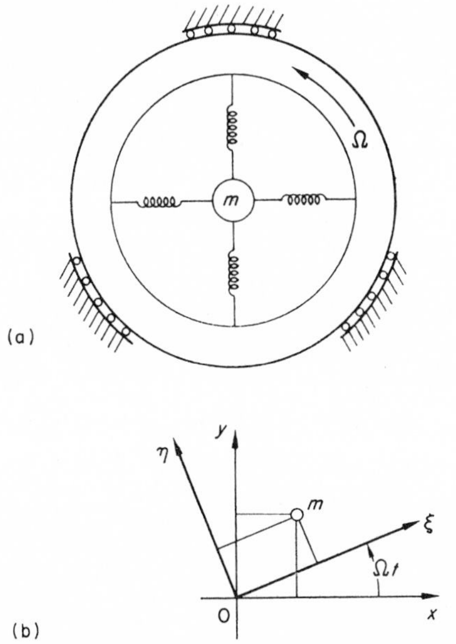

Figure 8. Model of whirling shaft. (a) Point mass m suspended by symmetrical springs from rotating frame, (b) stationary co-ordinates x, y and rotating co-ordinates ξ, η to locate mass point when displaced from origin.

Instability can be introduced by constraining the mass point to vibrate along a rotating diameter. For example, if a guide rail is installed along the ξ-axis so that η is constrained to vanish identically, equation (26) reduces to

$$m\ddot{\xi}+(k-m\Omega^2)\,\xi=0 \tag{27}$$

which predicts unbounded growth for ξ as soon as $\Omega^2>k/m=\omega_n^2$, i.e. as soon as the centrifugal field overpowers the elastic field.

Now let us return to Figure 8 with two degrees of freedom and examine the effect of introducing damping. Retaining circular symmetry we still have to distinguish whether the damping of the mass is with respect to motion relative to the stationary x, y system or with respect to motion relative to the rotating ξ, η system. This can be considered as a problem of ascertaining the spatial distribution of damping. There is, however, an interaction with the frequency dependence of the damping because a whirl which appears to have the frequency ω with respect to the x, y system will appear to have the frequency $\omega-\Omega$ to an observer stationed in the rotating co-ordinate system.

A formal procedure for introducing frequency-dependent dashpots is to write time-domain differential equations for constant parameter dashpots, transform to the frequency domain and *then* replace the constant parameters by frequency dependent parameters. Thus assuming resisting forces $c_s\dot{x}$ and $c_s\dot{y}$ parallel to the *stationary* x- and y-axes and resisting forces $c_r\dot{\xi}$ and $c_r\dot{\eta}$ parallel to the *rotating* ξ- and η-axes, we can extend the undamped equations of (25) (26) to the following equations for constant-parameter dashpots. In the stationary co-ordinates we have

$$m\begin{Bmatrix}\ddot{x}\\ \ddot{y}\end{Bmatrix}+(c_s+c_r)\begin{Bmatrix}\dot{x}\\ \dot{y}\end{Bmatrix}+k\begin{Bmatrix}x\\ y\end{Bmatrix}+c_r\begin{bmatrix}0 & \Omega\\ -\Omega & 0\end{bmatrix}\begin{Bmatrix}x\\ y\end{Bmatrix}=0, \tag{28}$$

while in the rotating co-ordinates we have

$$m\begin{Bmatrix}\ddot{\xi}\\ \ddot{\eta}\end{Bmatrix}-m\Omega^2\begin{Bmatrix}\xi\\ \eta\end{Bmatrix}+(c_s+c_r)\begin{Bmatrix}\dot{\xi}\\ \dot{\eta}\end{Bmatrix}+k\begin{Bmatrix}\xi\\ \eta\end{Bmatrix}+\begin{bmatrix}0 & -\Omega\\ \Omega & 0\end{bmatrix}\left(2m\begin{Bmatrix}\dot{\xi}\\ \dot{\eta}\end{Bmatrix}+c_s\begin{Bmatrix}\xi\\ \eta\end{Bmatrix}\right)=0 \tag{29}$$

as equations for free damped motions with constant parameter damping (c_s with respect to the stationary axes and c_r with respect to the rotating axes).

A discussion of general solutions of these is omitted and we proceed directly to the stability borderline where a steady whirl of amplitude a is maintained. If this whirl has the (unknown) frequency ω with respect to the stationary axes then

$$\begin{aligned} x &= a\cos\omega t,\\ y &= a\sin\omega t, \end{aligned} \tag{30}$$

and the stationary dashpot-parameter can be replaced by a frequency-dependent parameter (13) evaluated at frequency ω

$$c_s=\frac{k\eta_s(\omega)}{|\omega|}. \tag{31}$$

Viewed from the rotating axes the same whirl appears as

$$\begin{aligned} \xi &= a\cos(\omega-\Omega)t,\\ \eta &= a\sin(\omega-\Omega)t, \end{aligned} \tag{32}$$

and the rotating dashpot parameter can be replaced by a frequency dependent parameter (13) evaluated at the frequency $\omega-\Omega$

$$c_r=\frac{k\eta_r(\omega-\Omega)}{|\omega-\Omega|}. \tag{33}$$

It is now a routine algebraic task to draw out the conditions for neutral stability from either (28) or (29). We find that neutral stability will occur if $\omega^2=\omega_n^2$ and if the steady rotation Ω satisfies the following requirement:

$$\eta_s(\omega_n)\,\mathrm{sgn}\,\omega_n+\eta_r(\omega_n-\Omega)\,\mathrm{sgn}\,(\omega_n-\Omega)=0. \tag{34}$$

There is no solution to (34) if $0<\Omega<\omega_n$, but if $0<\omega_n<\Omega$ then a solution is possible if

$$\eta_s(\omega_n)=\eta_r(\Omega-\omega_n). \tag{35}$$

The two sides of equation (35) are sketched in Figure 9. Neutral equilibrium occurs when the two curves intersect, i.e. when the stationary loss factor (at frequency ω_n) just equals the rotational loss factor (at frequency $\Omega-\omega_n$). Further study shows that the system is unstable when the rotational loss factor is greater than the stationary loss factor.

This example illustrates the remarkable fact that a stable undamped system can be made unstable by adding a damping mechanism. Actually it is only the rotational damping which is destabilizing. Damping of motion with respect to the stationary axes always tends to stabilize the system. Damping of motion with respect to the rotating axes acts somewhat like (although not as effectively as) a diametral constraint in unleashing the centrifugal field. Viewed from the stationary axes the neutral whirl is at the natural frequency ω_n but the frequency of the whirl with respect to the rotating axes (and hence the rotation speed Ω) depends critically on the frequency dependence of the rotational damping loss factor.

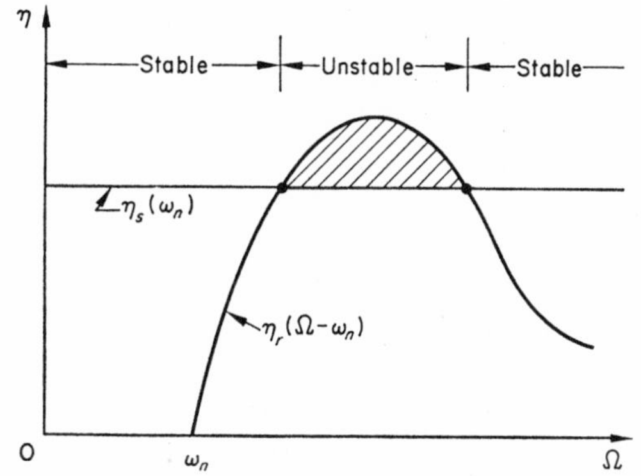

Figure 9. Location of stability borderlines according to equation (35). Neutral stability occurs at rotational speeds Ω for which rotational and stationary loss factors are equal.

The sensitivity of the location of a stability limit for small damping can be illustrated by considering what happens to Figure 9 when the damping approaches zero. A wide range of results is possible depending on the exact nature of the limiting process. For example, if η_r is allowed to vanish while η_s remains fixed, the system is *stable* for all speeds Ω and remains so when η_s is subsequently made to vanish. Alternatively if η_s is made to vanish first the system is *unstable* for all speeds Ω that are greater than ω_n. Furthermore this condition remains unchanged if subsequently the magnitude of η_r is decreased without limit. Any number of intermediate results is also possible. For example, if η_s and η_r are maintained in strict proportion as they are simultaneously brought to zero, then at any stage the unstable range will have exactly the same limits as in Figure 9.

6. FURTHER EXAMPLES

We briefly draw attention to the sensitive role of small damping in problems of internal and external flutter. Figure 10 shows an experimental set-up [34] in which a flexible cantilever tube transmits a steady flow of water. For low rates of flow the straight tube is stable with respect to small disturbances. If the flow rate is slowly increased a critical rate is reached and the pipe begins to flutter as shown in the strobe-lighted photo of Plate 2. This problem has been studied at some length [35, 36]. The sensitivity of the results to the type of damping assumed can be indicated by displaying some results from [34]. Experimental results for three different configurations are compared with analytical predictions in Figures 11 and 12. The parameter specifying the configuration of the tube-fluid system is the ratio $m_F/(m_F + m_T)$ of the fluid mass (per unit length of tube) to the total mass of fluid plus tube. At the stability borderline the important parameters are the flow velocity of the fluid V and the frequency ω of the neutral oscillation. In the experiments only the first mode loss factor was measured. The analytical results were based on a 4-mode Galerkin approximation. The first mode loss factor was put equal to the experimental value. Then two different hypotheses were made.

In the first, the loss factors of the higher modes were assumed to be *inversely* proportional to the modal natural frequencies (this would be the case if there were uniform viscous resistance to the transverse *velocity* of the tube). In the second the loss factors of the higher modes were assumed to be *directly* proportional to the modal natural frequencies (this would be the

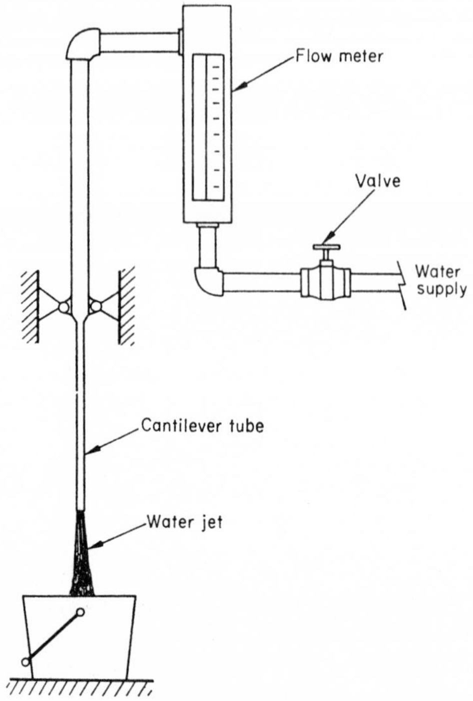

Figure 10. Flexible cantilever tube will "flutter" when flow rate surpasses critical value.

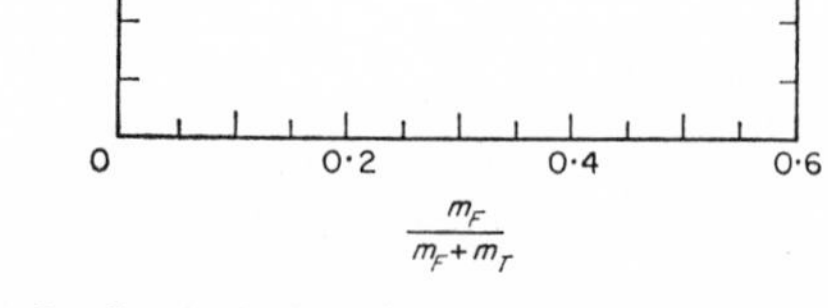

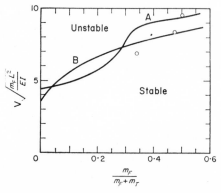

Figure 11. Stability borderlines for flow velocity. Comparison of experimental points (○) with prediction based on: A, loss factors inversely proportional to frequency; B, loss factors directly proportional to frequency. Adapted from [34].

case if resisting stresses were developed uniformly throughout the tube in proportion to the bending *strain*-rate).

In Figure 11 the predicted stability borderlines are compared with the three measured critical flow rates. In Figure 12 the measured frequencies of neutral oscillation at the stability borderline are compared with the analytical predictions. Note the wide divergence in the two analytical models (especially for the frequency) due only to the different assumptions concerning the damping in the higher modes.

Damping also plays an important role in external flutter. For example, the problem of panel flutter due to supersonic flow over the outside face of a panel has been widely studied

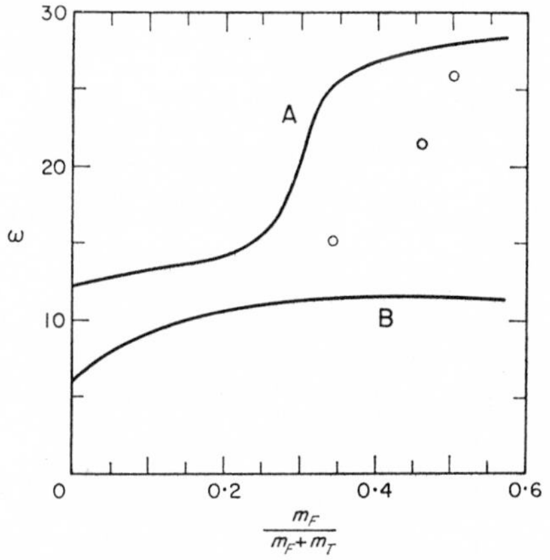

Figure 12. Frequency of neutral oscillation at stability borderline. Comparison of experimental points (○) with prediction based on: A, loss factors inversely proportional to frequency; B, loss factors directly propoıtional to frequency. Adapted from [34].

in recent years [37, 38]. The critical flow velocity depends on the aspect ratio of the panel, the panel boundary conditions, the magnitude of in-plane preloading and the aerodynamic damping as well as on the internal mechanical damping. The interplay between these many factors is complex. Several investigators [32, 39, 40] have noted the sensitivity of the results to the type of assumption made concerning the spatial distribution and frequency dependence of the mechanical damping.

ACKNOWLEDGMENTS

The author would like to express his thanks to L. E. Wittig for measuring the dynamic characteristics of the beam shown in Plate 1. Support for this research was provided by the National Aeronautics and Space Administration under Contract NAS 8-2504.

REFERENCES

1. C. Zener 1948 *Elasticity and Anelasticity of Metals.* Chicago: University of Chicago Press.
2. B. J. Lazan 1968 *Damping of Materials and Members in Structural Mechanics.* Oxford: Pergamon Press.
3. J. C. Heine 1966 *Thesis, Department of Mechanical Engineering, M.I.T.* The stress and frequency dependence of material damping in some engineering alloys.
4. N. Granick and J. E. Stern 1966 *Shock and Vibration Bulletin* **34**, 5, 177. Material damping of aluminum by a resonant dwell technique.
5. P. M. Morse and K. U. Ingard 1968 *Theoretical Acoustics.* New York: McGraw-Hill Book Co.
6. M. Uemura and M. Takehana 1953 *Tokyo University Report, Inst. Sci., and Tech.* **7**, 99. Damped vibration of thin beams (I. The effect of air resistance).
7. W. E. Baker and F. J. Allen 1957 *Aberdeen Proving Ground B.R.L. Report No.* 1033. The damping of thin beams in air.
8. W. E. Baker, W. E. Woolam and D. Young 1967 *Int. J. Mech. Sci.* **9**, 743. Air and internal damping of thin cantilever beams.
9. T. K. Caughey 1963 *Random Vibration*, vol. 2, chapter 3, ed. by S. H. Crandall, Cambridge, Mass.: The M.I.T. Press. Nonstationary random inputs and responses.

K. A. Foss 1958 *J. appl. Mech.* **25**, 361. Coordinates which uncouple the equations of motion of damped linear dynamic systems.
11. S. H. Crandall and R. B. McCalley 1961 *Shock and Vibration Handbook*, vol. 2, chapter 28, ed. by C. M. Harris and C. E. Crede, New York: McGraw-Hill Book Co. Numerical methods of analysis.
12. T. K. Caughey and M. E. J. O'Kelly 1961 *J. acoust. Soc. Amer.* **33**, 1458. Effect of damping on the natural frequencies of linear dynamic systems.
13. T. K. Caughey 1960 *J. appl. Mech.* **27**, 649. Random excitation of a system with bilinear hysteresis.
14. S. H. Crandall, G. R. Khabbaz and J. E. Manning 1964 *J. acoust. Soc. Amer.* **36**, 1330. Random vibration of an oscillator with nonlinear damping.
15. M. R. Torres and C. D. Mote, Jr. 1969 *ASME Paper No. 69-Vibr*-34. Expected equivalent damping under random excitation.
16. A. Papoulis 1962 *The Fourier Integral and Its Applications.* New York: McGraw-Hill Book Co.
17. W. W. Soroka 1949 *J. aeronaut. Sci.* **16**, 409. Note on the relations between viscous and structural damping coefficients.
18. W. Pinsker 1949 *J. aeronaut. Sci.* **16**, 694. Structural damping.
19. R. H. Scanlan and F. W. Rosenbaum 1951 *Introduction to the Study of Aircraft Vibration and Flutter.* New York: Macmillan.
20. N. O. Myklestad 1952 *J. appl. Mech.* **19**, 284. The concept of complex damping.
21. R. E. D. Bishop 1955 *Jl R. aeronaut. Soc.* **59**, 738. The treatment of damping forces in vibration theory.
22. T. J. Ried 1956 *Jl R. aeronaut. Soc.* **60**, 283. Free vibration and hysteretic damping.
23. S. Neumark 1957 *Aero Res. Council R & M* No. 3269. Concept of complex stiffness applied to problems of oscillations with viscous and hysteretic damping.
24. P. Lancaster 1960 *Jl R. aeronaut. Soc.* **64**, 229. Free vibration and hysteretic damping.
25. R. H. Scanlan and A. Mendelson 1963 *AIAA J.* **1**, 938. Structural damping.
26. E. Skudrzyk 1968 *Simple and Complex Vibratory Systems.* Univ. Park, Pennsylvania: Pennsylvania State University Press.
27. M. J. Lighthill 1962 *Fourier Analysis and Generalized Functions.* Cambridge University Press.
28. B. M. Fraeijs de Veubeke 1960 *AGARD Manual on Elasticity*, vol. 1, chapter 3. Influence of internal damping on aircraft resonance.
29. T. K. Caughey 1962 *Proc. Fourth U.S. natn. Congr. appl. Mech.* 87, New York: ASME. Vibration of dynamic systems with linear hysteretic damping.
30. S. H. Crandall 1963 *Air, Space, and Instruments, Draper Anniversary Volume*, p. 183, ed. by S. Lees, New York: McGraw-Hill Book Co. Dynamic response of systems with structural damping.
31. M. A. Biot 1958 *Proc. Third U.S. natn. Congr. appl. Mech.* 1. New York: ASME, Linear thermodynamics and mechanics of solids.
32. V. V. Bolotin 1963 *Nonconservative Problems of the Theory of Elastic Stability.* New York: Macmillan.
33. A. L. Kimball 1924 *Gen. Elec. Rev.* **27**, 244. Internal friction theory of shaft whirling.
34. A. S. Greenwald and J. Dugundji 1967 *M.I.T. Aeroelastic and Structures Research Laboratory Report ASRL TR* 134-3. Static and dynamic instabilities of a propellant line.
35. R. W. Gregory and M. P. Paidoussis 1966 *Proc. R. Soc.* **A293**, 512. Unstable oscillation of tubular cantilevers conveying fluid, Part 1—Theory, Part II—Experiments.
36. G. Herrman and S. Nemat-Nasser 1967 *Int. J. Solids and Structures* **3**, 39. Instability modes of cantilevered bars induced by fluid flow through attached pipes.
37. Y. C. Fung 1963 *AIAA J.* **1**, 898. Some recent contributions to panel flutter research.
38. J. Dugundji 1966 *AIAA J.* **4**, 1257. Theoretical considerations of panel flutter at high supersonic Mach numbers.
39. C. H. Ellen 1968 *AIAA J.* **6**, 2169. Influence of structural damping on panel flutter.
40. C. P. Shore 1969 *NASA TN D*-4990. Effects of structural damping on flutter of stressed panels.

Reprinted from *J. Sound Vibr.*, **26**(1), 141–154 (1973)

THE STATUS OF ENGINEERING KNOWLEDGE CONCERNING THE DAMPING OF BUILT-UP STRUCTURES

E. E. UNGAR

Bolt Beranek and Newman Inc.,
50 *Moulton Street,*
Cambridge, Massachusetts 02138, *U.S.A.*

(*Received* 31 *August* 1972)

The importance of the effects of structural joints on the damping of built-up structures is pointed out, and the energy dissipation mechanisms associated with squeezing, rocking, and shearing motions are discussed for simple joints that are dry, lubricated, or provided with viscoelastic inserts. The damping mechanisms and behaviors of built-up beams and of skin–stringer structures are discussed as far as they are currently understood, and available damping estimation methods are summarized. Difficulties in defining and measuring the damping of skin–stringer structures are indicated, and it is pointed out that particularly the high-frequency damping of built-up beams and the low-frequency damping of skin–stringer configurations require further investigation.

1. INTRODUCTION

The fuselages, wings, and control surfaces of modern aircraft typically are built up from beam, plate, and shell-like components, fastened together by means of rivets, bolts, welds, or adhesives. Design and test techniques relating to the static strength and to the gross dynamic responses of such structures are highly developed and have long been used throughout the aircraft industry. In recent years, however, with the advent of jet propulsion —and particularly with the current increased interest in short take-off and landing aircraft—it has become necessary to pay increasing attention to the higher-frequency motions of such structures, for example, in relation to their "sonic fatigue". As discussed below, these motions depend strongly on the structures' damping or capability for dissipation of vibratory energy—and it is to a review and analysis of the current engineering knowledge of this damping that this report is devoted.

It is well known that the steady-state response of a simple spring-and-mass system that is subjected to sinusoidal excitation is determined by its stiffness, if the excitation frequency is below the system's resonance frequency. Similarly, the system's high-frequency response is determined by its mass. Thus, the system's low-frequency response depends on its potential-energy storage characteristic (i.e., stiffness) and its high-frequency response depends on its kinetic-energy storage characteristics (i.e., mass). Its response at resonance, on the other hand, depends on its energy dissipation characteristics (i.e., damping) (see, e.g., reference [1]). Since the response of a simple system to broadband (random) excitation is known to be dominated by contributions at and near its resonance, its responses to random excitation also depend on its damping (see, e.g., reference [2]).

Since one may generally describe the response of a structure to a given excitation in terms of a superposition of the responses of the structure's modes, and since each mode behaves like a simple mass-spring-dashpot (single degree-of-freedom) system [1, 3] the statements in the foregoing paragraph apply also to structural modes. Thus, the responses of a structure to resonant or random excitation depend on the structure's damping, and one needs to know the (modal) damping of the structure before one can predict its resonant or random response—and such consequences of these responses as fatigue or excessive excursions.

A structure's inertia and stiffness properties usually can be predicted rather readily; at the very least they can be determined from relatively simple static measurements. On the other hand, the damping (energy dissipation) properties of practical structures usually are difficult to predict reliably, and their experimental determination requires comparatively complex dynamic measurements.

Unlike mass and stiffness, damping does not refer to a unique physical phenomenon; that is the reason damping generally is so much more difficult to predict. There are as many damping mechanisms as there are ways of converting (ordered) mechanical energy into (disordered) thermal energy. The most important of these mechanisms generally are interface friction, fluid viscosity, and mechanical hysteresis (also called internal friction or material damping), but fluid turbulence and acoustic radiation may also extract energy from a vibrating structure, and occasionally such electromagnetic effects as eddy currents and magnetic hysteresis may also play significant roles. In order to analyze or predict the damping of a given structure, one ideally should take into account all possible damping mechanisms; fortunately, in most practical cases, one or two mechanisms predominate, so that one may neglect the effects of all others.

Material damping and the closely related problem of the damping of monolithic structures (i.e., structures made of one continuous piece of the same material) have been studied rather extensively, and engineering procedures for the quantitative prediction of these types of damping are reasonably well in hand [4]; monolithic structures made of the common structural metals typically have loss factors† between 10^{-4} and 10^{-3} [4, 5]. On the other hand, built-up structures (i.e., structures made by joining together skins, stringers, frames, etc.) have been found to exhibit much higher damping; loss factors of the order of 10^{-2} are typical of common aircraft structures [6]. Because built-up structures differ from similar one-piece structures only in that the former incorporate joints, it is clear that the damping of built-up structures is dominated by the effects of the joints.

Thus, the designer of built-up structures that are to be exposed to broadband excitation requires at least a qualitative understanding of how structural joints dissipate energy and what parameters affect this damping. For design optimization, or for the prediction of structural responses, one of course also requires means for the quantitative prediction of this damping. It is the purpose of this report to review what is currently known about the damping of built-up structures, in order to provide the structural designer with the necessary insights into the important physical phenomena and to guide him to appropriate sources for quantitative information—to the extent this information exists.

Accordingly, the discussion in the following sections progresses from simple ideal joints to more complex joints and then to built-up beams and skin–stringer structures. The emphasis throughout is on realistic metal structures that are not designed primarily for high damping, but some aspects of designing structures for high damping are also considered.

† The loss factor η is a commonly used measure of damping, defined by $\eta = D/2\pi W$, where D denotes the energy dissipated in one cycle of the vibration and W represents the maximum energy stored in the system during the cycle. For small damping $\eta = 2c/c_c$, where c/c_c represents the fraction of critical damping [5–7].

2. DAMPING OF SIMPLE JOINTS UNDER IDEAL CONDITIONS

One may study the important mechanisms and parameters that affect joint damping most simply by dealing with simple joint geometries, with relative motions that are well defined, and with low frequencies. The low-frequency restriction permits one to neglect inertia effects and ensures that the vibratory wavelengths are considerably greater than all structural dimensions of concern (so that the relative motions are everywhere in phase).

2.1 SQUEEZING AND ROCKING MOTIONS

The damping produced by metallic interfaces that are pressed together by time-varying normal forces has been subject to a number of laboratory investigations. Because this damping generally is very small, the experimental parts of such studies generally involve "stacks" of interfaces, like that shown in Figure 1, to facilitate measurement. Such experimental measurements [8] have shown for dry (unlubricated) interfaces that (i) the dynamic stiffness increases with increasing normal preload and with increasing surface smoothness, and (ii) the damping is so small that it cannot be measured. There appears to be no effect of frequency over the range investigated (i.e., up to 300 Hz).

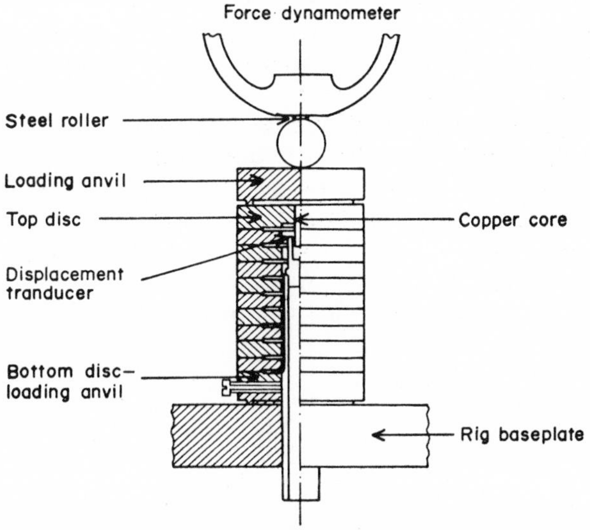

Figure 1. Stack of discs for measurement of joint damping associated with oscillatory normal forces. From) reference [8].)

Experiments with lubricants introduced between the interfaces indicate that the presence of such lubricants serves to increase the dynamic normal stiffness of the joint and produces significant amounts of damping. The stiffness then increases with increasing lubricant viscosity and increasing frequency, whereas the damping (loss factor) increases with the viscosity and quantity of lubricant, but varies relatively little with amplitude and frequency [8]. The same sort of behavior is also exhibited by joints that are subject to rocking motion—i.e., where the stack of experimental disks of Figure 1 is loaded by an oscillating bending moment, rather than by a varying normal force [9].†

These experimental findings tend to indicate that the action of dry joints subject to oscillatory normal forces is dominated by the elastic interaction of asperities of the two

† Note that with oscillatory moments or relative rocking motions at the interface, the local forces and motions are also normal to the interface.

contacting surfaces. On the other hand, it appears that the action of lubricated joints is largely due to the lubricant; stiffness increases occur as a result of the entrapment of oil, and damping occurs as a result of viscous losses produced in the oil as it is made to flow essentially parallel to the mating surfaces by the normal relative motions of these surfaces. Theoretical analyses and experiments [10] have shown that the damping associated with viscoelastic materials in the interspaces between facing surfaces is also dominated by these "pumping" effects. The fact that the presence of cavitation bubbles in the oil film reduces the joint stiffness and damping [11] serves as further evidence for the validity of the oil pumping model.

2.2. SHEARING MOTIONS

Although one may readily visualize Coulomb or "dry" friction to be the dominant mechanism responsible for energy dissipation at dry joints in which the relative motions of the mating surfaces are parallel to the interface, the details of the relative motions and the parametric dependences may be rather complex functions of the joint loading and geometry [12–14].

The action of dry lap joints which are held together by a uniformly distributed clamping pressure (Figure 2) has been analyzed rather thoroughly. If the overlapping parts of the joint are rigid, then relative motion occurs over the entire contact area at once, if there occurs any motion at all. Since the friction force here is proportional to the clamping pressure and to the coefficient of friction, the energy dissipation for a given motion increases as the clamping pressure and friction coefficient increase. Of course, if the externally applied force is less than the friction force, then there occurs no slippage at all, and therefore no energy dissipation.

If the overlapping parts of the joint are not rigid, then the interface area over which there occurs relative motion (and the associated energy dissipation) depends on the instantaneous external load and on the loading history. As indicated in the lower part of Figure 2, during the initial loading phase slippage occurs first at the outermost portions of the contact

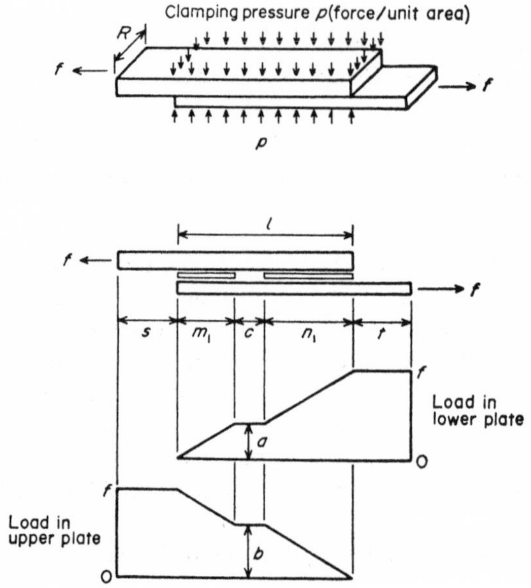

Figure 2. Load distribution in lap joint with dry friction, during initial loading phase. (From reference [14].)

area (nearest the joint edges), and the slipped area propagates inward from the edges as the externally applied load increases. During unloading, reverse slippage occurs, again beginning at the edges; the extent of the reversed slip areas then depends on the reduction in the load from the greatest value. Reloading then again produces slippage (of the reversed slip areas again beginning from the edges) in the initial directions. The energy D dissipated during each further cycle of loading and unloading, during which the external force varies between F_{max} and F_{min} is given by [14]

$$D = \frac{(F_{max} - F_{min})^3}{12R\mu p}\left(\frac{1}{A} + \frac{1}{B} - \frac{3}{A + B}\right)$$

where A is the extensional stiffness of the upper plate, B the extensional stiffness of the lower plate, R the width of the joint, μ the coefficient of friction and p the interface clamping pressure.

The dependence of the energy dissipation on the friction coefficient and interface pressure is of particular interest. Note [12] that for a fully slipped (essentially rigid overlap) joint, the energy dissipated is directly proportional to μp, whereas for a partially slipped (elastic overlap) joint, the energy dissipated is inversely proportional to μp.

A generalized analysis of the energy dissipation produced by partially slipped joints with Coulomb friction, taking into account non-uniform pressure and friction coefficient distributions (including circular distributions representative of conditions around bolts or rivets), appears in Appendix IV of reference [13]. This analysis is somewhat more complex than that for uniformly distributed pressures, but reveals no new phenomena. No analyses are available that take into account such realistic effects as constraints provided by bolt or rivet shanks extending through the jointed plates, or the reduced stiffness of the joined plates due to the presence of holes.

If a viscoelastic material, such as a plastic or rubbery substance, is present between the mating surfaces, then energy dissipation results essentially due to shearing of this material [10]. One may similarly expect that a lubricant introduced between the mating surfaces will provide energy dissipation by virtue of viscous losses associated with shear of the lubricant. Experimentally validated theories are available for bolt-less [10] and for bolted joints [15] with viscoelastic interlayers; however, in the latter the shear stiffness contribution provided by the bolts is obtained from a separate measurement, rather than from first principles. No similar investigations of lubricated joints appear to have been undertaken.

3. DAMPING OF BUILT-UP BEAMS

The concepts discussed in the previous section in relation to ideal joints in essence apply also to realistic joints at low frequencies, at which the important wavelengths are much greater than the characteristic dimensions of the joint, and at which local inertia effects are unimportant. However, realistic joints generally involve unknowns and complexities that usually cannot be treated purely analytically, on the basis of first principles.

3.1. INTERFACES WITH COULOMB FRICTION

Built-up beams, consisting of primary beams with narrow cover-plates fastened to their tops and bottoms by means of closely spaced bolts, behave much like lap-joints with uniform clamping pressure, provided that the bolts do not make contact with the cylindrical surfaces of the holes (Figure 3). The behavior of such a beam vibrating at or below its fundamental natural frequency is analyzed in reference [16], on the basis of the slipped areas that occur at any given stage of loading and on the basis of the shear force (per unit area) required to produce slippage. Experimental results were found to agree well with theoretical predictions based on separately measured values of the shear force at which slippage occurs (for various values of bolt torque).

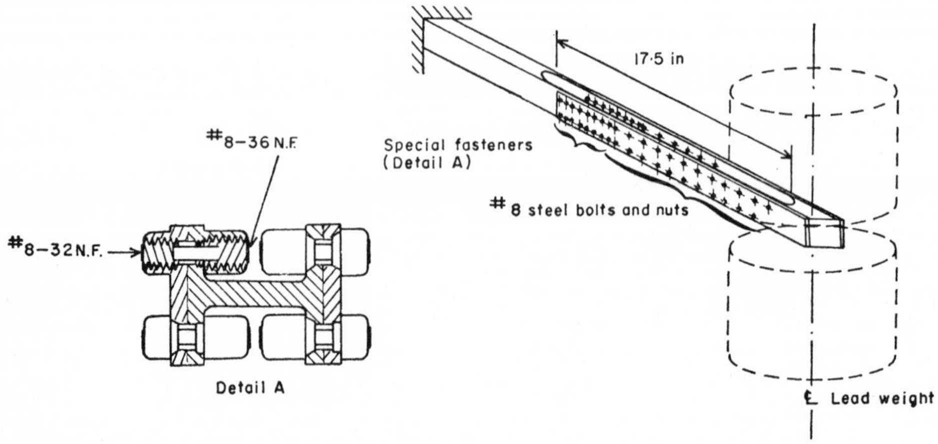

Figure 3. Experimental built-up beam. (From reference [17].)

In reference [17] the analysis of reference [16] is extended to take into account the restriction of the motion due to tight fit of the bolts, by considering the effect of the (closely spaced) bolts as that of a continuous shear joint. Again, experimental results were found to agree well with theoretical calculations based on values of shear joint stiffness and limiting friction force obtained on the basis of separate experimental measurements.

Although the approach taken in references [16] and [17] represents a useful engineering technique, it is applicable only to the lowest vibrational mode—or, at best, to cases for which the mode shape is fully known. This technique is also limited to cases where the fasteners (bolts or rivets) are uniformly tensioned and spaced so closely that the resulting interface pressure may be considered as uniform. However, the interface pressure produced by a tensioned bolt is very non-uniform; it decreases (approximately linearly with radius) from a maximum value at the bolt diameter to very nearly zero at three times the bolt diameter [18, 19]. Some experimental observations [20] also have found fretting to be confined to arcs of small annular regions around rivets, indicating that slip is likely to be limited to these regions. Thus, the method of references [16] and [17] is likely to be applicable in only a very limited number of practical cases, where the fasteners are very closely spaced.

The analysis of references [16] and [17] implies an amplitude-dependence of damping that differs considerably from that for linear (viscous) damping. The effects of non-uniform interface pressures around bolts or rivets would be expected to increase the deviation from linearity even further; however, measurements of the damping due to cover plates riveted onto partially cut-through beams [10] showed this damping to be surprisingly linear (i.e., to result in energy dissipation per cycle varying as the square of the amplitude) up to quite considerable amplitudes. Therefore, there exists some doubt that Coulomb friction indeed is the dominant mechanism, except perhaps at quite large amplitudes.

There also exist some further difficulties with applying the Coulomb friction model to analysis or design of realistic structures. In "dry friction" analyses, velocity effects generally are not considered, but at high frequencies the relative velocities of the mating surfaces may be great enough so that the actual velocity-dependences of the friction coefficients (e.g., see reference [21]) need to be considered. In practical situations, interface pressures are extremely difficult to measure or control, particularly for riveted joints, and furthermore this pressure and the resulting friction may be critically affected by local roughness, wear, oxidation, or other chemical contamination.

It appears that, as for simple joints, there occurs an optimum condition for which the damping is greatest. This optimum must occur between the condition where relative motions are greatly restricted (by the friction forces and/or fastener shear stiffnesses), and that where the friction force is so small that little energy is dissipated, even with rather large relative motions. Joints that have near optimum damping are likely to be relatively loose, however, and thus to be unacceptable from the basic practical structural standpoint.

3.2. INTERFACES WITH LUBRICANTS OR VISCOELASTIC INSERTS

The effect of lubricants on the damping of built-up beams has not been studied extensively; because lubrication of the interfaces of built-up structures may be expected to have the generally undesirable effect of reducing the stiffnesses of these structures, lubrication is not usually employed in practical structures. Because practical structures are likely to be "too tight" for optimum damping, in the sense of the foregoing paragraph, the "loosening" produced by the introduction of a lubricant may be expected to change the stiffness toward the optimum, and thus to result in an increase in damping. Measurements [22] made on cantilever aluminium strips whose ends were bolted to supporting brackets indeed showed that lubrication of the interface between the metal strips and supporting brackets resulted in a considerable increase in damping.

The effects of viscoelastic interlayers introduced into built-up structures also have not been investigated thoroughly—again, probably because viscoelastically damped structures tend to be less stiff than conventional ones, and thus less efficient from the basic static load-carrying capacity point of view. To a good first approximation, the action of a not-too-stiff viscoelastic insert may be expected to be much like that of an oil film: that is, it facilitates relative motion and dissipates energy by virtue of its viscosity. Thus, one would again expect such inserts to "loosen" practical structures and to increase their damping—and again, this type of behavior has been observed in measurements made on the previously mentioned end-supported aluminum strips [22].

Because joints with viscoelastic inserts can be made stiffer than lubricated joints (since viscoelastic materials are available with a wide range of stiffnesses) and because viscoelastic interlayers can be designed and controlled more precisely, they hold some practical promise. Therefore, simple bolted joints with viscoelastic interlayers have been studied quite thoroughly [15]; although such studies have not yet been performed for extended built-up structures, the corresponding analyses probably can be performed readily by means of rather well-established techniques.

4. DAMPING OF BUILT-UP STIFFENED-PLATE STRUCTURES

Aerospace structures typically consist of metal "skin" plating, attached to frames, and reinforced by stringers (Figure 4). The frames usually are rather massive and therefore experience deformations that are much smaller than those of the stringers and skin. It thus is reasonable to neglect frame deformations when one is concerned with the vibrations of skin structures, and to deal with one-dimensional arrays of beam-reinforced panels—in effect assuming that panels on opposite sides of a frame do not interact with each other.

The vibrations of such one-dimensional skin–stringer arrangements have been subject to considerable study for more than a decade, and are reasonably well understood. For example, reference [23] deals with the modes of an infinitely repeated array, reference [24] addresses the corresponding finite array problem, reference [25] presents a wave approach for calcualtion of the natural frequencies, and reference [26] summarizes response analysis in a way that gives good insight into the mode grouping that occurs in periodic structures. Related measurements have been made on both flat [27] and curved [28] panel rows.

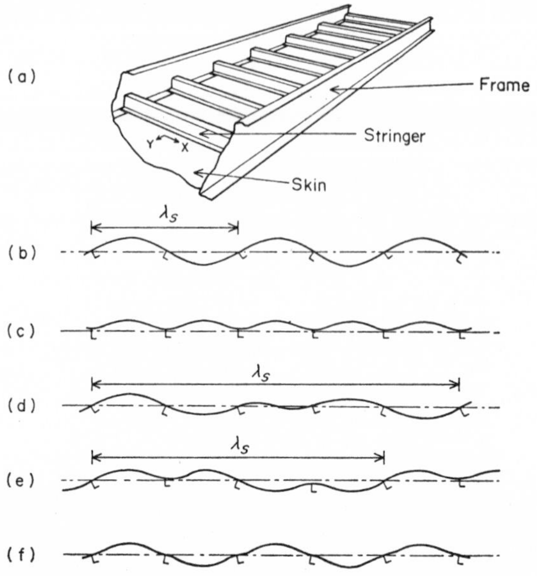

Figure 4. Skin–stringer structure and its typical modal deflections. (From reference [27].) (a) The structure. (b) The "stringer torsion" mode (flexible stringers). (c) The "stringer bending" mode. (d) An "intermediate" mode (I). (e) An "intermediate" mode (II). (f) The "stringer torsion" mode (stiffer stringers).

At the lowest natural frequency of such a skin–stringer panel row, adjacent panels are deflected in opposite directions and all the stringers deform essentially only in torsion (Figure 4(b)). If the stringers are torsionally very soft, then the natural frequency of this mode is very nearly that of the fundamental natural frequency of a rectangular skin panel that is simply supported on two opposite edges (at the stringers), with boundary conditions appropriate to the frame support at the other two edges.

Above this lowest natural frequency, there occurs a series of more complex "intermediate" modes, where the dominant motions of different stringers involve either torsion or bending (e.g., see Figures 4(d) and 4(e)). The highest natural frequency associated with simple (first mode) stringer deflections occurs when all stringers deform essentially only in flexure; then all panels deflect in the same direction at the same time (Figure 4(c)), and the associated natural frequency is very nearly that for a panel that is clamped at the two edges that correspond to the stringers. (Note that for typical aircraft structures the stringer bending displacements tend to be much smaller than the skin panel displacements.)

The lower-frequency motions, which involve torsion of stringers and generally opposite deflections of adjacent panels, tend to be affected considerably by the stringers' torsional stiffness, but not by curvature of the structure in the direction of the frames, which, however, affects modes involving stringer bending. At high frequencies, the motions of the skin panels tend to be so much greater than those of the stringers, that one may generally neglect the stringer deflections altogether.

4.1. DAMPING AT LOW FREQUENCIES

There exists considerable uncertainty concerning the mechanism that dominates the damping of skin–stringer structures at "low frequencies", at which the stringers share significantly in the vibratory energy. Classical acoustic radiation damping may be ruled out in general; calculated values for radiation damping tend to be much smaller than actually measured damping values. One may conjecture that energy dissipation occurs "at the rivets"—but just how the energy is dissipated, and what parameters this damping obeys, is largely unknown.

Experiments have shown that the damping of conventional (unlubricated) riveted skin–stringer structures varies considerably from mode to mode, with no easily recognizable trends. A study in which only the skin thickness was varied, while all other parameters were kept the same, showed that the damping of each of the lowest few modes varied with skin thickness in a complicated non-monotonic manner [27]. No attempt appears to have been made to determine the effects of such other parameters as stringer torsional stiffness, rivet size and spacing, and ambient air pressure.

Although it has been shown that appropriately chosen (different) constant damping values for stringer torsion and stringer bending, coupled with a detailed analysis of the skin and stringer motions, can reproduce experimentally observed values of skin–stringer system damping reasonably well, at least for one experimental configuration [29], there exists no rationale for selecting the stringer damping values *a priori*. The damping values that were found to lead to good matching of the experimental data are considerably greater than those associated with the bending or twisting of monolithic beams, leading one once again to conclude that some mechanism associated with the joints or interfaces must be at work.

In spite of the fact that the damping of conventional skin–stringer structures is poorly understood at best, considerable attention has been paid to increasing the damping of such structures by means of viscoelastic treatments. The damping increases resulting from the addition of viscoelastic coatings to the metal skin or from replacing the simple metal skin by a thin metal sandwich with a viscoelastic interlayer have been explored both theoretically and experimentally [27, 30], as have the effects of tuned dampers added to the skin panels [30]. It has also been shown [31] that viscoelastic materials added only to the portions of the stringers that are far from the skin (i.e., to the tops of the stringers of Figure 4) can produce significant damping increases with relatively small weight penalties. The effect of viscoelastic inserts between the skin and the stringers or of the use of stringers that are themselves highly damped (e.g., made of a metal sandwich with a viscoelastic mid-layer) have not been reported in the literature.

4.2. DAMPING AT HIGH FREQUENCIES

At the "high frequencies", where the stringers of typical skin–stringer structures deflect relatively little, and where the vibrations of each skin panel are little affected by those of adjacent panels, one need essentially consider only one panel at a time. In contrast to the low-frequency case, the dominant high-frequency damping mechanism has been identified, and reasonable engineering methods for its prediction have been proposed.

An extensive series of damping measurements [13] made on panels to which reinforcing beams (stringers) were bolted, riveted, or spot-welded showed that the damping of these structures is independent of interface pressure (bolt torque), surface finish, beam rigidity, and details of the fasteners; also, linear damping behavior was observed over a wide range of amplitudes. All these observations indicate that Coulomb friction could not be the dominant mechanism. On the other hand, the damping of these structures was found to depend on the frequency (*via* a dependence on the ratio of the plate bending wavelength to the bolt spacing), on the presence of lubricant at the interface, and on the length and width of the contact area between the beam and plate. It was also found that any stratagem (e.g., metal inserts, beads of glue, a second beam confining the plate between it and the first beam) that restricted the parts of the plate between fasteners from moving toward and away from the beam resulted in a decrease in the damping. Finally, it was demonstrated that the damping of lubricant-free beam-reinforced plates depended crucially on the ambient air; evacuation of the space around such structures was found to reduce their damping to that of monolithic structures.

Further experiments at various air pressures and in different gases, as well as good agreement between measured damping values and a corresponding theory [32], as illustrated in Figure 5, lend considerable support to a model that ascribes the high-frequency damping of beam-reinforced panels to the "pumping" of the fluid contained between the mating surfaces of the beams and plates. This pumping results as these surfaces move apart and together, alternately forcing the fluid into and out of the thin spaces between the beams and plates, between adjacent fasteners. Energy then is dissipated due to viscous losses in the fluid, which may be air, other gases, a lubricant purposely added at the joint, or even a soft visoelastic insert at the joint.

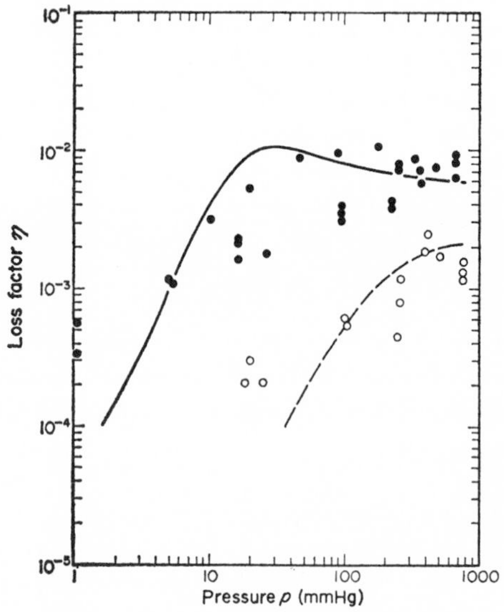

Figure 5. Sample comparison between experimentally measured and theoretically predicted gas-pumping loss factors, for test panel vibrating at two frequencies and various ambient air pressures. (From reference [33].) ——, Theory, 1000 Hz; ---, theory, 16,000 Hz; ●, Experiment, 1000 Hz; ○, experiment, 16,000 Hz.

A semi-empirical method has been proposed [7, 32, 33] for estimating the high-frequency damping of beam-reinforced panels (i.e., skin–stringer structures). This method makes use of the absorption coefficient γ of a reinforcing beam (or other discontinuity) on the panel; this coefficient is defined as the fraction of the panel bending-wave energy impinging on the beam that is not reflected back to the panel. The loss factor η of a panel of area S is given [33] by

$$\eta = \eta_0 + \frac{\lambda}{\pi^2 S} \sum_i \gamma_i L_i,$$

where η_0 represents the loss factor of the panel by itself, without any reinforcing beams or edge supports (i.e., essentially, the loss factor associated only with the plate material).

The symbol L_i refers to the effective length of the ith reinforcing beam; this length is equal to the actual length of the beam if it is located at an edge of the panel under consideration (or within one plate flexural wavelength λ of an edge); otherwise, it is equal to twice the actual beam length. γ_i denotes the absorption coefficient of the ith reinforcing beam, and λ represents the wavelength of bending waves on the panel at the frequency under consideration. For homogeneous isotropic panels of thickness h,

$$\lambda \approx 1{\cdot}35\sqrt{hc_L/f},$$

where f represents the frequency and c_L the longitudinal wave velocity in the plate material. ($c_L = \sqrt{E/\rho}$, where E denotes the Young's modulus and ρ the density of the material.)

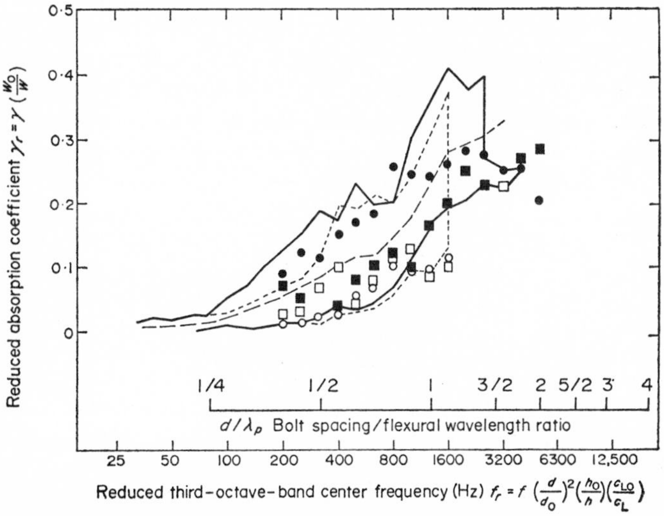

Figure 6. Summary of reduced absorption coefficient data, suggested for estimation of absorption coefficients of riveted, bolted, or spot-welded panel reinforcements. (From reference [7].) Plate thickness h (in), beamwidth w (in): □, $h = 1/16$, $w = 1/2$; ○, $h = 1/16$, $w = 2$; ■, $h = 1/32$, $w = 1/2$; ●, $h = 1/32$, $w = 2$. (all aluminium). ——, Enclosed region contains data points for $w = 1$ in and various bolt spacings. — — —, Approximate average (for estimation). - - -, Enclosed region contains data points for aluminium or steel beams on 1/32 in aluminium or steel plates ($w = 1$ in). Reference value for 1 atm ambient pressure: $h_0 = 1/32$ in $= 0{\cdot}078$ cm; $d_0 = 3$ in $= 7{\cdot}6$ cm; $c_{L0} = 2 \times 10^5$ in/s $= 5100$ m/s; $w_0 = 1$ in $= 2{\cdot}5$ cm.

Figure 6 is a plot of reduced absorption coefficient *versus* reduced frequency, based on extensive experimental data. These reduced parameters, whose definitions appear in the ordinate and abscissa labels of the figure, take into account the observed dependences of the absorption coefficient on beam width, fastener spacing, panel thickness, and panel material, and enable one to use Figure 6 for estimating γ for a wide variety of configurations.

Applicability of the absorption coefficient approach for loss factor estimation, it must be emphasized, is indeed limited to high frequencies, at which the lengths of the panel edges and of the reinforcing beams are at least several times as great as a wavelength λ on the panel. For shorter reinforcing beams (or lower frequencies), use of this approach generally leads to considerable overestimation of the damping (e.g., see reference [34]).

4.3. MEASUREMENT PROBLEMS

Since, as has been mentioned, the damping of skin–stringer structures at low frequencies depends markedly on the stringer deformations (and thus on the type of vibration mode), damping measurements at these frequencies must be confined to one mode at a time, if they are to be meaningful. For such measurements one generally needs to excite the structure at a single frequency or with a very narrow band of frequencies. If the structure is large and/or rather highly damped, one may need to excite it simultaneously at several points. Care must then be taken to match the phasing of the various excitations to the shape of the mode to be excited.

In relation to high-frequency damping measurements, one must take note of the fact that all available measurement methods (such as the decay-rate and the input power measurement techniques) in essence determine the rate at which energy disappears from a measurement point. Since these methods cannot distinguish energy that is actually dissipated (i.e., converted into heat) from energy that travels—still as mechanical energy—from the measurement point to other parts of the structure, localized damping measurements on extended structures (e.g., on one panel of a skin–stringer row) tend to yield erroneous values† of damping (if damping is defined in terms of energy dissipation only, as usually is done). In order to measure the damping of a single panel which is part of a larger structure, one thus needs to prevent energy from being transported away from the panel under consideration. Unfortunately, the only way this energy transport can be eliminated in practice is by disconnecting the panel from the adjacent structures; attempts at blocking energy flow by means of very stiff or very massive additive structures have proved fruitless [19].

It is a basic premise of the recently developed "statistical energy analysis" approach to dealing with the random vibrations of complex structures that the time-average rate of flow of vibratory energy (in a given not too narrow frequency band) from one structure to another is proportional to the difference between the average energies per mode of the two structures (e.g., see reference [35]). Thus, one may be able to prevent energy from flowing from the test panel to its neighbors by providing an excitation and control system that keeps equal the average modal energies of all panels adjacent to the test panel. However, such a system is likely to be complicated and difficult to implement, except for the case where all panels and stringers are alike (in which case one would merely need to excite all panels similarly: e.g., by means of a sound field).

Because of the above discussed difference between what is usually defined as damping and what one can readily measure on panels of skin–stringer structures, it is important to keep in mind the purpose for which one desires to know the damping in any given situation. For example, the response of a panel to localized excitation depends on the total effective damping (i.e., the damping one measures by any of the conventional techniques, if one takes no special precaution), which is due to energy transport away from the test panel, as well as energy dissipation. On the other hand, the response of an array of like panels to uniform excitation depends only on the energy dissipation of the individual panels; the average energy transport between the panels here is zero, because their vibratory energies are equal. Clearly, in these two cases one needs to carry out different types of measurements in order to determine "the damping".

5. CONCLUSION

The phenomenology of how energy is dissipated at simple ideal joints is reasonably well understood at present and provides some valuable insights into the mechanism that

† These values either may be too large or too small. Steady state power input measurements always yield damping values that are too high, because these measurements also indicate the energy that is conducted away. However, decay rate measurements may indicate values either larger or smaller than those associated with energy dissipation in the panel under consideration, depending on whether the net energy transport between that panel and its neighbors after removal of the excitation is away from or into the test panel. The direction of this energy transport depends on the distribution of energy in the system, on the distribution of the damping, and on the energy transmission characteristics of the boundaries between adjacent panels—all, at the frequency under consideration [35].

dominates the damping of built-up beams vibrating at low frequencies. On the other hand, little is known about the damping of such beams vibrating at frequencies considerably above their fundamental resonance frequencies, particularly for configurations where the fasteners are spaced rather far apart.

Although the often quite complex low-frequency modal motions of riveted skin–stringer structures have been subject to considerable investigation and are understood quite well, the same is not true of the damping associated with these motions. The amplitude and frequency dependence of this damping has been studied only briefly, and virtually nothing is known about the dominant mechanism. In contrast, the damping of the high-frequency motions, which essentially involve only the skin panels between frames and stringers, has been investigated rather thoroughly; "air pumping" has been shown to be the dominant mechanism, and a corresponding engineering estimation technique is in hand.

A considerable amount of further investigation is required to bring the state of engineering knowledge pertaining to the damping of built-up structures to the point where it sufficies for the prediction of structural responses and for the design of optimized structures; the greatest need for further work exists with respect to the higher-frequency vibrations of built-up beams and with respect to the lower-frequency modes of skin–stringer structures.

ACKNOWLEDGMENT

This paper is based in part on an invited paper presented orally in 1968 at the 76th Meeting of the Acoustical Society of America. Preparation of the present updated written version of this paper was undertaken under NASA Langley Research Center Contract NAS 1–9557–17. NASA's sponsorship is gratefully acknowledged.

REFERENCES

1. E. E. Ungar 1964 In *Mechanical Design and Systems Handbook* (H. A. Rothbart, ed.). New York: McGraw-Hill. Section 6: Mechanical vibrations.
2. D. C. Karnopp 1963 In *Random Vibration* (Vol. 2, S. H. Crandall, ed.). Cambridge Massachusetts: The M.I.T. Press. Basic theory of random vibration.
3. K. N. Tong 1960 *Theory of Mechanical Vibration.* New York: John Wiley & Sons.
4. B. J. Lazan 1968 *Damping of Materials and Members in Structural Mechanics.* New York: Pergamon Press.
5. L. Cremer and M. Heckl 1967 *Koerperschall.* Berlin: Springer-Verlag. Chapter III.
6. D. J. Mead 1968 In *Noise and Acoustic Fatigue in Aeronautics* (E. J. Richards and D. J. Mead, eds.). London: John Wiley. Chapter 18: The damping of jet-excited structures.
7. E. E. Ungar 1971 In *Noise and Vibration Control* (L. L. Beranek, ed.). New York: McGraw-Hill. Chapter 14: Damping of panels.
8. C. Andrew, J. A. Cockburn and A. E. Waring 1968 *Conference on Properties and Metrology of Surfaces.* London: Institution of Mechanical Engineers. Paper 22: Metal surfaces in contact under normal forces: some dynamic stiffness and damping characteristics.
9. D. N. Reshetov and Z. M. Levina 1956 *Vestnik Mashinostroyeniya* **3,** no. 12. Damping of oscillations in the couplings and components of machines.
10. T. J. Mentel 1967 *Journal of Engineering for Industry* **89,** 797–805. Joint interface layer damping.
11. S. Hother-Lushington and D. C. Johnson 1963 *Journal of Mechanical Engineering Science* **5,** 175–181. Damping properties of thin oil films subjected to high-frequency alternating loads.
12. H. Himelblau 1971 *Journal of the Acoustical Society of America* **50,** 96. Comparison of slip and slide damping for a friction-clamped beam excited by sinusoidal vibration at resonance.
13. E. E. Ungar 1964 *AFFDL–TDR*–64–98. Energy dissipation at structural joints; mechanisms and magnitudes.

14. A. F. METHERELL and S. V. DILLER 1968 *Journal of Applied Mechanics* **35,** 123. Instantaneous energy dissipation rate in a lap joint—uniform clamping pressure.
15. D. J. MEAD and D. G. C. EATON 1960 *A.A.S.U. Report No.* 153. Interface damping at riveted joints, Part I: Theoretical analysis. 1963 *A.A.S.U. Report No.* 241. Part II: Damping and fatigue measurements.
16. T. H. H. PIAN and F. C. HALLOWELL, JR. 1952 *Proceedings of the First U. S. National Congress of Applied Mechanics.* New York: American Society of Mechanical Engineers, pp. 97–102. Structural damping in a simple built-up beam.
17. T. H. H. PIAN 1957 *Journal of Applied Mechanics* **24,** 35–38. Structural damping of simple built-up beam with riveted joints in bending.
18. J. FERNLUND 1961 *Institute of Machine Elements Report No.* 11, *Chalmers University of Technology, Gothenberg, Sweden.* A method to calculate the pressure between bolted or riveted plates.
19. E. E. UNGAR and K. S. LEE 1967 *AFFDL–TR*–67–86. Considerations in the design of supports for panels in fatigue tests. See Appendix V.
20. D. J. MEAD 1961 *WADC–University of Minnesota Conference on Acoustic Fatigue, WADL–TR–*59–676 (W. J. Trapp and D. M. Forney, Jr.. eds.), pp. 235–261. The damping, stiffness and fatigue properties of joints and configurations representative of aircraft structures.
21. E. RABINOWICZ 1960 *Product Engineering* September 26, pp. 51–54. Practical approach to friction coefficients.
22. B. R. HANKS and D. G. STEPHENS 1967 *Shock and Vibration Bulletin* **36,** Part 4, 1–8. Mechanisms and scaling of damping in a practical structural joint.
23. Y. K. LIN 1960 *Journal of Applied Mechanics* **27,** 669. Free vibrations of continuous skin–stringer panels.
24. Y. K. LIN, I. D. BROWN, P. C. DEUTSCHLE 1964 *Journal of Sound and Vibration* **1,** 14–27. Free vibration of a finite row of continuous skin–stringer panels.
25. G. SEN GUPTA 1971 *Journal of Sound and Vibration* **16,** 567–580. Natural frequencies of periodic skin–stringer structures using a wave approach.
26. D. J. MEAD 1971 *Journal of Engineering for Industry* **93,** 783–792. Vibration response and wave propagation in periodic structures.
27. D. J. MEAD 1964 *Institute of Sound and Vibration Research Memorandum No.* 110, *University of Southampton.* The damping of stiffened plate structures.
28. B. L. CLARKSON and R. D. FORD 1962 *A.A.S.U. Report No.* 206. The response of a model structure to noise: Part II—Curved panel.
29. J. P. HENDERSON and A. D. NASHIF 1972 *Journal of Engineering for Industry* **94,** 159–166. The effect of stringer width and damping on the response of skin–stringer structures.
30. D. I. G. JONES and W. J. TRAPP 1971 *Journal of Sound and Vibration* **17,** 157–185. Influence of additive damping on resonance fatigue of structures.
31. F. CICCI 1971 *Journal of Sound and Vibration* **17,** 63–76. The damping of integrally stiffened structures.
32. E. E. UNGAR and G. MAIDANIK 1968 *Journal of the Acoustical Society of America* **44,** 292–295. High frequency plate damping due to gas pumping in riveted joints.
33. E. E. UNGAR and J. R. CARBONELL 1966 *American Institute of Aeronautics and Astronautics Journal* **4,** 1385–1390. On panel vibration damping due to structural joints.
34. F. J. FAHY and R. B. S. WEE 1968 *Journal of Sound and Vibration* **7,** 431–436. Some experiments with stiffened plates under acoustic excitation.
35. E. E. UNGAR and J. E. MANNING 1968 In *Dynamics of Structured Solids* (G. Herrmann, ed.). New York: American Society of Mechanical Engineers. Analysis of vibratory energy distributions in composite structures.

Reprinted from *J. Eng. Indus.*, **84,** 149–155 (1962)

Maximum Stresses in Beams and Plates Vibrating at Resonance

ERIC E. UNGAR
Senior Engineering Scientist,
Bolt Beranek and Newman, Inc.,
Cambridge, Mass. Assoc. Mem. ASME

Expressions are derived which relate the maximum stresses encountered in simply supported beams and rectangular plates and in clamped circular plates vibrating at resonance to modal displacements and modal loadings. Computation of modal loadings from time-wise harmonic or random pressures is discussed. It is shown that the resonant maximum stress may be reasonably approximated by a simple formula suitable for conservative design calculations for all types of beams and plates.

Introduction

In the past it has generally been possible to sidestep the difficulties associated with structural resonances either by designing the structures so that their resonant frequencies fall outside of the range of exciting frequencies, or by "red-lining" the design; i.e., by simply not operating at such troublesome frequencies. In recent years, however, there have occurred marked increases in the abundance of intense sources of broad-band excitation, such as rockets and jets. One usually cannot design equipment or structures that avoid the resonance problem in the presence of such sources. Since under such conditions red-lining is usually either impossible due to the nature of the source, or impractical in view of the many resonances that can occur in a given configuration, the design of many structures must take resonances into account.

Many of the recent advances made in the fields of material fatigue and of structural damping have probably been motivated to a large extent by the afore-mentioned resonance problem, and as a result of such advances the design of structures capable of long term operation in wide-band dynamic environments has now become feasible. The prediction of the maximum oscillatory stresses in vibrating structures is of prime importance in designing these to attain a given fatigue life. In view of this importance it is not surprising that a number of researchers have assigned themselves to related problems; for example Bisplinghoff, Isakson, and Pian [1][1] have outlined methods suitable for analysis of transient stresses in aircraft, and Eringen [2] has dealt with the motion and stress responses of viscously damped beams and plates subject to random loads. Although these investigators have outlined general methods of attack and presented valuable formal expressions for stress distributions and cross-correlations, it appears that no one has as yet attempted to obtain expressions by means of which the maximum stresses in simple structures vibrating at resonance may be estimated for design purposes. The present paper is intended as a step toward providing such expressions.

The following paragraphs deal in detail with rectangular plates simply supported on all edges, with circular plates clamped at their circumferences, and with simply supported beams. The two particular plate configurations were selected because they lead to relatively simple results, since the applicable boundary condition expressions in terms of the deflections do not depend on Poisson's ratio. Although the results of these analyses apply in the strictest sense only for the configurations considered, they may be expected to apply with reasonable accuracy also for the higher modes of configurations with other boundary conditions, since at these higher modes the boundary conditions are known to have little effect on the structure not immediately adjacent to the boundaries [3, 4].

In order to provide a framework for the subsequent analyses the ideas associated with normal modes are first reviewed briefly in the following section, and expressions are presented by means of which the modal loadings may be computed for pressures which are sinusoidal or random in time. Thereafter the principal stresses associated with the normal modes of the two afore-mentioned types of plates are calculated, and the maxima of these principal stresses are determined. Finally, the plate analyses are reinterpreted for beams, and results are summarized.

Basic Concepts and Relations

Normal Modes. The classical equation of undamped motion of uniform elastic plates may be written as [5]

$$D\nabla^4\bar{\eta} + \rho h \frac{\partial^2 \bar{\eta}}{\partial t^2} = \bar{p}(x, y, t) \tag{1}$$

where

$\bar{\eta}$ = deflection of plate out of its plane of equilibrium
D = $Eh^3/12(1 - \nu^2)$ = flexural rigidity
E = Young's modulus
ν = Poisson's ratio
ρ = density of plate material
h = plate thickness
$\bar{p}$ = forcing pressure distribution
x, y = Cartesian co-ordinates
t = time

If the forcing function is sinusoidal in time, then one may use the usual conventions of complex variable notation [5] to write

$$\bar{p}(x, y, t) = p(x, y)e^{i\omega t}, \quad \bar{\eta} = \eta(x, y)e^{i\omega t}, \tag{2}$$

and introduce damping by permitting the rigidity to take on complex values [5]. Then Eq. (1) becomes

$$D(1 + i\beta)\nabla^4\eta - \rho h\omega^2\eta = p(x, y), \tag{3}$$

where β is the "loss tangent" or damping factor [6] of the plate and is assumed to be known as a function of frequency.[2]

For a given finite plate one may in general obtain a set of functions $\psi_{mn}(x, y)$, each of which satisfies all the boundary conditions and a homogeneous equation

[1] Numbers in brackets designate References at end of paper.

Contributed by the Machine Design Division for presentation at the Summer Annual Meeting, Los Angeles, Calif., June 11–14, 1961, of The American Society of Mechanical Engineers. Manuscript received at ASME Headquarters, July 28, 1960. Paper No. 61—SA-14.

[2] This formulation is more general than the common treatment assuming viscous damping. The latter is included here as a special case; β is then related to the viscous damping coefficient c as $\beta = c\omega/D$.

$$D\nabla^4\psi_{mn} - \rho h\omega_{mn}^2\psi_{mn} = 0 \quad (4)$$

for a definite value ω_{mn}. The functions ψ_{mn} are called eigenfunctions, the corresponding values ω_{mn} are eigenvalues. For plates with clamped and/or simply supported edges the eigenfunctions ψ_{mn} are found to be orthogonal or normal, that is,

$$\int_A \psi_{mn}\psi_{kl}dA = 0 \quad \text{for } m, n \neq k, l, \quad (5)$$

where the indicated integration is to be carried out over the entire plate surface A. The ψ_{mn} may then be called "normal modes," and the ω_{mn} natural frequencies.

A given function, such as $p(x, y)$, may conveniently be expanded in a normal mode series

$$p(x, y) = \sum_{m,n=0}^{\infty} P_{mn}\psi_{mn}, \quad (6)$$

where, because of the orthogonality of the ψ's,

$$P_{mn} = \frac{\int_A p(x, y)\psi_{mn}dA}{\int_A \psi_{mn}^2 dA}. \quad (7)$$

If one postulates a similar expansion for the plate deflection, i.e.,

$$\eta(x, y) = \sum_{m,n=0}^{\infty} U_{mn}\psi_{mn}, \quad (8)$$

one finds by substitution into Eq. (3) that

$$\frac{P_{mn}}{U_{mn}} = D(1 + i\beta)\frac{\nabla^4\psi_{mn}}{\psi_{mn}} - \rho h\omega^2 = \rho h[(1 + i\beta)\omega_{mn}^2 - \omega^2] \quad (9)$$

This relation permits the response to be determined directly if the excitation is known.

If, for reasonably small values of damping, the driving frequency ω is nearly equal to the natural frequency ω_{mn}, then the corresponding deformation coefficient U_{mn} will take on a large value (provided the corresponding modal loading coefficient P_{mn} is finite). Since $\omega = \omega_{mn}$ corresponds to resonance of the mn mode, one may conclude that at resonance of the mn mode

$$\eta(x, y) \approx U_{mn}\psi_{mn}(x, y), \quad (10a)$$

$$U_{mn} \approx P_{mn}/i\beta\rho h\omega_{mn}^2, \quad (10b)$$

provided that damping is not so small that the deflection U_{mn} becomes large enough to make the originally assumed linear plate equation (1) totally inapplicable.

Evaluation of Modal Loading P_{mn}. It has already been pointed out that for pressures varying sinusoidally in time, as given in Eq. (2), the modal loading may be computed by means of Eq. (7). Randomly time-variant pressures, given by

$$\bar{P}(x, y, t) = p(x, y)h(t), \quad (11)$$

where h is a random function of time, result in time-dependent coefficients $\bar{P}_{mn}(t)$ when introduced into Eq. (7). These coefficients are related to the usual ones as

$$\bar{P}_{mn}(t) = P_{mn}h(t). \quad (12)$$

The mean square modal pressure may then be expressed as

$$\langle \bar{P}_{mn}^2 \rangle = P_{mn}^2 \langle h^2 \rangle = P_{mn}^2 \int_0^{\infty} S_p(f)df \quad (13)$$

where $P_{mn}^2S_p(f)$ is the modal power spectral density expressed (as usual) as a function of cyclic frequency $f = \omega/2\pi$. The plate response in the mn mode is affected essentially only by the pressure contributions at frequencies near the natural frequency f_{mn} and contained in a frequency band of width β. Thus [7]

$$\langle \bar{P}_{mn}^2 \rangle \approx P_{mn}^2S_p(f_{mn})\Delta f \approx P_{mn}^2S_p(f_{mn}) \cdot \beta\frac{\omega_{mn}}{2\pi}, \quad (14)$$

and in view of Eq. (10b) the root-mean-square deformation coefficient obeys

$$|U_{mn}|_{\text{rms}} \approx |\bar{P}_{mn}|_{\text{rms}}/\beta\rho h\omega_{mn}^2 \approx \frac{|P_{mn}|}{\rho h\omega_{mn}}\sqrt{\frac{S_p(f_{mn})}{\beta f_{mn}}}. \quad (15)$$

Stresses. For a given plate deformation $\eta(x, y)$ one may compute the stresses in the outermost fibers of the plate from [8]

$$\pm\frac{h^2}{6D}\sigma_x = \eta_{xx} + \nu\eta_{yy}, \qquad \pm\frac{h^2}{6D}\sigma_y = \nu\eta_{xx} + \eta_{yy},$$
$$\pm\frac{h^3}{6D}\tau_{xy} = (1 - \nu)\eta_{xy}, \quad (16)$$

where subscripts on η denote differentiation and the stress notation is that of Timoshenko [8]. These equations represent stresses acting on planes normal to the co-ordinate axes chosen, and hence not necessarily the maximum stresses. Thus at each point of the plate surface one must establish the value of the (larger) principal stress by means of the equation

$$\sigma_{\text{prin}} = \frac{|\sigma_x| + |\sigma_y|}{2} + \sqrt{\left[\frac{|\sigma_x| - |\sigma_y|}{2}\right]^2 + \tau_{xy}^2}. \quad (17)$$

Finally, in order to determine the maximum value of stress in the plate, one must determine the largest value of principal stress encountered anywhere on the plate.

Stresses Associated With Straight-Crested Flexural Waves

In many cases, the deformations of finite plates at "higher frequencies" in regions "far" from the boundaries may be approximated in terms of straight-crested flexural waves. If one assumes straight-crested flexural waves traveling in the x-direction (in regions where the boundaries have no significant effect) one may write the corresponding deflection as [5]

$$\eta = Ue^{i(\omega t - \gamma x)}, \quad \gamma^2 = \omega\sqrt{\rho h/D}. \quad (18)$$

The maximum stress due to such waves may be found from Eqs. (16) and (17) to obey

$$\sigma_{\text{prin max}} = |\sigma_x| = \frac{6D}{h^2}|\eta_{xx}| = \frac{6D}{h^2}\gamma^2 U \quad (19)$$

and the ratio R_u of stress amplitude to displacement amplitude may be written simply as

$$R_u \equiv \left|\frac{\sigma}{U}\right| = \frac{6\omega}{h^2}\sqrt{\rho hD} = \sqrt{3}\,\omega\rho c_L \quad (20)$$

where

$$c_L = [E/\rho(1 - \nu^2)]^{1/2} \quad (21)$$

is the velocity of propagation of quasi-longitudinal waves [9] in the plate and depends only on properties of the material.[3]

With the ratio of the modal pressure to the modal displacement amplitude at resonance as obtained from Eq. (10b), use of

[3] It is interesting to note that the maximum strain is given by

$$\epsilon_{\max} = \frac{\sqrt{3}}{1 - \nu^2}\frac{V}{c_L} \approx \sqrt{3}\frac{V}{c_L}, \quad (20a)$$

where $V = U\omega$ is the velocity amplitude, and that the maximum strain thus is proportional to the "Mach number" of the oscillation referred to the speed of sound in the plate material.

Eq. (20) results in the following simple ratio R_p between the maximum stress and modal loading:

$$R_p \equiv |\sigma_{max}/P_{mn}| = \sqrt{3}\, c_L/\beta h\omega_{mn} = 6\sqrt{D/\rho h}/\beta h^2\omega_{mn}. \quad (22)$$

The fact that the resonant displacement of a plate may be considered as a superposition of oppositely traveling waves not only justifies the use of Eq. (10*b*) for the foregoing deduction, but also makes the present result more generally applicable. Of course, Eq. (22) may produce a rather poor approximation for modes for which the superposed waves have strongly curved crests.

Modal Stresses in Simply Supported Rectangular Plates

The normal modes and natural frequencies for a simply supported rectangular plate of edge lengths a and b (in the x and y-directions, respectively) are [10]

$$\psi_{mn} = \sin(m\pi x/a)\sin(n\pi y/b); \quad \omega_{mn} = \sqrt{D/\rho h}\cdot[(m\pi/a)^2 + (n\pi/b)^2]. \quad (23)$$

The modal stresses on the plate surface may be determined directly from Eqs. (16), and thereafter the principal stress at every point on the surface may be computed from Eq. (17). One may then show that σ_{prin} attains its largest value for those values of x and y for which $\sin^2(m\pi x/a) = \sin^2(n\pi y/b) = 1$. Then one finds simply that the largest value of the normal stress obeys

$$\sigma_{prin\,max} = (6\pi^2 DU_{mn}/h^2)\cdot\max[m^2/a^2 + \nu n^2/b^2,\ \nu m^2/a^2 + n^2/b^2]$$
$$= \frac{6P_{mn}}{\pi^2 h^2\beta}\cdot\frac{\max[m^2/a^2 + \nu n^2/b^2,\ \nu m^2/a^2 + n^2/b^2]}{[m^2/a^2 + n^2/b^2]^2} \quad (24)$$

where the second form is obtained from the first by use of Eqs. (10) and (23) and the symbolism max $[\xi, \zeta]$ denotes the larger of the two values ξ, ζ.

Inspection of the previous result and use of Eqs. (21) and (23) lead to the following simple relation:

$$|\sigma_{prin\,max}/P_{mn}| < 6\sqrt{D/\rho h}/\beta h^2\omega_{mn} = R_p, \quad (25)$$

where R_p is defined in Eq. (22).

The foregoing relation may conveniently be used in conservative design calculations since it represents an upper bound to the stresses attained in resonant simply supported rectangular plates. That this upper bound corresponds to the straight-crested flexural wave approximation is evident from comparison of it with Eq. (22). This correspondence is not surprising, since the step leading from Eq. (24) to Eq. (25) amounts in effect to neglecting plate curvatures in the direction perpendicular to the direction of wave propagation—this is precisely the assumption leading to Eq. (22).

Modal Stresses in Clamped Circular Plates

1 Mode Shapes and Natural Frequencies. The eigenfunctions for circular plates clamped at their circumferences may be written[4] as [5]

$$\psi_{mn}(r,\phi) = \begin{bmatrix}\cos\\ \sin\end{bmatrix}(m\phi)\,[J_m(\gamma_m r) + B_{mn} I_m(\gamma_{mn} r)],\qquad m = 0, 1, 2, \ldots;\ n = 1, 2, 3, \ldots \quad (26)$$

[4] $\begin{bmatrix}\cos\\ \sin\end{bmatrix}(m\phi)$ means either $\sin(m\phi)$ or $\cos(m\phi)$. Eq. (26) implies two subsets of eigenfunctions, an even set involving cosines and an odd set involving sines. This double notation is retained throughout the subsequent analysis for the sake of brevity.

in terms of the usual polar co-ordinates r and ϕ. J_m and I_m denote the mth order ordinary and hyperbolic Bessel functions, respectively, and the "radial wave number" γ_{mn} is related to natural frequency ω_{mn} as

$$\gamma_{mn}^4 = \frac{\rho h}{D}\omega_{mn}^2. \quad (27)$$

B_{mn} is a constant for a given mode. Its value is dictated by the requirement of zero plate deflection at the clamped circumference and is given by

$$B_{mn} = -J_m(\gamma_{mn}a)/I_m(\gamma_{mn}a), \quad (28)$$

where a denotes the plate radius.

The natural frequencies may be found from the eigenvalues of the governing equation. These are the values of γ_{mn} which satisfy the relation

$$I_m[J_{m-1} - J_{m+1}] = J_m[I_{m-1} + I_{m+1}] \quad (29)$$

arising from the condition that the radial slope vanish at the boundary. The argument $(\gamma_{mn}a)$ is implied for all Bessel functions of Eq. (29). Values of $(\gamma_m a)$ which satisfy this relation are listed in Table 1. From this tabulation it is evident that for larger mode numbers the eigenvalues approach

$$\gamma_{mn}a \approx \left(n + \frac{m}{2}\right)\pi. \quad (30)$$

Table 1 [5]

Eigenvalues for clamped circular plates
[Values of $\gamma_{mn}a/\pi$ for which $\gamma_{mn}a$ satisfies Eq. (29)]

n \ m	0	1	2	$m > 2$
1	1.015	1.468	1.879	. . . 1 + $m/2$
2	2.007	2.483	2.992	. . . 2 + $m/2$
3	3.000	3.490	4.000	. . . 3 + $m/2$
. . .	. . .	. . .	. . .	. . .
$n>3$	n	$n + 1/2$	$n + 1$	. . . $n + m/2$

2 General Stress Relations. The stresses in the outermost plate fibers may be expressed in polar co-ordinates as [8]

$$\pm h^2\sigma_r/6D = \eta_{rr} + \nu\eta_{\phi\phi}/r^2; \quad \pm h^2\sigma_\phi/6D = \nu\eta_{rr} + \eta_{\phi\phi}/r^2;$$
$$\pm h^2\tau_{r\phi}/6D = (1-\nu)\eta_{r\phi}/r, \quad (31)$$

whence the stresses obtained at resonance may be computed by substitution of Eqs. (10*a*) and (26). One finds

$$\frac{h^2}{6DU_{mn}}\sigma_r = \begin{bmatrix}\cos\\ \sin\end{bmatrix}(m\phi)\left[J''_m + B_{mn}I''_m - \frac{\nu m^2}{r^2}(J_m + B_{mn}I_m)\right]$$

$$\frac{h^2}{6DU_{mn}}\sigma_\phi = \begin{bmatrix}\cos\\ \sin\end{bmatrix}(m\phi)\left[\nu(J''_m + B_{mn}I''_m) - \frac{m^2}{r^2}(J_m + B_{mn}I_m)\right] \quad (32)$$

$$\frac{h^2}{6DU_{mn}}\tau_{r\phi} = (1-\nu)\frac{m}{r}\begin{bmatrix}-\sin\\ \cos\end{bmatrix}(m\phi)\,[J'_m + B_{mn}I'_m]$$

where the primes denote differentiation with respect to r and the argument $(\gamma_{mn}r)$ is implied for all Bessel functions. The resulting expressions suffice, in conjunction with Eq. (17), to establish the maximum normal stress experienced at any position on the plate for a given resonance. [For the present analysis one must, of course, replace the x, y subscripts of Eq. (17) by r, ϕ, respectively.] Determination of the positions at which the greatest stress occurs is much more difficult here than in the previously discussed case of rectangular plates.

From Eqs. (17) and (32) one may demonstrate that ϕ-wise relative maxima of σ_{prin} occur at those angles for which $\begin{bmatrix} \cos \\ \sin \end{bmatrix}(m\phi) = 1$. At these angles the shear stress is found to vanish, so that σ_r and σ_ϕ are principal stresses. If one may ignore other possible relative maxima (as is reasonable for estimates of the maximum stress), then the maximum normal stress corresponds to the highest of the r-wise maxima of $|\sigma_r|$ and $|\sigma_\phi|$ that occur with $\begin{bmatrix} \cos \\ \sin \end{bmatrix}(m\phi)$ set equal to unity.

The evaluation of the afore-mentioned maxima presents a considerable amount of difficulty, because of the Bessel form of the eigenfunctions ψ_{mn}. Although it is possible to calculate the maximum stresses associated with a given mode shape exactly by numerical means, such a procedure would not shed much light on the over-all problem and is not likely to lead to rules useful for design purposes. Hence, one must be content with some rather rough approximations if one desires to obtain some general results in tractable form.

In the following subsections the maximum stresses encountered at the circumference and center of the plate and at those positions where the Bessel functions attain relative maxima in the plate interior are determined. Then these stresses are compared so that the greatest may be selected.

3 Stresses at Circumference. The stresses at the clamped boundary may be determined with relative ease. Since B_{mn} is chosen according to Eq. (28) so as to make $\psi_{mn}(a, \phi) = 0$, the maximum radial and tangential stresses at $r = a$ are, in view of Eqs. (32), given by

$$\sigma_{\phi\,\max}]_{r=a} = \nu\sigma_{r\,\max}]_{r=a} < \sigma_{r\,\max}]_{r=a};$$

$$\frac{h^2}{6DU_{mn}}\sigma_{r\,\max}]_{r=a} = \frac{d^2}{dr^2}[J_m(r\gamma_{mn}) + B_{mn}I_m(r\gamma_{mn})]_{r=a} \quad (33)$$

$$= \frac{\gamma_{mn}^2}{4}[J_{m-2} - 2J_m + J_{m+2} + B_{mn}(I_{m-2} + 2I_m + I_{m+2})]$$

where the argument $(a\gamma_{mn})$ is implied in the last line and the expression is obtained by use of the well-known differentiation properties of Bessel functions [11]. For large values of the argument, the Bessel functions approach [11]

$$J_m(z) \approx \sqrt{\frac{2}{\pi z}}\cos\left[z - \frac{\pi}{2}\left(m + \frac{1}{2}\right)\right], \quad I_m(z) \approx e^z/\sqrt{2\pi z} \quad (34)$$

so that in view of Eq. (30)

$$\frac{h^2}{6DU_{mn}}\sigma_{\max}\Big]_{r=a} \approx \gamma_{mn}^2(J_m - B_{mn}I_m) = 2\gamma_{mn}^2 J_m(a\gamma_{mn}), \quad (35)$$

at least for the higher modes.

In order to determine the error involved in using Eq. (35) for the lower modes, the more exact expression of Eq. (33) was evaluated for several values of m and n. A comparison of the more exact stress values (to slide rule accuracy) with those obtained by the approximation appears in Table 2. The approximation is seen to be quite good for $n > 1$, and the errors are found not to exceed 15 per cent even for $n = 1$.

Table 2 Ratio of "exact" stresses [Eq. (33)] to approximate values [Eq. (35)]

$n \backslash m$	0	1	2
1	1.035	0.919	0.872
2	1.008	0.982	1.000
3	0.980	1.030	0.993

4 Stresses at Plate Center. The stresses at the center of the plate may also be evaluated without a great deal of difficulty. From Eq. (32) and by use of small argument approximations [11][5] for J_m and I_m, one may establish that

$$\frac{h^2}{6DU_{mn}}|\sigma_r|\Big]_{r=0} \approx \begin{cases} \frac{1}{2}\gamma_{0n}^2(1 + B_{0n}), & m = 0 \\ \frac{1}{4}\gamma_{2n}^2(1 - 2\nu)(1 + B_{2n}), & m = 2 \\ 0, & m = 1, m > 2 \end{cases} \quad (36)$$

$$|\sigma_r|]_{r=0} \geq |\sigma_\phi|]_{r=0}, \qquad m = 0, 1, 2, 3, \ldots$$

One may show by the use of the large argument asymptotic expansions for J_m and I_m and of the approximation of Eq. (30) for $a\gamma_{mn}$ that for larger values[6] of $a\gamma_{mn}$,

$$|B_{mn}| \approx \sqrt{2}\,e^{-\pi(n+m/2)} \ll 1. \quad (37)$$

Equations (36) may thus be simplified further by neglecting the B terms.

5 Stresses Where Bessel Functions Have Maxima. Determination of the maximum stresses at only the plate center and edge may be insufficient. It is conceivable, in view of the oscillatory nature of Bessel functions, that high stress levels occur at other positions. To obtain an estimate of these levels one may note from the derivative properties of the Bessel functions that for large values of γ_{mn}

$$(d^2/dr^2)[J_m + B_{mn}I_m] \approx -\gamma_{mn}^2[J_m(\gamma_{mn}r) + J_m(\gamma_{mn}a)I_m(\gamma_{mn}r)/I_m(\gamma_{mn}a)]. \quad (38)$$

Since $I_m(z)$ increases monotonically and nearly exponentially with z, one may observe that for values of r not too near a the $B_{mn}I_m$ term contributes relatively little to the value of the second derivative of $(J_m + B_{mn}I_m)$ or to the value of the expression itself, provided that $J_m(\gamma_{mn}r)$ is not small. In order to evaluate maximum stresses approximately one should, therefore, deal with the maxima of $J_m(\gamma_{mn}r)$.

By setting $\begin{bmatrix} \cos \\ \sin \end{bmatrix}(m\phi) = 1$ and also setting the cos term of Eq. (34) equal to unity, one may find from Eq. (32) that

$$\frac{h^2}{6DU_{mn}}|\sigma_r|_{\max} \approx \left|J''_m - \frac{\nu m^2}{r_m^2}J_m\right|$$

$$\approx \sqrt{\frac{2}{\pi\gamma_{mn}r_m}}\left(\gamma_{mn}^2 + \frac{\nu m^2}{r_m^2}\right), \quad (39)$$

where r_m denotes the smallest value of r which corresponds to a maximum of J_m. By referring to contours of constant $J_m(z)$ plotted on the m-z plane, as by Jahnke and Emde [12], one may establish that the first maximum in $J_m(z)$ occurs very nearly at $z \approx 1 + m$, for $m > 2$. By setting $\gamma_{mn}r_m = 1 + m$ in Eq. (39), one finds

$$\frac{h^2}{6DU_{mn}}|\sigma_{\max}| \approx \sqrt{\frac{2}{\pi(1 + m)}}\left[1 + \nu\cdot\left(\frac{m}{1 + m}\right)^2\right]\gamma_{mn}^2, \quad (40)$$

since $|\sigma_r|_{\max} \approx |\sigma_\phi|_{\max}$ in view of the forms of Eqs. (32).

Summary and Comparison of Stresses. Collecting Eqs. (35), (36), and (40), one finds that one may write these results as

[5] For $|z| \ll 1$, $J_m(z) \approx I_m(z) \approx \frac{1}{m!}\left(\frac{z}{2}\right)^m$.

[6] Since it was shown that "large value" approximations are not unreasonable even for the lower modes, these approximations are used henceforth for the sake of obtaining results in simple closed form.

$$\left|\frac{\sigma_{\max}}{U_{mn}}\right| = \frac{6D}{h^2}\gamma_{mn}{}^2 g_{mn} = R_u g_{mn}, \quad \left|\frac{\sigma_{\max}}{P_{mn}}\right| = R_p g_{mn} \qquad (41)$$

where R_u and R_p are similar ratios of stress to displacement and to modal pressure corresponding to straight-crested waves and are defined in Eqs. (20) and (22). The function g_{mn} is given by

$$g_{mn} \approx \begin{cases} 2J_m(a\gamma_{mn}) \approx 2/\pi\sqrt{n + m/2} & \text{for } r = a \\ \left.\begin{array}{ll} 1/2, & m = 0 \\ (1 - 2\nu)/4, & m = 2 \\ 0, & \text{otherwise} \end{array}\right\} & \text{for } r = 0 \\ \left[1 + \nu\left(\frac{m}{1+m}\right)^2\right]\left[\frac{2}{\pi(1+m)}\right]^{1/2}, \ m > 2, & \text{for } r = r_m. \end{cases} \qquad (42)$$

By comparing the previous expressions (and noting that the last expression does not hold for $m < 2$) one finds that the maximum stress for the fundamental mode ($m = 0$, $n = 1$) occurs at the plate circumference and is given by Eq. (41) with $g_{mn} \approx 2/\pi$. Similarly, one may determine the other entries of Table 3, which shows the largest values of g_{mn} and the corresponding locations on the plate.

Table 3 Maximum values of g_{mn} and locations of maximum stress

$m \backslash n$	1	2	3	>3
0	0.636 $r = a$	0.500 $r = 0$	0.500 $r = 0$	0.500 $r = 0$
1	0.520 $r = 0$	0.520 $r = 0$	0.520 $r = 0$	0.520 $r = 0$
2	0.450 $r = a$	0.367 $r = a$	0.318 $r = a$	see below
3	←——0.400(1 + 9ν/16)——→ ←—— $r = r_m$ ——→			

For $m = 2$, $n > 3$: $g_{mn} \approx 2\pi\sqrt{n+1}$ [at $r = a$], until for very large n, $g_{mn} \approx (1 - 2\nu)/4$ [at $r = 0$].

For $m > 2$: $g_{mn} \approx \left[1 + \nu\left(\frac{m}{m+1}\right)^2\right]\sqrt{\frac{2}{\pi(m+1)}}$ [at $r = r_m$].

For large m: $g_{mn} \approx (1 + \nu)\sqrt{2/m\pi}$ [at $r = r_m$].

$$\frac{r_m}{a} \approx \frac{2}{\pi}\left(\frac{1+m}{2n+m}\right).$$

Modal Stresses in Beams

The general treatment of plates embodied by Eqs. (1) to (15) may be extended to apply also to beams. One need merely to replace the various plate parameters by the corresponding beam parameters. That is, one must replace the plate flexural rigidity D by the beam rigidity $B = EI$, and the plate surface density ρh by the beam linear density ρS; where I and S denote the centroidal moment of inertia and the area of the beam cross section, respectively. In addition, one must delete all y-dependences, all n-subscripts, and all summations on n in the afore-mentioned equations; also, all integrations carried out over the plate surface A must be reinterpreted as taken over the beam length L.

The displacement of a beam resonating in its mth mode then is found to be given by[7]

$$\bar{\eta}(x, t) = \eta(x)e^{i\omega_m t}; \quad \eta(x) \approx U_m\psi_m(x) \qquad (10a')$$

where

$$U_m = P_m/i\beta\rho S\omega_m{}^2, \qquad (10b')$$

[7] The beam equations of this section bear the same numbers as the analogous plate equations, but are distinguished by primes.

$$P_m = \left[\int_L P(x)\psi_m(x)dx\right]\Big/\left[\int_L \psi_m{}^2(x)dx\right], \qquad (7')$$

and for time-wise random loading,

$$\langle \bar{P}_m{}^2\rangle \approx P_m{}^2 S_p(f_m)\cdot\beta\omega_m/2\pi \qquad (14')$$

Unlike in plates, however, the flexural stresses in beams are independent of Poisson's ratio. If C denotes the distance of the extreme fiber from the beam's neutral surface, the stress in this fiber is given by the well-known relation

$$\sigma_x = CE\eta_{xx}, \qquad (16')$$

so that the maximum stress can easily be found if $\eta(x)$ is known.

The flexural motions of simply supported beams may be described by Eq. (18) if the previously outlined notational modifications are made. One then finds that the maximum stress $\sigma_{\max}$ is related to modal displacement U_m as

$$R_u \equiv |\sigma_{\max}/U_m| = C\omega_m\sqrt{E\rho S/I} = (C/r)\omega_m\rho c_A \qquad (20')$$

where

$$r = \sqrt{I/S}$$

is the radius of gyration of the cross section, and

$$c_A = \sqrt{E/\rho} \qquad (21')$$

is the velocity of pure longitudinal waves (i.e., the sound velocity) in the beam material. The ratio of maximum stress to modal pressure may similarly be written as

$$R_p = |\sigma_{\max}/P_m| = (C/r)(c_A/\beta S\omega_m), \qquad (22')$$

and one may note that the maximum strain is proportional to the "Mach number" of the flexural oscillation, as for plates[8]:

$$\epsilon_{\max} = C|\eta_{xx}|_{\max} = (C/r)(V/c_A). \qquad (20a')$$

$V = U_m\omega_m$ is the maximum transverse velocity of the beam.

The mode shapes ψ_m and natural frequencies ω_m for simply supported beams, namely [10]

$$\psi_m(x) = \sin(m\pi x/L); \quad \omega_m = (m\pi/L)^2\sqrt{EI/\rho S} = (m\pi/L)^2 rc_A \qquad (23')$$

were implied when Eq. (18) was introduced. The foregoing results thus apply in the strictest sense only to simply supported beams; however, they may be expected also to provide good approximations to conditions at locations far from the supports in beams with other boundary conditions (for resonant vibrations at the higher modes).

Table 4 was constructed in order to provide an estimate of the maximum errors one may expect if one applies Eq. (20a′) to beams with other than simple supports and includes boundary effects. This table, computed from the mode shape tabulations of Bishop and Johnson [13], lists the numbers by which one must multiply the right-hand side of Eq. (20a′) in order to obtain exact values for the maximum strain. Since Eq. (20a′) without the correction factors applies to the maximum stresses and velocities far from the beam ends, the factors of Table 4 may also be considered as "end correction" factors; they essentially account for end stress and/or velocity deviations from the "interior" nearly sinusoidal conditions.

Summary of Results

For the clamped circular and simply supported rectangular plates considered, the ratios of maximum modal stress to modal

[8] For beams of rectangular cross section, $C/r = \sqrt{3}$. The differences between beam and plate results are due only to Poisson effects.

Table 4 "End correction" factors g_m applicable to strain-Mach number, Equation (20a′), for beams with various end conditions

End conditions	Mode number m					Location of	
	1	2	3	4	5	Maximum stress	Maximum deflection
Pinned—Pinned	1.00	1.00	1.00	1.00	1.00	All antinodes	All antinodes
Clamped—Clamped	1.26	1.33	1.33	1.32	1.33	Clamps	Interior antinodes
Free—Free	0.79	0.75	0.75	0.76	0.75	Interior antinodes	Free ends
Clamped—Free	1.00	1.00	1.00	1.00	1.00	Clamp	Free end
Clamped—Pinned	1.33	1.32	1.32	1.33	1.32	Clamp	Interior antinodes
Pinned—Free	0.76	0.76	0.76	0.75	0.76	Interior antinodes	Free end

displacement U_{mn} and to modal loading P_{mn} may be expressed as

$$\left|\frac{\sigma_{\max}}{U_{mn}}\right| = R_u g_{mn}, \qquad \left|\frac{\sigma_{\max}}{P_{mn}}\right| = R_p g_{mn}, \tag{41}$$

where R_u and R_p denote the corresponding ratios obtained for straight-crested waves (of sinusoidal shape) and are given by Eqs. (20) and (22). The symbol g_{mn} denotes a numerical factor that depends on plate geometry, boundary conditions, and mode number. Since it was demonstrated that $g_{mn} \le 1$ in all cases considered here, one need merely set $g_{mn} = 1$ in Eqs. (41) to obtain a set of relations suitable for simple conservative design calculations. The same relations may be expected to hold also for other plate configurations at regions a few wave lengths from the boundaries.

Eqs. (41) may be applied also to beams with various end conditions, if the subscripts n are deleted. If one is concerned only with stresses and deflections at points not too near the ends of beams vibrating in their higher modes, one may set $g_m = 1$ for all practical purposes. For cantilevers and simply supported beams $g_m = 1$ for all conditions as indicated in Table 4. However, if end effects are included, then $g_m \approx 1.33$ for clamped-clamped or clamped-pinned beams and $g_m \approx 0.75$ for free-free or pinned-free beams.

Thus, one may use the following relations for designing plates or beams that can oscillate at resonance:

$$\epsilon_{\max} = \left(\frac{C}{r}\right)\frac{|V_{mn}|}{c}\cdot g_{mn} \tag{20a″}$$

$$\sigma_{\max} = \left(\frac{C}{r}\right)|V_{mn}|\,\rho c\cdot g_{mn} \tag{20″}$$

$$|V_{mn}| = |P_{mn}|/\beta d\omega_{mn} = |U_{mn}\omega_{mn}| \tag{10b″}$$

For beams the subscript n must be disregarded, and c and d have slightly different meanings for beams and plates:

For beams:

$$c = c_A = \sqrt{E/\rho}, \qquad d = \rho S;$$

For plates:

$$c = c_L = c_A/\sqrt{1-\nu^2}, \qquad d = \rho h.$$

The value of (C/r) depends only on the cross-sectional shape; for plates and for beams with rectangular cross sections $(C/r) = \sqrt{3}$.

The factor g_{mn} (or g_m) may be set equal to unity for:

1 All cases where only conditions at more than a few wave lengths from the boundaries are of interest; or for

2 Over-all analyses of:
   Rectangular plates on simple supports
   Circular plates with clamped circumferences
   Beams on simple supports
   Cantilever beams.

The value g_{mn} (or g_m) ≈ 0.75 should be used for over-all analyses (i.e., including boundary stresses and displacements) of beams and rectangular plates with free boundaries, and of pinned-free beams; g_{mn} (or g_m) ≈ 1.33 should be used for beams and rectangular plates having one or more clamped boundaries, except for cantilever arrangements. Although no detailed calculations have been presented that apply to circular plates with other than clamped boundaries, it appears likely that $g_{mn} = 1$ will result in conservative designs for such plates. Similarly, it appears that use of a value $g_{mn} = 1.33$ will result in conservative designs for all types of resonant beams and plates.

Concluding Remarks

The previously summarized results should, hopefully, serve as convenient sign-posts for the design of beam-like and plate-like structures in which resonances cannot be avoided. The aforementioned equations demonstrate that beam cross sections with small C/r values that are desirable for static designs are desirable also for resonant designs. Eq. (10b″) shows the utility of increased damping β in reducing stresses and deflections at resonance, and indicates the effects of changes in plate thickness h and beam cross-sectional area S on these.

Once the variation of modal loading P_{mn} with frequency has been established for a given structure, the foregoing equations may be used as a basis for delineating the amounts of damping required at the various frequencies. If $|P_{mn}/\omega_{mn}|$ decreases with increasing frequency, as one often finds, then the greatest modal stress occurs for the lowest resonance that falls within the range of exciting frequencies, provided the damping β does not decrease with frequency. This is seldom the case in structures in which no special damping provisions are included; hence, the rule of thumb that the maximum stress occurs at the fundamental is not unreasonable for such structures.

The effect of damping on the modal stresses is evident from Eqs. (20″) and (10b″). For loads that vary purely sinusoidally with time P_{mn} does not depend on damping, and the modal stress varies inversely as the damping factor β. However, for loads that vary randomly with time the root-mean-square P_{mn} is proportional to $\sqrt{\beta}$, so that the root-mean-square stress varies inversely as only $\sqrt{\beta}$.

From Eq. (10b″) one may also note that the modal stress varies inversely as plate thickness h or as beam cross-sectional area S, provided frequency ω_{mn} is held constant. However, ω_{mn} is proportional to h for plates, and ω_m is proportional to r for beams. Thus, plate resonance stress varies as h^{-2} at a given mode, but only as h^{-1} at a given frequency; beam resonance stress varies as $(rS)^{-1}$ at a given mode, and as S^{-1} at a given frequency. Thus, for example, if one doubles the thickness of a given plate which developed a stress σ at resonance at frequency f, the new plate would develop a stress of $\sigma/2$ at frequency f (if this frequency corresponds to a resonance of the thicker plate and if the modal pressure is the same). If the new plate has at f' a mode shape which is the same as that which the original plate had at f, then the new plate will experience a stress $\sigma/4$ at f'. Similar remarks apply also to beams, but for these one must account for changes in both section area and shape.

Acknowledgment

The author is indebted to Drs. Peter A. Franken and Ira Dyer, both of Bolt Beranek and Newman, Inc., for their many helpful suggestions. In particular, Dr. Franken indicated the need for the present study and Dr. Dyer pointed out the strain-Mach number relation and some of the conclusions. The results presented in this paper grew out of a study sponsored by the Army Ballistic Missile Agency under contract No. DA-19-020-ORD-5038, and in part were conceived in the course of analyses sponsored by the U. S. Air Force, Wright Air Development Division, under contract No. AF 18(600)-1857.

References

1 R. L. Bisplinghoff, G. Isakson, and T. H. H. Pian, "Methods in Transient Stress Analysis," *Journal of Aeronautical Sciences*, vol. 17, 1950, pp. 259–270.

2 A. C. Eringen, "Response of Plates and Beams to Random Loads," *Journal of Applied Mechanics*, vol. 24, Trans. ASME, vol. 79, 1957, pp. 46–52.

3 Lord Rayleigh, "The Theory of Sound," Dover Publications, New York, N. Y., 1945, vol. 1.

4 E. E. Ungar, "Free Oscillations of Edge-Connected Simply Supported Plate Systems," ASME Paper No. 60—WA-112, Winter Annual Meeting, New York, Nov. 28–Dec. 2, 1960.

5 P. M. Morse, "Vibration and Sound," McGraw-Hill Book Company, Inc., New York, N. Y., third edition, 1948, chapter V.

6 J. E. Ruzicka, Ed., "Structural Damping," The American Society of Mechanical Engineers, New York, N. Y., 1959, Sections I, III.

7 S. Crandall, "Random Vibration," Technology Press, Cambridge, Mass., 1958, chapters 1, 4.

8 S. Timoshenko, "Theory of Plates and Shells," McGraw-Hill Book Company, Inc., New York, N. Y., 1940.

9 L. Cremer, "The Propagation of Structure-Borne Sound," Sponsored Research (Germany) Report No. 1 (Series B) (British) Dept. of Scientific and Industrial Research, circa 1948.

10 S. Timoshenko, "Vibration Problems in Engineering," D. Van Nostrand Co., Inc., New York, N. Y., second edition, 1937, pp. 420–425.

11 P. M. Morse and H. Feshbach, "Methods of Theoretical Physics," McGraw-Hill Book Company, Inc., New York, N. Y., 1953, part II, chapter 10.

12 E. Jahnke and F. Emde, "Tables of Functions," Dover Publications, New York, N. Y., fourth edition, 1954.

13 R. E. D. Bishop and D. C. Johnson, "Vibration Analysis Tables," Cambridge University Press, Cambridge, England, 1956.

AUTHOR CITATION INDEX

SUBJECT INDEX

Biographical Notes on Reprint Authors

J. D. ACHENBACH (1935–   ) is Professor of Civil Engineering at Northwestern University. His areas of interest include wave propagation and continuum mechanics, and he recently published a treatise on stress wave propagation.

J. R. AIREY (1868–1937) was Mathematician and Principal of the City of Leeds Training College. He is known for his work on toroidal functions, elliptic integrals, and Bessel functions.

R. N. ARNOLD (1908–1963) was Regius Professor of Engineering at the University of Edinburgh. His publications included work on the embrittlement of steels, the cutting of metals, impact, mechanical vibration, and gyrodynamics.

B. A. BOLEY (1924–   ) is Dean of the Technological Institute of Northwestern University. During his years at Columbia and Cornell he has become widely known for his work on structural theory, thermoelasticity, and related thermal conductivity and melting problems.

B. BUDIANSKY (1925–   ) is Gordon McKay Professor of Applied Mechanics at Harvard University. He has pioneered in contributions to the elastic–plastic behavior of structures, plasticity theory, dynamic response of structures, and buckling and postbuckling of elastic structures.

G. F. CARRIER (1918–   ) is T. Jefferson Coolidge Professor of Applied Mathematics at Harvard University. He has written extensively on many aspects of applied mathematics and mechanics, and is well known for his work on perturbation theory, complex variables, and the mathematical modeling of mechanics problems.

H. -N. CHU (1923–   ) is a Research Fellow at the California Institute of Technology. He has contributed to the understanding of nonlinear shell dynamics and, after many years of industrial experience, is currently working on environmental problems.

S. H. CRANDALL (1920–   ) is Ford Professor of Mechanical Engineering at the Massachusetts Institute of Technology. He has stressed heavily the role of random processes in vibration theory, and has applied his work in acoustics and earthquake engineering.

H. DERESIEWICZ (1925–) is Professor of Mechanical Engineering at Columbia University. His work includes contributions to the theory of crystal vibrations, thermoelasticity, and elastic contact problems.

C. L. DYM (1942–) is with Bolt Beranek and Newman Inc. He has worked on vibration and stability of shells and shell-type structures, and on problems of sound transmission and acoustics.

B. R. EL-ZAOUK (1946–) is a consulting engineer in Beirut, Lebanon. He received his Ph.D. in Civil Engineering from Carnegie-Mellon University in 1973.

D. A. EVENSEN (1937–) is with the J. H. Williams Company in Redondo Beach, California. Since his graduation from the California Institute of Technology, he has been a leading investigator of the nonlinear vibrations of shells.

D. FEIT (1937–) is with the Naval Ship Research and Development Command. He has contributed heavily to the theory of fluid–solid interaction, being particularly interested in the acoustic response of structures submerged in water.

K. FORSBERG (1934–) has been with the Lockheed Missiles and Space Company since 1956. He has worked on static and dynamic behavior of shell structures and applications of computer graphics to engineering research and design.

D. C. GAZIS (1930–) is Consultant to the Director of IBM Research at the IBM Thomas J. Watson Research Center. He has published widely on the subjects of dynamics of plates and shells, traffic theory, and applications of computers to environmental and social problems.

R. GREIF (1938–) is Associate Professor at Tufts University. He has worked on elasticity, structural design, response to dynamic loading, and stress waves.

G. HERRMANN (1921–) is Professor in and Chairman of the Department of Applied Mechanics at Stanford University. He has written, translated, and edited many works on shells, structural dynamics, stability, and dynamic stability.

D. E. JOHNSON (1933–) is Senior Staff Scientist and Group Leader in the Structures Department at Avco Systems Division. He has worked on thermoelasticity, shells, dynamic response, and numerical methods.

A. KALNINS (1931–) is Professor of Mechanics at Lehigh University. He has worked on many aspects of static and dynamic behavior of thin shells and numerical solutions of boundary value problems.

H. KRAUS (1932–) is Professor at the Hartford Graduate Center of Rensselaer Polytechnic Institute. He has worked on structural analysis of industrial equipment, shells, elasticity, thermal stresses, vibration of solids.

H. LAMB (1849–1934) began his career at Cambridge, but spent most of his life at Manchester. He was knighted in 1931 for his research in fluid mechanics, acoustics, and wave propagation.

R. W. LEONARD (1926–) is Associate Chief of the Structures and Dynamics Division at the NASA Langley Research Center. He has done research in the areas of structural behavior, shell theory, and on the dynamics and stability of structures.

A. E. H. LOVE (1863–1940), a well-known theoretical mechanician, is best known through his landmark volume *Treatise on the Mathematical Theory of Elasticity.* He contributed significantly to the understanding of the elasticity of solids, particularly to the behavior of the earth crust. He was the discoverer of distortional surface waves that now bear the name Love Waves.

M. A. MEDICK (1927–) is Professor of Mechanical Engineering at Michigan State University. His research has been concerned with wave motion and impacts in solids, and, more recently, biomechanics.

J. MIKLOWITZ (1919–) is Professor of Applied Mechanics at the California Institute of Technology. The bulk of his work has been in the area of wave propagation in elastic and inelastic media, although he has also investigated various yield phenomena in materials.

R. D. MINDLIN (1906–) is Professor Emeritus of Civil Engineering and Engineering Mechanics at Columbia University. His contributions over the years to photoelasticity, crystal vibrations, and elasticity theory have won him many, many honors and awards.

M. P. MORTELL (1941–) is Lecturer in the Department of Mathematical Physics at the University of Cork, Ireland. His publications include papers on dynamics of shells, finite amplitude waves, and waves in viscoelastic media.

J. L. NOWINSKI (1905–) is Professor of Mechanics and Aerospace Engineering at the University of Delaware. He has contributed to the theory of nonlinear vibrations; analysis of composite materials; and, more recently, problems in biomechanics.

LORD RAYLEIGH (J. W. STRUTT) (1842–1919) is known to acousticians from his several volumes *Treatise on Theory of Sound.* He has contributed also to other fields, such as the physics of light and color, the dynamics of resonance, and the vibration of gases.

E. REISSNER (1913–   ) is Professor of Applied Mechanics and Mathematics at University of California, San Diego. He has been a dominant figure in theoretical mechanics for the last 25 years. His contributions to the elastic behavior of solids, especially elastic plates and shells, have earned him much recognition and many honors. He is perhaps most widely known for a plate theory that bears his name.

E. W. ROSS, JR. (1925–   ) is Mathematician at the U.S. Army Natick Laboratories. He has published papers in mathematical plasticity, vibrations of thin shells, and stress and flexibility analysis of tents and parachutes.

J. C. SNOWDON (1932–   ) is Professor of Engineering Research at the Ordinance Research Laboratory of the Pennsylvania State University. He has contributed extensively to the shock and vibration literature, particularly emphasizing the role of damping, and effectively exposing the concepts of impedance and transmissibility.

W. R. SPILLERS (1934–   ) is Professor of Civil Engineering at Columbia University. He has worked in continuum mechanics and applications of computers to engineering problems.

C. -T. SUN (1938–   ) is Associate Professor of Aeronautics and Astronautics at Purdue University. He has worked mostly in the areas of the mechanics of composite materials and stree-wave propagation.

S. P. TIMOSHENKO (1878–1972) may, without exaggeration, be considered the father of engineering mechanics in the United States. He influenced at the highest level the development of mechanics in this country for most of the first half of this century.

E. E. UNGAR (1926–   ) is a Principal Engineer at Bolt Beranek and Newman Inc. He has been active both as a researcher and a consultant, and is well known for his work on damping.

G. B. WARBURTON (1924–   ) is Professor of Applied Mechanics at the University of Nottingham. He has made significant contributions to the area of vibration of solids.

J. P. D. WILKINSON (1938–   ) is Manager of the Solid Mechanics Program at General Electric Company. He has worked on radiation of submerged structures, transducers, jet engine noise, and many applications of structural dynamics.

D. YOUNG (1904–   ) is Senior Vice President of the Southwest Research Institute. He has contributed heavily to the applications of vibration theory in structural dynamics and shock isolation.

About the Editors

ARTURS KALNINS is Professor of Mechanics at Lehigh University and teaches courses in static and dynamic behavior of thin plates and shells. Professor Kalnins received the B.S. in 1955, the M.S. in 1956, and the Ph.D. in 1960, in Engineering Mechanics from the University of Michigan. He was Assistant Professor at Yale University from 1960 to 1965.

He has conducted research on many aspects of the behavior of shells, in particular of shells of revolution, and has developed a method of solution for shells of revolution that is now being used widely by analysts in many countries of the world.

Dr. Kalnins is a Fellow of the Acoustical Society of America and a member of the American Society of Mechanical Engineers. Presently he serves as associate editor of the Journal of the Acoustical Society of America.

CLIVE L. DYM is presently Manager of the Dynamics and Structures Group at Bolt Beranek and Newman Inc., where he has been a Supervisory Consultant since 1974. Prior to this appointment he served on the faculties of the State University of New York at Buffalo and Carnegie-Mellon University, and held visiting appointments at the TECHNION-Israel Institute of Technology and at the Institute for Sound and Vibration Research at the University of Southampton.

Dr. Dym completed his undergraduate work at the Cooper Union, and received the M.S. degree from the Polytechnic Institute of Brooklyn and the Ph.D. degree from Stanford University. Since completing his studies, Dr. Dym has done research on a variety of problems in structural dynamics and stability and on some aspects of transmission loss of sandwich panels. The results of these and other researches have been published in some sixty journal articles and technical reports, and in three books, the latest of which is *Stability Theory and Its Applications to Structural Mechanics* (Noordhoff, 1974).

Dr. Dym is a Fellow of the Acoustical Society of America, a founding member of the American Academy of Mechanics, a member of the American Society of Civil Engineers and the American Society of Mechanical Engineers, and has served on a number of technical committees of these societies. He has also been elected to a number of honorary societies, and is listed in *Who's Who in the East.* Dr. Dym holds professional engineering registration in a number of states, and is affiliated with the Institute of Noise Control Engineering.